U0802496

2011

CHINA ENERGY CONSERVATION AND EMISSION REDUCTION DEVELOPMENT REPORT

中国节能减排发展报告

从“十一五”到“十二五”

中国工业节能与清洁生产协会
中国节能环保集团公司 ●编

中国经济出版社
CHINA ECONOMIC PUBLISHING HOUSE
北 京

图书在版编目（CIP）数据

2011 中国节能减排发展报告/中国工业节能与清洁协会，中国节能环保集团公司编.
北京：中国经济出版社，2011. 11
ISBN 978 - 7 - 5136 - 1209 - 8
Ⅰ. ①2… Ⅱ. ①中… ②中… Ⅲ. ①节能—研究报告—中国—2011 Ⅳ. ①TK01

中国版本图书馆 CIP 数据核字（2011）第 221427 号

责任编辑　吴航斌　陈　佳
责任审读　贺　静
责任印制　张江虹
封面设计　白朝文

出版发行　中国经济出版社
印 刷 者　北京市人民文学印刷厂
经 销 者　各地新华书店
开　　本　210mm × 285mm
印　　张　33. 75
字　　数　860 千字
版　　次　2011 年 11 月第 1 版
印　　次　2011 年 11 月第 1 次
书　　号　ISBN 978 - 7 - 5136 - 1209 - 8/Z · 987
定　　价　160. 00 元

中国经济出版社 **网址** www. economyph. com **社址** 北京市西城区百万庄北街 3 号 **邮编** 100037
本版图书如存在印装质量问题，请与本社发行中心联系调换（联系电话：010 - 68319116）

版权所有　盗版必究（举报电话：010 - 68359418　010 - 68319282）
国家版权局反盗版举报中心（举报电话：12390）　　　服务热线：010 - 68344225　88386794

《2011中国节能减排发展报告》

编　委　会

总　顾　问　李毅中　解振华　汪光焘

顾　　　问　（以姓氏笔画为序）

王茂林　王显政　冯之浚　曲格平　许宪春

柴松岳　濮洪九

编委会主任　周　建　翟立功　王小康

编委会副主任　周长益　赵世堂　文兼武

主　　　编　王小康

副　主　编　陈津恩　余海龙　余红辉

编　　　委　（以姓氏笔画为序）

马　骏　刘文强　安　宜　闫长乐　余　薇

张永泽　张　超　李　杰　李龙生　李新创

朱继明　杨义华　杨家义　邹结富　陈曙光

郑起宇　孟合合　郁　聪　赵志林　智　慧

雷仲敏　戴彦德

撰　稿　人　（以姓氏笔画为序）

马　骏　王小兵　王　毅　刘　丹　刘志亭　刘昊明

安治永　许　泓　闫长乐　任俊涛　宋丽娟　张永泽

张　雷　李长胜　李瑾梁　林　燕　赵友民　胡正鸣

曾　武　雷仲敏　霍中和

主办单位　中国工业节能与清洁生产协会

中国节能环保集团公司

前　言

《2011中国节能减排发展报告》是我们对中国节能减排整体状况进行观察、描绘、分析的第三本年度报告，该报告将继续秉承“内容详实、资料丰富、数据准确、观点鲜明”的指导思想对节能减排工作进行持续跟踪，为产业界、政府部门、投资界、学术界的决策者、研究者以及实践者提供全面、权威、客观、多元的参考意见。

“十一五”时期，我国节能减排取得显著成效。全国单位国内生产总值能耗降低19.1%，二氧化硫、化学需氧量排放总量分别下降14.29%和12.45%，基本实现了“十一五”规划纲要确定的约束性目标，扭转了“十五”后期单位国内生产总值能耗和主要污染物排放总量大幅上升的趋势，为保持经济平稳较快发展提供了有力支撑，为应对全球气候变化作出了重要贡献。

“十二五”时期，我国节能减排形势仍然十分严峻，任务十分艰巨。特别是我国节能减排工作还存在责任落实不到位、推进难度增大、激励约束机制不健全、基础工作薄弱、能力建设滞后、监管不力等问题。这种状况如不及时改变，不但“十二五”节能减排目标难以实现，还将严重影响经济结构调整和经济发展方式转变。为此，国务院总理温家宝在部署“十二五”节能减排工作时强调，要把节能减排作为促进科学发展的“硬任务”，转变经济发展方式的“硬举措”，考核各级干部的“硬指标”。

站在“十二五”开端这一时间节点之上，《2011中国节能减排发展报告》基于全球气候、能源、经济及政治大背景，真实记录国家总体、各地区、重点耗能行业节能减排目标完成情况，以及节能环保、新能源等战略新兴产业的发展，深入分析国际国内环境保护、节能减排政策及产业发展趋势，介绍关于节能减排的不同观点，为我国“十二五”期间开展节能减排工作提供判断、预测和展望，为中国节能减排事业发展作出贡献。

我们希望把《中国节能减排发展报告》塑造成高水平、高质量的中国节能减排发展蓝皮书，为节能减排事业奉献绵薄之力。为使以后的工作做得更好，希望您提出宝贵的批评和建议，也期待更多的专家给予斧正。

主编：王小康

2011年11月

前 言

[illegible]

Introduction of the *Report*

China Energy Conservation and Emission Reduction Development Report (hereafter referred to as the" *Report*") is a major annual blue paper developed by experts and scholars in the relevant field, organized by China Industry Energy Conservation and Clean Production Association and China Energy Conservation and Emission Reduction Group. The paper aims at an in - depth research into and analysis of the energy conservation and emission reduction sector in China and presents the status quo and development trend for this sector in the area of policy, market, technology, investment opportunities, etc. In a word, the *Report* is a systematic document about the sector in discussion.

The *Report* has received support from a lot of senior officials. General counsels to the writing process include such figures as Li Yizhong, Former Minister of Industry and Information Technology, Xie Zhenhua, Deputy Director of the National Development and Reform Commission, and Wang Guangtao, Chairman of the Environment and Resource Protection Committee in the 11th National People's Congress. Names on the Editorial Board for the *Report* are Zhou Jian, Minister of Environmental Protection, Zhai Ligong, Chairman of the Supervising Committee for Major Enterprises, the State Council, and Wang Xiaokang, President of CECEP and Head of China Industry Energy Conservation and Clean Production Association. In addition to the support from officials, the *Report* has enjoyed support and help from numerous government agencies, industry associations and research institutions, such as National Development and Reform Commission, State Assets Administration Committee, Ministry of Industry and Information Technology, China National Coal Association, China Electricity Council, China Academy of Building Research, China Energy Conservation Association, Petroleum and Chemical Planning Institute, Institute of Metallurgy Planning, China Non-ferrous Metals Industry Association, China Green Lights Association, China Building Materials Industry Association, Tsinghua University and Beijing University of Posts and Telecommunications.

The annual blue paper contains not only a snapshot of energy conservation and emission reduction in China, but also realities for different industrial sectors and regions in the sector. Moreover, the *Report* goes beyond mere summarization of the development for the sector. It researches into and explores development trends and investment opportunities, judging the future and looking into the prospect of the field in discussion.

The *Report* sums up the research subject, target readers, basic framework and major conclusions from reports in the last three years. In this way, we want to tell readers our research subjects, aim, approach of research and main conclusions.

1 Research Subject

The sector of energy conservation and emission reduction, or our research subject, refers to a comprehensive industry cluster that pursues the three major goals of low energy consumption, pollution and emission, has under it three sub – industries, namely, energy conservation, environmental protection and new energy, and covers all links of the industrial chain, including upstream materials, equipment providers, midstream processing, manufacturers and downstream service providers.

There are two common ways for categorizing the energy conservation and emission reduction sector: the first divides it into two, energy conservation and environmental protection, and new energy; the second into three, energy conservation, environmental protection, and new energy. The second approach is explained by the fact of relevantly three independent industrial sections within the sector due to different industrial needs. Currently, we need to look at the inherent relations of the three branches so that they fit into an integral whole. Energy (in the sense of traditional fossil energy) conservation represents the best way of environmental protection. Introduction of new energy counts as the best way of saving conventional energy. Hence, the three fields are inseparable. An industrial plan that neglects their internal links means we do not have a clear mind of their relations.

2 Target Readers

Target readers for our report include:

- Decision – makers and managers of energy conservation, environmental protection, new energy and industrial energy – depleting businesses;
- Decision – makers and managers of the functional departments for energy conservation and emission reduction both in the central and local governments;
- Investors in energy conservation and emission reduction, environmental protection, new energy, low carbon economy and circular economy;
- Researchers on energy conservation and emission reduction, environmental protection, new energy, low carbon economy and circular economy.

For decision – makers and managers of energy conservation, environmental protection and new energy industrial energy – depleting businesses, we have included in the *Report* relevant forward – looking market opportunities, analysis of policy, economic, social and technology environment and industrial competitors, and research on the business model for the sector.

For decision – makers and managers of industrial energy – depleting businesses, we have included in the *Report* benchmarking progress for key indicators in the sector, road – map for energy conservation and emission reduction in different sectors, and specific guidelines for the relevant businesses as to how industries can conserve energy and reduce emission through energy efficiency auditing and clean production verification, and how can they achieve relevant goals through Energy Performance Contracting.

For decision – makers and managers of the functional departments for energy conservation and emission reduction both in the central and local governments, we have included in the *Report* contents as to how to monitor the process of energy conservation and emission reduction, assess the implementation effect of government control policies and develop future control targets and relevant policies.

For investors in energy conservation and emission reduction, environmental protection, new energy, low carbon economy and circular economy, we have included in the *Report* snapshot of the sector they are about to enter, must – know analysis of the sector trends for a to – be – investor, fundamentals, key performance and financial indicators of major players in the field, investment and financing mode, legal and policy environment, and the status quo and trend for investment and financing needs.

In addition, for researchers on energy conservation and emission reduction, environmental protection, new energy, low carbon economy and circular economy, we have included in the *Report* relevant sector data, the latest ideas and representative case studies.

3 Basic Framework

To deliver the afore – mentioned promises, we have designed a basic framework for the *Report* as shown in the following table.

Number	Name	Contents	Written in the year(s) of
1	Snapshot	A general view, main institutions and latest progress on major issues in the sector of energy conservation and emission reduction in China in the global context	2009/2010/2011
2	International practice	Bring in advanced practices in the area of low carbon economy and energy conservation and emission reduction from government agencies and established businesses in developed countries. Their practices can be learned by domestic government agencies and businesses	2010
3	Progress across sectors	Track the progress in energy conservation and emission reduction across the main energy – intensive sectors, such as iron & steel, petro – chemical, non – ferrous metals and electricity	2009/2011
4	Emerging energy	Introduce the industrialization of new sources of energy, such as solar energy, wind power, nuclear power and biomass energy	2009/2010/2011
5	Environmental protection	Introduce the status quo and industrialization process of environmental protection sectors, such as waste water, air pollution and solid waste treatment	2009/2011
6	Regional comparison	Introduce the status quo of and make comparison among energy conservation and emission reduction across regions, such as the east, central, northeastern parts of China, divided by administration or economic development level	2009/2011

续表

Number	Name	Contents	Written in the year(s) of
7	Technology	Track the development of major and advanced technologies in the sector and analyze their applicability in the Chinese context	2010
8	Green businesses	Analyze the fundamentals, typical experiences and investment values of businesses in the area of low carbon economy and energy conservation and emission reduction	2009/2010/2011
9	Major events	Annual list of major events in the sector of low carbon economy and energy conservation and emission reduction	2009/2010/2011
10	Relevant data	Annual authorized data concerning low carbon economy and energy conservation and emission reduction	2010/2011

As made clear by the table, the 2009 *Report* gives a general and comprehensive view of energy conservation and emission reduction in China and introduces the situation of the sector in that year. The 2010 and 2011 reports can be discussed together, although each report has its own focuses. Both the 2010 and 2011 reports have included in them the parts about snapshot, green businesses, major events and annual data. As to focuses, the 2010 *Report* looks more at international practices, new sources of energy and technology, whereas the 2011 one gives more attention to progress across sectors, environmental protection and regional comparison.

4 Main Conclusions

4.1 General background

In the global context, research on energy conservation and emission reduction is attributable to the issue of climate change. Yet climate change is now not just a matter of climate, but one that concerns politics, diplomacy and economy, grabbing the centre place in the international arena. With the introduction of "low carbon economy", the fight against climate change will shape people's production mode, lifestyle, values, national interests, and even human civilization.

In the context of China, the gravity on climate change is accompanied by a focus on development. China is a developing country, characterized by the drastic expansion of its economic output in the last two decades, giving rise to surging aggregate needs of resources. In addition, in the process of speeding up its industrialization, China faces intensive resource and energy consumption. Genuine efforts to conserve energy and protect the environment are entailed by China's basic national conditions, large aggregate economic output and economic development level. Also, energy – saving and emission reduction is needed to restructure its economy, transform its development mode and create a new pole for growth to develop the economy and fight against the financial crisis. In a word, a balance between energy conservation and emission reduction and economic development is in the following decades a core issue to be confronted and addressed in China.

4.2 From political consensus to industrial consensus

After 2009, new trends in global energy conservation and emission reduction have emerged. A major one is that despite controversy among scientists, climate change has been accepted as a political consensus; and low carbon development, an economic one. With accumulated advantages in industries, capital and technologies over time, developed countries are seeking the lead in developing low carbon economy and have produced initial fruits in advancing the sector of energy conservation and emission reduction. However, as a developing country, China faces a different set of challenges from its developed counterparts. To be specific, China has to balance energy conservation and emission reduction with its economic development.

According to the *Report*, the sector of energy conservation and emission reduction will be at the core for China to realize its goals concerning low carbon development and promote a balanced focus between economic development and environmental protection. The Chinese government has promised to the world that, by the year 2020, CO_2 emission per unit of GDP will be reduced by 40%—45% and clean energy will take up 15% of the total energy consumption. There are three approaches to deliver these promises. First, compulsory goals set by the government concerning energy conservation and emission reduction. Second, active and voluntary efforts on the part of businesses and the society at large. Third, developing a market - driven sector of energy conservation and emission reduction. Administrative efforts on the part of the government can be effective in the short term, but like heavy doses, they are only transitory and cannot work over the long term. Voluntary efforts by businesses and the society have their limitations since they are constrained by the willingness and consciousness of environmental protection in the minds of customers; such efforts are like "health products" that do not get to the core of an illness. As to a market - driven energy conservation and emission reduction sector, it is a sound combination of policy guidance and intrinsic needs for industrial development; it is like the overall long - term "therapeutic plan". Expectedly, for China, whose carbon emission leads most of the countries in the world, energy conservation and emission reduction will in the future contribute more than 30% of the fruits in the development of a low carbon economy.

Also according to the *Report*, 2010, as the ending year for China's 11th Five - year Plan of Energy Conservation and Emission Reduction and the beginning year for China's 12th Five - year Plan, marks the transition from "energy conservation and emission reduction dominated by administrative measures" to "one dominated by market mechanism".

In 2010, driven by economic recovery, global energy consumption has gone through a new round of rebound. As a result, the decline in global energy consumption as of the year of 2009 fails to develop into a medium to long - term trend. As regards further projection, global energy demand shall increase by a big margin by either 2035 or 2050. Among this, energy demand driven by rapid economic growth of the developing countries shall take a lion's share with the energy demand of China in particular.

In order to curb the increase in carbon emission due to energy consumption, the development of renewable energy shall help to improve the global energy consumption structure. Being a mainstay in promotiong the transition of global enegy structure to sustainable energy, China shall play a leading role in the development of and investment into nuclear power and renewable energy. In the meanwhile, development of renewable energy shall boost economic recovery of the developed countries who are deeply affected by the financial crisis. However, to our concern, while becoming the engine for economic growth, 'low – carbon' is also constructed as green trade barrier under disguise, resulting in new international economic rules.

4.3 Energy conservation and emission reduction sector in China has to learn from international practices and be based on our national conditions

Many developed countries have placed green and low carbon economy at the core of their national strategies and at the pinnacle of global competition. In other words, developing green and low carbon economy is a new international race initiated by the developed world, a competition of a higher level and more intense nature. In the sense of competition, developed nations and their businesses, with accumulated advantages in industries, capital and technologies, are our competitors, but also role models.

The *Report* has looked at the features of developing low carbon economy across the developed world: the UK is the initiator, advocate and pioneering paver in the global development of green economy; the US prides itself on its exploration of clean energy and clean application of conventional energy and hopes that such advantages can maintain its leading position in a world with new economic environment; Japan is trying to build a comprehensive "low carbon society", in the hope that such a development mode will get out the shackles of its limited resources and energy; European countries such as Germany, France and Denmark boast of their localized practices, showing the diversity in developing low carbon economy and the need for such an economy to fit national conditions.

As the *Report* suggests, even amidst competition among countries, China can learn from its competitors. For example, we should learn from the UK for its accurate understanding of new trends in the new economy; and from the US, for its determination to lead the global economy by improving energy efficiency and applying clean energy. Based upon these international practices, we have to develop the sector of energy conservation and emission reduction conforming to China's basic national conditions. For example, coal is the main resource of energy and China is in the process of speeding up industrialization.

A case in point in the *Report* is the part about new sources of energy. In developing new energy as a step of restructuring its energy mix, China has considered its national conditions by adopting a "double track" strategy of supporting the development of new sources of energy and promoting reform in conventional energy. New energy refers only to new and renewable energy, such as nuclear, wind, solar and biomass power. New sources of energy cover a broader area, including clean application of conventional energy, such as smart grid and clean coal. In fact, development of renewable energy and clean application of conventional energy will become two driving forces for the low carbon course in China.

4. 4 Initial fruits in energy conservation and emission reduction in energy – intensive industries

Main energy – intensive industries refer to industries that consume a lot of energy, such as iron & steel, non – ferrous metals, electricity, petro – chemical, building materials, transportation, and industries with intensive energy needs and high emission and discharge, such as paper – making, spinning, printing and dyeing and building materials. These industries take up more than 64% of the total end industrial energy consumption, and almost half of the total national energy consumption.

The output and scale of many industrial sectors in China occupies the world front rows. The output of many industrial products such as steel, coal, cement, fertilizer and etc. all rank No. 1 in the world. In the 11th Five – year Plan, China still registered an apparent trend of heavy industrialization. The growth rate of heavy industry outspeeded that of light industry. For this reason, the industrial sector remains to be main energy consumer and main source of pollution emission. The priority and difficulty of energy conservation and emission reduction all lie in the industrial sector.

During the 11th Five – year Plan period, China has made remarkable achievements in industrial energy conservation and emission reduction, which contributes greatly to the attainment of energy conservation and emission reduction goals for the 11th Five – year Plan. For example, during the 11th Five – year Plan period, energy consumption for power supply has decreased from 367g/kw in 2006 to 335g/kw. Average energy consumption per ton steel output of key iron and steel manufacturers has declined from 694kg to 607kg in 2010 with a total of 46. 11 million tce saved. In the non – ferrous sector, energy consumption of major products has declinced with technique and economic performance improved remarkably.

However, looking into the future, industrialization and urbanization in China shall continue to develop. The 12th Five – year Plan shall be a critical period for China's industrialization. People's consumption pattern shall be upgraded, resulting in increased demand for industrial products. The ratio of heavey industry in China shall remain relatively high. Demand for energy, transportation and raw materials shall increase further. Energy consumption shall be intensified. Consequently, the resource constraints on industrial development shall become more and more severe.

4.5 Environmental protection shall be more than just end pollution treatment

The extensive growth pattern exerts great pressure on the environment and the environmental issue is becoming increasingly serious. Both the central government and ordinary people have come to realize the urgency of this issue. To this end, the government has put into place various policies to curb the emission of industrial pollutants and made the environmental issue an important economic one to be solved in the national development planning. The central government has made the construction of resource – saving and environment – friendly society as a top priority for China's economic and social development.

Hence, China should speed up its efforts in development energy – saving and environment – friendly industries and strategic sunrise industries. Investments and policies for such industries should be enhanced. With a strategic position in national economic development, these strategic sunrise industries are of great and profound significance to economic development and national security. They possess the potentiality to become the pillar industry for future economic development of a country.

However, energy conservation and environmental protection industry in China still lies in the medium and lower stream of international labor chain and takes a weak position in international competition. With a low industrial concentration rate, such industry has failed to form an industrial chain featurhing labor division nand coordinated development. Currently, guidance to the development of energy conservation and environmental protection industry is mainly implemented through lauching of 'large projects' or subsidies for end products. In a bid to boost the development of energy conservation and environmental protection industry, we should take effective measures to stimulate both its external as well as endogenous growth.

With economic and social progress and the advancement of industrial restructuring in China, the sector of environmental protection will play an ever more important role in its national economy, and become a sector that speeds up economic development, improves its quality and brings into the economy more technologies. Future branches of the sector might include clean technologies, products and environmental protection services. In addition, the dominant concept in the sector will be "environment industry" or "green industry".

4.6 Renewable energy production and clean application of conventional energy will bring further the restructuring of energy mix

New sources of energy and new energy are similar in expression, but different in their contents. New energy refers only to new and renewable energy, such as nuclear, wind, solar power and biomass. New sources of energy cover a broader area, including clean application of conventional energy, such as smart grid and clean coal. The national "New Energy Industry Plan" has been renamed as the "New Sources of Energy Industry Plan", a move signaling the "double track" strategy of supporting the development of new sources of energy and promoting reform in conventional energy in restructuring the energy mix in China.

In other words, the renaming implies a quiet shift in China's energy industry plan in China. The revitalizing plan for new energy industry that focuses on wind, solar and biomass power before will be replaced by a new plan on new sources of energy that covers the whole industrial chain. According to the National Energy Bureau, the revitalizing plan for new sources of energy that encompasses nuclear, wind, solar, and biomass power, clean coal, smart grid, distributed energy and new energy for vehicles has been submitted to the State Council. By October, 2020, an additional investment of ¥5 trillion on these sources of energy is expected to have been made.

4.7 Energy conservation and emission reduction will pressure regions for economic restructuring

Comparatively speaking, China has limited supply of resources and energy, which is vastly but unevenly distributed. In other words, specific resources and energy are concentrated in certain regions. For example, the eastern part is rich in offshore oil, gas and wind; the central part produces a lot of coal; the western part is the base for producing new sources of energy; the northeastern part is the base for producing conventional energy and rich in oil. Yet energy distribution cannot match the development needs of localities, the mismatch that entails the constant transportation of resources and energy from one place to another to meet the needs in economic development.

In terms of regional environment, the three kinds of pollution (water, air and solid waste pollution) have shown different trends. Generally speaking, water pollution has been getting better over the years, whereas air and solid pollution are getting worse. In this context, energy conservation and emission reduction has become a compulsory choice for different regions to address their shortage in resources, transform their economic development model and restructure their regional economy.

After the unveiling and implementation of measures concerning energy conservation and emission reduction by the Chinese government, localities have stricken major progress in this regard and significantly increased energy efficiency. Despite of such exciting news, there are still problems across regions. For example, industrial structure needs further improvement; energy – intensive industries are continuing their fast expansion. Particularly after the introduction of policies by the government to stimulate domestic consumption and promote a sound and fast economic development, some localities began again to give green light to energy – intensive and high – polluting projects; some green businesses suffered decreasing profits, undermining the implementation of projects on energy conservation and emission reduction. All these problems represent new challenges for the green sector.

During the 11th Five – year Plan period, national polices on energy conservation and emission reduction focus more on the general arrangement of relevant requirements, without taking enough consideration of uneven social and economic development levels across regions. Hence, while implementing national polices, local governments should take stock of local social and economic conditions and come up with specific policies that fit local realities.

During the 11th Five - year Plan period, energy consumption and emission reduction in China is mainly driven by governments at various levels. It is expected that during the 12th Five - year Plan, this mode shall gradually evolve into a market - oriented one. However, the basic pattern of government dominance shall remain unchanged. That being the case, it is of realistic significance to analyze energy conservation and emission reduction in China by regions.

In order to comprehensively compare and analyze the overall effect of energy conservation and emission reduction by various regions in the 11th Five - year Plan period, in this *Report*, we construct three comprehensive evaluation indexes for regional energy conservation and emission reduction, namely emission reduction result, degree of effort - making and potentiality tapping. In line with such indexes, we divide various provinces into four categories.

According to our study, the first category mainly focuses on eastern China, including Beijing, Shanghai, Jiangsu, Guangdong and Zhejiang specifically. This indicates that among various reigons, eastern China has done the best in energy conservation and emission reduction. The second category mainly incudes Tianjin, Hebei, Fujian, Shandong and Hainan in eastern China, Anhui, Jiangxi, Henan, Hubei and Hunan in the middle part of China, Chongqing, Sichuan, Shannxi and Yunan in western China and Heilongjiang, Liaoning and Jilin in northeastern China. The third catergory includes Shanxi in the middle part of China, Guangxi, Gansu and Inner Mongolia in weastern China. The fourth category mainly focuses on western China, including Guizhou, Qinghai and Ningxia specifically.

4.8 Identify investment values on energy conservation and emission reduction in businesses

As the practitioners of energy conservation and emission reduction, businesses have on them clear responsibilities and obligations. The recent policy orientation in China has clearly emphasized the diversification of participants in developing low carbon economy and the co - investment of private and state - owned capital. Weak basis in low carbon development, imperfect policy and legal environment and lacking of supporting currency and financial polices all demonstrate the bumpy road ahead for the sector of energy conservation and emission reduction in the country. But excitingly, as a major sector for economic transition, it has received ever more attention. More consensuses have been reached. Particularly encouraging is the solid improvement in the aspect of financing, which is characterized by the combination of indirect financing that depends mainly on bank loans and direct financing from the market in the form of equities.

There are already listed companies in the energy conservation and emission reduction plate of the stock market, a financing market. Since the introduction of Growth Enterprise Market (GEM) Board in China's capital market in 2009, nearly one hundred small – and – medium – sized enterprises, supported mainly by private capital, have entered the A – share market. By July, 2010, about 28 of these businesses were involved in different areas of low carbon economy, either in terms of their main business, or their future orientation. Altogether, they have received a finance of about ¥12.867 billion. To be specific, the industries they are in and their scope of business cover such areas as new energy, energy conservation and environmental protection, smart grid and new materials. These businesses also have their counterparts in the energy conservation and emission reduction plates on the main boards of Shanghai Stock Exchange and Shenzhen Stock Exchange. These small businesses have succeeded in utilizing both the industrial and financial capital, boast sound market performance and have attracted capital from all corners.

To analyze within the framework of industrial features and investment values, the *Report* has adopted the following sub – industry categorization based on the nature of industrial chains: new sources of energy (wind power, photovoltaic solar power, nuclear power and biomass), energy conservation (energy conservation from frequency conversion and using waste heat and excess pressure), environmental protection (air pollution treatment, waste water treatment and utilization of water resources and solid waste treatment). In each of the category, there are case studies of listed companies in the relevant field. The *Report* has chosen such an approach to explore opportunities for technology development and investment in energy conservation and emission reduction in the country, and to judge the future and expect the prospect of the sector.

In a nutshell, the annual *Report* on the development of energy conservation and emission reduction sets the sector itself as its research subject, gives a general view of the sector, analyzes the status quo for the sector and environmental protection, and tries to present the realities on environmental protection and new energy across industries and regions. We hope that this *Report* can become the most authoritative major annual blue paper in the sector of energy conservation and emission reduction.

特约专稿

低碳发展　绿色发展　和谐发展

冯之浚

“十二五”规划明确提出，我国要以科学发展为主题，以加快转变经济发展方式为主线，并把加快转变经济社会发展方式贯穿于经济社会发展的全过程和各领域。一言以蔽之，目前中国发展的主题可以概括为一个“转”字，即转变、转型、转轨、转制。转变，就是转变经济发展方式，创新发展模式，破解发展难题，提高发展质量；转轨，就是不能再走拼人力物力、拼资源能源、先污染后治理的路子，而是走出一条有别于先期工业化国家的资源节约型、环境友好型的新路；转型，就是政府要实现由计划经济向市场经济的角色调整。当政府在市场经济中担当主角时，政府更关心财政收入，对居民收入重视不够，更关心生产性投资，对公共需求投入重视不够，更关心效率，对公平的重视不够，更关心国有企业，对民营经济重视不够。我们要界定好政府与市场的关系，更加充分地发挥市场“无形之手”的基础性作用；转制，就是要大力推进经济体制改革，加快推进文化体制、社会体制改革，不断完善社会主义市场经济体制，为科学发展和社会和谐提供有力保障。

“转”的核心是转变经济发展方式。改革开放30年多来，中国的经济发展取得了令人瞩目的成就，年均经济增长保持了近10%的快速水平，国内生产总值居世界前列。但同时长期形成的粗放型增长方式尚未根本改变，工业化、城镇化快速发展同能源资源和生态环境的矛盾日趋突出，影响科学发展的体制机制障碍依然存在，转变经济发展方式迫在眉睫。

一、低碳技术在世界范围内方兴未艾

当前，世界经济走出国际金融危机最困难的时期并出现复苏，但同时影响世界经济全面复苏的不确定因素依然较多。面对国际金融危机冲击，尽管我国迅速采取一系列重大应对措施，实现了经济总体回升向好，但长期形成的粗放型增长方式尚未根本改变，工业化、城镇化快速发展同能源资源和生态环境的矛盾日趋突出，影响科学发展的体制机制障碍依然存在。因此，科学应对国际金融危机的影响，不仅仅是提高经济增长速度的问题，更是转变经济发展方式的问题。目前，我国要实现经济发展方式的根本性转变，一要依靠科技进步；二要依靠绿色发展。而发展低碳技术，则是兼具了两条路径的一项共同选择。

历史发展的经验表明，每一次大的经济危机常常伴随着一场新的科技革命。从增长周期角度看，世界经济史上存在着由重大科技创新所主导的周期性经济波动，依靠重大技术创新的形成，催生集群式产业发展，方能走出经济衰退的困局。信息技术自上个世纪引发了重大变革之后，生命科学、生物技术不断取得重大突破，相关的新产品、新产业快速发展，可再生能源、绿色能源正处在替代化石能源的前期，循环经济、低碳技术方兴未艾。在未来一段时期内，这些领域的技术将成为推动全球发展、转变世界经济发展方式的新的动力。

事实上，目前世界上主要国家已普遍意识到金融危机将催生新一轮科技革命。英美欧等发达国家已经在这些领域投入了巨大的力量。就在许多国家受全球金融危机影响纷纷转移精力、削减投入甚而放松减排要求的情况下，英国却宣布启动了一项“绿色振兴计划”，尝试以低碳发展模式从衰退中复苏。目前，英国已初步形成了以市场为基础，以政府为主导，以全体企业、公共部门和居民为主体的互动体系，从低碳技术研发推广、政策发挥建设到国民认知姿态等诸多方面，都处在了世界领先位置。美国总统奥巴马上台以来，在能源政策方面采取了一系列措施，诸如计划在未来五年准备投入1500亿，用于能源新技术方面的大规模投资建设，这是继IT革命之后，美国技术储备的又一个主要方向。2007年底，欧盟提出了战略能源技术计划，该计划是包括欧洲风能、太阳能、生物能和CO_2捕集、运送和贮存以及电网、核裂变等的综合性启动计划。同时，欧盟采取了一系列有力措施推进低碳产业发展，无论从目标制定到科研投入、机制建设、标准与立法、项目投资，都走在世界前列。

可以相信，蓄势待发的低碳技术等新经济形态很有可能会改变未来的世界经济版图。对于我国来说，能否抓住这次历史变革的机遇，将对我国的未来产生极为深远的影响。

二、推行低碳技术是我国转变发展方式的当务之急

作为发展中的大国，我国全力贯彻落实科学发展观，努力建设资源节约型与环境友好型社会，大力倡导循环经济，在中央文件和领导人讲话中，多次提出要将节能减排、推行低碳技术作为国家发展的重要任务。

在我国，推行低碳技术是实现经济社会可持续发展的当务之急。研究指出，2007年，我国消费煤炭约23亿吨，碳基燃料共排放出CO_2达到54.3亿吨，居全球第二。在2007年，我国每建成1平方米的房屋，约释放出0.8吨CO_2；每生产1度电，要释放1公斤CO_2；每燃烧1升汽油，要释放出2.2公斤CO_2。这些数字表明，中国的能源消费处于“高碳消耗”状态，加上中国的化石能源占总能源数量的92%，其中煤炭占68%，电力生产中的78%依赖燃煤发电，而能源、汽车、钢铁、交通、化工、建材等六大高耗能产业的加速发展，使得中国成为“高碳”的典型代表。如果不推行低碳技术，实现经济发展方式的转变，我国的资源能源将难以支撑，生态环境将难以承受，国家竞争力将难以提升，国家安全也将难以保障。这就要求我们必须把推行低碳技术提到国家战略层面上加以思考。

（1）资源能源难以支撑。我国的资源总量和人均资源都严重不足，而资源消耗的增长速度却十分惊人。在资源总量方面，我国石油储量仅占世界1.8%，天然气占0.7%，铁矿石不足9%，铜矿不足5%，铝土矿不足2%。在人均资源量方面，我国人均45种主要矿产资源为世界平均水平的1/2，人均耕地、草地资源为1/3，人均水资源为1/4，人均森林资源为1/5，人均石油占有量仅为1/10。我国的国内资源已难以支撑传统工业文明的持续增长，现实的状况要求我们必须通过加快转变经济发展方式，缓解经济增长中环境代价太大的问题。

（2）生态环境难以承受。当前，我国所面临的环境形势十分严峻。我国现有荒漠化土地面积已占国土总面积的27.9%，而且每年仍在增加1万多平方公里；我国七大江河水系，劣五类水质占27%，75%的湖泊出现不同程度的富营养化；我国600多座城市中有400多座供水不足，

其中100多个城市严重缺水；我国尚有3.6亿农村人口喝不上符合卫生标准的水；2010年环保重点城市空气污染物监测中，有近三成的城市空气质量劣于国家二级标准；全国酸雨面积约占国土面积的12.6%，人民身体健康受到一定损害。我国的生态环境状况已临近生态阈值，难以支撑当前的高污染、高消耗、低效益生产方式的技术扩张。

（3）国家竞争力难以提升。产业要在国际上开拓市场，一般需要经过三个阶段：首先靠劳动力、土地资源、政策优惠发挥“成本优势”；获得一定的资本积累进行研发产生“技术优势”；然后自创名牌，在国际竞争中发挥“品牌优势”。从目前情况看，我国仍处在“成本优势”为主的阶段，必须加强原始自主创新，尽快向“技术优势”和“品牌优势”过渡。若不如此，我们的“打工经济”在生产过程中消耗的大量能源和造成的环境污染等隐形损失，与所获得利润相比将得不偿失。我们生产1台电脑只能赚10个苹果的钱；生产1个鼠标，外国公司拿走了大部分利润，留给我们的只有7.5%；在我国生产的世界名牌领带产量占世界的75%，而利润却只有6%左右；由于不掌握核心技术，我们不得不将每部国产手机售价的20%，数控机床售价的20%—40%用于向国外支付专利费。我国的发展成本高于世界平均水平，严重影响了国家竞争力。特别是我国加入世贸组织以后，国际上各式各样严格的法规和标准接踵而来，如何在日趋激烈的国际竞争中占有一席之地是急需探讨的重大问题。资源环境因素在国际贸易中的作用日益突显，形成了事实上的贸易壁垒，这种“绿色壁垒”成为我国扩大出口面临最多也是最难突破的问题，有的已对我国产品在国际市场的竞争力造成了严重影响。

（4）国家安全难以保障。目前，我国对外技术依存度高达50%，美国、日本仅为5%左右；我国的设备投资有60%以上要靠进口。更何况涉及战略需求的核心技术，就是花再多的钱别人也不会卖给我们。同时，我国重要资源对外依存度也不断上升。目前，我国先进技术和主要资源的对外依赖度已经达到了一个相当高的水平，国家的能源安全、经济安全受到严重的威胁。

当然，我们应该看到，低碳概念是发达国家在保护全球气候的名义下有利于后工业化国家的策略，其用意是十分深刻的。一方面，发达国家通过将高碳模式转换到低碳模式中，在这一新的平台上和新的游戏规则下，利用其低碳技术方面的领先地位重新拉大与发展中国家的经济竞争优势，成为西方国家的一个巧妙战略；另一方面，发达国家以低碳为名，实施碳关税、技术性贸易壁垒和市场准入等贸易保护主义的倾向愈演愈烈，限制发展中国家发展的图谋已暴露无遗。发达国家虽有环境保护、温室减排之意，但其更深刻的目的却是要强行将发展中国家绑架到他们精心设计的轨道上去。

因此，将“低碳”这一名词引入我国，在实施中必须注意与中国的实际相结合，既吸收有关新能源、技术创新等积极内容，又要充分体现我国的国情、发展阶段和利益。目前，我国所处的发展环境可以概括为以下四个方面：

第一，随着经济全球化的深入发展，国际产业分工加快重组，国际产业结构不断调整，我国由于经济整体素质不高，在国际产业链中，处于不利的分工地位。我国在国际产业分工与竞争中，缺乏自主知识产权，缺少世界知名品牌，出口商品中90%是贴牌产品，不得不依靠廉价劳动力比较优势换来微薄的利润，成为低端产品的“世界工厂”，承受随之而来的资源、环境代价。

第二，我国进入工业化中期，这一时期产业结构的特点是“重化工业”加速发展，导致资源能源消费增加。先期实现工业化的国家，如英国、美国大都走过一条大量消耗资源能源、先污染后治理的路子，能源和重要金属消费增长超过经济增长，出现非常严重的环境污染。当年日本也是如此，1960—1970 年日本在加速工业化过程中，能源消费弹性系数为 1.21，是日本环境问题最严重的时期。我国已开始实行新型工业化的政策，坚持走出一条有别于先期工业化国家的资源节约型、环境友好型的新路。我们必须重视重化工业的层次提升和节能减排的结构调整与技术进步，避免少数地方盲目发展和引进重化工业中的高能耗、高污染项目，造成资源能源的双重压力。

第三，中国特色社会主义道路要求工业化与城镇化同时并举。西方发达国家，在实现现代化的过程中，工业化与城镇化之间，是有一段间隔的。我们根据中国国情，更好地解决“三农”问题，实现城乡统筹发展，需要两者兼顾。工业化进程的加速，如果不提高科技含量，调整产业结构，势必导致资源能源消费量迅速增长，城镇化进程加快，城市人口增加，基础设施规模加大，人均生活能源消费量较快增长，加之，生活模式明显改变，消费结构不断提升，以住房和汽车消费为标志的能源消费升级，是导致能源消费强度上升的主要原因之一。

第四，政府职能的改革与市场经济的完善。如何配置资源合理利用能源，减少环境污染既是政府的职能，也是市场经济完善的表征之一，应避免政府失灵和市场失灵并存的现象。我们过去迷信计划经济，认为计划经济能够有力地解决经济增长方式的转变，随着计划经济色彩的逐渐淡化，政府职能如何转变成为改革中的重要问题。不能过于迷信市场，市场并非万能的，它对自动配置资源、合理利用能源的调节是有限的，因此，对市场经济必须要加以引导和完善。

在我国谈论低碳概念和低碳技术不应以单纯的碳减排为核心，应具有更广泛的内涵，要以节能减排为核心，重点解决提高能源利用效率，发展可再生能源，提高环境质量，为中国的经济转型和结构调整打好基础，避免走发达国家曾经走过的道路，而是要走出一条具有中国自身特色的低碳发展之路。我们在推进新能源等技术研发的同时，必须兼顾传统工业生产能效与资源利用效率的提高，对其加以绿色改造，开发适合我国国情的能源和环境新技术、新工艺、重大装备，为发展绿色经济提供技术支撑。

三、走中国特色的低碳发展之路

中国特色的低碳发展之路必须要以科学发展观为指导，坚持全面、统筹、可持续的原则。具体来说包括以下四个方面：

一是倡导以可持续发展为核心的绿色发展。在绿色经济的框架下，推动包括循环经济、低碳技术、生态经济、合理消费等在内各领域的发展。把低碳技术的概念落在节能减排、环保、能源效率提高上，落在传统产业的改造和提升上，落在产业结构的调整和升级上。

二是中国的低碳技术要建立在科技创新的基础上。我们应根据我国的经济社会发展状况和国际社会未来应对气候变化的趋势，形成具有自主知识产权的先进技术、适用技术的层次体系，做到各种技术之间互相协调。不仅要加大适用技术的推广应用力度，也要着力加强新能源技术、清洁煤技术、节能降耗等战略性技术的研发和储备。

三是通过低碳技术和产业政策，扬长避短，改造提升传统工业，培育战略性新兴产业。推动低碳技术涉及国民经济各个部门和各个行业，关系到经济结构的调整，传统产业的改造提升和战略性新兴产业的培育，关系到国际的科技竞争和市场竞争。低碳技术必须“魂要附体”，而传统产业以及新兴产业正是这个“体”。

四是为应对西方国家所可能采取的措施有所准备。一方面我们要自主实现节能减排，服务我们自身的发展；另一方面，也应充分考虑对碳关税等贸易保护措施的反制措施，在全球变化经济博弈中不受损失或少受损失。

（冯之浚：国务院参事，中国可持续发展研究会名誉理事长）

大力推进节能减排　加快建设“两型”社会

杨　洁

节能减排是事关中华民族长远发展的大计和当前一项十分紧迫的任务，是贯彻落实科学发展观、推动经济结构调整和发展方式转变的重要抓手。改革开放30多年来，我国经济发展保持年均10%左右的增长速度，综合国力和人民生活水平不断提升，但也付出了巨大的资源和环境代价，经济发展与资源环境的矛盾日趋尖锐，环境污染问题正在成为影响社会和谐的重要制约因素。如果节能减排和环境保护问题不能妥善解决，必然使资源难以支撑，环境难以容纳，社会难以承受，发展难以持续，中国经济不但走不好，也走不远。为解决资源环境约束日益加重的问题，“十一五”规划《纲要》提出了“十一五”期末单位国内生产总值能源消耗比“十五”期末降低20%左右、主要污染物排放总量减少10%的节能减排目标，并首次将其作为国民经济和社会发展的约束性指标。这一重要目标的设定，突出体现了建设资源节约型、环境友好型社会和实现可持续发展的要求。面对日趋强化的资源环境约束，“十二五”规划《纲要》进一步明确指出：未来五年，要坚持把建设资源节约型、环境友好型社会作为加快转变经济发展方式的重要着力点，以节能减排为重点，健全激励与约束机制，加快构建资源节约、环境友好的生产方式和消费模式，增强可持续发展能力，提高生态文明水平。

一、立足节约资源保护环境推动发展，“十一五”节能减排成效显著

节约资源和保护环境是我国的基本国策。“十一五”期间，国务院成立了节能减排工作领导小组，发布了加强节能工作的决定和节能减排综合性工作方案，制定了促进节能减排的一系列政策措施，确保实现节能减排约束性指标，推动经济社会又好又快发展。各地区、各部门认真贯彻落实党中央、国务院的决策部署，把节能减排作为调整经济结构、转变经济发展方式、推动科学发展的重要抓手和突破口，作为加强和改善宏观调控，促进经济社会又好又快发展的重要目标，严格责任考核，加大资金投入，完善政策机制，加强综合协调，扎实推进节能减排、生态建设和环境保护。

（1）落实目标责任，强化监督检查。为把节能降耗和污染减排两大约束性指标落到实处，国务院组织制定了《“十一五”各地区万元GDP能耗降低指标分解方案》和《“十一五”期间全国主要污染物排放总量控制计划》，并将“十一五”节能减排目标分解落实到省级人民政府、千家高耗能企业和部分中央企业。各省（区、市）也将“十一五”节能减排目标进一步分解落实到市（地）县和重点排污单位。2008年以来，组织开展了省级政府和千家企业年度节能目标责任评价考核、省级政府主要污染物总量减排核查核算，并向社会公告了考核结果。国务院发

布了节能减排统计监测及考核实施方案和办法，将节能减排完成情况作为政府领导干部综合考核评价和企业负责人业绩考核的重要内容。目前，责任清晰、目标明确、逐级考核的节能减排目标责任制基本形成。为确保“十一五”节能减排目标按时完成，国务院派出督查组对18个重点地区进行节能减排专项督查，全国人大组成检查组对重点地区组织开展了节约能源法执法检查、节能减排专项督查和环保专项行动。

（2）增加资金投入，实施重点工程。“十一五”期间，中央政府投入资金2000多亿元，全国节能环保投入约达1.6万亿元，比“十五”期间增加了70%，大力支持节能减排。其中，安排中央专项资金，支持十大重点节能工程、节能产品惠民工程、城镇污水处理设施及配套管网建设、城市矿产示范基地建设和半导体照明应用示范工程；安排中央财政补贴资金，推广节能灯、高效节能空调、节能汽车以及高效节能电机等先进节能技术和产品；中央财政采取“以奖代补”、“以奖促治”、转移支付等方式支持节能减排重点项目。

（3）加大政策支持，完善法规标准。一方面，强化产业政策和项目管理，提高节能环保准入门槛，严把土地、环保、信贷等关口。近年来，采取的经济政策和措施主要有：出台《固定资产投资项目节能评估和审查暂行办法》，将节能评估和审查作为项目审批、核准以及开工建设的前置条件；调整或取消产品出口退税及配额，清理和纠正高耗能行业优惠电价，严格控制高耗能、高排放行业和产能过剩行业扩大产能项目；实施成品油价格和税费改革，实行烟气脱硫机组上网电价政策，对限制类、淘汰类的高耗能产品实施差别电价政策；对节能节水环保设备给予税收优惠；在部分省市开展排污权有偿使用和交易试点；扩大能效标识、节能产品认证和政府采购实施范围；加快淘汰落后产能，“十一五”关停小火电机组7000多万千瓦、炼铁产能超过1亿吨、水泥产能超过2.6亿吨，节能约1.3亿吨标准煤。另一方面，修订颁布了节约能源法、水污染防治法，制定了《民用建筑节能条例》和《公共机构节能条例》。为切实降低高耗能行业能源消耗、推动企业提高能源利用水平，“十一五”期间，发布近30项高耗能产品能耗限额强制性国家标准、40余项主要终端用能产品强制性能效标准和20多项污染物排放标准，实行主要用能产品能耗限额标准；制定发布了节能汽车推广目录和国家重点节能技术推广目录、国家重点行业清洁生产技术导向目录和评价指标体系，引导企业采用先进的清洁生产工艺和技术，积极防治工业污染。

（4）加强生态建设和污染防治，积极应对气候变化。“十一五”期间，继续推进天然林资源保护、京津风沙源治理、重点防护林体系建设，以及三江源自然保护区生态保护和建设工程，巩固退耕还林、退牧还草等重点生态建设工程成果。全面推进太湖流域水环境综合治理和渤海、三峡库区、丹江口库区等重点流域环境治理，以及尾矿库隐患治理、铬渣污染综合整治和节能环保能力建设。“十一五”期间，国家大力推动火电脱硫发展，以燃煤电厂安装脱硫设施为重点，在新上机组必须安装脱硫设施的同时，组织实施火电行业老机组脱硫改造。截至2010年底，全国已投运烟气脱硫机组超过5.6亿千瓦，约占全国煤电机组容量的86%，火电脱硫成效明显。与此同时，把推进节能减排和应对气候变化紧密结合起来，实施应对气候变化国家方案，采取优化能源结构、发展可再生能源、增加森林碳汇、开展低碳试点等一系列措施，努力控制温室气体排放。目前，我国水电装机容量、风电新增装机容量、太阳能热水器集热面积均居世界第一，非化石能源占一次能源消费的比重不断提高。“十一五”通过节能减排提高能效少消耗能源6.3亿吨标准煤，减少二氧化碳排放14.6亿吨，为应对全球气候变化作出了重要贡献，彰显了中国负责任的大国形象。

“十一五”期间，我国节能减排力度之大世界少有，成效之显著居各国前列：全国单位国内生产总值能耗下降了19.1%，全国二氧化硫和化学需氧量排放总量分别下降了14.29%和12.45%，实现了“十一五”规划《纲要》确定的约束性目标；以能源消费年均6.6%的增速支撑了国民经济年均11.2%的增速，能源消费弹性系数由“十五”时期的1.04下降到0.59；实施十大重点节能工程，形成节能能力2.6亿吨标准煤；开展千家企业节能行动，节能1.5亿吨标准煤；关停小火电机组7000多万千瓦、淘汰炼铁产能超过1亿吨、水泥产能超过2.6亿吨；推广节能灯3.6亿只以上，高效节能空调、高效电机、节能汽车都在大力推广之中。节能减排目标的完成，不仅缓解了能源供需矛盾，改善了环境质量，更重要的是促进了产业结构升级和节能减排技术进步，扭转了我国工业化、城镇化加快发展阶段能源消耗强度和污染物排放大幅上升的势头，为保持经济平稳较快发展提供了有力支撑，也为实现“十二五”节能减排目标奠定了坚实基础。

近年来的实践充分表明，节能减排目标的完成，是各方面共同努力、多种举措共同作用的结果。实现节能减排约束性目标，首先，要坚持科学发展，更加注重节约资源和保护环境，加快经济结构优化和企业转型升级，发挥节能减排对优化经济结构、推动节能技术进步的促进作用，不断提高发展的质量和效益，增强产业竞争力和可持续发展能力。其次，要加快构建以企业为主体、政府为主导、市场有效驱动、全社会共同参与，适应社会主义市场经济要求的节能减排长效机制。既要切实发挥政府主导作用，综合运用经济、法律、技术和必要的行政手段，将落实责任、加强法制与完善政策有机结合起来，形成有效的激励和约束机制；更要进一步发挥市场机制作用，加大节能减排市场化机制推广力度，真正把节能减排转化为企业和各类社会主体的内在要求。最后，要在全社会营造重视资源节约和环境保护的舆论氛围，倡导文明、节约、绿色、低碳消费理念，使资源节约、环境保护和发展循环经济的理念深入人心，推动形成与我国国情相适应的绿色生活方式和消费模式。

二、坚持不懈地推进节能减排，着力提高生态文明水平

“十二五”时期，随着工业化、城镇化进程的进一步加快和消费结构的持续升级，我国能源需求呈刚性增长，加之受国内资源保障能力和环境容量制约以及全球性能源安全和应对气候变化影响，经济增长的资源环境约束日趋强化，节能减排形势依然十分严峻。要实现“十二五”规划《纲要》确定的单位国内生产总值能耗降低16%，二氧化碳排放降低17%，化学需氧量、氨氮、二氧化硫、氮氧化物四种主要污染物下降8%—10%的约束性指标的目标，任务非常艰巨。特别是我国节能减排工作还存在责任落实不到位、推进难度增大、激励约束机制不健全、基础工作薄弱、能力建设滞后、监管不力等问题。这种状况如不及时加以改变，不但“十二五”节能减排目标难以如期实现，经济结构调整和经济发展方式转变也难以取得实质性进展。因此，做好“十二五”时期的节能减排工作，关键是要真正把思想和行动统一到中央的决策部署上来，树立绿色低碳发展理念，进一步把节能减排作为落实科学发展观、加快转变经济发展方式的重要抓手，作为检验经济是否实现又好又快发展的重要标准，下更大决心，用更大气力，采取更加有力的政策措施，大力推进节能减排，加快形成资源节约、环境友好的生产方式和消费模式，增强可持续发展能力。

第一，强化节能减排目标责任考核，健全节能减排激励约束机制。建立健全科学的节能减

排统计、监测和考核体系。综合考虑地区经济发展水平、产业结构、节能潜力、环境容量等因素，科学确定节能减排五年目标与年度目标，并合理分解落实到各地区、各行业。明确企业的节能减排主体责任，进一步完善节能减排考核办法，健全奖惩制度，强化监督检查，严格考核问责。完善节能减排政策体系，健全激励机制。深化资源性产品价格改革，健全污水垃圾处理费征收和管理制度，加强对节能环保项目的财政支持和税收优惠。

第二，加强节能减排重点工程建设，加快节能减排技术开发和推广应用。加快实施重点节能改造工程、节能技术产业化示范工程、节能产品惠民工程、合同能源管理推广工程、污染物减排重点工程和循环经济重点工程。当前，重点推进城镇污水处理设施及配套管网建设，加强重点流域区域污染综合治理；实施脱硫脱硝工程，推动燃煤电厂、钢铁行业烧结机脱硫工程和节能减排重大技术与装备产业化工程，以及资源综合利用、废旧商品回收体系、“城市矿产”示范基地、再制造产业化、餐厨废弃物资源化、产业园区循环化改造、资源循环利用技术示范推广等循环经济重点工程。

第三，强化节能减排管理，提高节能增效整体水平。建立能源消费总量控制目标分解落实机制，合理控制能源消费总量。完善节能法规和标准，制定完善并严格执行主要耗能产品能耗限额和产品能效标准，严格执行固定资产投资项目节能评估审查制度。开展能耗总量控制和节能量交易试点。加快推行合同能源管理和电力需求侧管理，完善能效标识、节能产品认证和节能产品政府强制采购制度。推广先进节能技术和产品。加强节能能力建设。突出抓好工业、建筑、交通运输、公共机构等领域的用能管理和重点耗能企业的节能管理，开展万家企业节能低碳行动，实施重点行业能效水平对标，公告能源利用状况。推行产品“领跑者”能效标准、能效标识等市场化节能管理机制。

第四，大力发展循环经济，加快构建覆盖全社会的资源循环利用体系。健全中央财政支持循环经济发展的政策体系，推行循环型生产方式。深化循环经济示范试点，推广循环经济典型模式。完善循环经济指标体系，推动建立循环经济统计评价制度。发布重点行业清洁生产推行方案，实施清洁生产示范工程，推广应用清洁生产技术，全面推行清洁生产。推进大宗固体废物综合利用和海水淡化，加快资源再生利用产业化、再制造产业化，促进垃圾资源化利用。

第五，强化污染物减排和生态治理，积极应对气候变化。实施主要污染物排放总量控制。实行严格的饮用水水源地保护制度，提高集中式饮用水水源地水质达标率。推进重点流域、重点区域、重点城市环境保护和生态治理，改善人居环境。加大高污染行业二氧化硫和氮氧化物治理力度。建立健全区域大气污染联防联控机制，控制区域复合型大气污染。加强天然林保护、退耕还林、退牧还草、水土流失治理、湿地保护、荒漠化石漠化治理等生态工程建设，强化对水源、土地、森林、草原、海洋、生物等自然资源，以及自然保护区、重要生态功能区、海岸带等的生态保护，改善生态环境。抓好低碳省市试点和重点行业低碳技术产业化，开展碳排放交易试点和低碳产品认证标识试点，研究制定碳排放行业标准。

第六，深入开展节能减排全民行动，推进全社会节能减排。加强节能减排宣传教育，进一步增强全社会的能源忧患意识、资源节约和环境保护意识，倡导健康文明、节约环保、绿色低碳的消费模式和生活习惯，形成全社会共同参与、共同促进节能减排的良好氛围。

（杨洁：国家发展改革委政策研究室副主任）

目录

第1篇　综合篇

第2篇　行业篇

第3篇　区域篇

第4篇　环保篇

第5篇 新兴能源篇

第6篇　技术篇

第7篇　数据篇

Contents

第1篇　综合篇

“十一五”期间，我国基本实现了在五年前制定的节能减排目标，经济增长在逐步摆脱对于能源的过度依赖，为应对全球气候变化作出了重要贡献。这主要得益于一系列的政府强力推动政策，例如大力淘汰落后产能、开展十大重点节能工程、千家重点能耗企业节能行动等。

“十二五”时期，我国发展仍处于可以大有作为的重要战略机遇期。随着工业化、城镇化进程加快和消费结构持续升级，我国能源需求呈刚性增长，受国内资源保障能力和环境容量制约以及全球性能源安全和应对气候变化影响，资源环境约束日趋强化，“十二五”时期节能减排形势仍然十分严峻，任务十分艰巨。特别是我国节能减排工作还存在责任落实不到位、推进难度增大、激励约束机制不健全、基础工作薄弱、能力建设滞后、监管不力等问题。这种状况如不及时改变，不但“十二五”节能减排目标难以实现，还将严重影响经济结构调整和经济发展方式转变。

与此同时，全球2010年能源消费在经济复苏的推动下经历了一波反弹。这使得2009年全球能源消费的降低并未能成为中长期趋势。就更进一步的预测而言，全球能源需求总量不论是到2035年，还是到2050年，均会有大幅度增长。其中，发展中国家强劲经济增长带来的能源需求将占据主要位置。

本篇我们将描述中国节能减排整体状况，主要对“十一五”节能减排指标实现的历程进行回顾、评估，对“十二五”节能减排的目标进行展望，希望我国各级政府、部门能够严格落实节能减排目标责任，进一步形成政府为主导、企业为主体、市场有效驱动、全社会共同参与的推进节能减排工作格局。同时，向读者介绍2010年度全球能源消费以及低碳经济、气候谈判等的最新进展。

1 中国“十一五”节能减排回顾

1.1 “十一五”节能指标实现情况及历程

“十一五”期间，我国的节能主要目标是到2010年，万元国内生产总值能耗由2005年的1.22吨标准煤下降到1吨标准煤以下，降低20%左右。随着国家和各地区节能降耗工作力度的不断加大，各项政策措施逐步深入落实，节能降耗取得明显成效。

“十一五”期间，我国能源消费及能源节约方面的主要数据如下：

- 为应对全球气候变化作出了重要贡献。“十一五”期间通过节能提高能效少消耗能源6.3亿吨标准煤，减少二氧化碳排放14.6亿吨。
- 统计数据显示，2010年，全国能源消费总量达32.5亿吨标准煤，同比增长6%。相对于2005年消费23.6亿吨标准煤而言，增加能源消费10.8亿吨标准煤，增加46%。按照2011年同比继续增长6%来估算，2011年能源消费总量将达到34.4亿吨标准煤，如图1－1所示。
- 我国在“十一五”期间的经济增长已经逐步摆脱对于能源的过度依赖，2006—2010年的经济平均增速是11.2%，能源消费平均增速是6.6%（如图1－2），能源消费弹性系数由最高点2004年的1.6已经降至2010年的0.57，如图1－3所示。

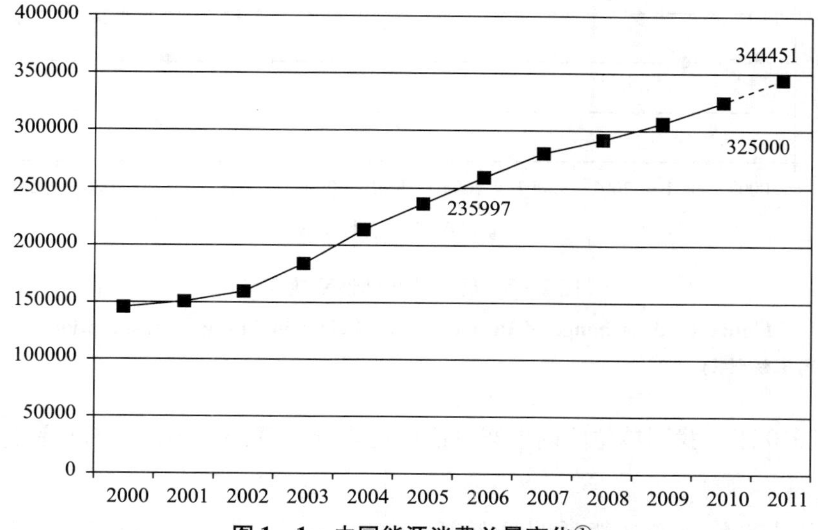

图1－1 中国能源消费总量变化①

Figure 1－1 Change in China's Total Energy Consumption

资料来源：国家统计局

---

① 本书图中的虚线部分数值为估计值，后文不再逐一说明。

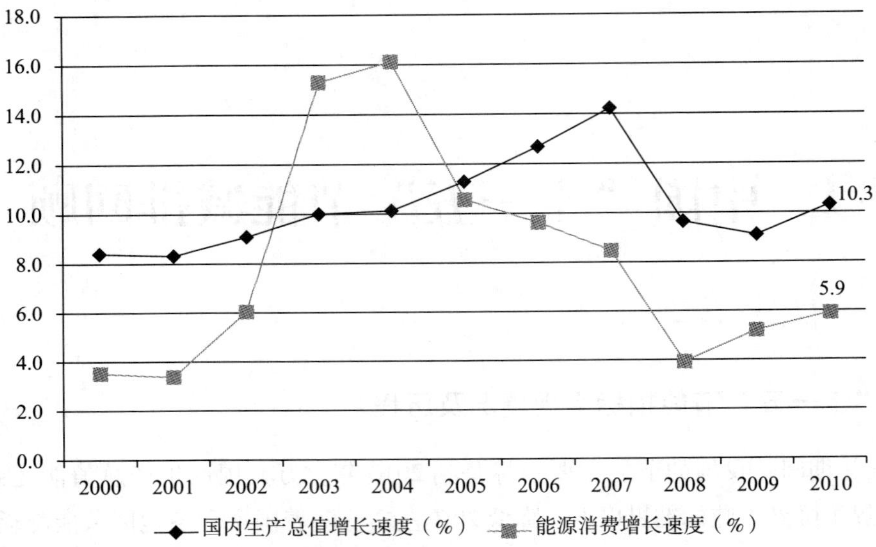

图 1-2 国内生产总值增速与能源消费增速比较

Figure 1-2 GDP Growth vs. Energy Consumption Growth

资料来源：国家统计局

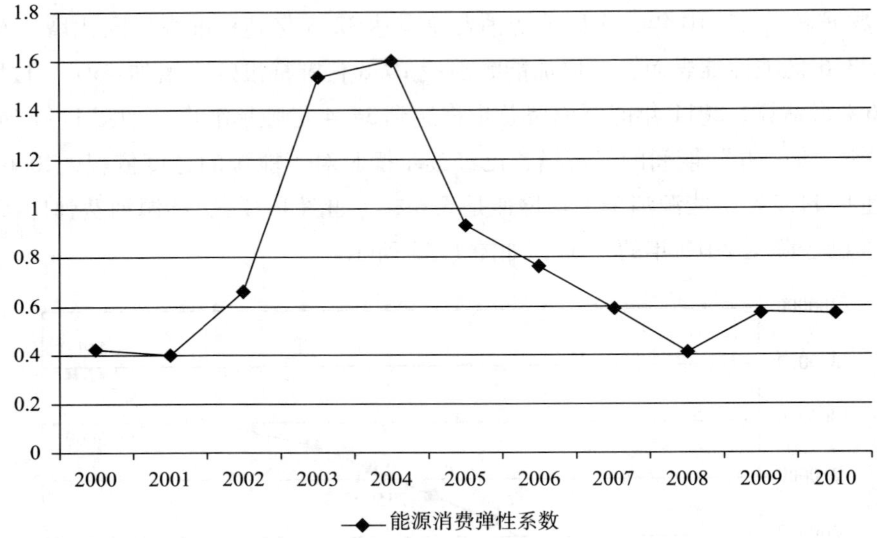

图 1-3 能源消费弹性系数变化

Figure 1-3 Change of Resilience Coefficient in Energy Consumption

资料来源：国家统计局

- 2006—2010 年，我国单位国内生产总值能耗累计下降 19.06%，基本完成“十一五”节能降耗目标。

主要耗能产品的单位产品能耗明显下降。“十一五”期间，单位铜冶炼综合能耗下降 35.9%，单位烧碱生产综合能耗下降 34.8%，吨水泥综合能耗下降 28.6%，原油加工单位综合能耗下降 28.4%，电厂火力发电标准煤耗下降 16.1%，吨钢综合能耗下降 12.1%，单位电解铝综合能耗下降 12.0%，单位乙烯生产综合能耗下降 11.5%。

促进了结构优化升级。重点行业先进生产能力比重明显提高，大型、高效装备得到推广应用。2009 年与 2005 年相比，电力行业 300 兆瓦以上火电机组占火电装机容量比重由 47% 上升到 69%，钢铁行业 1000 立方米以上大型高炉比重由 21% 上升到 34%，电解铝行业大型预焙槽产量比重由 80% 上升到 90%，建材行业新型干法水泥熟料产量比重由 56.4% 上升到 72.2%。

但是，节能的历程并不是一帆风顺，期间经历了诸多波折与往复。以下按时间顺序回顾我国单位 GDP 能耗指标的实现历程①：

2006 年是推进节能减排工作的起始年，也是阻力最大的一年。这种阻力主要来自我国粗放式经济发展所形成的偏重高耗能投资的巨大惯性，使得产业结构调整进展缓慢。同时，在各项支撑条件没有完全到位、基础工作和管理力量薄弱的前提下，节能减排工作并未得到地方的足够重视。因此，2006 年单位 GDP 能耗仅下降了 2.74%，4% 的既定目标未能达到。

随后，中央进一步强化了节能减排法律和政策支撑，九大措施的开展特别是落实节能减排目标责任制，使得 2007 年的单位 GDP 能耗调控开始渐入正轨，当年单位 GDP 能耗下降了 5.04%。

2008 年是调控单位 GDP 能耗的关键之年，全年单位 GDP 能耗下降了 5.2%，但 2008 年为应对国际金融危机出台的产业振兴计划和经济刺激计划又对后来的节能减排工作产生了一定影响。

由于经济振兴计划主要针对制造业，而经济刺激计划中的 4 万亿元投资也基本用于基础建设，这使得单位 GDP 能耗的下降幅度自 2009 年起开始逐季度减缓：2009 年一季度下降 2.89%，上半年累计下降 3.35%，2009 年全年仅下降 3.61%。

进入 2010 年以后，高耗能产品需求持续旺盛，落后产能复产，这些都使得 2010 年一季度单位 GDP 能耗不降反升 3.2%，虽然在二季度升幅得到了控制，但是，2010 年上半年仍然同比上升了 0.09%，导致了 2010 年下半年需要完成超过全年的降耗任务。

1.2 “十一五”节能领域重要举措及效果

“十一五”期间，我国能够基本实现期初制定的节能目标，主要在于采取了一系列针对性的措施，主要举措及效果如下：

1.2.1 淘汰高排放的落后产能成效突出

“十一五”期间，全国累计淘汰炼铁落后产能约 11172 万吨，炼钢落后产能约 6683 万吨，焦炭落后产能约 10538 万吨，铁合金落后产能约 663 万吨。主要政策措施如下：

一是分解落实目标任务，加大奖励惩罚力度。每年将目标任务分解到市、县和具体企业。对完成淘汰落后产能任务较好的地区和企业，在资金、土地、融资等方面给予倾斜，中央财政设立专项资金对经济欠发达地区淘汰落后产能给予奖励。对未按期完成淘汰落后产能任务的地区，暂停对该地区项目的环评、核准和审批；对未按要求淘汰落后产能的企业，依据有关法律法规责令停产或予以关闭。

二是完善相关技术标准，坚决淘汰落后产能。制定实施主要用能产品能耗限额标准，对达不到能耗限额标准的落后产能坚决予以淘汰。切实加强对落后产能企业执行环境保护标准、产品质量标准、能耗限额标准和安全生产规定的监督检查。

---

① 宏观经济研究院国地所课题组卢伟《“十一五”我国节能减排工作回顾及“十二五”政策建议》。

三是充分发挥市场机制作用，提高落后产能企业成本。推进资源性产品价格改革，提高土地使用价格，实行差别电价，提高差别电价加价标准。对超过限额标准的，实行惩罚性价格政策。

四是确保社会稳定，高度重视职工安置工作。研究制定职工安置措施，认真落实安置政策，通过多种渠道、多种手段，促进淘汰落后产能企业职工再就业。

1.2.2 十大重点节能工程取得了良好的经济和社会效益

十大重点节能工程[①]是实现“十一五”单位GDP能耗降低20%左右目标的一项重要的工程技术措施，目标是在“十一五”期间实现节能2.4亿吨标准煤。“十一五”期间，十大重点节能工程的实施取得了良好的经济和社会效益：

一是大幅度提高了能源利用效率。2010年与2005年相比，火电供电煤耗由370克标准煤/千瓦时降到333克标准煤/千瓦时，下降了10.0%；吨钢综合能耗由694千克标准煤降到605千克标准煤，下降了12.8%；水泥综合能耗下降了24.6%；乙烯综合能耗下降了11.6%；合成氨综合能耗下降了14.3%。

二是促进了先进节能技术的推广应用。2010年与2005年相比，钢铁行业干熄焦技术普及率由不足30%提高到80%以上，水泥行业低温余热回收发电技术由开始起步提高到55%，烧碱行业离子膜法烧碱比重由29.5%提高到84.3%。新型阴极铝电解槽、高压变频、稀土永磁电机、等离子无油点火等一大批高效节能技术和产品得到普遍应用。

三是促进了节能环保产业发展。我国高效照明产品、家用电器、电机、新型节能墙材等节能设备和产品的市场规模得到大幅度提升，节能环保装备的研发和制造水平显著提高。

1.2.3 千家企业节能行动[②]节能1.5亿吨标准煤

“十一五”期间，千家企业单位氧化铝综合能耗、乙烯生产综合能耗、烧碱生产综合能耗等指标下降了30%以上，单位原油加工综合能耗、电解铝综合能耗、水泥综合能耗等指标下降了10%以上，供电煤耗下降近10%，部分企业的指标达到了国际先进水平。“十一五”以来，围绕千家企业节能行动，国家主要采取以下措施：

一是对千家企业节能行动全面部署。国家发展改革委等部门印发了实施方案，明确了工作要求和保障措施，公布了千家企业名单。

二是开展目标责任考核。国家发展改革委与地方政府签订了千家企业节能目标责任书，各地分别与本地所属千家企业签订了节能目标责任书，每年组织考核，结果向社会公告。

三是开展企业能源审计和编制节能规划。千家企业均按要求编制实施了能源审计和节能规

---

① 十大重点节能工程包括：燃煤工业锅炉（窑炉）改造工程、区域热电联产工程、余热余压利用工程、节约和替代石油工程、电机系统节能工程、能量系统优化工程、建筑节能工程、绿色照明工程、政府机构节能工程、节能监测和技术服务体系建设工程。

② 千家企业节能行动是指国家对年综合能源消费量18万吨标准煤以上约一千家企业加强节能管理，其能源消费量占工业能源消费量的一半，占全国能源消费量的1/3。2006年4月，国家发展改革委会同有关部门启动了千家企业节能行动，目标是“十一五”期间节能1亿吨标准煤左右。通过对千家中953家企业节能目标完成情况和节能措施落实情况考核的综合评分，391家考核结果为超额完成等级，占41%；456家考核结果为完成等级，占47.8%；32家考核结果为基本完成等级，占3.4%；74家考核结果为未完成等级，占7.8%。“千家企业节能减排行动”成为推动全国节能减排工作的重要抓手，在全国起到了良好的示范作用。

划，各地组织专家对能源审计报告集中进行审核，对不合格的报告要求企业进行整改。

四是实施能源利用状况报告制度。研究制定了《重点用能单位能源利用状况报告制度实施方案》，要求重点用能单位每年报送能源利用状况报告。发布了《千家企业能源利用状况公报》。

五是开展能效水平对标活动。印发了《重点耗能企业能效水平对标活动实施方案》，部署实施能效水平对标活动，在钢铁、化工、水泥三个高耗能行业开展对标试点，编制了《重点耗能行业能效对标指南》。

六是积极推进千家企业节能技术改造。国家加大了对千家企业节能技术改造项目的支持力度，通过节能技术改造，千家企业单位产品综合能耗指标大幅度下降。

七是强化节能管理。95%以上的千家企业建立了专门的能源管理机构，配备了相关能源管理人员，完善三级计量仪器、仪表配备，在部分企业开展能源管理师试点。

1.2.4　重点行业节能减排成果显著

“十一五”期间，许多传统的重点耗能行业节能减排成果显著，不少行业提前、超额完成本行业的“十一五”规划约束目标。

电力行业于2009年底提前完成“十一五”节能减排规划目标，为全国节能减排目标的实现作出了重要贡献。2009年与2005年相比，火电供电煤耗由370克/千瓦时降到340克/千瓦时，下降了8.11%。

“十一五”前四年，煤炭行业少消耗了4.9亿吨标准煤，减少二氧化碳排放11.3亿吨。

2009年与2005年相比，钢铁行业的吨钢综合能耗由694千克标准煤降到615千克标准煤，下降了11.4%。

1.2.5　节能服务产业快速发展

“十一五”期间，我国节能服务产业迅速发展，专业化节能服务公司迅速壮大、产业规模大幅增长、服务范围不断扩展、服务水平显著提高，节能服务公司已成为我国节能战线上一支重要力量。2010年与2005年相比，节能服务公司从80多家增加到800多家，从业人员从1.6万人增加到18万人，节能服务产业规模从47亿元增加到840亿元，合同能源管理项目投资从13亿元增加到290亿元，形成年节能能力从60多万吨标准煤增加到1300多万吨标准煤。

“十一五”期间，节能服务产业拉动社会投资累计超过1800亿元。为加快节能服务产业发展，2010年4月，国务院制定了《关于加快推行合同能源管理促进节能服务产业发展的意见》，为合同能源管理推广创造了良好的政策和体制环境。

第一，加大资金支持力度。将合同能源管理项目纳入中央预算内投资和中央财政节能减排专项资金支持范围。出台了《合同能源管理财政奖励资金管理暂行办法》，2010年中央财政共安排20亿元，支持节能服务公司利用合同能源管理方式实施节能改造。

第二，实行税收扶持政策。节能服务公司实施的合同能源管理项目，暂免征收营业税，所得税三免三减半，对合同期满后节能服务公司无偿转让给用能单位因实施合同能源管理项目形成的资产，免征增值税。用能企业按照合同支付给节能服务公司的合理支出，均可以在计算当期应纳税所得额时扣除。

第三，完善相关会计制度。各级政府机构采用合同能源管理方式实施节能改造，按照合同支付给节能服务公司的支出视同能源费用进行列支。事业单位采用合同能源管理方式实施节能改造，按照合同支付给节能服务公司的支出计入相关支出；企业采用合同能源管理方式实施节能改造，按照合同支付给节能服务公司的支出作为费用列支。

第四，进一步改善金融服务。鼓励银行等金融机构根据节能服务公司的融资需求特点，创新信贷产品，拓宽担保品范围，简化申请和审批手续，为节能服务公司提供项目融资、保理等金融服务。节能服务公司实施合同能源管理项目投入的固定资产可按有关规定向银行申请抵押贷款。

1.2.6 “节能产品惠民工程”取得明显成效

为扩大消费需求，提高能源效率，促进产业升级，2009 年 6 月，国家发展改革委、财政部组织实施了“节能产品惠民工程”，以财政补贴方式推广高效节能空调、节能汽车、节能灯、三相异步电动机和稀土永磁电动机等高效节能产品，目前已形成家用电器、交通工具、照明产品、工业设备等四大类高效节能产品推广体系。

1.3 “十一五”减排目标实现情况及历程

“十一五”期间，我国的节能减排主要目标是：到 2010 年，全国化学需氧量和二氧化硫排放量分别比 2005 年下降 10%，即化学需氧量排放量由 2005 年的 1414.2 万吨减少到 1272.8 万吨，二氧化硫排放量由 2005 年的 2549.4 万吨减少到 2294.4 万吨。截至 2010 年，化学需氧量和二氧化硫排放量分别比 2005 年下降 12.45% 和 14.29%，双双超额完成“十一五”减排任务。

其他环境方面主要数据如下：

• 扭转了我国工业化、城镇化加快发展阶段能源消耗强度和污染物排放大幅上升的势头。“十五”后三年全国单位 GDP 能耗上升了 9.8%，全国二氧化硫和化学需氧量排放总量分别上升了 32.3% 和 3.5%；“十一五”期间，全国单位 GDP 能耗下降了 19.1%，全国二氧化硫和化学需氧量排放总量分别下降了 14.29% 和 12.45%。

• “十一五”期间，环境质量持续改善。2010 年，七大水系的水质监测断面中，Ⅰ～Ⅲ类水质断面比例占 59.6%，比 2005 年提高 18.6 个百分点。

• 在监测城市中空气质量达到二级以上（含二级）标准的城市占监测城市数的 82.7%，比 2005 年提高 22.4 个百分点。

• 截至 2010 年末，城市污水处理厂日处理能力达 10262 万立方米，比 2005 年末增长 79.2%；城市污水处理率达到 76.9%，提高 24.9 个百分点。

另外，根据《2010 年中国环境状况公报》，我国总体环境方面主要状况如下：

（1）淡水环境。全国地表水污染依然较重。七大水系总体为轻度污染，浙闽区河流和西南诸河水质良好，西北诸河水质为优。但是，湖泊（水库）富营养化问题突出，如图 1－4 所示。

废水和主要污染物排放量。2010 年，全国废水排放总量为 617.3 亿吨，比上年增加 4.7%；化学需氧量排放量为 1238.1 万吨，比上年下降 3.1%；氨氮排放量为 120.3 万吨，比上年下降 1.9%，如表 1－1 所示。

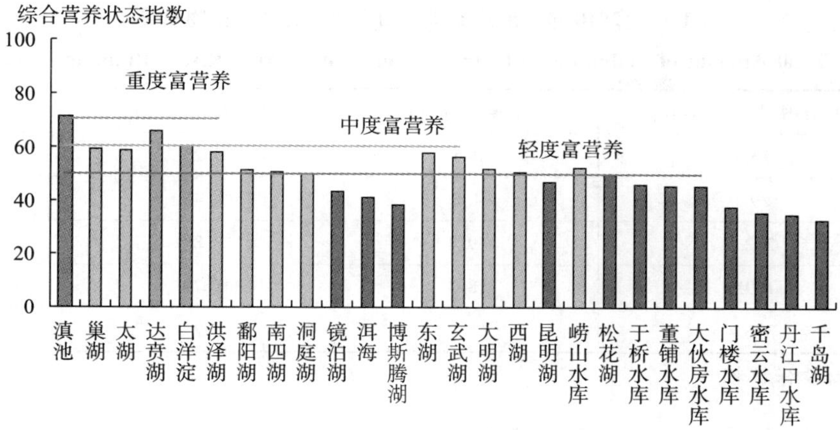

图1－4　2010年重点湖库综合营养状态指数

Figure 1－4　Comprehensive Trophic State Index of Key Rivers and Lakes in 2010

资料来源：环境保护部

表1－1　全国废水和主要污染物排放量年际变化

Table 1　Inter－annual Variation of Waste Water and Major Pollutants Emissions in China

项目 年度	废水排放量（亿吨）			化学需氧量排放量（万吨）			氨氮排放量（万吨）		
	合计	工业	生活	合计	工业	生活	合计	工业	生活
2006	536.8	240.2	296.6	1428.2	541.5	886.7	141.3	42.5	98.8
2007	556.8	246.6	310.2	1381.8	511.1	870.8	132.3	34.1	98.3
2008	572.0	241.9	330.1	1320.7	457.6	863.1	127.0	29.7	97.3
2009	589.2	234.4	354.8	1277.5	439.7	837.8	122.6	27.3	95.3
2010	617.3	237.5	379.8	1238.1	434.8	803.3	120.3	27.3	93.0

资料来源：环境保护部

（2）海水水质。全国近岸海域水质总体为轻度污染。2010年，近岸海域监测面积共279225平方千米，其中一、二类海水面积177825平方千米；三类海水面积44614平方千米；四类、劣四类海水面积56786平方千米。

陆源污染物入海状况。192个入海河流监测断面水质总体较差（如表1－2），河流污染物入海量大于直排海污染源污染物入海量。192个入海河流断面主要污染物排海总量约为：高锰酸盐指数396.4万吨，氨氮65.7万吨，石油类5.21万吨，总磷23.67万吨。东海的河流污染物入海总量远高于其他海区，如表1－3所示。

表1－2　2010年入海河流监测断面水质类别

Table 1－2　Water Quality Grades of the Monitoring Sections of Rivers Running into Sea in 2010

海区	永质类别（个）						
	Ⅰ	Ⅱ	Ⅲ	Ⅳ	Ⅴ	劣Ⅴ	合计
渤海	0	0	9	4	7	28	48
黄海	1	3	15	16	5	13	53
东海	0	0	5	7	6	7	25
南海	0	10	21	20	8	7	66
合计	1	13	50	47	26	55	192

资料来源：环境保护部

表1-3 2010年入海河流排入四大海区各项污染物总量

Table 1-3 Total Amount of Pollutants in Four Key Sea Waters from Rivers Running into Sea in 2010

海区	高锰酸盐指数（万吨）	氨氮（万吨）	石油类（万吨）	总磷（万吨）
渤海	15.6	3.5	0.23	0.36
黄海	29.0	4.5	0.40	0.84
东海	242.7	39.4	2.88	18.90
南海	109.1	18.3	1.70	3.56
合计	396.4	65.7	5.21	23.67

资料来源：环境保护部

直排海污染源。461个日排污水量大于100吨的直排海工业污染源、生活污染源和综合排污口的污水排放总量为50.92亿吨，各项污染物排放总量为：化学需氧量21.94万吨，石油类1215吨，氨氮22870吨，总磷2901吨，汞244.6千克，六价铬1076千克，铅1064千克，镉392千克，如表1-4、表1-5所示。

表1-4 2010年各类直排海污染源排放情况

Table 1-4 Emissions of Pollutants Directly Running into the Sea in 2010

污染源类别	废水量（亿吨）	化学需氧量（万吨）	石油类（吨）	氨氮（吨）	总磷（吨）	汞（千克）	六价铬（千克）	铅（千克）	镉（千克）
工业	14.98	2.82	91	1182	67	3.49	603	553	157
生活	8.45	4.48	373	5248	796	22.11	359	130	180
综合	27.48	14.64	751	16440	2038	219	114	381	55
合计	50.92	21.94	1215	22870	2901	244.6	1076	1064	392

资料来源：环境保护部

表1-5 2010年四大海区受纳直排海污染源污染物情况

Table 1-5 Direct Sources of Pollutants Discharged into Four Key Sea Waters in 2010

海区	废水量（亿吨）	化学需氧量（万吨）	氨氮（万吨）	石油类（吨）	总磷（吨）
渤海	1.81	0.79	0.11	74.7	66.9
黄海	8.78	4.5	0.53	81.5	774.3
东海	29.78	11.87	1.12	598	1068
南海	10.55	4.77	0.53	460.8	991.4

资料来源：环境保护部

（3）大气环境。全国城市空气质量总体良好，比上年有所提高，但部分城市污染仍较重；全国酸雨分布区域保持稳定，但酸雨污染仍较重。

2010年，全国471个县级及以上城市开展环境空气质量监测，监测项目为二氧化硫、二氧化氮和可吸入颗粒物。其中3.6%的城市达到一级标准，79.2%的城市达到二级标准，15.5%的城市达到三级标准，1.7%的城市劣于三级标准。全国县级城市的达标比例为85.5%，略高于地级及以上城市的达标比例。

废气中主要污染物排放量。2010年，二氧化硫排放量为2185.1万吨，烟尘排放量为829.1

万吨，工业粉尘排放量为448.7万吨，分别比上年下降1.3%、2.2%、14.3%。全国废气中主要污染物排放量年际变化情况如表1－6所示。

表1－6　全国废气中主要污染物排放量年际变化

Table 1－6　Inter－annual Variation of Major Air Pollutants Emissions in China

项目／年度	二氧化硫排放量（万吨）			烟尘排放量（万吨）			工业粉尘排放量（万吨）
	合计	工业	生活	合计	工业	生活	
2006	2588.8	2234.8	354.0	1088.8	864.5	224.3	808.4
2007	2468.1	2140.0	328.1	986.6	771.1	215.5	698.7
2008	2321.2	1991.3	329.9	901.6	670.7	230.9	584.9
2009	2214.4	1866.1	348.3	847.2	603.9	243.3	523.6
2010	2185.1	1864.4	320.7	829.1	603.2	225.9	448.7

资料来源：环境保护部

（4）固体废物。2010年固体废物主要数据如下：

- 全国工业固体废物产生量为240943.5万吨，比上年增加18.1%。
- 排放量为498.2万吨，比上年减少29.9%。
- 综合利用量（含利用往年贮存量）、贮存量、处置量分别为161772.0万吨、23918.3万吨、57263.8万吨，分别占产生量的67.1%、9.9%、23.8%。
- 危险废物产生量为1586.8万吨，综合利用量（含利用往年贮存量）、贮存量、处置量分别为976.8万吨、166.3万吨、512.7万吨。2010年全国工业固体废物产生及处理情况如表1－7所示。

表1－7　2010年全国工业固体废物产生及处理情况

Table 1－7　Production and Treatment of Industrial Solid Waste in China in 2010

产生量（万吨）		综合利用量（万吨）		贮存量（万吨）		处置量（万吨）	
合计	危险废物	合计	危险废物	合计	危险废物	合计	危险废物
240943.5	1586.8	161772.0	976.8	23918.3	166.3	57263.8	512.7

资料来源：环境保护部

（5）自然生态。全国部分生态系统功能有所改善，但总体生态系统服务功能不强，生物多样性下降趋势尚未得到有效遏制，遗传资源不断丧失和流失，外来入侵物种危害严重。特别是近年来受全球气候变暖等自然因素影响，加之人为开垦、超载过牧、破坏生态用地等影响，主要生态环境问题依然突出。

（6）农村环境状况。农村环境问题日益显现，农业源污染物排放总量较大，局部地区形势有所好转，但总体形势仍十分严峻。突出表现为畜禽养殖污染物排放量巨大，农业面源污染形势严峻，农村生活污染局部增加，农村工矿污染凸显，城市污染向农村转移有加速趋势，农村生态退化尚未得到有效遏制。

（7）森林资源概况。根据第七次全国森林资源清查（2004—2008年）结果，全国森林面积19545.22万公顷，森林覆盖率20.36%，活立木总蓄积149.13亿立方米，森林蓄积137.21亿立方米。乔木林平均每公顷蓄积量85.88立方米。中国森林面积列世界第五位，森林蓄积列世界第六位，人工林面积居世界首位。

（8）气候状况。2010年，中国气候形势复杂异常，极端天气气候事件频繁发生，气象灾害造成的损失为本世纪以来之最。全国降水偏多，但季节和区域分布不均，旱涝灾害交替发生。年平均气温偏高，季节转换总体偏晚，高温日数创历史新高。2010年，中国年平均气温9.5℃，较常年偏高0.7℃，为1961年以来第10高值，也是1997年以来连续第14个偏高年份，如图1－5所示。

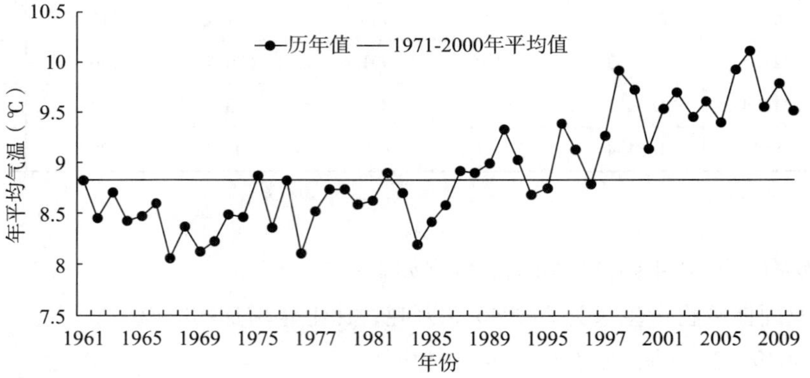

图1－5 1961—2010年中国年平均气温变化曲线图（℃）

Figure 1－5 Curve of China's Annual Mean Temperature Change from 1961—2010（℃）

资料来源：环境保护部

与节能类似，减排的历程较为轻松，其目标实现也比较早，但期间也经历了诸多波折与往复。以下按时间顺序回顾我国COD排放量和二氧化硫指标的实现历程①：

（1）COD排放量。2006年，我国COD排放量不降反升，较2005年增长1%，远未达到下降2%的目标。虽然中央加大了对COD排放量的减排力度，但由于“十一五”时期前两年经济增长仍超预期，“两高一资”行业仍发展迅速，城镇化进程逐渐加速等原因，直至2007年第三季度，COD排放量才首次转增为降。由于4万亿元投资的产能所属行业并非控制COD排放的重点行业（造纸、农副产品加工业和化学原料及化学品制造），COD减排工作并未受到4万亿元刺激计划较大的影响。2008—2009年，COD排放量分别下降了4.42%和3.27%，2009年COD排放量较2005年累计下降9.66%，距离“十一五”减排目标已经非常接近。

（2）二氧化硫排放量。2006年，我国二氧化硫减排同样未能实现下降2%的既定目标，反而上升了1.55%。2007年推出的减排指标考核统计体系，遏制了地方虚报、瞒报污染物排放量数据的行为。同时，三大减排措施（工程减排、结构减排和管理减排）逐步推进，污染物减排工作逐渐走上正轨。通过工程减排，燃煤脱硫机组装机容量不断增加；通过结构减排，小火电装机容量不断降低，落后产能不断被淘汰；通过管理减排，废气重点污染源达标排放率得到了很大的提高。2007—2009年，各年二氧化硫排放量同比下降4.66%、5.95%和4.60%。2009年二氧化硫排放量与2005年相比下降了13.14%，提前一年实现“十一五”减排目标。

① 宏观经济研究院国地所课题组卢伟的《“十一五”我国节能减排工作回顾及“十二五”政策建议》。

1.4 “十一五”减排领域重要举措及效果

“十一五”期间，工程减排、结构减排和管理减排三大措施稳步推进，全面发挥减排效力。

（1）工程减排。“十一五”主要数据如下：

• 全国新增燃煤脱硫机组装机容量1.07亿千瓦，火电脱硫机组装机容量达到5.78亿千瓦，占全部火电机组的比例从2005年的12%提高到82.6%。

• 新增城市污水日处理能力1900万立方米，污水日处理能力达到1.25亿立方米，城市污水处理率由2005年的52%提高到75%以上。

• 钢铁烧结机烟气脱硫设施累计建成运行170台，占烧结机台数的比例由2005年的0提高到2010年的15.6%。

（2）结构减排。“十一五”主要数据如下：

• 累计关停小火电机组7210万千瓦，提前一年半完成关停5000万千瓦的任务。

• 钢铁、水泥、焦化及造纸、酒精、味精等高耗能高排放行业淘汰落后产能均超额完成任务。

• 2010年，全国电力行业30万千瓦以上火电机组占火电装机容量比重从2005年的47%提高到70%以上，火电供电煤耗下降9.5%。

• 造纸行业单位产品化学需氧量排污负荷下降45%。

（3）管理减排。“十一五”主要数据如下：

•“十一五”中央财政投入100多亿元，用于支持全国污染减排“三大体系”和环保监管能力建设，建成污染源监控中心343个，对1.5万家企业实施自动监控，配备监测执法设备10万多台（套），环境监管能力显著增强。

• 南方电网公司和多个省份开展节能减排发电调度，对燃煤脱硫机组实行投运率考核并扣减脱硫电价，投运率由2005年不足60%提高到2010年的95%以上。

• 国控重点污染源二氧化硫和化学需氧量达标率分别为92%和94%，较2005年分别提高22和34个百分点。

1.5 “十一五”节能减排政策推动历程①

为实现节能减排目标，2006—2010年，我国政府出台了一系列政策措施，以前所未有的力度，综合运用经济、法律乃至必要的行政手段，推进节能减排工作。

1. 深化能源价格改革

2008年12月，我国出台了成品油价格和税费改革方案，为实现国内成品油价格与国际市场接轨奠定了必要的制度基础。

2009年5月，我国又出台了《石油价格管理办法》，使国内成品油价格监管制度更加科学、规范、透明。电价改革不断推进，需求侧电价管理制度逐步完善。实行促进节约用水的水价制度，鼓励使用再生水。

2. 实行有利于节能的税收政策

第一，将汽车排量与消费税税率挂钩，鼓励购买小排量车型；扩大消费税征收范围，将石

---

① 刘慧娴．“十一五”节能减排攻坚战［J］．中国财政，2010，（20）．

脑油、润滑油、溶剂油、航空煤油、燃料油等成品油纳入其中。

第二，制定了环境保护专用设备、节能节水专用设备、资源综合利用等企业所得税优惠目录。

第三，提高了部分有色金属矿产品的资源税税额标准，调整了有色金属、铁矿石的资源税优惠政策。

第四，对引进国内尚不能生产并在规定范围内的节能环保设备及技术，免征进口关税和进口环节增值税，鼓励先进节能减排技术和设备进口等。

3. 建立生态环境补偿机制

围绕水资源、水环境保护和水污染防治，省际间以签订合同或协议形式，建立起下游对上游水资源、水环境保护的补偿和上游对下游超标排污或环境责任事故赔偿的双向责任机制，逐步建立起以省际横向补偿或赔偿为主、中央财政引导或奖励为辅的利益补偿制度。

2008 年，财政部联合相关部门开始实施国家重点生态功能区转移支付试点，将天然林保护、青海三江源和南水北调等重大生态功能区所辖 230 个县纳入转移支付范围。

4. 完善矿产资源有偿使用制度改革

2006 年 9 月，经国务院批复同意，财政部联合相关部委在山西、内蒙古等 8 个煤炭主产省区率先开展了煤炭资源有偿使用制度改革试点。

截至 2007 年底，试点省区 9963 个煤炭矿业权中，75% 的矿业权实现了有偿取得，山西、河南二省矿业权有偿处置率已达到 90% 以上；煤炭矿业权有偿出让收入共计 322 亿元；煤炭企业累计提取矿山环境治理恢复保证金 10 亿元，为加强矿山环境治理和生态恢复提供了稳定的资金来源。

5. 推进排污权有偿取得和交易试点

2008 年，先在江苏省电力行业和太湖流域试点，随后又扩大到天津市滨海新区、浙江省电力行业及太湖流域杭嘉湖地区和钱塘江流域。

截至 2009 年底，已交易二氧化硫 33120 吨，交易金额 1.4 亿元；化学需氧量 2509 吨，交易金额 0.9 亿元。

6. 建立政府强制采购节能产品制度

2007 年，我国建立了政府强制采购节能产品制度，并定期发布“节能产品政府采购清单”。

截至 2009 年，清单中节能产品种类已由原来的 18 类 4770 种扩大到 33 类 15087 种，其中空调机、双端荧光灯和自镇流荧光灯、电视机、电热水器、计算机等九类产品为政府强制采购节能产品。

7. 鼓励节能型消费

2009 年，财政部、国家发改委启动“节能惠民工程”，通过财政补贴方式对能效等级一级或二级以上的空调、冰箱、平板电视、洗衣机、电机等十大类高效节能产品予以推广应用，可使高效节能产品市场份额提高 10—20 个百分点，达到 30% 以上；每年可节电 750 亿千瓦时，减排二氧化碳 7500 万吨。当年，中央财政已兑付高效节能空调补贴近 40 亿元，推广高效节能空调近 1000 万台。

同年，财政部、国家发改委又开始实行汽车、家电“以旧换新”政策，中央财政安排了 70 亿元资金，用于鼓励汽车、家电“以旧换新”9 个试点省市共回收五大类旧家电 200 多万台，销售五大类新家电200多万台，销售额达80多亿元。

2010年是“十一五”规划的收官之年，也是实现“十一五”节能减排指标的最后一年，形势十分严峻。要完成“十一五”期间单位GDP能耗比2005年降低20%的目标，2010年单位GDP能耗至少要降低5.62%。此外，为应对全球气候变化，我国政府承诺到2020年单位国内生产总值二氧化碳排放要比2005年下降40%—45%，节能提高能效的贡献率要达到85%以上，这也给节能减排工作带来巨大挑战。

在这一严峻的形势下，政府领导指出“无论面临多大的困难，我们的承诺不能改变，决心不能动摇，工作不能减弱”，“必须采取铁的手腕，不折不扣地按期完成”。到“十一五”末，要对节能减排目标完成情况算总账，实行严格的问责制。

在此背景下，政府各部门密集出台各项政策措施，以确保实现“十一五”节能减排目标。

5月26日，财政部等三部门下发了《“节能产品惠民工程”节能汽车（1.6升及以下乘用车）推广实施细则》，对消费者购买发动机排量在1.6升及以下的燃用汽油、柴油的乘用车（含混合动力汽车和双燃料汽车）给予一次性定额补助，补助标准为3000元/辆。

5月27日，在全国工业系统淘汰落后产能工作会议上，中央18部委联席向31个省市地区下军令状，将淘汰落后产能任务指标分解下达各地。当天，国家发改委有关负责人表示，2010年，我国将对省级政府和重点企业进行节能目标责任评价考核，并向社会公告考核结果，实行奖惩问责，到年底要对责任单位实行问责制。

5月31日，财政部、国家发改委联合下发了《关于印发“节能产品惠民工程”高效电机推广实施细则的通知》，对能效等级为一级和二级、额定功率为0.55千瓦（含）至315千瓦（含）的低压三相异步电机给予每千瓦15元—40元补贴；对损耗比现行准入标准低20%、额定功率为355千瓦（含）至25000千瓦（含）的高压三相异步电机给予每千瓦12元补贴；对额定功率为0.55千瓦（含）至315千瓦（含）的稀土永磁电机给予每千瓦40元—60元补贴。

5月31日，财政部等四部门联合下发《关于扩大公共服务领域节能与新能源汽车示范推广有关工作的通知》，进一步扩大公共服务领域节能与新能源汽车示范推广工作，在现有13个试点城市的基础上，增加天津、海口、郑州、厦门、苏州、唐山、广州等7个试点城市。

5月31日，财政部等四部门联合下发《关于开展私人购买新能源汽车补贴试点的通知》，自6月1日起，在上海、长春、深圳、杭州、合肥5个城市启动私人购买新能源汽车补贴试点工作，加快启动新能源汽车私人消费市场。中央财政对试点城市私人购买、登记注册和使用的插电式混合动力乘用车和纯电动乘用车给予一次性补贴。

6月1日，讨论多年的资源税改革方案浮出水面，新疆原油、天然气资源税改革王式启动，原油、天然气资源税实行从价计征，税率为5%。为此，财政部、国家税务总局联合下发了《新疆原油、天然气资源税改革若干问题的规定》，对新疆原油、天然气资源税改革的纳税人、纳税方式、纳税额计算公式以及减免税情形等内容予以明确。

6月1日，国家发改委等六部门开始在全国范围内开展节能减排电力价格大检查，重点检查地方政府及其有关部门违反国家政策规定，不执行或推迟执行对电解铝、电石、铁合金等高耗能行业差别电价政策，擅自制定并实施对高耗能企业或产业园区优惠电价行为。六部门组成6个联合督查组，分赴6个大区进行重点督查。

为加快推行合同能源管理，促进节能服务产业发展，中央财政决定2010年安排20亿元，用于支持节能服务公司采取合同能源管理方式在工业、建筑、交通等领域以及公共机构实行节能改造。为规范和加强财政奖励资金管理，提高资金使用效益，6月3日，财政部、国家发改委联合出台了《关于印发合同能源管理财政奖励资金管理暂行办法的通知》。

6月18日，财政部等三部门联合下发通知，宣布将汽车"以旧换新"政策实施期限由2010年5月31日延长至2010年12月31日。

6月22日，财政部、国家税务总局发出通知，自2010年7月15日起，取消部分钢材、有色金属加工材、农药、医药、化工产品、塑料及制品、橡胶及制品、玻璃及制品的出口退税。取消出口退税的商品共有400多种，大部分是高污染、高耗能和资源类产品。这是自2008年下半年以来，出口退税政策首次进行反向调整，力度之大显示出国家加快产业结构调整的决心。

7月6日，为加快启动私人购买新能源汽车试点工作，并确保达到预期目标，财政部等四部委在深圳召开私人购买新能源汽车试点会议，对私人购买新能源汽车试点工作进行动员和部署，并举行了深圳市私人购买新能源汽车试点启动仪式。

8月8日，18个工业行业淘汰落后产能企业名单公布，这些企业的落后产能必须在2010年9月底前关闭，否则，将吊销企业排污许可证；银行业金融机构不再提供任何形式的新增授信支持；投资管理部门不予审批和核准新的投资项目；国土资源管理部门不予批准新增用地；相关管理部门不予办理生产许可，已颁发生产许可证、安全生产许可证的要依法撤回。必要时，政府相关部门可要求电力供应企业依法对落后产能企业停止供电。

1.6 "十一五"节能减排工作启示[①]

1.6.1 节能减排考评制度的建立能够产生以点带面效应

1. 地方节能减排考评制度的建立，为我国发展观和政绩观的转型奠定了基础

自上世纪90年代以来，中央一直是以资源节约和环境保护作为实现经济增长方式转变的重要抓手，并围绕其进行了一系列的制度安排。但在当前阶段，我国的资源节约和环境保护措施实施的有效程度，取决于地方政府对中央政策的配合程度。

由于资源节约和环境保护事业的公益性强和经济效益低，许多政策并未得到有效执行。"十一五"时期，围绕节能减排所形成的制度安排特别是地方节能减排考评制度，为实现中央和地方在转变发展观和政绩观上达成一致奠定了基础。同时，这种制度安排也将对我国环境保护和经济发展产生深远的影响。

2. 节能减排约束性指标有效促进了地方经济转型和产业结构调整

我国首次将节能减排指标作为约束性指标写入了"十一五"规划纲要中。在围绕节能减排强制性的指标所采取的两大措施中，环保措施的实施逐渐完善了我国环境保护和污染治理的底层结构，而节能措施的实施则是以能源结构调整为突破口，有效促进了地方产业结构的调整和经济的升级转型。

3. 节能减排技术的推行，为我国实现能源安全、应对气候变化等作出了重要贡献

能源问题是当前我国经济社会发展所面临的最重要的问题之一。节能减排工作所遵循的

---

① 宏观经济研究院国地所课题组卢伟的《"十一五"我国节能减排工作回顾及"十二五"政策建议》。

“提升旧能源使用效率，提高新能源使用比率”的思路，不断推动我国在能源领域的技术创新。这种技术创新所形成的能源产品替代，逐步完善了我国能源安全战略框架。由能源替代所带来的温室气体排放量减少、能源消耗所产生污染物排放量的降低等一系列连锁反应，使得节能减排工作成为我国应对气候变化和环境保护的又一突破。

1.6.2　节能减排的推进过程具有自上而下的特征

节能减排工作的推进特征，一方面，源于我国中央与地方的权力架构；另一方面，源于其本身的全局和公益特征。事实上，节能减排的公益特征与许多地方政府的发展思路是不相符的。同时，我国正处在后金融危机时代的战略机遇期，如果仅通过开展试点试探性地开展节能减排工作，那就会使我国错失提升总体经济发展质量，提高经济系统对抗国际风险能力，把握新兴经济增长点，增强企业和产品国际竞争力的最佳时期。因此，自上而下全面铺开式的节能减排工作是迫切而必要的，事实证明，也是非常有成效的。

1.6.3　节能减排的指标设置具有全过程控制雏形

从当前我国节能减排的指标设定上看，两个环保指标（SO_2 排放量和 COD 排放量）都是末端治理指标，而节能指标则为清洁生产指标。这种设定，使得节能减排工作有助于构建由末端到前端全过程控制的环境管理体系，体现了未来节能减排向全过程控制方向发展的必然趋势，但如何使得以生产过程控制为主的节能工作和以末端治理为主的环保工作在技术和管理上有机对接，是需要探索的问题。

1.6.4　节能减排在具体执行上存在变相达标风险

节能减排制度安排的一个最大特点是其强制性。由于问责制和“一票否决”制，使地方政府对确保节能减排目标完成的重视，在某种程度上已超过了确保 GDP 增长率的目标，但这种制度安排又产生了变相达标的风险：当通过常规措施难以达到节能减排目标时，地方政府可能采取一些非常规的措施来强行达到目标，这就可能会对正常的经济运行产生影响。事实上，这种变相达标风险产生的根本原因，是地方政府对行政手段的过度使用和市场调节力量的缺失。

1.6.5　我国政策目标的设定存在效益悖反现象

在我国政策目标的设定上，短期政策目标以及应对突发性事件政策目标和较长期政策目标往往会存在矛盾，从而易产生“效益悖反”的现象。节能减排目标是一个五年目标，属于中长期政策目标；而我国为应对 2008 年发生的全球金融危机而提出的 2009 年的“保 8”目标，是一个短期目标。一方面，政府需要通过保证 GDP 的增长目标来复苏经济；另一方面，复苏经济就增加了节能减排目标实现的难度。同时，由于不同目标实现先后顺序的不断变化，政策的执行成本也会进一步增加。2009 年，“保增长”被放在主要的位置，而从 2010 年初开始，随着形势的变化，“保增长”又被放在相对次要的位置，而调结构和节能减排成为更重要的任务。但由于为实现保增长目标而进一步加大了节能减排目标实现的难度，使得政府为完成节能减排目标所需支付的额外成本大幅增加。因此，如何最大程度地消除我国不同政策目标的“效益悖反”现象，是需要深入思考的问题。

2 中国“十二五”节能减排的趋向

2.1 “十二五”节能减排的整体目标

我国“十二五”节能减排主要目标如下：

（1）节能目标。到2015年，全国万元国内生产总值能耗下降到0.869吨标准煤（按2005年价格计算），比2010年的1.034吨标准煤下降16%，比2005年的1.276吨标准煤下降32%（如图2－1）；“十二五”期间，实现节约能源6.7亿吨标准煤。

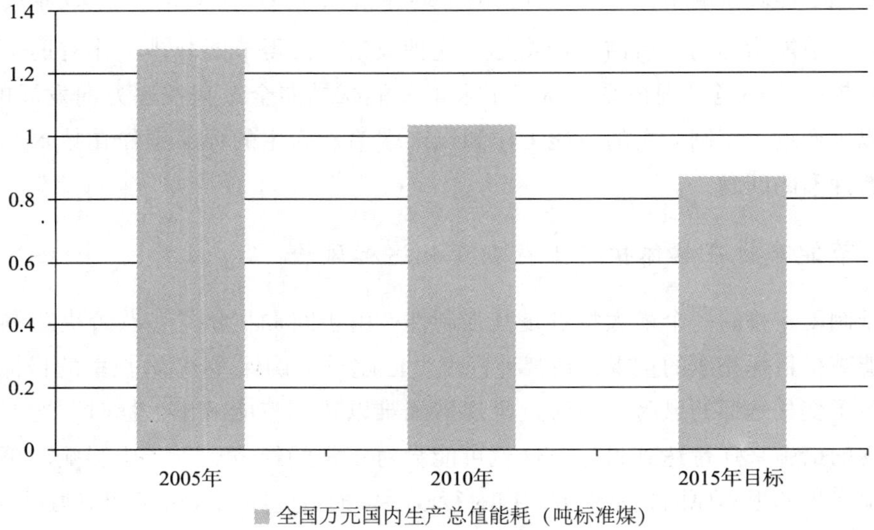

图2－1 中国“十二五”节能目标

Figure 2－1 Energy Conservation Targets of China's 12th Five－year Plan

资料来源：《“十二五”节能减排综合性工作方案》

（2）减排目标。提出2015年，全国化学需氧量和二氧化硫排放总量分别控制在2347.6万吨、2086.4万吨，比2010年的2551.7万吨、2267.8万吨分别下降8%；全国氨氮和氮氧化物排放总量分别控制在238.0万吨、2046.2万吨，比2010年的264.4万吨、2273.6万吨分别下降10%，如图2－2所示。

2.2 “十二五”节能减排目标责任分解

在综合考虑经济发展水平、产业结构、节能潜力、环境容量及国家产业布局等因素基础之上，全国节能减排目标被分解到各地区。

（1）节能目标分解。天津、上海、江苏、浙江、广东下降18%；北京、河北、辽宁、山东

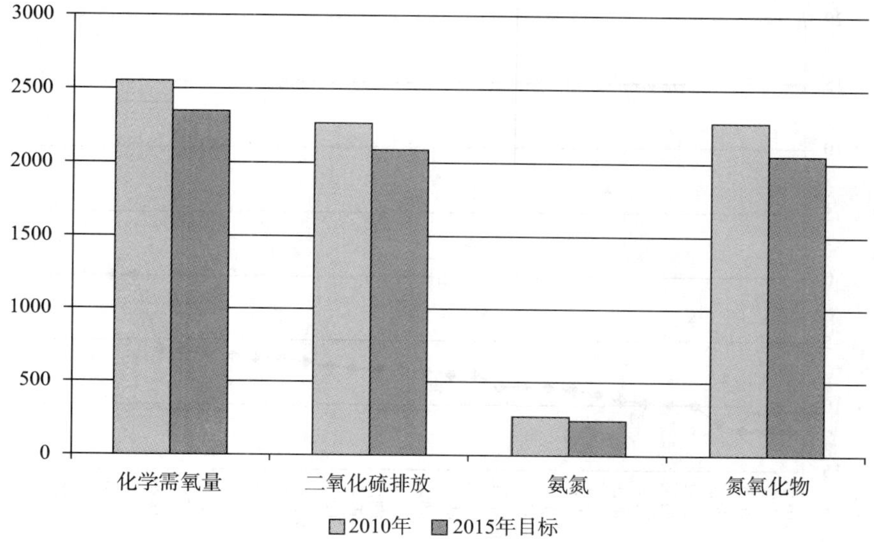

图 2-2 中国“十二五”减排目标

Figure 2-2 Emission Reduction Targets of China's 12th Five-year Plan

资料来源：《“十二五”节能减排综合性工作方案》

下降17%；山西、吉林、黑龙江、安徽、福建、江西、河南、湖北、湖南、重庆、四川、陕西下降16%；内蒙古、广西、贵州、云南、甘肃、宁夏下降15%；海南、西藏、青海、新疆下降10%，如图2-3所示。

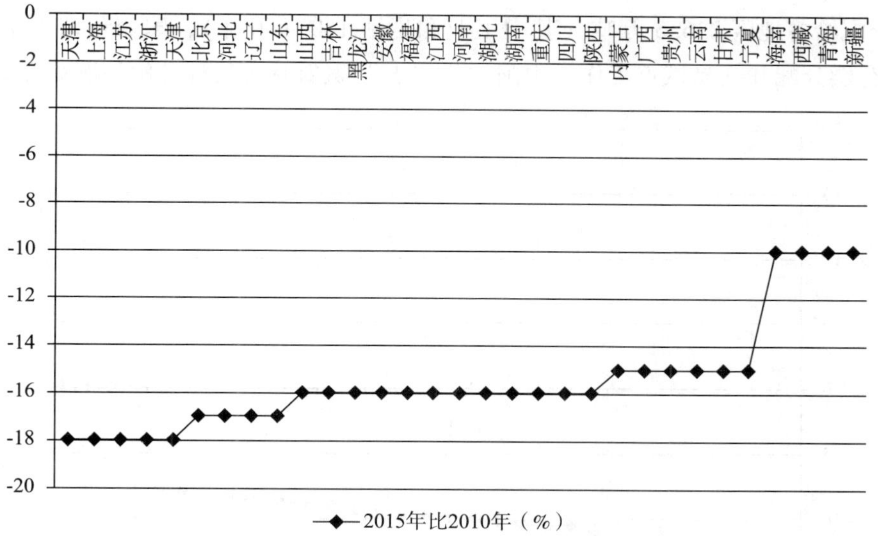

图 2-3 中国“十二五”节能目标分解

Figure 2-3 Targets Decomposition of Energy Conservation in China's 12th Five-year Plan

资料来源：《“十二五”节能减排综合性工作方案》

（2）化学需氧量减排目标分解。全国化学需氧量排放量削减8%的总量控制目标为2347.6万吨（其中工业和生活1221.9万吨），实际分配给各地区2335.2万吨（其中工业和生活1214.6万吨），国家预留12.4万吨，用于化学需氧量排污权有偿分配和交易试点工作。中国“十二五”化学需氧量减排目标分解情况如图2-4所示。

（3）全国氨氮排放量削减10%的总量控制目标为238.0万吨（其中工业和生活161.5万

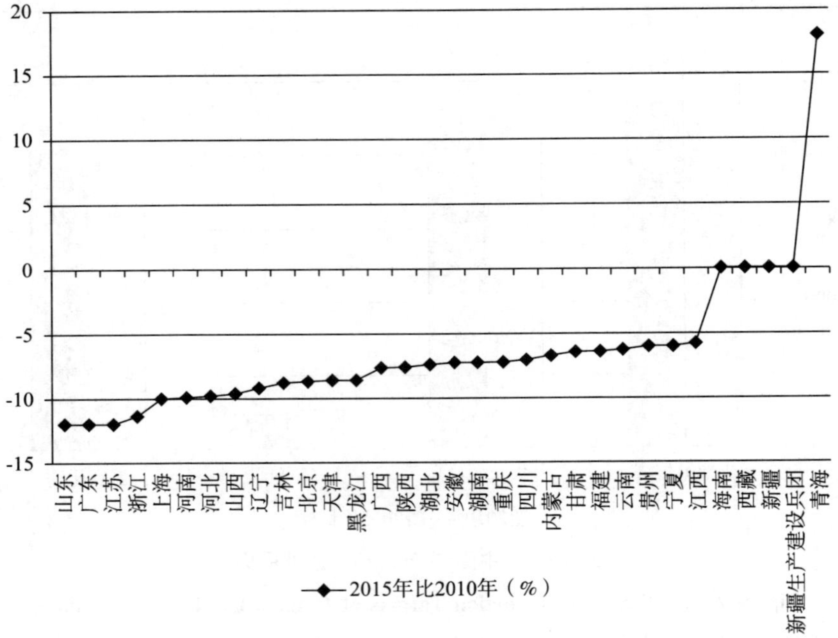

图2－4　中国“十二五”化学需氧量减排目标分解

Figure 2－4　Targets Decomposition of Chemical Oxygen Demand Emission Reduction in China's 12th Five－year Plan

资料来源：《“十二五”节能减排综合性工作方案》

吨)，实际分配给各地区236.6万吨（其中工业和生活160.4万吨），国家预留1.4万吨，用于氨氮排污权有偿分配和交易试点工作。中国“十二五”氨氮减排目标分解情况如图2－5所示。

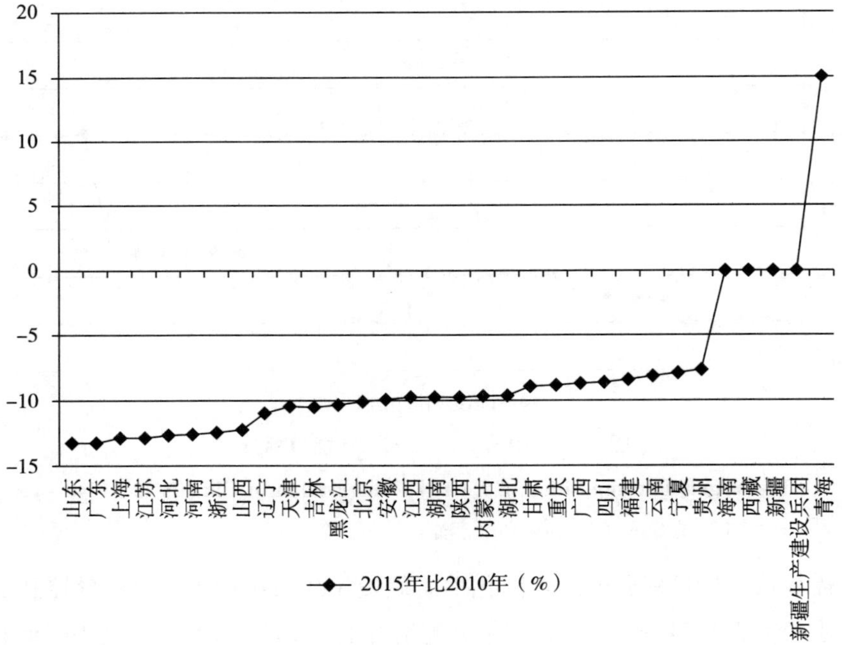

图2－5　中国“十二五”氨氮减排目标分解

Figure 2－5　Targets Decomposition of Ammonia Nitrogen Emission Reduction in China's 12th Five－year Plan

资料来源：《“十二五”节能减排综合性工作方案》

（4）全国二氧化硫排放量削减 8% 的总量控制目标为 2086.4 万吨，实际分配给各地区 2067.4 万吨，国家预留 19.0 万吨，用于二氧化硫排污权有偿分配和交易试点工作。中国“十二五”二氧化硫减排目标分解情况如图 2－6 所示。

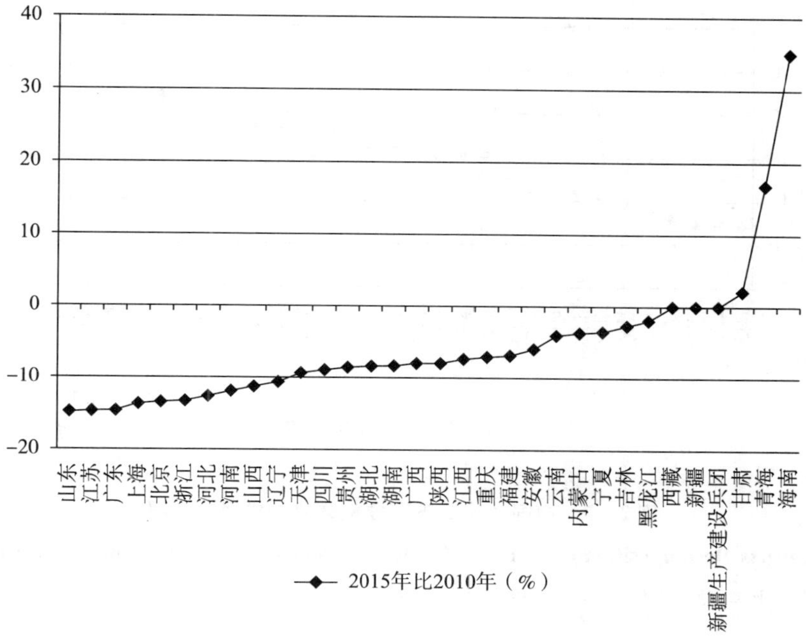

图 2－6 中国“十二五”二氧化硫减排目标分解

Figure 2－6 Targets Decomposition of Sulfur Dioxide Emission Reduction in China's 12th Five－year Plan

资料来源：《“十二五”节能减排综合性工作方案》

（5）全国氮氧化物排放量削减 10% 的总量控制目标为 2046.2 万吨，实际分配给各地区 2021.6 万吨，国家预留 24.6 万吨，用于氮氧化物排污权有偿分配和交易试点工作。中国“十二五”氮氧化物减排目标分解情况如图 2－7 所示。

2.3 “十二五”节能减排的可能趋向

“十一五”规划将单位 GDP 能耗五年降低 20% 左右作为约束性指标，对节约能源起到了很好的作用，增强了大家的节能意识。但是，只控制单位 GDP 能耗，而不控制能源消费总量，难以约束地方的超高速发展和重化工业的扩张冲动，难以从根本上扭转敞开口消耗能源的状态①。单位 GDP 能耗取决于分子和分母的相对关系，虽然我国单位 GDP 能耗逐渐下降，但能源消耗总量的增长速度太快，按这个速度下去，国内自有资源和国外进口资源都是不够用的②。在单纯考核能耗强度的情况下，地方政府可以通过做大分母（GDP）的方式来稀释能耗强度。

“十一五”期间，虽然单位 GDP 能耗指标基本完成，但能源消费总量指标执行情况并不算好。能源消费总量指标执行情况如下：

- 2004 年，国家发展改革委制定并经国务院批准的我国能源中长期发展规划纲要规定，

① 房维中．能源消费总量指标应列入“十二五”规划［J］．宏观经济管理，2010，（10）．
② 刘雅鸣，王志，李江涛．“十二五”规划的政策期待［J］．瞭望，2010．

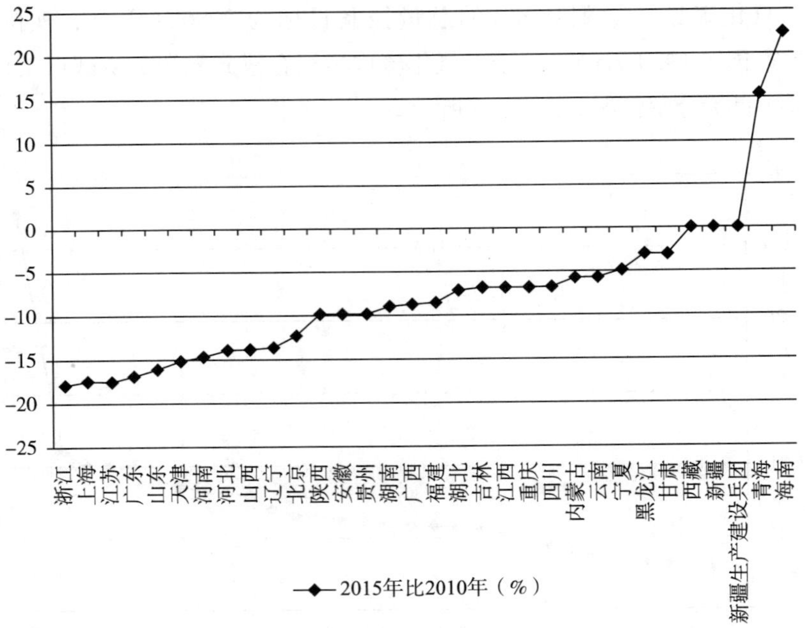

图 2－7　中国“十二五”氮氧化物减排目标分解

Figure 2－7　Targets Decomposition of Nitrogen Oxide Emission Reduction in China's 12th Five－year Plan

资料来源：《“十二五”节能减排综合性工作方案》

2010 年能源消费总量控制在 24 亿吨标准煤以内，2020 年控制在 30 亿吨左右标准煤。

• 2007 年国家发展改革委制定的“十一五”能源专项规划规定，2010 年我国能源消费总量控制在 27 亿吨左右标准煤。

• 2010 年达到的 32.5 亿吨标准煤，比 2000 年的 14.5 亿吨增加 1.2 倍，10 年翻了一番还多，超出 2004 年规定的国家控制目标 8.5 亿吨、2007 年规定的国家控制目标 5.5 亿吨。

经济增长不能再以高能耗为代价，而应以能源环境的可承受力决定我国适当的经济增速。如果单位能耗降低的多，GDP 的增长速度就可以快一些；若单位能耗降低的少，GDP 的增长速度就要减慢。规定能源消费总量，特别是化石能源的消费总量，对内对外具有双重意义：

第一，对内可以约束国内地方政府的扩张冲动。有了单位 GDP 能耗降低指标作为约束性指标，又有能源消费总量作为约束性指标，就可防止地方增速的过分偏离。

第二，履行国际上所承担的控制二氧化碳排放的共同而有区别的责任。如果不控制能源消费总量，让经济增长速度超过国家的预期目标 1—2 个百分点，即使保持每个五年单耗各降低 20%，2020 年二氧化碳排放量也将达到 92 亿吨—100 亿吨，比 2005 年的 52 亿吨，增长 75% 到将近 1 倍，这必将受到国际上的沉重压力。

《“十二五”节能减排综合性工作方案》提出，要合理控制能源消费总量。从而，“十一五”对能耗强度的单一控制升级为“十二五”对能耗强度与能源消费总量的综合控制。如果要使控制达到较好的效果，还需要重点研究以下几个方面[①]。

第一，研究如何确定能源总量控制的目标。能源是支持经济增长的重要保障，同时经济增长需要以一定数量的能源来支持。要维持现阶段比较高的经济增长，能源供应的总量将随之增

① 林伯强．如何实行能源消费总量控制，获取地址：http://finance.ifeng.com/news/industry/20110519/4044518.shtml。

长，需要研究阶段性能源增长与GDP之间的关系。制定一个既能保证一定经济增长速度，又能较好实现能源总量控制的目标十分重要。

第二，研究能源总量控制目标的分配问题。能源总量控制要求各个省份控制甚至减少能源消费数量，这相当于给地方经济增长加了一个紧箍咒。地方政府出于对经济发展的担忧，可能在能源总量控制目标的分配上斤斤计较。需要根据各地情况，在经济增长和能源消费之间寻找平衡，合理制定能源总量控制目标。

第三，研究转变经济发展方式对能源总量控制的意义。与“十一五”不同，“十二五”的节能需要改变经济发展方式。各地政府以节能环保产业、新信息技术产业、生物产业、高端设备制造和新材料为重点，培育战略性新兴产业。其他方面包括能源节约和资源的综合利用，加快清洁能源的发展速度。

第四，研究地方政府控制能源总量的动力。能源总量控制对GDP增长是比较大的障碍，如何引导地方政府真心真意完成控制目标，是能源消费总量控制的一个关键点。目前地方政府的“十二五”规划，把经济增长的速度定得很高，甚至是按照翻番的目标来制定的，可能对能源消费总量控制消极应对。现阶段的能源消费刚性问题将导致很难进行有效的能源消费总量控制。

第五，研究地方政府完成能源总量控制目标的方式。“十一五”末，各级政府为完成节能减排目标，采用拉闸限电，强制性减少能源使用，给企业生产和居民生活带来了不便。如何将节能手段与能源消费总量控制有效结合，也是完成能源消费总量控制的关键。除了关闭落后产能，调整经济结构，各级政府可以加强消费引导，对公共交通的发展提供支持政策，降低交通能耗；积极推广节能住宅，发展“小户大区”的居住模式，以此降低居住能耗；加强节能宣传，普及节能知识，推广节能产品，以此减少家用电器和燃具耗能。

第六，研究能源消费总量控制的评估体系。能源消费总量控制目标具有行政命令性强、可监控性差的特点，实际操作起来不容易，只有解决如何监测地方政府对控制目标完成情况的问题，能源消费总量控制才可以实际操作。

第七，研究如何应对可能的能源价格上涨。目前，国际原油价格已经飙升超过100美元一桶，不断升高的国际油价，不仅阻碍了全球的经济复苏，也增加了我国能源进口的成本。此外，近几年异常天气和能源生产安全事故对能源行业的影响也越来越大，能源的平稳运行和持续稳定供应将面临挑战。一旦实施能源供应总量控制，能源供求关系可能失衡，能源价格可能上涨。在老百姓日益关注油价、电价、水价等价格的情况下，政府要考虑如何解决能源价格上涨所带来的问题。

如上所述，虽然国家在节能减排方面提出了整体推进方案，但是能否取得预期目标，还需要继续观察。这既取决于各级政府的决心，也取决于各类企业内生动力的激发。2011年，我国单位GDP能耗下降目标为3.5%，然而上半年这一指标仅下降0.8%，因此“十一五”开端的节能减排形势非常严峻。这说明，如何避免“十二五”节能减排工作“前松后紧”，仍是一个亟待破解的难题。

3 全球的节能与减排

3.1 能源消费和碳排放

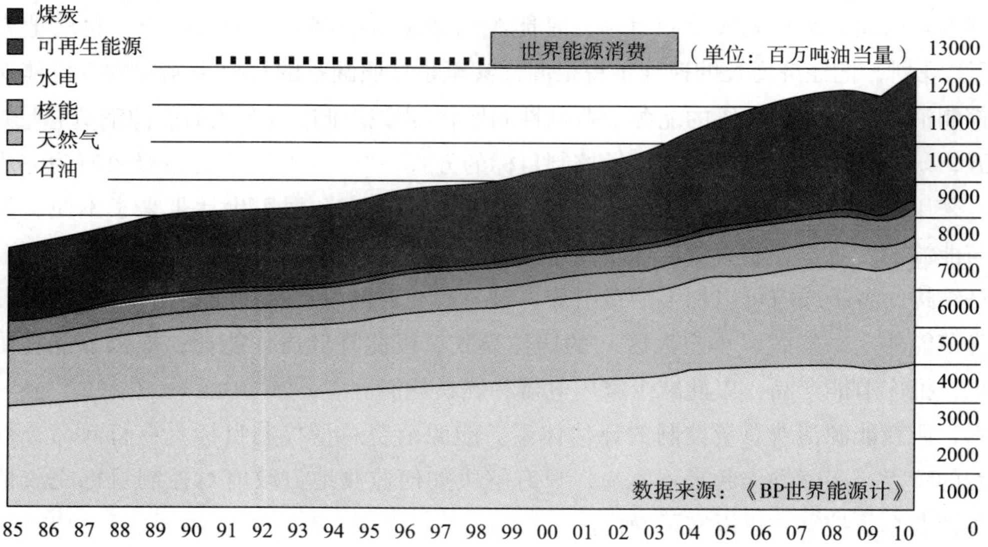

图 3－1 全球一次能源消费总量

Figure 3－1 Global Total Primary Energy Consumption

资料来源：《BP 世界能源统计》

根据《BP 世界能源统计》，2010 年全球能源消费在经济复苏的推动下，经历了一波强劲的反弹。2010 年全球主要能源数据如下：

- 全球一次能源消费总量为 12002.4 百万吨油当量，比上年增长了 5.6%，一举超过了 2008 年经济衰退前的峰值，达到了 1973 年以来的最高水平（如图 3－1）。与之相对应的是全球二氧化碳排放量为 33158.4 百万吨，比上一年增长了 5.8%，如图 3－2 所示。

- 2010 年，全球电力消费总量中有 1.3% 是来自可再生能源，这一比例比 2000 年增加了 0.7%。

- 石油。全球石油消费量在连续两年下降后，2010 年每天增加了 270 万桶，达到创纪录的每天 8740 万桶，增幅为 3.1%。与此同时，全球的石油生产也有 2.2% 的总体增幅，每天增产大约 180 万桶。

- 天然气。全球天然气消费增长了 7.4%，为自 1984 年以来的最快增长。全球天然气产量增加了 7.3%，如图 3－3 所示。

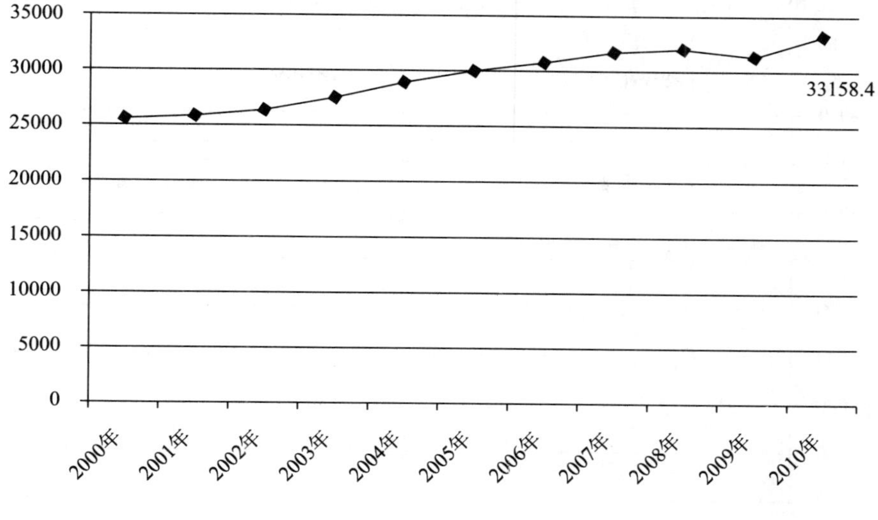

图 3－2　全球二氧化碳排放量　单位：百万吨

Figure 3－2　Global Carbon Dioxide Emissions　Unit：Million Tons

资料来源：《BP 世界能源统计》

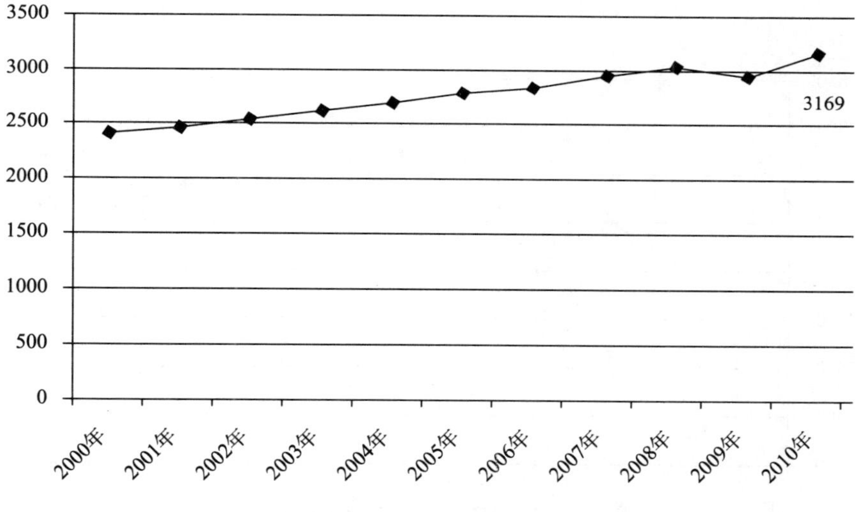

图 3－3　全球天然气消费量　单位：10 亿立方米

Figure 3－3　Global Natural Gas Consumption　Unit：1 Billion m^3

资料来源：《BP 世界能源统计》

• 煤炭。全球煤炭消费增长了 7.6%，是 2003 年以来最快的增速。目前，煤炭已经占到全球能源消费总量的 29.6%。

• 水电。全球水电消费量实现连续七年增长，总消费量增加了大约 5.3%，也打破了以往的消费纪录（如图 3－4）。其中，消费增长最为强劲的国家分别为：葡萄牙增长 88.2%；西班牙增长 60.9%；比利时增长 59.4%。

• 核电。全球核电在连续三年下滑之后，2010 年仍有 2% 的增长。其中，约 3/4 的增长来自经合组织国家，如图 3－5 所示。

• 其他可再生能源。2010 年，全球其他可再生能源也持续快速增长。其中，全球生物燃料的产量增加了13.8%，相当于每天增长24万桶油当量，成为全球液态燃料中产量增长最多的一项，如图3－6所示。

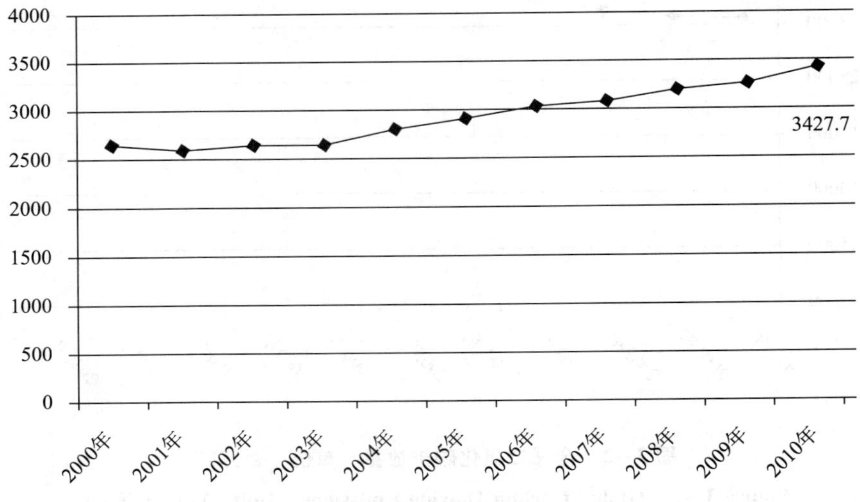

图3－4　水电消费量　单位：太瓦小时

Figure 3－4　Water and Power Consumption　Unit：TWh

资料来源：《BP世界能源统计》

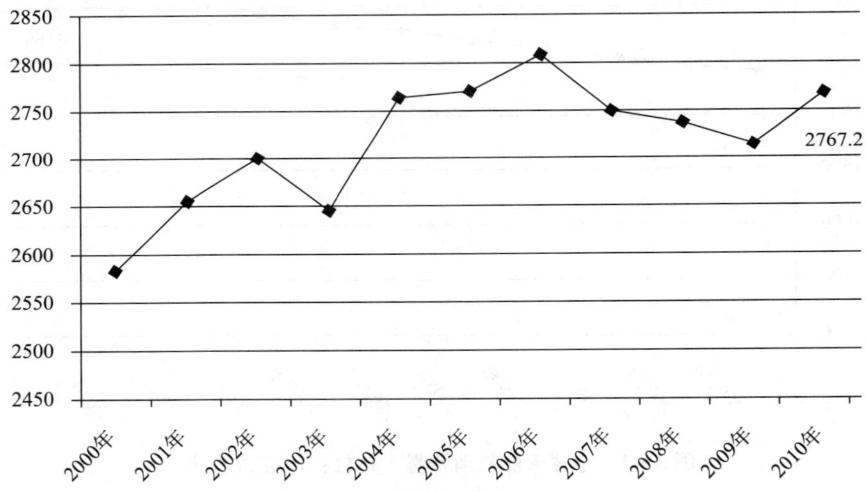

图3－5　核电消费量　单位：太瓦小时

Figure 3－5　Nuclear Power Consumption　Unit：TWh

资料来源：《BP世界能源统计》

报告对不同国家和地区的能源数据进行对比，有如下主要结论与数据：

- 经合组织国家仍是全球可再生能源电力的主要生产国，占全球总产量的77.5%。
- 美国和巴西仍然是全球生物燃料生产的领军国家。美国生物燃料2010年的产量增长了17%，巴西则增长了11.5%。
- 法国的核电产量增长最多，增幅达到4.4%。
- 中国是全球最大的水电生产国，占全球水力发电量的21%。其他份额超过5%的国家包括巴西、加拿大、美国，其水电产量分别占全球的11.6%、10.7%和7.6%；煤炭消费量增加了

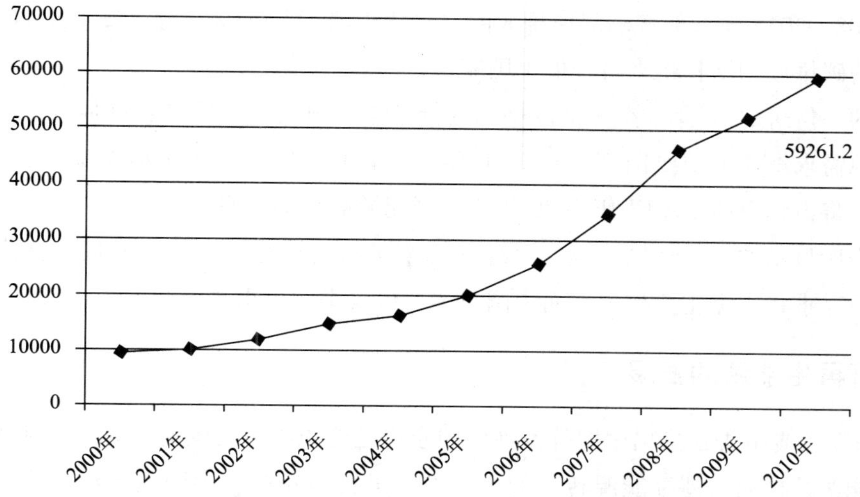

图 3－6　生物质能源生产量　单位：千吨油当量

Figure 3－6　Production Capacity of Biomass Energy　Unit: Thousand Tons of Oil Equivalent

资料来源：《BP 世界能源统计》

10.1%，大约占全球增幅的2/3。不过，去年中国的煤炭产量也增加了9%，在全球煤炭产量6.3%的增幅中也占了2/3左右。

- 美国和中国的风力发电增长最为显著，占到全球总增长的近70%。

同时，华盛顿的独立研究机构世界观察研究所在一份研究报告中表示，全球能源强度，即全球能源消耗量与全球总产值的比例，在过去两年的增长速度超过了全球经济的增速。在2010年，全球能源强度增加了1.35%。而在过去30年的时间里，全球能源强度一直呈下降的趋势。根据世界观察研究所的统计，在1981—2010年期间，全球能源强度下降了20.5%，即每年的降幅为0.8%。该报告认为，全球能源强度之所以增加，主要原因是“在全球经济低迷的情况下，很多国家都采取了刺激经济的促进措施，而促进措施中有很大一部分是基础设施的投资，像中国一样投资很多铁路、高速公路的建设，从而刺激像钢铁这些行业的发展。当然，美国也差不多，它的经济刺激计划也有很多基础建设的项目，而基建项目很多都是高耗能的。”

两份报告显示出了全球能源消耗量及能源强度均出现了较为显著的增长。这也印证了国际能源署《世界能源展望2009》中的分析结论，各国政府如果保持现有政策和发展模式不变的情况下，一旦经济复苏，能源消费将很快回升。因此，2009年全球一次能源消费虽然下降了1.1%，但未能形成持续下降趋势。

那么，未来将会怎样？美国能源部发布的《2011国际能源展望报告》给出的分析显示，全球能源需求总量到2035年将从2008年水平再增长53%，其中，中印两国将占据增长比例的一半之多。中印两国占全球能源总消耗的比例将从2008年的21%升至2035年的31%。到2035年，中国能源需求将超过美国能源需求68%。

该报告其他主要观点及数据如下：

- 从2008—2035年，全球天然气消耗总量预计将增长52%，至169000亿立方米，同时，天然气需求的增长速度将赶超其他有机能源需求的增长速度。
- 从2008—2035年，能源需求总量将平均增长1.6%。发展中国家强劲的经济增长将促使能源消耗增长速度的加快。

• 从2008—2035年，能源消耗所排放的二氧化碳将增长43%，至432亿公吨。报告指出，未来大部分的碳排放物增长将来自发展中国家。

麦肯锡的一份对2050年前全球能源格局进行考察和预判的报告也着重强调：今后发达国家不是推动能源需求增长的主要国家，未来40年中，新兴国家将占到全球能源需求增长的90%。这之中，中国将占到30%，相当于2030年前全球能源需求的26%。

上述报告所显示的一致性结论是：全球能源需求将持续增长，其中发展中国家强劲经济增长带来的能源需求将占据主要位置，而中国是主要位置中的最重要一个。

3.2 可再生能源的发展

3.1节的麦肯锡报告在指出中国等发展中国家将是能源消费大国的同时，也侧重指出：中国是推动全球能源结构向可持续能源转型的主力军。早在两年前，中国就已经成为全球最大的可再生能源投资国，其投资额远远高于排名第二及第三的美国和英国。本节我们将集中呈现全球领域对于可再生能源的主要观点及数据。

麦肯锡报告认为，中国未来20年将引领全球核能和可再生能源的发展和投资。该报告数据显示：

• 今后20年，中国或将占到全球新增可再生能源发电能力的23%。在此期间，中国将占到全球新增水力发电容量的32%、风力发电容量的26%和核能发电容量的47%。

• 从2010—2030年，中国平均每年将投入1300亿美元建造约110万兆瓦电力的新电厂，相当于目前全球最大电力生产国美国整个装机容量。

• 在全球每年4900亿美元总投资中，中国占到27%。在新增发电能力中，将近一半即47万兆瓦的电力，为风力、水力和太阳能等可再生能源，相当于目前德国发电总量的3倍以上。

与麦肯锡的报告视角较为接近，罗兰·贝格管理咨询公司发布的《2010—2011全球及中国新能源产业发展报告》侧重分析了光伏、风能、核能等新能源行业。主要结论如下：

• 到2012年，全球光伏市场每年将增长35%，总体规模将达到19GWp，欧洲依然是主要市场，预计将占50%的市场份额。但是美国、中国和印度等新市场发展潜力较大，如图3－7所示。

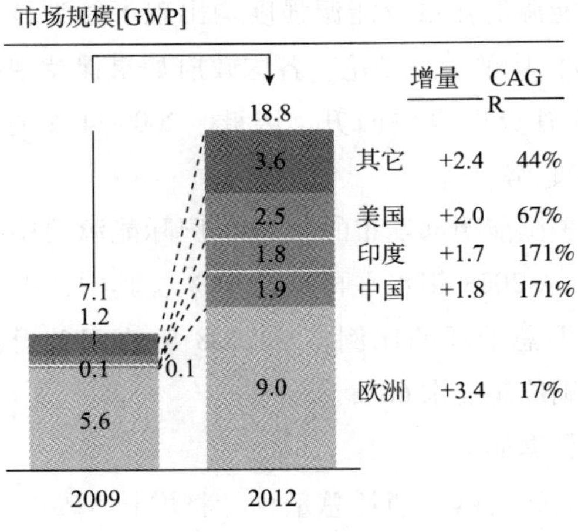

图3－7 全球光伏市场规模

Figure 3－7 Global PV Market Scale

资料来源：罗兰·贝格

• 在中短期内，风电行业对全球增长的预期调低。在未来几年，中国和美国将是最大的市场。海上风电的发展已经不会逆转，但仍存在不确定性。全球风电新增产能预测情况如图 3－8 所示。

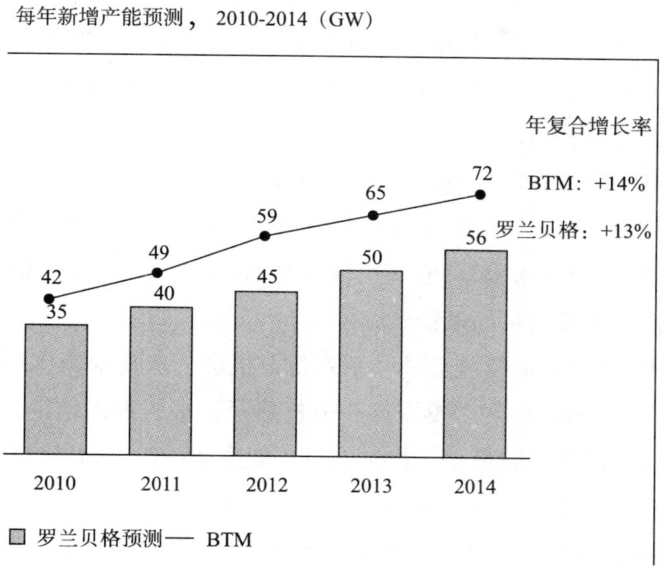

图 3－8 全球风电新增产能预测

Figure 3－8 Forecasting on Global Wind Power New Capacity

资料来源：罗兰·贝格

从 1954 年第一台核电站投运以来，全球核电产业的发展经历了多个发展阶段，未来 20 年内将迎来新一轮增长高峰。（该报告发布时间早于日本福岛地震引发的核泄漏事件，因此可能未能考虑此项因素导致的政策影响。）

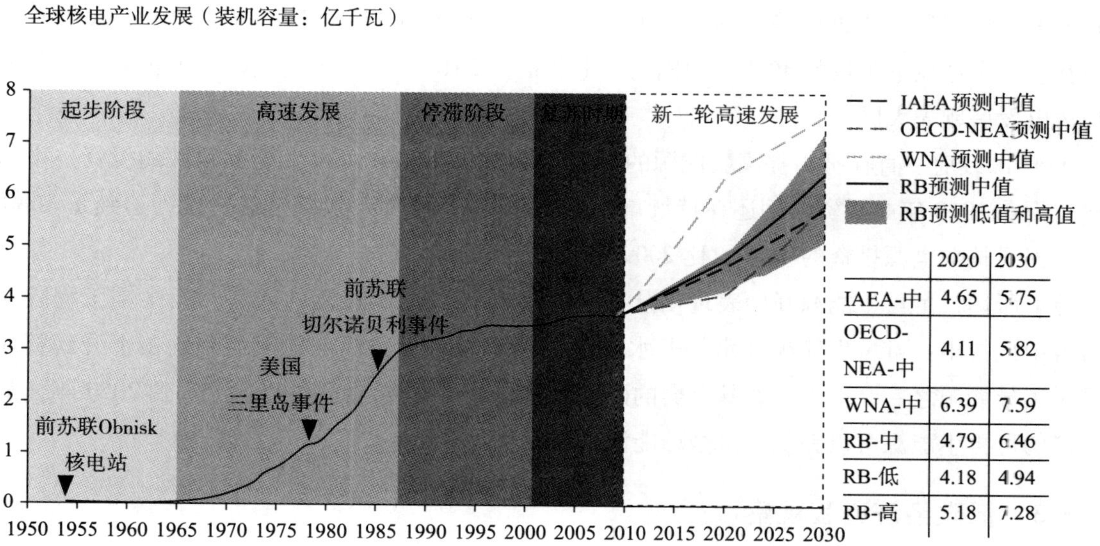

	2020	2030
IAEA-中	4.65	5.75
OECD-NEA-中	4.11	5.82
WNA-中	6.39	7.59
RB-中	4.79	6.46
RB-低	4.18	4.94
RB-高	5.18	7.28

图 3－9 全球核电产业发展预测

Figure 3－9 Forecasting on Global Nuclear Power Industry Development

资料来源：罗兰·贝格

新能源、可再生能源一方面是调整能源结构的手段，可以利于环境保护，减少二氧化碳排放；另一方面，不论是麦肯锡，还是罗兰·贝格，都把其视为一个产业来进行分析。目前，诸多的研究报告呈现了这一视点。美国皮尤慈善信托基金会（Pew Charitable Trusts）发布的《全球清洁能源的未来：价值2万3千亿美元的机会》的研究报告更是把新能源视作“一个利润可观的普通产业”。该报告指出，如果执行强硬新能源政策，未来10年20国集团对新能源项目的累积投资将达23000亿美元，比执行现行政策多出5460亿美元。到2020年，20国集团的年投资总额将达到3370亿美元，比2010年增长161%。其中，“美国、印度、英国将是最大受益国。”

皮尤慈善信托基金会的报告同时指出，新能源政策对气候影响、电价变化产生的积极影响也是报告着力强调的方面。“只有这样才能确保全球在2015年前达到温室气体排放峰值，使全球升温不超过2摄氏度。”报告指出，“它还可使未来10年20国集团的累积发电量增加1180千兆瓦，相当于目前清洁能源发电量的4倍。”

“清洁能源行业的发展沿供应链创造了一系列就业机会—从技术操作到运输—市场的扩张将惠及全球的工人和企业。”通过市场—投资这一分析路径，报告得出结论，“哪个国家采取的政策框架最强硬，哪里吸引的资金就最多，得到的包括创造就业在内的经济利益也就更多。”

国际环保组织绿色和平发布的《全球电力市场发展研究报告》宣称，一场全球可持续能源革命已经悄然发生。绿色和平呼吁中国政府抓住此机遇，大力推进可再生能源发展、摆脱对煤炭的依赖，跟上全球能源转型的时代步伐。研究报告主要数据显示：

- 全球可再生能源在2010年增长了8700万千瓦，第一次超过了煤电的发展。其中，风电在2010年一年就了3500万千瓦的新增装机，而核电却花了10年才达到这一规模。
- 同时，煤电增长在中国以外的地区正在逐渐放缓，全球煤电在2010年一共增加了6300万千瓦，其中88%发生在中国。在过去10年间，中国新增3.75亿千瓦煤电，占全球新增4.75亿千瓦煤电装机的79%，这10年修的煤电厂到本世纪中叶前能排放高达55亿吨的二氧化碳。
- 虽然中国仍然在大规模的上马火电，但中国的可再生能源发展势头迅猛。中国的风电的增长速度举世瞩目。从2003年开始，中国的新增风电装机量年年翻番，在2010年达到了1800万千瓦，占全球风电市场的49%。中国的风电装机在2010年已经超过美国位列世界第一，成为全球风电发展领先大国。
- 中国政府一直在努力提高新能源的投资，2009年中国在可再生能源领域的投资一举超过煤电，而在2010年此投资比例还在持续增加，2010年中国在水电与风电的投资达到了1680亿元，占据了整个电源投资的46%，比2009年的比例增加了4.8%。

综上所述，可再生能源开始表现出从补充能源向替代能源过渡的趋势，而中国工程院在一份报告中也指出：可再生能源（水和非水）的战略地位将由目前的补充能源逐步上升到替代能源乃至主导能源之一。世界自然基金会的研究报告则表明了更为乐观的观点：到2050年，清洁和经济的可再生能源可以完全满足全球能源的需求。

3.3 低碳经济进程观察

3.2节提到的新能源、可再生能源是目前主流认识中低碳经济的核心范畴，也是节能减排（从传统化石能源节约，二氧化碳排放降低的角度而言）的核心范畴。进入21世纪后，低碳经济作为一种以低能耗、低污染、低排放为基础的新经济模式日益受到世界各国的青睐。该概念的提出是循环经济、绿色经济这两个提法在21世纪气候变化新背景下的延续。低碳经济的概念缘起如图3-10所示。

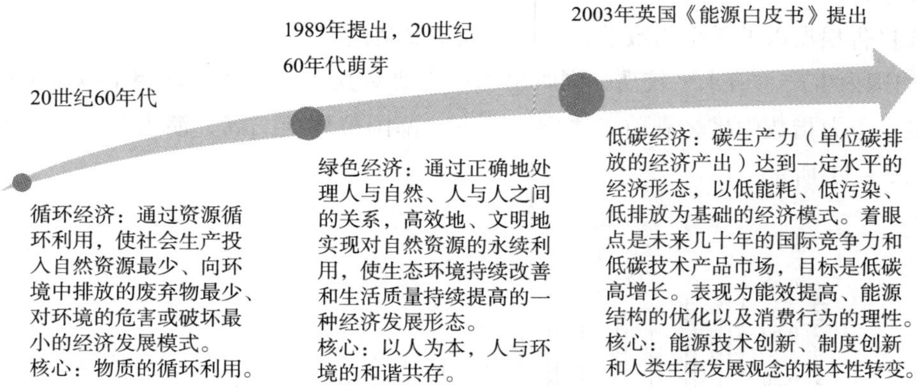

图3-10 低碳经济的概念缘起

Figure 3-10 Origin of the Concept of Low Carbon Economy

有分析认为，英国成为全球低碳经济的积极倡导者和先行者，英国已初步形成了以市场为基础，以政府为主导，以全体企业、公共部门和居民为主体的互动体系，从低碳技术研发推广、政策发挥建设到国民认知姿态等诸多方面，都处在了世界领先位置；低碳经济或为欧盟就业出路，低碳经济给欧盟带来的不仅是维持欧盟在环保领域的优势地位，提高竞争力，还可以大大缓解失业带来的巨大社会压力，保持欧盟发展模式的生命力；日本强化低碳经济旨在实现低碳社会，日本期望通过低碳革命和引领世界二氧化碳低排放革命来建设健康长寿社会并发挥日本魅力。

3.3.1 G20的低碳竞争力

进一步而言，第三代环保主义机构”（E3G）推出的一份题为《20国集团[①]低碳竞争力》的报告，对20国集团中的19个国家的低碳竞争力进行了更为具体和量化的评估。因为传统的竞争力研究方法无法对各国应对限制碳排放这一新情况所带来的机遇与成本进行评估，所以这份报告是通过数据对比对各国在目前和将来实施这一转变过程中所取得的进展进行评估的。

评估整体低碳竞争力主要通过三个方面的因素：该国目前的情况、改变的速度以及改变的规模。因此，这份报告按照以下三个指标对20国集团成员国的表现加以对比：

（1）低碳竞争力指数。根据每个国家当前的政策环境和政策指标，衡量各自在低碳世界中的竞争力，以及为其国民带来物质繁荣的能力。

（2）低碳提升指数。各国所展现出来的随着发展提高碳竞争力的潜力。对于人均国内生产总值很低的国家来说，他们或许并没有很强的碳竞争力，但是却可能在提高碳竞争力方面取得长足进步。

（3）低碳缺口指数。该国的实际改进速度和要求达到的改进速度之间的差额；要求达到的

---

① 20国集团由19个国家和欧盟组成，其成员为：阿根廷（ARG）、澳大利亚（AUS）、巴西（BRA）、加拿大（CAN）、中国（CHIN）、欧盟（EU）、法国（FRA）、德国（GER）、印度（IND）、印度尼西亚（IDN）、意大利（ITA）、日本（JPN）、墨西哥（MEX）、俄罗斯（RUS）、沙特阿拉伯（SAU）、南非（ZAF）、韩国（KOR）、土耳其（TUR）、英国（GBR）和美国（USA）。其中，欧盟作为一个整体在低碳竞争力方面的表现不在此报告的研究范围内。

改进速度是指，按照该国预计的经济增长率计算，为了使温室气体的大气浓度稳定在450ppm（每百万含量）二氧化碳当量，该国应该将碳排放降低到规定的程度，为了实现这一目标，该国应达到的改进速度。

进而该报告根据以上三个指数的比较，形成主要结论如下：

• 人均国内生产总值水平较高，同时也承认需要对其经济发展进行调整以降低碳排放的国家，在低碳竞争力指数中排名靠前。法国、日本、德国位列前三名，澳大利亚、南非等国居于后列，如图3-11所示。

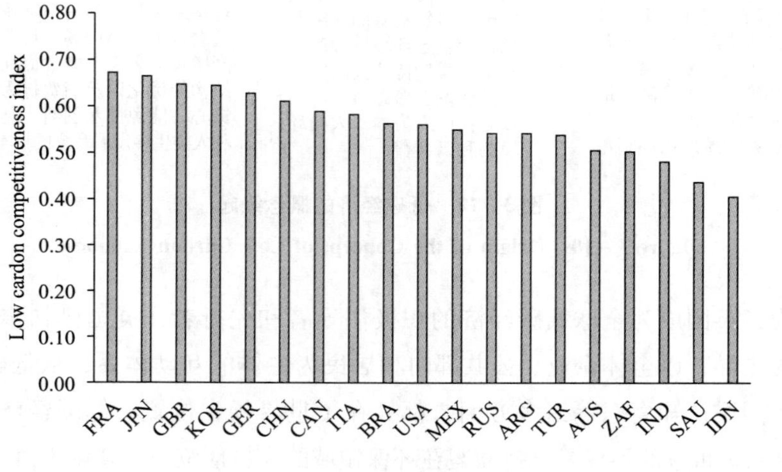

图3-11 G20低碳竞争力排名

Figure 3-11 Low Carbon Comperitiveness Ranking in G20

资料来源：E3G

• 虽然德国在低碳提升上位居首位，但是许多中等收入国家却以比某些发达国家更快的速度提高他们的碳生产力，尤其是位于第二位和第三位的南非和墨西哥。在过去几年中，这三个国家证明了他们有能力在保持经济增长的同时，还使每吨二氧化碳排放所创造出的国内生产总值大幅提高，如图3-12所示。

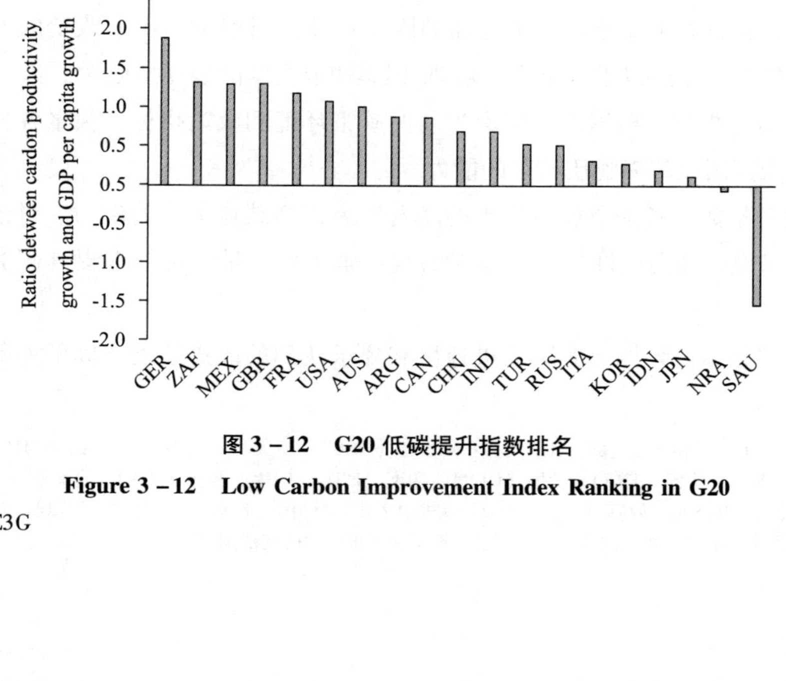

图3-12 G20低碳提升指数排名

Figure 3-12 Low Carbon Improvement Index Ranking in G20

资料来源：E3G

● 在各自既定的人口增长率和 GDP 增长率下，目前有两个国家的碳生产率达到了全球减排目标的标准。这两个国家就是墨西哥和阿根廷。作为非附件一国家（未被列入气候公约“附件一”中的各工业化国家，即发展中国家），他们通过提高碳效率及降低排放双管齐下的方法，使其碳生产力保持持续增长。中国、南非以及德国正逐步靠近正确的发展轨道。其他国家则需要提高他们的碳生产率。而澳大利亚，土耳其、俄罗斯以及沙特阿拉伯则必须实现碳生产力的大幅度扭转。低碳缺口指数排名如图 3－13 所示。

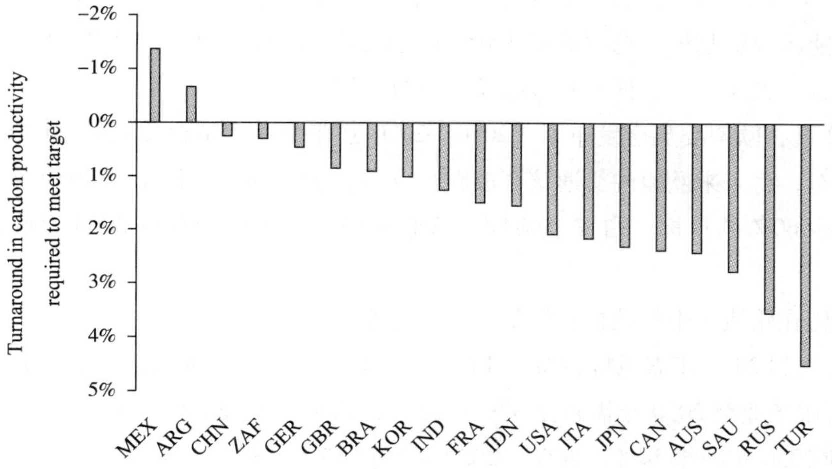

图 3－13 低碳缺口指数排名

Figure 3－13 Low Carbon Gap Index Ranking

资料来源：E3G

总体而言，由于 20 世纪 90 年代的经济结构转型，欧洲国家（法国、英国和德国）目前成为向低碳世界转变的领头者，但一些发展中国家正迅速赶上。该报告显示，20 国集团各成员国在低碳竞争力方面存在着巨大差异。这些差异虽然反映了他们各自不同的出发点和目标水平。但是，通过报告，我们还能看到，各国存在着提高并向低碳经济转型的潜力。

提高碳生产力不一定需要以经济增长为代价，这不仅对发达国家如此，对发展中国家同样如此。德国、南非、以及墨西哥三国存在着巨大的不同。但是，在最近几年，他们均显示出使经济增长与碳排放脱钩的实力。

下面我们将继续从低碳经济角度对欧盟、美国等国的最新动向进行介绍，主要方法是从近期上述国家和地区最新的动态性事件入手，来捕捉各方观点，以力图揭示深层立场和可能趋向。

3.3.2 欧盟

欧盟需要关注的事情主要有两个：其一是 2011 年 3 月份发布的《欧盟 2050 低碳经济路线图》；其二是欧盟拟征收航空碳排放费。

1. 欧盟 2050 低碳经济路线图

《欧盟 2050 低碳经济路线图》表示，欧盟将在 2020 年实现 CO_2 减排 25%，同时将现有能效提高 20%。其最终目标是：相较 1990 年的排放值，2050 年实现温室气体减排 80%—90%。

为了达到欧盟 2050 年减排 80%—90% 的目标，路线图规划如下：

● 欧盟希望 2030 年达到温室气体减排 40%。

- 2040年减排60%。

根据路线图要求，以1990年排放值为基准：

- 2020年之前，年减排目标应每年递增1%。
- 从2020—2030年，年减排目标应每年递增1.5%。
- 从2030—2050年，年减排目标应每年递增2%。

为了满足如上条件，欧盟出台了一系列政策：

- 2011年1月31日，欧盟委员会于布鲁塞尔发布了其《面向2020年——新能源计划》，包括计划将新能源投资翻倍，总数额将达到700亿欧元。同时，加大利用多样化金融工具，包括公共债券、风投、夹层融资、证券基金以及保险投资等。
- 2月22日，欧盟委员会发布了《2010—2020欧盟交通政策白皮书》，主要内容包括加大使用新能源汽车，将未来欧盟的交通核心放在公共运输上。同时，提出了2050年交通方面的温室气体排放减少60%的目标。白皮书预计，欧盟未来将每年至少减少5000万吨的二氧化碳排放量。

同时，欧盟在路线图中将减排目标细分到了行业。

- 电力部门到2030年需减排34%—40%，到2050年则需实现减排93%—99%。
- 住宅与服务业至2030年需减排37%—53%，2050年减排88%—91%。
- 工业到2030年减排34%—40%，2050年减排83%—87%。
- 农业的非二氧化碳温室气体减排，其目标为2030年实现减排36%—37%，2050年达到42%—49%。

根据路线图的预计，若全面实施此计划，欧盟委员会预计未来10年将需要至少增加500亿欧元的研发和示范资金。欧盟委员会提示，成员国应充分利用拍卖市场和欧盟的区域经济援助。

目前，欧洲在低碳经济上的投资占总GDP的19%，其中包括：智能电网、CCS、交通电力化、被动式建筑。欧盟委员会预计，至少需要未来40年达到每年2700亿欧元的总投资。

有观点认为，欧盟上述决策的背后是不可或缺的经济驱动。该观点可以从下述事例及观点得到支撑：

- 德国环境部2月21日发布的工作报告称，如果将2020年的减排目标定于30%，欧盟GDP每年将会额外增加0.6%，同时增加600万工作机会。
- 报告主要起草者Carlo C. Jaeger在报告中认为："低碳经济的投资将会在未来带来18%—22%的GDP增长，所有27个成员国都可以享受到这样的经济成果，同时取得失业率的下降和更多的投资机会。"
- 欧盟委员会也认为：根据路线图的轨道发展，减排将会给碳市场带来大量密集的投资。稍后，碳价格必然会得到一定提高。更为重要的是，这将会使得在低碳领域的总投资大于总成本，低碳经济将得以可持续发展。

总体而言，欧盟认为，低碳经济的投资将会在未来带来更大回报。

2. 欧盟拟征收航空碳排放费

另一动向性事件是欧盟拟征收航空碳排放税。欧盟在2011年3月7日发布的规定将欧盟航空业总排放许可量上限规定为2004—2006年三年平均排放量的97%，而到2013年必须降至95%。5月13日，欧盟宣布从2012年1月份起将航空业正式纳入欧盟碳排放交易系统，飞经欧盟的航空公司可免费获得碳排放配额的82%，剩余份额中的15%将由航空公司通过拃卖的方式获取，其余的3%的排放量将被分配给高速成长中的航空公司以及行业后来者。欧盟希望通过这样的机制来保证欧盟航空业的二氧化碳排放不再增长并且可以逐年下降。

对此，中国航空运输协会与欧盟会晤并展开对话后声明，不承认欧盟碳排放交易。根据中国民航局节能减排办公室的介绍，一旦欧盟根据计划于2012年起征收该笔费用，中国民航业界明年预计将为此支付8亿元人民币，而至2020年，则可能将支付超过30亿元人民币，九年间累计支出约176亿元。

中航协具体表明了对欧盟碳排放交易的质疑，指出其不合理之处在于：

其一，非直接减排手段，没有对节能减排产生直接和实际的效果，而是单方面建立的一种以节能减排名义实施的迂回的具有金融性质的制度设计，将阻碍、抑制广大发展中国家航空运输业的发展。

其二，没有对发展中国家的航空运输业发展需要作出合理的安排，违背了国际社会在气候变化领域普遍遵循的“共同但有区别的责任”原则。

其三，违背了《芝加哥公约》和国际民航组织对航空减排的基本立场。

其四，实施范围超越了欧盟的管辖权，违反了国际法和国际惯例并触犯了国家主权。

其五，实施对象颠倒，没有首先针对飞机及发动机的制造商，减排没有直接从源头抓起。

美国航空运输协会亦称，将欧盟的法律体系“应用于国际航空业违反了国际法，这是一种很不好的政策”。美国航空运营商认为，欧盟这项法律违反了《京都议定书》和《国际民用航空公约》的规定。

厦门大学林伯强认为：虽然说的是排放交易系统，实质上已经是对航空业征收碳关税了，同时也揭开了“低碳全球化”导致的贸易格局大变动和贸易博弈的序幕。无论出于何种动机，发达国家显然想把低碳全球化做成推动其经济增长的机会，关键是今后的贸易保护可以打着低碳发展的旗号。

具体而言，发达国家可能通过贸易（碳关税）和其他非贸易手段（减少能源补贴），迫使发展中国家减少排放。发达国家可以先利用资金和技术优势，创造低碳技术和管理的制高点，然后在气候谈判中施加温室气体减排压力，并向外输出“绿色技术和管理模式”；同时，提高进入市场产品的环保标准，制造“绿色壁垒”。

3.3.3　美国

从上文可以看出，在全球低碳经济发展的背景下，国际贸易摩擦及贸易壁垒也更多的与“低碳”结合起来。2010年10月，美国贸易代表办公室宣布，应美国钢铁工人联合会申请，启动对华清洁能源有关政策和措施的301调查。据悉，该调查范围涉及中国的风能、太阳能、高级电池及节能汽车等行业。

美国钢铁工人联合会按照《1974年美国贸易法案》第301条款规定，向美国贸易代表办公室提交了长达5800页的诉状。起诉文件对中国绿色产业的政策和措施提出多项指责，包括中方控制关键原材料、为国内厂商提供大量补贴、歧视外国公司和产品、要求外国投资者转让技术等。申诉书主要观点如下：

- 中国政府为其风能、太阳能、电池及节能汽车等产品提供了不公平的支持，导致美国相关企业的利益受到损害、美国的贸易逆差增大、美国在此类行业的就业机会减少。

• 中方控制关键原材料、要求外国投资者转让技术、歧视外国公司和产品、为国内厂商提供大量补贴等。在重点领域之一的风电行业中，美方认为中国政府为国内风机制造商提供补贴、优惠贷款等，并在风电特许权招标中歧视国外公司，导致美国企业向中国出口的风机整机和关键零部件大幅下滑。

• 中国政府自2007年起对光伏产业实施了多项政策，如对用户侧光伏系统补贴50%及偏远地区补贴70%的“金太阳计划”、财政部的《太阳能光伏建筑应用财政补助资金管理暂行办法》、山东省能源基金等。

对此，中国政府积极回应。前国家发展和改革委员会副主任、国家能源局局长张国宝表示，美国的调查只会将其对本国新能源产业的补助更多地暴露在世界面前。国家能源局和商务部表述了如下的观点：

• 中国在各个方面对新能源企业的补贴总额是微乎其微的。有关资料显示，美国2010年1—9月份，已经通过现金支付方式给美国的新能源企业提供了46亿美元补助，美国的风电企业获得了其中30亿美元的补贴。

• 截至2010年10月，中国仅向美国出口了3台风机，容量还不到1万千瓦；而美国通用电气公司2005—2009年累计向中国出口的风机容量达到113万千瓦。

• 美方此举向外界发出了贸易保护主义的错误信号。

另外，在“碳关税”尚未正式通过生效之时，欧美等国家关于节能标准、碳评级、碳标识等碳壁垒的诸多细小动作，都会引来包括多数发展中国家在内的普遍关注。

3.4 德班会议

《联合国气候变化框架公约（UNFCCC）》第17次缔约方大会（COP17）暨《京都议定书》第七次缔约方大会（CMP7）（以下简称“德班大会”）于2011年11月28日—12月9日在南非德班召开。该会议的核心焦点是《京都议定书》第二承诺的走向。南非环境部长埃德娜·莫莱瓦4月20日在南非议会发表讲话时表示，作为主办国，南非不希望《京都议定书》在此次会议上“死亡”，恰恰相反，南非政府认为延长《京都议定书》是至关重要的。

但是，7月3日—4日，来自35个国家的环境部长和谈判代表在德国彼得堡为2011年11月28日开始的南非德班国际气候变化谈判会议作准备，在会议中，默克尔表示：“目前的二氧化碳自愿减排远不充分，排放令人担忧，谈判速度迄今却仍以蜗牛的速度前进。”为此，德国和欧盟认为需要一个单一的、具有法律约束力的《京都议定书》后继者。但是，该后继者直接触碰到了发展中国家谈判的底线，即发展中国家坚持的“双轨制”以及“共同但有区别的责任”是否会被抹杀。

《联合国气候变化框架公约》执行秘书菲格雷斯认为德班会议如果能够解决以下四个关键问题，国际社会应对气候变化的能力将会迈出重大步伐。

（1）有关《京都议定书》第二承诺期的政治问题。削减温室气体排放量不仅仅只需要发达国家的努力，发达国家正在发挥领导作用，同时还需要全球的集体努力，我们要解决如何以透明的方式来推进全球集体努力，并且随着时间的推移不断扩大削减目标。

（2）在去年的坎昆会议上，国际社会承诺从2013年开始审议各国采取的减排措施是否能够达到控制气候变化的科学要求，有必要在此之前确定审议规则。

（3）明确应对气候变化的资金问题，一方面确定明确的第一个气候融资截止时间；另一方面要对到2020年之前每年提供1000亿美元资金的承诺制订明确的计划。

（4）确定新技术机制和适应气候变化委员会的运作细节。

对于气候谈判，不同的国家及利益集团持有不同的立场。我们对基础四国、欧盟的立场进行简单介绍如下。

基础四国是由中国、印度、巴西和南非四个主要发展中国家组成的气候谈判集团，作为由巴西、南非、印度和中国等主要发展中国家组成的气候谈判集团，各成员间在应对气候变化中既有共同的利益追求，由此决定了基础四国在后京都气候谈判中重要谈判议题上的立场具有一定的共同点。其主要立场有以下四点：

（1）后京都气候条约的法律地位。后京都气候条约的法律地位一直是谈判中分歧颇大的议题之一。2009年底的联合国哥本哈根气候会议虽达成了《哥本哈根协议》，但《公约》缔约方会议（COP）在大会最终决议中仅表示注意到（take note of）它的存在，但并未确立其在联合国多边气候谈判中的位置，更未解决未来气候条约与《公约》以及《议定书》的关系问题。在后哥本哈根气候谈判中，基础四国强调《公约》渠道是气候谈判的唯一合法平台，未来的气候谈判应继续坚持“双轨制”。

（2）发达国家的减排承诺。2012年后发达国家的进一步减排是构建后京都气候机制的关键。基础四国认为，未来全球升温与工业革命前相比不应该超过2℃。据此，基础四国坚持发达国家应承担涵盖所有经济部门的、量化的、有法律约束力和能够实现“可衡量、可报告和可核实”的减排义务。对于发达国家的量化减排目标，基础四国认为空谈发达国家的长期目标徒劳无益，应先规定发达国家的中期减排目标，并且把2020年作为确定中期减排目标的时间点。

（3）发展中国家的适当减排行为。根据“巴厘岛路线图”，在后京都时代，发展中国家在发达国家的支持下应当采取适当的国家减排行动。对于发展中国家应承担的义务，基础四国一方面反对发达国家给发展中国家整体温室气体排放设限，另一方面反对将发展中国家区别对待，更反对通过渐进的“晋级制”（Graduation Process）使他们承担新的义务。对于发展中国家适当减排行为的性质，基础四国认为，适当国家减排行为和发达国家的减排行为有着本质的区别，应是发展中国家根据国情采取的自愿行为，不具有强制性和法律约束力，并且其实施是以发达国家的减排行为和支持为前提的。

（4）发达国家的资金和技术支持。发达国家的支持是发展中国家应对气候变化的重要保证，发达国家向发展中国家的资金提供和技术转让仍是双方矛盾的焦点之一。除了坚持发达国家的支持要做到可衡量、可报告和可核查外，基础四国详尽地阐明了其在发达国家资金和技术支持上的立场。在资金支持上，对于资金的来源，基础四国均强调发达国家应提供新的、额外的、充足的和可预期的资金，来自发达国家政府的公共资金应发挥主导性的作用，私营投资和全球碳市场可作为有益的补充。

2009年1月28日，欧盟委员会发布《哥本哈根气候变化综合协议》文件，阐明了欧盟在气候谈判上的基本立场。此后，欧盟通过参加G8峰会、主要温室气体排放国会议、联合国气候变化峰会、G20金融峰会以及联合国哥本哈根气候会议的机会，进一步完善了其在气候谈判上的立场。

（1）发达国家的减排承诺。认为发达国家应该率先减排，作为一个整体到2020年将其温室气体排放量在1990年的基础上减少30%，进而实现全球排放总量在2020年达到峰值，到2050年减少为1990年的50%。

（2）发展中国家的适当减排行为。欧盟认为发展中国家的温室气体排放正在迅速增加，倘若发展中国家不参与全球减排，将抵消发达国家所做出的减排努力。因此，发展中国家应采取适当的减排行为，作为一个整体到2020年将其温室气体排放量的增长与基准排放相比减少15%—30%。

（3）发达国家资金和技术支持的力度。资金支持在欧盟理事会的讨论中是颇具争议的议题。从目前欧盟内部协调的结果来看，欧盟认为资金应来自三个方面：发展中国家的自有资金（包括公私部门）、国际碳市场和国际公共资金（来自发达国家公共部门）。与此同时，欧盟强调私人资金应发挥主导作用，公共资金发挥辅助作用。

（4）发达国家资金和技术支持的管理结构。对于资金的管理，欧盟认为应遵循透明性、责任性、一致性、辅助性和可预测性等基本原则。总体管理应是分权型，国家主导以及自下而上的透明结构，从而确保对发展中国家支持的有效性。

相对于欧盟立场较为中立、温和而言，美国、加拿大、日本等国家持有和基础四国等发展中国家较为对立的立场。

（1）美国。美国政府于1997年在《京都议定书》上签字，但美国参议院没有核准。美国是唯一一个退出《京都议定书》的国家。从1997年以来，美国的二氧化碳排放既没有下降也没有减缓增长，而是快速增加，其幅度在发达国家中名列前茅。在气候变化谈判中，美国常常以未签署《京都议定书》为由，自起炉灶，我行我素。即使《京都议定书》有许多好的实施经验和方法，美国也常常拒之千里之外。

（2）加拿大。加拿大保守党政府在气候变化问题上，以美国马首是瞻，亦步亦趋。加拿大是哥本哈根会议后，第一个将承诺目标降低的国家。哥本哈根会议之前，加拿大政府提交的承诺目标是：2020年在2006年的基础上降低20%，等效于2020年在1990年的基础上降低3%。现在的目标是：2020年在2005年的基础上降低17%，等效于2020年在1990年的基础上增加3%。这两个目标之间相差6个百分点，而且基准年改为2005年，与美国的基准年相同。

（3）日本。在波恩会议上，日本代表团成员一改以往闪烁其辞的回答，而是重复申明，日本政府拒绝《京都议定书》第二承诺期的立场是明确和坚定的。尽管日本首相菅直人于今年6月1日在国会中肯定要坚守2020年在1990年基础上下降25%的目标，但在日本有一些团体试图在给国会提交的气候提案中抹掉25%的目标。日本政府在福岛核电站灾难事故之后，更加快了对《京都议定书》的抛弃。

从上文可以看到，在长期的气候变化谈判中，基本上形成了发达国家和发展中国家两大阵营，巨大的分歧很难在短时间磨合。如何在现有的政治运作模式中，加强沟通交流、协商和合作，是对谈判政治的一大考验。而且，2010年的全世界温室气体总排放量中，发达国家与发展中国家的比例大致相当。但在对2030—2050年的预测中，发展中国家的排放总量将逐渐地上升到排放总量的70%以上，但从人均碳排放量的角度来看，发达国家仍然比发展中国家高出几倍以上。这种动态变化的趋势，使气候变化谈判的基础和政治生态环境相应的发生变化。

第2篇 行业篇

工业、交通、建筑是我国能源消耗和污染物排放的重要领域，节能减排重点与难点体现在上述领域。“十一五”期间，我国在上述领域的节能减排取得了积极成效，为确保完成“十一五”节能减排目标作出了贡献。

但是，从未来发展看，工业化、城市化将进一步发展，“十二五”仍是我国工业化发展的关键阶段，居民消费结构继续升级，对工业品的需求也将继续增加，我国重化工业所占的比重仍将保持较高水平，对能源、交通、原材料等需求还会进一步增大，资源消耗总量将进一步增加，我国经济与社会发展面临的资源约束将进一步加大。

本篇我们重点分析工业领域，尤其是电力、钢铁、有色金属、石化、煤炭、建材、纺织以及交通、建筑等高耗能、高污染、高产值行业的转型与节能减排之路。

4　行业节能减排概述

4.1　行业节能减排“十一五”总结

工业、交通、建筑是我国能源消耗和污染物排放的重要领域，节能减排重点与难点均在上述领域。“十一五”期间的主要数据如下：

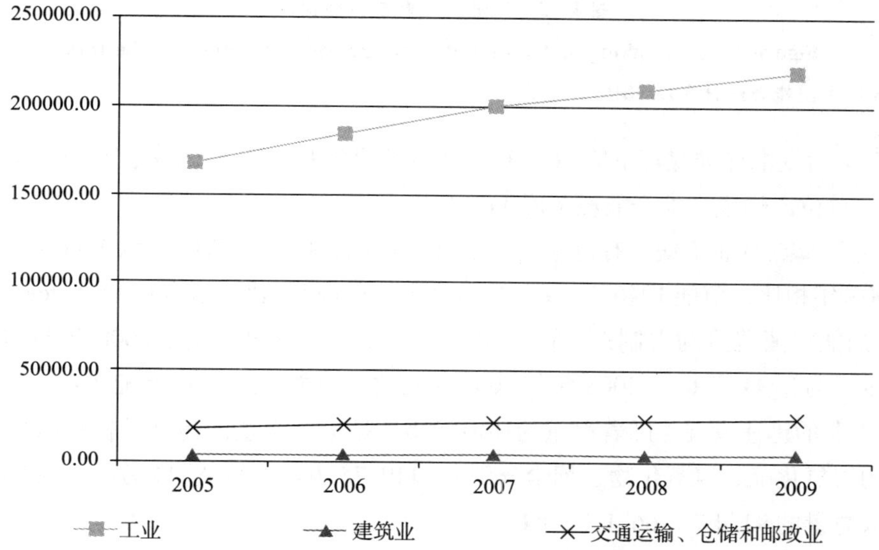

图 4－1　工业、交通、建筑总能耗增长趋势

Figure 4－1　Total Energy Consumption Growth of Industry, Transport and Construction

资料来源：中国能源统计年鉴 2010

“十一五”期间，工业总能耗由 2005 年 16.87 亿吨标准煤增加到 2009 年 21.9 亿吨标准煤，交通领域总能耗由 2005 年的 1.83 亿吨标准煤增加到 2009 年的 2.37 亿吨标准煤，建筑业总能耗由 2005 年的 0.34 亿吨标准煤增加到 2009 年的 0.46 亿吨标准煤，如图 4－1 所示。

2009 年工业领域主要耗能行业依次是黑色金属冶炼及压延加工业、化学原料及化学制品制造业、非金属矿物制品业、电力、热力的生产和供应业、石油加工、炼焦及核燃料加工业、有色金属冶炼及压延加工业、煤炭开采和洗选业、纺织业、造纸及纸制品业、石油和天然气开采业，上述行业能耗总量占工业能耗总量的 83.5%，如图 4－2 所示。

- 工业化学需氧量和二氧化硫排放量分别下降 25% 和 20% 左右。
- 火电供电煤耗由 370 克/千瓦时降到 335 克/千瓦时，单位铜冶炼综合能耗下降 35.9%，单位烧碱生产综合能耗下降 34.8%，吨水泥综合能耗下降 28.6%，原油加工单位综合能耗下降

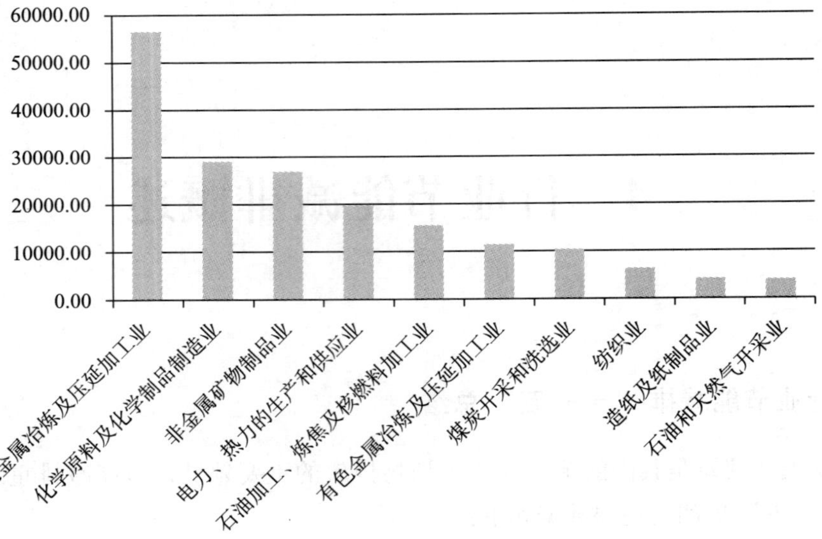

图 4－2　工业主要耗能行业排序

Figure 4－2　Ranking of Major Industrial Energy Consumers by Sectors

资料来源：中国能源统计年鉴 2010

28.4%，电厂火力发电标准煤耗下降 16.1%，吨钢综合能耗下降 12.1%，单位电解铝综合能耗下降 12.0%，单位乙烯生产综合能耗下降 11.5%。

- 工业部门重点行业钢铁、有色金属、建材、石化、化工和造纸等六个行业的 12 种产品，2008 年与 2007 年相比，节能 1946 万 tce，2008 年与 2005 年相比，节能 6816 万 tce。这六个行业的 12 种产品的能源消费量约占制造业能源消费量的 65%。按此推算，2008 年与 2007 年相比，制造业节能量约为 2994 万 tce；2008 年与 2005 年相比，节能量约为 1.05 亿 tce。
- 我国工业的迅速发展伴随着严重的环境污染。据第一次全国污染源普查结果，2007 年工业部门排放的二氧化硫、氮氧化物、烟尘分别为 2119.75 万吨、1188.44 万吨、982.01 万吨，分别占全国排放总量的 91.4%、66.1%、84.2%。

2009 年交通领域（交通运输、仓储和邮政业）的能源消费总量为 23691 万吨标准煤，2010 年估计为 24493 万吨标准煤。能源消费总量占全国能源消费总量的比重一直稳定在 7.7% 左右。

到 2010 年底，全国城镇新建建筑设计阶段执行节能强制性标准的比例为 99.5%，施工阶段执行节能强制性标准的比例为 95.4%，分别比 2005 年提高了 42 个百分点和 71 个百分点，完成了国务院提出的“新建建筑施工阶段执行节能强制性标准的比例达到 95% 以上”的工作目标。

4.2　行业节能减排“十二五”展望

“十二五”时期是我国由中等收入国家向中等发达国家迈进的重要阶段，工业化和城镇化进程将进一步加快，能源需求呈刚性增长，资源环境约束突出，传统增长模式面临新挑战。“十二五”时期，我国节能减排的主要目标是：

（1）到 2015 年，全国万元国内生产总值能耗下降到 0.869 吨标准煤（按 2005 年价格计算），比 2010 年的 1.034 吨标准煤下降 16%，比 2005 年的 1.276 吨标准煤下降 32%；“十二五”期间，实现节约能源 6.7 亿吨标准煤。

（2）2015 年，全国化学需氧量和二氧化硫排放总量分别控制在 2347.6 万吨、2086.4 万吨，比 2010 年的 2551.7 万吨、2267.8 万吨分别下降 8%；全国氨氮和氮氧化物排放总量分别控制在 238.0 万吨、2046.2 万吨，比 2010 年的 264.4 万吨、2273.6 万吨分别下降 10%。

1. 工业节能减排

对于工业节能减排，《“十二五”节能减排综合性工作方案》中主要安排如下：

一是重点推进电力、煤炭、钢铁、有色金属、石油石化、化工、建材、造纸、纺织、印染、食品加工等行业节能减排，明确目标任务，加强行业指导，推动技术进步，强化监督管理。

二是发展热电联产，推广分布式能源。开展智能电网试点。

三是推广煤炭清洁利用，提高原煤入洗比例，加快煤层气开发利用。

四是实施工业和信息产业能效提升计划。推动信息数据中心、通信机房和基站节能改造。

五是实行电力、钢铁、造纸、印染等行业主要污染物排放总量控制。

六是加强脱硫脱硝设施的强制安装及配套安装。

- 新建燃煤机组全部安装脱硫脱硝设施，现役燃煤机组必须安装脱硫设施，不能稳定达标排放的要进行更新改造，烟气脱硫设施要按照规定取消烟气旁路。
- 单机容量 30 万千瓦及以上燃煤机组全部加装脱硝设施。
- 钢铁行业全面实施烧结机烟气脱硫，新建烧结机配套安装脱硫脱硝设施。
- 石油石化、有色金属、建材等重点行业实施脱硫改造。
- 新型干法水泥窑实施低氮燃烧技术改造，配套建设脱硝设施。

七是加强重点区域、重点行业和重点企业重金属污染防治，以湘江流域为重点开展重金属污染治理与修复试点示范。

2. 交通领域节能减排

对于交通领域节能减排，《“十二五”节能减排综合性工作方案》中主要安排如下：

一是加快构建综合交通运输体系，优化交通运输结构。积极发展城市公共交通，科学合理配置城市各种交通资源，有序推进城市轨道交通建设。提高铁路电气化比重。

二是实施低碳交通运输体系建设城市试点，深入开展“车船路港”千家企业低碳交通运输专项行动，推广公路甩挂运输，全面推行不停车收费系统，实施内河船型标准化，优化航路航线，推进航空、远洋运输业节能减排。开展机场、码头、车站节能改造。加速淘汰老旧汽车、机车、船舶，基本淘汰 2005 年以前注册运营的“黄标车”，加快提升车用燃油品质。

三是实施第四阶段机动车排放标准，在有条件的重点城市和地区逐步实施第五阶段排放标准。全面推行机动车环保标志管理，探索城市调控机动车保有总量，积极推广节能与新能源汽车。

3. 建筑领域节能减排

对于建筑领域节能减排，《“十二五”节能减排综合性工作方案》中主要安排如下：

一是制定并实施绿色建筑行动方案，从规划、法规、技术、标准、设计等方面全面推进建筑节能。新建建筑严格执行建筑节能标准，提高标准执行率。

二是推进北方采暖地区既有建筑供热计量和节能改造，实施“节能暖房”工程，改造供热老旧管网，实行供热计量收费和能耗定额管理。做好夏热冬冷地区建筑节能改造。

三是推动可再生能源与建筑一体化应用，推广使用新型节能建材和再生建材，继续推广散装水泥。

四是加强公共建筑节能监管体系建设，完善能源审计、能效公示，推动节能改造与运行管理。研究建立建筑使用全寿命周期管理制度，严格建筑拆除管理。

五是加强城市照明管理，严格防止和纠正过度装饰和亮化。

从未来发展看，工业化、城市化将进一步发展，“十二五”仍是我国工业化发展的关键阶段，居民消费结构继续升级，对工业品的需求也将继续增加，我国重化工业所占的比重仍将保持较高水平，对能源、交通、原材料等需求还会进一步增大，资源消耗总量将进一步增加，我国经济与社会发展面临的资源约束将进一步加大。

5　电力行业的节能减排[①]

5.1　电力行业“十一五”运行综述

“十一五”期间，我国电力工业持续快速发展，取得了重大成就。

（1）电力供应能力进一步增强。全国全口径发电量42280亿千瓦时，同比增长14.85%，比上年提高8.18个百分点。其中，水电发电量6863亿千瓦时，同比增长20.05%；火电发电量34145亿千瓦时，同比增长13.38%；核电发电量768亿千瓦时，同比增长9.66%；风电发电量501亿千瓦时，同比增长81.41%，如图5－1所示。

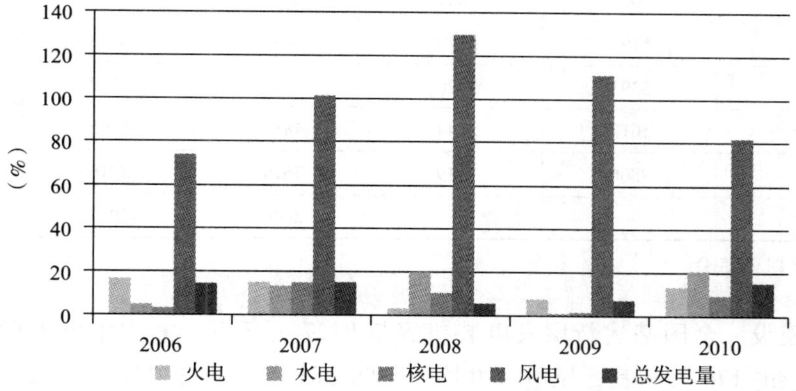

图5－1　“十一五”发电量增速情况

Figure 5－1　Growth of Power Energy Production during China's 11th Five－year Plan Period

资料来源：中国电监会

（2）电力需求总体旺盛。全社会用电量为41923亿千瓦时，同比增长14.56%，比上年提高8.12个百分点。

（3）发电设备利用小时略有回升。全国6000千瓦及以上电厂累计平均设备利用小时数为4660小时，同比增加114小时，实现了自2004年持续下降后的首次回升。其中，水电设备利用小时数为3429小时，同比增加101小时；火电设备利用小时数为5031小时，同比增加166小时；核电设备利用小时数为7924小时，同比提高208小时；风电设备利用小时数为2097小时，同比增加20小时，如表5－1所示。

① 本章资料及数据主要来源于电监会《电力年度监管公报2010》、《电力企业节能减排情况通报2009》。

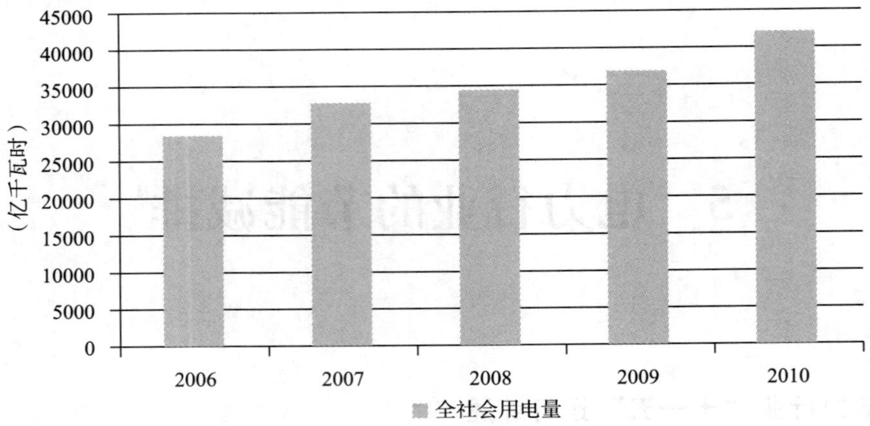

图5－2 “十一五”全国用电量情况

Figure 5－2 National Power Energy Consumption during China's 11th Five－year Plan Period

资料来源：中国电监会

表5－1 “十一五”发电设备利用情况

Table 5－1 Utilization State of Power Generating Facilities during China's 11th Five－year Plan Period

年份	2006	2007	2008	2009	2010
平均发电小时	5198	5020	4648	4546	4660
水电小时数	3393	3520	3589	3328	3429
火电小时数	5612	5344	4885	4865	5031
核电小时数	7806	7719	7679	7716	7924
风电小时数	—	—	2046	2077	2097

资料来源：中国电监会

（4）电力建设。全国基建新增发电装机容量9127万千瓦，其中水电1661万千瓦，火电5872万千瓦，核电174万千瓦，风电（并网）1399万千瓦。截至2010年底，全国发电设备容量达96219万千瓦，同比增长10.08%。其中，水电21340万千瓦，同比增长8.72%；火电70663万千瓦，同比增长8.53%；核电1082万千瓦，同比增长19.16%；风电并网总容量3107万千瓦，同比增长76.53%，如图5－3所示。

（5）电源结构调整取得一定成效，全国新增发电装机9127万千瓦，其中，水电1661万千瓦，火电5872万千瓦，核电174万千瓦，风电1399万千瓦。水电、核电和风电等非化石能源装机容量比重由上年度的25.5%提升至26.5%，如图5－4所示。

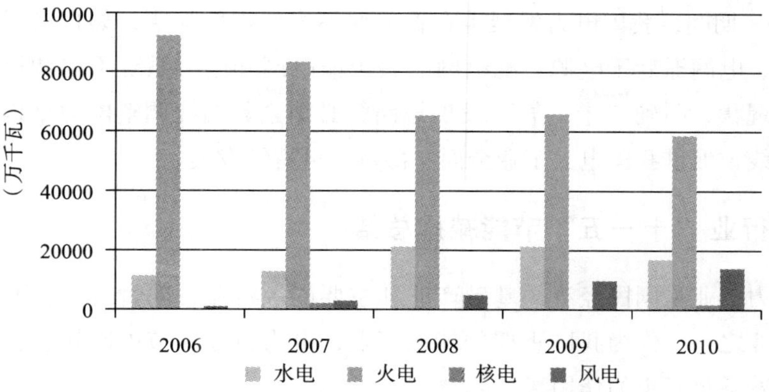

图5-3 “十一五”基建新增装机容量情况

Figure 5-3 New Installed Capacity of Infrastructure Construction during China's 11th Five-year Plan Period

资料来源：中国电监会

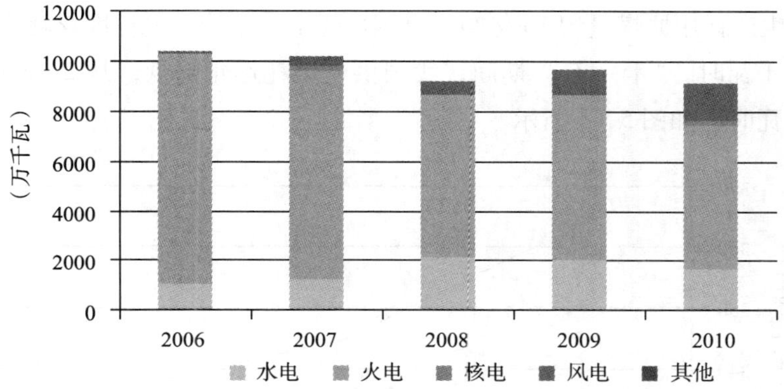

图5-4 “十一五”新增发电设备的结构变化情况

Figure 5-4 Structure Change of New Power Generating Facilities during China's 11th Five-year Plan Period

资料来源：中国电监会

（6）从投资结构看，电源工程建设投资持续向非化石能源发电领域倾斜。2010年全国电源工程建设完成投资3641亿元，其中，水电791亿元，火电1311亿元，核电629亿元、风电891亿元。水电、核电、风电等非化石能源发电投资占电源投资的比重从2005年的29%持续提高到2010年的64%，如表5-2所示。

表5-2 “十一五”各类电源投资同比增速情况

Table 5-2 Year-on-year Growth of Power Sources Investment during China's 11th Five-year Plan Period

	2006	2007	2008	2009	2010
合计	-1.01	0.97	5.61	11.62	-4.26
水电	-9.07	9.60	-1.18	2.16	-8.79
火电	-1.82	-10.08	-16.26	-8.04	-15.06
核电	179.21	73.89	101.39	77.25	7.67
风电	37.76	173.54	208.27	48.27	13.97

资料来源：中国电监会

但“十一五”期间，我国电力发展也存在一些不容忽视的问题，如盲目无序发展，导致电力整体结构失衡、电网滞后于电源、配电网滞后于输电网和电力系统不协调等问题。客观认识和分析电力发展现状，明确“十二五”发展思路，必须紧紧围绕调整电力结构、加快电力发展方式转变这条主线，促进我国电力工业全面、协调、可持续发展。

5.2 电力行业“十一五”节能减排总结

在我国，电力工业是国民经济的基础产业和主要能源行业，同时也是主要的能源资源消耗和污染物排放行业之一。作为我国重要的能源行业，电力行业在节能减排、实现低碳发展和应对气候变化中起着举足轻重的作用。

“十一五”期间，电力行业充分认识到节能减排的重要性，按照国家统一要求与部署，积极推进“上大压小”，优化电源结构，加强电网建设，加大二氧化硫治理和节能减排管理力度，提前完成“十一五”节能减排规划目标，为全国节能减排目标的实现作出了重要贡献。主要指标性数据如下：

• 全国发电生产耗用原煤量约16亿吨。全国燃煤电厂平均供电标准煤耗为335克/千瓦时，比上年降低5克/千瓦时。“十一五”期间，我国供电煤耗逐年降低，从2006年367克/千瓦时下降到335克/千瓦时，如图5－5所示。

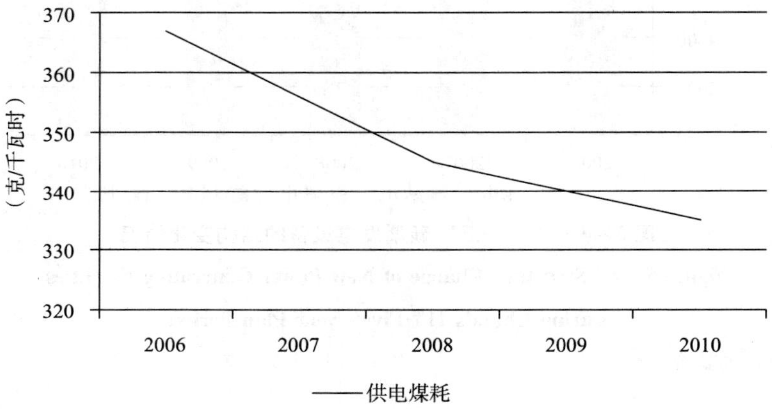

图5－5 “十一五”供电煤耗变化情况

Figure 5－5 Change of Coal Consumption for Power Supply during China’s 11th Five－year Plan Period

资料来源：中国电监会

• 全国电网输电线路损失率为6.49%，比上年下降0.14个百分点。“十一五”期间，我国电网输电线路损失率总体呈下降态势，从7.04%下降到6.49%，如图5－6所示。

• 电力二氧化硫排放量继续下降。

① 2009年全国电力二氧化硫排放量约939.3万吨，比2005年下降了29.3%，提前一年达到“十一五”末电力二氧化硫排放951.7万吨的目标。

② 全国电力二氧化硫排放量占全国二氧化硫排放量的比例由2005年的51.0%下降到42.4%，减少8.6个百分点。

③ 火电二氧化硫排放绩效值由2005年的6.4克/千瓦时下降到3.1克/千瓦时，减少3.3克/千瓦时。

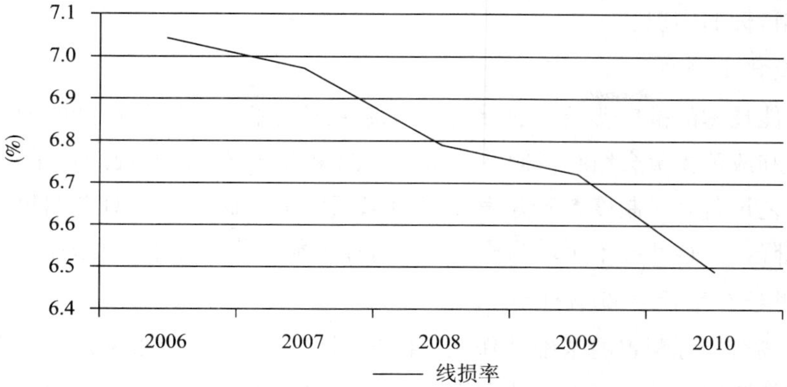

图 5-6 “十一五”线损率变化情况

Figure 5-6 Change of Line Loss Rate during China's 11th Five-year Plan Period

资料来源：中国电监会

④ 从 2006—2009 年，电力二氧化硫排放量共下降 388.7 万吨，为全国二氧化硫总量减排作出了重要贡献。

• 截至 2009 年底，全国燃煤电厂烟气脱硫机组容量 4.7 亿千瓦，比上年增长 29.5%；烟气脱硫机组占煤电机组的比例约为 76%，比上年增加 12 个百分点。2005—2009 年全国燃煤电厂烟气脱硫机组发展情如图 5-7 所示。

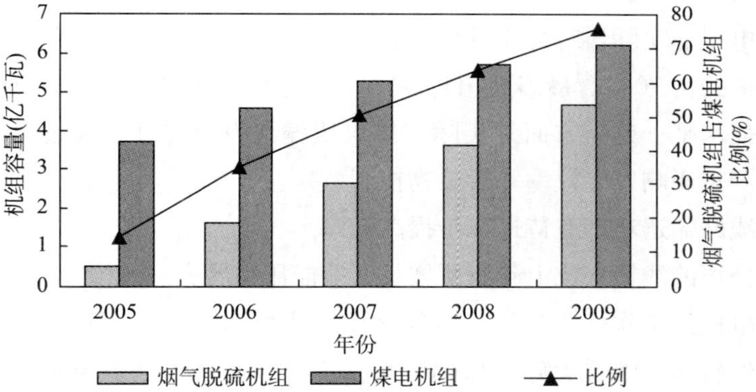

图 5-7 2005—2009 年全国燃煤电厂烟气脱硫机组发展情况

Figure 5-7 Development of Fume De-Sulphur Device in Coal Power Plant across China from 2005—2009

资料来源：中国电监会

“十一五”期间电力行业节能减排工作主要措施如下：

1. 出台了相关法规和政策措施

电力行业企业在国家法律政策基础上，建立了具有支持性的标准和管理办法。在政策措施上，建立节能减排目标责任制，层层分解责任，强化考核，按年度向社会发布电力企业节能减排情况通报，向社会公布各省（区、市）和五大电力集团主要污染物总量减排考核结果，对问题电厂责令限期整改，并进行处罚。按照国务院要求，各地区节能减排考核结果将作为地方党政领导班子和领导干部综合评价考核的重要依据。不断完善的法规体系和政策措施有力地保障了电力节能减排目标的实现。

2. 加快节能技术进步

加快重点节能技术的推广普及、引导先进节能工艺技术的采用是提高能源利用效率的重要手段。国家发展和改革委员会组织编制并公布了《国家重点节能技术推广目录》，建立国家鼓励的脱硫、脱硝工艺设备名录制度，新建发电机组采用脱硫脱硝设备，对老旧机组安装脱硫脱硝设施，促进节能降耗。这些技术手段为推动电力行业节能工作发挥了重要的作用。

3. 推动采用经济手段和市场机制

充分发挥市场对资源配置的基础性作用，因地制宜，调整电价水平，疏导电价矛盾，完善电价结构，对包括脱硫电价、可再生能源电价在内的电价水平进行了适当调整，使电力成本真正体现资源和环境成本。开展发电权、排污权交易，针对水电比重较大的区域或省份，探索利用市场机制发挥水火互济作用，优化资源配置。开展烟气脱硫特许经营。市场机制推动了节能减排健康有序开展。

5.3 电力行业“十二五”节能减排背景

“十一五”期间，电力行业在节能减排方面取得了显著成绩，在经济增速和能源消费总量均超过规划预期的情况下，我国二氧化硫减排目标仍提前一年实现。对此，电力行业功不可没。但是，在“十一五”期间，电力行业节能减排仍存在如下问题：

1. 节能减排潜力有待进一步挖掘

一是大部分发电企业实际煤耗高于设计值。如国家电监会在2009专项督查机组中发现实际供电煤耗高于设计值的占70%，最多超出近30克/千瓦时。其主要原因一方面是机组偏离经济工况运行，影响煤耗水平；另一方面是实际燃煤煤质偏离设计煤种，超过设计值1倍甚至数倍的现象时有发生，严重影响发电效率和污染物控制效果。

二是部分输电线路输送功率有待进一步提高。

三是部分电网公司的线损率比上年有所增长，节能压力增大。

四是关停小火电机组工作任重道远。尽管淘汰小火电机组取得了巨大成绩，但能耗高、污染重的火电装机依然较多。随着关停工作的不断深入，人员安置任务更加繁重、债务处理难度更大，资产关系更加复杂，关停小火电工作仍然任重道远。

2. 电力企业节能减排管理水平有待进一步提高

一是节能减排统计系统有待进一步完善。部分电厂脱硫系统台账不完整，历史记录不完整，物料不平衡。少数电网企业线损统计分析管理信息系统尚未建立，部分电网企业线损率统计不规范。

二是节能减排运行管理人员素质参差不齐，经验水平有待加强。

三是电力企业节能减排考核和评估机制有待进一步完善。

3. 运用经济手段推动节能减排的作用有待进一步发挥

一是需求侧管理机制有待完善。目前，需求侧管理缺乏财政、税务、物价等相关的配套激励政策。

二是部分机组脱硫成本超出脱硫电价补偿。由于近年来电厂燃煤质量变化波动大，硫分和灰分增加，再加上煤炭、石灰石等原材料的涨价，电价、水价的上涨，机组利用小时数的下降，对企业脱硫成本影响较大。

4. 脱硫设施运行管理水平仍有待提高

“十一五”前四年，燃煤电厂烟气脱硫装置大规模快速建设，平均年投运容量约1亿千瓦，开创了世界的先河。

在烟气脱硫装置快速建设的同时，也暴露出部分工程设计参数选取不合理，工程质量、设备质量问题突出，运行管理水平不高，脱硫装置建成投运不久即面临技术改造等问题，导致部分发电企业脱硫设施运行不正常，投运率不高，二氧化硫超标排放，在一定程度上影响了二氧化硫控制水平。

例如，山西忻州广宇煤电有限公司、广西方元电力股份有限公司来宾电厂、广西来宾法资发电有限公司、广西柳州发电有限责任公司、四川嘉陵电力有限公司、宜宾天原集团股份有限公司等6家电厂存在脱硫设施运行不正常或烟气在线监测数据作假问题。专项督查中发现部分企业自备电厂未同步安装脱硫装置或脱硫装置尚未验收，但其发电利用小时数却明显高于公用电厂利用小时数。

5. 烟气排放连续监测系统作用有待进一步发挥

从烟气排放连续监测系统（CEMS）建设情况看，全国燃煤电厂基本均安装了CEMS，江苏、贵州、山东、京津冀等地区已实现了对脱硫机组的在线监测，其他地区的联网工作正在逐步推进中。

但CEMS的使用效果不尽人意。由于CEMS招投标不规范，前期质量把关不严，部分项目售后服务不完善，造成CEMS由于本身质量问题不能正常运行，且由于在维护过程中受到时间、技术等条件限制，造成部分仪表出现数据漂移、损坏、腐蚀等问题，对有效实施脱硫设施的运行管理和准确监管脱硫电价造成一定影响。

6. 脱硫石膏综合利用存在地区差异

从全国平均水平来看，2009年脱硫石膏综合利用率约56%，但受地域和经济发展水平等诸多因素影响，各地脱硫石膏综合利用情况不平衡。

例如，专项督查的河北省11家采用石灰石—石膏湿法脱硫的电厂中，有10家电厂脱硫石膏全部实现利用，而四川省、贵州省脱硫石膏平均综合利用率不到30%，部分电厂出现了脱硫石膏无处消纳的问题。

石膏堆放和抛弃不仅占用大量土地，增加灰场的投资，而且处理不好可能会对周围环境造成二次污染。脱硫石膏产量逐年提高，脱硫石膏综合利用问题不容忽视。

电力行业在节能减排和应对气候变化中起着举足轻重的作用。因为应对气候变化的核心是减少温室气体的排放和增加碳汇，其中，减少能源生产、转换、消费过程中产生的二氧化碳是重大措施之一。电力行业在能源转换过程中排放的二氧化碳约占全国排放总量的50%，是二氧化碳减排的关键部门之一。同时，电力行业也是实现全社会、各行各业减排的重要桥梁。通过电力结构调整、节能降耗和建设应用智能电力系统，可以在转变电力发展方式的同时，促进工业和民用领域加快应用低碳技术和全社会节能减排。但是，“十一五”期间，电力企业已经将减排潜力和减排阈度几乎用完，面对国家更坚定的减排决心和更严厉的约束措施，被称为二氧化硫和氮氧化物排放大户的电力企业“十二五”期间如何应对环境压力？要想更好地完成新的节能减排目标，转变电力发展方式、提高发展质量将是中国电力行业节能减排、应对气候变化的战略任务。

5.4 电力行业“十二五”节能减排展望

中电联的《电力工业“十二五”规划研究报告》中对促进电力工业绿色和谐发展提出了规划目标：通过发展非化石能源、降低供电煤耗和线损等途径，实现2015年比2010年节约标准煤2.64亿吨，减排二氧化碳6.55亿吨，减排二氧化硫565万吨，减排氮氧化物248万吨。

为实现规划目标，电力行业应有效采取应对措施。

1. 进一步挖掘电力节能减排潜力

一是积极推进电源结构调整。加强电力规划，加快电源结构由过度依赖煤电向提高非化石能源比重转变。优先开发水电、优化发展煤电、大力发展核电、积极推进风电、太阳能等新能源发电，不断提高清洁能源比重。

二是积极建设大型高效环保煤电机组。进一步发展超临界、超超临界、热电联产等高效火电机组，不断提高高参数、高效率、低消耗的火电机组比重和能源利用效率。

三是继续做好小火电机组关停工作。电力工业要把保增长与调结构紧密结合起来，要通过发展逐步解决结构的问题，要在发展中突出结构调整，要把淘汰小火电作为电力工业结构调整的一项重要任务，持之以恒地开展下去，总结经验，完善政策，不断优化火电结构，确保电力工业长期、稳定和可持续发展。

四是注重电网结构优化和运行优化。第一，加强城乡电网规划与建设，优化电网结构，解决电网运行中瓶颈问题和利用效率偏低的问题，提高负荷中心的电源支撑能力，提高城乡电网吸纳小水电等清洁能源能力；第二，优先安排节能、环保、高效机组发电，限制高耗能、高污染机组发电，提高电力工业整体效率和效益；第三，在保证电网安全的基础上，实现电力系统的经济运行，提高发电机组的负荷率，避免机组长时间在低负荷下运行，发挥高参数、大容量的超（超）临界机组节能减排的潜力。

五是继续推行节能发电调度，降低煤炭消耗。目前，我国多数地区仍普遍实行平均分配发电利用小时数的调度方式，这导致大容量高参数机组节能优势难以发挥。为改变这种状况，我国已在河南、四川、江苏、广东和贵州五省进行了节能发电调度试点，并取得了一定的成绩。建议根据试点情况完善有关配套政策，继续推行节能发电调度。

2. 进一步提高节能减排管理水平

一是加强电煤管理。第一，加强对煤炭的调控力度，努力稳定煤炭价格、保证煤炭供应，提高煤炭质量，控制国家明令限采的煤炭进入市场；第二，加强煤炭在生产流通领域、检测市场等环节的质量监管，规范煤炭市场秩序，避免掺杂使假现象；第三，推进区域性大型配煤中心建设，提高煤炭洗选比例，重点区域内未配备脱硫设施的企业，禁止直接燃用含硫量超过0.5%的煤炭。

二是继续改进和完善企业内部节能降耗管理工作。发电企业要进一步强化内部节能降耗管理工作，通过加强用煤管理，加强机组设备的运行监测及故障诊断，建立企业内部节煤、节电企业管理制度等措施，不断提高生产管理和运行水平，促进煤耗的进一步降低。

三是强化煤耗计量、统计工作的监督和检查。建立发电企业供电煤耗统计制度，加强统计工作，定期公布发电企业煤耗水平；建立发电企业供电煤耗评价考核机制，包括建立供电煤耗指标体系、评估体系、核查制度等，定期组织开展发电企业供电煤耗监督和检查，公布检查结果，确保节能降耗工作各项目标的落实。

四是加强火电厂烟气脱硫工程后评估，将后评估结果作为脱硫电价调整、脱硫装置运行效果考核等的主要依据之一。

五是组织开展现役机组火电厂节能技术评估，分析生产各环节的节能状况和潜力。通过节能减排评估，掌握现役机组节能减排可挖潜的空间，提出改进措施，推广采用成熟可靠的新技术、新工艺，最大限度地提高现役机组节能减排的能力。

3. 充分利用市场手段，建立健全以市场为导向的节能减排新机制

一是推广实施已取得实践经验的发电权交易机制，扩大发电权交易范围及交易主体，积极稳妥地推进以水代火、以核代火、以风代火、以大代小、以高效代低效、以先进代落后等手段促进节能发电调度工作。

二是通过电价引导，通过峰谷/差别电价促进用户开展节能减排，鼓励用户改善用电特性，削峰填谷。

三是大力推进节能服务产业的发展，培育节能服务体系，逐步推广合同能源管理。

四是加快节能新技术、新产品、新设备的推广应用。

五是运用价格机制，严格差别电价政策，提高需求侧管理水平。

六是继续完善脱硫电价补偿机制。对供热电厂的供热部分、老电厂、硫份高的电厂以及由于客观条件导致脱硫成本高的特殊电厂（例如，煤质很差的坑口电厂）继续合理补偿脱硫成本。

4. 加大减排力度，提高脱硫装置运行水平，提高监管有效性

一是加大减排力度，鼓励发电企业加快建设脱硫设施，采取有效手段减少温室气体排放。

二是规范招投标行为，防止以牺牲质量的低价中标。

三是开展脱硫装置可靠性研究，加强设备的可靠性管理。

四是加强脱硫装置的运行管理，强化脱硫系统的优化调整。

五是加强脱硫装置技术改造力度，提高对劣质煤炭的适应能力。

六是全面评估CEMS的建设、运行和维护情况以及数据的使用情况等，使监测数据能够真实反映污染物排放情况。

5. 高度重视脱硫石膏综合利用，防止二次污染，从政策上加大对脱硫石膏综合利用的支持力度

一是要进一步研究出台支持利用脱硫石膏的财税、金融、价格等相关经济政策，例如，适时出台对开采天然石膏增加资源税的政策。同时，要出台、完善相关政策推动二氧化硫治理技术（例如，氨法脱硫、镁法脱硫等）发展。

二是电力企业要加强对脱硫装置的运行管理，提高脱硫石膏品质，同时加强与建材企业合作，为综合利用脱硫石膏创造条件。

三是加大脱硫石膏综合利用技术研发力度与经验交流，针对我国脱硫石膏的实际品质，开发脱硫石膏大规模资源化利用技术，提升脱硫石膏在应用过程中的品质，努力开拓大宗脱硫石膏综合利用的新领域。

四是建立区域脱硫石膏综合利用示范工程。

5.5 典型电力企业的节能减排动向

表 5-3 2009年主要大型发电集团公司供电煤耗

Table 5-3 Coal Consumption for Power Supply of Major Large-scale Power Providers in 2009

2009年主要大型发电集团公司供电煤耗				
序号	发电公司名称	火电装机容量（万千瓦）	火电发电量（亿千瓦时）	供电煤耗（克/千瓦时）
1	中国华能集团公司	9309	3924	327.7
2	中国大唐集团公司	8222	3397	328.41
3	中国国电集团公司	7025	3307	331.8
4	中国华电集团公司	6239	2733	331.68
5	中国电力投资集团公司	4331	2074	343.23
6	神华国华电力有限公司	2591	1186	321
7	华润电力控股有限公司	1890	948	334
8	浙江省能源集团有限公司	1775	879	328.5
9	广东粤电集团有限公司	1755	947	338.18
10	国投电力公司	1196	484	332.37

注：按2009年发电公司火电装机容量排序。

资料来源：中国电监会

表 5-4 2009年主要大型发电集团公司二氧化硫排放一览表

Table 5-4 List of Sulfur Dioxide Emissions of Major Large-scale Power Providers in 2009

2009年主要大型发电集团公司二氧化硫排放一览表				
序号	发电公司名称	火电装机容量（万千瓦）	火电发电量（亿千瓦时）	二氧化硫排放量（万吨）
1	中国华能集团公司	9309	3924	91.5
2	中国大唐集团公司	8222	3397	92.2
3	中国国电集团公司	7025	3307	98.0
4	中国华电集团公司	6239	2733	89.0
5	中国电力投资集团公司	4331	2074	76.9

注：1. 按2009年发电公司火电装机容量排序；
2. 五大电力集团公司资料来源于环境保护部公告。

资料来源：中国电监会

6 钢铁行业的节能减排

6.1 钢铁行业“十一五”运行综述

“十一五”期间以及2010年，中国钢铁行业的主要数据如下：

- 中国2010年的粗钢产量为6.267亿吨，同比增长9.3%，如按此速率继续增长，2011年将达到6.9166亿吨（如图6－1）。同期，2010年全球粗钢产量达到14.14亿吨，同比增长15%，创历史新高。主要产钢国和地区的增长速度都基本达到两位数，由于2009年的基数较低，美国和欧盟地区的增幅较高，而亚洲和独联体的增速相对较低。

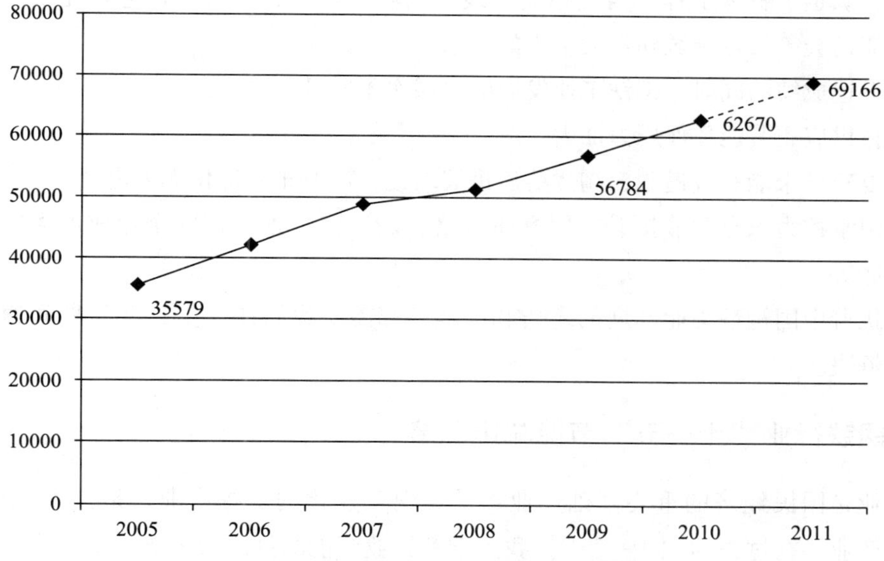

图6－1 中国的粗钢产量（2011年为估计值）

Figure 6－1 Production Capacity of Crude Steel in China (2011 Based)

资料来源：中国钢铁工业协会

- 数据显示，中国是粗钢生产第一大国，占全球总产量的44.3%。日本是继中国之后的全球第二大产钢国，产量为1.096亿吨，之后是美国（8060万吨）、俄罗斯（6700万吨）。印度是全球第五大产钢国（6680万吨），韩国排名第六（5850万吨），如图6－2所示。
- 钢铁行业装备取得重大发展，1000立方米以上大型高炉由2005年的88座，增加到2010年底的260座，已占全国炼铁总产能的52%。100公称吨位以上转炉由2005年80座，增加到2010年底的200多座，50吨级以上电炉80多座，占全国炼钢总产能的比重上升到51%。
- 产品结构进一步改善，产品质量大幅提高。

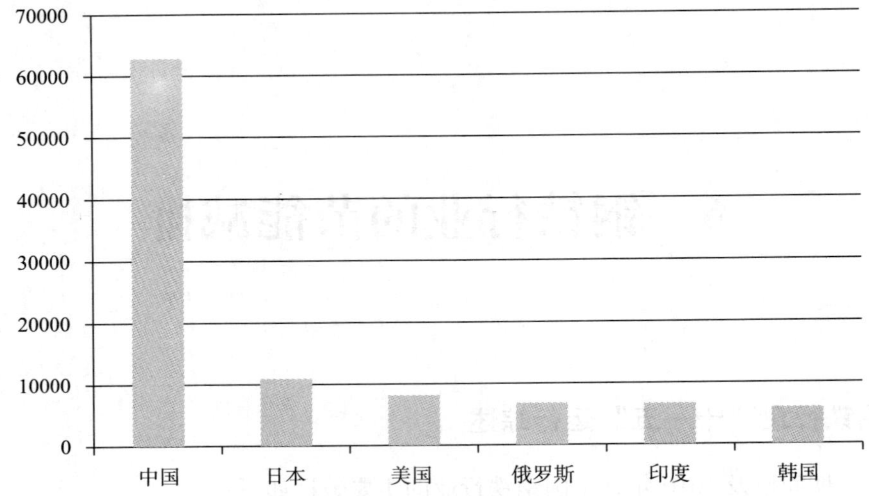

图 6－2　2010 年粗钢主要生产国

Figure 6－2：Major Crude Steel Producers in 2010

资料来源：国际钢铁协会

• 过去的五年，钢铁工业紧随下游行业快速发展和优化升级的步伐，通过产品结构调整和新产品开发，满足了奥运工程、西气东输二线、三峡水电站、高铁及核电站建设，以及上海世博会、广州亚运会等重点领域和重大工程的要求。

在取得上述成绩的同时，钢铁工业发展的深层次矛盾日益显现：

一是原料供应、节能环保压力加大。

二是重复建设未能有效遏制，淘汰落后进展缓慢，产业布局优化无大进展。

三是资源掌控力未有明显提升，原燃料价格不断上涨，推动钢铁企业成本大幅上升，行业盈利水平过底等。

这些困扰着中国钢铁工业发展的矛盾和问题，都需要我们在“十二五”期间通过转变发展方式等加以解决。

6.2　钢铁行业“十一五”节能减排总结

钢铁产业是国民经济的重要基础产业，是实现工业化的支撑产业，是技术、资金、资源、能源密集型产业。在过去的“十一五”，我国钢铁行业节能减排成效显著，能源利用效率明显提高。重点大中型钢铁企业的吨钢综合能耗大幅降低，“十一五”期间累计完成淘汰落后炼铁、炼钢产能预定目标的 117% 和 126%。行业节能对国家 GDP 节能贡献较大。“十一五”期间钢铁行业能源消耗量变化主要情况如下：

• 钢铁行业用能总量在 2009 年达到 35149 万吨标准煤（如图 6－3），比 2005 年的 24692 万吨标准煤增加了 42.34%，支持了钢产量 59.6% 的增长，如图 6－4 所示。

• 由于钢产量增幅大于总用能量增幅，钢铁行业的能源消耗强度仍保持下降趋势。与 2005 年相比，重点统计钢铁企业 2010 年平均吨钢综合能耗预计由 694 千克标准煤降至 607 千克标准煤，实现节能总量约 4611 万吨标准煤，如图 6－5 所示。

• 按钢铁行业工业增加值占国家 GDP 的 3.25% 计算，钢铁行业为国家 GDP 能耗降低的贡献率约为 0.88%，对国家 GDP 节能贡献率百分比为 4.4%。

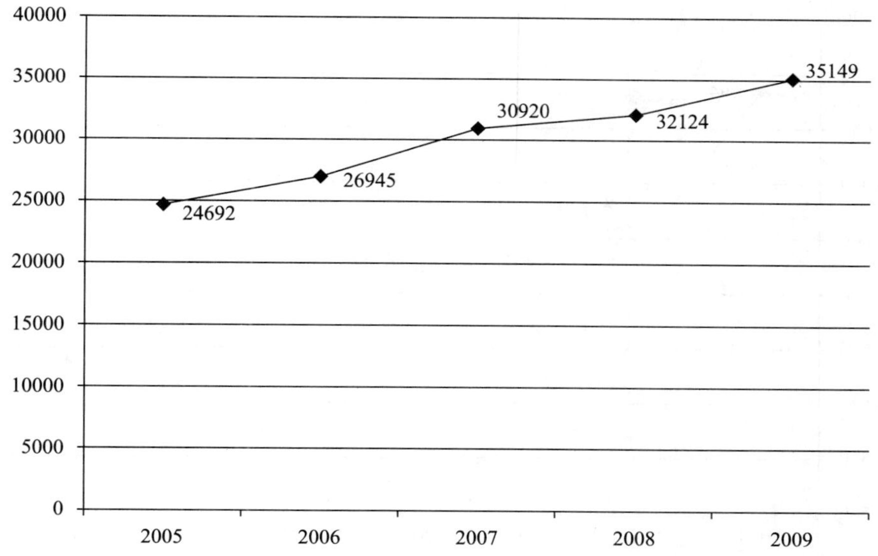

图 6－3 钢铁行业用能总量 单位：万吨标准煤

Figure 6－3 Total Energy Consumption by Steel Sector Unit：10000 Tons of Standard Coal Equivalent

资料来源：中国钢铁工业协会

总体而言，钢铁行业在“十一五”期间通过加大节能减排力度，为国家GDP能耗下降20%目标，作出了应有贡献。

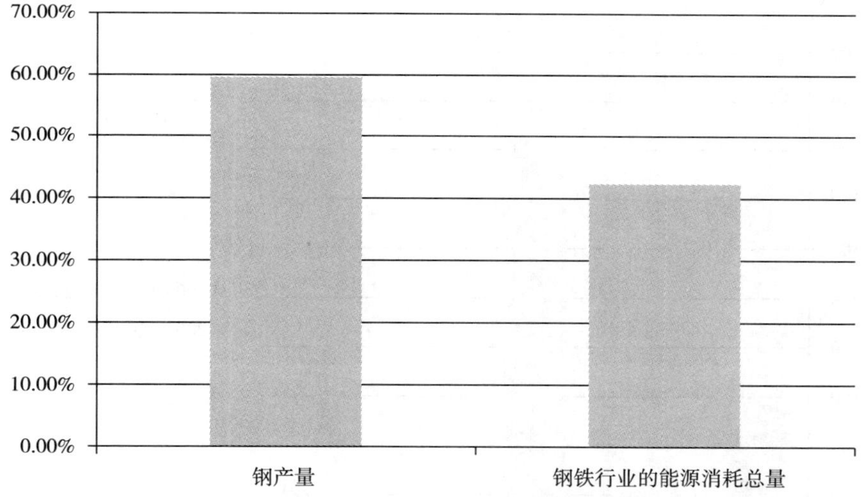

图 6－4 2005—2009 钢产量与能源消耗总量增长率对比

Figure 6－4 Comparison of Growth Rate in Steel Production and Total Energy Consumption from 2005—2009

资料来源：中国钢铁工业协会

“十一五”期间钢铁行业主要生产工序能耗[①]稳步下降，为降低吨钢综合能耗奠定了基础，如图6－6所示。

2009年与2005年相比，焦化工序能耗企业达标率提高了18.6%，烧结工序能耗企业达标率

① 2005年各工序对标的电力折标系数为0.404Kgce/KWh；2009年各工序对标的电力折标系数为0.1229Kgce/KWh。

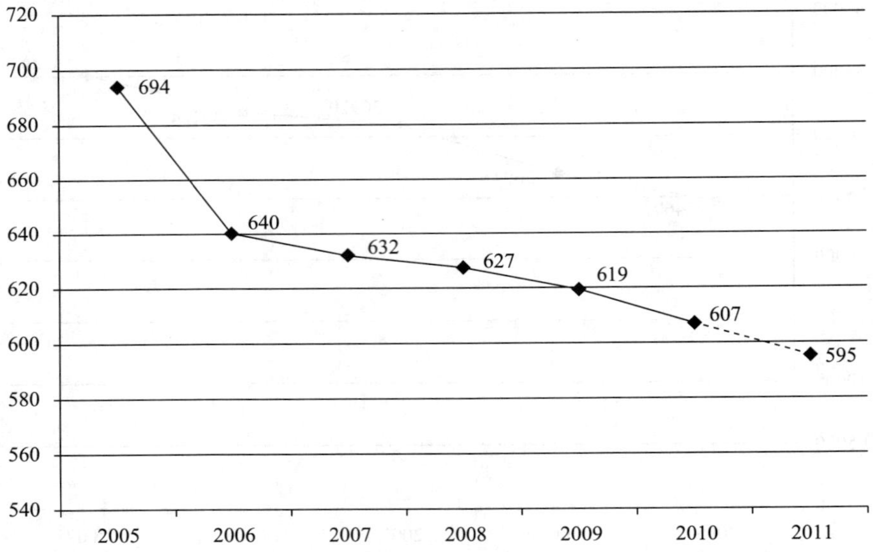

图 6－5 吨钢综合能耗

Figure 6－5 Comprehensive Energy Consumption of Per Ton of Steel

资料来源：中国钢铁工业协会

提高了9.7%，炼铁工序能耗企业达标率提高了32.7%，转炉工序能耗企业达标率提高了31.9%，电炉工序能耗达标率提高了20.8%（如图6－7）。这说明，"十一五"期间我国钢铁企业各生产工序节能措施得当，节能效果明显，取得了显著成绩。

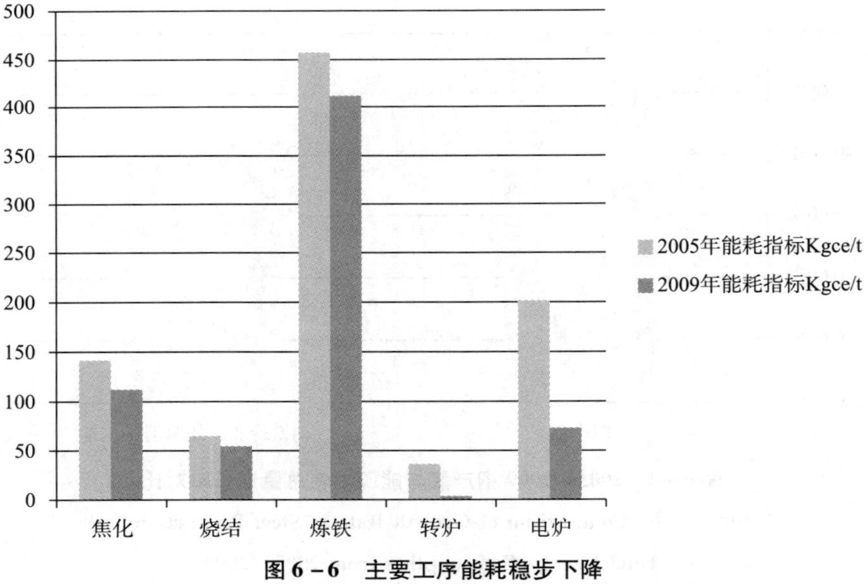

图 6－6 主要工序能耗稳步下降

Figure 6－6 Steady Reduction of Energy Consumption in Major Processes

资料来源：中国钢铁工业协会

（1）二次能源回收利用水平提高。2005—2009年间，占钢铁企业二次能源回收总量70%的高、焦、转炉煤气的利用量逐年提高，损失率逐年降低。2009年与2005年相比，焦炉煤气回收利用率由95.48%提高到98.16%，提高了2.67个百分点；高炉煤气回收利用率由90.74%提高到95.01%，提高了4.27个百分点；企业自发电比例由平均27.44%提高到32.9%，企业自发电

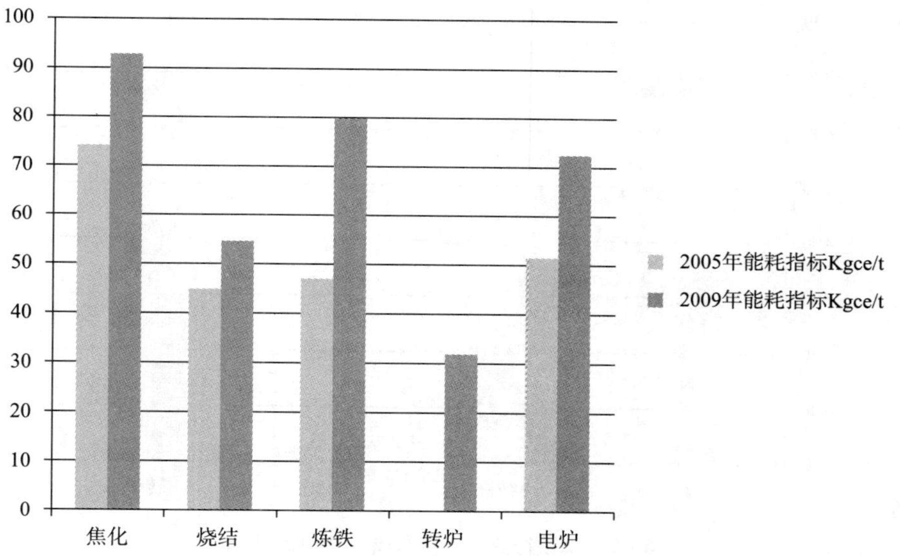

图 6 –7 主要工序达标率增长

Figure 6 –7 Growth of Attainment Rate in Major Processes

资料来源：中国钢铁工业协会

量占生产总用电量比例提高了 5. 46 个百分点（如图 6 –8）。同时，吨钢转炉煤气回收量由 47. 5 立方米/吨提高到 76. 5 立方米/吨。这些成果说明，钢铁行业近几年在二次能源利用和节能管理方面有了很大提高。

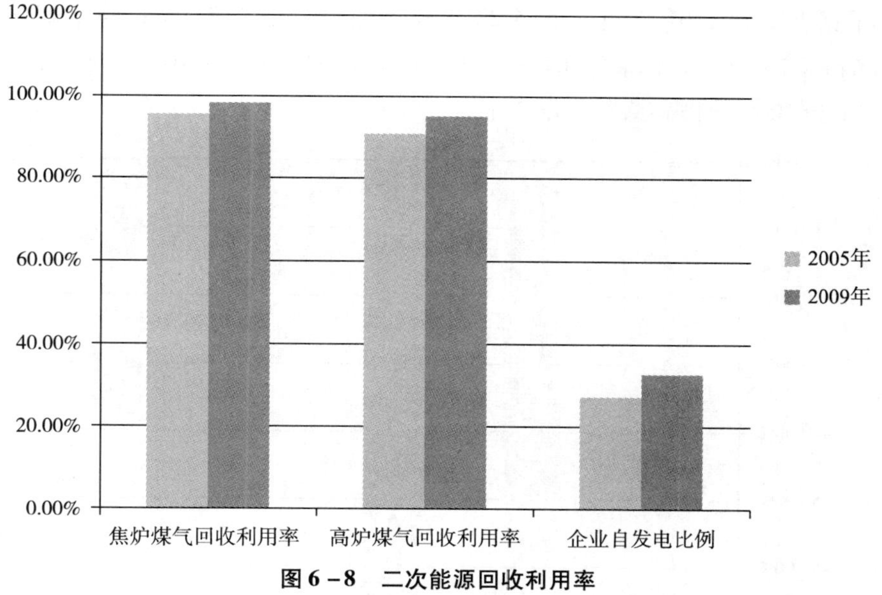

图 6 –8 二次能源回收利用率

Figure 6 –8 Recycling Rate of Secondary Energy

资料来源：中国钢铁工业协会

（2）主工艺装备水平大幅提高。"十一五" 期间钢铁行业积极贯彻国家钢铁产业政策，以大型化、现代化和采用先进节能技术，使各主体设施的工序能耗、吨钢综合能耗不断下降。2009 年底，仅重点大中型钢铁企业就拥有各类机械化焦炉约 340 座，拥有 456 台烧结机，拥有高炉 560 座，拥有转炉 689 座，如图 6 –9 所示。

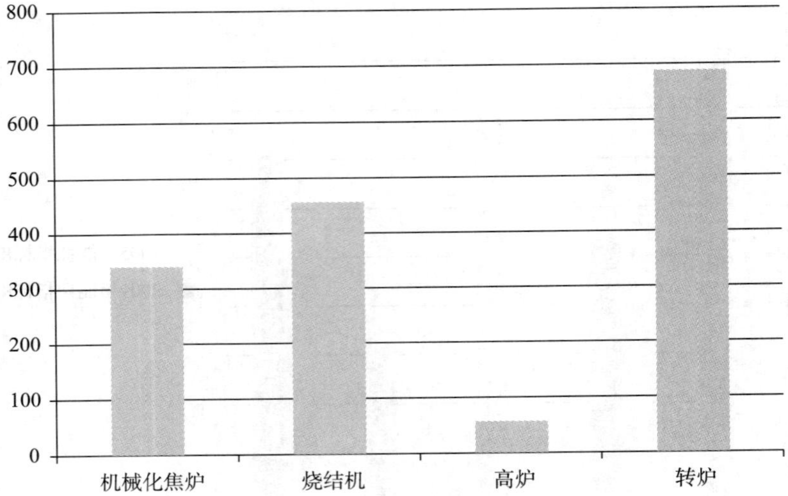

图 6－9　钢铁行业主工艺装备水平（2009 年）

Figure 6－9　Equipments for Major Processes of Steel & Iron Sector（2009）

资料来源：中国钢铁工业协会

（3）吨钢耗新水及主要污染物单位排放量明显降低。与 2005 年相比，2009 年重点统计钢铁企业吨钢耗新水由 8.6 吨下降到 4.5 吨，下降 47.7%；水重复利用率由 94.04% 提高到 97.04%；吨钢外排废水量由 4.89 吨下降到 2.06 吨，下降 57.9%，外排废水总量下降 35.3%；吨钢化学耗氧量排放量由 0.254 千克下降到 0.09 千克，下降 64.6%，总量下降 46.1%；吨钢二氧化硫排放量由 2.83 千克下降到 2.01 千克，下降 28.97%，但总量增长 7.1%；吨钢烟粉尘排放量由 2.18 千克下降到 1.35 千克，下降 38.07%，总量下降 6.3%。图 6－10 为 2009 年吨钢排放下降情况，图 6－11 为 2009 年排放总量下降情况。

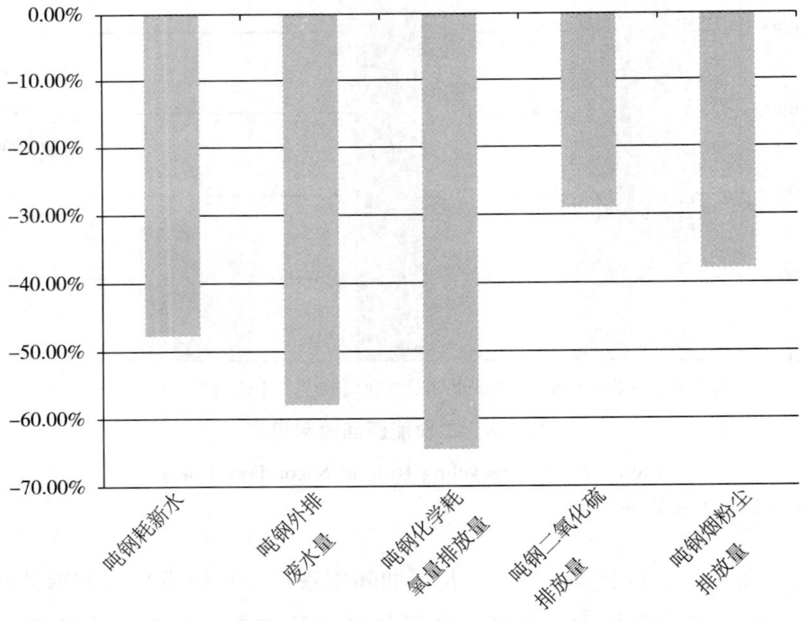

图 6－10　吨钢排放下降情况（2009）

Figure 6－10　Emission Reduction of Per Ton of Steel（2009）

资料来源：中国钢铁工业协会

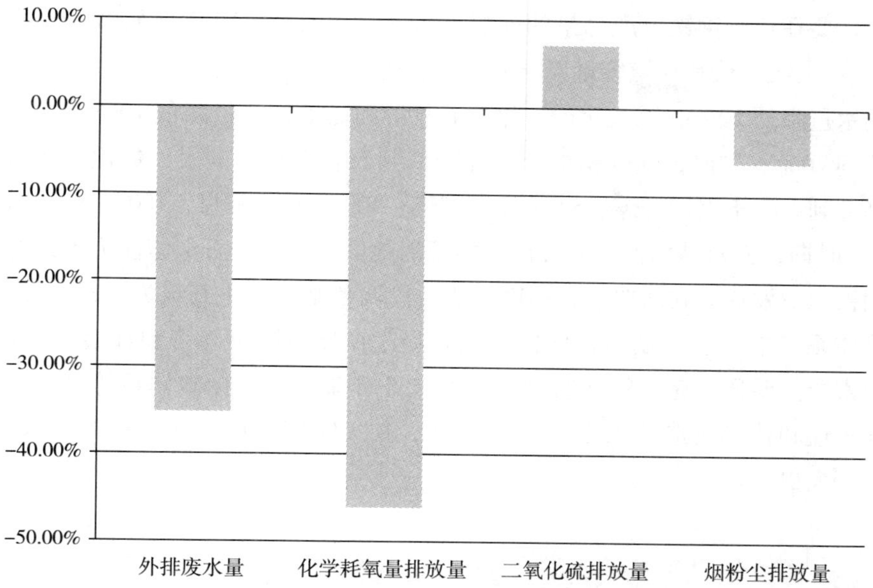

图 6－11 排放总量下降情况（2009）

Figure 6－11 Reduction of Total Emissions（2009）

资料来源：中国钢铁工业协会

6.3 钢铁业“十二五”节能减排形势

我国钢铁工业经过了 10 多年的高速发展，发展的同时造成了生产能力上的重复扩张，能源消耗高，污染物排放量大，污染治理欠账较多。按照目前增长率预测，按目前消耗和治理水平的趋势发展下去，到 2015 年，粗钢产量将达到 10 亿吨以上，钢铁工业每年的能源消耗将超过 6 亿吨标准煤（如图 6－12）。这将大大加剧我国资源和能源紧张的局势，大气污染压力仍十分严峻，应对气候变化工作十分艰巨。

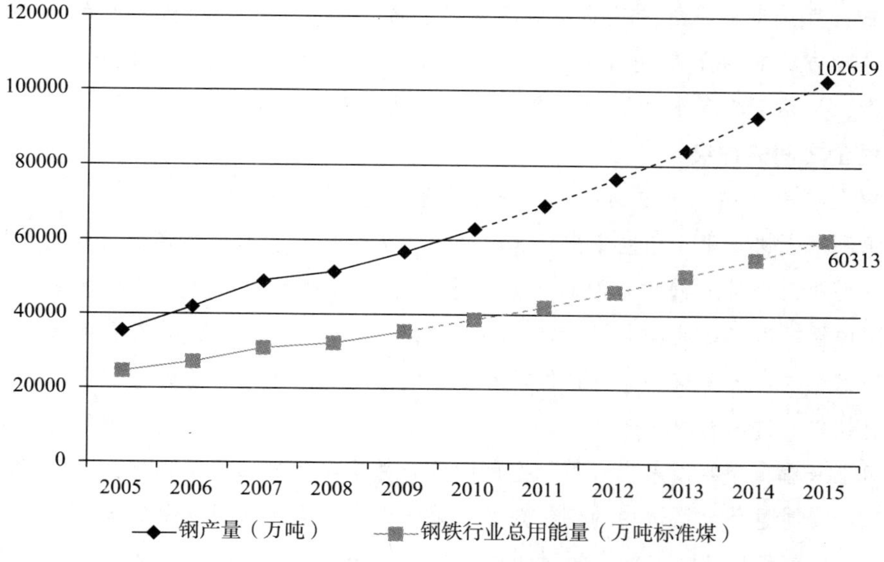

图 6－12 钢铁产量及用能量的预测

Figure 6－12 Forecasting on Steel & Iron Production and Its Consumption

资料来源：中国钢铁工业协会

今后五年，要在控制钢铁生产总量的前提下，通过普及推广已成熟的节能减排技术，达到国家对2015年污染物排放和环境容量要求的目标，例如二氧化碳排放强度降低16%以上，单位工业增加值用水量降低20%，二氧化硫、化学需氧量和氨氮排放量降低10%，工业固体废弃物综合利用率达到76%；同时，使吨钢综合能耗降低到590千克标准煤，焦化、烧结、炼铁各工序能耗分别降低到108千克标准煤、52千克标准煤、400千克标准煤，100%实现负能炼钢。

“十二五”时期，面对结构性矛盾突出、能源消耗量大、铁矿石保障能力不足等深层次问题和加快转变钢铁工业发展方式的要求，钢铁工业要以调整结构优化升级为主线，加强节能减排，完善资源保障体系，推动技术创新，优化产业布局，在保障国民经济和社会发展需求的同时，着力提高产业素质。要通过完善行业管理、落实产业政策、健全标准体系等措施，突出抓好淘汰落后、兼并重组和技术改造等工作，力争“十二五”时期钢铁行业结构调整取得显著进展，核心竞争力有较大提高。

专栏6-1

钢铁产业结构调整中的矛盾与冲突

众所周知，我国钢铁行业产业结构调整的口号已经喊了很多年，而且也调整了很多年，但是进度非常缓慢，甚至出现了不如不调整的声音。业内专家认为，在产业结构调整中所面临的矛盾和冲突依然非常突出，而这些冲突是否能够得到妥善的解决，将对钢铁行业产业结构调整的效果产生决定性的影响。

中央与地方的冲突

这种矛盾应该说由来已久，主要表现在地方政府的权力偏大，对中央的很多政策会结合地方的利益进行落实，这就必然会造成不利于地方的政策在落实的过程中被大打折扣。

地方政府为了推进城市化进程，为了地方的经济发展，甚至是为了政府的业绩，开始向银行大量借贷，直接造成了地方政府负债率明显偏高。而作为这些大大小小的钢铁企业来讲，在不同级别的地方都算得上是纳税大户，淘汰他们就直接断了地方的“财路”了，这就使得节能减排和淘汰落后产能等相关政策的贯彻和落实更加难上加难。

国企与民企之间的冲突

这个冲突主要表现在两个方面：一是资源利用；二是合并重组。

先从资源利用上讲，上半年国家出台相关政策，大力整合国内矿产资源，并且资源配置优先考虑大型钢铁企业，除了沙钢和日照等个别钢企之外，其他的民营钢企基本上属于中小钢铁企业，所以，国内铁矿石资源的优先权肯定轮不到他们。另外，从进口铁矿石资源来看，每年参加铁矿石谈判的都是国内大型钢企，进口铁矿石长协价格也只适用于这些大型钢企，大部分中小钢企根本就没有铁矿石进出口权，即便长协价格谈得很低，他们也享受不到这种优惠。再从钢企之间的合并重组上来看，今年以来，大型钢企之间的合并重组明显加快了步伐，跨区域的整合先不必说，地方政府更愿意整合区域内的钢铁企业，以便形成更大的区域竞争优势。在完成中央制定的将800多家钢企重组成200家钢企的过程中，这种“大鱼吃小鱼”的现象将会一直存在。

钢企与钢贸商之间的冲突

这种冲突随着钢贸企业的不断增多以及市场信息的不断透明，正在变得日益严重。

首先，钢贸商的下游市场部分被钢厂抢占，钢厂利用自身的优势一直在向下游终端渗透，原本下游很多由钢贸商供货的大客户，都变成了钢厂的直供户，钢贸商们的生存空间被大幅压缩。另外，最近在钢材圈闹得沸沸扬扬、影响比较恶劣的一件事就是，28 家钢贸商联名致信一大型钢企领导，表达对其子公司的不满，这在业内引起了很大的反响。究其原因就是钢厂在行情好的的时候，只部分或者延期供给钢贸商的订货量，而在行情不好的时候，把以前没有供足的量全都塞给钢贸商，这就导致了钢贸商在该赚钱的时候赚不到钱，在赔钱的时候亏损会更加严重。其实这种现象由来已久，只是随着钢铁行业的利润逐渐压缩愈演愈烈而已，根本原因在于"钢老大"的思想根深蒂固，但是他们却不知道"水能载舟亦能覆舟"这个道理，这也是钢铁行业在产业结构调整中，必须要解决的实际问题。

其实矛盾和冲突还有很多，像钢企与矿山之间的矛盾，进口矿与国内矿之间的冲突，以及国内钢企在国外投资或者合并重组过程中所造成的冲突等等，这些矛盾都或多或少地影响着钢铁行业产业结构调整的进程和效果。那么如何正确、全面、稳妥地处理这些矛盾，变成了这次改革的当务之急。虽说欲速则不达，但是"钝刀割肉"我们已经尝试了很多年了，很显然没有任何效果，怎么样快速、高效地完成钢铁产业结构调整的重担，将是考验政府的决心和智慧的关键，也是维持钢铁行业健康、稳定、可持续发展的必要前提。

资料来源：宝钢新闻中心

6.4 钢铁业"十二五"节能减排展望

"十二五"时期是中国加快转变经济发展方式的攻坚时期，也将是中国钢铁行业进一步推进兼并重组、调整结构、转型升级的攻坚时期。2011 年是"十二五"开局之年，我国钢铁行业面临的节能减排形势严峻，任务艰巨。

钢铁行业"十二五"规划的初稿之中，要求淘汰 400 立方米及以下高炉和 30 吨及以下转炉；在节能减排标准方面，重点大中型企业吨钢综合能耗不得超过 580 千克标准煤，而此前，2010 年底前要求的标准仅为淘汰 300 立方米及以下高炉，淘汰 20 吨及以下转炉、电炉。

"十二五"期间中国钢铁业节能减排的目标已确立：

（1）普及推广已成熟的节能减排技术，到 2015 年国内大中型钢铁企业的干熄焦率达到 100%。

（2）按照国家相关要求，中钢协会员企业焦化、烧结、炼铁、炼钢的工序能耗达标率要达到 100%。

（3）凡没有通过国家建设项目环境影响评价的企业，要努力完善环保措施，补办手续，获得审批。

作为产能过剩行业，钢铁行业"十二五"发展无疑要以转方式、调结构为根本，国家提出了淘汰落后，兼并重组、技术改造相结合的思路，为了确保目标的实现，"十二五"期间钢铁行业必将在多个方面继续提高与完善。随着社会发展对环境质量要求越来越高，钢铁生产必须与社会、环境相和谐，必须是低碳、绿色制造的过程，国家颁布的一系列关于单位产品耗能限额标准，也给钢铁工业带来了巨大的压力，据权威统计，2009 年，重点统计钢铁企业平均工序能耗基本达到标准限额值的要求，但其中烧结工序能耗未达标企业比例仍达 45.5%，炼铁工序未达标企业比例仍达 20.3%（如图 6-7），而"十二五"初期，即将出台的新排放标准又将会给企业带来更多挑战。

为了实现生产工序的节能，就要求钢铁企业从结构节能、技术节能、管理节能等多个方面对标挖潜，在钢铁生产各环节制定详细的节能减排措施。

- 原材料方面，降低原燃料灰分、硫分，提高入炉矿品位等措施，可能增加采购成本，但有利于节能减排，仍可降低制造成本。
- 技术创新方面，加快推广节能减排新技术，例如焦化煤调湿技术、降低烧结漏风率技术、烧结余热发电技术、高炉脱湿鼓风技术、高炉喷吹焦炉煤气技术等。
- 加强循环利用，提高效率。例如，与供热企业合作，扩大余热利用范围；利用高炉或焦炉消纳社会废塑料等，发挥钢铁企业的社会功能。
- 与电力企业合作，开展“共同火力”发电；副产煤气的资源化高效利用；与建材企业合作，提高冶金渣利用附加值等。

同时，必须继续加大淘汰落后产能的力度。《国务院关于进一步加强淘汰落后产能工作的通知》明确2011年前，钢铁企业高炉有效容积400立方米以下，转炉或电炉公称容量30吨以下（不含特殊钢电炉）产能将被全部淘汰。相应淘汰落后炼铁能力7200万吨、炼钢能力2500万吨。同时，国务院要求，实施淘汰落后、建设钢铁大厂的地区和其他有条件的地区，要将淘汰落后产能标准提高到1000立方米以下高炉及相应的炼钢产能。此举意味着，中小钢铁企业将大量出局，其中，为数众多的民营钢铁企业将不得不遭遇生死劫。若这些淘汰落后的措施能得以切实有效的落实，不仅于节能减排非常有利，而且对于缓解我国钢铁产能增长过快，部分产品供大于求的矛盾也将产生实际的作用。

另外，深化节能减排，必须把钢铁与其他相关产业深度融合，发展循环经济，是进一步提高二次能源和资源利用效率的根本出路。

（1）与供热企业合作，扩大余热利用范围。有条件的钢铁企业可考虑与供热企业联合为周边城市居民供热等。

（2）利用高炉或焦炉消纳社会废塑料等，发挥钢铁企业的社会功能。处理大宗废塑料，包括通过高炉风口喷吹或热压处理后装入焦炉，实现废弃物再资源化利用。在高炉风口喷吹1吨废塑料，相当于喷入1吨油的热量。利用焦炉将废塑料和焦煤压块处理可以处理大量的废塑料，据目前焦炉消纳废塑料先进技术，废塑料能添加到焦炭产量的2%左右，折合吨钢废塑料消纳量为8千克。另外，钢厂处理废轮胎和罐类的循环利用技术也趋于成熟、稳定。

（3）与电力企业合作，开展“共同火力”发电。建立“共同火力”产融模式是实现钢铁企业煤气“零排放”的重要途径之一。共同火力整合了钢厂和电厂的各自优势，充分利用钢厂的剩余资源（副产煤气）和火力发电厂现有设备（锅炉及大容量的发电机组），只需增加煤气输送管道和对锅炉进行改造即可进行发电，在投入较少的情况下就可以获得较好效益。国家应出台相应激励政策。

（4）副产煤气的资源化高效利用。副产煤气资源化利用可提高煤气利用的附加值，可生产氢、甲醇、二甲醚（CH_3OCH_3）等。利用焦炉煤气吸附制氢是目前比较成熟的技术，重点是研究焦炉煤气吸附制氢的大型煤气压缩机等关键技术，焦炉煤气制氢将比直接使用较贵的天然气和煤炭等制氢更加经济，可提供石化行业、氢气动力汽车及氢冶金等，是大规模、高效、低成本地生产廉价氢气的有效途径。

（5）与建材企业合作，提高冶金渣利用附加值。促进钢渣为代表的固废资源综合利用产业融合。利用高炉渣生产水泥已成为普及最好的固体废弃物利用技术，目前高炉渣水泥中高炉渣的配比可达到40%左右，相应可节省资源、能源消耗约40%，降低 CO_2 排放约45%。钢渣也可以用来生产钢渣水泥或混凝土等，取代部分水泥熟料。虽然由于钢渣的水硬活性较高炉渣低，目前在水泥和混凝土中的掺量仍较低。

（6）系统集成脱盐、城市污水利用、深度处理回用综合水利用技术，发挥钢铁企业的城市功能。难降解、高污染物浓度的废水处理一直是钢铁企业水处理的重点、难点，“十二五”期间应继续加大投入，努力对其进行深度处理、科学减量、合理消纳，为废水的“零排放”创造条件。重点是：浓盐水处理及减量化，焦化废水的深度处理与回用，市政污水再生水回用。

综合上文分析，对钢铁行业“十二五”节能减排提出如下建议。

1. 加快淘汰落后生产能力和实现产业结构及布局合理化

解决生产能力过剩问题的关键是加快淘汰落后生产能力。使用进口铁矿超过60%的内地企业应不再新增炼铁能力。通过联合重组在沿海有条件的地区建设现代化钢厂，为今后若干年产业布局调整创造条件。

2. 加快推进企业联合重组，提高产业集中度

政府要加快解决企业联合重组的体制机制问题，国有资产管理部门对中央和地方分权管理的国有或国有控股企业，要指定跨地区联合重组的规划，落实《钢铁产业发展政策》中关于提高产业集中度的目标要求，避免高水平重复建设，只有提高产业集中度，才能加快以企业为主体的自主创新体系建设，提高国际竞争力。

3. 把节能降耗、改善环保作为企业生存发展的前提，加大技改资金的投入

现阶段，借鉴国外先进的节能减排技术是迅速缩小我国与世界钢铁行业先进水平的主要途径，但要真正使我国从“钢铁大国”转变为“钢铁强国”，自主知识的节能减排技术开发和运用是不可或缺的。为此，针对我国的钢铁企业小而分散的特点，钢铁协会和主要钢铁企业，应走联合开发的道路，针对我国钢铁行业的特点，有预见性地加大资金和人才的投入，研究和开发具有自主知识产权的钢铁节能减排新技术。

4. 转变经营方式，重视环境经营

随着国家制定和实施一系列与环境问题相关的法律法规及消费者环境保护意识的增强，实行以降低资源消耗、减少环境污染为目标的环境管理在企业经营活动中占越来越重要的位置，它与企业的生死存亡密切相关。我国钢铁企业必须顺应趋势，转变经营方式，从获取竞争优势的战略层面重视环境管理，确保企业持续、快速、健康发展。

5. 国家应制定和实施相关鼓励政策

制定鼓励钢铁行业资源、能源利用和环境友好的政策，促进发展循环经济；完善废钢进出口政策；重视废钢和废塑料的回收、分类、处理和供应；对废塑料回收和处理在技术开发上给予资金和政策支持；鼓励钢铁渣利用和余热发电上网等政策。

此外，研究制定钢渣产品进入建筑材料市场的扶持政策，完善衔接生产企业和钢渣产品市场的供销机制，形成从生产现场回收、厂外集中处理加工和综合利用的规模化钢渣资源综合利用产业；建立建筑材料用钢渣标准，尽快完善建筑领域工程建设标准体系，研究出台建筑材料用钢渣标准，进一步完善钢渣粉用于混凝土掺合料和钢渣砖等建筑用钢渣产品的生产技术标准，

以利于开展钢渣综合处理工艺设计和生产。

"十二五"时期，我国钢铁工业节能减排、环境保护工作将面对低碳经济及更严格的污染物排放标准等新形势、新压力，由单一技术、单一工序节能减排技术向系统集成优化转变，力争在重点技术领域有所突破，真正从末端治理向源头治理、过程控制转变，持续深化节能减排工作。

6.5 典型钢铁企业节能减排动态

6.5.1 宝钢的"十二五"节能减排规划

在"十二五"期间，宝钢将在继续提升能源环保管理体系能力的基础上，对已有产线、新投产线和收购产线，结合各企业行业特点、产品结构、产线装备水平、所处环境条件等，从持续改进、科技创新、升级改造等各方面，加大节能减排技术改造力度。宝钢计划投入101亿元，实施重点项目83个，争取节能环保工作有新突破。

宝钢"十二五"节能减排规划期目标为：坚持减量化优先策略，不断提高资源与能源利用效率；持续提高各制造单元的清洁生产水平，开展清洁生产达标和绩效持续提升行动；不断提高副产资源返生产利用率，提升副产资源社会化利用的价值，发展节能环保工艺技术与装备制造技术产业，提升节能环保服务产业；建立与完善循环经济管理组织体系、指标体系，建立环境管理会计制度与评价体系；履行企业社会责任，促进社区和谐。

一是对已有产线，宝钢将采取持续改进与增加投入并重的措施。在钢铁单元全面开展清洁生产达标和绩效提升行动，要求三年内主要指标上一个台阶，未列入国家标准的工序指标和能源系统转换指标提高3个百分点，不断提高基础管理水平，加强节能环保设施运行精度管理，提高余能回收与利用率，确保环保设施的环保效果，以推行清洁生产达到持续改进。

二是对新投产线，将依据市场条件，有效开展"四达"管理。为严控增量，宝钢将特别加强基本建设项目"四达"管理。将把能源环保指标作为项目验收的主要门槛，改变以往重生产达标，轻消耗排放达标的项目管理。

三是对新收购产线，将抓合规性整改，以升级改造促淘汰落后。在收购活动中，宝钢将开展能源环保尽职调查工作，对收购对象从能源环保管理、技术装备、外部资源条件等进行细致调研，提出对策。对于新并入集团的企业，宝钢在加强管理监督的同时，将更加积极主动地组织协同提升工作，制订切实可行的改造方案，协调集团内部资源，推动实施，通过带动落后企业进步，促进共同发展。

6.5.2 新首钢循环经济发展模式建立①

新首钢位于河北唐山渤海之滨的曹妃甸工业区，该区是国务院批准设立的循环经济示范区。2005年，首钢与唐钢联合组建首钢京唐钢铁公司即新首钢后，开始实施搬迁工程。随着奥运会的临近，北京及周边地区钢厂的减产、停产工作也已经启动。首钢为迎接奥运，推动节能减排工作的顺利进行，积极采取各种有效措施，促进了新首钢循环经济发展模式的建立。

---

① 新首钢创造循环经济节能减排初见成效。

首钢有效地调整产业结构。在奥运会期间将暂停在京的主要钢铁生产，之后逐步压缩在京的800万吨年生产能力，至2010年全部停产；河北省也因组建新首钢将关停35家污染严重的小型钢铁企业共730万吨的年生产能力。如此一来，每年可减少北京市约1.8万吨的可吸入颗粒物，消减唐山地区2.3万吨烟粉尘、2.4万吨二氧化硫，对加速钢铁产业结构调整，推动节能减排，确保资源循环再利用，有效改善北京和唐山地区环境质量、减少渤海海域的环境压力发挥了重要作用。

以创新为发展动力，广泛采用先进工艺，从技术和设备上改善新首钢节能减排局面。在设备方面新首钢集中采用220项大型装备及先进工艺技术，现拥有我国目前最大的高炉、亚洲最大的焦炉、世界上第一个自主集成的铁水罐加盖保温并“一罐到底”技术等一批成果，确保了各项节能减排指标都能达到国际领先水平，从源头上节能减排、控制污染。

此外，新首钢充分利用临海建厂、曹妃甸深水大港、铁路公路便利等优势，采用大型船舶运输并直接在厂区码头装卸，实现了原料输入、企业生产、产品输出的高度集约化及生产过程中各环节的最短运距，吨钢占地仅0.9平方米，达到国际先进水平，每年可减少铁路公路运输压力1800万吨，不仅充分发挥了新首钢临海布局的优势，还节约了土地资源和交通资源的大量社会资源。

污水和废渣是否能够合理处理是衡量一个钢铁企业是否能够达到真正节能减排的重要标准。新首钢采用先进的污水处理技术将雨水、生产生活污水处理后全部回用于高炉水冲渣、钢渣冷却、厂区绿化等；海水淡化工程每年淡化海水3000万立方米，可解决全厂40%的用水量，吨钢耗用新水仅3.84吨，仅为钢铁行业现平均水平的约1/3，水循环率可达97.5%，基本实现污水“零排放”，对渤海海区的海洋环境保护起到示范作用。对炼钢所产生的废渣分门别类进行再利用，将含铁金属部分全部回收继续炼钢，其余部分则分别作为水泥、建材等行业的原料，每年可减少水泥行业石灰石开采量约320万吨、二氧化碳排放约220万吨、标准煤消耗约22万吨、粉尘排放约7万吨，有效降低了下游相关产业的能源消耗、排污压力和生产成本，也为新首钢创造了可贵的社会效益和经济效益。

与此同时，新首钢利用余热淡化海水，剩余的高含盐水又大幅缩短下游制盐行业的生产过程。炼钢过程产生的各类煤气全部回收，每年可利用余压、余气自发电47.5亿千瓦时，为新首钢的正常生产提供了可靠的能源保障。

新首钢在建设过程中，始终贯彻循环经济理念，资源的循环再利用，也成为新首钢“发家致富”的新法宝，在争取创造良好经济效益的同时，也将为新首钢节能降耗和保护环境带来实际效益。

7 有色金属行业的节能减排

7.1 有色金属行业“十一五”运行综述

有色金属是国民经济的重要基础原材料产业，在经济建设、国防建设和社会发展中发挥着重要作用。有色金属工业是工业领域六个重点行业之一，能耗占全国能源消耗总量的4.3%，其中，电力消耗占全国电力消耗量的6.6%，是推进节能降耗的重点行业。我国有色金属工业能源消耗主要集中在矿山、冶炼和加工三大领域，整体能耗仍在较高基数下保持增长，因此有色金属工业面临巨大节能降耗政策压力。

“十一五”期间，我国有色金属工业持续快速发展，已成为世界有色金属生产大国和消费大国，成为全球有色金属产业发展的重要因素。相关主要数据如下：

- 至2010年，包括铝、铜、铅、锌等在内的10种有色金属工业产量达到3092万吨（如图7-1），同比增长17%，“十一五”期间平均增长13.8%，产量消费量占世界1/3。

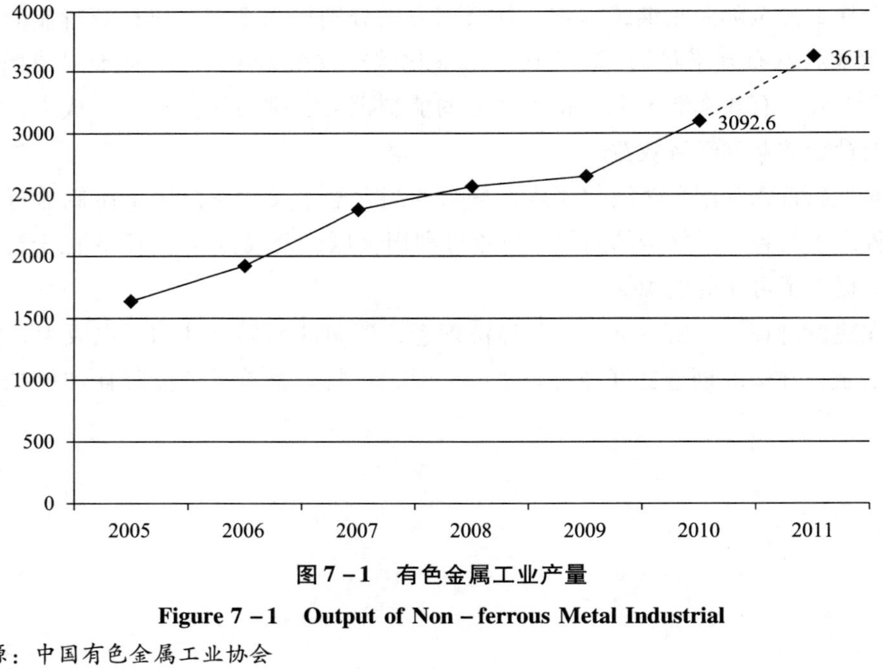

图7-1 有色金属工业产量

Figure 7-1 Output of Non-ferrous Metal Industrial

资料来源：中国有色金属工业协会

- 在“十一五”期间，中国有色金属行业投资年均增长32.7%，2010年有色金属工业累计完成固定资产投资3627.9亿元，同比增长33.5%。
- 2010年我国金属工业实现主营业务收入约3万亿元，主营业务收入增长29.4%，完成增加值按可比价格计算比上年增长12.7%，“十一五”期间年均增长16.2%，占GDP比重由2005年的1.19%增加到2010年的1.99%。新开工项目投资总额为3684.5亿元，同比增长28.8%。

- 2010年我国有色金属进出口贸易总额达到1203.4亿美元，同比增长43.7%。“十一五”期间，我国有色金属产品进出口贸易额年均增长20.8%。

7.2　有色金属行业“十一五”节能减排总结

“十一五”时期，我国有色金属工业节能减排成效显著，成为世界有色金属产业节能减排领导者和领头羊。目前，全行业的经济规模不断扩大，综合实力明显提高，在国际同行业中的影响力和竞争力日益增强。进一步推进有色金属工业的节能减排，是贯彻落实科学发展观、转变发展方式的紧迫任务，也是我国应对全球气候变化的实际行动。“十一五”期间钢铁行业能源消耗量变化主要情况如下：

（1）能耗总量。根据中国有色金属工业协会统计，2009年我国有色金属工业年消费标准煤12741万吨（如图7－2），有色金属工业电力消耗200亿千瓦时（如图7－3），分别比2005年增长56%和69%。

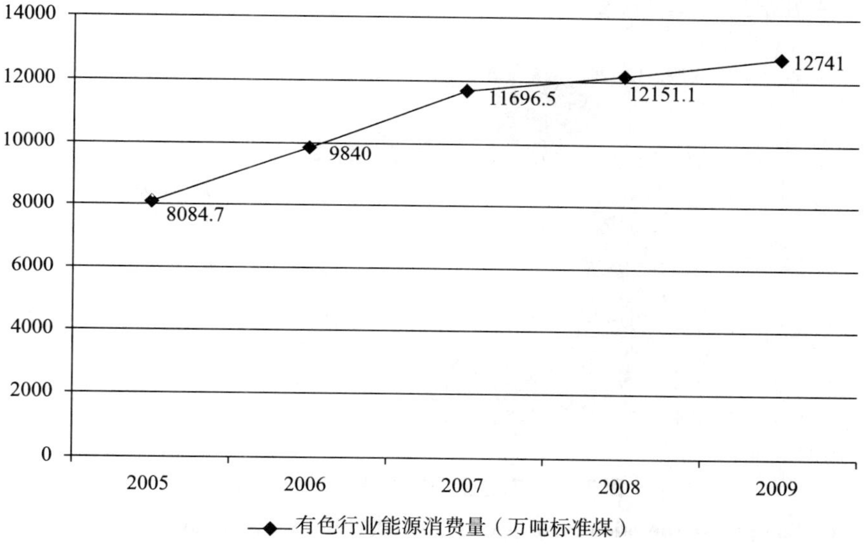

图7－2　有色金属行业能源消费量

Figure 7－2　Energy Consumption by Non－ferrous Metal Sector

资料来源：中国有色金属工业协会

（2）能耗结构。2008年我国有色金属工业消费电力2259亿千瓦时（折标准煤2776万吨）、煤炭4378万吨（折标准煤3127万吨）、焦炭234万吨、燃料油73.5万吨、柴油54.7万吨、汽油5.8万吨、天然气8.7亿立方米以及其他燃料折99万吨标准煤。其中，煤炭消费占比最高为46.23%，电力消费占比为41.05%，其他能源消费如焦炭、燃料油、柴油等合计占比为12.72%。有色金属行业能源消费结构情况如图7－4所示。

（3）主要产品能耗下降，技术经济指标明显提高。2010年，铝锭综合交流电耗为13979千瓦时/吨；氧化铝综合能耗为632.4千克标准煤/吨；铜冶炼综合能耗为360.3千克标准煤/吨；铅冶炼综合能耗降到453.5千克标准煤/吨；电锌综合能耗为946.6千克标准煤/吨，如

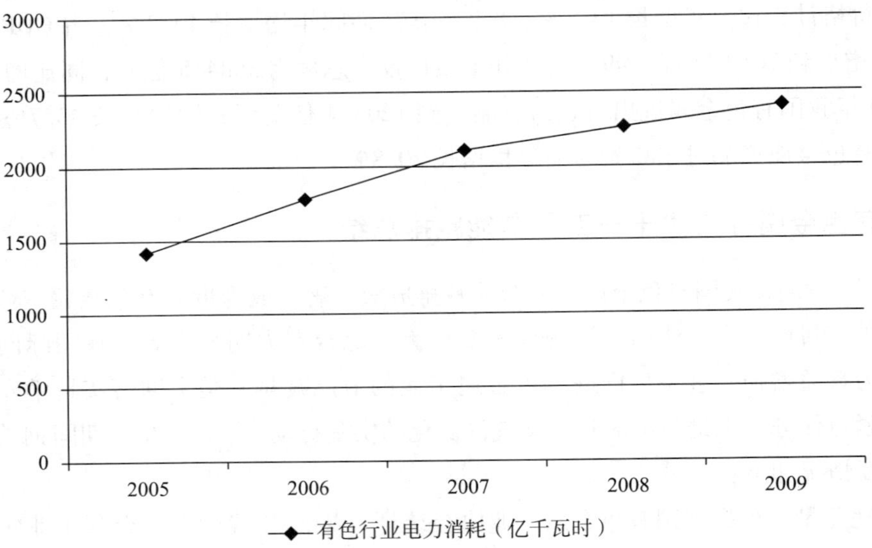

图 7－3　有色金属工业电力消费

Figure 7－3　Power Consumption by Non－ferrous Metal Sector

资料来源：中国有色金属工业协会

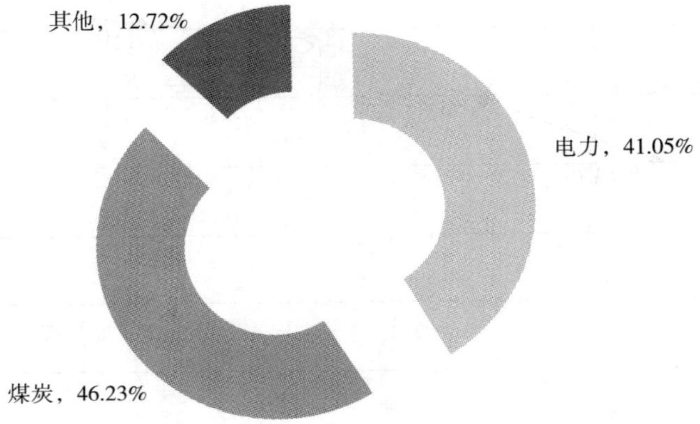

图 7－4　有色金属行业能源消费结构

Figure 7－4　Energy Consumption Structure in Non－ferrous Metal Sector

资料来源：中国有色金属工业协会

图 7－5 所示。

（4）落后产能淘汰进展。2006 年，电解铝落后的自焙槽已全部淘汰，电解铝技术装备水平大大提高，能耗大幅下降。目前，广泛采用的 300KA 大型预焙槽与原先采用的 60KA 自焙槽相比，吨铝电耗可降低2000KWh以上，氟化物排放由 20Kg/t 下降到 1Kg/t 以下。

2008 年末，有色金属工业落后产能仍占 25%。应淘汰的电解铝小型预焙槽产能有 80 万吨，铜冶炼反射炉和鼓风炉产能 30 万吨，铅冶炼烧结机产能 150 万吨，锌冶炼落后产能 40 万吨。

2009 年淘汰落后铜冶炼产能 30 万吨、铅冶炼产能 60 万吨、锌冶炼产能 40 万吨；2010 年 9 月底已关停铜冶炼、电解铝、铅冶炼和锌冶炼的 93 家企业，其中铜冶炼产能 14.49 万吨、电解铝 37.1 万吨、铅冶炼 26.58 万吨、锌冶炼 29.37 万吨。

2011 年底前，计划淘汰电解铝、烧结锅炼铅等产业的大量落后工艺及设备，实现重点骨干

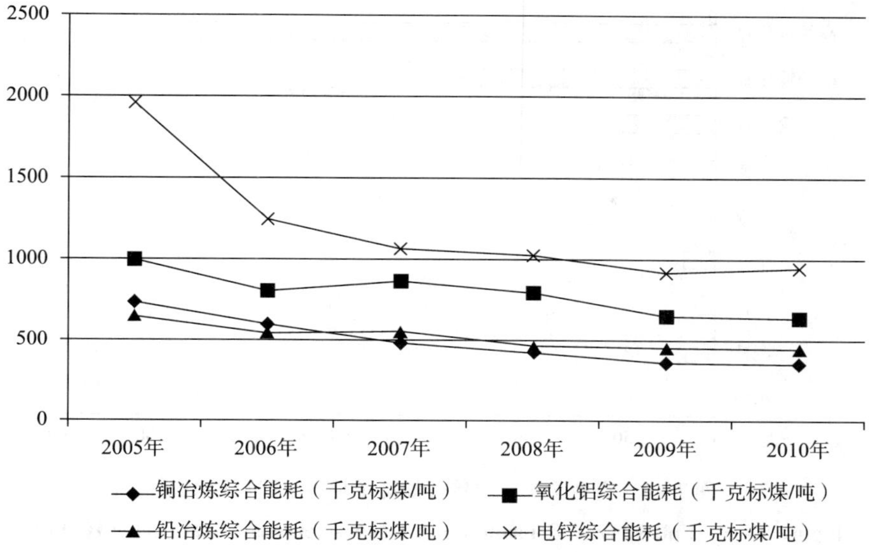

图 7-5 2005—2009 年主要产品能耗下降情况

Figure 7-5 Energy Consumption Reduction of Major Products from 2005—2009

资料来源：中国有色金属工业协会

电解铝厂吨铝直流电耗下降到 12500 千瓦时以下，粗铅冶炼综合能耗低于每吨 380 千克标准煤，余热基本 100% 回收利用。

（5）产业结构调整。2008 年，全国氧化铝企业平均规模达到 110 万吨，在 25 个氧化铝企业中，8 个年产 100 万吨以上的企业的产量占全国的 74.2%。目前，全国电解铝企业平均规模已超过 10 万吨，而 2000 年仅 3 万吨。

2008 年，年产 10 万吨以上的电解铝企业有 16 家，产量占总产量的 70%。精炼铜企业通过整合重组，年产 10 万吨以上的企业由 3 家增至 6 家，这 6 家企业的产量占全国产量的 63%。铅年产 10 万吨以上的有 7 家，产量占总产量的 31.5%。锌年产 20 万吨以上的有 4 家，产量占总产量的 30.5%。

（6）加快企业兼并重组，产业集中度有所提高。大型骨干企业在节能减排中发挥了骨干作用，树立了行业典范。中国铝业公司氧化铝单位能耗与 2005 年相比下降 33.2%。江西铜业通过新增的 6 台阳极炉余热锅炉、4 台硫酸余热锅炉和 1 套电解车间蒸汽压差能量回收发电装置，全面进行综合回收利用，2010 年可发电 1.1 亿千瓦时以上，大约节约 3.5 万吨标准煤。

但是，整体而言有色金属行业的单位能耗水平还是远高于水泥和钢铁。有色金属中，镍生铁的吨能源消耗量最高，其次为含硫镍和铝。铜、铅的单位能耗水平最低，如图 7-6 所示。

（7）再生金属利用。据测算，与原生金属生产相比，每吨再生铜、再生铝、再生铅分别相当于节能 1054 千克、3443 千克、659 千克标准煤，节水 395 立方米、22 立方米、235 立方米，减少固体废物排放 380 吨、20 吨、128 吨，每吨再生铜、再生铅分别相当于少排放二氧化硫 0.137 吨、0.03 吨。“十一五”前四年，我国再生有色金属产业与生产等量原生金属相比，相当于节能 4650 万吨标准煤，节水 18.8 亿立方米，减少固废排放 34.5 亿吨，减少二氧化硫排放 112 万吨①。

① 资料来源：再生有色金属产业发展推进计划。

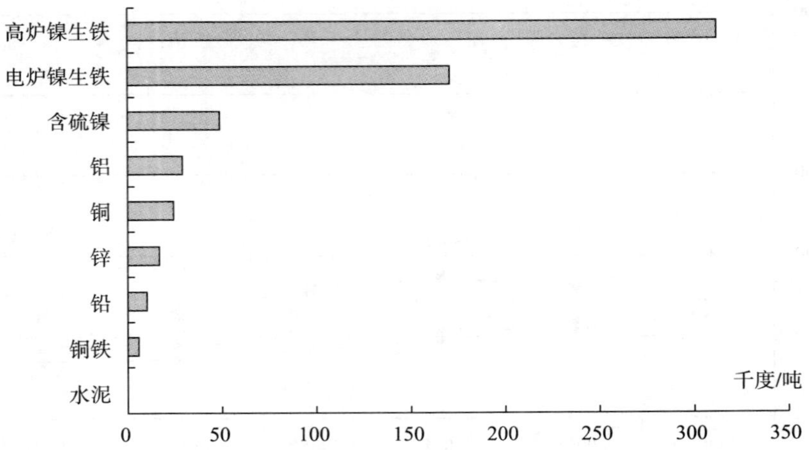

图7－6 有色金属行业单位能耗水平高于钢铁和水泥

Figure 7－6 Unit Energy Consumption by Non－ferrous Metal Sector Higher Than That of Steel & Iron and Cement Sectors

资料来源：中金公司

“十一五”以来，我国有色金属工业虽然在节能减排方面取得较好的成效，但仍存在一系列问题需要去面对。

1. 部分产品单耗与世界先进水平仍存在一定的差距

受我国铝土矿为一水硬铝石的影响，我国氧化铝的生产能耗高于国外采用三水软铝石铝土矿的企业，即使同为拜耳法，我国的溶出温度也较国外企业高一倍左右。当前国内采用国产铝土矿生产氧化铝能力达2250万吨，尚有700万吨在建和拟建产能，恰恰也是利用耗能较高的国产铝土矿生产。国内采用三水铝石铝土矿的企业氧化铝综合能耗400千克标准煤/吨氧化铝，已经达到世界先进水平，2008年我国铅冶炼综合能耗463.31千克标准煤/吨铅，与国外的先进水平300千克标准煤/吨铅仍存在一定的差距。我国火法炼锌综合能耗为1888.36千克标准煤/吨，虽然接近国外1.7吨标准煤/吨锌的先进水平，仍有一定差距。

2. 国内企业间能耗水平相差悬殊

2008年电解铝综合交流电耗下降到14283千瓦时/吨，国内先进水平达到了13500千瓦时/吨铝，已处于世界先进水平，但是国内铝电解企业之间差距较大，最好的企业为13000千瓦时/吨铝左右，最差企业为16054千瓦时/吨铝，相差3000千瓦时/吨铝以上。氧化铝生产企业之间差距也较大，采用一水硬铝石的部分企业氧化铝综合能耗仍有1200多千克标准煤/吨氧化铝，比全国氧化铝综合能耗平均水平794.4千克标准煤/吨氧化铝，都高出400多千克标准煤/吨氧化铝。

3. 淘汰落后工艺任务艰巨

虽然淘汰落后生产能力在有色金属工业局部取得重要进展，但从整体观察，能源消耗高、环境污染大、劳动强度高的落后生产能力在有色金属工业中仍占相当比例。截至2008年底，国内能耗高、污染重的低于80KA及以下的落后预焙槽产能为102万吨，100KA及以下的预焙槽产能为167万吨；反射炉及鼓风炉铜冶炼产能有30万吨/年；烧结锅铅冶炼和落后锌冶炼产能分别有40万吨/年、60万吨/年。此外，还有较为落后的烧结机铅冶炼能力150万吨/年。有色金属工业淘汰落后产能的任务仍十分艰巨，尤其是铅锌产业企业集中度低，中小企业居多，很多企业位于经济发展相对落后的地区，淘汰落后生产能力难度很大。

4. 矿产资源品质下降形成能耗上升的压力

我国有色金属矿产资源经过多年开采后，原矿和精矿品位出现下降趋势。2008 年国内铜矿坑采出矿品位为 0.81%，而 2000 年为 1.0%；铜矿露天开采出矿品位为 0.51%，而 2000 年为 0.52%。2000 年国内铅坑采出矿品位为 3.64%，2008 年下降到 3.39%；锌露天开采出矿品位 2000 年为 9.10%，2008 年下降到 7.50%。矿山原矿品位的下降，必然导致能源消耗上升。这是行业进一步推进节能减排面临的一个重要障碍。

7.3 有色金属行业“十二五”节能减排背景

1. 有色金属需求持续增长

我国是近年来全球有色金属需求增长最快的国家，但人均消费量与发达国家相比还有很大增长空间，经济发展对有色金属的需求仍将处于增长阶段。不断增加的社会蓄积量为有色金属循环利用奠定了良好基础。截至 2009 年底，消费领域蓄积有色金属资源超过 2 亿吨。

2. 有色金属原生矿产资源约束不断加剧

我国有色金属矿产资源相对短缺，资源消耗量持续增加，重要资源对外依存度逐年攀升，目前铜原料约 65%、铝原料约 55%、铅锌原料约 30% 以上依靠进口，并且还有进一步扩大的趋势。大力发展再生有色金属产业是缓解资源约束的有效途径，有利于解决国内自然矿产资源不足与有色金属需求增长之间的突出矛盾。

3. 有色金属产业面临的节能环保压力日益加大

有色金属作为传统高耗能行业，节能减排任务艰巨。环境保护对产业发展形成“绿色屏障”。排放标准将越来越严格；重金属污染治理任重道远；固体废弃物，例如赤泥、尾矿等利用率难题仍未有大的突破。

4. 再生资源利用水平较低

除少数骨干企业外，多数企业技术装备水平低，金属回收率低，冶炼过程污染比较严重，以再生铅最为突出。

5. 淘汰落后产能任务艰巨

按照国务院国发［2010］7 号文件要求，到 2011 年底，要淘汰 100 千安及以下电解铝小预焙槽；鼓风炉炼铜，烧结锅、无尾气吸收装置炼铅，小竖罐炼锌等落后生产工艺及设备。

6. 产业结构调整缓慢，产品结构有待提高

当前，我国有色金属冶炼规模持续扩张，已经超出国内资源、能源、环境支撑能力对上游的矿产开采和下游的深精加工环节的重视与投入不足，导致产业结构性矛盾突出，不利于行业产业链的发展与完善。同时，由于产业退出机制不完善，高能耗的落后生产能力淘汰难度大。此外，我国有色金属工业在常规有色金属产品方面已能够基本满足国民经济发展的需要，但我国具有资源优势的有色金属，目前还都只能加工成初级矿产品或初级冶炼产品，除少量国内应用外，极大部分出口外国，资源优势尚未变成经济优势。

7. 能源消耗总量持续上升，大气环境污染压力增大

有色金属工业是高耗能产业，尽管近年部分产品单位能耗略有降低，但有色金属工业总能耗仍呈持续增长势头。目前，我国有色金属工业产品单位平均能耗比国际先进水平高 15% 左右。

同时，尽管近年来不断加大环境治理的力度，但是有色金属企业污染问题还远没有彻底解决，特别是中小地方企业和民营企业对环境的污染尤为严重。

8. 水资源消耗大，水环境压力增大

目前，我国有色金属工业用水量较大。2005 年有色金属工业用水量为 57.5 亿立方米，占整个工业行业用水量的 4.47%，2007 年有色金属工业用水量有所下降，但仍较高，为 69 亿立方米，占整个工业行业用水量的 4.89%。据预测，尽管单位工业增加值用水量持续下降，但未来有色金属工业的总用水量仍将上升，特别是在 2010—2015 年期间，由于我国有色金属行业的污染生产量所占比例较高，水资源消耗增加较快。

9. 环境污染压力持续增大

据测算，未来 5 年—10 年，随着有色金属工业进一步发展，不论是废水生产量、重金属生产量、大气污染物生产量，还是固体废物生产量，都将持续上升，特别是在 2010—2015 年期间，我国有色金属行业的污染物所占比例较高，是特别需要重视的时期。

10. 要达到预定的污染控制目标，治理投入将进一步加大

据预测，未来 5 年—10 年，由于我国有色金属工业废水生产量、大气污染物生产量和固体废物生产量持续上升，为了减轻对环境的压力，达到污染控制目标，需要进一步加大投入，特别是在 2010—2015 年期间，无论是治理投资，还是运行费用的投入都将进一步加大。

7.4 有色金属行业“十二五”节能减排展望

2011 年是“十二五”开局之年，为有效控制冶炼产能的过快增长，做强有色金属产业，“十二五”规划提出控制有色金属产量增速，10 种有色金属产量年均增幅将要控制在 8% 左右。进一步推进有色金属工业的节能减排，是贯彻落实科学发展观、转变发展方式的紧迫任务，也是我国应对全球气候变化的实际行动。建设资源节约型、环境友好型产业，要大力发展循环经济、绿色经济和低碳技术，研发节能减排、清洁生产共性技术，突破制约行业发展的能源和环境瓶颈。加强资源综合利用，提高资源利用效率，发展再生有色金属工业。

根据有色金属行业“十二五”规划草案，“十二五”期间，我国有色金属行业的主要任务将集中在有效控制产能、努力提高资源保障程度、提高技术水平、优化产业结构、加快兼并重组、开展节能减排等 10 个方面。

7.4.1 总体目标

（1）产能控制目标。根据草案，计划到 2015 年，粗铜冶炼控制在 500 万吨以内，电解铜控制在 650 万吨—700 万吨之间，氧化铝控制在 4100 万吨以内，电解铝控制在 2000 万吨以内，铅冶炼控制在 550 万吨以内，锌冶炼控制在 670 万吨以内。从数字看，未来有色金属冶炼总产能扩张的空间将相当有限。

（2）产量控制目标。2015 年 10 种有色金属产量控制在 4200 万吨以内，其中铜 650 万吨（矿产铜 500 万吨），电解铝 2000 万吨（氧化铝 4100 万吨），铅 500 万吨（矿产铅 350 万吨），锌 680 万吨，镁 100 万吨，如图 7 - 7 所示。

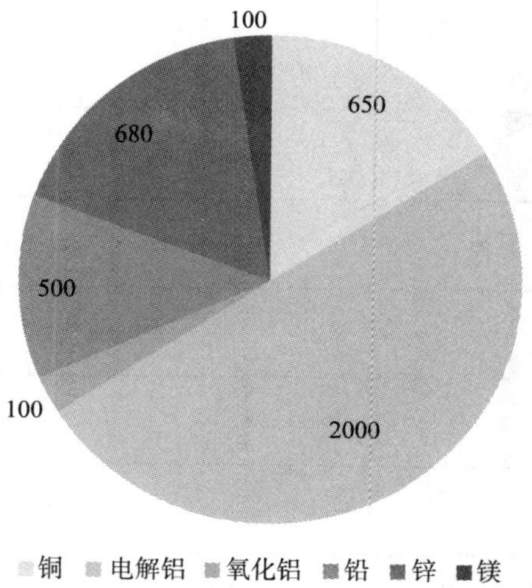

图 7 - 7 有色金属产品的产量控制目标

Figure 7 - 7 Goal of Production Capacity Control for Non - ferrous Metals

资料来源：《有色金属行业“十二五”规划草案》

（1）电解铝。2015 年，平均每吨铝综合交流电耗为 13670KWh，与 2005 年比，吨铝降低电耗 905KWh。图 7 - 8 为电解铝节能目标。

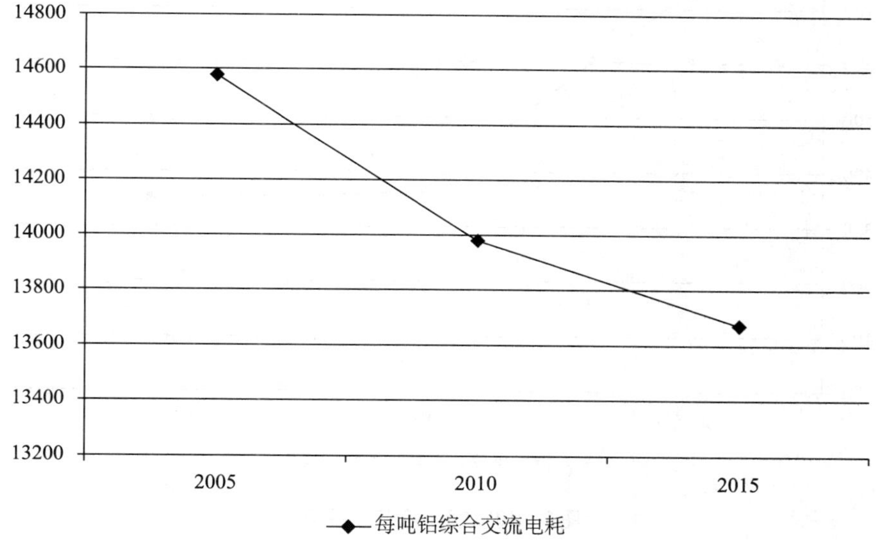

图 7 - 8 电解铝节能目标

Figure 7 - 8 Energy Conservation Goal of Electrolytic Aluminium

资料来源：《有色金属行业“十二五”规划草案》

（2）氧化铝。2015 年氧化铝平均综合能耗降为 650 千克标准煤，与 2005 年比，将降低能耗 348.2 千克标准煤。图 7 - 9 为氧化铝节能目标。

（3）铜冶炼。2015 年，平均每吨精炼铜综合能耗为 350 千克标准煤，与 2005 年比，吨铜综合能耗降低 102 千克标准煤。图 7 - 10 为铜冶炼节能目标。

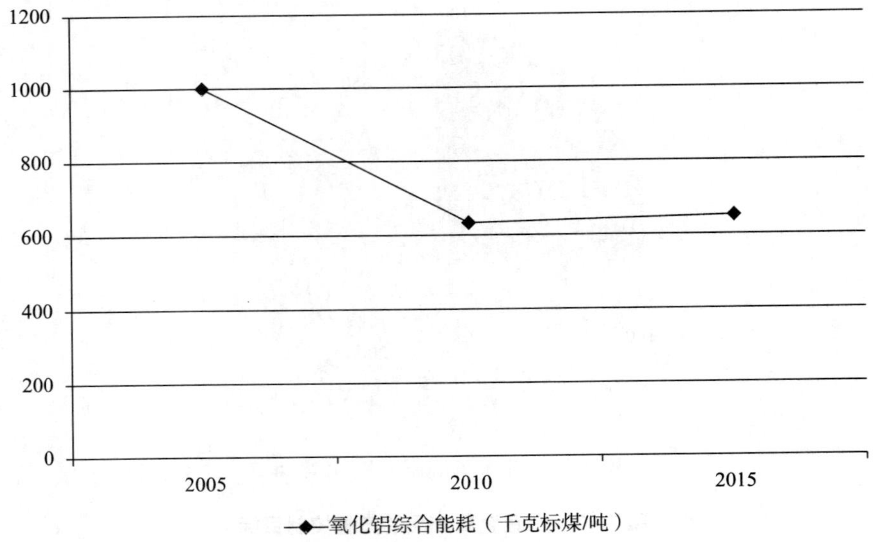

图 7－9　氧化铝节能目标

Figure 7－9　Energy Conservation Goal of Aluminum Oxide

资料来源：《有色金属行业“十二五”规划草案》

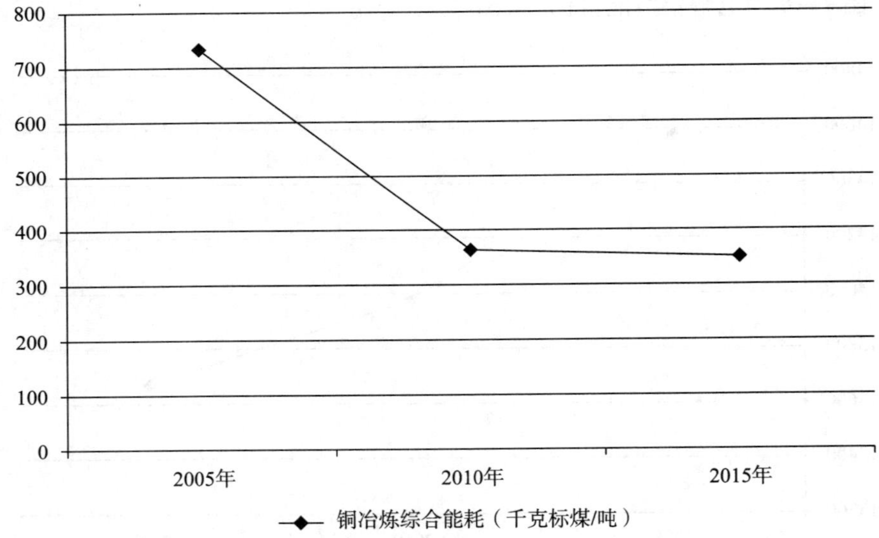

图 7－10　铜冶炼节能目标

Figure 7－10　Energy Conservation Goal of Copper Metallurgy

资料来源：《有色金属行业“十二五”规划草案》

（4）铅（锌）冶炼。2015 年，平均每吨铅冶炼综合能耗为 430 千克标准煤，与 2005 年比，吨铅降低能耗 225 千克标准煤。图 7－11 为铅冶炼综合能耗目标。

（5）镁冶炼。到 2015 年，吨镁综合能耗下降到 4000 千克标准煤，与 2005 年比，吨镁降低能耗 4635 千克标准煤。

（6）其他。根据测算，其它有色金属品种（包括镍、锡、锑、钛等其他金属冶炼、铜铝加工及矿山采选等）2015 年节能量为 250 万吨。

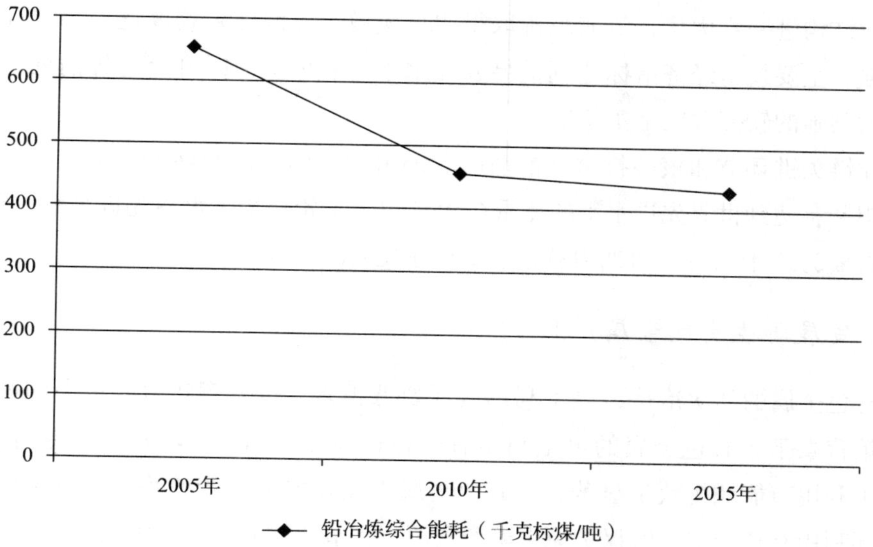

图7-11 铅冶炼综合能耗目标

Figure 7-11 Energy Consumption Goal of Lead Metallurgy

资料来源：《有色金属行业“十二五”规划草案》

7.4.2 提升集中度，优化产业结构挑战

在“有色金属十二五规划”中，规划要求到2015年，铜、铝、铅、锌排名前十位企业产量占全国总产量比例分别达到90%、90%、70%和70%。建议国家继续鼓励地方大型企业集团的发展。

“十二五”加强有色行业兼并重组促进产业升级、未来5年—10年，是我国有色产业发展的关键期。兼并重组将会使我国有色产业发展的格局产生重大变化，将会定格我国有色产业若干年的发展方向，但是，在迎来“兼并重组、产业升级”这一潮流时，我国有色产业也会遇到一系列问题。

目前，我国有色金属各品种企业和发达国家相比较仍然是数量多、规模小、分布散、无法形成一致的竞争力。无疑产业政策应该推动企业走向规模化、集约化经营，才能适应国际市场的激烈竞争，企业才能提高技术装备水平和研发能力，才有力量解决环保治理、节能和实现规模效益。小型的甚至是作坊式的有色金属企业，很难实现产业升级和节能减排。

与此同时，更重要的是应加强有色金属上下游企业或产品的结合，不断优化产业链结构，形成最合理的资源配置。要以勘探为基础，以采矿为核心，以冶炼为中心，以地域关联、资金连接为纽带，通过参股、联合等多种形式，实施从勘探、采矿、选矿、冶炼到产品深加工的联盟战略，将有色金属企业整合为一个市场化、现代化的战略联盟，提高整体的经济效益，进而实现“十二五”实现节能减排目标。

“十二五”有色金属工业发展规划的主要目标有：着力调整产业结构，加快转变发展方式，产业组织和布局结构要明显改善；促进产业改造提升，坚持控制总量与优化结构相结合、兼并重组与优化布局相结合、自主创新能力要显著增强；自主创新与技术改造相结合、节能减排要取得显著成效；资源保障能力要明显提高；开发国内资源与国外资源相结合，资源开发与节约利用相结合的原则。提高发展质量和效益，有效控制冶炼产能过快增长。力争到2015年，有色金属工业产业结构进一步优化，节能减排取得明显成效，资源保障程度进一步提高，技术创新能力显著增强，主要技术经济指标达到世界领先水平，总体实力跃升至世界前列，为实现有色金属工业由大到强的转变奠定坚实基础。

掌控产业链关键环节和核心技术，推动产业链由低端向中高端转变，提升核心竞争力。铜冶炼、电解铝装备达到世界先进水平的比重在95%以上，铅锌冶炼世界先进水平达到85%以上，调整和整合将成为“十二五”时期内有色金属行业规划的重点。

7.4.3 发展再生有色金属产业

相对于有色金属的初次冶炼，再生利用能耗要低很多，也就意味着有更高的利润空间。在节能降耗政策背景下，有色金属的再生利用有望在较高经济利益的驱动下，驰向发展快车道。有色金属再生利用节能减排效果显著，是有色金属工业发展的重要趋势。发展再生有色金属产业，多次循环利用有色金属，既保护原生矿产资源，又节约能源、减少污染。

近年来，有色金属再生利用得到快速发展，生产和消费规模不断扩大，产业比重逐步提高，技术装备水平不断提升，再生有色金属产业已成为我国有色金属工业的重要组成部分。

1. 产业规模快速扩大

本世纪以来，再生有色金属产量连续10年保持快速增长，再生铜、再生铝、再生铅等主要再生有色金属产量年均增长27%，从2000年的72万吨增加到2009年的633万吨。2009年主要再生有色金属产量占当年10种有色金属产量的24.3%，相当于10年前全国10种有色金属总产量，再生有色金属产业已形成一定规模。

2. 产业集中度逐步提高

已建成一批年产5万吨以上再生有色金属企业，其中最大的再生铝企业产能达65万吨，再生铜企业产能超过40万吨，再生铅企业产能超过20万吨。珠江三角洲、长江三角洲、环渤海经济圈和成渝经济区等逐步形成再生有色金属产业集群，一批进口再生资源加工园区和国内回收交易市场，以及规模化再生有色金属利用工程正在建设。

3. 技术水平不断提升

再生有色金属技术装备和清洁生产水平持续进步，金属熔炼回收率不断提高，产品结构不断优化。一批原生矿产冶炼龙头企业加快进入再生有色金属领域，快速拉升产业整体发展水平。

4. 社会效益日益显现

再生有色金属产业是典型的劳动密集型产业，其回收、分类、拆解、冶炼各环节需要大量劳动力资源，目前，行业从业人员达到150万人以上，为缓解就业压力、促进社会稳定发挥了重要作用。

当前，我国再生有色金属产业也面临着突出的矛盾：

1. 产业集中度低，亟待建立行业准入制度

多数企业生产规模小，全行业产业集中度普遍较低。据不完全统计，全国有300多家再生铅企业，2009年平均产能仅为4100吨；年产量超过10万吨的再生铜企业只有2家，多数企业年产量低于3万吨；大型再生铝企业年产量达到30万吨以上，小企业年产量仅有几百吨。目前，行业缺乏准入管理，发展水平参差不齐，市场竞争无序，亟待加以规范。

2. 技术装备水平落后，环保形势严峻

综合能耗、污染物排放、资源回收利用率等关键指标与发达国家差距明显。再生铜行业，大部分中小企业仍采用落后的传统固定式阳极炉；再生铅行业，小企业产能占50%，大多采用人工拆解废铅酸蓄电池，废铅酸液随意倾倒，冶炼工艺及设备落后，铅膏、铅栅未实现分类熔炼，带来极大环境污染隐患。

3. 标准政策体系有待完善，先进产能竞争力弱

我国废旧有色金属回收、拆解及利用环节标准规范较为薄弱，政策法规体系不完善，不利于形成公平的行业竞争环境。规模化、规范化企业节能环保投入大，生产成本相对较高，在废旧有色金属原料采购竞争中处于劣势地位，生产经营困难，产能开工不足。整个行业呈现出“规模经济不出效益”、“环保科技不出效益”、“先进产能吃不饱”等不正常状态。

4. 加工园区和交易市场有待进一步规范

许多地方未充分结合资源条件、环境形势和供需市场，纷纷投资建设进口再生资源加工园区、交易市场或产业集群（以下简称“加工园区”）。加工园区建设缺乏科学规划，造成无序竞争和资源浪费，不利于产业健康发展。加工园区内部尚未形成覆盖回收、拆解和深加工的产业链。

再生有色金属能耗目标是：到2015年，再生铜熔炼（杂铜—阴极铜）能耗低于290千克标准煤/吨，再生铜熔炼金属回收率达到96%以上；再生铝熔炼能耗低于140千克标准煤/吨，再生铝熔炼金属回收率达到95%以上；再生铅熔炼能耗低于130千克标准煤/吨，废铅渣100%无害化处置，再生铅熔炼金属回收率达到95%以上。

7.5　典型有色金属企业节能减排动态

7.5.1　中国铝业公司

中铝公司“十一五”节能减排的总体目标是：与2005年相比，2010年公司二氧化硫排放总量减少10%，万元工业增加值能耗下降20%，万元增加值取水量下降30%，矿产资源综合利用率提高5个百分点，工业固体废物综合利用率提高5个百分点，工业水重复利用率达到90%。

同时，必须完成国资委下达的“十一五”节能减排具体指标任务和2009年目标：到“十一五”期末，与2005年相比，万元增加值综合能耗下降20%，低于5.3吨标准煤；吨氧化铝综合能耗低于拜耳法生产线528千克标准煤/吨，烧结法生产线1150千克标准煤/吨，混联法生产线930千克标准煤/吨，串联法生产线1100千克标准煤/吨；铝锭综合交流电耗达到14300千瓦时/吨；再生铝冶炼回收率达到国际先进水平。到2009年，公司的万元增加值综合能耗下降16%以上，低于5.54吨标准煤；吨氧化铝综合能耗低于拜耳法生产线538千克标准煤/吨，烧结法生产线1180千克标准煤/吨，混联法生产线940千克标准煤/吨，串联法生产线1115千克标准煤/吨；铝锭综合交流电耗下降2%。

1. 固体废弃物循环利用

公司坚持“废弃物就是放错位置的资源”的理念，以最小化、无害化、资源化为方向，加大废弃物循环利用，实现废物资源化和环境友好化。公司通过加大投入，采用固体废弃物综合处理工艺，以沉降分离、蒸发提浓、铝酸钠结晶析出等实现了系统循环，高效回收利用氧化铝

和氧化钠。

山东分公司拥有国内首创、国际先进的拜尔法赤泥资源化综合利用技术。2010年，山东分公司综合利用赤泥55万吨，拜耳法氧化铝赤泥选铁、除砂和除钒生产不断优化，日产铁粉1500吨左右，年增效益近2亿元，被授予“全国氧化铝赤泥综合利用示范企业”。

山西铝厂利用粉煤灰生产新型墙体材料和抗震节能材料，形成年产3亿块粉煤灰砖、60万平方米节能建筑板材、30万平方米新型环保地板、60万吨新型干法水泥、2万吨赤泥粉煤灰保温材料的生产规模，年综合利用废弃物61万吨，被列入山西省政府采购产品明录，被授予“全国资源综合利用节能建筑环保建材示范基地”。

兰州分公司将47万吨粉煤灰、15立方米废炉渣等固体废弃物外销制作水泥和空心砖等建筑材料，实现了废弃物再利用，减少了环境污染。

2. 减少废气排放

2010年，公司加大脱硫改造力度，山东分公司自备电厂环保节能改造如期完成，中州分公司1#—5#锅炉脱硫工程通过河南省环保竣工验收并投入试运行；公司铝土矿综合利用试验基地对豫西、长通等含硫1%—8%的高硫铝土矿进行环保改造工业试验。公司二氧化硫排放量比2005年下降15.2%。

3. 加强废水治理

按照“控制废水产生、完善废水处理、开发利用再生水”的治水原则，公司加大污水治理力度。2010年，山东企业氯碱生产完成了电解岗位所有泵冷却纯水回收和综合利用，工业废水排放达标率由91.4%提高到100%。公司氧化铝、电解铝生产已基本实现工业废水“零”排放，工业废水重复利用率达到93.09%，超额完成国资委“十一五”下达的90%的计划目标。

8 石化行业的节能减排

8.1 石化行业“十一五”运行综述

石油和化学工业是我国国民经济重要的能源和基础原材料工业，也是国民经济的支柱性产业，经过多年发展，形成了包括油气开采、炼油、基础化学原料、化肥、农药、专用化学品、橡胶制品等约50个重要子行业，可生产6万多个（种）产品，涉及国民经济各领域的完整工业体系，成为世界最大的石油和化工产品生产和消费国之一。

“十一五”时期，我国石油和化工行业成功应对国际金融危机的巨大冲击和特大地震等自然灾害带来的严重挑战，实现了全行业平稳较快发展，全面完成了“十一五”规划的各项主要经济指标。主要数据如下：

- 2010年，全行业规模以上企业达3.66万家，从业人员约690万人，实现总产值8.88万亿元（现行价格），仅次于美国，位居世界第二。其中，化学工业产值达5.23万亿元，超越美国，跃居世界第一。
- “十一五”时期，石油和化工行业总产值、利润、销售收入和资产年均增长率分别为21.3%、13.2%、21.5%和21.4%，是历史发展最快的时期之一。图8-1为石化行业经济规模增长情况。

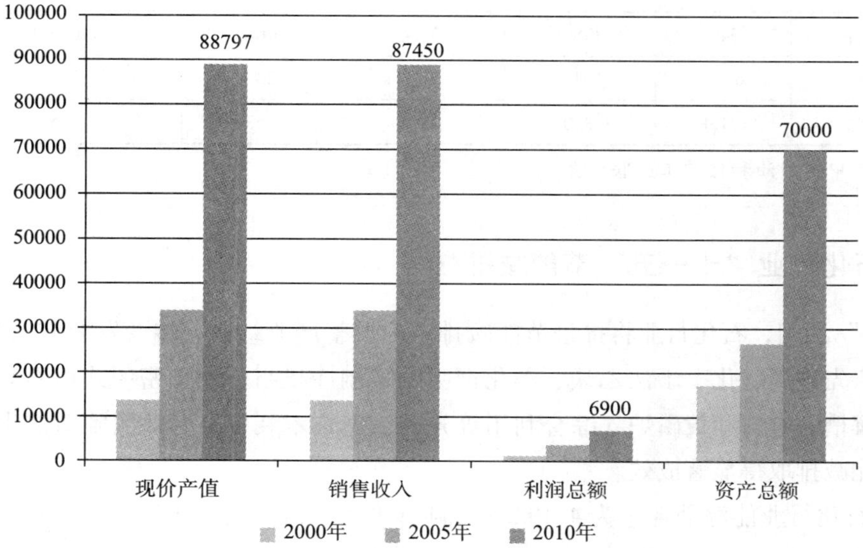

图8-1 石化行业经济规模增长 单位：亿元

Figure 8-1 Economic Growth in Petrochemical Industry Unit: 100 Million RMB

资料来源：中国石油和化学工业联合会

• 2010年，全行业有20多种大宗产品产量位居世界前列。其中，氮肥、磷肥、纯碱、烧碱、硫酸、电石、农药、染料、轮胎、甲醇、合成树脂、合成橡胶、合成纤维等排名世界第一；原油加工量、乙烯等排名世界第二；原油产量达2.03亿吨，排名世界第四；天然气产量接近1000亿立方米，居世界第五，如表8-1所示。

表8-1 石化大宗产品产量位居世界前列

Table 8-1 Production Capacity of Staple Petrochemical Products Tops the World

序号	产品	2000年	2005年	“十五”年均增幅%	2010年	“十一五”年均增幅%	世界排名
1	原油	16262	18083.9	2.1	20301.4	2.3	4
2	天然气	277.3	499.5	12.5	944.8	13.6	5
3	原油加工量	21062	28622	6.3	42286.8	8.1	2
4	成品油	12086.6	17455.5	7.6	25277.4	7.7	
5	化肥	3185.7	4888	8.9	6619.8	6.3	1
6	氮肥	2398	3576	8.3	4521.1	4.8	1
7	磷肥	663	1075	10.1	1701.4	9.6	1
8	钾肥	124.6	232.7	13.3	396.8	11.3	4
9	硫酸	2365	4462	13.5	7060.1	9.6	1
10	纯碱	834	1421	11.2	2029.3	7.4	1
11	烧碱	667.9	1240	13.2	2086.7	11	1
12	甲醇	199	536	21.9	1574.3	24	1
13	乙烯	470	770	10.4	1418.9	13	2
14	PTA	202	556	22.4	1455.4	21.2	1
15	合成橡胶	83.6	163.2	14.3	310	13.7	1
16	合成树脂	1079.45	2141.9	14.7	4360.9	15.3	1
17	PVC	240	649	22	1130.1	11.7	1
18	电石	340	895	21.4	1462.3	10.3	1
19	农药	64.8	104	9.9	234.2	17.6	1
20	轮胎	12158	31820	21.2	77634.4	19.5	1

资料来源：中国石油和化学工业联合会

8.2 石化行业“十一五”节能减排总结

“十一五”期间，石化行业将推进节能减排作为调整产业结构、转变发展方式的重要手段，通过大力推广先进煤气化、干法乙炔、溴化钾吸收式制冷机组、大型密闭式电石炉、新型膜极距离子膜电解槽、电石和黄磷炉气综合利用等先进工艺技术装备，有效实施清洁生产和循环经济，行业节能减排取得显著成效。

2009年石化行业能源消耗量为47192.5万吨标准煤，占全国能源消耗总量的15.2%；占工业能源消耗量的22.6%。

2009年全国石油和化学工业万元增加值能耗为3.65吨标准煤，比2005年下降13.5%，重点耗能产品单耗明显下降。其中，原油加工每吨耗标油72.82千克，下降7%；乙烯每吨耗标油

637.08千克，下降7.7%；合成氨每吨耗标准煤1425.98千克，下降5.1%；烧碱每吨耗标准煤534.29千克，下降17.9%；电石每吨耗标准煤1018.11千克，下降14%，如图8-2所示。

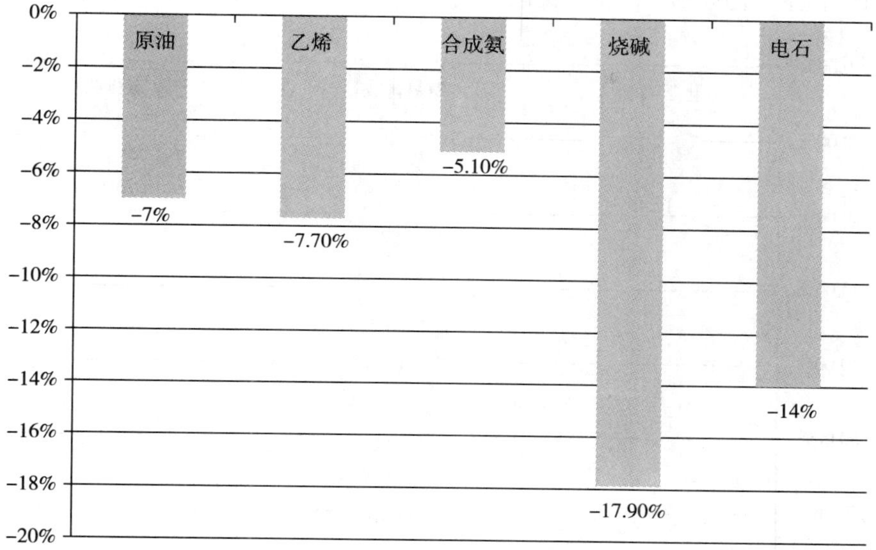

图8-2 石化主要产品单位综合能耗下降率

Figure 8-2 Reduction Rate of Energy Consumption Per Unit of Major Petrochemical Products

资料来源：中国石油和化学工业联合会

“十一五”前四年，石油和化工行业万元工业增加值能耗累计下降13.5%，其中原油加工下降24.1%、乙烯下降8.7%、合成氨下降13.7%、烧碱下降19.1%、电石下降14.9%、纯碱下降28.9%，如图8-3所示。

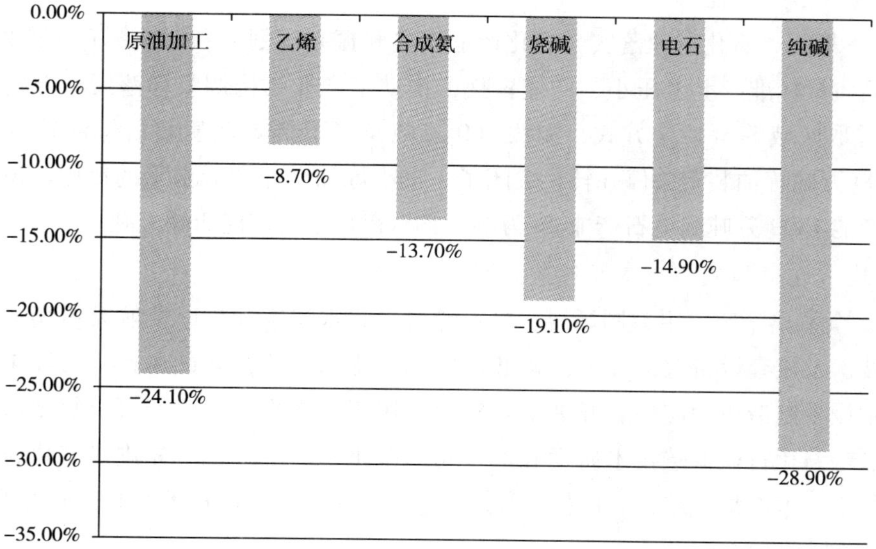

图8-3 主要产品的工业增加值能耗下降率

Figure 8-3 Reduction Rate of Energy Consumption Industrial Added Value of Major Products

资料来源：中国石油和化学工业联合会

行业环保治理水平明显提高，污染物排放得到有效控制。2009 年全行业化学需氧量排放量 65.2 万吨，比 2005 年下降 16.4%；氨氮排放量 10.1 万吨，下降 58.1%；二氧化硫排放量 177.7 万吨，下降 14.1%；主要污染物排放量均提前完成了“十一五”国家规划目标，如图 8－4 所示。

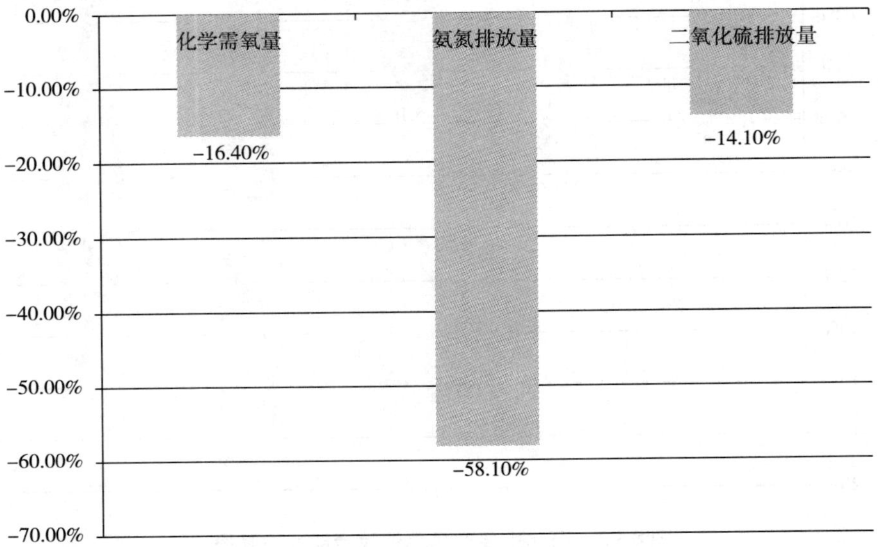

图 8－4　主要污染物下降率

Figure 8－4　Decrease Rate of Key Pollutants

资料来源：中国石油和化学工业联合会

“十一五”期间，围绕节能减排目标，石化行业在淘汰落后产能、技术进步推广，以及内部能源管理等方面做了大量工作。

1. 积极淘汰落后产能

“十一五”期间，石化行业落实《石化产业调整和振兴规划》，通过修订产业政策、提高准入门槛和产品质量标准、上大压小、产能置换，推动了产业结构调整和落后产能淘汰工作。从 2009 年起，采取区域等量替代方式，淘汰 100 万吨及以下低效低质落后炼油装置，积极引导 100 万吨—200 万吨炼油装置关停并转；关闭了一批小炼油企业和低水平的氮肥、磷肥企业，淘汰落后焦炭产能 1809 万吨和电石产能 46 万吨；淘汰落后电石产能 360 万吨。

2. 行业技术进步加快，技术节能效果显著

2010 年，全氟离子膜、甲醇制烯烃、多喷嘴对置式水煤浆气化以及航天粉煤加压气化技术等一批关键技术及成套设备取得突破，相继建成工业化示范装置并试运行，乙烯裂解气压缩机等国产化关键设备制造成功应用。开展了二氧化碳用于三次采油、生产可降解塑料、绿藻生物转化制取油品等石化行业低碳技术研发和产业化示范工程，继续推广先进煤气化技术、大型密闭式电石炉技术、电石炉气综合利用技术等，一些重点领域的关键技术产业化取得重大突破。

3. 加强企业能源管理

开展重点能耗企业能源审计。在石油和化工企业中，节能减排已经变成自觉行动。中国化工集团公司制定和印发了《中国化工集团公司节能减排管理暂行办法》；四川天华股份有限公司等企业率先在石化行业开展了能源管理体系认证。

4. 开展重点耗能产品能效对标

做好重点耗能产品能效对标工作，研究制定包括尿素等18个重点耗能产品的能源消耗限额强制性国家标准，加强对已出台的烧碱、电石、合成氨、黄磷等4个产品的能源消耗限额强制性国家标准的宣传和贯彻落实，保障"十一五"节能目标的实现。氮肥、磷肥、氯碱及石油石化等行业先后推行了废水零排放、废气和废渣综合治理等技术，污水回用、废气回收、废渣治理在石油和化工企业成为主攻方向。

8.3 石化行业"十二五"节能减排背景

石化行业能源消费集中度较高，为行业节能降耗的研究锁定了目标。石化行业是国民经济中能耗较高的产业部门，其能耗占工业能耗接近1/5，占全国总能耗的14%左右。其中，原油加工、氮肥制造、有机化学原料制造等八个子行业的年能源消费量均超过1000万吨标准煤，其能源消费量之和则超过行业总能耗量的85%。其中，原油加工和氮肥制造两个子行业能源消费比重最大，这两个行业能耗量之和已超过全行业能源消费量的一半以上。而整个行业中另外二三十个子行业的用能量却相对较少，总和不足总量的15%。因此，行业内的低端、高耗能产品都集中在这八个子行业领域，这就为行业节能降耗锁定了重点领域。石化耗能子行业的能源消费比重情况如图8-5所示。

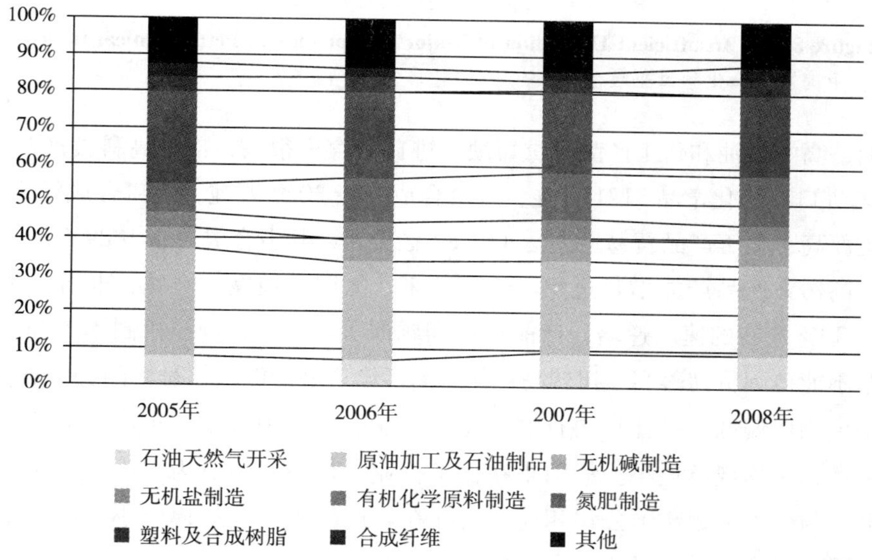

图8-5 石化耗能子行业的能源消费比重

Figure 8-5 Energy Consumption Ratio of Petrochemical Sub-industries

资料来源：中国石油和化学工业联合会

与此同时，石化行业结构性过剩矛盾日益突出。"十一五"期间，石油和化工行业大多数产品产能大幅增长，产品供给由"整体数量短缺"转变为"结构性过剩"。传统大宗石化产品的总产能明显超过国内市场需求，装置开工负荷较低，资源浪费严重。据行业协会统计，2010年主要装置开率数据（如图8-6）如下：

- 我国甲醇产能约为3840万吨/年，装置平均开工率不足45%。
- 尿素总产能达6606万吨/年（实物量），超过国内需求20%以上，装置平均开工率不足75%。

- 烧碱产能达3021万吨/年，装置平均开工率约为72%。
- 聚氯乙烯产能2043万吨/年，装置平均开工率仅为54%。

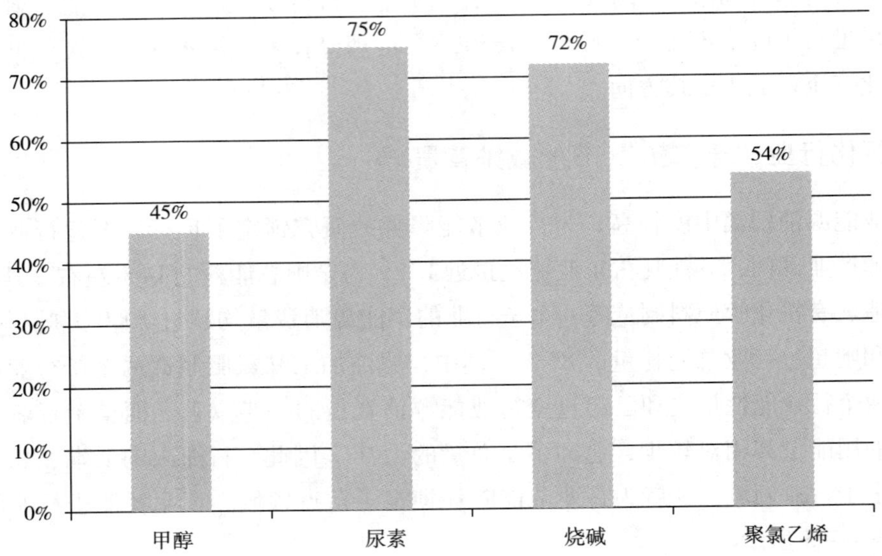

图8-6 石化行业主要装置开工率不足

Figure 8-6 Insufficient Utilization of Major Equipments in Petrochemical Industry

资料来源：中国石油和化学工业联合会

另一方面，高端石油和化工产品严重短缺，进口依存度很高，部分高科技产品尚处于空白。2010年，我国进口有机化学品3121万吨，进口合成树脂3069万吨，进口合成纤维单体1439万吨，均创历史新高；化工产品贸易逆差达433.82亿美元，其中合成树脂占89.9%。

高耗能、高污染产品产能增长过快。近几年来，氮肥、纯碱、烧碱、电石、黄磷等产业的投资增长均以30%左右的速度递增，产能过剩问题严重。同时，调控机制不力和地方保护，使国家产业政策不能得到贯彻落实，扭转高耗能、高污染产品产能增长过快的势头难度很大。

产业集中度仍然偏低。尽管我国石油和化学工业产业集中度不断提高，但从整体上看，企业仍然过多过散，平均规模很小。特别是在化工行业，这一问题更为突出。即使是我国化工行业的龙头企业，与国外大型跨国公司相比，也存在着产品单系列规模偏小、市场竞争力不强的问题。2010年我国主要石化行业企业集中度数据如下：

- 化肥行业规模以上企业达2789家，平均产量2.34万吨。
- 合成氨生产企业423家，平均产量11.73万吨。
- 农药行业规模企业985家，平均原药产量仅为0.24万吨。
- 无机化工原料制造业规模以上企业多达2480家，平均产值只有1.39亿元。

而产业集中度偏低，企业规模太小，是影响行业竞争力提升的重要因素之一，也是“十二五”产业结构调整的难点和重点。

随着经济快速发展，石油和化学工业面临的能源和环境约束不断加大。煤炭、石油、天然气等既是行业发展必须的能源，又是生产不可或缺的原料，是能源消耗量较大的行业。2009年全行业综合能源消费量达3.8亿吨标准煤，占工业能源总消费量的17.4%，位居所有工业部门的第二位。石油和化学工业也是污染物排放大户，废水、废气、废固等污染物排放在全国工业领域分别居第一、第四和第五位。我国政府向国际社会做出承诺，到2020年，单位GDP二氧化碳排放降低40%—45%。国家“十二五”规划纲要也对节能减排提出了更加严格的约束性指标，石油和化工行业节能减排工作任重道远。

8.4　石化行业“十二五”节能减排展望

石化行业“十二五”节能减排的主要目标是：

（1）2015年，万元工业增加值能源消耗和二氧化碳排放量均比“十一五”末下降15%。

（2）COD和氮氧化物排放总量均减少10%，氨氮排放总量减少12%，二氧化硫排放总量减少8%，废水达标排放。

（3）化工固体废物综合利用率达到75%、有效处置率达到100%。

（4）全行业清洁生产达到历史最好水平。

为此，中国石油和化学工业联合会最新发布的《中国石油和化工行业经济运行报告——2010年回顾与2011年市场预测》明确提出“十二五”期间的发展思路。

一是要以转变经济发展方式为主线，大力调整产业结构，在保障国内能源供应的情况下，积极布局新兴能源产业。改造升级传统产业，继续加速淘汰高能耗、高污染的落后生产能力和设施，进一步发挥传统产业的比较优势，推进“三废”处理和综合利用，提高产品附加值和市场竞争力，有序推进大型炼油、化工生产基地的建设，为实现结构调整、节能减排的战略目标提供强有力的支撑。

二是要增强自主创新能力，推进技术改革，通过技术变革来提高企业节约资源、减少环境污染的能力；广泛应用和开发清洁生产工艺以及资源循环利用技术。重点治理存在污染隐患的生产装置和企业，例如，实施火炬熄灭计划，解决燃料气放空问题；提高燃煤锅炉等耗能设备的效率，降低单位能源消费的二氧化碳排放强度；气代油；对排放浓度大且集中的温室气体逐步进行回收处理，尽量资源化。

三是要积极发展高端石化产品，改善品种质量，推动节能减排，提高资源能源利用效率；众所周知，节能减排工作在石油化工行业任务艰巨。一方面，我国石油和化工行业目前仍以生产基础原材料产品为主，精细化工产业比重不大，能源消耗较高、污染物排放量较大的特点一时还难以彻底改变；另一方面，近几年我国石油和化工产业年均增长率均在20%以上，这相当于几年就再造一个同等规模的石化产业。这样一个产业结构，这样一个发展速度，要实现能耗下降20%、排放减少10%的目标，是十分艰难的。

四是要通过建立有效的生产体系和管理体系，提高企业在安全环保以及节约能源方面的能力，加快促进企业走低碳经济发展的道路。加快催化裂化再生烟气脱硫、加热炉烟气、工艺排气及电站排气中二氧化硫和氮氧化物处理技术的创新和应用，开发合成氨副产气回收技术、甲醇装置二氧化碳回收制尿素原料技术；研发低成本二氧化碳的高效捕集、回收及化工利用技术。大力发展多联产设施，提高资源和能源综合利用率。加强油气资源的低碳、高附加值转化技术的开发利用。

而石油和化学工业“十二五”发展指南也指出大力促进节能减排的重要工作内容如下：

一是加强重点用能企业的节能管理，推进企业能源管理中心建设。

二是完善主要耗能产品能源消耗限额标准体系。

三是推广先进节能技术和设备，做好炼油、乙烯、氮肥、氯碱、纯碱、电石、黄磷等高耗能行业的节能技术改造工作。

四是全面推行清洁生产，建立主要行业清洁生产指标体系、清洁生产评价标准体系、污染物排放标准体系、环境保护管理体系，做好农药、染料、涂料等行业的清洁生产示范项目，创建一批“化工清洁生产绿色园区”。

五是抓好农药、染料、铬盐等高污染行业的结构调整，采用先进环保技术，加快改造落后生产装置。

六是大力发展可再生资源和过程排放控制技术，强化源头和过程管理，减少“三废”和二氧化碳排放。

七是鼓励发展二氧化碳的捕集、封存和综合利用新产品、新技术和新设备。

总体而言，循环经济、低碳经济将引领石化行业未来经济发展方向，战略性新兴产业将成为行业经济新的增长点。应该说，循环经济、低碳经济和战略性新兴产业也正是石化行业未来发展的着力点。“十二五”期间，石化行业要一如既往地发展循环经济和低碳经济，构建从原材料采购、运输、存储、生产至包装、流通加工、配送、销售、废弃物回收利用全过程的循环经济体系，以此作为推动行业经济结构调整和发展方式转变的根本途径，走出一条科技含量高、经济效益好、能源资源消耗低、环境污染少的新型工业化道路，努力建设低投入、高产出、低消耗、少排放、能循环、可持续的石油和化学工业发展模式。

8.5 典型石化企业节能减排动态

8.5.1 中国石化

2010年，中国石化万元产值综合能耗为0.77吨标准煤，与2005年相比下降15.4%，累计节约1444万吨标准煤，超额完成“十一五”节能目标任务。各板块也都“八仙过海，各显其能”，节能降耗成效显著。

2010年，中国石化油田板块万元产值综合能耗为0.41吨标准煤，比上年下降1.77%。油气综合能耗100.40千克标准煤/吨，比计划低1.07千克标准煤/吨，比上年降低1.98%；原油商品率96.66%，比计划高0.01个百分点，比上年提高0.02个百分点；天然气商品率为88.40%，比计划高3.03个百分点，比上年提高3.07个百分点。

2010年，中国石化炼油板块万元产值综合能耗为0.34吨标准煤，比上年下降5.86%。炼油综合能耗58.25千克标准油/吨，比计划低2.55千克标准油/吨，比上年降低5.04%；轻油收率75.70%，比计划高0.4个百分点，比上年上升0.28个百分点；综合商品率94.81%，比计划高0.51个百分点，比上年上升0.29个百分点；原油加工损失率0.57%，比计划低0.05个百分点，比上年降低0.08个百分点。

2010年，中国石化化工板块万元产值综合能耗为1.62吨标准煤，比上年下降3.33%。乙烯装置燃动能耗609.28千克标准油/吨，比计划低14.92千克标准油/吨，比上年下降2.14%；乙烯收率31.66%，比计划高0.14个百分点，比上年上升0.16个百分点；双烯收率46.56%，比计划高0.15个百分点，比上年升高0.12个百分点；乙烯装置损失率0.50%，比计划高0.07个百分点，比上年上升0.14个百分点。芳烃装置三苯收率完成50.62%，比计划高1.62个百分点，比上年上升1.64个百分点；综合能耗完成719.33千克标准油/吨，比计划低80.67千克标准油/吨，比上年下降56.47个单位。

2010年，中国石化油品销售板块分环节油品损耗率均控制在3‰以内；成品油吨油运杂费68.9元，比计划降低4.4元/吨。油田未上市企业万元产值综合能耗为0.72吨标准煤，比上年下降1.88%。资产管理公司万元产值综合能耗为1.68吨标准煤，比上年下降1.07%。

2010年，中国石化废水排放量48000万吨，COD排放量3.16万吨、下降18.1%；二氧化硫排放量36.7万吨、下降35.7%，均超额完成“十一五”减排目标任务。

“十一五”期间，中国石化加大燃煤燃焦锅炉脱硫治理，累计投资30亿元，对全系统19家自备电站、107台燃煤锅炉实施烟气脱硫改造。二氧化硫排放量由2005年的57.1万吨下降到2010年的36.7万吨，降幅35.7%。在污水排放指标日益严格的情况下，各个企业对老设施实行技术改造，增加深度处理，得以使COD浓度大大降低，总量削减18.1%。

“十一五”期间，中国石化共投入492亿元用于油品质量升级。2009年前后，中国石化所属炼油企业抓紧落实汽油质量升级方案和装置改造，所产车用汽油于2010年5月1日全部达到国Ⅲ及以上标准。国Ⅲ柴油质量升级项目正加紧推进，预计2012年12月全部建成投用。这些项目对减少消费环节二氧化硫排放量无疑具有重要意义。

8.5.2 中国石油

五年间累计投入61亿元，实施污染减排项目273个，相继启动减排“十大工程”和节能“十大工程”，提前完成国资委下达的“十一五”二氧化硫、COD的考核任务和节能目标。这一系列成绩的取得，折射出中国石油对环境保护与节能减排工作的力度不断加大。

环境保护工作在“十一五”期间得到中国石油前所未有的重视。集团公司成立以蒋洁敏总经理为组长的节能减排工作领导小组，制定《节能减排工作实施方案》，建立以HSE管理体系为基础，以清洁生产为技术政策，以ISO14000认证为促进的管理体系和污染减排统计、监测与考核约束机制。

中国石油在所有炼油化工企业全部建立水污染应急三级防控体系。“十一五”末，中国石油二氧化硫排放量比2005年下降18%，COD排放量比2005年下降36%，超额完成减排工作目标，获得国资委颁发的“节能减排特别贡献奖”。

为实现企业与环境的和谐，中国石油开展了冀东油田绿色油田建设、大连石化生态工业园建设和兰州石化创建环境友好企业等试点工作。西气东输管道工程、新疆石西油田开发工程和塔里木油田沙漠公路绿化工程相继被评为“国家环境友好工程”。

以实际行动推行清洁生产，为减缓全球气候变化做出积极努力是中国石油的庄严承诺。“十一五”期间，中国石油积极研发、应用清洁工艺，努力实现生产过程清洁化。吉林油田二氧化碳驱油及埋存关键技术研究重大科技专项攻关取得进展，初步形成了二氧化碳驱油优化设计和注采工程配套技术，为二氧化碳减排拓展了思路。

中国石油努力生产和投放清洁产品，努力实现产品消耗过程清洁化。通过技术创新，保持天然气储量持续增长，初步建成横跨东西、纵贯南北、覆盖全国、连通海外的油气管网体系，满足油品清洁输送需求，同时完成对21家炼化企业的升级改造，提前转产供应国Ⅲ标准汽油，硫含量指标降低70%。

温室气体控制工作也于“十一五”期间全面展开。2008年9月，中国石油参资设立天津排放权交易所，实施生物固碳的碳汇项目，推动清洁发展机制（CDM）项目，已实施的2个CDM项目每年可减排1270万吨碳指标，环境风险防范和污染事故控制能力明显增强。

“十一五”期间，中国石油采取进一步落实节能目标责任书，将万元工业产值综合能耗考核指标扩大为四项主要业务的考核指标，首次将销售企业纳入节能节水型企业的考核评选范围，推进企业低碳发展。

通过调整优化结构、淘汰落后产能、实施重点节能工程、推广应用成熟适用节能技术和强化节能目标责任等多重举措，中国石油提前完成“十一五”节能目标，预计完成节能量860万吨标准煤，比“十一五”规划目标多完成200万吨标准煤。

“十一五”期间，中国石油信息化与工业化融合推动节能减排。炼油与化工生产运行管理系统上线应用，有效对生产运行过程进行全面控制，避免装置生产波动的同时降低能耗，仅兰州石化2009年清油收率就比上年增长2.1%。管道生产管理系统将4万公里管道和5万个数据采集与监控点纳入系统，有效促进了油气储运环节的节能降耗。其中，西气东输管道年输损率下降1.55%，相当于减少天然气损耗2700万立方米。

9　煤炭行业的节能减排

9.1　煤炭行业“十一五”运行综述

“十一五”是我国煤炭工业史上发展最快的五年。国家煤炭政策法规体系不断健全，市场化改革步伐加快，行业发展环境显著改善，企业效益稳步提高；煤炭投资规模扩大，煤炭产能快速增长；科技进步速度加快，一大批关键技术和重大装备攻关取得成效；结构调整力度加大，产业集中度大幅提高；企业主体地位不断提升，市场竞争能力明显增强，对外开放稳步推进，在有力地保障了国家煤炭稳定供应的同时，煤炭工业面貌发生了巨大变化。行业主要数据如下：

- “十一五”期间，全国累计生产煤炭160亿吨。中国煤炭生产消费量占全世界比重均超过45%，多年位居世界第一。2010年全国煤炭产量达到32.4亿吨，比2005年的23.5亿吨，增长8.9亿吨，年均增长1.78亿吨，如图9－1所示。

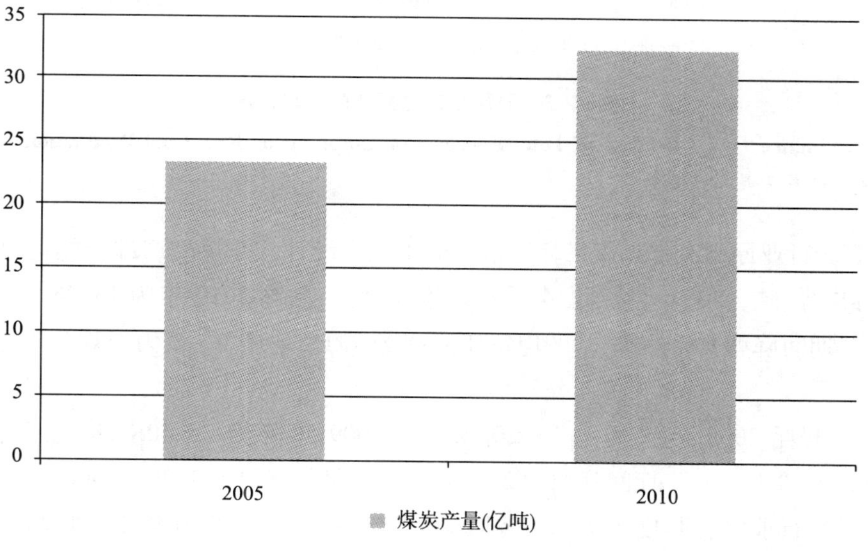

图9－1　煤炭产量增长

Figure 9－1　Growth in Coal Production

资料来源：中国煤碳工业协会

- “十一五”煤炭行业累计完成固定资产投资1.25万亿元人民币，是“十五”投资总额的5倍多，是新中国成立55年投资总和的2.8倍。全国有16个行业投资开办煤矿。全国共建成安全高效煤矿359处，合计产量10.2亿吨，占全国总量的1/3。

• 2010 年主要港口煤炭发运完成 5.56 亿吨，同比增加 9854 万吨，增长 21.5%。2010 年全年煤炭进口 1.65 亿吨，出口 1903 万吨，净进口 1.46 亿吨；2005 年我国进口煤炭 2622.57 万吨，出口煤炭 8462.74 万吨，煤炭净出口量 5840.17 万吨。

9.2 煤炭行业“十一五”节能减排总结

“十一五”以来，煤炭企业依靠科技进步，调整产业、产品结构，延伸产业链，加强资源综合利用，发展循环经济，节能减排工作取得了很大进展。

在消化了原煤生产能耗年均自然增长 4% 情况下，原煤生产主要能耗指标均有下降，并且煤炭行业节能减排指标完成进度快于全国。

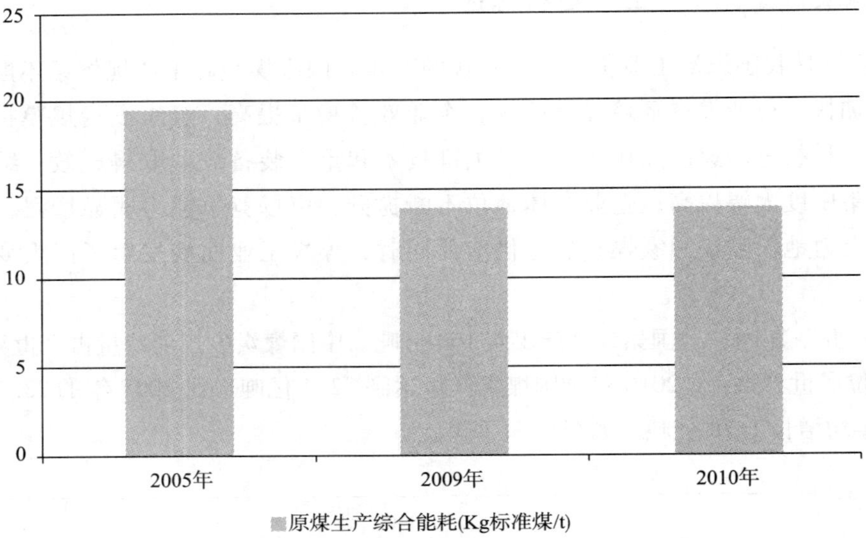

图 9-2 原煤生产总额能耗下降情况

Figure 9-2 Decease of Total Energy Consumption for Raw Coal Production

资料来源：中国煤碳工业协会

根据对煤炭行业原国有重点煤矿耗能情况的不完全统计，原煤生产综合能耗 2005 年平均为 19.50 千克标准煤/吨，2009 年降为 14 千克标准煤/吨，预测 2010 年为 13.96 千克标准煤/吨，即“十一五”期间降幅约 28.4%，平均每年降幅为 7.1%。图 9-2 为原煤生产总额能耗下降情况。

原煤生产电耗 2005 年平均为 33.20KWh/t，2009 年降为 25.40KWh/t，预测 2010 年为 25.39KWh/t，即“十一五”期间降幅约 23.5%，平均每年降幅为 5.9%，如图 9-3 所示。

2009 年煤炭行业能耗强度约为 1.13 吨标准煤/万元，相比 2006 年 2.2 吨标准煤/万元下降了 48.6%。

煤炭节能重点企业成绩明显。“十一五”期间，国家开展了千家企业节能行动，参与该行动的 64 家煤炭企业目前重组为 61 家。这些企业在 2006 年、2007 年、2008 年、2009 年分别完成节能量为 160.15 万吨标准煤、441.97 万吨标准煤、261.46 万吨标准煤、304.8 万吨标准煤，即前四年已经提前超额完成“十一五”煤炭企业节能量为 639.26 万吨标准煤的节能目标，如图 9-4 所示。

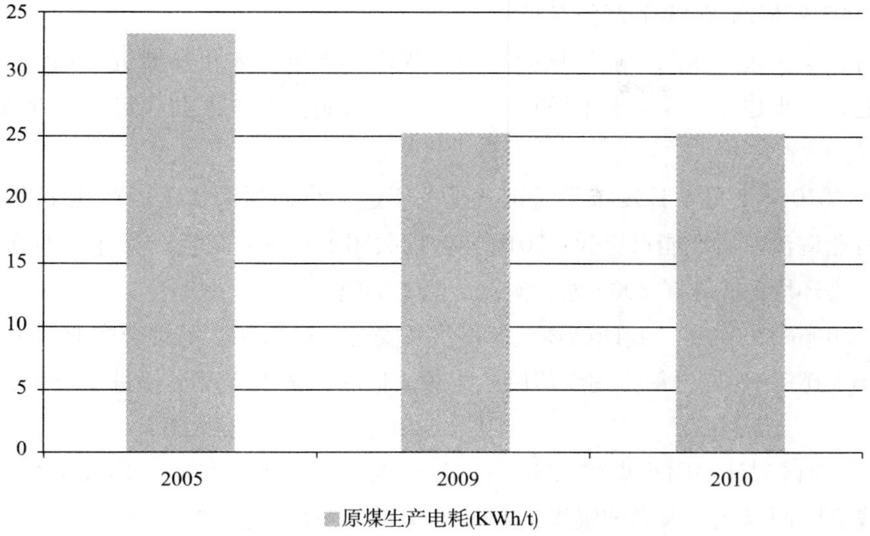

图 9－3 原煤生产电耗下降情况

Figure 9－3 Decrease of Power Energy Consumption for Raw Coal Production

资料来源：中国煤碳工业协会

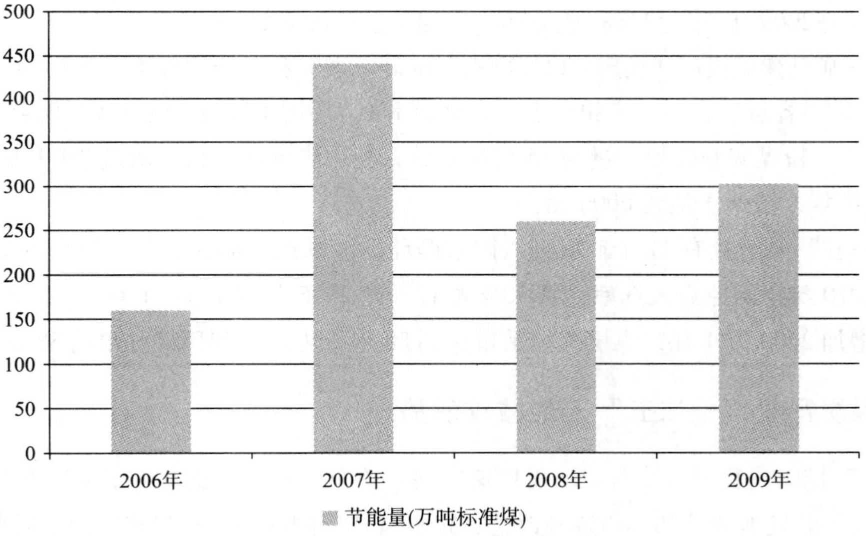

图 9－4 煤企的年度节能量

Figure 9－4 Yearly Energy Consumption by Coal Companies

资料来源：中国煤碳工业协会

煤炭企业取得上述成绩的重要原因是结构调整以及落后产能淘汰工作的成绩显著。

- 到 2009 年底，全国煤矿数量约 1.5 万处，比 2005 年减少了 40%。
- 平均单井规模达 20 万吨，比 2005 年增加了 108%。
- 大型煤炭基地产量达到 26 亿吨，占全国的 87%。
- 年产量超过千万吨的企业 43 家、产量 17.28 亿吨、占全国总产量的 58.12%，比 2005 年提高了 22 个百分点。
- 年产 120 万吨以上的大型矿井 434 处、核定能力 12.63 亿吨，占全国煤矿总核定能力的

50.6%，比2005年提高了13个百分点。

与此同时，煤炭企业多元产业发展格局初具规模，煤电一体化发展进程加快，新型煤化工产业逐渐兴起，初步建立了煤炭上下游产业联合发展机制，多数大型煤炭企业非煤产业产值已超过50%以上。

国务院于2010年2月6日发布了《国务院关于进一步加强淘汰落后产能工作的通知》，明确提出煤炭行业淘汰落后产能目标为：2010年底前关闭不具备安全生产条件、不符合产业政策、浪费资源、污染环境的小煤矿8000处，淘汰产能2亿吨。

据统计，2006—2009年，我国已经关闭不具备安全生产条件、不符合产业政策、浪费资源、污染环境的小煤矿7466处，每年淘汰落后产能约3亿吨，很大程度上促进了整个行业节能工作的开展。

“十一五”全国累计关闭小煤矿9234处，淘汰落后产能近5亿吨，形成了神华、中煤、山西焦煤、同煤和陕西煤化5家亿吨级煤炭企业和10家5000万吨级煤炭企业，原煤产量约占全国总产量的44%。

煤炭行业积极发展循环经济，延长以原煤和共伴生矿物为基础的产业链，通过与电力、化工、建材等产业联营，利用煤矸石发电、煤矸石制建材等方式，充分利用煤矸石中的热能。同时取消小锅炉，利用园区内的余热资源，增加了园区的经济效益，并且产生了巨大的社会效益和环境效益。到2009年底，煤炭行业循环经济相关主要数据如下：

- 全国煤矿已建成投产和在建的以煤矸石、洗矸、煤泥和部分中煤为主要燃料的低热值燃料电厂400座，装机容量（包括已建和在建）2500万KW，燃用煤矸石、洗矸和煤泥合计约2亿吨。
- 煤矸石、粉煤灰制砖年产量约85亿块标砖，利用煤矸石、粉煤灰逾2600万t/a，节约能源80万吨标准煤，节约土地约6000亩。
- “十一五”全行业有11个矿区列入国家循环经济试点，形成了各具特色的矿区循环经济典型模式。2010年全国原煤入选率达到50%左右，比2005年提高16个百分点。全国煤矸石电厂装机容量增加2600万千瓦，煤层气抽采量达到88亿立方米，利用矿井水约38亿立方米。

9.3 煤炭行业“十二五”节能减排形势

“十一五”期间，我国煤炭行业取得快速发展，整个行业的面貌发生了巨大变化。在新的历史时期，煤炭行业发展的外部环境逐渐改善，仍具有较广阔的发展空间和新的发展机遇。

但是，我们必须注意到，煤炭行业仍然存在结构性过剩压力增大、安全生产与生态环境的压力增大、转变发展方式的任务艰巨、行业管理体制有待完善等挑战。

1. 煤炭结构性过剩压力增大

近年来，在市场需求的拉动下，煤炭产量持续以年均2.0亿多吨的速度增长。且煤炭生产越来越向晋陕蒙宁等西部地区集中，煤炭生产供应区与主要消费区的距离增加，煤炭结构性过剩的压力加大。

2. 安全生产与生态环境的压力增大

我国煤炭资源赋存条件与主要产煤国家相比相对较差，煤层埋藏深、自然灾害多，且随着强度开采，深度逐年增加，条件越来越复杂，煤矿安全生产的压力加大。我国是世界上最大的煤炭消费国，在全球气候变化加剧，国际社会碳减排压力加大的背景下，高强度的生产和消费煤炭，生态环境和国际社会舆论的压力增加。2007 年，全球排放 CO_2 总量为 208 亿吨，其中，中国排放 53 亿吨，美国排放 59 亿吨。我国碳减排压力增大。同时，高强度开发煤炭资源，富煤地区环境承载能力面临挑战。

3. 煤炭工业转变发展方式的任务依然艰巨

近年来，虽然煤炭行业经济效益提高，多元产业格局初步形成，但煤炭行业发展多元产业仍然面临技术、资金、人才等诸多因素制约，结构调整、转变发展方式的任务依然艰巨。

在“十二五”期间，我国煤炭行业应该进一步认清形势，增强机遇意识和忧患意识，以科学发展为主题，以转变发展方式为主线，努力破解发展难题，构建新型煤炭工业体系，提升煤炭行业科学化发展水平，科学把握好总量与结构、生产与消费、环境与产量、大基地建设与中小骨干矿之间的关系，实现由产量速度型向质量效益型转变，实现由单一煤炭生产向煤炭资源综合利用、深加工方向转变，实现由资源环境制约向生态环境友好型转变。坚持循环经济发展理念，推动资源综合利用和节能减排工作，加快科技创新和新技术研发，推进煤矿绿色开采，建立矿区生态环境恢复与治理机制、以最少的资源和环境消耗，支撑国民经济又好又快发展。

9.4 煤炭行业“十二五”节能减排展望

按照目前的形势，到“十二五”末的 2015 年，我国煤炭总需求量或将达到 38 亿吨，到 2015 年，中国需要的一次能源消费量将达到 42 亿吨标准煤。其中，水电可提供 2.6 亿吨标准煤的供应能力，风电可提供 0.57 亿吨标准煤供应能力。其他方面，生物质能有 2000 万吨标准煤，核电有 6800 万吨标准煤，石油天然气有 2 亿吨标准煤。这些非煤炭的能源共能提供 5.14 亿吨标准煤的能源供应能力，其余的还有 26.8 亿吨标准煤的能源供应需要煤炭提供，折合成原煤量约为 38 亿吨。考虑到其他能源形式在今后几年仍处于起步阶段，即使尽其所能地发展，其提供的供应能力有限。因此，煤炭在一次能源消费中的主体地位仍无法改变，产量的增加也势在必行。煤炭行业节能减排主要从以下几个方面努力：

1. 提高采区回采率

煤炭行业各种所有制、各种规模的企业并存，这些企业建设、生产、经营、管理、技术水平参差不齐、差别巨大，这直接或间接导致煤炭行业整体生产效率低、回采率低、资源浪费严重、矿区环境恶化、污染严重等问题。目前全国拥有大中小型煤矿共计 1.1 万个，年均生产能力不足 30 万吨，资源回采率达 35%。

2. 加大煤炭洗选能力，为煤炭用户节能创造条件，促节能减排

2009 年我国原煤入洗率为 43%，动力煤入洗率为 20%，分别比美国、南非、俄罗斯等煤炭行业发达的国家低 12 和 20 个百分点。如果将动力煤入洗率提高到 40%，每年将节约 3400 万吨标准煤，节约运输耗煤近 1000 万吨，节能潜力巨大。因此，“十二五”期间，国家将鼓励企业多建洗煤厂，把优质煤运输出矿区。另外，我国煤矸石、煤矿瓦斯、矿井水综合利用率为 40%，如果将这些二次资源的利用率提高到 70%，每年可节约 2000 万吨标准煤，如果将所有矿井的灯都换成节能灯，每年将节约 6000 亿度电。因此，这些节约能源的形式未来也将成为节能减排的重要推动力。

3. 实施煤炭清洁化利用，发展煤化工产业

以发展煤化工为代表的煤炭清洁化利用，近年来成为政府和企业促进煤炭行业可持续发展

的重要着力点之一。我国煤炭资源与水资源逆向分布的现实，发展煤化工的耗水问题不容忽视。相较之下，发展煤转化为电更经济合理。在欧美发达国家，煤电产业占据整个行业的80%—90%。因此，“十二五”期间，煤化工将采取适度发展的原则，国家鼓励大型煤炭企业搞煤炭、港口、铁路等一体化经营。

4. 大力发展循环经济，从源头促进节能减排

目前，全国已建成了一批以大同塔山煤矿循环经济园区为代表的循环经济产业园，这些园区发展模式代表了煤炭结构调整、促进产业升级、协调发展的成功案例，引领了煤炭行业节约发展、清洁发展、安全发展和可持续发展的方向。“十二五”期间，一是要统筹考虑大型煤炭基地建设、大型坑口电厂建设、煤焦化工产业发展，建立具有产业链发展特点的循环经济工业园区；二是煤炭生产规划要与下游产业发展规划有效衔接，合理布局，促进资源消耗“减量化”，废物利用“资源化”，提高煤炭资源综合利用水平；三是鼓励煤气共采、煤矸石和矿井水利用，实现煤与共伴生资源在产业链条上“再利用、再循环”，促进节能减排。

5. 促进煤炭企业节电

煤炭企业电力消耗占煤矿总耗能的70%—80%，节电潜力较大。一是加强用电管理，减少设备空负荷运转；二是逐步淘汰国家明令禁止的高耗能设备，减少大马拉小车用电设备；三是加强用电计量管理，制定用电消耗定额管理和节奖超罚激励措施；四是合理匹配用电负荷，积极推广应用节电技术。

6. 加强技术创新

科技进步的发展，为煤炭节能提供了技术支撑，提供了节能的空间。据统计，全国煤矿相当部分耗能设备属于落后淘汰设备，许多电机设备并未在最佳工况下运行，原国有重点煤矿主通风机、主排水泵的平均效率和系统运行效率均低于国外先进水平。

9.5 典型煤炭企业节能减排动态

9.5.1 神华集团

神华集团把节能减排作为促进企业科学发展和转变经济发展方式的重要抓手，打出了一套节能降耗的“组合拳”，超额完成了节能减排总目标。截至2009年，神华集团万元产值综合能耗为1.84吨标准煤/万元，较2005年下降了20.3%；所属的8家重点企业共节能769384吨标准煤。2009年，神华集团二氧化硫、COD排放量分别下降33%和36.6%，均超额完成国家下达的“十一五”节能减排考核指标。

1. 健全节能减排管理体制

一是完善了组织机构，规范了管理。神华集团公司成立了节能减排领导小组，制定了节能减排工作管理暂行办法等规章制度，建立监测统计体系和指标考核奖惩体系，促进了节能减排工作制度化、规范化。

二是组建了节能环保专业化技术服务公司，开展技术咨询和基于合同能源管理模式的技术服务。开展的合同能源管理项目实现年节约标准煤1.6万吨，年减排二氧化碳4.5万吨。

三是积极开展节能减排战略研究，对煤炭、电力、运输、煤制油与煤化工四个板块制定中长期节能减排目标。

四是强化项目管理，设立节能减排重点工程项目，确保节能减排目标的实现。2007年以来，神华集团组织实施了包括工艺系统节能改造、节能减排计量系统优化、矿井水综合利用、电厂/锅炉脱硫改造以及固废综合治理等共计70多个节能减排项目，有效提高了能源与资源的综合利用水平。

2. 加大节能减排投入

2010年神华集团节能减排计划投入6.3亿元，比2009年增长7.5%。神华神东煤炭集团投入5655万元，完成煤矿生产设备提效节能系统等16个项目的建设，发挥了良好的效益。

神华国华电力投资4亿多元完成绥中电厂火力发电机组烟气脱硫改造工程，每年减少二氧化硫排放31867万吨，实现全公司发电机组脱硫装置配置率100%。

同时，神华集团积极利用国家的节能减排财政支持政策，神华乌海能源公司、神华国华电力公司、神华黄骅港务公司等单位的一批项目获得了国家资金支持。

3. 增强节能减排科技创新能力

一是加强自主创新，以神东7米大采高液压支架项目和国华沧电日产1.2万吨海水淡化项目为代表，一批具有自主知识产权的节能减排项目投产，取得了良好的经济和社会效益。

二是推动技术改造和内部挖潜，“火电厂节水发电技术研究”、“黄骅电厂海水淡化项目产业化及大型化研究”、“神东矿区1.7米以下薄煤层开采技术方案研究”等项目，为挖掘新的价值增长点奠定了基础。神华国华电力在进行机组节能技术改造的同时，提高精益管理水平，采用对标评比方式，在供电煤耗国内领先的情况下，又显著降低4.1克/千瓦时。

三是提升综合利用水平，矿区企业积极规划建设矿井水处理与综合利用、瓦斯发电、煤矸石发电、高温烟气发电与煤气化工品合成等项目。全集团在建煤矸石发电机组全部投产后预计每年可利用煤矸石超过1千万吨。煤脱水干燥技术、CO_2捕集和封存关键技术集成与示范、400MW机组IGCC多联产、粉煤灰综合利用工业化技术系统研发等多个项目正在积极推进。

4. 建立节能减排考核体系

一是将节能减排评价体系与神华“五型企业”绩效评价体系有机结合，构建符合神华实际，具有神华特色的节能减排工作考核机制。

二是积极推进能源审计和能耗监测，2009年完成了神华乌海能源等20家单位能源审计工作，为查找重点耗能环节，挖掘节能潜力，推进精细化管理，合理安排资金，以及科学考核提供了依据。

三是电力板块大型火力发电机组全部安装了烟气在线监测仪，实时对污染物排放情况进行监测，强化了对污染物排放量的控制。

10 建材行业的节能减排

10.1 建材行业“十一五”运行综述

“十一五”期间，在满足市场需求强劲增长的同时，全行业结构调整、技术进步和工艺装备水平、节能减排等各方面都取得了长足进步。“十一五”是建国以来我国建材工业发展最快的时期之一，也是发展水平提升、发展质量提高和经济效益最好的一个五年期。建材工业“十一五”主要数据指标完成情况如表10-1所示。

• 目前，我国已经是世界上最大的建筑材料生产国和消费国，主要建材产品水泥、平板玻璃、建筑卫生陶瓷、石材和墙体材料等产量多年居世界首位。

• 2010年水泥产量18.8亿吨，平板玻璃产量6.6亿重量箱，建筑陶瓷产量80.8亿平方米，卫生陶瓷产量1.6亿件，其产量年均增长分别为11.7%、10.3%、14.2%、21.3%。我国已成为名副其实的世界建材生产大国。

• 2010年规模以上建材工业企业实现销售收入2.7万亿元，“十一五”期间年均增长29.5%；2010年实现利润总额2000亿元，“十一五”期间年均增长44.2%。全年水泥产销平稳。

• “十一五”期间建材商品出口总体保持了较快增长态势，年均增长速度为17.3%。以具有国际先进水平的大型新型干法水泥成套技术为依托，大企业积极参与国际工程建设领域的竞争，并由此带动了成套技术装备和劳务的出口。

• 玻璃纤维行业大企业启动实施了海外投资战略，瞄准海外市场，“走出去”积极参与国际市场竞争，提升了行业在国际竞争中的影响力。

表10-1 建材工业“十一五”指标完成情况

Table 10-1 Targets Fulfilled by Building Materials Sector in China's 11th Five-year Plan

指标	2005年	“十一五”指标	2010年 完成指标
建材工业增加值年均增速（%）	16.4	8-10	27.9
水泥产量（亿吨）	10.6	12.5	18.8
新型干法熟料比重（%）	40	70	81
前10家水泥企业生产集中度（%）	15.4	30	25
累计淘汰水泥落后生产能力（亿吨）	—	[2.5]	[3.4]
浮法玻璃比重（%）	79	>90	87
前10家玻璃企业生产集中度（%）	47	70	50

续表

指标	2005 年	“十一五”指标	2010 年 完成指标
新型墙体材料比重（%）	42	60	55
累计淘汰玻璃落后生产能力（万重箱）			6000
建材万元增加值能耗（吨标准煤/万元）	6.38	-20%	3
全行业年利用工业废渣量（亿吨）	4.6	>5	>6
新型干法水泥采用余热发电生产线比例（%）		40	55
烟气粉尘排放量减少（%）			[46]
二氧化硫排放量减少（%）			[12]
平板玻璃综合能耗下降（%）		[20]	[20]
建筑陶瓷单位综合能耗下降（%）		[20]	[20]
卫生陶瓷单位综合能耗下降（%）		[20]	[25]

注：[ ] 内指标为累计完成指标
资料来源：《建筑材料工业“十二五”发展指导意见》

10.2　建材行业“十一五”节能减排总结

建材工业“十一五”规划纲要明确提出了节能的主要目标任务：“十一五”期间，建材工业将按照建设资源节约型、环境友好型社会的要求和循环经济的发展模式，继续加快产业结构调整，加大节能型、环保型技术、装备和产品的开发力度，积极推进企业和行业信息化建设，增强自主创新能力和竞争力，提高经济效益和社会效益，大力发展循环经济，将建材工业建成与经济、社会、环境协调发展的资源节约型和环境友好型产业。

具体节能目标如下：

（1）总目标。力争使建材工业万元增加值能耗降低 20%。

（2）主要产品的综合单耗有较大幅度下降。

• 水泥：新型干法水泥能耗指标达到中等发达国家水平。日产 4000 吨以上大型新型干法生产线熟料烧成热耗小于 740 千卡/千克，吨水泥综合电耗小于 95 千瓦时，新型干法水泥采用余热发电生产线比例达到 40%。

• 平板玻璃：平板玻璃综合能耗达到 16.5 千克标准煤/重箱，下降 20%。浮法玻璃平均能耗降低到 6500 千焦/千克以下。

• 建筑卫生陶瓷：建筑瓷砖综合能耗达到 300 千克标准煤/吨，卫生陶瓷单位综合能耗达到 700 千克标准煤/吨，下降约 20%。

• 墙体材料：每万块标砖耗标准煤下降 15% 以上。

（3）资源节约及综合利用目标。全行业年利用工业废渣（含粉煤灰、高炉矿渣等）量达 5 亿吨以上，新型干法水泥生产资源利用率达到中等发展国家水平。

上述目标完成情况如下：

（1）截至 2009 年，主要行业能源消费总量如下：

• 2009 年水泥全行业能源消耗总量为 1.76 亿吨左右，如图 10-1 所示。

- 2009 年平板玻璃行业能源消耗量约为 1100 万吨标准煤，如图 10－2 所示。
- 2009 年建筑卫生陶瓷行业全年能耗总量约为 3800 多万吨标准煤，如图 10－3 所示。
- 2009 年墙体材料行业年耗能总量约为 7300 吨标准煤，如图 10－4 所示。

（2）2010 年建材工业万元增加值综合能耗为 3 吨标准煤，比 2005 年下降了 52.6%，超额完成了“十一五”计划目标。

（3）“十一五”期间，由于产业结构优化和节能技术的推广，建材主要产品单位综合能耗进一步下降。其中：

- 2009 年水泥行业吨水泥能耗较“十五”末期下降了 18%，如图 10－5 所示。
- 2009 年平板玻璃单位产品能耗较“十五”末期下降了 18%，如图 10－6 所示。

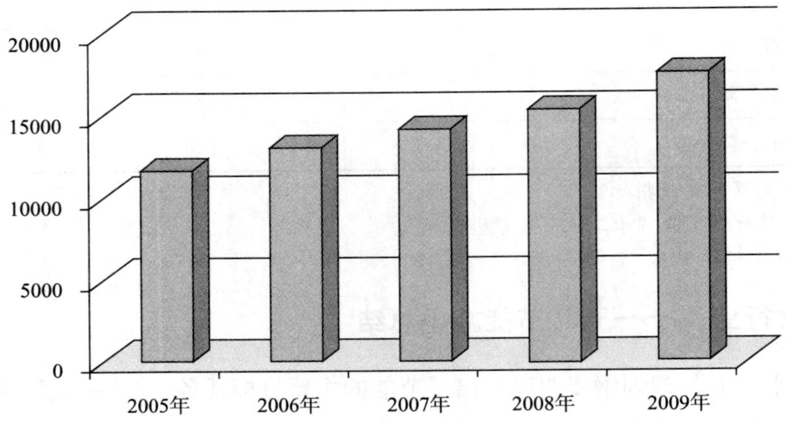

图 10－1　水泥行业能源消耗总量（2005—2009）

Figure 10－1　Total Energy Consumption by Cement Industry（2005—2009）

资料来源：《建筑材料“十二五”发展指导意见》

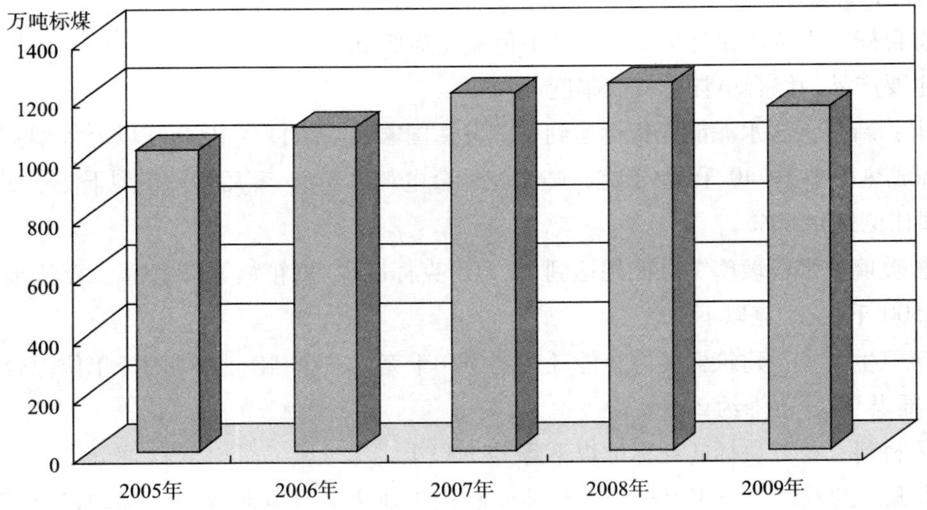

图 10－2　平板玻璃能源消耗总量（2005—2009）

Figure 10－2　Total Energy Consumption by Plate Glass Sector（2005—2009）

资料来源：《建筑材料“十二五”发展指导意见》

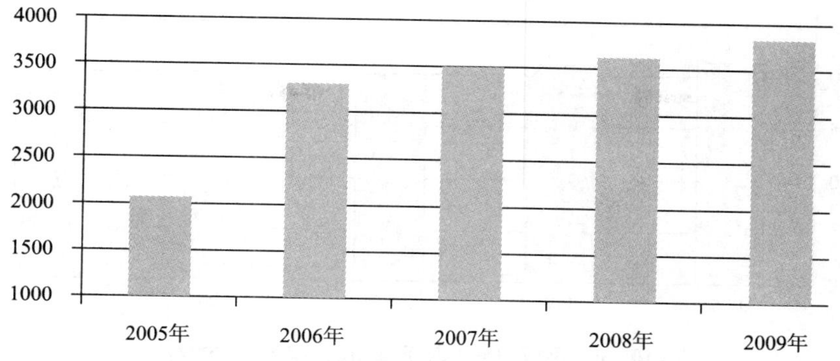

图 10－3　建筑卫生陶瓷行业能源消耗总量（2005—2009）

Figure 10－3　Total Energy Consumption by Architectural and Sanitary Ceramics Sector（2005—2009）

资料来源：《建筑材料“十二五”发展指导意见》

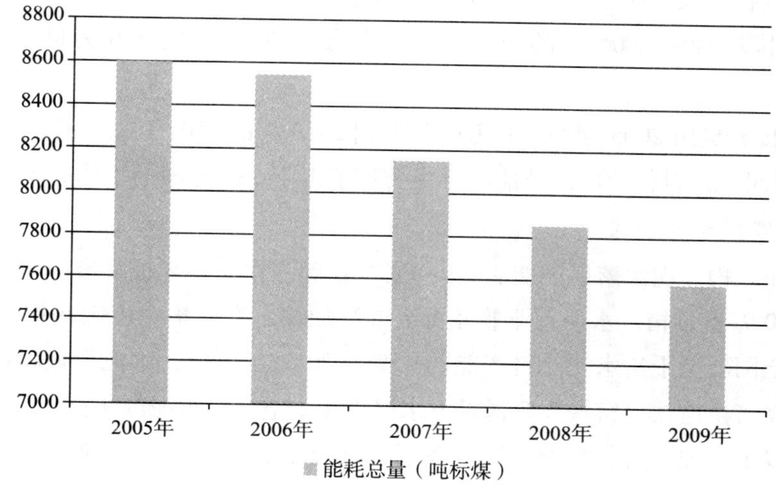

图 10－4　墙体材料行业能源消耗总量（2005—2009）

Figure 10－4　Total Energy Consumption by Walling Materials Sector（2005—2009）

资料来源：《建筑材料“十二五”发展指导意见》

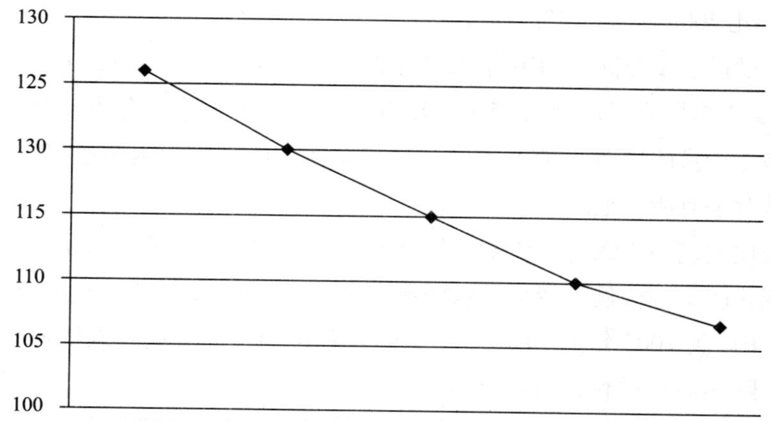

图 10－5　水泥能耗下降情况（2005—2009）

Figure 10－5　Reduction of Energy Consumption Per Unit by Cement Sector（2005—2009）

资料来源：《建筑材料“十二五”发展指导意见》

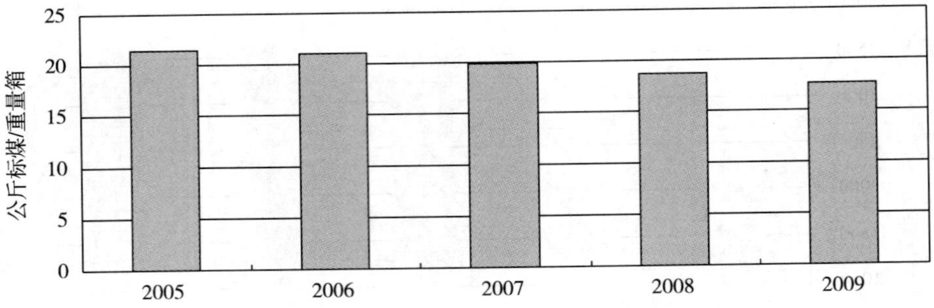

图 10－6 平板玻璃能耗下降情况（2005—2009）

Figure 10－6 Reduction of Energy Consumption Per Unit by Plate Glass Sector（2005—2009）

资料来源：《建筑材料“十二五”发展指导意见》

（4）主要污染物排放总体呈明显下降趋势，其中：

• 烟气粉尘排放量由 2005 年的 702 万吨，降低到 2010 年的 380 万吨，比 2005 年减少了 46%。

• 二氧化硫排放量由 2005 年的 170 万吨降低到 2010 年的 150 万吨，下降了 12%。

“十一五”期间，建材行业围绕节能减排主要作了以下三个方面的工作。

1. 结构调整取得重大进展

一是优化产业结构，淘汰落后产能。“十一五”期间，水泥、平板玻璃分别累计淘汰落后产能 3.4 亿吨、6000 万重量箱。水泥行业兼并重组继续加快，产业集中度大幅提高，10 家水泥企业熟料年产能占全国新型干法水泥熟料产能的 45%。2012 年以前，水泥行业将淘汰窑径 3.0 米以下水泥机械化立窑生产线、窑径 2.5 米以下水泥干法中空窑、水泥湿法窑生产线、直径 3.0 米以下的水泥磨机以及水泥土窑、普通立窑等落后水泥产能；玻璃行业平拉工艺平板玻璃生产线将全部淘汰。

二是技术和产品结构不断优化。2010 年新型干法水泥熟料产量比重达到 81%，比 2005 年提高 41 个百分点；浮法玻璃产量比重达到 87%，比 2005 年提高 8 个百分点，单线最大规模达到日熔化量 1000 吨；池窑玻璃纤维产量比重达到 85%，比 2005 年提高 16 个百分点；新型墙体材料比重达到 55%，比 2005 年提高 13 个百分点。

三是产业集中度进一步提高。2010 年水泥行业前 10 家企业产量 4.7 亿吨（其中有 2 家企业水泥产能过亿吨），占全国水泥产量的 25%，较 2005 年增长 9.6 个百分点；全国前 10 家平板玻璃企业产量占全国总产量的 57%；有 8 家大型建材企业集团进入中国企业 500 强，市场供需秩序有所改变，区域竞争有所规范。

四是产业结构比例进一步优化。传统的能耗高的产业增加值比重下降是“十一五”时期建材工业结构发生变化的又一个显著特征。水泥制造业工业增加值占建材工业增加值总量的比重从 2002 年的 43% 下降为 2005 年的 34%，2010 年下降到 24%；玻璃纤维增强塑料、建筑用石、云母和石棉制品、隔热隔音材料、防水材料、土砂石开采、技术玻璃、水泥制品等行业的发展速度远超过了传统的能耗高的行业增长速度。2010 年低能耗产业工业增加值比重已经达到建材工业增加值总量的 43%，其对建材工业增长的贡献率，已接近传统的能耗高的建材产业。

五是产业布局进一步改善。在国家实施西部大开发战略和产业转移等政策引领下，中西部地区建材工业的发展速度和主要建材产品生产能力增长明显快于东部地区。“十一五”时期中西部地区建材工业增加值年均增长32.7%，高于东部地区年均增长10.7个百分点。

2. 节能减排、综合利废成效显著

2010年，建材工业利用各类工业固体废弃物6亿吨，其中粉煤灰的综合利用量占到全国利用总量的30%以上，煤矸石的利用量占全国利用总量的50%以上。工业副产石膏也得到有效利用。

水泥行业余热利用技术得到普遍认可和推广应用，到2010年底，累计约有700条生产线建成余热发电站，总装机容量达到4800兆瓦；玻璃熔窑余热发电技术也已得到开发应用。

产品单位能耗显著下降。水泥熟料单位煤耗2010年下降到120千克标准煤，比2005年下降17.7%，“十一五”期间水泥工业因熟料单位煤耗的下降，共节约煤炭消耗1.16亿吨。“十一五”后期，1/4的日产4000吨及以上生产线吨水泥熟料标准煤耗在103千克标准煤以内，在大型水泥企业中，亚东、华润、海螺、冀东、塔牌集团吨熟料标准煤耗整体水平已低于110千克标准煤。

水泥行业已基本掌握了利用水泥窑无害化处置工业废弃物的关键技术。利用水泥窑协同处置工业废弃物、危险废弃物、城市生活垃圾、污泥等综合利用工程陆续启动。同时，以可燃性废弃物替代燃料的研究与实施也在积极推进之中。

3. 提升生产工艺，发展循环经济

2009年日产4000吨及以上生产线能力占新型干法能力比重超过50%，2010年达到60%以上。利用粉煤灰、煤矸石等固体废弃物大力发展新型墙体材料，目前占墙体材料的比例已超过50%。浮法玻璃比重超过80%，水泥和玻璃生产技术和装备水平都已经接近或者达到国际先进水平。水泥、玻璃行业低温余热发电得到推广；水泥行业处置城市污泥、工业危废物、生活垃圾技术得到逐渐普及。

10.3 建材行业“十二五”节能减排形势

虽然“十一五”期间通过加快产业技术结构调整、推广节能新技术、新工艺和新装备，建材工业取得了较为明显的节能效果，然而，与发达国家相比，在主要能耗指标和能源综合利用等方面仍然存在突出的问题，成为制约行业节能发展的重要因素。主要问题分析如下：

1. 产业结构不合理，产品结构有待优化

当前，我国建材行业产业结构不合理的状况仍然较为突出，虽然新型干法水泥的比重不断提高，但水泥产量的45%左右仍为落后的立窑等生产工艺，产业集中度较低，大企业比较少，而分散的小企业仍然很多，导致资源能源消耗高，污染物排放高。

同时，水泥制品、玻璃制品等后期污染轻、附加值高的行业比重仍非常低，造成整个建材行业产业结构的不合理。另外，水泥产品的质量不高，目前主要是低标号水泥，而高标号水泥的产量较少，导致水泥需求量增大，间接造成能源消耗与污染排放的增加。

2. 主要能耗指标与国际先进水平存在较大差距

我国大型水泥企业新型干法工艺的热耗指标已接近世界先进水平，但水泥电耗与世界先进水平尚存在较大差距。例如，世界先进水平水泥综合电耗仅为80千瓦时/吨，而我国平均水平为90千瓦时/吨左右。

现有的平板玻璃生产工艺已与国际接轨，但能耗与国际水平仍存在一定差距。就浮法生产而言，我国现有生产线平均能耗比发达国家平均水平高20%，600吨/天和800吨/天浮法玻璃生产线平均熔化能耗发达国家分别为6500千焦/千克玻璃液和5700千焦/千克玻璃液。

3. 能源综合利用未得到广泛推广和应用

工业能源综合利用与发达国家存在较大差距。例如，发达国家在水泥窑处置工业废弃物、城市垃圾和污泥等方面已积累了丰富的经验，并得到了广泛的推广和应用，而我国尚处于探索和试验阶段，我国吨水泥废弃物利用率（不含商品砼用量）仅为日本的一半左右。这不仅存在技术方面的原因，政策体系的不健全以及政府的推广力度不足也影响了行业能源综合利用的进展。

4. 碳绝对排放量还在持续增长

我国正处于工业化和城市化的中期阶段，建材行业作为基础工业始终处于先行发展的状态。目前，虽然主要建材产品都已超过世界总产量的一半以上，且部分地区主要建材产品出现产能过剩，但由于近年来全社会固定资产投资持续高速增长，基础设施建设和房屋建设规模的不断扩大，拉动了建材工业主要产品产量仍在快速增长（如表10-2），由此导致全行业的能耗总量和温室气体排放总量还在持续增长。

表10-2 2015年主要建材产品国内需求量预测

Table 10-2 Forecasting on Domestic Demand of Key Building Materials in 2015

产品	2015年国内需求量
水泥（亿吨）	22
平板玻璃（亿重量箱）	7.5
镀膜玻璃（亿平方米）	1.4
太阳能光伏玻璃（亿平方米）	1.6
汽车安全玻璃（亿平方米）	2.2
超薄屏显基板玻璃（亿平方米）	1.2
建筑陶瓷（亿平方米）	90
卫生陶瓷（亿件）	1.6
玻璃纤维（万吨）	200
复合材料（万吨）	600
新型墙体材料（亿块标砖）	5700
纸面石膏板（亿平方米）	28
建筑防水材料（亿平方米）	16.6
保温绝热材料（万吨）	1000
花岗石板材（亿平方米）	6.6
大理石板材（万平方米）	9000

资料来源：《建筑材料“十二五”发展指导意见》

10.4 建材行业“十二五”节能减排展望

“十二五”期间，建材行业的节能减排及循环经济发展目标：到“十二五”末，建材主要行业能耗、二氧化碳和污染物排放量均达到国家规定，万元增加值能耗比2010年降低20%，万元增加值二氧化碳排放比2010年降低18%，如表10-3所示。

其中，主要产品的综合单耗有较大幅度下降。具体如下：

- 水泥：平均每千克熟料烧成热耗小于740千卡，吨水泥综合电耗小于85千瓦时。
- 平板玻璃：综合能源耗小于15千克标准煤/重箱。
- 建筑陶瓷：单位产品综合能耗为4.59千克标准煤/平方米。
- 卫生陶瓷：单位产品综合能耗为655千克标准煤/吨。
- 墙体材料：单位产品能耗为800千克标准煤/万块标砖。

表10-3 “十二五”时期建材工业主要发展目标

Table 10-3 Major Targets Set for Building Materials Sector in China's 12th Five-year Plan

类别		指标	2010年	2015年	年均增长（%）
经济发展目标		工业增加值年均增长（%）	26		>10
		利润总额年均增长（%）	44		>10
		总资产贡献率（%）	17	>20	
结构调整目标	产业结构调整目标	低能耗产业增加值比重（%）	43	>50	
	技术结构调整目标	新型干法熟料比重（%）	81	>95	
		浮法玻璃的比重（%）	87	>90	
		平板玻璃深加工率（%）	35	45	[10]
		新型墙体材料产量比重（%）	55	65	[10]
	组织结构调整目标	前10家水泥企业生产集中度（%）	25	35	
		前10家玻璃企业生产集中度（%）	57	75	
		进入世界500强企业数（家）		1-2	
	淘汰落后目标	淘汰落后水泥产能（亿吨）			[2.5]
		淘汰落后平板玻璃产能（万重量箱）			[5000]
节能减排和循环经济目标		万元工业增加值能耗降低（%）	3吨标准煤		[18-20]
		万元工业增加值 CO_2 排放量降低（%）			[18]
		氮氧化物排放总量减少（%）			[10]
		二氧化硫排放总量减少（%）			[8]
		综合利用和协同处置废弃物总量比2010年增加（%）	6亿吨		[20]
技术进步目标		规模以上企业科技投入占企业销售收入的比重（%）		1.5	

注：1. [ ] 内为五年累计数；

2. 低能耗产业：建材低能耗产业指万元增加值能耗低于万元国内生产总值能耗的产业；

3. 前10家企业生产集中度：均指产量占比。

资料来源：《建筑材料“十二五”发展指导意见》

“十二五”建材行业节能减排发展的重点如下：

1. 提高生产节能水平

推广先进节能技术与装备，加快对现有生产线实施以节能为中心的技术改造，强化能耗管理，全面提高建材产品生产的能效水平。

2. 实施清洁生产

推广应用高效除尘技术，进一步降低建材产品生产过程中的烟粉尘排放量。推广应用先进、实用的烟气脱硫脱硝技术、降低噪声污染的技术。加强生产过程中粉尘无组织排放的控制，积极开展清洁生产审核，进一步完善清洁生产评价体系。

3. 发展循环经济

以水泥、墙体材料、水泥混凝土及其制品、石材等行业为重点，继续鼓励企业对矿渣、煤矸石、粉煤灰、尾矿、工业副产石膏、建筑垃圾、碎石粉等大宗工业废弃物的综合利用，发展循环经济。鼓励水泥企业利用废弃物、生物质燃料替代原燃料，推广利用水泥窑协同处置城市生活垃圾、城市污泥和工业废弃物。推广应用包括纯低温余热发电在内的建材工业窑炉余热梯级利用技术。

4. 技术进步和技术改造重点

第一，关于重点关键技术的研发，应该重点突破制约建材工业节能减排的脱硫脱硝一体化、二氧化碳减排关键技术以及无机非金属新材料加工制造核心技术、产业升级的新工艺、新技术、新装备。

第二，在推进企业技术改造方面，应充分发挥科技对产业升级的支撑作用，支持符合国家产业政策和规划布局的企业，运用高新技术和先进适用技术，推进“两化”融合，以提高质量、发展高端产品、推进节能降耗、环境保护和改善装备、安全生产等为重点，进行技术改造，实现产业升级。

第三，着重提升和完善建材标准。围绕自主创新、节能减排、安全生产、战略性新兴产业等行业发展急需或缺失的领域，加强建材标准的制定和修订，进一步完善建材标准体系，加强与国际标准对标及参与或主持国际标准制定工作。

5. 工程重点

一是节能改造工程。通过技术改造降低单位产品能耗水平，推动行业节能减排。到2015年水泥、平板玻璃、建筑卫生陶瓷、玻璃纤维等行业能耗水平接近或达到世界同期先进水平。

二是淘汰落后工程。2015年，基本淘汰水泥、平板玻璃和建筑卫生陶瓷等落后产能，新型干法水泥比重达到95%以上、浮法玻璃比重达到90%以上。

三是环保产业工程。利用水泥窑协同处置城市生活垃圾、污泥和工业危废物，改善城市生态环境，使城市周边水泥企业成为保护生态环境，促进工业化和城镇化发展不可或缺的环保企业。

四是新材料工程。开发新型高性能、多功能和结构材料及制品，为战略性新兴产业和其他相关产业的发展提供支撑，发展无机非金属新材料及其制品业。

五是绿色节能建材工程。为绿色节能建筑提供绿色节能材料及产品，推动房屋建筑用绿色建材及制品业发展。

10.5 典型建材企业节能减排动态

10.5.1 中国建筑材料集团有限公司①

中国建筑材料集团有限公司（以下简称“集团”）的能源消耗在产业板块中所占比重各不相

① 资料来源：中国建筑材料集团有限公司《2010年社会责任报告》。

同，水泥产业能源消耗占集团能源消耗总量的90.3%，其次为石膏板、玻璃和玻璃纤维。直接能源消耗以煤炭为主，占集团综合能源消费量的86.4%，其次为电力。通过技术手段，实现综合利用煤矸石、污泥、城市垃圾等可燃废物替代部分化石能源；回收利用废气中的余热资源发电，替代部分外购电力。

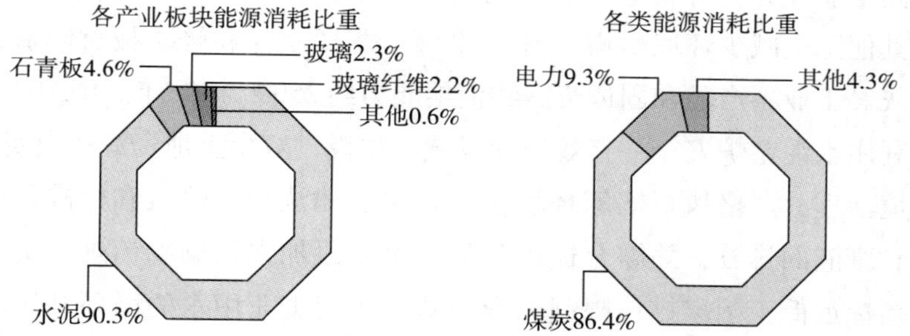

图10－7 能源消耗比重

Figure 10－7 The Proportion of Energy Consumption

资料来源：中国建筑材料集团有限公司

集团在制造过程中向环境排放的主要污染物根据各产业板块不同也各不相同，水泥产业主要排放包括氮氧化物、二氧化碳、粉尘和二氧化硫等；玻璃产业主要为二氧化硫；石膏板、玻璃纤维和复合材料产业主要为废水中的COD，各产业自产固体废弃物基本都可实现自回收利用。

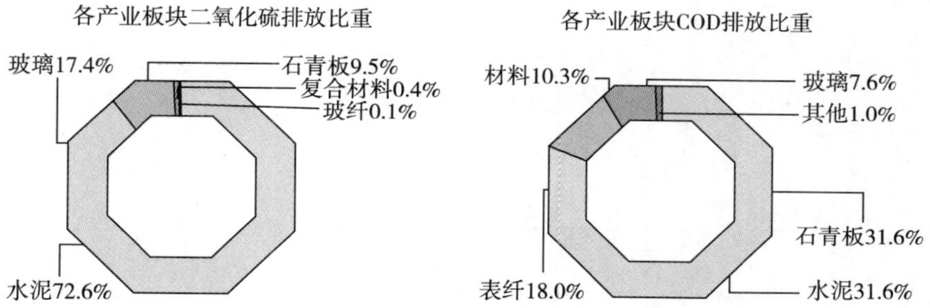

图10－8 排放比重

Figure 10－8 Emission Proportion

资料来源：中国建筑材料集团有限公司

集团及时淘汰落后的生产技术和设备，采用先进的工艺，选用高效环保的装备制造产品。重视节能技术改造，努力提高能源利用效率，例如在玻璃和玻璃纤维生产线实施全（富）氧燃烧技术，提高燃烧效率；对大型用电设备实施变频改造，提高用电效率等，努力实现每一单位产品的消耗和排放水平不断降低。

“十一五”期间，集团累计淘汰水泥（熟料）落后产能51台套，产能合计680万吨/年；淘汰浮法玻璃产能1台套，产能231万重量箱/年；以节能减排为目标的产业结构调整、生产工艺升级改造、既有生产设备节能减排技术改造投入占集团固定资产投入总额的比重均保持在80%左右。

集团努力减少生产过程中化石能源的使用。通过技术改造，实现生产线综合利用煤矸石、污泥、城市垃圾等可燃废物替代煤炭等化石能源；在符合条件的水泥和玻璃生产线上全部配套建设余热发电系统，回收利用生产过程中的废热转化为电能，回供到生产线上使用，减少外购电力的使用量。截至2010年底，已建和在建水泥余热发电系统装机容量近900MW，形成发电能力约44亿千瓦时/年，年可节约标准煤140万吨，相当于减排二氧化碳350万吨。

集团发挥工艺优势，循环利用工业和城市废弃资源。通过技术革新，在确保质量和安全的前提下，不断提高工业废弃资源在原材料中的使用比重，提高脱硫石膏、矿渣、钢渣的消纳和利用能力，协助其他工业减少环境影响。到2010年，集团水泥和轻质板材制造板块形成脱硫石膏、矿渣、粉煤灰等工业副产品和固体废弃物消纳能力约2500万吨/年。中国建筑材料集团有限公司还积极试点利用水泥窑炉安全、有效地处置城市垃圾，减少土地资源的浪费和环境污染。

集团重视环境治理，严格执行国家环境保护相关法律法规、政策和标准，努力减少生产过程中废水、废气和废渣的排放，并综合运用各种技术手段加强污染物治理。集团绝大部分企业各项污染物监测指标远低于国家排放限制标准，基本实现工业用水的循环利用和自产固体废弃物100%回收利用。

集团通过采购清洁能源，提高燃烧系统中氧气含量，减少废气中的二氧化硫和氮氧化物等污染物的形成和排放；通过使用袋除尘等除尘技术，有效降低废气中的粉尘排放量；通过建设污水处理和循环使用系统，努力实现废水“零”排放。

11　纺织行业的节能减排

11.1　“十一五”纺织行业运行综述

总体看来，在“十一五”期间，我国纺织行业保持了较快的增长势头。“十一五”行业主要数据如下：

• 2010年纺织行业总产值达到47650亿元，较2005年增长1.31倍，“十一五”期间平均增速为18.22%。

• 2010年纺织行业实现主营业务收入46900亿元，较2005年增长1.33倍；2010年纺织行业利润额2550亿元，较2005年增长2.60倍。

• 主要产品产量增幅明显：

① 2010年国内化纤产量达到3100万吨，较2005年增长了86.21%，“十一五”期间年均增速为13.24%。

② 纤维加工总量达4130万吨，较2005年增长了60.70%。

③ 纱产量达到2730万吨，较2005年增长88.21%，“十一五”期间的年均增速为13.48%。

④ 布产量达到790亿米，较2005年增长63.09%，“十一五”期间年均增速为10.28%。2005—2010年，产销衔接一直稳定在97%—98%区间。

• 我国纺织品服装的出口竞争力在不断提升，占全球的纺织品服装贸易比重仍在继续提升。据我国海关数据，2010年纺织服装出口总额2065亿美元，较2005年增长75.72%，年均增长11.93%。

同时，“十一五”期间，我国纺织工业经济运行效益和质量进一步提升，科技含量和创新能力大幅提高，大量科技成果的产业化成为行业在新时期增长的重要支撑，较大地提高了纺织行业的运行质量和效益水平。

此外，围绕“十一五”《纺织工业科技进步发展纲要》部署的目标和任务，纺织行业科技进步速度加快，自主创新能力迅速提升，开展了大量关键技术攻关和成果产业化推广工作，以自主创新的工艺、技术、装备为主体的发展速度加快，新产品开发能力和品牌创建能力明显增强，节能减排取得较大进步，有效支撑了产业结构调整和产业升级，为行业由大变强的发展奠定了坚实基础。

11.2　“十一五”纺织行业节能减排总结

“十一五”期间，我国纺织工业的重点工作是加大结构调整力度，提高科技贡献率和品牌贡献率，加速产业升级。其中，节能降耗、保护环境、实现行业可持续发展是结构调整工作的重要内容。为此，《纺织工业“十一五”发展纲要》明确提出了相关约束性指标：

（1）节能指标为吨纤维耗电量比2005年降低10%。

（2）降耗指标为单位产值的纤维使用量比2005年降低20%。

（3）节水指标为吨纤维耗水量比2005年降低20%。

（4）环保指标为单位产值的污水排放量比2005年降低22%。

“十一五”期间，纺织行业的节能减排主要完成情况如下：

• 根据中国纺织工业协会公布的数据：“十一五”期间，按可比价计算，纺织工业单位增加值综合能耗累计下降约40%，污水排放量累计下降幅度超过40%；印染行业百米印染布生产新鲜水取水量由4吨下降到2.5吨，累计减少37.5%。

• 纺织行业能源消耗总量呈现了一个先升后降的过程，从2005年的6867万吨标准煤升至2007年的顶点8438万吨标准煤后逐步下降至2010年的7876万吨标准煤（估计值），如图11－1所示。

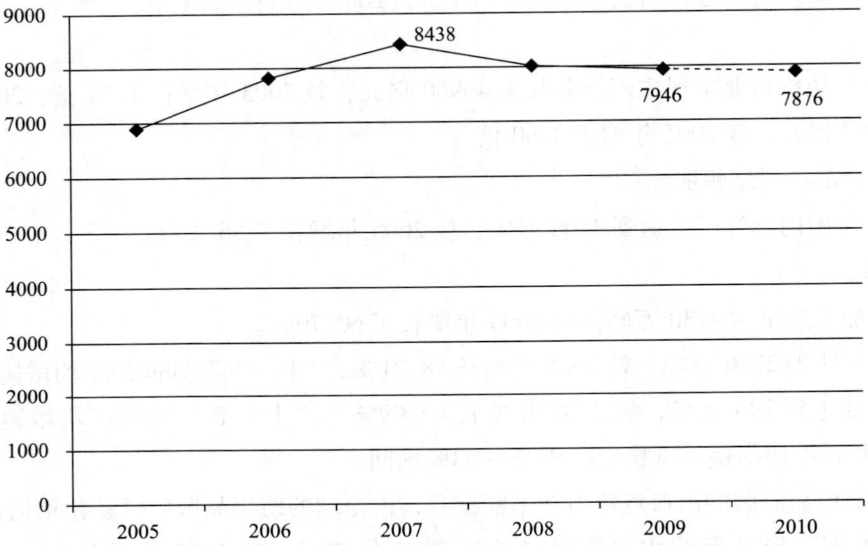

图11－1 “十一五”纺织行业能源消耗总量变化

Figure 11－1 Change of Total Energy Consumption by Textile Sector during China's 11th Five－year Plan Period

资料来源：《纺织工业“十一五”发展纲要》

• 纺织行业能耗比重逐步下降。纺织工业能耗所占工业能耗比重在逐年下降，从2005年的4.34%下降到2009年的3.60%，下降了0.74个百分点，如图11－2所示。

• 以较低的能耗增长速度支撑了较高的工业增加值增长。工业增加值从2005年的5145.36亿元增加到2009年的10351.88亿元，上升了101.18%，能耗从2005年的6866.82万吨标准煤增加到2009年的7945.65万吨标准煤，上升了15.71%。

• 纺织工业万元产值能耗由2005年的0.34吨标准煤降至2009年的0.21吨标准煤，累计降低38.24%，按可比价计算，年均节能率7.65%，如图11－3所示。

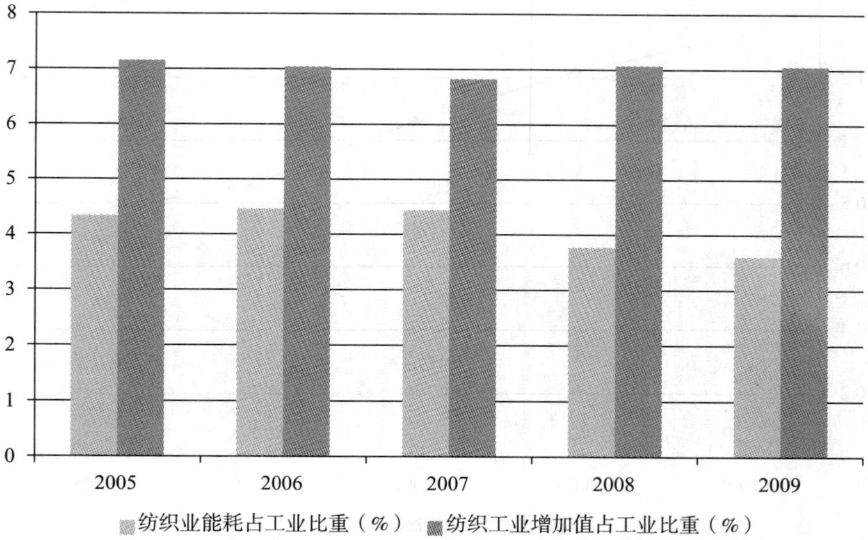

图 11-2 纺织业能耗及增加值占工业比重

Figure 11-2 Ratios of Energy Consumption and Added Value of Textile Sector to the Industry

资料来源：《纺织工业“十一五”发展纲要》

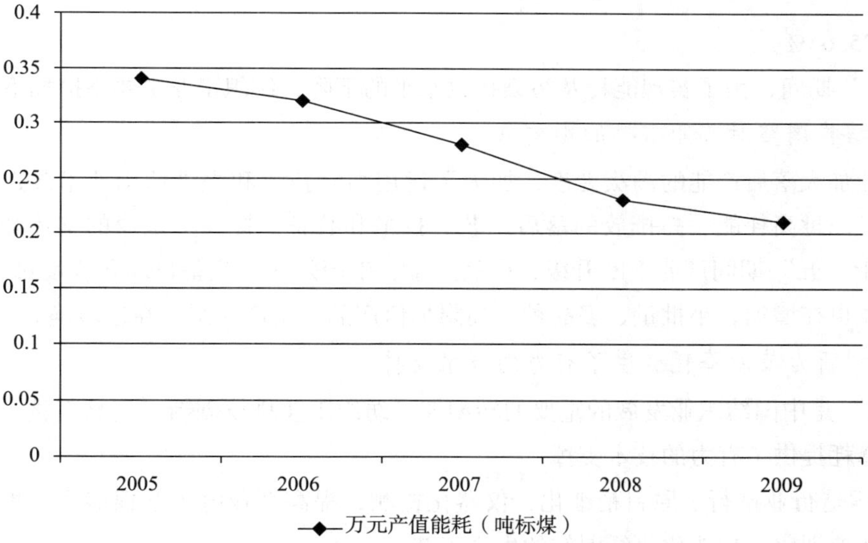

图 11-3 “十一五”纺织工业万元产值能耗变化

Figure 11-3 Change of Energy Consumption Per RMB 10000 Added Values by Textile Sector during China's 11th Five-year Plan Period

资料来源：《纺织工业“十一五”发展纲要》

• 纺织工业万元增加值能耗由2005年的1.33吨标准煤降至2009年的0.77吨标准煤，如图11-4所示。

• 2008年工业污水排放量为2173775万吨，纺织行业污水排放量为293693万吨，纺织污水排放占工业污水排放的13.51%，从2005年到2008年上升了2.86个百分点。

• “十一五”以来，纺织工业增加值污水排放持续下降。从2005的44.69吨降至2008年的31.44吨，下降了29.65%。纺织工业万元产值排污由2005年的11.35吨降至2009年的8.44

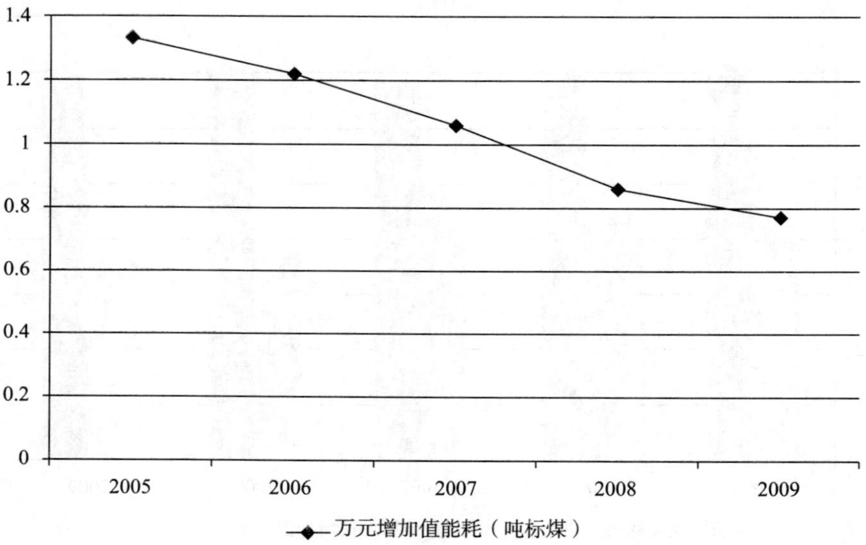

图 11－4 “十一五”纺织工业万元增加值能耗变化

Figure 11－4 Change of Energy Consumption Per RMB 10000 Added Values by Textile Sector during China's 11th Five－year Plan Period

资料来源：《纺织工业“十一五”发展纲要》

吨，降低了25.63%。

“十一五”期间，为了实现能耗及污染排放水平的下降，纺织工业主要举措如下：

1. 产业结构调整对节能减排的影响

纺织行业加大落后产能的淘汰力度，加大了运用高新技术和先进适用技术改造传统产业的步伐，淘汰了一批高耗能、高排放的落后工艺、技术和装备，提高了工业的整体能耗水平。纺织工业在“十一五”期间产业不断升级，产品产量、生产效率、装备自动化程度都在不断提高，产品平均幅宽也在增加，小批量、多品种、高附加值产品的生产体系正在逐步确立。

2. 科技创新为节能降耗提供了有力的技术支撑

“十一五”是中国纺织业发展的重要的转型期，纺织工业科技创新、用高新技术改造传统产业，为节能降耗提供了有力的技术支撑。

- 纺纱织造行业推行了原料精细化、仪器化检测，提高企业电子配棉能力，推广了多种纤维混纺纱线和差别化、功能化化纤混纺的生产工艺。
- 化纤工业大力发展高新技术纤维，用高新技术改造提升传统化纤产业。
- 棉纺行业推广采用节能电机、空调自动控制等技术，其中空调自动控制技术可降低空调能耗10%—15%。
- 化纤行业推广差别化直纺技术、新型纺丝冷却技术等实用节能型加工技术，其中新型熔体直纺热媒加热系统可减少燃料消耗近1/3。
- 印染行业节能降耗的新工艺技术研发和推广成效显著，其中高效短流程前处理技术可节约电、汽消耗30%以上，已经应用于各类棉及其混纺织物；冷轧堆染色可节约蒸汽40%，已在中厚型织物上应用。

3.“十一五”期间，污染物减排及治理技术明显进步

- 印染行业开发了对废水分质分流进行深度处理及回用的新技术，实现废水处理稳定达标，

同时使印染布生产水回用率由2005年的7%提高到2010年的15%，大幅减少了污水排放。

- 丝绸行业研发了缫丝生产废水深度净化循环技术，缫丝废水循环使用率可达90%以上，基本实现污水零排放，目前已在大中型缫丝企业中推广应用。

- 化纤行业采用膜技术处理化纤废水，采用长网洗浆机、连续打浆机和漂白自控系统等装置进行粘胶浆粕黑液治理，采用活性碳吸附法、废气制硫酸装置等治理粘胶废气，有效减少了液体、气体污染物的排放，提高了行业的清洁生产水平。

4. 投资力度加大是节能的重要保证

燃煤工业锅炉改造、区域热电联产、余热余压利用、电机系统节能、能量系统优化、以及节能监测和技术服务体系建设等十大重点工程，在“十一五”前四年得到了超过1200亿元的国家资金支持。其中，中央预算内投资480亿元、中央财政资金805亿元。国家实施的十大节能工程及绩效购买性质的投资，企业及社会投资等，改进了重点行业的技术、装备和工艺水平，直接增加了企业的节能能力，从而产生了节能效果。

5. 责任制起到了重要的导向作用

建立健全责任制，特别是领导问责制，实行“节能减排一票否决制”，所起的作用显而易见。中国纺织工业将“十一五”节能目标进行了有效的分解，是纺织各行业节能减排的导向，企业领导重视是工业节能的最大动力，考核则是企业领导或管理层重视节能的标尺。

6. 采用节能降耗新技术与加强管理相结合，发挥了最大的节能效果

把各项节能措施落实，纺织企业加强管理入手做到责任落实、维护精细、调控科学、成效显著。建立了能耗考核体系，建立能源统计、计量管理制度，把考核指标分解落实到车间和个人。

7. “十一五”期间纺织行业采取的节能减排技术

纺织工业在“十一五”期间，推广利用了一系列的先进技术、工艺，大大提高了设备技术水平，降低了单位产品能耗。

- 行业智能化控制系统研究取得实质进展。例如，企业采用数据在线采集控制系统，不但可以进行生产数据的查询追踪，还可对生产各环节进行精细化控制。这不但提高了企业产品质量，还有效控制了企业能源消耗，减少了企业各项生产成本，增加了企业信息反馈速度，提高了企业生产效率，使企业各项奖惩制度的贯彻落实做到有据可依。

- 棉纺织行业节能减排新技术。在棉纺行业，进行能量系统优化技术，推广使用热量平衡风量平衡技术，降低了纺纱车间和织造车间采暖和制冷的能耗。采用空调自动控制技术，提高车间温湿度的控制精度，提高了产品质量，降低了能量消耗，减少了设备维护费用。

- 印染行业节能减排新技术。在染整行业主要采用的节能技术和装备有：① 高效前处理技术；② 无水等离子体染整前处理技术；③ 无水涂料染色、印花新工艺新技术；④ 少水、高效、环保、节能染色关键技术；⑤ 高效、环保纺织印染助剂开发应用技术；⑥ 数码喷射印花技术；⑦ 气流染色设备。

- 化纤行业节能减排新技术。化纤行业主要采用的节能技术和装备有：聚酯聚合节能组合技术；高效节能的热媒系统技术，采用差别化直纺新技术。新型纺丝冷却装置。氨纶行业主要采用的节能技术：连续聚合、高速纺丝、环保溶剂。腈纶行业改用新型溶剂DMAC等。

- 服装行业节能减排新技术。“十一五”期间服装企业推广“蒸汽节能器”节能率在20%

以上，采用“工业缝纫机电子调速电机”节电30%。

- 纺织行业通用节能减排技术。纺织企业推广绿色照明，采用三基色T8－21W代替40W荧光灯可节省照明用电40%，采用LED照明替代传统荧光灯。太阳能集热技术充分利用了纺织企业厂房集中平整的优势，在厂房屋面上建造太阳能集热器，将冷水加热到设定的温度（70℃左右）后进入保温储热水箱，再将水箱的热水通过泵输送至用水设备，再利用蒸汽加热至90℃—120℃，太阳能集热技术解决了纺织行业对化石能源的过分依赖程度，从而达到节能的目的。

11.3 “十二五”纺织行业节能减排背景

“十二五”纺织行业节能减排仍面临如下问题：

1. 企业规模小，生产方式落后，增长方式粗放

目前，我国纺织企业，特别是印染企业仍以中小型为主，增长方式粗放，多数缺乏高科技含量，产品平均价格较低，多数以量取胜。同时，许多企业是来样加工生产，产品设计创新不足，市场中低档产品较多，高附加值产品少。这样的生产格局造成了纺织企业生产过程中的高投入、高消耗、高能耗和高污染，水环境不堪重负。

2. 治理投入严重不足

虽然政府和企业花了大力气治理废水，但治理速度仍赶不上污染发展的速度。为了节省投资，尽量压缩环保投资，有的企业污水集中处理厂没有接管标准或者接管标准放得很宽。在场内环保预处理设施投资上，有些纺织小企业也是“能省就省”，环保经费、技术、装备常常缺位，也是增加污染物排放的重要原因。

3. 行业准入标准不严，产业转移显现突出

纺织行业是劳动密集型产业，是污染较重的产业，也是比较容易转移的产业。在发达国家劳动力成本和环境成本都较高。自改革开放以来，发达国家及我国台湾、港澳地区的纺织企业陆续迁往我国沿海地带，其主要因素：一是当地劳动力便宜；二是在中国设厂环保成本低，环境准入标准不严。特别是发达国家环保要求很高，违法排放将严重受罚，因此，他们看中我国沿海地区，只要交通发达，当地环保允许，就举厂搬迁。

4. 环保法制观念淡薄，重经济效益、轻环境效益

由于我国环境和资源成本不计入企业生产成本，水资源费是象征性收费，不少企业实质是靠牺牲环境来获得利润。虽然国家制定了不少环保法规，开展过大力宣传，但少数企业环保意识仍然淡薄。分析原因，粗放型的纺织印染加工企业，产品大都为中、低档产品，利润很低，加工1米布仅0.4元—0.5元，有的企业还不到0.2元，甚至出现亏损。对于这类粗放型企业，污染治理成本可能高于所获利润，因此，置环保法制而不顾，环保设施开开停停，或者千方百计偷排废水，毕竟对于他们来说偷排了废水就是获得利润。

5. 对新工艺、新原料、新产品的应用缺乏环境效应评估

随着科学技术的不断发展，纺织印染行业新工艺、新原料、新产品不断出现，这些产品无疑会推动传统纺织业发展与进步。但是否每一项工艺，每个新原材料都是绿色环保的，进入自然界后的环境效应如何，常常被人忽略，对这些工艺造成的严重水污染认识不足，进而缺乏水污染防范的有效措施。

11.4 纺织行业“十二五”节能减排展望

“十二五”期间的纺织工业发展，在节能环保方面也制定了具体目标：在节能减排全面达到国家强制执行标准要求、完成国家下达的节能减排任务的基础上，大规模实现清洁生产，基本建立低碳、绿色、循环经济体系。

为了实现这一目标，未来五年纺织行业必将加快研发和推广绿色环保技术，资源循环利用技术，高性能、高效率、节能减排的先进适用工艺、技术和装备，淘汰落后产能，全面调整产业结构，大幅提高技术水平，并重点推出一些节能减排工程。

按照“十二五”规划纲要提出的约束性指标要求，未来几年纺织行业的节能减排任务仍相当艰巨。纺织行业需要更加突出节能工作在促进产业升级中的抓手作用，通过淘汰落后，提高行业准入门槛，遏制行业产能过剩的现象，同时重点发展技术含量高、附加值高、资源消耗低的产品，提升全行业的竞争力。其重要举措如下：

1. 积极参加“万家企业节能行动”

作为“十一五”千家企业节能行动的扩展和延续，国家将在“十二五”期间开展“万家企业节能行动”，纺织行业也将参与其中。这一行动将结合行业特点，开展以下几个方面的工作：

- 加强能源计量、能耗统计等基础性工作，建立能源利用状况动态监测体系。
- 实施能源平衡测试，开展能源审计，编制节能规划。
- 广泛开展能效对标，向国际先进水平看齐。
- 大力推广适宜的节能技术、工艺，实施节能技术改造。
- 设立能源管理岗位，配备专职能源管理人员，加强能源管理体系建设等。

2. 纺织行业能效对标工程

“十二五”期间通过能效对标活动，纺织品单位能耗必须以每年1%速度下降，才能达到国家“十二五”节能减排总目标。

3. 在“十二五”期间逐步开展棉纺织企业能效对标活动、印染企业能效对标活动、化纤企业能效对标活动、服装企业能效对标活动

该计划具体实施方案如下：

- 2011年建立分行业能效对标体系，投入8000万建立4个能效对标平台，对行业内5%重点企业进行能效对标试验，对标企业产品单耗平均下降5%以上。节能300万吨以上标准煤。
- 2012年对行业内15%企业实施能效对标，节能600万吨以上标准煤。
- 2013年对行业内30%企业实施能效对标，节能800万吨以上标准煤。
- 2014年对行业内60%企业实施能效对标，节能900万吨以上标准煤。
- 2015年对行业内90%企业实施能效对标，节能1000万吨以上标准煤。

4. 绿色照明工程

“十二五”期间每年推广507万支节能灯，用T8－21代替T5－40。每年节约标准煤9.55万吨，每年需投入改造资金2.54亿元。“十二五”期间共向纺织行业推出2535万支节能灯。在“十二五”期间逐步开展纺织企业绿色照明工程（潜力分析见表13－6）。其实施计划如下：

- 2011年投入2.54亿元用于改造纺织行业照明系统，对全行业推广高效节能灯，实现节能9.55万吨以上标准煤。

- 2012 年对全行业推广高效节能灯，节能 19.1 万吨以上标准煤。
- 2013 年对全行业推广高效节能灯，节能 28.6 万吨以上标准煤。
- 2014 年对全行业推广高效节能灯，节能 57.2 万吨以上标准煤。
- 2015 年对全行业推广高效节能灯，节能 114.5 万吨以上标准煤。

5. 太阳能集热工程

“十二五”期间每年推广 60 万平米太阳能集热系统，每平米每年可产生 55 度热水 18.24 立方米。每年共产生热水 1094.5 万立方米，每年节约标准煤 24.2 万吨，每年需投入改造资金 18 亿元。“十二五”期间共推广 300 万平方米太阳能集热系统。其实施计划如下：

- 2011 年投入 18 亿元用于推广太阳能集热系统，对行业内企业推广 60 万平米太阳能集热系统，实现节能 24.2 万吨以上标准煤。
- 2012 年对行业内推广 60 万平米太阳能集热系统，节能 48.4 万吨以上标准煤。
- 2013 年对行业内推广 60 万平米太阳能集热系统，节能 72.6 万吨以上标准煤。
- 2014 年对行业内推广 60 万平米太阳能集热系统，节能 145.2 万吨以上标准煤。
- 2015 年对行业内推广 60 万平米太阳能集热系统，节能 290.4 万吨以上标准煤。

6. 能耗在线监控系统

能耗在线监测系统可通过总线，与企业中的电、水、气能耗计量以及在线称量设备直接连通，解决了计量表计分散的问题。能耗监测中心稳定地接收实时能耗数据，并实现对能耗测控设备的集中管理，数据库完成对能耗实时数据的实时采集、实时分析计算（包括折标、统计和报警计算）和支持能耗实时数据的高速查询，实现海量的能耗数据存储，基于实时数据应用平台构建的工业能耗在线监测系统，将实时数据库中的能耗实时信息、统计分析信息和报警信息以图形化方式直观地呈现给政府能耗工作人员，并以明晰准确的实时图表为政府领导提供决策支撑。

7. 纺织节能公共信息服务平台

以信息化促进节能减排，建设纺织行业节能公共服务平台，建立节能信息发布制度，利用节能公共服务平台，及时发布国内外能效信息、先进的节能新技术、新工艺、新设备及先进的管理经验，引导企业挖潜改造，提高能效，提高国际竞争力。

12 交通行业的节能减排

12.1 交通行业“十一五”运行综述

“十一五”时期是我国交通运输事业发展最快、成绩最突出的五年。不论是交通设施总量、规模，还是运输能力供给以及运输质量等方面都取得了巨大成就，安全监管和应急处置能力明显加强，对国民经济发展的支撑作用明显增强。“十一五”期间主要数据如下：

（1）交通运输基础设施建设进一步加快，网络更趋完善，结构更加优化。

- 全国公路网总里程五年新增63.9万公里，高速公路新增3.3万公里。
- “五纵七横”12条国道主干线提前13年全部建成，西部开发8条省际通道基本贯通。
- 全国农村公路五年新增里程52.7万公里，全国所有具备条件的乡镇都通了公路。
- 沿海港口五年建成深水泊位661个，新增通过能力30亿吨。新增及改善内河航道4181公里，初步形成了国家高等级航道网络。
- 民航定期航班机场达到176个，五年新增35个。
- 邮政基础设施补建空白乡镇邮政局所2500多个，改造西部及农村邮政局所7000多个。

（2）运输量快速增长，运输效率明显提高，服务能力显著提升。

- 各种运输方式的总里程，从2005年的558.64万公里增加到2010年的704.27万公里，增长26.1%，年均增长4.7%。
- 全社会主要运输方式完成客运量由2005年的184.70亿人增加到2010年的327.91亿人，增长77.5%，年均增长12.2%。
- 旅客周转量由17466.7亿人公里增加到27779.2亿人公里，增长59.0%，年均增长9.7%。
- 货运量由186.20亿吨增加到320.30亿吨，增长72.0%，年均增长11.5%。
- 货物周转量由80258.1亿吨公里增加到137329.0亿吨公里，增长71.1%，年均增长11.3%。
- 港口和集装箱吞吐量连续7年保持世界第一。
- 民航客货运量比五年前增加近一倍，运输总周转量跃居世界第二。
- 邮政业务总量完成1975亿元，快递业务量日均突破1000万件。

12.2 交通行业“十一五”节能减排总结

交通运输行业“十一五”节能减排的主要目标是：与2005年相比，到2010年和2020年，营运货车单位运输量能耗分别下降5%和16%，营运船舶单位运输量能耗分别下降10%和20%

（其中海运船舶分别下降11%和20%，内河船舶分别下降8%和20%），港口生产单位吞吐量综合能耗分别下降5%和10%。

"十一五"期间，交通行业节能减排目标完成情况如下：

• 全国营运车船单位运输周转量能耗、港口生产综合单耗分别比"十五"末下降了5%、7%①。

• 2009年交通运输、仓储和邮政业的能源消费总量为23691万吨标准煤，2010年估计为24493万吨标准煤（如图12－1）。能源消费总量占全国能源消费总量的比重一直稳定在7.7%左右。

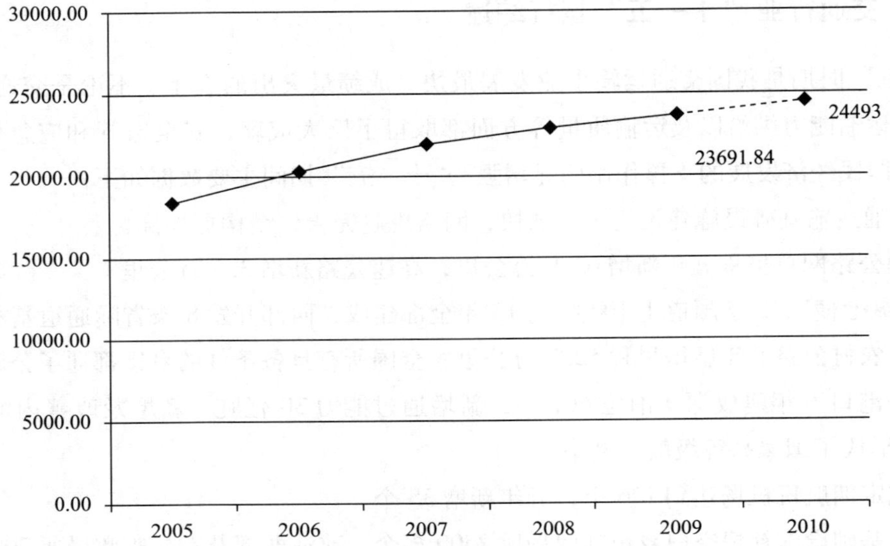

图12－1 "十一五"期间交通行业能源消费总量

Figure 12－1 Total Energy Consumption by Transportation Sector during China's 11th Five－year Plan Period

资料来源：交通运输部

"十一五"期间，交通行业节能减排的主要工作如下：

1. 优化交通运输结构，节能减排的集约效应日益显现

一是大力推进综合运输体系建设，运输结构进一步优化，特别是内河航运、城市公共交通等节能环保的比较优势日益显现。

二是着力优化交通基础设施网络布局、改善技术状况，2010年底全国二级及二级以上公路、有铺装路面和简易铺装路面公路、万吨级以上泊位和五级以上内河航道所占比重分别比"十五"末提高了1.3、20.2、2.3和1.1个百分点，为交通运输节能减排提供了良好的物质基础。

三是努力改善车船运力和工程机械装备结构，交通运输装备的大型化、专业化和标准化趋势明显，2010年底营运货车平均吨位、营运船舶平均净载重量分别比"十五"末提高了36.1%、105.9%，整体能效水平显著提高。

---

① 高宏峰副部长在"车、船、路、港"千家企业低碳交通运输专项行动总结会暨低碳交通运输体系城市试点启动会上的讲话。

四是加快调整交通运输企业组织结构，企业的集约化程度和用能管理水平有效提升。交通运输结构的不断优化，使节能减排的网络效应、规模效应和集约效应得到充分发挥，大大提升了交通运输系统节能减排的整体水平。

2. 加强科技创新与成果推广，节能减排技术基础明显增强

一是组织开展了资源节约型环境友好型交通发展模式、低碳交通运输体系建设等一批重大战略规划与政策研究，为行业科学决策提供了有力支撑。

二是推进信息化建设，大力推广应用不停车收费（ETC）、智能交通系统（ITS）、物流公共信息平台、公众出行信息服务系统、无线射频识别技术（RFID）、全球导航卫星系统（GNSS）等现代信息技术，交通运输生产效率和服务水平有效提升。

三是大力推进采用先进运输装备技术，开展了货运汽车及汽车列车推荐车型、客运车辆等级评定和内河船型标准化工作，配合开展了“十城千辆”节能与新能源汽车示范推广，积极开展港口轮胎式集装箱门式起重机（RTG）“油改电”，探索应用靠港船舶使用岸电技术等。

四是发布了两批共40项重点推广在用车船节能产品（技术）目录，大力推广应用燃油添加剂、节油器等先进适用技术与产品。

五是在交通基础设施建设养护中积极采用新结构、新工艺和新材料，推广应用隧道节能照明、路面材料再生、温拌沥青等新技术，探索应用太阳能、风能等可再生能源。通过这些科技创新与推广应用，交通运输节能减排的技术基础和保障能力不断增强。

3. 注重提升节能减排监管能力，节能减排管理水平明显提高

一是建立健全节能减排管理机构。交通运输部成立了新的节能减排工作领导小组，并下设节能减排与应对气候变化工作办公室，各地交通运输主管部门、大型交通运输企业也普遍设立了相应的节能减排工作管理机构或岗位，为做好节能减排工作提供了有力的组织保障。

二是组织开展了交通运输行业能源统计体系研究，公路运输、水路运输和港口生产能源统计指标初步纳入国家统计；着手研究构建行业节能减排监测考核体系，并在山东等地开展监测考核试点。

三是积极推动实施了燃油消费税改革，实施了营运车辆燃料消耗量准入、实载率低于70%的客运线路不得新增运力等政策，强化交通运输市场监管，改进行业能源利用水平。

四是积极引导企业规模化、集约化发展，组织开展甩挂运输、江海直达、多式联运等试点工作，不断提升交通运输系统运行效率。

12.3　交通行业“十二五”节能减排形势

尽管“十一五”期间交通运输行业节能减排工作取得了一定成绩，但与国外先进水平相比，与全面落实科学发展观、发展绿色低碳经济的更高要求相比，还存在一定的差距与不足，主要体现在：

1. 交通运输结构性矛盾尚未根本解决

- 内河航运比较优势尚未充分发挥，综合交通枢纽建设滞后，综合运输组合效率尚未充分显现。
- 城市公共交通服务能力和质量不高，吸引力不强。
- 断头路、局部瓶颈等已成为影响基础设施网络效应的重要制约因素，部分沿海港口进出

港航道能力不足。

- 运输装备结构不尽合理，大型化、专业化车船比重不高，老旧车船比重偏高。
- 道路运输规模化、集约化程度还比较低；替代能源、可再生能源比重亟待提高。

2. 交通运输节能减排技术创新与服务体系仍需健全

- 节能减排科技研发投入不足，节能减排技术、产品推广应用进展较为缓慢。
- 现代信息技术应用推广还比较滞后。
- 交通运输节能减排技术服务体系尚未建立，节能减排技术产品和服务市场还有待进一步规范。

3. 交通运输节能减排监管能力还有待提升

- 节能减排意识有待进一步增强，专职管理机构和人员缺乏，体制机制性障碍尚未根本消除。
- 引导性资金投入明显不足，节能减排长效机制尚未形成。
- 政策法规和标准规范体系还不完善。
- 节能减排统计计量、检测监测与考核评价等基础性工作薄弱。

12.4 交通行业“十二五”节能减排展望

“十二五”期间，交通行业的节能减排相关发展目标如下（如表12－1）：

（1）能源强度指标。与2005年相比，营运车辆单位运输周转量能耗下降10%，其中营运客车、营运货车分别下降6%和12%；营运船舶单位运输周转量能耗下降15%，其中海洋和内河船舶分别下降16%和14%；港口生产单位吞吐量综合能耗下降8%。

（2）CO_2排放强度指标。与2005年相比，营运车辆单位运输周转量CO_2排放下降11%，其中营运客车、营运货车分别下降7%和13%；营运船舶单位运输周转量CO_2排放下降16%，其中海洋和内河船舶分别下降17%和15%；港口生产单位吞吐量CO_2排放下降10%。

（3）环境保护力度进一步加强。重大交通工程生态修复取得明显进展，主要污染物排放强度进一步降低，力争行业总悬浮颗粒物（TSP）和化学需氧量（COD）等主要污染物排放强度比“十一五”末降低20%。

（4）资源集约利用程度进一步提高。国省道单位行驶量用地面积下降5%，沿海港口单位长度码头岸线通过能力提高5%。

（5）港口、公路服务区等生产、生活污水的循环利用水平，路面废弃材料等资源的再生利用水平显著提高。

表12－1 “十二五”期间交通行业节能减排目标

Table 12－1 Targets of Energy Conservation and Emission Reduction Set for Transportation Sector during China's 12th Five－year Plan Period

营运车辆单位运输周转量能耗和二氧化碳排放下降率（%，基年：2005）	10、11
营运船舶单位运输周转量能耗和二氧化碳排放下降率（%，基年：2005）	15、16
营运客、货车单位运输周转量能耗下降率（%，基年：2005）	6、12
海洋和内河货运船舶单位运输周转量能耗下降率（%，基年：2005）	16、14

续表

港口生产单位吞吐量综合能耗下降率（%，基年：2005）	8
民航运输吨公里能耗和二氧化碳排放下降率（%，基年：2010）	>3
国省道单位行驶量用地面积下降率（%，基年：2010）	5
沿海港口单位长度码头岸线通过能力提高率（%，基年：2010）	5
总悬浮颗粒物（TSP）和化学需氧量（COD）等主要污染物排放强度（吨/亿吨公里）下降率（%，基年：2010）	20

资料来源：交通运输部

为了实现上述目标，交通行业节能减排的主要举措如下：

1. 结构性节能减排

充分发挥各种运输方式比较优势，优化交通运输资源配置，发挥综合运输的整体优势和组合效率，降低能源消耗强度；加快发展城市公共交通、水运等低能耗运输方式，倡导低碳型交通消费模式和出行方式。

大力优化公路网结构，提高路网通行能力和效率，提升公路技术等级和路面等级，调整公路运输运力结构；推进港口结构调整，发展大型化、专业化港口；提升航道技术等级，加快形成以高等级航道为主体的内河航道网。推进交通能源消费结构优化，鼓励替代能源在营运车船中的应用。

2. 技术性节能减排

积极采用混合动力汽车、替代燃料车等节能环保型营运车辆以及双尾船、新能源动力船等节能环保型营运船舶，推广应用自重轻、载重量大的运输装备。对营运车船设置能耗和排放限制标准，提高准入门槛，淘汰低标准及老旧车船。鼓励使用天然气动力和电动车等节能环保型城市公交车，开展混合动力、电能出租汽车试点工作。与2005年相比，力争“十二五”末营运客车、货车单位运输周转量能耗分别下降6%和12%，海洋和内河营运船舶单位运输周转量能耗分别下降16%和14%。

大力研发推广隧道智能通风照明控制技术，推行隧道“绿色照明工程”。加快发展高能效电力驱动港口装卸设备，研发推广电能回馈、储能回用等新工艺新技术，基本完成轮胎式集装箱门式起重机“油改电”技术改造，推进船舶靠泊使用岸电技术改造。逐步推广港口太阳能、地源及海水源能、潮汐能、风能等新能源利用技术。与2005年相比，力争“十二五”末港口生产单位吞吐量综合能耗下降8%。

推广高速公路不停车收费（ETC）系统，推广内河船舶免停靠报港信息服务系统，完善公众出行信息服务系统，促进客货运输市场的电子化、网络化，实现信息共享，提高运输效率，降低能源消耗，实现节能减排。

3. 管理性节能减排

发展先进公路运输组织方式，加强货运组织和运力调配，利用回程运力，降低车辆空驶率。鼓励厢式运输、集装箱运输等专业化运输方式，发挥甩挂运输效率高和减排效果好的优势。合理安排客运线路，完善道路客运信息监测、分析和发布制度，提高客车实载率。推动建立绿色汽车维修体制机制，建立较完善的驾培行业节能减排体系。加强水路运输组织管理，鼓励航运企业联合经营，发挥规模优势，提高运输组织化程度。发展大宗散货专业化运输、多式联运等现代运输组织方式，全面提升船舶营运组织效率和节能减排水平。优化港口机械作业和港内运输组织管理。鼓励道路运输企业及港口企业完善能源管理体系，推广能源合同管理。

发挥港口、公路枢纽站场在物流发展中的节点作用，引导运输企业向依托港口、公路货运枢纽的物流园区集聚，提高不同运输方式间货物换装效率，促进节能减排。

加强民航业节能减排，以航空公司、机场、空管为主体，政府主导与市场调节相结合，加强监督检查和综合协调，控制航空排放。

4. 示范推广工程

营运车船燃料消耗量准入与退出工程：实施营运车辆燃料消耗量限值标准，加快淘汰高能耗、高污染的运输车辆和船舶。

节能与新能源车辆示范推广工程：促进混合动力、纯电动、天然气等新能源和清洁燃料车辆在公共汽车和出租车领域的示范推广应用，在城际客货运输和城市物流配送车辆中试点推广新能源和天然气车辆。

绿色驾驶与维修工程：大力推广绿色节能驾驶技术，组织实施绿色维修工程。

智能交通节能减排工程：推广电子不停车收费技术、内河船舶免停靠报港信息服务系统，建设物流公共信息平台、公众出行信息服务系统。

公路建设和运营节能减排技术推广工程：推广应用温拌沥青铺路技术、交通建设材料循环利用技术，实施公路隧道通风照明智能控制、高速公路服务区、收费站等节能减排技术改造，大力推进太阳能、风能等可再生能源利用，建设低碳服务区等一批试点工程。

绿色港航建设工程：加快港口集装箱码头轮胎式集装箱起重机“油改电”和船舶靠港使用岸电技术改造，积极推广太阳能、地热能等再生能源利用。

合同能源管理推广工程：逐步使合同能源管理成为交通运输行业节能技术服务市场的重要机制。

船舶能效管理体系与数据库建设工程：参照国际上在船舶能效改进方面的先进做法和经验，积极推动航运企业将船舶能效纳入体系管理。

13　建筑行业的节能减排

13.1　建筑行业“十一五”运行综述

“十一五”时期，我国国民经济保持了平稳快速发展，固定资产投资规模不断扩大，为建筑业的发展提供了良好的市场环境。

1. 产业规模创历史新高

• 2010 年，全国具有资质等级的总承包和专业承包建筑业企业完成建筑业总产值 95206 亿元，全社会建筑业实现增加值 26451 亿元。

• 全国工程勘察设计企业营业收入 9547 亿元。

• 全国工程监理企业营业收入 1196 亿元。

•“十一五”期间，建筑业增加值年均增长 20.6%，全国工程勘察设计企业营业收入年均增长 26.5%，全国工程监理企业营业收入年均增长 33.7%，均超过“十一五”规划的发展目标。

2. 在国民经济中的支柱地位不断加强

•“十一五”期间，建筑业增加值占国内生产总值的比重保持在 6% 左右，2010 年达到 6.6%。

• 建筑业全社会从业人员达到 4000 万人以上，成为大量吸纳农村富余劳动力就业、拉动国民经济发展的重要产业，在国民经济中的支柱地位不断加强。

3. 国际市场开拓取得新进展

•“十一五”期间，建筑企业积极开拓国际市场，对外承包工程营业额年均增长 30% 以上。

• 2010 年对外承包工程完成营业额 922 亿美元，新签合同额 1344 亿美元。

4. 技术进步和创新成效明显

•“十一五”以来，许多大型工程勘察设计企业和建筑施工企业加大科技投入，建立企业技术开发中心和管理体系，重视工程技术标准规范的研究，突出核心技术攻关，设计、建造能力显著提高。

• 超高层大跨度房屋建筑、大型工业设施设计建造与安装、大跨径长距离桥梁建造、高速铁路、大体积混凝土筑坝、钢结构施工、特高压输电等领域技术达到国际领先或先进水平。

13.2　建筑行业“十一五”节能减排总结

建筑行业“十一五”节能减排主要任务完成情况如下：

1. 新建建筑执行节能强制性标准成效显著

到 2010 年底，全国城镇新建建筑设计阶段执行节能强制性标准的比例为 99.5%，施工阶段执行节能强制性标准的比例为95.4%，分别比2005年提高了42个百分点和71个百分点，完成了国务院提出的“新建建筑施工阶段执行节能强制性标准的比例达到95%以上”的工作目标。

2010年新增节能建筑面积12.2亿平方米，可形成1150万吨标准煤的节能能力。

“十一五”期间累计建成节能建筑面积48.57亿平方米，共形成4600万吨标准煤的节能能力。

全国城镇节能建筑占既有建筑面积的比例为23.1%，比例超过30%的省市有北京、天津、上海、重庆、河北、吉林、辽宁、江苏、宁夏、青海、新疆等省（区、市）。

2. 北方采暖地区既有居住建筑供热计量及节能改造任务超额完成

截至2010年底，北方采暖地区15省市共完成改造面积1.82亿平方米，其中2010年完成改造面积8623万平方米，超额完成了国务院确定的1.5亿平方米改造任务。河北、吉林、山东、内蒙古、新疆、新疆兵团、北京等省（区、市）超额完成任务10%以上。

据测算，完成节能改造的项目可形成年节约200万吨标准煤的能力，减排二氧化碳520万吨，减排二氧化硫40万吨。

改造后同步实行按用热量计量收费，平均节省采暖费用10%以上，室内热舒适度明显提高，并有效解决老旧房屋渗水、噪音等问题。

部分地区将节能改造与保障性住房建设、旧城区综合整治等民生工程统筹进行，综合效益突出。

3. 可再生能源建筑应用呈现快速发展的良好态势

截至2010年底，财政部会同住房城乡建设部共实施了371个可再生能源建筑应用示范项目、210个太阳能光电建筑应用示范项目、47个可再生能源建筑应用城市、98个示范县。

山东、江苏、海南等省已经开始强制推广太阳能热水系统。全国太阳能光热应用面积14.8亿平方米，浅层地能应用面积2.27亿平方米，分别比2009年增长25.5%、63.3%，光电建筑应用已建成及正在建设的装机容量达850.6兆瓦，实现突破性增长，形成年替代传统能源2000万吨标准煤能力。

4. 绿色建筑与绿色生态城区建设稳步推进

各地把推广绿色建筑、推进绿色生态城区建设作为促进城乡建设模式转变的重要抓手，加大绿色建筑示范工程和绿色建筑评价标识推进力度，截至2010年底，全国有112个项目获得了绿色建筑评价标识，建筑面积超过1300万平方米，上海、苏州、深圳、杭州、北京、天津等市获得标识项目较多。

全国实施了217个绿色建筑示范工程，建筑面积超过4000万平方米。天津市滨海新区、深圳市光明新区、河北省唐山市曹妃甸新区、江苏省苏州市工业园区、湖南长株谭和湖北武汉资源节约环境友好配套改革试验区等正在进行绿色生态城区建设实践，对引导我国城市建设走绿色生态可持续方向发展道路，具有重要意义。

5. 国家机关办公建筑和大型公共建筑节能监管体系建设继续深入

能耗统计、能源审计、能效公示工作全面开展，截至2010年底，全国共完成国家机关办公建筑和大型公共建筑能耗统计33000栋，完成能源审计4850栋，公示了近6000栋建筑的能耗状况，已对1500余栋建筑的能耗进行了动态监测。在北京、天津、深圳、江苏、重庆、内蒙古、上海、浙江、贵州等九省市开展能耗动态监测平台建设试点工作，共启动了72所节约型校园建设试点。通过节能监管体系建设，全面掌握了公共建筑的能耗水平及耗能特点，带动了节能运行与改造的积极性，有力地促进了节能潜力向现实节能的转化。

"十一五"期间，主要的做法包括：

1. 加强领导，健全机构，增强建筑节能管理能力

一是加强建筑节能组织领导，各省（区、市）住房城乡建设主管部门均成立了主要领导或分管领导任组长的建筑节能领导小组，北京、天津、上海、山西、内蒙古、黑龙江、吉林、江苏、浙江、广东、广西等省（区、市）成立了政府分管领导任组长，相关部门参加的建筑节能工作领导小组，形成了各部门联动、齐抓共管的局面。

二是建筑节能管理机构能力进一步加强。部分省市住房城乡建设部门借机构改革契机，设置建筑节能专门处室，加强了职能，充实了管理力量。山西省省、市两级都建立了建筑节能监管机构，专职管理人员146人。全国有20个省（区，市）墙体材料革新工作和建筑节能工作统一由住房城乡建设主管部门负责，以墙改为抓手推进建筑节能，效果明显。

2. 完善法规，强化激励，建立建筑节能规范与引导机制

一是在法律法规方面。各地积极制定本地区的节能行政法规，河北、陕西、山西、湖北、湖南、上海、重庆、青岛、深圳、武汉、乌鲁木齐等地出台了建筑节能条例。有15个省（区、市）出台了资源节约及墙体材料革新等相关法规，24个省（区、市）出台了相关政府令，建筑节能法律体系初步建立。

二是在经济政策方面。"十一五"期间，中央财政共计安排资金152亿元，用于支持北方采暖地区既有居住建筑供热计量及节能改造、可再生能源建筑应用、国家机关办公建筑和大型公共建筑节能监管体系建设等方面。北京、上海、内蒙古、山西、江苏、深圳等地对建筑节能的财政支持力度较大，安排了专项资金。据不完全统计，"十一五"期间，省级财政共安排69亿元建筑节能专项资金，地级及以上城市市级财政安排65亿元建筑节能专项资金工作，为建筑节能提供了良好的政策环境和财力保障。

3. 健全标准，科技支撑，提升建筑节能技术水平

一是在技术标准方面，建筑节能标准体系不断完善，基本涵盖了设计、施工、验收、运行管理等各个环节。各地结合地区实际，对国家标准进行了细化，部分地区执行了更高水平的标准。及时把先进成熟的技术产品编入工程技术标准和标准图，通过标准引导技术进步。上海、天津、重庆、广西、深圳等地制定了绿色生态示范城区及绿色建筑等具有前瞻性的评价标准，发挥了标准的引导和规范作用。

二是在科研开发方面，"十一五"期间，国家科技支撑计划把建筑节能、绿色建筑、可再生能源建筑应用等作为重大项目，对一批共性关键技术进行研究攻关，取得了明显成效。各地围绕建筑节能工作发展需要，结合地区实际，积极筹措资金，安排科研项目，为建筑节能深入发展做好科技储备。

三是在示范推广方面，各地以建筑节能示范工程为载体，一方面积极申报国家级的示范项目，另一方面结合本地实际，不断丰富示范类型，提高示范水平。通过示范既推广了建筑节能技术和产品，又引导地区建筑节能的发展方向。

4. 加强监管，强化考核，落实建筑节能各项目标任务

一是在质量监管方面，各地充分利用现有法律法规确定的许可和制度，建立建筑节能专项

设计审查、节能工程施工质量监督、建筑节能专项验收、建筑能效测评标识、建筑节能信息公示等制度，实现了从设计、施工图审查、施工、竣工验收备案到销售和使用的全过程监管机制，效果明显。各省市都组织开展了建筑节能专项检查，对违反建筑节能有关法律法规及强制性标准的行为进行了处罚。

二是在目标考核方面，各省市强化建筑节能目标责任考核，对新建建筑执行建筑节能标准、既有建筑节能改造、可再生能源建筑推广等工作，采取签订目标协议等方式，逐级进行分解，按年度进行考核，确保建筑节能各项目标得到落实。

13.3 建筑行业“十二五”节能减排形势

建筑行业“十一五”节能减排各项工作仍存在一些问题，使得“十二五”期间的节能减排会遭遇各类阻碍。

1. 部分地方政府对建筑节能工作的认识不到位

一是对建筑节能的考核没有纳入政府层面，部分省（区、市）对建筑节能的考核评价仍局限在住房城乡建设系统内部，没有纳入本地区单位GDP能耗下降目标考核体系，使相关部门难以形成合力，相应的政策、资金难以落实。

二是对建筑节能能力建设重视不够，部分省级住房城乡建设主管部门建筑节能管理人员只有1—2人，没有专门的管理和执行机构，各项政策制度的落实大打折扣。

2. 建筑节能法规与经济支持政策仍需完善

一是落实《节约能源法》、《民用建筑节能管理条例》各项法律制度所需的部门规章、地方行政法规的制定工作仍然滞后。

二是各地对建筑节能的经济支持力度远远不够，尤其是中央财政大力投入的北方采暖地区既有居住建筑供热计量及节能改造、可再生能源建筑应用、公共建筑节能监管体系建设等方面，大部分地区没有落实配套资金，影响了中央财政支持政策的落实效果。

3. 新建建筑执行节能标准水平仍不平衡

执行建筑节能标准，施工阶段比设计阶段差，中小城市比大城市差，经济欠发达地区比经济发达地区差。

建筑节能工程施工过程中，外墙、门窗等保温工程施工工艺不过关，管理不规范，存在质量与火险隐患。

各地尤其是地级以下城市普遍缺乏建筑节能材料、产品、部品的节能性能检测能力，造成政府监管缺位。

4. 北方地区既有建筑节能改造工作任重道远

一是改造任务重。2000年以前我国建成的建筑大多为非节能建筑，占城镇建筑面积的80%，民用建筑外墙平均保温水平仅为欧洲同纬度发达国家的1/3，建筑能耗高出2—3倍，北方地区有超过20亿平米的既有建筑急需节能改造。

二是改造资金筹措压力大。节能改造成本在200元/平方米以上，再考虑热源改造，资金投入需求更大。但北方多数地区经济欠发达，地方政府财力投入有限，市场融资能力较弱。

三是供热计量改革滞后。热计量收费是运用市场机制促进行为节能最有效手段，但这项改革一直进展缓慢，目前北方地区130多个地级市，出台供热计量收费办法地级市仅有40余个，制约了节能效果与企业居民投资改造的积极性。

5. 可再生能源建筑应用推广任务依然繁重

我国在建筑领域推广应用可再生能源总体上仍处于起步阶段，据测算，目前可再生能源消费占建筑用能比重在2%左右，这与我国丰富的资源禀赋、快速增长的建筑用能需求、调整用能结构的迫切要求相比都有很大的差距。

可再生能源建筑应用相关技术标准体系还不完善，产业支撑能力不足，核心技术仍没有掌握，系统集成、工程咨询、运行管理等配套产业能力不强。

6. 农村建筑节能工作尚未正式启动

目前，我国广大农村地区的建筑节能工作尚未开展。随着农村生活水平的不断改善，使用商品能源和用能水平将不断提高，需采取措施，引导其科学发展。

13.4 建筑行业“十二五”节能减排展望

建筑行业“十二五”发展的基本原则之一是坚持节能减排与科技创新相结合：发展绿色建筑，加强工程建设全过程的节能减排，实现低耗、环保、高效生产；大力推进建筑业技术创新、管理创新，推进绿色施工，发展现代工业化生产方式，使节能减排成为建筑业发展新的增长点。

此原则之下的建筑节能目标如下：

（1）绿色建筑、绿色施工评价体系基本确立。

（2）建筑产品施工过程的单位增加值能耗下降10%，C60以上的混凝土用量达到总用量10%，HRB400以上钢筋用量达到总用量45%，钢结构工程比例增加。

（3）新建工程的工程设计符合国家建筑节能标准要达到100%，新建工程的建筑施工符合国家建筑节能标准要求。

（4）全行业对资源节约型社会的贡献率明显提高。

为实现上述目标，“十二五”期间建筑行业的相关举措如下：

1. 加强建筑节能体制机制建设

一是继续完善建筑节能法规体系。落实《节约能源法》、《民用建筑节能条例》确定的基本法律制度，研究制定配套的部门规章和政策措施。

二是制定建筑节能“十二五”专项规划，明确建筑节能工作目标、思路、重点工作任务及保障措施，指导各地编制本地区建筑节能“十二五”专项规划。

三是继续完善建筑节能标准体系，指导有条件地区制定并实施绿色建筑强制性标准。

四是研究完善经济制度，在发展绿色建筑、既有建筑节能改造、可再生能源建筑应用等方面，制定出台财税政策。

五是继续推动建筑节能科技进步，以国家科技支撑计划项目为依托，努力实现建筑节能关键技术的突破，并通过发布技术、产品推广限制禁止目录等方式，加快科技成果转化。

2. 严格履行节能减排责任

一是政府部门履行建筑执行节能标准的监管责任，着力抓好设计、施工阶段执行节能标准的监管和稽查。

二是各类企业履行节能减排社会责任，严格执行国家、地方的各项节能减排标准，确保节能减排标准落实到位。

3. 继续抓好新建建筑节能

一是继续强化新建建筑执行节能标准的监管力度，着力抓好施工阶段等薄弱环节以及中小城市等薄弱地区执行标准的监管力度，做好北方采暖地区以及夏热冬冷地区新颁布建筑节能标准的贯彻实施工作。

二是全面推行民用建筑能效测评标识、绿色建筑评价标识、民用建筑节能信息公示等制度。进一步加强建筑节能材料、产品、设备在生产、流通和使用环节的质量监管，严格工程准入。

4. 大力推动绿色建筑发展

一是制定绿色建筑发展专项规划。

二是继续完善绿色建筑标准体系，加大绿色建筑评价标识实施力度。

三是推动有条件地区开展强制性推广绿色建筑试点。

四是加快绿色建筑相关共性关键技术研究开发及推广力度。

五是研究支持绿色建筑发展的财税政策。

六是启动绿色建筑区域推广示范。

5. 加大北方采暖地区既有居住建筑供热计量及节能改造力度

一是贯彻落实《关于进一步深入开展北方采暖地区既有居住建筑供热计量及节能改造工作的通知》（财建［2011］12 号），住房与城乡建设部与各地签订“十二五”既有居住建筑供热计量及节能改造协议，明确改造目标。

二是在重点市县推动实施“节能暖房工程”。总结天津、吉林、内蒙古、唐山等地的改造经验，指导各地因地制宜确定改造模式，多渠道筹措改造资金，全面推进供热计量改革，力争 2015 年北方地区完成具备改造价值的老旧住宅的供热计量及节能改造面积的 35% 以上。

6. 加强国家机关办公建筑和大型公共建筑节能管理

一是深入开展针对公共建筑，特别是国家机关办公建筑及大型公共建筑的能耗统计、能源审计及能效公示工作，进一步扩大能耗动态监测平台试点范围，力争将所有重点建筑纳入监测范围，基本建成部、省、市三级构架的能耗传输及分析平台。

二是在公共建筑节能监管体系已经建立、具备节能量核查手段的城市，开展公共建筑节能改造示范。指导各地制定本地区公共建筑能耗限额标准，引导和约束用能单位的用能行为。继续抓好“节约型高等学校”建设工作。

7. 抓好可再生能源建筑一体化成规模应用

一是推进可再生能源建筑应用示范深入发展，继续组织可再生能源建筑应用城市示范和农村地区县级示范，实施“太阳能屋顶计划”。

二是适时开展新建建筑强制性应用可再生能源试点，在可再生能源资源条件好、技术成熟度和社会接受程度比较高、法规政策比较配套的地区，开展强制性应用可再生能源试点。

三是进一步完善修订现有标准，同时加快研究制定不同类型可再生能源建筑应用技术在设计、施工、能效检测等各环节的国家和地方标准规范。

四是加大对太阳能采暖制冷、太阳能与浅层地能耦合利用、城镇生活垃圾、污泥沼气利用技术、工业余热、污水热泵技术、深层地热能梯级利用等新技术的推广力度，拓展应用领域。

五是支持可再生能源建筑应用产品、设备性能检测机构、建筑应用效果检测评估机构等公共服务平台建设。

8. 促进建筑节能相关产业发展

一是做好新型建筑节能材料的推广应用。编制新型建筑节能技术、材料、产品推广目录，引导产业发展。支持建筑垃圾等建筑废弃物的再生利用。推广高强钢和高性能混凝土等高性能、低材耗建筑材料的利用。

二是大力培育建筑节能服务产业，在北方地区既有居住建筑供热计量及节能改造、公共建筑节能改造、可再生能源建筑应用等工作中，鼓励采用合同能源管理、能效交易、建设运行一体化、区域能源系统特许经营等市场化推进方式。大力培育与建筑节能直接相关的专业设计施工、工程咨询、系统集成、节能评估等配套产业，切实增强产业支撑能力。

9. 建立健全建筑节能统计、监测及考核评价体系

一是切实履行在建筑节能领域政府的公共管理职责，促成各地人民政府把建筑节能纳入本地单位 GDP 能耗下降的总体目标，明确任务，建立目标责任制，完善配套措施，落实经济激励政策，进行考核评价。

二是各级住房城乡建设主管部门要按照相关的法律制度和强制性节能标准，组织开展建筑节能检查，对节能目标落实情况、贯彻管理制度和执行节能强制性标准等，进行考核评价，落实节能目标责任制和问责制。

10. 采用先进的节能减排技术和材料

一是建立有利于建筑业低碳发展的激励机制，鼓励先进成熟的节能减排技术、工艺、工法、产品向工程建设标准、应用转化，降低碳排放量大的建材产品使用，逐步提高高强度、高性能建材使用比例。

二是推动建筑垃圾有效处理和再利用，控制建筑过程噪声、水污染，降低建筑物建造过程对环境的不良影响。

三是开展绿色施工示范工程等节能减排技术集成项目试点，全面建立房屋建筑的绿色标识制度。

13.5 典型项目

根据《全国绿色建筑创新奖管理办法》、《全国绿色建筑创新奖实施细则》和《全国绿色建筑创新奖评审标准》，住房和城乡建设部组织完成了“2011 年全国绿色建筑创新奖”申报项目的评审和公示。经审定，“深圳市建科大厦”等 16 个项目获得“2011 年全国绿色建筑创新奖”。以下对“深圳市建科大厦”的项目概况进行介绍①。

深圳建科大楼是深圳市建筑科学研究院（以下简称 IBR）科研办公楼，该项目以探索低成本和软技术为核心的绿色建筑实现模式为宗旨，以实现建筑全寿命周期内最大限度节约和高效利用资源、保护环境、减少污染为目标，建科大楼融入了深圳市建科院多年实践中取得的研究成果、专利技术，承载了实践绿色生活、绿色办公方式的梦想，期望成为一座荟萃地域特色、绿色科技和建筑艺术的绿色科研办公建筑。

建科大楼位于深圳市福田区北部梅林片区，总建筑面积 1.8 万平方米，地上 12 层，地下 2 层，建筑功能包括实验、研发、办公、学术交流、地下停车、休闲及生活辅助设施等。建筑设

① 以下内容选自《实践平民化的绿色建筑》一文，原作者袁小宜等。

计采用功能立体叠加的方式，将各功能块根据性质、空间需求和流线组织，分别安排在不同的竖向空间体块中，附以针对不同需求的建筑外围护构造，从而形成由内而外自然生成的独特建筑形态。

建科大楼已实现了最初的建设目标，其以4300元/平方米的工程单方造价，达到了国家绿色建筑评价标准三星级和美国LEED金级的要求，取得了较为突出的社会效益。经初步测算分析，1.8万平方米规模的整座大楼每年可减少运行费用约150万元，其中相对常规建筑节约电费145万元，节约水费5.4万元，节约标准煤610吨，每年可减排二氧化碳21600吨。

建筑设计过程是个共享参与权的过程，设计的全过程体现了权利和资源的共享，关系人共同参与设计；同时，建筑本身是一个共享平台，设计结果要实现建筑本身为人与人、人与自然、物质与精神共享提供一个有效、经济平台的目的。

从方案创作开始，整个过程都定量验证，并大量应用新的设计技术，利用计算机对能耗、通风、采光、噪声、太阳能等进行模拟分析。楼体的竖向布局与功能相关联，材料、通风、自然采光、外墙构造、立面及开窗形式等各个方面的确立也都经过优化组合。在设计中，为了确定一种技术方式，往往会研究十几种技术路线的贡献率后才选定最佳平衡值。

从设计到建设，建科大楼采用了一系列适宜技术共有40多项（其中被动、低成本和管理技术占68%左右），每一项技术都是建科大楼这一整合运用平台上“血肉相连”的一部分。它们并非机械地对应于绿色建筑的某单项指标，而是在机理上响应绿色建筑的总体诉求，是对节能、节地、节水、节材诸环节进行整体考虑并能够满足人们舒适健康需求的综合性措施。

第3篇　区域篇

“十一五”期间，我国的节能减排是以各级政府推进为主导模式。“十二五”时期，预计这种模式会以向市场力量推进的模式来演化。但是，其政府为主导的基本工作格局不会改变。因此，从行政区域的角度来分析中国的节能减排具有现实意义。

为了全面对比分析各地区“十一五”期间节能减排总体效果，本报告构建了涵盖减排效果、努力程度和潜力实现三个方面的区域节能减排综合评价指数，并根据此指数将各区域省份划分为四类。

根据我们的研究，第一类地区全部分布在东部地区，包含的省市有北京市、上海市、江苏省、广东省、浙江省，反映了东部地区在各区域中节能减排效果整体最好。第二类地区主要包括东部的天津市、河北省、福建省、山东省和海南省，中部的安徽省、江西省、河南省、湖北省和湖南省，西部的重庆市、四川省、陕西省和云南省，东北的黑龙江省、辽宁省和吉林省。第三类地区包括中部的山西省，西部的广西、甘肃和内蒙古。第四类地区全部集中在西部地区，包括贵州、青海省和宁夏。

基于上述基本研究，本篇我们重点分析东部、中部、西部、东北部地区“十一五”期间在节能减排方面的目标实现、经验教训，继而对各区域及省份“十二五”节能减排的战略目标及发展趋向进行分析。

14 我国区域节能减排“十一五”发展评述及“十二五”展望

14.1 “十一五”期间区域节能减排综述

14.1.1 区域经济发展概述

“十一五”时期，我国西部大开发战略、振兴东北地区等老工业基地、促进中部地区崛起、鼓励东部地区率先发展的发展区域总体战略进一步向纵深推进，特别是一系列国家综合配套改革试验区①、战略经济区②和跨区域战略规划③的相继实施，使得国家区域发展总体战略得到了进一步深化和细化，以西部大开发、东北振兴、中部崛起、东部加快发展为主体的区域经济发展格局日益清晰。“十一五”末，针对区域经济发展不平衡、区域经济发展过程中环境承载压力日益增大以及区域经济发展行政条块分割严重等突出问题，我国又提出了《全国主体功能区规划》，该规划将与区域发展总体战略相辅相成，共同构成了我国区域发展在国土空间开发完整的战略格局，确保我国区域经济协调可持续发展。

1. 区域发展的协调性明显增强

“十一五”时期，中西部及东北地区加快发展，经济总量和投资占全国的比重持续上升，产业结构进一步优化，城市化水平进一步提高，呈现出沿海与内陆协调发展的良好势头。东中西及东北地区经济实力显著增强，经济结构进一步优化，城镇化水平持续提高，特别是广大中西部地区经济增速普遍提高，区域发展增速差距趋于缩小，区域间、城乡间发展的协调性进一步增强，区域发展呈现了良性互动协调发展推进的新局面。

“十一五”时期，东部地区 GDP 平均增长速度比“十五”时期加快 0.1 个百分点；中部地区加快 2.1 个百分点；西部地区加快 2.4 个百分点；东北地区加快 2.6 个百分点。2010 年，东部地区生产总值占全国的比重为 53.0%，比 2005 年降低了 2.5 个百分点；中部地区生产总值占全国的比重为 19.7%，比 2005 年提高了 0.9 个百分点；西部地区生产总值占全国的比重为 18.7%，比 2005 年提高了 1.6 个百分点；东北地区生产总值占全国的比重为 8.6%，与 2005 年

---

① 国家综合配套改革试验区：上海浦东新区综合配套改革试点、天津滨海新区综合配套改革试验区、重庆和成都统筹城乡综改区、汉城市圈和长株潭城市群开展综合配套改革试区、深圳综合配套改革试点、沈阳经济区国家新型工业化综合配套改革试验区、山西省国家资源型经济综合配套改革试验区。

② 战略经济区：广西北部湾经济区、海峡西岸经济区、江苏沿海地区、横琴新区、关中—天水经济区、辽宁沿海经济带、长吉图开发开放先导区、黄河三角洲高效生态经济区、鄱阳湖生态经济区、甘肃循环经济示范区、海南国际旅游岛、皖江城市带、青海柴达木循环经济试验区、沈阳经济区、两江新区。

③ 跨区域战略规划：东北地区振兴规划、珠三角地区改革发展规划、中部崛起规划、长江三角洲区域规划。

基本持平。图 14－1、14－2 分别为 2005 年、2010 年各地区 GDP 占全国的比重情况。

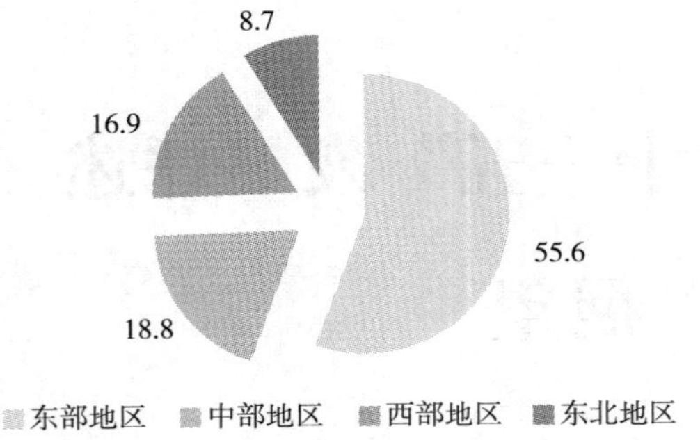

图 14－1 2005 年各地区 GDP 占全国的比重（%）

Figure 14－1 Proportions of Regional GDP in National Total in 2005

资料来源：《"十一五"经济社会发展成就系列报告》

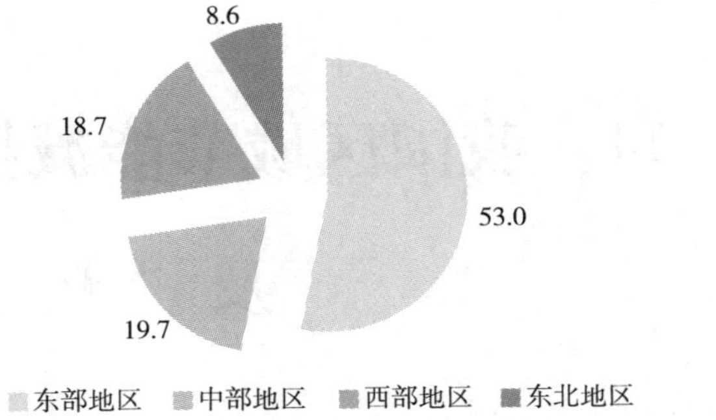

图 14－2 2010 年各地区 GDP 占全国的比重（%）

Figure 14－2 Proportions of Regional GDP in National Total in 2010

资料来源：《"十一五"经济社会发展成就系列报告》

2. 区域投资布局进一步优化

随着各项区域性政策的逐步落实，中部、西部和东北三省投资增长加快，比重提高，区域投资结构呈现了东、中、西、东北地区投资协同发展的局面。西部地区固定资产投资快速增长，五年累计完成投资 197758 亿元，年均增长 28.2%，比同期全国固定资产投资增速高 2.7 个百分点。2010 年，西部地区固定资产投资占全国的比重为 22.2%，比 2005 年提高 2.3 个百分点。中部地区五年累计完成投资 198084 亿元，年均增长 31.6%，增速比"十五"时期高 10.2 个百分点，比同期全国投资增速高 6.1 个百分点；中部地区投资占全国投资的比重为 22.6%，比 2005 年提高 4.4 个百分点。东北地区五年累计完成投资 97613 亿元，年均增长 32.9%，增速比"十五"时期平均增速高 13.1 个百分点，比同期全国投资增速高 7.4 个百分点；东北地区投资占全国投资的比重为 11%，比 2005 年提高 2.4 个百分点。东部地区五年累计完成投资 408767 亿元，年均增长 20.1%，增速比"十五"时期高 0.3 个百分点，比同期全国固定资产投资增速低 5.4 个百分点；东部地区完成投资占全国投资的比重为 42.7%，比 2005 年回落 9.7 个百分点。

3. 区域产业结构持续改善

"十一五"时期，我国三次产业全面快速发展，服务业水平及比重明显提高，区域经济结构优化升级取得新进展。

从国家层面的三次产业结构看，2010 年与 2005 年相比，第一产业所占比重明显下降，第二产业所占比重略有下降，第三产业所占比重上升。其中，第一产业所占的比重从 2005 年的 12.6% 下降到 2010 年的 9.4%，下降了 3.2 个百分点；第二产业所占比重由 47.5% 上升为 50.6%，上升了 2.9 个百分点；第三产业所占比重由 39.9 上升为 40.0%，上升 0.1 个百分点。图 14－3、14－4 分别为 2005 年、2010 年各地区投资占全国的比重情况。

从各个区域的产业结构看，2010 年与 2005 年相比，各个区域的第一产业所占的比重均呈现了不同幅度的下降，东部、中部、西部和东部分别下从 2005 年的 7.9%、16.7%、17.7%、12.8% 分别下降到 6.4%、13.2%、13.2% 和 10.7%，西部地区降幅最大达到 4.44 个百分点；第二产业所占比重除了东部地区均呈现了不同幅度的增加，其中东部地区从 2005 年的 51.6% 下降到 2010 年的 49.8%，中部、西部和东北地区第二产业结构比重分别从 2005 年的 42.8%、50.1%

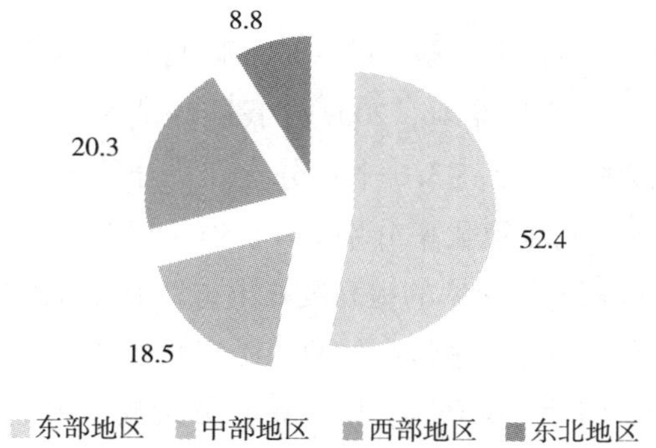

图 14－3 2005 年各地区投资占全国的比重（%）

Figure 14－3 Proportions of Regional Investment in National Total in 2005

资料来源：《“十一五”经济社会发展成就系列报告》

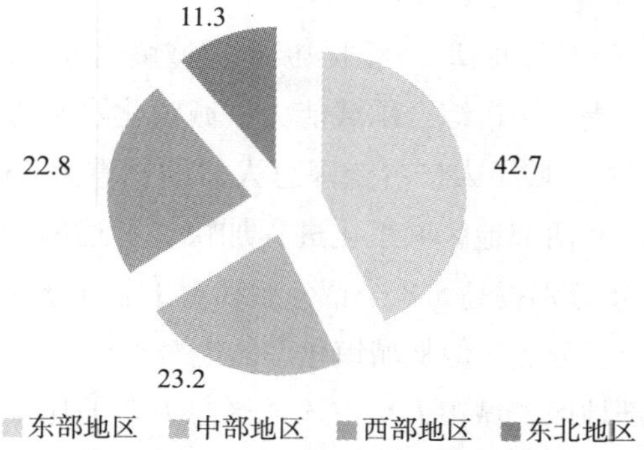

图 14－4 2010 年各地区投资占全国的比重（%）

Figure 14－4 Proportions of Regional Investment in National Total in 2010

资料来源：《“十一五”经济社会发展成就系列报告》

和 49.6% 增加到 2010 年的 52.7%、50.1% 和 52.3%，西部地区的增幅最大达到 7.3 个百分点，其次是中部地区为 5.96 个百分点；各地区第三产业结构变化与第二产业结构变化正好相反，除了东部外，其余地区第三产业结构均出现了不同幅度的降低，其中，东部从 2005 年的 40.5% 上升到 43.9%，增长了 3.4 百分点，中部、西部和东北地区第三产业结构比重分别从 2005 年的 36.6%、39.5% 和 37.6% 下降到 2010 年的 34.1%、36.7% 和 37%，西部地区的降幅最大达到 2.9 个百分点。中部、西部和东北地区产业结构呈现了明显的工业化倾向，产业结构调整还需进一步加强。“十一五”期间各地区产业结构变化情况如图 14－5 所示。

4. 区域间收入差距有所缩小

“十一五”时期，国家坚持实施推进西部大开发，振兴东北地区老工业基地，促进中部地区崛起，鼓励东部地区率先发展的区域发展总体战略，有效带动了各地区居民收入的协调增长，一定程度上缩小了地区间的收入差距。2010 年全国城镇居民人均可支配收入达到 19109 元，比 2005 年的 10493 元增长 82.1%，扣除价格因素，实际增长 59.0%，年均实际增长 9.7%。从分地区的城镇居民收入情况看，2010 年东部、中部、西部和东北地区城镇居民人均可支配收入分别比 2005 年增长 74.0%、81.2%、80.0% 和 82.6%，东部、中部、西部和东北地区收入比值从 2005 年的 152.3∶100.3∶100∶99.4 变为 2010 年的 147.2∶101.1∶100∶100.9，东部地区与其他地区间的收入差距有所缩小。

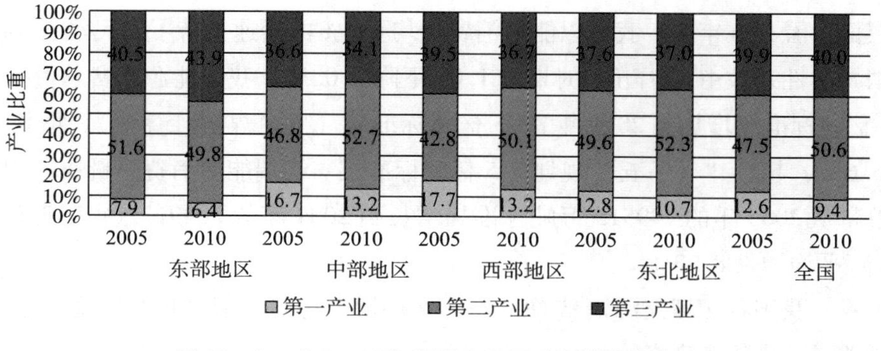

图 14－5 “十一五”期间各地区产业结构变化图

Figure 14－5 Industrial Restructuring of Different Regions during China's 11th Five－year Plan Period

资料来源：《中国统计摘要 2011》

5. 城镇化水平持续提高

“十一五”时期，我国积极推进城镇化进程，引导人口合理有序流动，进一步加强城市基础设施建设，提升城市综合承载能力，城镇化水平进一步提高。从全国来看，2009 年我国城镇人口达 62186 万人，城镇人口占全国总人口的比重为 46.6%，比 2005 年提高 3.6 个百分点。从分地区来看，中部和西部地区城镇化进程加快，上述两个地区 2009 年底城镇化率分别为 42.3% 和 39.4%，比 2005 年分别提高 5.8 个百分点和 4.8 个百分点，分别比国家城镇化率的提高度高出 2.2 个百分点和 1.2 个百分点。但从城镇化率的绝对水平来看，2009 年底城镇化水平最高的地区是东北地区，其次是东部地区，城镇人口占该地区总人口的比重分别为 56.9% 和 56.7%，比 2005 年分别提高 1.7 个百分点和 3.9 个百分点。“十一五”期间各地区城镇化率变化情况如图 14－6 所示。

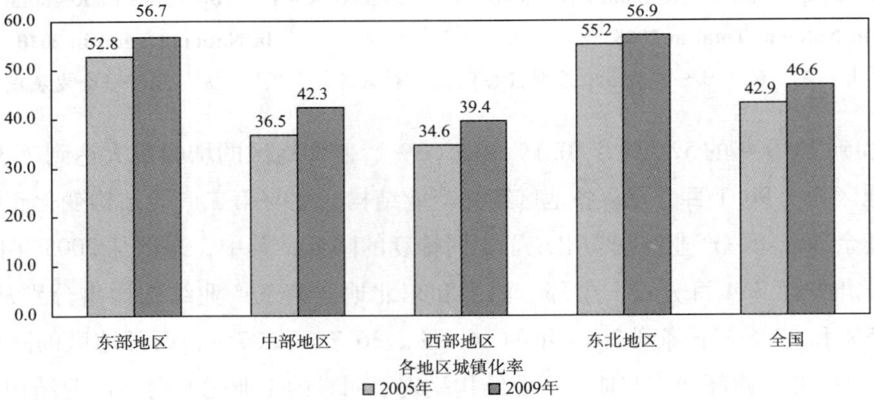

图 14－6 “十一五”期间各地区城镇化率变化图

Figure 14－6 Urbanization Progress of Different Regions during China's 11th Five－year Plan Period

资料来源：《中国统计摘要 2011》

14.1.2 “十一五”时期区域节能评述

1. 区域能源消费评述

“十一五”期间，我国克服了世界金融危机的影响，实现了国民经济平稳较快发展，在资源环境、节能减排目标的约束下，能源消费增速稳中有降。

（1）能源消费弹性系数①进一步降低，但能源消费总量持续快速增长。

从国家层面来看，五年来，我国以能源消费年均 6.6% 的增速支持了国民经济年均 11.2% 的增速，能源消费弹性系数由“十五”时期的 1.04 下降到 0.59。我国能源消费总量持续增长，从 2005 年 23.6 亿吨标准煤增加到 2010 年 32.5 亿吨标准煤（发电煤耗计算法），增长了 29.94%，年平均增长 6.6%，超过“十一五”规划 5.5 亿吨标准煤。终端能源消费总量快速增长，全国终端能源消费总量从 2005 年的 149515 万吨标准煤增长到 2009 年的 204912.14 万吨标准煤，增长了 36.85%，年平均增长 8.73%。

分区域来看，我国各区域能源消费总量一直保持增长态势，但各区域能源增长的速度呈现出明显的区域差异，具体表现在：

① 能源消费弹性：反映能源消费增长速度与国民经济增长速度之间比例关系的指标。

• 从东部地区来看，能源消费从2005年115913万吨标准煤的增加到2009年154730万吨标准煤，增长了33.49%，年平均增长7.49%；终端能源消费从2005年的65673.11万吨标准煤增长到2009年的88961.72万吨标准煤，增长了35.46%，年平均增长7.88%。能源消费量和终端能源消费量占全国的比重有所降低，分别从2005年的44.00%、43.92%下降到2009年的43.31%、43.31%。

• 从中部地区来看，能源消费从2005年57958万吨标准煤的增加到2009年77074万吨标准煤，增长了32.98%，年平均增长7.39%；终端能源消费从2005年的33208.47万吨标准煤增长到2009年的45073.07万吨标准煤，增长了35.73%，年平均增长7.94%。能源消费量和终端能源消费量占全国的比重有所降低，分别从2005年的22.00%、22.21%下降到2009年的21.58%、22.03%。

• 从西部地区看，随着我国西部大开发的持续推进，西部地区经济加快发展，带动了能源消费总量的快速增加，在全国能源消费总量中的比重有所上升。西部地区能源消费总量分别从2005年62611万吨标准煤增加到2009年的88155万吨标准煤，增长了40.80%，年平均增长8.93%，能源消费量占全国的比重从2005年的23.77%上升到2009年的24.68%；终端能源消费量分别从2005年的35183.26万吨标准煤增长到2009年的49179.11万吨标准煤，增长了39.78%，年平均增长率分别为8.73%。终端能源消费量占全国的比重从2005年的23.53%、10.33%上升到2009年的24.04%。

• 从东北地区来看，随着振兴东部老工业基地战略的持续推进，东北地区经济加快发展，带动了能源需求的增长。东北地区能源消费总量增速加快，在全国的比重有所上升。东部地区能源消费总量分别从2005年26976万吨标准煤增加到2009年的37276万吨标准煤，增长了38.18%，年平均增长8.42%，能源消费量占全国的比重从2005年的10.24%上升到2009年的10.43%；终端能源消费量从2005年的15450.51万吨标准煤增长到2009年的21398.24万吨标准煤，增长了38.5%，年平均增长率分别为8.48%。终端能源消费量占全国的比重从2005年的10.33%上升到2009年的10.46%。表14-1为各区域能源消费量和终端能源消费量情况，表14-2为各区域2005年与2009年终端能源消费量增长对比情况。

表14-1 各区域能源消费量和终端能源消费量

Table 1 Energy Consumption and Final Energy Consumption of Different Regions

	2005年与2009年能源消费量①对比（各省相加总）					
	能源消费总量（万吨标准煤）		比重（%）		增长（%）	
	2005年	2009年	比重		增长率	年平均增长率
东部地区	115913	154730	44.00	43.31	33.49	7.49
中部地区	57958	77074	22.00	21.58	32.98	7.39
西部地区	62611	88155	23.77	24.68	40.80	8.93
东北地区	26976	37276	10.24	10.43	38.18	8.42
	2005年与2009年能源终端消费量对比					
	终端消费总量（万吨标准煤）		比重（%）		增长（%）	
	2005年	2009年	2005年	2009年	增长率	年平均增长率
东部地区	65673.11	88961.72	43.92	43.48	35.46	7.88

① 由于折算系数的不同，故各地区能源消费量相加数与全国能源消费数不等。终端能源消费总量为各省终端能源消费总量加总。

续表

	2005年与2009年能源终端消费量对比					
	终端消费总量（万吨标准煤）		比重（%）		增长（%）	
中部地区	33208.47	45073.07	22.21	22.03	35.73	7.94
西部地区	35183.26	49179.11	23.53	24.04	39.78	8.73
东北地区	15450.51	21398.24	10.33	10.46	38.5	8.48

资料来源：《中国能源统计年鉴2005》、《中国能源统计年鉴2010》

表14-2　各区域2005年与2009年终端能源消费量增长对比

Tabel 14-2　Increases in Final Energy Consumption of Different Regions from 2005 to 2009

	2005年与2009年各品种能源终端消费量对比					
	煤及煤制品（万吨标准煤）			石油及制品（万吨标准煤）		
	2005年	2009年	增长率	2005年	2009年	增长率
全国	87800.54	116975.02	33.23	43823.08	58071.69	32.51
东部地区	34084.27	44247.69	29.82	24447.79	31445.87	28.62
中部地区	24526.96	31414.32	28.08	6304.12	8872.01	40.73
西部地区	22301.31	31099.51	39.45	7296.85	10606.75	45.36
东北地区	6888.00	10213.52	48.28	5774.32	7147.06	23.77
	天然气（万吨标准煤）			电力及其他能源（万吨标准煤）		
	2005年	2009年	增长率	2005年	2009年	增长率
全国	5063.04	9306.61	83.81	12828.69	20258.81	57.92
东部地区	1186.23	3652.74	207.93	5954.83	9615.43	61.47
中部地区	403.79	1079.40	167.32	1973.60	3707.34	87.85
西部地区	2981.59	3879.15	30.10	2603.50	3593.70	38.03
东北地区	491.44	695.32	41.49	2296.76	3342.34	45.52

资料来源：《中国能源统计年鉴2005》、《中国能源统计年鉴2010》

（2）区域能源消费结构调整不均衡，优质能源消费比重有所上升。

我国能源资源中煤炭资源相对丰富，决定了我国以煤炭为主的能源消费结构，“十一五”期间，随着我国节能减排战略的实施，我国加大了产业结构调整力度，进而带动了能源消费结构进一步优化，但我国由于不同区域资源状况不同，以及各地区经济发展的不平衡依然存在，因此，各地区能源结构仍然存在很大差异。

从全国来看，终端能源消费结构进一步优化。煤及煤制品、石油及制品、天然气、电力（其他能源）在终端能源消费中的比例从2005年的58.72%、29.31%、8.58%和3.39%调整到2009年的57.17%、28.38%、9.90%和4.45%，2009年终端能源消费中煤及煤制品、石油及制品的比重比2005年分别降低1.55个百分点和0.93个百分点，天然气、电力（其他能源）在终端能源消费中的比重分别上升1.06个百分点和1.32个百分点，石油、天然气等较为高效清洁能源的比重比2005年提高了1.55个百分点。

分区域来看，虽然区域能源消费结构得到了一定的优化，但煤炭依然在我国区域能源消费结构中占有主导地位，特别是中部和西部地区，终端能源消费中煤及煤制品的比重均在60%以上，比重最低的东北地区的比重依然达到47.73%，改善以煤为主的能源消费结构任重而道远。具体而言：

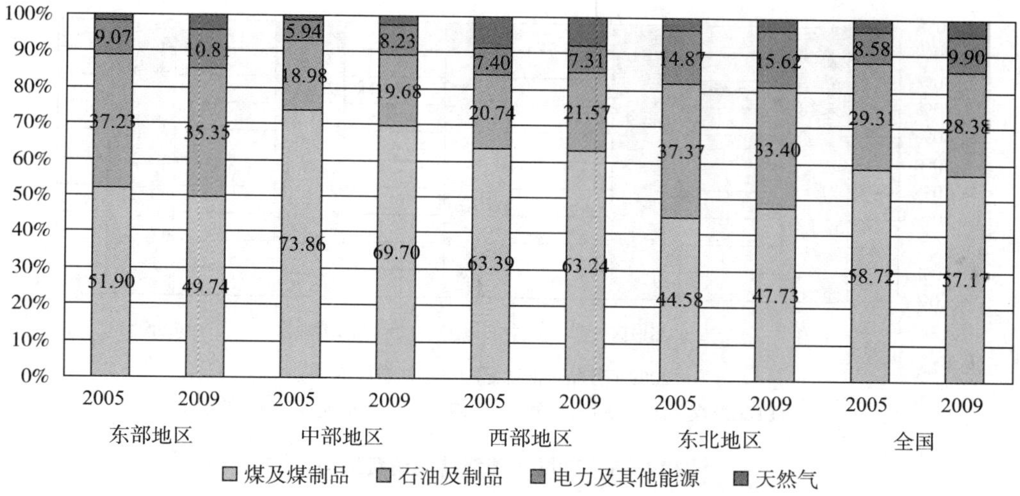

图 14－7 “十一五”期间能源终端消费结构变化图

Figure 14－7 Final Energy Consumption Restructuring during China's 11th Five－year Plan Period

资料来源：《中国能源统计年鉴 2005》、《中国能源统计年鉴 2010》

• 东部和东北地区终端能源消费结构在各区域中相对合理，消费结构进一步优化。2009 年东部地区和东北地区终端能源消费结构中煤及煤制品的比重分别为 49.74% 和 47.73%；石油及制品的比重分别为 35.35% 和 33.40%；电力（其他能源）的比重分别为 4.11% 和 3.25%；天然气的比重分别 10.81% 和 15.62%。东部地区，天然气和电力（其他能源）在终端能源消费中的比重有所上升，分别比 2005 年上升了 2.30 个百分点和 1.73 个百分点；但东北地区终端能源消费结构中煤及煤制品的比重有所上升，比 2005 年上升 2.15 个百分点；而石油及制品的比重有所降低，下降了 3.97 个百分点；其他形式的能源的比重基本维持不变。

• 中部地区终端能源消费对煤及煤制品的依赖显著降低，结构优化明显。2009 年煤及煤制品、石油及制品、电力（其他能源）和天然气的比重分别为 69.70%、19.68%、2.39% 和 8.23%，煤及煤制品的比重显著降低，分别比 2005 年的 73.86% 下降 4.16 个百分点；石油及制品、天然气和电力（其他能源）在终端能源消费中的比重有所上升，分别比 2005 年的 18.98%、1.18% 和 9.07 上升了 0.70、1.18 和 2.28 个百分点。

• 西部地区终端能源消费结构有所改善，但变化改善不太明显。2009 年煤及煤制品、石油及制品、电力（其他能源）和天然气的比重分别为 63.24%、21.57%、7.89% 和 7.31%，与 2005 年相比，各种能源在终端能源消费结构的比重分别变化－0.15、0.83、－0.59、－0.09 个百分点。但天然气占终端能源消费结构的比重在各区域中最高。图 14－7 为“十一五”期间能源终端消费结构变化图，图 14－8 为 2009 年能源终端消费结构情况。

2. 区域节能概述

为调整经济结构、转变发展方式、应对气候变化、推动科学发展，我国提出“十一五”单位 GDP 能耗降低 20% 左右的约束性指标，随着一系列强有力的措施的制定和执行，“十一五”期间，全国单位 GDP 能耗下降 19.1%，如期完成了“十一五”规划《纲要》确定的节能目标任务。

（1）各区域节能分析。经过五年的节能工作推进，各区域平均能耗指标都出现了下降，但由于经济发展的不平衡性，各区域节能效果存在一定的差异，具体表现如下：

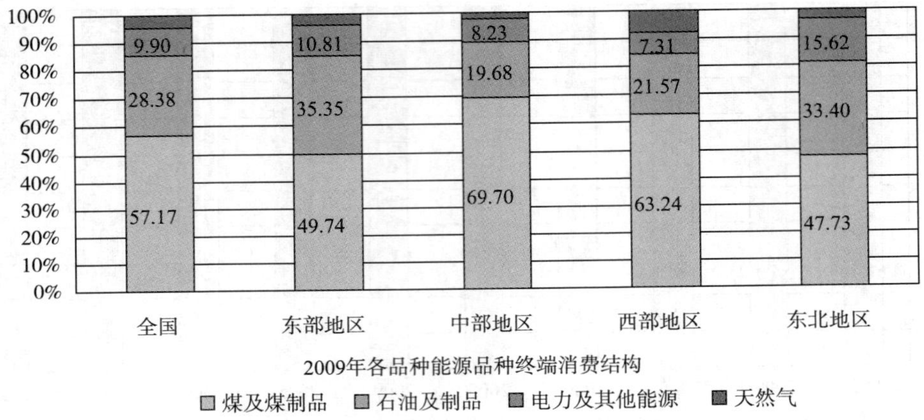

图 14－8 2009 年能源终端消费结构

Figure 14－8 Final Energy Consumption Structure of 2009

资料来源：《中国能源统计年鉴 2005》、《中国能源统计年鉴 2010》

表 14－3 "十一五"期间各区域平均能耗指标对比表

Table 14－3 Comparison of Average Energy－consuming Indexes between Different Regions during China's 11th Five－year Plan Period

区域	2006 年			2007 年			2008 年			2009 年			2010 年		
	单位地区生产总值能耗（等价值）	单位工业增加值能耗（规模以上，当量值）	单位地区生产总值电耗（等价值）	单位地区生产总值能耗（等价值）	单位工业增加值能耗（规模以上，当量值）	单位地区生产总值电耗（等价值）	单位地区生产总值能耗（等价值）	单位工业增加值能耗（规模以上，当量值）	单位地区生产总值电耗（等价值）	单位地区生产总值能耗（等价值）	单位工业增加值能耗（规模以上，当量值）	单位地区生产总值电耗（等价值）	单位地区生产总值能耗（等价值）	单位工业增加值能耗（规模以上，当量值）	单位地区生产总值电耗（等价值）
	（吨标准煤/万元）	（吨标准煤/万元）	（千瓦小时/万元）	（吨标准煤/万元）	（吨标准煤/万元）	（千瓦小时/万元）	（吨标准煤/万元）	（吨标准煤/万元）	（千瓦小时/万元）	（吨标准煤/万元）	（吨标准煤/万元）	（千瓦小时/万元）	（吨标准煤/万元）	（吨标准煤/万元）	（千瓦小时/万元）
东部地区	1.017	1.863	1112.110	0.978	1.690	1111.970	0.926	1.517	1052.286	0.873	1.412	989.343	0.843		
中部地区	1.539	3.553	1319.550	1.478	3.223	1368.062	1.386	2.818	1280.577	1.308	2.492	1180.160	1.253		
西部地区	2.210	4.054	2229.309	2.131	3.773	2287.146	2.021	3.428	2160.652	1.867	3.126	2030.388	1.783		
东北地区	1.593	2.650	1109.800	1.526	2.371	1066.144	1.450	2.100	991.880	1.287	1.753	909.263	1.227		

资料来源：《"十一五"各地区节能目标完成情况表》

表 14-4 各区域平均能耗指标 2009 年与 2005 年相比的降幅

Table 14-4 Reduction of Average Energy-consuming Indexes of Different Regions from 2005 to 2009

	2009 比 2005 年降幅		
	单位地区生产总值能耗（等价值）	单位工业增加值能耗（规模以上，当量值）	单位地区生产总值电耗（等价值）
	（%）	（%）	（%）
东部地区	11.18	19.53	11.49
中部地区	12.53	26.18	11.51
西部地区	14.16	20.16	13.39
东北地区	16.51	30.02	15.38

资料来源：由表 14-3 计算

• 东部地区能源效率整体最高，与其他地区相比平均能耗指标降幅较小。2009 年东部地区平均单位 GDP 能耗、平均单位工业增加值能耗、平均单位 GDP 电耗分别为 0.873 吨标准煤/万元、1.412 吨标准煤/万元和 989.34 千瓦小时/万元，分别比 2005 年下降 11.18%、19.53% 和 11.49%，单位 GDP 能耗、单位工业增加值能耗为各区域最低。该区域中，单位 GDP 能耗北京最低，为 0.606 吨标准煤/万元，较 2005 年降低 15.85%，降幅最高的是上海，降幅为 18.55%；单位工业增加值能耗最低的广东，为 0.809 吨标准煤/万元，单位工业增加值能耗降幅最大的是天津市，降幅为 25.56%；单位 GDP 电耗最高的河北省为 1449.94 千瓦小时/万元，最低的北京市为 681.85 千瓦小时/万元，降幅最高的天津市，为 22.3%。

• 中部地区能源效率居中，平均单位工业增加值能耗降幅明显。2009 年中部地区平均单位 GDP 能耗、平均单位工业增加值能耗、平均单位 GDP 电耗分别为 1.308 吨标准煤/万元、2.492 吨标准煤/万元和 1180.16 千瓦小时/万元，分别比 2005 年下降 12.53%、26.18% 和 11.51%；该区域中，单位 GDP 能耗最低江西省，为 0.880 吨标准煤/万元，降幅最高的是山西省为 16.5%；单位工业增加值能耗最低的湖南省，为 1.57 吨标准煤/万元，同时也是单位工业增加值能耗降幅最大，降幅为 40.08%；单位 GDP 电耗最低的湖南省，为 911.00 千瓦小时/万元，降幅最高的山西省为 21.10%。

• 西部地区节能显著，各种能耗指标下降明显，但能耗指标绝对值依然较高。2009 年西部地区平均单位 GDP 能耗、平均单位工业增加值能耗、平均单位 GDP 电耗分别为 1.867 吨标准煤/万元、3.126 吨标准煤/万元和 2030.388 千瓦小时/万元，分别比 2005 年下降 14.16%、20.16% 和 11.51%；该区域中，单位 GDP 能耗最低广西，为 1.057 吨标准煤/万元，降幅最高的是贵州省，为 24.96%；单位工业增加值能耗最低的陕西省，为 1.367 吨标准煤/万元，同时也是单位工业增加值能耗降幅最大省，降幅为 40.19%；单位 GDP 电耗最低的重庆市为 894.27 千瓦小时/万元，同时也是降幅最高省市，为 26.91%。

• 东北地区节能成效最为显著，各种能耗指标下降均十分明显，2009 年东北地区平均单位 GDP 能耗、平均单位工业增加值能耗、平均单位 GDP 电耗分别为 1.287 吨标准煤/万元、1.753 吨标准煤/万元和 909.263 千瓦小时/万元，分别比 2005 年下降 16.51%、30.02% 和 15.38%，平均单位 GDP 电耗在各区域中最小，同时上述三个能效指标降幅均为各区之最；该区域中，单位 GDP 能耗最低吉林，为 1.209 吨标准煤/万元同时也是降幅最大的省份，降幅为 21.4%；单位工业增加值能耗最低的黑龙江省，为1.382吨标准煤/万元，单位工业增加值能耗降幅最大的是吉林，降幅为39.44%；单位GDP电耗最低的黑龙江省，为798.67千瓦小时/万元，降幅最高的辽宁为17.57%。表14－3为“十一五”期间各区域平均能耗指标对比情况，表14－4为各区域平均能耗指标2009年与2005年相比的降幅情况。

（2）各省区节能情况分析。“十一五”时期，各地区、各部门认真落实党中央、国务院的决策部署，把节能作为调整经济结构、转变发展方式的重要抓手和突破口，放在更加突出的位置，采取了一系列强有力政策措施，取得了显著成效。除对新疆另行考核外，全国其他地区均完成了“十一五”国家下达的节能目标任务，有28个地区超额完成了“十一五”节能目标任务，超额完成目标较多的10个地区分别为：北京（超额32.95%，下同）、天津（5%）、山西（3%）、内蒙古（2.82%）、黑龙江（3.95%）、福建（2.81%）、湖北（8.35%）、广东（2.63%）、重庆（4.75%）、云南（2.41%），其中北京、湖北、天津分别超出目标6.59、1.67、1个百分点。表14－5为“十一五”各地区节能目标完成情况。

表14－5 “十一五”各地区节能目标完成情况表

Table 14－5 Energy Conservation Performance of Different Regions during China's 11th Five－year Plan Period

地区	2005年		2010年		超额完成（%）
	单位GDP能耗（吨标准煤/万元）	“十一五”时期计划降低（%）	单位GDP能耗（吨标准煤/万元）	“十一五”时期计划降低（%）	
北京	0.792	－20	0.582	－26.59	32.95
天津	1.046	－20	0.826	－21	5.00
河北	1.981	－20	1.583	－20.11	0.55
山西	2.89	－22	2.235	－22.66	3.00
内蒙古	2.475	－22	1.915	－22.62	2.82
辽宁	1.726	－20	1.38	－20.01	0.05
吉林	1.468	－22	1.145	－22.04	0.18
黑龙江	1.46	－20	1.156	－20.79	3.95
上海	0.889	－20	0.712	－20	0.00
江苏	0.92	－20	0.734	－20.45	2.25
浙江	0.897	－20	0.717	－20.01	0.05
安徽	1.216	－20	0.969	－20.36	1.80
福建	0.937	－16	0.783	－16.45	2.81
江西	1.057	－20	0.845	－20.04	0.20
山东	1.316	－22	1.025	－22.09	0.41
河南	1.396	－20	1.115	－20.12	0.60
湖北	1.51	－20	1.183	－21.67	8.35
湖南	1.472	－20	1.17	－20.43	2.15
广东	0.794	－16	0.664	－16.42	2.63
广西	1.222	－15	1.036	－15.22	1.47
海南	0.92	－12	0.808	－12.14	1.17
重庆	1.425	－20	1.127	－20.95	4.75
四川	1.6	－20	1.275	－20.31	1.55

续表

地　区	2005 年		2010 年		超额完成（%）
	单位 GDP 能耗（吨标准煤/万元）	“十一五”时期计划降低（%）	单位 GDP 能耗（吨标准煤/万元）	“十一五”时期计划降低（%）	
贵　州	2.813	-20	2.248	-20.06	0.30
云　南	1.74	-17	1.438	-17.41	2.41
陕　西	1.416	-20	1.129	-20.25	1.25
甘　肃	2.26	-20	1.801	-20.26	1.30
青　海	3.074	-17	2.55	-17.04	0.24
宁　夏	4.14	-20	3.308	-20.09	0.45

资料来源：《“十一五”各地区节能目标完成情况表》

14.1.3 区域减排评述

《国民经济和社会发展第十一个五年规划纲要》提出了“十一五”期间主要污染物排放总量减少 10% 的约束性指标。要求到 2010 年，全国化学需氧量和二氧化硫排放量分别比 2005 年下降 10%，即化学需氧量排放量由 2005 年的 1414.2 万吨减少到 1272.8 万吨，二氧化硫排放量由 2005 年的 2549.4 万吨减少到 2294.4 万吨。“十一五”期间，在一系列减排政策的推动下，主要污染物排放总量明显下降，2010 年中国化学需氧量实际排放总量 1238.1 万吨，二氧化硫排放实际总量 2185.1 万吨，与 2005 年相比，化学需氧量和二氧化硫排放总量分别下降 12.45% 和 14.29%，均超额完成 10% 的减排任务。表 14-6 为“十一五”期间主要污染物排放情况。

表 14-6　“十一五”期间主要污染物排放情况

Table 14-6　Emission of Major Pollutants during China's 11th Five-year Plan Period

	化学需氧量			二氧化硫		
	2005 年排放量（万吨）	2010 年排放量（万吨）	“十一五”降低幅度（%）	2005 年排放量（万吨）	2010 年排放量（万吨）	“十一五”降低幅度（%）
东部地区	510.50	421.05	17.52	847.80	671.41	20.81
中部地区	352.00	316.63	10.05	596.10	511.18	14.25
西部地区	367.70	334.32	9.08	844.70	758.26	10.23
东北地区	155.50	133.85	13.92	208.70	186.87	10.46
全国	1414.2	1238.1	12.45	2185.1	2294.4	14.29

资料来源：《2010 年度及“十一五”全国主要污染物总量减排情况考核结果》整理

1. 区域减排分析

由于我国经济发展的不平衡性和承担减排任务的差异性，各区域总体减排效果存在一定的差异，具体表现如下：

• 东部地区主要污染物减排成效最为显著。2010 年东部地区化学需氧量排放量和二氧化硫排放量分别为 421.05 万吨和 671.41 万吨，分别比 2005 年下降 17.52% 和 20.81%，化学需氧量排放量和二氧化硫排放量降幅为各区域最大。该区域中，北京市化学需氧量排放量和二氧化硫排放量降幅最高，较 2005 年分别降低 20.69% 和 39.74%；化学需氧量排放量降幅最小的是海南省，降幅为 2.84%；二氧化硫排放量降幅最小的是福建省，较 2005 年降低 8.03%。

• 中部地区：二氧化硫降幅较为明显。2010 年中部地区化学需氧量排放量和二氧化硫排放量分别为316.63万吨和511.81万吨，分别比2005年下降10.05%和14.25%，二氧化硫排放量降幅仅次于东部地区。该区域中，化学需氧量排放量降幅最高的是河南省，较2005年降低14.02%，降幅最小的是湖北省，降幅为7.14%；二氧化硫排放量降幅最高的是山西省，较2005年降低17.60%，降幅最小的是安徽省，降幅为6.72%。

• 西部地区：由于该地区部分省份“十一五”期间没有承担减排任务，与其他区域相比，总体上减排效果不太明显，2010年西部地区化学需氧量排放量和二氧化硫排放量分别为334.32万吨和758.26万吨，分别比2005年下降9.08%和10.23%。考虑到西部地区特定的经济发展阶段，取得上述减排成绩实属不易。该区域中，化学需氧量排放量降幅最高的是宁夏，较2005年降低14.09%，降幅最小的是四川省为5.43%；二氧化硫排放量减排比例最高的是陕西省，较2005年降低15.55%，降幅最小的是云南省，降幅为4.08%。青海省“十一五”期间没有承担减排任务，其化学需氧量排放量和二氧化硫排放量呈现了一定程度的上升，与2005年上相比分别升了15.42%和15.65%。

• 东北地区：2010年西部地区化学需氧量排放量和二氧化硫排放量分别为133.85万吨和186.87万吨，分别比2005年下降13.92%和10.46%，化学需氧量排放量降幅在各区域中仅次于东部地区。该区域中，辽宁省的化学需氧量排放量降幅和二氧化硫排放量的降幅最高，较2005年分别降低15.90%和14.60%；黑龙江省的降幅最小，与2005年相比分别降低11.77%和3.52%。表14－7为“十一五”期间各区域主要污染物排放情况。

表14－7 “十一五”期间各区域主要污染物排放情况

Table 14－7 Emission of Major Pollutants in Different Regions during China's 11th Five－year Plan Period

区域	化学需氧量排放量（万吨）					二氧化硫排放量（万吨）				
	2006	2007	2008	2009	2010	2006	2007	2008	2009	2010
东部地区	506.60	488.28	461.71	441.50	447.32	835.59	789.83	730.38	686.59	721.98
中部地区	358.35	349.29	335.83	325.41	325.36	600.13	575.55	540.95	518.13	515.72
西部地区	404.33	391.06	378.14	370.55	370.76	925.13	885.85	846.38	817.01	861.32
东北地区	155.50	151.57	143.44	138.53	141.68	218.50	214.82	201.45	190.47	188.26

资料来源：根据《2010年度及“十一五”全国主要污染物总量减排情况考核结果》整理

2. 省市减排分析

“十一五”时期，各地区认真落实党中央、国务院的决策部署，把节能减排作为调整经济结构、转变发展方式的重要抓手和突破口，放在更加突出的位置，采取了区域限批、挂牌督办及经济处罚等一系列处理措施，主要污染物减排取得了显著成效。

从表14－8可以看出，全国各地区均完成了“十一五”国家下达的主要污染物减排目标任务，大部分地区超额完成了“十一五”主要污染物减排目标任务，化学需氧量超额完成目标较多的10个地区分别为：上海（超额87.14%，下同）、湖北（42.86%）、北京（40.75%）、吉林（30.68%）、山东（30.20%）、河南（29.81%）、广东（25.83%）、辽宁（23.26%）、江苏（22.52%）、陕西（20.86%），其中，上海、北京、山东分别超出目标12.90、5.99、4.50个百分点。二氧化硫超额完成目标较多的10个地区分别为：北京（94.80%）、黑龙江（76.00%）、安徽（68.00%）、湖北（51.09%）、吉林（42.77%）、湖南（42.33%）、浙江（40.85%）、江苏（30.56%）、江西（30.43%）、陕西（29.58%），其中，北京、浙江、江苏分别超出目标 19.34、6.13、5.5 个百分点。

表 14－8　各区域主要污染物“十一五”减排目标完成情况

Table 14－8　Performance in Emission Reduction of Major Pollutants in Different Regions during China's 11th Five－year Plan Period

区域	省份	化学需氧量					二氧化硫				
		2005 年排放量（万吨）	“十一五”降低目标（%）	2010 年排放量（万吨）	“十一五”实际降幅（%）	超额完成幅度（%）	2005 年排放量（万吨）	“十一五”降低目标（%）	2010 年排放量（万吨）	“十一五”实际降幅（%）	超额完成幅度（%）
东部地区	北　京	11.6	－14.70	9.20	－20.69	40.75	19.1	－20.4	11.51	－39.74	94.80
	天　津	14.6	－9.60	13.20	－9.60	0.00	26.5	－9.4	23.52	－11.25	19.63
	河　北	66.1	－15.10	54.64	－17.34	14.83	149.6	－15.0	123.38	－17.53	16.87
	上　海	30.4	－14.80	21.98	－27.70	87.14	51.3	－25.9	35.81	－30.19	16.58
	江　苏	96.6	－15.10	78.73	－18.50	22.52	137.3	－18.0	105.03	－23.50	30.56
	浙　江	59.5	－15.10	48.68	－18.18	20.43	86	－15.0	67.83	－21.13	40.85
	福　建	39.4	－4.80	37.50	－4.82	0.47	46.1	－8.0	42.40	－8.03	0.33
	山　东	77	－14.90	62.06	－19.40	30.20	200.3	－20.0	154.03	－23.10	15.50
	广　东	105.8	－15.00	85.83	－18.88	25.83	129.4	－15.0	105.06	－18.81	25.40
	海　南	9.5	0.00	9.23	－2.84	/	2.2	0.0	2.84	－35.45	/
中部地区	山　西	38.7	－13.20	33.31	－13.93	5.51	151.6	－14.0	124.92	－17.60	25.71
	安　徽	44.4	－6.50	41.11	－7.41	14.00	57.1	－4.0	53.26	－6.72	68.00
	江　西	45.7	－5.00	43.11	－5.67	13.35	61.3	－7.0	55.70	－9.13	30.43
	河　南	72.1	－10.80	61.99	－14.02	29.81	162.5	－14.0	133.92	－17.59	25.64
	湖　北	61.6	－5.00	57.20	－7.14	42.86	71.7	－7.8	63.25	－11.79	51.09
	湖　南	89.5	－10.10	79.91	－10.72	6.14	91.9	－9.0	80.13	－12.81	42.33
西部地区	广　西	107	－12.10	93.69	－12.44	2.80	102.3	－9.9	90.38	－11.65	17.70
	重　庆	26.9	－11.20	23.45	－12.82	14.46	83.7	－11.9	71.95	－14.04	17.98
	四　川	78.3	－5.00	74.05	－5.43	8.60	129.9	－11.9	113.10	－12.93	8.68
	贵　州	22.6	－7.10	20.78	－8.05	13.42	135.8	－15.0	114.89	－15.40	2.65
	云　南	28.5	－4.90	26.83	－5.86	19.58	52.2	－4.0	50.07	－4.08	2.01
	陕　西	35	－10.00	30.77	－12.09	20.86	92.2	－12.0	77.86	－15.55	29.58
	甘　肃	18.2	－7.70	16.76	－7.91	2.75	56.3	0.0	55.18	－1.99	/
	青　海	7.2	0.00	8.31	15.42	/	12.4	0.0	14.34	15.65	/
	宁　夏	14.3	－14.70	12.17	－14.90	1.33	34.3	－9.3	31.08	－9.39	0.94
	内蒙古	29.7	－6.70	27.51	－7.37	10.06	145.6	－3.8	139.41	－4.25	11.88
东北地区	辽　宁	64.4	－12.90	54.16	－15.90	23.26	119.7	－12.0	102.22	－14.60	21.69
	吉　林	40.7	－10.30	35.22	－13.46	30.68	38.2	－4.7	35.64	－6.71	42.77
	黑龙江	50.4	－10.30	44.47	－11.77	14.27	50.8	－2.0	49.01	－3.52	76.00

资料来源：《2010 年度及“十一五”全国主要污染物总量减排情况考核结果》整理

14.1.4 区域节能减排综合指数分析①

为了全面对比分析各地区“十一五”期间节能减排总体效果，本报告构建了涵盖减排效果、努力程度和潜力实现三个方面的区域节能减排综合评价指数，根据节能减排综合评价指数对各区域省份进行分类，把节能减排综合指数在［2.00－3.00）之间的为第一类地区，在［1.50－2.00）之间的为第二类地区，在［1.00－1.50）之间的为第三类地区，［0.00－1.00）在之间的为第四类地区。全国各地区节能减排综合评价指数的分布情况如图14－9所示。

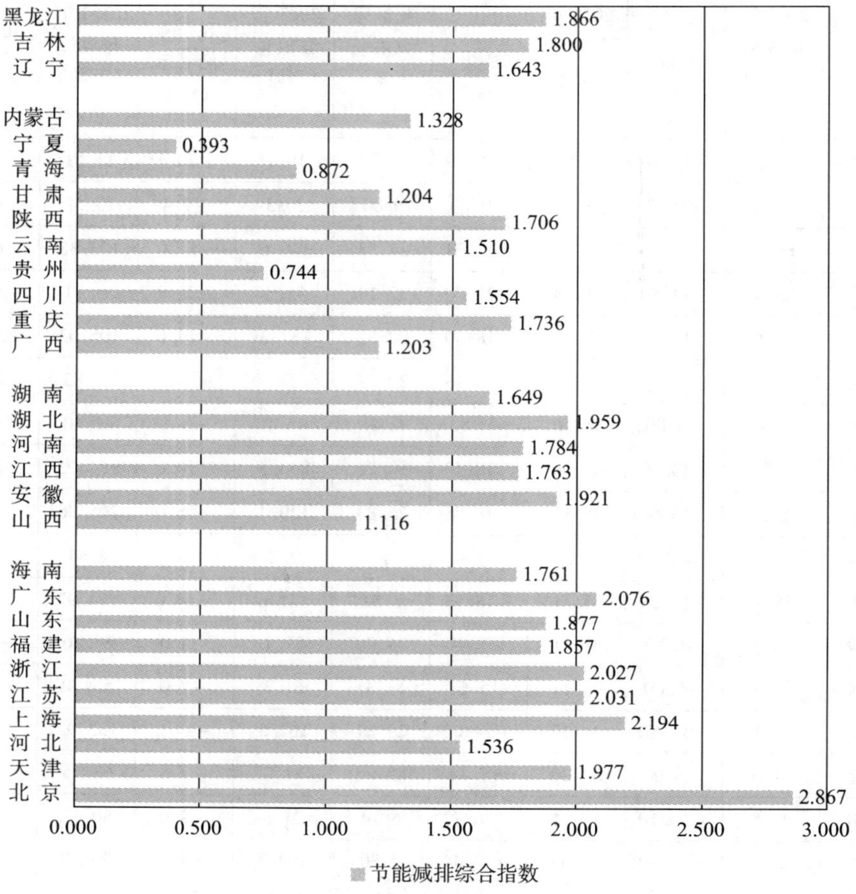

图14－9 各地区节能减排综合指数柱状图

Figure 14－9 Bargraph of Energy Conservation and Pollution Reduction Composite Index in Different Regions

资料来源：《中国能源统计年鉴2005》、《中国能源统计年鉴2010》、《2010年度及“十一五”全国主要污染物总量减排情况考核结果》、《“十一五”各地区节能目标完成情况表》

① 节能减排综合指数定义为：反映该地区节能减排取得的实际效果、努力程度和潜力实现指数综合指数，它涵盖三个二级指标：效果指数、努力程度指数和潜力实现指数。其中，效果指数反映各地区节能减取得实际效果；努力程度指数反映各地节能减排工作的努力程度；潜力实现指数反映各地区节能减排与最好的地区相比存在的差距。该指数的具体算法：将效果指数、努力程度指数和潜力实现指数按0.5、0.25和0.25权重求和得出，其中效果指数是将2010年单位GDP能耗、单位GDP化学需氧量排放量、单位GDP二氧化硫排放量经过数据标准化后按按0.5、0.25和0.25权重进行求和；努力程度指数是将“十一五”时期单位GDP能耗下降、化学需氧量排放量减排和二氧化硫排放量减排目标的超额完成百分比经过数据标准化后，再按0.5、0.25和0.25权重进行求和；潜力实现指数是将各地区2010年单位GDP能耗、单位GDP化学需氧量排放量及单位GDP二氧化硫排放量分别与各指标最小的地区进行对标，并将对标后的数据标准化后按0.5、0.25和0.25权重进行求和，潜力实现指数越大表示该地区与最好的地区的差距越小，节能减排越好。

第一类地区全部分布在东部地区，包含的省市有北京市、上海市、江苏省、广东省、浙江省，反映了东部地区在各区域中节能减排效果整体最好。第一类地区中，节能减排综合指数得分分别为：2.867、2.194、2.076、2.031 和 2.027；从效果指数、努力程度指数和潜力实现指数来看，北京市各项得分分别为 1.000、0.867 和 1.000，各项指数得分均为各地区最高。从单位 GDP 能耗的绝对量和降幅来看，2010 年，北京市 GDP 能耗最低为 0.582 吨标准煤/万元，比 2005 年降低 26.59%，降幅在全国最高；广东的 GDP 能耗最高为 0.734 吨标准煤/万元，比 2005 年降低 20.45%。从主要污染物减排的绝对量和降幅来看，2010 年，北京市化学需氧量排放量和二氧化硫排放量最少，分别为 9.20 万吨和 11.51 万吨，与 2005 年相比分别降低 20.69% 和 39.74%；化学需氧量排放量和二氧化硫排放量最到的是广东省，分别为 85.83 万吨和 105.03 万吨，与 2005 年相比分别降低 18.88% 和 18.81%；从超额完成“十一五”节能减排目标的幅度来看，单位 GDP 能耗、二氧化硫排放超额幅度最高的是北京市，分别超额完成“十一五”目标的 32.95% 和 94.80%；化学需氧量排放超额幅度最高的是上海市，超额完成“十一五”目标的 87.14%。

第二类地区主要包括东部的天津市、河北省、福建省、山东省和海南省，中部的安徽省、江西省、河南省、湖北省和湖南省，西部的重庆市、四川省、陕西省和云南省，东北的黑龙江省、辽宁省和吉林省，因为海南省不承担“十一五”减排任务，故分析不包含海南省。该类地区节能减排综合指数最高的是东部地区的天津市为 1.977，最小的是西部地区的云南省为 1.510。从效果指数、努力程度指数和潜力实现指数来看，天津市的效果指数和潜力实现指数最高，分别为 0.925 和 0.925，湖北省的努力程度指数最高，为 0.384；从节能减排的绝对量和降幅来看，2010 年，单位 GDP 能耗绝对值最低是东部地区的福建省，为 0.783 吨标准煤/万元，与 2005 年相比降低 16.45%；化学需氧量排放量和二氧化硫排放量最低的是天津市，分别为 13.2 万吨和 23.52 万吨，分别比 2005 年降低 9.6% 和 11.25%；从降幅来看，2010 年与 2005 年相比，单位 GDP 能耗降幅、化学需氧量降幅和二氧化硫降幅最大的都是东部地区的山东省，降幅分别为 22.09%、19.40% 和 23.10%。从超额完成“十一五”节能减排目标的幅度来看，单位 GDP 能耗超额完成最多的是中部的湖北省，超额完成 8.35%；化学需氧量减排超额完成最多的是中部的湖北省，超额完成 42.86%；二氧化硫排放量减排超额完成最多的是东部地区的黑龙江省，超额完成 76%。

第三类地区包括中部的山西省，西部的广西、甘肃和内蒙古。从节能减排综合指数来看，最高的是西部地区的内蒙古，为 1.328，最小的是中部地区的山西省，为 1.116；从效果指数、努力程度指数和潜力实现指数来看，内蒙古的效果指数和潜力实现指数最高，分别为 0.613 和 0.613，努力程度指数最高的是山西省，为 0.129；从节能减排的绝对量和降幅来看，2010 年单位 GDP 能耗绝对值最低的省份是西部地区的广西，为 1.036 吨标准煤/万元，与 2005 年相比降低 15.22%，化学需氧量排放量和二氧化硫排放量最低是西部地区的甘肃省，分别为 16.76 万吨和 55.18 万吨，与 2005 年相比分别下降了 7.91% 和 1.99%；从降幅来看，2010 年与 2005 年相比，单位 GDP 能耗降幅、化学需氧量降幅和二氧化硫排放量降幅最大都是中部地区的山西省，降幅分别为 22.66%、13.93% 和 17.60%；从超额完成“十一五”节能减排目标的幅度来看，单位 GDP 能耗超额完成最多的中部的山西省，超额完成 3.00%，化学需氧量减排超额完成最多的内蒙古，超额完成 10.06%；二氧化硫排放量减排超额完成最多的是山西省，超额完成25.71%。

第四类地区全部集中在西部地区，包括贵州、青海省和宁夏，这类地区的节能减排总体效果较差。从节能减排综合指数来看，最高的青海省，为0.872，最小的是宁夏，为0.393；从效果指数、努力程度指数和潜力实现指数来看，效果指数和潜力实现指数青海省最高，分别为0.434和0.435，努力程度指数最高的是贵州省，为0.049；从节能减排的绝对量和降幅来看，2010年该三个省份的单位GDP能耗分别为2.248吨标准煤/万元、2.55吨标准煤/万元和3.308吨标准煤/万元，较2005年分别降低20.06%、17.04%和20.09%；贵州省、青海省和宁夏三个省份的化学需氧量排放量分别为20.78万吨、8.31万吨和12.17万吨，与2005年相比分别降低8.05%、-15.42%和14.90%；三个省份二氧化硫排放总量分别为114.89万吨、14.34万吨和31.08万吨，与2005年相比分别降低15.40%、-15.65%和9.39%。

从超额完成“十一五”节能减排目标的幅度来看，单位GDP能耗超额完成最多的宁夏，降幅为20.09%，因青海和宁夏没有承担“十一五”节能减排任务，不再分析该三个省份主要污染物超额完成情况。

表14-9 各地区节能减排综合指数及排名

Table 14-9 Energy Conservation and Pollution Reduction Composite Index in Different Regions and the Ranking

	综合指数及排名		二级指数					
	节能减排综合指数	排名	效果指数	排名	努力程度指数	排名	潜力实现指数	排名
北 京	2.867	1	1.000	1	0.867	1	1.000	1
上 海	2.194	2	0.951	2	0.293	4	0.950	2
广 东	2.076	3	0.948	3	0.181	7	0.948	3
江 苏	2.031	4	0.927	5	0.179	8	0.925	5
浙 江	2.027	5	0.930	4	0.167	9	0.930	4
天 津	1.977	6	0.925	6	0.127	16	0.925	6
湖 北	1.959	7	0.788	15	0.384	2	0.788	15
安 徽	1.921	8	0.837	10	0.247	5	0.838	10
山 东	1.877	9	0.871	8	0.133	14	0.873	8
黑龙江	1.866	10	0.782	16	0.301	3	0.783	16
福 建	1.857	11	0.906	7	0.044	25	0.908	7
吉 林	1.800	12	0.799	13	0.203	6	0.798	13
河 南	1.784	13	0.812	12	0.162	10	0.810	12
江 西	1.763	14	0.822	11	0.121	18	0.820	11
海 南	1.761	15	0.871	9	0.017	27	0.873	9
重 庆	1.736	16	0.788	14	0.161	12	0.788	14
陕 西	1.706	17	0.775	17	0.156	13	0.775	17
湖 南	1.649	18	0.745	19	0.162	11	0.743	19
辽 宁	1.643	19	0.759	18	0.124	17	0.760	18
四 川	1.554	20	0.741	20	0.071	23	0.743	20
河 北	1.536	21	0.722	21	0.095	21	0.720	21

续表

	综合指数及排名		二级指数					
	节能减排综合指数	排名	效果指数	排名	努力程度指数	排名	潜力实现指数	排名
云　南	1.510	22	0.706	22	0.097	20	0.708	22
内蒙古	1.328	23	0.613	23	0.102	19	0.613	23
甘　肃	1.204	24	0.589	24	0.027	26	0.588	24
广　西	1.203	25	0.564	25	0.077	22	0.563	25
山　西	1.116	26	0.495	26	0.129	15	0.493	26
青　海	0.872	27	0.434	27	0.003	29	0.435	27
贵　州	0.744	28	0.347	28	0.049	24	0.348	28
宁　夏	0.393	29	0.190	29	0.013	28	0.190	29

资料来源：《中国能源统计年鉴2005》、《中国能源统计年鉴2010》、《2010年度及“十一五”全国主要污染物总量减排情况考核结果》、《“十一五”各地区节能目标完成情况表》

14.1.5 “十一五”区域节能减排的主要经验及存在不足

1. 主要经验

第一，健全地方政府和主要国有企业的政绩考核体系，将节能减排考核纳入地方政府政绩考核是节能减排取得显著成效的直接原因。中央政府于2007年开始把节能减排指标完成情况纳入各地经济社会发展综合评价体系，作为政府领导干部综合考核评价和企业负责人业绩考核的重要内容，实行“一票否决”制，并着手建立健全节能减排工作责任制和问责制，地方各级人民政府对本行政区域节能减排负总责，政府主要领导是第一责任人。同时，节能减排各项工作目标和任务逐级分解到各市（地）、县和重点企业，各地方政府纷纷建立健全节能减排工作责任制和问责制。甘肃不仅对不达标的地区实行一票否决制，还要对重点排污企业恶意排污行为实行重罚，严重的还要追究刑事责任。实践表明，将节能减排考核纳入地方政府政绩考核迅速地扭转了节能减排工作的不利局面，是节能减排取得显著成效的最直接的原因。

第二，节能减排领导执行机构的建立为节能减排顺利开展提供了组织保障。全国各省（市、区）都成立了由政府主要领导任组长的节能减排工作领导小组和节能减排办公室，协调解决当地节能减排工作中的重大问题，并要求各地市也建立相应领导小组。节能减排办公室对节能减排工作履行牵头抓总的职责，综合协调节能减排日常工作，负责全社会节能减排的监管和考核评价，重点企业节能减排的监督检查。节能减排领导执行机构的建立为节能减排顺利开展提供了组织保障。

第三，进一步调整优化产业结构，严格控制高耗能高污染行业过快增长，加快淘汰落后产能是实现节能减排目标的重要推动力。“十一五”期间，各地区、有关部门采取了一系列强有力的政策措施，综合运用法律、经济、技术及必要的行政手段，大力推动落后产能淘汰工作，上大压小、关停小火电机组7200万千瓦，淘汰落后炼铁产能12172万吨、炼钢产能6969万吨、水泥产能3.3亿吨等。主要政策措施：

一是分解落实目标任务，加大奖励惩罚力度。每年将目标任务分解到市、县和具体企业。对完成淘汰落后产能任务较好的地区和企业，在资金、土地、融资等方面给予倾斜，中央财政设立专项资金对经济欠发达地区淘汰落后产能给予奖励。对未按期完成淘汰落后产能任务的地区，暂停对该地区项目的环评、核准和审批；对未按要求淘汰落后产能的企业，依据有关法律法规责令停产或予以关闭。

二是完善相关技术标准，坚决淘汰落后产能。制定实施主要用能产品能耗限额标准，对达不到能耗限额标准的落后产能坚决予以淘汰。切实加强对落后产能企业执行环境保护标准、产品质量标准、能耗限额标准和安全生产规定的监督检查。

三是充分发挥市场机制作用，提高落后产能企业成本。推进资源性产品价格改革，提高土地使用价格，实行差别电价，提高差别电价加价标准。对超过限额标准的，实行惩罚性价格政策。

第四，重点企业和重点工程节能减排成为节能减排工作的重要抓手。千家企业节能行动取得了显著成效，能源利用效率大幅度提高，“十一五”期间，千家企业单位氧化铝综合能耗、乙烯生产综合能耗、烧碱生产综合能耗等指标下降了30%以上，单位原油加工综合能耗、电解铝综合能耗、水泥综合能耗等指标下降了10%以上，供电煤耗下降近10%，部分企业的指标达到了国际先进水平。

同时，国家制定发布了十大重点节能工程实施意见，中央和省级地方财政都设立了节能专项资金，对节能改造实行投资补助和财政奖励，十大重点节能工程共形成节能能力3.4亿吨标准煤。

第五，加强节能减排法规制度建设，有力地保证了节能减排工作的顺利开展。广东省建立和完善节能减排工作评价、考核制度，实施单位生产总值能耗和主要污染物排放总量公报制度，定期向社会公布辖区的节能减排主要指标，接受社会监督。同时，加快研究制订和完善重点产品的能耗限额标准、建筑业的能耗标准和节能设计规范，加快制定适合广东省实际的地方环保标准，建立完善有利于节能减排的评估审查制度，逐步建立有利于节能减排的利益导向机制。

第六，增加节能减排财政投入，制定激励企业节能减排的政策措施。2007—2009年，河北、山西、内蒙古、辽宁、吉林、黑龙江、江苏、浙江、安徽、福建、山东、河南、湖北、湖南、广东、广西、重庆、四川、贵州、陕西20个省、自治区、直辖市20个省累计投入节能减排专项资金1241.87亿元（不含新能源类资金），其中，中央财政918.28亿元，省级财政323.59亿元。这些资金主要投向了城市污水处理厂及其配套管网（占33.2%）、重点流域水污染防治（占25.3%）、十大重点节能工程（占15.1%）和淘汰落后产能（占11.6%）等方面，有力地支持了重点节能减排项目的顺利实施。

同时，各地区建立和完善有利于节能减排的财税政策，建立政府引导、企业为主和社会参与的节能减排投入机制。积极推进合同能源管理机制，鼓励和支持节能服务公司投资帮助企业实施节能技术改造。2009年，全国节能服务公司约502家，共实施节能项目4000多个，总投资280亿元，完成总产值580多亿元，形成年节能能力1350万吨标准煤。

2. 存在的问题

尽管“十一五”节能减排取得了显著成绩，节能工作如期完成了《国民经济和社会发展第十一个五年规划纲要》提出的“十一五”期间单位国内生产总值能耗降低20%左右的目标，主要污染物排放总量明显下降，超额完成了10%的减排任务，但从节能减排目标实施过程看，节能减排仍存在一些问题。

第一，各级政府部门和部分中央企业对节能减排的认识有待进一步提高。一些地方政府仍存在将经济增长与节能减排目标相互对立，片面追求经济增长。特别是在受金融危机影响、经济出现下滑的时候，认为经济增长必须靠投资拉动，盲目投资高耗能产业。

第二，各地区节能减排指标的分配缺乏充分的科学依据。各省节能减排指标分配的公平性有待提高。“十一五”确定全国单位GDP能耗强度降低20%，各省（区）目标均大致按此水平设定。这种做法有其合理性，但未能充分考虑到各地的资源禀赋、发展水平和技术上的能力，而是通过“自上而下”确定的节能减排目标，进行年度分解，各地区节能减排目标分解的随意性较强，同时由于能耗强度在经济发展过程中呈现倒U型变化规律，而那些欠发达省份大多正处在能耗强度上升的重化工业阶段，单位GDP能耗强度下降潜力较小，节能减排和经济发展之间矛盾较为突出。

第三，节能减排的推进过分依赖政府行政手段。尽管可以通过政府行政手段实现就节能减排目标的层层分解，具有较高的执行力度，能够确保完成节能目标的功效，但也带来了“一刀切”，前松后紧，临时突击，治标不治本等问题。

第四，中央政府对地方政府节能考核方式有待进一步完善。在“十一五”的节能减排目标考核中，政府基本上采用到最后一年整体考核是否完成节能减排目标，即采取“算总账”方法。如果在起初各年节能率低，五年算总账时，尽管能源强度下降的指示达到了，但能源消费总量大幅上涨。因此在考虑强度节能目标的同时，应考虑各年度的节能总量目标，将年度节能量目标和五年总的节能目标相结合，以达到实现强度节能目标的同时，降低能源消费总量。

第五，实施节能减排的配套政策体系急需完善。“十一五”期间随着就节能减排的推进，相关配套政策滞后在一定程度上制约了节能减排工作的深入推进，并且已出台的配套政策的协调性上有待进一步加强。

第六，节能减排各利益相关方参与机制不健全。节能减排关系到社会经济全局，政府在推动节能减排中发挥主导作用固然重要，但地方政府、企业和民众等利益相关者参与综合决策的渠道和机制还显不足。

第七，节能减排所需能源和排放的统计、报告、核实制度体系尚不健全。

14.2 “十二五”时期区域节能减排展望

14.2.1 “十二五”时期节能减排战略背景

1.“十二五”区域经济发展展望

依据国家《国民经济和社会发展第十二个五年规划纲要》，“十二五”期间，我国继续推进区域发展总体战略和实施主体功能区战略，逐步构筑区域经济优势互补、主体功能定位清晰、国土空间高效利用、人与自然和谐相处的区域发展格局，充分发挥不同地区比较优势，促进生产要素合理流动，深化区域合作，推进区域良性互动发展，区域发展差距将逐步缩小。

“十二五”期间各区域发展重点：推进新一轮西部大开发，把深入实施西部大开发战略放在区域发展总体战略优先位置，坚持以线串点、以点带面，推进重庆、成都、西安区域战略合作，推动呼包鄂榆、广西北部湾、成渝、黔中、滇中、藏中南、关中—天水、兰州—西宁、宁夏沿黄、天山北坡等经济区加快发展，培育新的经济增长极；全面振兴东北地区等老工业基地，统筹推进全国老工业基地调整改造，重点推进辽宁沿海经济带和沈阳经济区、长吉图经济区、哈大齐和牡绥地区等区域发展；大力促进中部地区崛起，重点推进太原城市群、皖江城市带、鄱阳湖生态经济区、中原经济区、武汉城市圈、环长株潭城市群等区域发展；积极支持东部地区率先发展，推进京津冀、长江三角洲、珠江三角洲地区区域经济一体化发展，打造首都经济圈，重点推进河北沿海地区、江苏沿海地区、浙江舟山群岛新区、海峡西岸经济区、山东半岛蓝色经济区等区域发展，建设海南国际旅游岛。按照全国经济合理布局的要求，规范开发秩序，控制开发强度，实施主体功能区战略，形成高效、协调、可持续的国土空间开发格局。

2. 节能减排面临的机遇和挑战

（1）国家层面。“十二五”时期是我国转变经济发展方式、加快经济结构战略性调整的关键时期。随着工业化、城镇化进程加快，消费结构升级，能源需求持续增长，资源环境约束日趋强化，对节能减排提出了更高要求。加上一些投入少、见效快的措施已经实施，“十二五”时期如果没有大的投入和技术突破，实现节能减排目标的难度很大。部分地方片面追求经济增长，对节能减排工作的重要性的认识有待进一步提高。同时，节能减排还存在责任落实不到位、推进难度增大、激励约束机制不健全、能力建设跟不上节能减排形势需要、监管不力等问题，这些问题解决的好坏对“十二五”节能减排目标的实现具有重要影响。

（2）区域节能减排挑战和机遇

• 东部地区：“十一五”时期承担了较大的节能减排任务，且大部分省市超额完成了目标，“十二五”时期进一步节能减排的难度加大。从单位GDP能耗来看，除了广东省、福建省、海南省外，该地区大部分省市“十一五”单位GDP能耗降低的任务都在20%或更高（山东省22%）；从主要污染物减排任务来看，除了天津市、福建省和海南省，该地区大部分省市“十一五”主要污染物减排任务明显高于国家10%的目标；按照节能减排综合指数划分的第一类省份全部在该区域，表明该区域节能减排总体效果在各区域表现最好；从效果指数及排名来看，该地区大部分省市均排在全国各省市的前列，表明单位GDP能耗、单位GDP化学需氧量排放量、单位GDP二氧化量排放量均处于全国较低水平，绝对减排空间较小；从潜力实现指数来看，该地区大部分省市也基本上排在全国各省市的前列，单位GDP能耗、单位GDP化学需氧量排放量、单位GDP二氧化量排放量与全国较低水平差距较小，相对减排空间减小；从努力程度指数及排名来看，有一半省市的排名在全国前十，超额完成了节能减排任务目标，特别是北京市和上海市，超额完成“十一五”节能减排指标幅度很大，北京市单位GDP能耗降低超额完成32.95%；上海市、北京市和山东省的化学需氧量减排超额完成幅度都在30%以上，其中，上海市超额幅度更是达到了87.14%；北京市、浙江省和江苏省二氧化硫减排超额完成幅度均在30%以上，其中，北京市超额幅度更是达到了94.8%。因此，见效快、难度低的节能减排政策所剩的空间已经不大。

“十一五”期间，我国节能取得的成效从具体实现途径来看，主要依靠淘汰高耗能落后产能、实施十大重点节能工程、节能产品惠民工程以及千家企业节能行动来实现的。过去五年，全国上大压小、关停小火电机组7200万千瓦，淘汰落后炼铁产能12172万吨、炼钢产能6969万吨、水泥产能3.3亿吨等；十大重点节能工程的实施，共形成节能能力3.4亿吨标准煤；千家企业可节能1.5亿吨标准煤。这些政策效果突显的同时，我们也必须看到这些见效快、难度相对较低的减排政策所剩空间已经大大被压缩，特别是东部地区，各省市产业结构调整升级力度较大，这些政策的空间更小，“十二五”期间东部地区节能必须依靠长效机制来实现。

从减排实现的政策来看，东部地区经济社会较为发达，城市污水处理、燃煤脱硫的基础设施相对完善，工程减排的潜力已经不大，通过快速关停、淘汰落后产能的结构减排空间也大大压缩。“十二五”期间，减排应该加大监督减排力度和通过产业优化升级，实现产业结构的高级化进而实现低排放。

“十二五”期间，东部地区仍然面临这较重的节能减排任务。从表14－10可以看出，除了福建省和海南省之外，东部地区其他省份的“十二五”节能目标均在17%或18%，其中，目标定在18%的省份全部在东部地区；从主要污染物减排任务来看，东部地区仍然承担着主要的减排任务，除了福建省和海南省之外，东部地区其他省份的“十二五”减排目标均在国家目标之上，考虑到福建省和海南省在该区域发展的特殊性，福建省的各项减排任务均低于国家的平均水平，而海南省不仅不承担化学需氧量和氨氮的减排任务，还允许其他两个减排指标一定幅度的增长，这也充分考虑到了地区间发展的不平衡性。减排任务在该区域各地区的分配充分体现了东部高、西部低，中部和东北地区持平的节能减排任务分配原则，也要该地区节能减排承载能力相适应。较高的节能减排任务必将加快该区域经济增长方式的转变步伐，提高经济增长质量和效率，加快推动该区域经济的低碳化进程。

● 中部地区：“十二五”期间进一步节能难度加大，但部分省市减排仍有一定空间。“十一五”期间，该地区承担了相对重的节能任务，单位GDP能耗降低的任务都在20%或更高（山西省22%）；“十一五”期间承担减排任务中，山西省、河南省和湖南省承担的化学需氧量任务分别为13.2%、10.8%和10.1%，高于国家整体10%的目标；而安徽省、江西省和湖北省分别承担了6.5%、5%和5%减排任务；二氧化硫减排任务只有山西省和河南省高于国家整体10%的目标，其余省市二氧化硫减排任务均在国家整体10%的目标以下。

从节能减排目标的完成情况来看，单位GDP能耗均小幅超额完成，化学需氧量减排幅度在10%以上只有山西省、河南省和湖南省；除了安徽省和江西省，其余各省的二氧化硫减排幅度均在10%以上。“十一五”期间承担相对较小的减排任务为“十二五”减排预留一定的空间。

从节能减排综合指数来看，该区域各省市排名在中等位置，个别省份的排名较靠后，表明节能减排效果在全国处于中游水平；从效果指数及排名来看，除了山西省外，其余各省的得分相对较高，表明这些省市的单位GDP能耗、单位GDP化学需氧量排放量、单位GDP二氧化量排放量均处于全国相对较低水平，绝对减排空间不太大；从潜力实现指数得分及排名来看，这些省市的单位GDP能耗、单位GDP化学需氧量排放量、单位GDP二氧化量排放量与全国最好的省份存在一定差距，表明存在一定的相对减排空间；从努力程度指数得分及排名来看，除了山西省外，各省的排名比较靠前，这表明“十一五”期间这些省份节能减排超额完成了任务目标。

“十二五”期间，中部地区将承担与全国平均水平相当的节能减排任务。从“十二五”区域节能减排目标整体趋势来看，中部地区将承担与全国平均水平相当的节能减排任务。从表14－10可以看出，该区域各省的“十二五”节能目标均在16%；从主要污染物减排任务来看，中部地区其他省份的“十二五”减排目标总体上与国家目标相当，但各省减排目标存在一定差异，山西省和河南省的各项减排目标均高于国家相应的平均降幅，而湖南省和湖北省只有二氧化硫的减排目标略高于国家的平均降幅。减排任务在该区域各地区的分配基本上与该区域经济发展阶段相适应，也为该地区转变经济增长方式和加快经济发展预留一定的空间。

• 西部地区："十二五"期间节能减排空间较大，但由于该区域经济社会发展的特特殊性，在承担"十二五"节能减排任务的上将会得到国家的照顾，承担的节能减排任务低于全国平均水平。"十一五"期间，国家在节能减排任务分解时已经考虑到西部地区特定的经济社会发展阶段，该区域大部分省市的减排的目标均低于国家平均水平。化学需氧量减排任务只有广西、重庆、陕西和宁夏的减排比例高于国家平均10%的水平，二氧化硫减排任务只有重庆、四川、贵州和陕西的减排比例高于国家平均10%的水平；单位GDP能耗降低目标中，广西、云南和甘肃的目标分别比国家平均水平低5、3和3个百分点。

从实际完成目标情况来看，化学需氧量实际减排水平只有广西、重庆、陕西和宁夏的减排比例高于国家平均10%的水平，二氧化硫减排实际完成情况要好于化学需氧量减排，但仍有云南、宁夏和内蒙古的减排比例低于国家平均10%的水平，广西、云南和甘肃的单位GDP能耗实际降低比例仍未达到国家平均20%的水平。

从节能减排综合指数来看，该区域各省市排名比较靠后，排名最好的是重庆也仅排在第16位，表明"十一五"期间节能减排整体效果一般；从效果指数及其排名来看，排名最高的是重庆市第14位，效果指数得分普遍比较低，表明单位GDP能耗、单位GDP化学需氧量排放量、单位GDP二氧化量排放量绝对值较高，减排绝对空间较大；从潜力实现指数得分及排名来看，这些省市的单位GDP能耗、单位GDP化学需氧量排放量、单位GDP二氧化量排放量与全国最好的省份存在较大差距，减排的相对空间较大；从努力程度指数得分及排名来看，该地区各省市超额完成"十一五"节能减排目标的幅度不大。

由于西部地区经济社会发展的特殊性，"十二五"承担的节能减排任务低于全国平均水平，从表14－10可以看出，该区域"十二五"节能目标，重庆、四川和陕西为16%，内蒙古、广西、贵州、云南、甘肃和宁夏为15%，西藏、青海和新疆为10%；从主要污染物减排任务来看，除了内蒙古的和贵州省的二氧化硫减排目标高于国家8%的平均降幅外，西部地区其他省份的"十二五"减排目标均低于国家平均水平。因考虑到青海省和新疆发展的特殊情况，"十二五"期间，新疆不承担减排任务，青海省的还允许各项减排指标不同程度的增长。减排任务在西部各地区的分配充分体现了各地区之间发展的差异性，为西部地区在"十二五"期间加快经济发展提供相对较大的发展空间。

• 东北地区：该地区情况与中部地区类似，"十二五"期间节能减排的难度进一步加大。"十一五"期间该地区承担了相对重的节能任务，单位GDP能耗降低的任务高于国家平均20%的水平（吉林22%），化学需氧量减排任务均高于国家平均10%的水平，二氧化硫减排任务相对较低，只有辽宁的减排任务高于国家平均10%的水平。

从实际完成目标情况来看，该地区化学需氧量实际减排大幅超过"十一五"计划目标，辽宁、吉林和黑龙江分别超额完成23.26%、30.68%和14.27%；二氧化硫实际减排也大幅超过"十一五"计划目标，辽宁、吉林和黑龙江分别超额完成21.69%、42.77%和76%；单位GDP能耗均小幅超额完成"十一五"计划目标。

从节能减排综合指数来看，该区域各省市排名在全国中间位置，排名最好的是黑龙江，排在第16位，表明"十一五"期间该区域节能减排整体效果优于全国平均水平；从效果指数及其排名来看，排名最高的是吉林第13位，表明单位GDP能耗、单位GDP化学需氧量排放量、单位GDP二氧化量排放量绝对值偏高，具有一定的减排绝对空间；从潜力实现指数得分及排名来看，这该区域三省排名居全国中等偏后位置，表明单位 GDP 能耗、单位 GDP 化学需氧量排放量、单位 GDP 二氧化量排放量与全国最好的省份存在一定差距，具有一定的相对减排空间；该区域各省努力程度指数得分及排名较为靠前，其中黑龙江排在第三位，表明该地区各省市超额完成“十一五”节能减排目标的幅度较大。

“十二五”期间，东北地区将承担与全国平均水平相当的节能减排任务，个别省份的节能减排任务可能高于全国平均水平。从表 14－10 可以看出，该区域中辽宁的“十二五”承担的节能目标是 17%，其余二省均为 16%；从主要污染物减排任务来看，该地区的辽宁省各项减排任务均高于国家的平均降幅，吉林和黑龙江二省的减排目标约低于国家的平均降幅，特别是二氧化硫的目标分别是下降 2.7% 和 2%，远低于国家 10% 的平均降幅。

14.2.2 “十二五”区域节能减排的战略目标

1. 国家目标

根据国家《国民经济和社会发展第十二个五年规划纲要》，我国提出了“十二五”期间节能减排的国家战略目标是：非化石能源占一次能源消费比重达到 11.4%，单位国内生产总值能源消耗降低 16%，单位国内生产总值二氧化碳排放降低 17%，主要污染物排放总量显著减少，化学需氧量、二氧化硫排放分别减少 8%，氨氮、氮氧化物排放分别减少 10%。

2. 地区目标

（1）节能目标。根据 2011 年 9 月 7 国务院《关于印发“十二五”节能减排综合性工作方案的通知》（国发〔2011〕26 号），《通知》提出了“十二五”期间的节能主要目标：到 2015 年，全国万元国内生产总值能耗下降到 0.869 吨标准煤（按 2005 年价格计算），比 2010 年的 1.034 吨标准煤下降 16%，比 2005 年的 1.276 吨标准煤下降 32%；“十二五”期间，实现节约能源 6.7 亿吨标准煤。各地区节能任务分解如表 14－10 所示。

表 14－10 “十二五”各地区节能目标

Table 14－10 Energy Conservation Targets of Different Regions for China’s 12th Five－year Plan

地区	国内生产总值能耗①			
	2010 年单位 GDP 能耗	2015 年控制量	2015 年比 2010 年下降	2006—2015 年累计下降
	（吨标准煤/万元）	（吨标准煤/万元）	（%）	（%）
全国	1.034	0.869	16	32.01
北京	0.582	0.483	17	39.07
天津	0.826	0.677	18	35.22
河北	1.583	1.314	17	33.69
山西	2.235	1.877	16	35.03
内蒙古	1.915	1.628	15	34.23
辽宁	1.38	1.145	17	33.61
吉林	1.145	0.962	16	34.51

① 该表 2010 年单位 GDP 能耗和 2015 年单位 GDP 能耗控制量是根据《国家“十一五”节能减排方案》和国家统计局公布的《“十一五”各地区节能目标完成情况表》测算的。

续表

地区	国内生产总值能耗			
	2010 年单位 GDP 能耗	2015 年控制量	2015 年比 2010 年下降	2006—2015 年累计下降
黑龙江	1. 156	0. 971	16	33. 46
上海	0. 712	0. 584	18	34. 4
江苏	0. 734	0. 602	18	34. 77
浙江	0. 717	0. 588	18	34. 41
安徽	0. 969	0. 814	16	33. 1
福建	0. 783	0. 658	16	29. 82
江西	0. 845	0. 710	16	32. 83
山东	1. 025	0. 851	17	35. 33
河南	1. 115	0. 937	16	32. 9
湖北	1. 183	0. 994	16	34. 2
湖南	1. 17	0. 983	16	33. 16
广东	0. 664	0. 544	18	31. 46
广西	1. 036	0. 881	15	27. 94
海南	0. 808	0. 727	10	20. 93
重庆	1. 127	0. 947	16	33. 6
四川	1. 275	1. 071	16	33. 06
贵州	2. 248	1. 911	15	32. 05
云南	1. 438	1. 222	15	29. 8
西藏	1. 276	1. 148	10	20. 8
陕西	1. 129	0. 948	16	33. 01
甘肃	1. 801	1. 531	15	32. 22
青海	2. 55	2. 295	10	25. 34
宁夏	3. 308	2. 812	15	32. 08
新疆	3. 771	3. 394	10	18. 02
新疆生产建设兵团	/	/	/	

资料来源：《“十二五”节能减排综合性工作方案》

（2）减排目标。根据2011 年9 月7 国务院关于印发“十二五”节能减排综合性工作方案的通知”（国发〔2011〕26 号），通知提出了“十二五”期间的减排主要目标：2015 年，全国化学需氧量和二氧化硫排放总量分别控制在 2347. 6 万吨、2086. 4 万吨，比 2010 年的 2551. 7 万吨、2267. 8 万吨分别下降 8%；全国氨氮和氮氧化物排放总量分别控制在 238. 0 万吨、2046. 2 万吨，比 2010 年的 264. 4 万吨、2273. 6 万吨分别下降 10%。各地区减排目标分解如表 14 - 11 所示。

表 14-11 “十二五”各地区减排目标

Table 14-11 Pollution Reduction Targets of Different Regions for China's 12th Five-year Plan

地区	化学需氧量排放总量控制计划			氨氮排放总量控制计划			二氧化硫排放总量控制计划			氮氧化物排放总量控制计划		
	2010年排放量（万吨）	2015年控制量（万吨）	2015年比2010年下降（%）	2010年排放量（万吨）	2015年控制量（万吨）	2015年比2010年下降（%）	2010年排放量（万吨）	2015年控制量（万吨）	2015年比2010年下降（%）	2010年排放量（万吨）	2015年控制量（万吨）	2015年比2010年下降（%）
全国	2551.7	2335.2	-8.5	264.4	236.6	-10.5	2267.8	2067.4	-8.8	2273.6	2021.6	-11.1
北京	20	18.3	-8.7	2.2	1.98	-10.1	10.4	9.0	-13.4	19.8	17.4	-12.3
天津	23.8	21.8	-8.6	2.79	2.5	-10.5	23.8	21.6	-9.4	34	28.8	-15.2
河北	142.2	128.3	-9.8	11.61	10.14	-12.7	143.8	125.5	-12.7	171.3	147.5	-13.9
山西	50.7	45.8	-9.6	5.93	5.21	-12.2	143.8	127.6	-11.3	124.1	106.9	-13.9
内蒙古	92.1	85.9	-6.7	5.45	4.92	-9.7	139.7	134.4	-3.8	131.4	123.8	-5.8
辽宁	137.3	124.7	-9.2	11.25	10.01	-11	117.2	104.7	-10.7	102	88	-13.7
吉林	83.4	76.1	-8.8	5.87	5.25	-10.5	41.7	40.6	-2.7	58.2	54.2	-6.9
黑龙江	161.2	147.3	-8.6	9.45	8.47	-10.4	51.3	50.3	-2	75.3	73	-3.1
上海	26.6	23.9	-10	5.21	4.54	-12.9	25.5	22	-13.7	44.3	36.5	-17.5
江苏	128	112.8	-11.9	16.12	14.04	-12.9	108.6	92.5	-14.8	147.2	121.4	-17.5
浙江	84.2	74.6	-11.4	11.84	10.36	-12.5	68.4	59.3	-13.3	85.3	69.9	-18
安徽	97.3	90.3	-7.2	11.2	10.09	-9.9	53.8	50.5	-6.1	90.9	82	-9.8
福建	69.6	65.2	-6.3	9.72	8.9	-8.4	39.3	36.5	-7	44.8	40.9	-8.6
江西	77.7	73.2	-5.8	9.45	8.52	-9.8	59.4	54.9	-7.5	58.2	54.2	-6.9
山东	201.6	177.4	-12	17.64	15.29	-13.3	188.1	160.1	-14.9	174	146	-16.1
河南	148.2	133.5	-9.9	15.57	13.61	-12.6	144	126.9	-11.9	159	135.6	-14.7
湖北	112.4	104.1	-7.4	13.29	12	-9.7	69.5	63.7	-8.3	63.1	58.6	-7.2
湖南	134.1	124.4	-7.2	16.95	15.29	-9.8	71	65.1	-8.3	60.4	55	-9
广东	193.3	170.1	-12	23.52	20.39	-13.3	83.9	71.5	-14.8	132.3	109.9	-16.9
广西	80.7	74.6	-7.6	8.45	7.71	-8.7	57.2	52.7	-7.9	45.1	41.1	-8.8
海南	20.4	20.4	0	2.29	2.29	0	3.1	4.2	34.9	8	9.8	22.3
重庆	42.6	39.5	-7.2	5.59	5.1	-8.8	60.9	56.6	-7.1	38.2	35.6	-6.9
四川	132.4	123.1	-7	14.56	13.31	-8.6	92.7	84.4	-9	62	57.7	-6.9
贵州	34.8	32.7	-6	4.03	3.72	-7.7	116.2	106.2	-8.6	49.3	44.5	-9.8
云南	56.4	52.9	-6.2	6	5.51	-8.1	70.4	67.6	-4	52	49	-5.8
西藏	2.7	2.7	0	0.33	0.33	0	0.4	0.4	0	3.8	3.8	0
陕西	57	52.7	-7.6	6.44	5.81	-9.8	94.8	87.3	-7.9	76.6	69	-9.9
甘肃	40.2	37.6	-6.4	4.33	3.94	-8.9	62.2	63.4	2	42	40.7	-3.1
青海	10.4	12.3	18	0.96	1.1	15	15.7	18.3	16.7	11.6	13.4	15.3

续表

地区	化学需氧量排放总量控制计划			氨氮排放总量控制计划			二氧化硫排放总量控制计划			氮氧化物排放总量控制计划		
	2010年排放量（万吨）	2015年控制量（万吨）	2015年比2010年下降（%）	2010年排放量（万吨）	2015年控制量（万吨）	2015年比2010年下降（%）	2010年排放量（万吨）	2015年控制量（万吨）	2015年比2010年下降（%）	2010年排放量（万吨）	2015年控制量（万吨）	2015年比2010年下降（%）
宁夏	24	22.6	-6	1.82	1.67	-8	38.3	36.9	-3.6	41.8	39.8	-4.9
新疆	56.9	56.9	0	4.06	4.06	0	63.1	63.1	0	58.8	58.8	0
新疆生产建设兵团	9.5	9.5	0	0.51	0.51	0	9.6	9.6	0	8.8	8.8	0

注：全国化学需氧量排放量削减8%的总量控制目标为2347.6万吨（其中工业和生活1221.9万吨），实际分配给各地区2335.2万吨（其中工业和生活1214.6万吨），国家预留12.4万吨，用于化学需氧量排污权有偿分配和交易试点工作；全国氨氮排放量削减10%的总量控制目标为238.0万吨（其中工业和生活161.5万吨），实际分配给各地区236.6万吨（其中工业和生活160.4万吨），国家预留1.4万吨，用于氨氮排污权有偿分配和交易试点工作；全国二氧化硫排放量削减8%的总量控制目标为2086.4万吨，实际分配给各地区2067.4万吨，国家预留19.0万吨，用于二氧化硫排污权有偿分配和交易试点工作；全国氮氧化物排放量削减10%的总量控制目标为2046.2万吨，实际分配给各地区2021.6万吨，国家预留24.6万吨，用于氮氧化物排污权有偿分配和交易试点工作。

资料来源：《“十二五”节能减排综合性工作方案》

14.2.3 “十二五”时期区域节能减排的战略选择

由于我国各区域经济处于不同的发展阶段和“十二五”节能减排目标的差异性，各区域为完成“十二五”节能减排目标所采取的战略侧重点略有差异。

1. 东部地区

（1）节能战略。第一，由侧重结构调整向技术节能、结构节能、管理节能统筹推进转变。结构节能的重点应由淘汰高耗能低端产能节能向实现产业高端化节能转变，重点发展节能环保产业、高端装备制造产业、新材料产业、新能源产业、新一代信息技术产业等新兴战略产业，通过降低产业的整体能耗实现结构节能。第二，节能战略重点将由强调工业节能的同时向更注重挖掘建筑、交通等第三产业节能潜力转变。工业节能以重点行业、重点园区、重点企业、重点产品为着力点，推动工业节能降耗，构建产业结构优化、产业链完备、科技含量高、资源消耗小、污染排放少、可持续发展的工业体系。第三，节能手段将由政府主导向政府主导、市场推动、全民参与转变。

（2）减排战略。第一，污染物的削减将继续通过结构减排、工程减排和管理减排来完成。结构减排的重点应由淘汰高耗能低端产能减排向实现产业高端化减少排放转变；工程减排的工程布局重点由省市向县域转变，全面启动县域污水处理设施建设，提高污水处理率和负荷率，继续加强燃煤电厂脱硫脱硝，严格控制机动车的尾气排放，进一步提高造纸、纺织、皮革、化工等行业主要污染物的排放标准；加强减排工程和设施运行的监督和检查，提高管理减排的效果。第二，循环经济发展侧重点将由推行循环型生产向推进生产、流通、消费各环节循环经济转变，把生活消费领域的循环经济发展放在更加重要的位置。通过鼓励消费者购买使用再生产品，减少使用一次性用品，限制过度包装，推行绿色采购等工作，形成绿色生活方式和消费模式。第三，在推动清洁工业生产的同时，更加注重建筑、交通和商贸领域。发展绿色生态和低碳建筑，积极采用清洁施工技术和措施。交通领域采用清洁、高效的运输方式和工具，减轻交通拥堵压力，减少能源消耗。商贸服务领域要减少流通环节、提高流通效率，大力推进节能设备、高效照明产品使用，鼓励采购和销售节能、节水、节材产品。第四，控制能源消费总量，特别是煤炭的消费总量，能源结构调整重点放在了进一步增加天然气和电力供应，提高清洁能源消费比例，大力发展核电，根据资源禀赋发展风能和太阳能。

2. 中部地区和东北地区

（1）节能战略。第一，坚持以结构调整为根本，更加注重技术进步和强化管理挖掘节能减排潜力。继续把结构节能减排作为今后一段时期的治本之策，应用高新技术和先进适用技术改造传统产业，通过控制高耗能产业进入和淘汰落后产能，减少经济增长对能源需求的压力，加大技术推广力度，大力支持先进节能技术产业化、节能技术改造、节能产品惠民工程。第二，把工业节能放在节能战略更加突出位置，围绕工业生产源头、过程、产品三个重点，实施能效提升计划；加强工业节能管理，鼓励企业建立节能计量、台账和统计制度，实施电力需求侧管理、能效标识、政府节能采购等管理方式；加强工业节能技术的研究开发和技术储备，加快资源综合利用与清洁生产关键技术产业化，推进先进高效技术、装备、产品的推广应用，支撑工业发展方式转变。

（2）减排战略。第一，污染物的削减将继续通过结构减排、工程减排和管理减排来完成。结构减排的重点应由淘汰高耗能低端产能减排向实现产业优化、减少排放强度转变；进一步加大工程减排力度，逐步启动县域减排工程建设，加强减排工程和设施运行的监督和检查，提高管理减排的效果。第二，循环经济发展重点是推行循环型生产，推行清洁生产、提高资源综合利用水平、建设改造各类产业园区等方面入手，实现土地集约利用、废物交换利用、能量梯级利用、废水循环利用和污染物集中处理，统筹园区、产业基地和产业带的空间布局，构筑上下链接、相互循环的农业、工业和服务业循环利用体系，实现产业循环式组合。第三，加大工业清洁生产推广力度，积极推行工业领域生态设计，从全生命周期减少资源消耗和污染物排放，加大重点行业清洁生产审核和清洁生产改造力度，加强清洁生产重大技术的研发和推广应用。第四，能源消费结构的重点是积极推进煤炭清洁高效化利用技术，降低煤炭在终端消费中的比例，提高煤炭消费用于发电的比重。

3. 西部地区

（1）节能战略。第一，坚持以结构节能为主，加大对高耗能产业的技术改造和产品升级，严格控制高耗能产业在区域的新的布局，坚决淘汰技术落后、浪费资源、污染严重的产品和产业，按照科技含量高、经济效益好、资源消耗低、环境污染少的新型工业化发展要求，加快工业结构的调整，逐步形成有利于资源节约和环境保护的工业体系，降低单位产出的能源消耗和排放。第二，加大工业节能力度，注重产品节能。加大重点行业包括钢铁、有色、石化等高耗能、高污染行业工业节能力度，重点发展余热余压利用、工业锅炉（窑炉）改造、热电联产等，提高工业能源利用率效率。提高财政补贴力度，加大家用电器、交通工具、照明产品等高效节能产品在西部区域的推广应用

（2）减排战略。第一，加大工程减排在污染物的削减的作用，继续完善城市污水、工业废气的处理设施和废水深度治理工程；提高规模以上燃煤脱硫、脱硝机组、钢铁烧结机烟气脱硫

设施的比重。第二，推进资源综合利用，实现循环发展。围绕能源、大宗短缺、稀缺金属等重要矿产资源，推进综合开发及共伴生矿综合利用；以尾矿、赤泥等大宗工业固废为重点，开展固废综合利用示范基地建设。第三，加快构建企业为主、政府指导、市场驱动、全民参与的清洁生产工作格局；通过抓好清洁生产审核、清洁生产评价，在重点区域、重点领域实施清洁生产重大工程，带动企业实施清洁生产改造。第四，积极调整优化能源消费结构，降低煤炭在终端消费中的比重，继续推进"上大压小"，大力发展大容量、高参数机组火电，不断提高一次能源的加工转化比重，特别是提高煤炭消费用于发电的比重。

14.3 "十二五"时期区域节能减排政策措施展望

《我国国民经济和社会发展第十二个五年规划纲要》指出：面对日趋强化的资源环境约束，必须增强危机意识，树立绿色、低碳发展理念，以节能减排为重点，健全激励和约束机制，加快构建资源节约型、环境友好型的生产方式和消费模式，增强可持续发展能力。从《纲要》有关节能减排政策的论述上，可以对未来五年我国区域节能减排政策的大致取向做如下判断：

14.3.1 国家层面节能减排政策取向

1. 合理控制能源消费总量，将成为产业结构调整的重要手段

控制能源消费总量，提高产业准入门槛，从源头抑制高耗能产业过快增长，通过严控增量，优化存量，切实改变敞开口子供应能源、无节制使用能源的现象，能源消费总量控制将成为产业结构调整的重要手段。同时，加快发展服务业和高技术产业，积极发展节能环保、新能源等战略性新兴产业，加快淘汰落后产能，改造提升传统产业，促进产业结构优化升级。环保部已公布"十二五"在东部六省市实施煤炭消费总量控制试点，这些政策将成为转变经济发展方式，调整产业结构的重要手段和推动力。

2. 深化节能减排考核力度，增加节能减排政策的执行力

将节能减排绩效纳入各级政府及主要国有企业领导人的政绩考核内容，极大地推动我国节能减排工作，但目前各地方政府政绩考核中仍偏重于经济增长速度的考核，"十二五"期间节能减排考核将在考核指标的分解、奖惩力度的确定、健全节能减排监督机制等方面进一步深化。

3. 节能减排的实现手段将呈现多样化

"十二五"期间，我国将逐步改变以政府行政手段推进节能减排的实现方式，市场化手段将在未来实现节能减排目标发挥更加重要的作用。一方面，将健全节能市场化机制，加快推行合同能源管理；另一方面，建立完善温室气体排放和节能减排统计监测制度，逐步建立碳排放交易市场。

4. 重点节能工程、重点领域节能仍然是实现我国节能目标的重要措施

"十一五"期间，十大重点节能工程的实施，大幅度提高了能源利用效率，促进了先进节能技术的推广应用，促进了节能环保产业发展，取得了显著的节能效益，共形成节能能力3.4亿吨标准煤。"十二五"期间，我国会继续实施重点节能工程，推广先进节能技术和产品，重点抓好工业、建筑、交通运输等重点领域节能。

5. 进一步完善节能减排法规和企业激励约束机制

在认真落实《节约能源法》、《民用建筑节能条例》、《公共机构节能条例》等法律法规，完

善配套法规和标准，加强节能执法监督检查，确保各项制度落实，制止各种浪费能源行为的基础上，有关节能减排的法律法规有望陆续出台，与我国节能减排相适合的法律法规体系将进一步建立健全。加大财政、税收、金融对企业节能的支持力度，完善企业节能减排的政策激励机制。全面改革资源税，推动资源性产品价格改革，形成反映资源稀缺程度、供求关系和环境成本的价格形成机制。开展节能量交易试点，推行产品“领跑者”能效标准、能效标识等市场化节能管理机制，通过市场手段合理引导企业节能减排。

6. 调整能源消费结构，增加非化石能源比重

加快新能源开发，推进传统能源清洁高效利用，在保护生态的前提下积极发展水电，在确保安全的基础上高效发展核电，加强电网建设，发展智能电网，完善油气管网，扩大油气战略储备。

14.3.2　区域节能减排政策取向

1. 东部地区

东部地区是我国经济社会发展较为发达的地区，节能减排工作整体上效果较好，按照节能减排综合指数该区域能省市大体可以分为三类，不同类型省份之间应区别对待，节能减排政策略有差异。

（1）大力优化调整产业结构。第一类型省份的北京市、天津市、江苏省，均是我国重要的经济省份。该类型的省份产业结构调整优化的重点要充分利用区位优势、人才优势和政策优势，优先布局和发展节能环保、新一代信息技术、生物、高端装备制造、新能源、新材料、新能源汽车等战略性新兴产业，通过战略性新兴产业的发展实现产业结构升级；加快先发展现代服务业，特别是金融、信息、文化创意、交通物流、商贸服务等高端服务业，不断提高第三产业的比例和产业水平；合理控制能源消费总量，特别是严格控制煤炭的消费总量；能源结构调整的重点放在了进一步增加天然气和电力供应，提高清洁能源消费比例，根据自身条件大力发展核电、风电和太阳能光伏发电。

第二类省份产业结构调整的重点，是加大工业产业结构调整力度，着力培育产业竞争新优势，加快发展战略性新兴产业、现代服务业和先进制造业。加快传统产业的淘汰步伐，坚决淘汰钢铁、水泥、电力等高耗能行业落后生产能力，适时制定高耗能产业的退出机制，不断提升产业整体水平；能源结构调整的重点降低煤炭的消费比例，加大优质高效能源的消费比例，加快发展太阳能、风能、核能等可再生能源；控制能源消费总量，不断减少煤炭的消费量。按照减量化、再利用、再循环的要求，积极推进已建工业园的改造，形成资源流、能源流、信息流和工业设施、基础设施、服务设施，通过废物交换循环利用、清洁生产等手段，减少污染物排放，提高资源利用效率。

第三类省份产业结构调整的重点，大力发展低耗能高附加值产业，按照“存量调强、增量条优”的方针，改造提升传统优势产业，坚决淘汰落后产能，加速推进高耗能产业结构的调整，实现传统产业的新型化；探索发展循环经济的有效途径，大力发展循环经济；能源结构调整的重点是以电力为中心，增加优质消费比例，减少原煤消费量，积极开放可再生能源，加快发展生物质能、太阳能、风能等可再生能源。

（2）节能减排技术政策。第一类型的省份在注重工业节能的同时，节能的重点应放在加大

对节能减排先进技术的研发和投入，重点推进交通运输节能和建筑节能，一方面建立大型公共节能监管体系，加大对新建工程设计、施工环保执行建筑节能设计标准、施工验收规范的监督检查力度，加快对既有非节能建筑和用能设市的改造步伐，进一步优化建筑用能管理，优化供热结构；另一方面优先发展城市公共交通，加快城市快速公交和轨道交通建设，确立公共交通在城市交通的中心地位，提高交通工具的能效标准，控制交通耗能的过快增长。

第二类型和第三类的省份主要为我国重要的工业强省，节能的重点应放在工业节能，加快推进工业技术创新和循环经济建设，支持能耗高的产业开展余热余压利用、燃煤锅炉改造、节能新设备的推广；同时加大推进能效标识和节能节水产品认证制度。

（3）加强防治力度，大力削减污染物排放。第一类省份削减污染物排放重点通过加强对各地减排统计监测和执法监管能力，通过环保部门污染源在线监控系统保证相关设施的正常运转，提高企业达标排放水平。在大气污染物排放方面，重点防治机动车排气污染、深化工业污染治理，全面削减大气污染物排放。加快固体废弃物处置设施的建设，推进固体废弃物源头削减和循环利用，提高固体废弃物污染防治能力和水平。

第二类省份削减污染物排放重点在加大重点节能减排工程建设力度，完善城镇污水处理厂及配套设施建设的同时，加大高耗能产业的淘汰步伐，即通过产业结构调整淘汰一批高耗能、高污染的底端产能，然后再通过产业优化升级实现产业结构的高级化，从而实现低排放、低污染的目标。在大气污染物排放方面，重点加大推进煤炭的洁净化利用，支持煤炭清洁利用关键技术的研发，大力发展IGCC（整体煤气化联合循环）、PFBC（增压流化床燃烧）等先进发电技术，大力发展煤直接液化、煤气化、煤间接液化、煤制天然气等煤炭高效低碳清洁转化技术。

（4）完善节能减排的激励约束机制。加大对节能设备产品生产企业和淘汰高耗能设备、大宗采购节能产品的单位给予扶持的政策体系，尽快落实有利于发展循环经济的价格政策、财政政策、产业调整政策、技术政策，加快完善节能节水、利用可再生资源的优惠政策；利用价格机制，适时研究调整资源产品价格，加大差别电价、水价的实施力度，形成节约资源的良好氛围；对节能环保产品，政府采购要优先给予采购。

（5）充分利用市场手段，提高节能减排经济效率。节能减排措施除了采用行政手段以外，可以多考虑一些更具基础性作用的市场手段，建立健全政府为主导、企业为主体，适应社会主义市场经济要求的节能减排长效机制。具体措施包括：第一，探讨形成一套基于市场手段符合经济规律的节能量交易的机制，采用上限与交易机制，允许节能量超额主体与节能量未超额主体之间进行配额单位交易，促使一些参与主体使用节能技术，并提高管理效率。第二，在条件较好的地区积极推进“碳排放权交易体系”试点建设，可以参照《京都议定书》具体的排放贸易机制，联合履行机制和清洁发展机制，通过上述机制分担发达省份的减排责任，同时发挥欠发达省份的减排优势，推动各种减排资源的优化配置，提高节能减排的整体经济效率。

2. 中部地区和东北地区

中部地区和东北地区从整体上来看，产业结构相似，第二产业和第三产业的比例相当。按照节能减排指数，第二类省份包括中部地区的安徽省、江西省、河南省、湖北省、湖南省和东北地区的辽宁省和吉林省；第三类省份包括中部地区的山西省以及东北地区的黑龙江省。

（1）产业结构调整优化政策措施。第二类省份和第三类省份产业调整重点完善现代产业体系，推动装备制造、原材料、汽车、农产品深加工等优势产业升级，大力发展金融、物流、旅游以及软件和服务外包等服务业。同时，积极该地区推进工业生产的清洁化，加快构建企业为主、政府指导、市场驱动、全民参与的清洁生产工作格局，可在重点区域、重点领域实施清洁生产重大工程，带动企业实施清洁生产改造，应用高新技术和先进适用技术改造传统产业，降低资源型经济在经济总量的比例，进一步优化产业结构和产品结构。

（2）节能减排的技术政策。第二类省份和第三类省份主要为该区域的经济大省，第二产业比重较高，第二产业正处于由传统工业向现代制造业过渡的阶段，工业节能潜力很大，应大力开展节能技术创新活动，在本地区重点行业推广一批潜力大、应用广的重大节能减排技术；同时加大建筑节能和交通运输节能工作，加大节能技术在建筑领域的推广和应用，积极推广各种燃料油节约、替代技术和清洁燃料汽车，制定鼓励发展城市职能环保型交通工具，加快淘汰能耗高、尾气排放不达标的机动车辆。对北方采暖地区既有居住建筑供热剂量及节能改造，农村大力发展省柴节煤的炉、窖、炕灶及新型燃料技术。

（3）大力削减污染物排放政策。对于正处于经济结构由传统工业向现代制造业转型阶段的省份，积极探索多元化投资渠道，大力建设城市污水处理工程、规模以上燃煤脱硫、脱硝机组以及建成一批废水深度治理工程、钢铁烧结机烟气脱硫设施等工程，通过这些治理措施达到主要污染物减排目标。

（4）激励引导政策。积极深化资源性产品价格改革，进一步理顺煤炭价格成本构成机制，建立煤热价格联动机制，推进成品油、天然气价格改革，对利用可再生能源发电及利用余热余压、煤矸石和城市垃圾发电，实行相应的电价政策。进一步完善节能减排的财政政策和税收政策，加大对节能环保领域的金融服务，鼓励和引导金融机构按照货币信贷政策与节能环保政策协调配合，进一步加大对循环经济、环境保护及节能减排技术改造项目的信贷支持；出台强制性政策或经济鼓励政策，促进煤炭的洁净化利用，对应用洁净煤技术的企业，给予补助、奖励、优惠电价、税收减免等优惠政策，采用先进发电技术的电厂可优先调度上网。

3. 西部地区

西部地区幅员广阔，区域内差异较大，部分地区生态环境脆弱，因此该地区的节能减排政策应与生态环境建设相匹配。第二类省份包括重庆市、四川省、陕西省、云南省和内蒙古；第三类省份包括云南省和甘肃省；第四类省份全部在西部地区，包括广西、贵州、青海省和宁夏。

（1）加大产业结构调整力度，努力转变经济增长方式。第二类和第三地区产业调整的重点，发挥资源优势，实施以市场为导向的优势资源转化战略，在资源富集地区布局一批资源开发及深加工项目，建设国家重要能源、战略资源接续地和产业集聚区，发展特色农业、旅游等优势产业。培育发展循环经济，推进新型工业化进程，促进经济增长方式由高耗能、高污染向资源节约型和环境友好型转变，提高经济增长质量和效益。

第四类地区调整优化经济结构的重点是培育发展生态经济，促进经济增长方式由传统工业化向资源节约型和环境友好型的生态经济转变，实现生态保护与地区可持续发展。首先，发挥西部地区丰富的自然和文化旅游资源的优势，大力发展旅游业等第三产业。其次，调整优化工业结构，进一步制定、实施资源能源利用效率和污染物排放限制等方面的准入制度，严格限制资源能源效率低、污染物排放强度高的产业发展，坚决淘汰技术落后、浪费资源、污染严重的产品和产业。

（2）调整优化能源结构，提高清洁能源消费比例。一方面优化一次能源供应结构，大力发展坑口电厂，降低煤炭在一次能源供应总量中的比重。在妥善处理生态环境等保护和移民安置等基础上大力发展水电，积极发展风电，鼓励发展生物质能源、太阳能等可再生能源；另一方面优化能源转换结构，发展清洁高效、大容量燃煤机组，优先发展大中城市、工业园区热电联产机组，以及大型坑口燃煤电站和煤矸石等综合利用电站，不断提高一次能源的加工转化比重，特别是提高煤炭消费用于发电的比重。

（3）完善节能减排管理体系。完善节能减排法律法规和相关政策，加大节能减排的宣传力度，不断提高广大人民群众的节能意识、环保意识、责任意识和参与意识，积极支持各类环保民间组织和环保热心人士开展工作，形成全民参与监督的机制。

15 东部地区的节能减排

15.1 “十一五”期间东部地区节能减排综述

15.1.1 “十一五”时期东部地区经济发展评述

东部地区一直是我国经济发展的龙头，我国著名的长江三角洲、珠江三角洲和京津唐城市带三大经济引擎都处于东部地区，更兼国家“主体功能区划”和“综合配套改革实验区”深入推进的任务。“十一五”期间，东部经济发展强劲，主要经济指标都保持良好增长势头。国内生产总值由2005年的110528.22亿元增加到2010年的229384.50亿元，按不变价计算增长81%，年均增长12.6%，快于“十五”时期。2010年东部地区生产总值占全国的比重为53.0%，比2005年降低了2.5个百分点。“十一五”期间东部地区生产总值增长情况如图15－1所示。

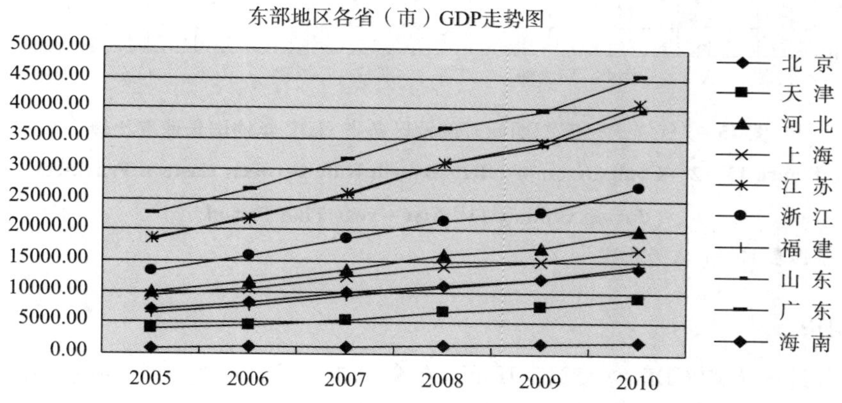

图15－1 “十一五”期间东部地区生产总值增长情况

Figure 15－1 GDP Growth in East China during the 11th Five－year Plan Period

资料来源：《中国统计年鉴》（历年）、《中国统计摘要2011》

1. 增长速度

东部地区内部各省年均GDP增长都在11%以上，从速度上看比较均衡。增长速度最快的是天津市，其五年的增长幅度达到111.09%，翻了一番还多，年均增长速度达到了16.12%，这主要得益于天津滨海新区的开发与建设；其次是福建省，受海西经济区开发的推动，福建省“十一五”经济规模增长了90.98%，接近翻番，年均增长也达到13.81%；经济增长速度较慢的是上海、北京，其五年增长幅度分别为69.36%和71.42%，年均增长速度分别为11.11%和11.38%。“十一五”期间东部地区各省经济增长情况如表15－1所示，“十一五”期间东部地区各省GDP年均增长速度比较情况如图15－2所示。

表 15－1 “十一五”期间东部地区各省经济增长

Table 15－1 GDP Growth in Eastern Provinces during China's 11th Five－year Plan Period

省（市）	五年增长幅度（%）	年均增长速度（%）
北 京	71.42	11.38
天 津	111.09	16.12
河 北	73.82	11.69
上 海	69.36	11.11
江 苏	88.31	13.49
浙 江	75.12	11.86
福 建	90.98	13.81
山 东	85.18	13.11
广 东	79.24	12.38
海 南	87.02	13.34

资料来源：《中国统计年鉴》（历年）、《中国统计摘要 2011》

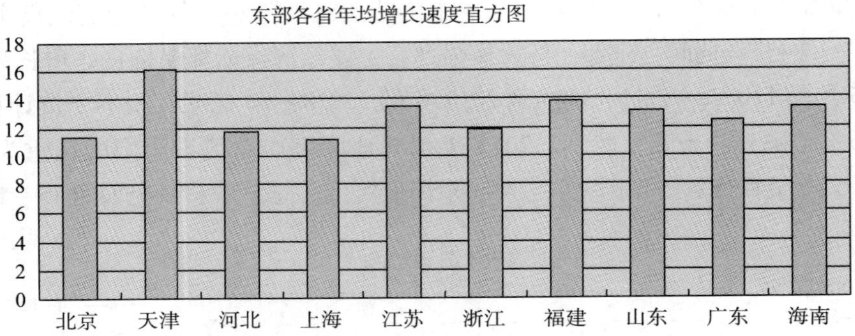

图 15－2 “十一五”期间东部地区各省 GDP 年均增长速度比较

Figure 15－2 Comparison of GDP Growth Rate between Eastern Provinces during China's 11th Five－year Plan Period

资料来源：根据表 15－1 数据整理

2. 人均 GDP

2010 年东部地区人均 GDP 为 45317 万元，在全国四个地区中处于第一位。在区域内部，三大直辖市居前，人均 GDP 均超过 70000 元。上海人均 GDP 最高，为 73297 元；天津次之，为 70402 元；北京第三，为 70251 元。人均 GDP 最低的是海南，为 23665 元；次低的是河北，为 28108 元。数据表明东部地区尽管经济发展总体水平比较高，但各省之间经济发展也存在较大差距。2010 年东部地区各省人均 GDP 情况如表 15－2 所示。

表 15－2 2010 年东部地区各省人均 GDP 情况

Table 15－2 Per Capita GDP of Eastern Provinces in 2010

省（市）	人均 GDP（元）
北 京	70251
天 津	70402
河 北	28108
上 海	73297

续表

省（市）	人均 GDP（元）
江　苏	52000
浙　江	50025
福　建	338914
山　东	41147
广　东	43597
海　南	23665
平　均	45317

资料来源：《中国统计摘要 2011》

3. 产业结构

东部地区各省一、二、三次产业结构呈现出较大的差别。在第一产业方面，上海、北京和天津都只占很少一点比重，而海南省的第一产业比重还很高，达到26%，河北、福建和山东也都在9%以上；而第二产业还是东部的主要经济支柱，除北京、海南和上海外，其他各省的第二产业比重都在50%以上，相差不多；而北京和上海的经济支柱是第三产业，比重分别为75%和57%，已基本完成了产业转型，其他省份都不很高，这说明东部地区第三产业还有有很大发展潜力。2010 年东部地区各省三次产业结构比重情况如表 15－3 所示。

表 15－3　2010 年东部地区各省三次产业结构比重

Table 15－3　Industrial Structure of Eastern Provinces in 2010

省（市）	第一产业（%）	第二产业（%）	第三产业（%）
北　京	0.9	24.1	75.0
天　津	1.6	53.1	45.3
河　北	12.7	53.0	34.3
上　海	0.7	42.3	57.0
江　苏	6.2	53.2	40.6
浙　江	5.0	51.9	43.1
福　建	9.5	51.3	39.2
山　东	9.1	54.3	36.6
广　东	5.0	50.4	44.6
海　南	26.3	27.6	46.1

资料来源：《中国统计摘要 2011》

4. 城镇化水平

“十一五”时期，随着我国城镇化进程的加快，东部地区也进一步加强了城市基础设施建设，提升了城市综合承载能力，致使城镇化水平得到进一步提高。

东部地区各省的城镇化水平差别也较大，上海、北京、天津分别达到88.6%、85%和78.01%，基本达到了发达国家水平；而河北、山东和海南三省的城镇化率还不足50%，总体东部地区的城镇化上还有很大潜力和空间。2009 年东部地区各省城镇化水平如表

15－4所示。

表15－4　2009年东部地区各省城镇化水平

Table 15－4　Urbanization Level of Eastern Provinces in 2009

省（市）	人口总数（千万）	城镇人口（千万）	城镇人口比重（%）
北　京	1755	1492	85.00
天　津	1228	958	78.01
河　北	7034	3025	43.00
上　海	1921	1702	88.60
江　苏	7725	4295	55.60
浙　江	5180	2999	57.90
福　建	3627	1864	51.40
山　东	9470	4576	48.32
广　东	9638	6110	63.40
海　南	864	424	49.13

资料来源：《中国统计年鉴2010》

15.1.2　"十一五"时期东部地区节能减排评述

伴随经济的快速增长，东部地区的能源消费总量也有了较大幅度增长，从2005年115913万吨标准煤的增加到2009年154730万吨标准煤，增长了33.49%，年均增长7.49%；终端能源消费从2005年的65673.11万吨标准煤增长到2009年的88961.72万吨标准煤，增长了35.46%，年均增长7.88%。但能源消费量和终端能源消费量占全国的比重有所降低，分别从2005年的44.00%、43.92%下降到2009年的43.31%、43.31%，整体呈现向好的趋势。

1. 节能评述

在单位产出能耗和节能方面，2009年东部地区平均单位GDP能耗、平均单位工业增加值能耗、平均单位GDP电耗分别为0.873吨标准煤/万元、1.412吨标准煤/万元和989.34千瓦小时/万元，分别比2005年下降11.18%、19.53%和11.49%，单位GDP能耗、单位工业增加值能耗为各区域最低。在东部地区各省市中，单位GDP能耗北京最低，为0.606吨标准煤/万元，较2005年降低15.85%，降幅最高的是上海，降幅为18.55%；单位工业增加值能耗最低的广东，为0.809吨标准煤/万元，单位工业增加值能耗降幅最大的是天津市，降幅为25.56%；单位GDP电耗最高的河北省为1449.94千瓦小时/万元，最低的北京市为681.85千瓦小时/万元，降幅最高的天津市，为22.3%。总体来说，东部地区的能源利用效率全国最高，但平均能耗指标降幅较小。2009年东部地区各省单位能耗情况如表15－5所示。

表15－5　2009年东部地区各省单位能耗情况

Table15－5 Unit Energy Consumption of Eastern Provinces in 2009

省份	单位GDP能耗（等价值）		单位工业增加值能耗（规模以上，当量值）		单位GDP电耗	
	2009年值（吨标准煤/万元）	较2005年上升（±%）	2009年值（吨标准煤/万元）	较2005年上升（±%）	2009年值（千瓦小时/万元）	较2005年上升（±%）
北　京	0.606	－15.85	0.909	－24.01	681.85	－8.79

续表

省份	单位 GDP 能耗（等价值）		单位工业增加值能耗（规模以上，当量值）		单位 GDP 电耗	
	2009 年值（吨标准煤/万元）	较 2005 年上升（±%）	2009 年值（吨标准煤/万元）	较 2005 年上升（±%）	2009 年值（千瓦小时/万元）	较 2005 年上升（±%）
天　津	0.836	-18.55	0.911	-25.56	782.88	-22.30
河　北	1.640	-10.70	2.999	-24.19	1449.94	-5.13
上　海	0.727	-13.52	0.957	-15.16	808.49	-12.53
江　苏	0.761	-11.49	1.107	-23.92	1064.25	-14.49
浙　江	0.741	-11.11	1.123	-19.95	1176.50	-7.59
福　建	0.811	-7.63	1.150	-11.36	1032.05	-10.09
山　东	1.072	-9.80	1.543	-19.00	972.49	-12.15
广　东	0.684	-8.61	0.809	-19.84	1002.09	-12.80
海　南	0.850	-4.97	2.613	-12.16	922.89	-10.57

资料来源：《中国能源统计年鉴》（历年）

在总体节能方面，单位 GDP 能耗小于 1.0 吨标准煤/万元的省份有北京、广东、上海、浙江、江苏、福建、天津、海南六个省市；单位 GDP 能耗在 1.0—2.0 吨标准煤/万元之间的省份有山东和河北；没有单位 GDP 能耗超过 2.0 吨标准煤/万元的省份，说明东部各省总体经济结构较好，节能减排也下了很大功夫。

在工业节能方面，单位工业增加值能耗小于 1.0 吨标准煤/万元的省份有广东、北京、天津、上海四个省市；单位工业增加值能耗在 1.0—2.0 吨标准煤/万元之间的省份有江苏、浙江、福建、山东四省；单位工业增加值能耗大于 2.0 吨标准煤/万元之间的省份有海南和河北，说明东部地区只有这两个省的工业节能技术还比较落后，需要加强改进。

在电力消耗方面，单位 GDP 电耗小于 1000 千瓦小时/万元的省份有北京、天津、上海、海南和山东五省，其他五省该项指标也都低于 1500 千瓦小时/万元，这至少说明两点：第一，东部地区的节电效果是不错的；第二，东部地区电力供应仍然缺乏。

在能源消费结构方面，东部地区终端能源消费结构在各区域中相对比较合理，并在逐步优化中。2009 年东部地区终端能源消费结构中煤及煤制品的比重为 49.74%，石油及制品的比重为 35.35%，电力（其他能源）的比重为 4.11%，天然气的比重 10.81%。天然气和电力（其他能源）在终端能源消费中的比重有所上升，分别比 2005 年上升了 2.30 个百分点和 1.73 个百分点。

2. 减排评述

“十一五”时期，随着东部地区经济发展水平的提升和技术进步加快，人们环保意识的增强，东部地区的环境保护水准有了很大提高，主要污染物减排成效非常显著。2010 年东部地区化学需氧量排放量和二氧化硫排放量分别为 421.05 万吨和 671.41 万吨，分别比 2005 年下降 17.52% 和 20.81%，化学需氧量排放量和二氧化硫排放量降幅是各地区域最大的。其中，北京市化学需氧量排放量和二氧化硫排放量降幅最高，较 2005 年分别降低 20.69% 和 39.74%；化学需氧量排放量降幅最小的是海南省，降幅为 2.84%；二氧化硫排放量降幅最小的是福建省，较

2005 年降低 8.03%。“十一五”末东部地区各省减排完成情况如表 15－6 所示。

表 15－6　“十一五”末东部地区各省减排完成情况

Table 15－6　Pollution Reduction Performance of Eastern Provinces at the End of China's 11th Five－year Plan Period

省份	化学需氧量			二氧化硫		
	“十一五”降低目标	“十一五”实际降幅	超额完成幅度（%）	“十一五”降低目标	“十一五”实际降幅	超额完成幅度（%）
北　京	－14.70	－20.69	40.75	－20.4	－39.74	94.80
天　津	－9.60	－9.60	0.00	－9.4	－11.25	19.63
河　北	－15.10	－17.34	14.83	－15.0	－17.53	16.87
上　海	－14.80	－27.70	87.14	－25.9	－30.19	16.58
江　苏	－15.10	－18.50	22.52	－18.0	－23.50	30.56
浙　江	－15.10	－18.18	20.43	－15.0	－21.13	40.85
福　建	－4.80	－4.82	0.47	－8.0	－8.03	0.33
山　东	－14.90	－19.40	30.20	－20.0	－23.10	15.50
广　东	－15.00	－18.88	25.83	－15.0	－18.81	25.40
海　南	0.00	－2.84	/	0.0	－35.45	/

资料来源：根据《“十一五”期间全国化学需氧量排放总量控制计划表》、《“十一五”期间全国二氧化硫排放总量控制计划表》整理

如果按单位 GDP 减排或人均减排分析，东部地区 2010 年与 2005 年相比，化学需氧量排放量和二氧化硫排放量都有较大幅度的降低。按单位 GDP 测算，东部地区的化学需氧量平均减排幅度达到 51.80%，二氧化硫减排幅度达到 53.02%。其中，山东的单位 GDP 化学需氧量减排幅度达到 56.30%，福建的单位 GDP 二氧化硫减排幅度达到 58.98%，分别是减排幅度最大的。2010 年东部地区各省比 2005 年单位减排情况如表 15－7 所示。

从单位 GDP 排放绝对值来看，北京的两项排放值都是最低的，分别为 8.19 吨/万元和 11.75 吨/万元；其次是上海、天津，排放值相对都较低；河北的两项排放值都比较高，尤其是二氧化硫排放值达到 83.40 吨/万元；海南的单位 GDP 化学需氧量排放是东部最高的，达到 51.92 吨/万元，其单位 GDP 二氧化硫排放量却很低，为 14.41 吨/万元，仅高于北京。

按人均排放测算，东部地区的人均化学需氧量平均减排幅度为 20.68%，人均二氧化硫平均减排幅度达到 22.55%。其中上海的人均化学需氧量减排幅度达到 34.70%，北京的人均二氧化硫减排幅度达到 42.51%，分别是减排幅度最大的，这两个城市的另两项人均减排降幅指标也名列前茅；人均化学需氧量减排幅度最小的是福建，仅为 4.95%；人均二氧化硫减排幅度最小的是河北，为 7.59%。值得注意的是，海南的人均二氧化硫不但没有减少，而且增加了 4.82%。

从人均排放绝对值来看，仍然是北京的两项排放值最低，分别为 49.87 吨/万人和 71.49 吨/万人；其次是山东、河北的人均化学需氧量排放为 64.51 吨/万人、74.26 吨/万人，福建、广东的人均二氧化硫排放量分别为 97.68 吨/万人、99.02 吨/万人，都是比较低的。在人均化学需氧量排放方面，有六个省（市）超过 100 吨/万人，但没有非常高的，最高的是上海的 111.65 吨/万人；在人均二氧化硫排放方面，排放值最高的是河北的 201.99 吨/万人，天津、山东、上海也都超过了 170 吨/万人。

表 15 - 7 2010 年东部地区各省比 2005 年单位减排情况

Table 15 - 7 Pollution Reduction of Eastern Provinces from 2005 to 2010

省份	按单位 GDP（吨/万元）				按人均（吨/万人）			
	化学需氧量排放量		二氧化硫排放量		化学需氧量排放量		二氧化硫排放量	
	2010 年	减排（%）	2010 年	减排（%）	2010 年	减排（%）	2010 年	减排（%）
北 京	8.19	-51.40	11.735	-57.69	49.87	-33.97	71.49	-42.51
天 津	17.44	-55.83	30.08	-58.03	111.14	-20.64	191.68	-24.59
河 北	30.66	-53.17	83.40	-43.72	74.26	-23.11	201.99	-7.59
上 海	16.41	-50.59	25.76	-54.04	111.65	-34.70	175.25	-39.26
江 苏	24.86	-52.89	30.71	-59.05	110.68	-14.44	136.74	-25.62
浙 江	23.47	-47.00	33.58	-47.53	101.31	-16.67	144.97	-17.50
福 建	31.25	-47.90	28.79	-58.98	106.03	-4.95	97.68	-25.16
山 东	18.17	-56.30	49.83	-53.93	64.51	-22.59	176.92	-18.39
广 东	22.82	-51.76	25.54	-55.85	88.45	-23.21	99.02	-29.71
海 南	51.92	-51.11	14.41	-41.41	100.56	-12.53	27.907	4.82
地区平均	24.52	-51.80	33.38	-53.02	91.85	-20.68	132.36	-22.55

资料来源：根据《"十一五"期间全国化学需氧量排放总量控制计划表》、《"十一五"期间全国二氧化硫排放总量控制计划表》整理

15.1.3 节能减排指标完成情况

1. 节能指标

尽管各省的产业结构不同、能源消费结构不同、科技进步水平不同，但是由于"十一五"节能目标带有强制性，而且是一项重要的考核目标，所以各省都想方设法完成了任务。完成目标最好的是北京市，任务目标是 2010 年比 2005 年单位 GDP 能耗降低 20%，实际上降低了 26.59%，超额完成比例 32.95%，在原本基数就很低的情况下实属不易；其次是天津市，任务目标是 20%，实际完成 21%，超额完成比例 5%；然后依次是福建、广东、江苏，超额完成比例都在 2% 以上，但福建和广东因为任务目标是降幅 16%，相对比较容易完成，而江苏任务目标是降幅 20%，而且 2005 年基数也比较低，也超额完成了任务；尤其值得一提的是山东，任务目标降幅 22%，是东部地区最高的，但仍然超额完成了任务。另外，河北 2010 年的单位 GDP 能耗仍然高达 1.583 吨标准煤/万元，节能潜力很大。"十一五"时期东部地区节能目标完成情况如表 15 - 8 所示。

表 15 - 8 "十一五"时期东部地区节能目标完成情况表

Table 15 - 8 Energy Conservation Performance of Eastern Provinces during China's 11th Five - year Plan Period

省份	2005 年单位 GDP 能耗（吨标准煤/万元）	"十一五"时期降低任务目标（%）	2010 年单位 GDP 能耗（吨标准煤/万元）	2010 年比 2005 年实际降低幅度（%）	超额完成比例（%）
北 京	0.792	-20	0.582	-26.59	32.95
天 津	1.046	-20	0.826	-21	5.00
河 北	1.981	-20	1.583	-20.11	0.55
上 海	0.889	-20	0.712	-20	0

续表

省份	2005 年单位 GDP 能耗（吨标准煤/万元）	"十一五"时期降低任务目标（%）	2010 年单位 GDP 能耗（吨标准煤/万元）	2010 年比 2005 年实际降低幅度（%）	超额完成比例（%）
江　苏	0.92	-20	0.734	-20.45	2.25
浙　江	0.897	-20	0.717	-20.01	0.05
福　建	0.937	-16	0.783	-16.45	2.81
山　东	1.316	-22	1.025	-22.09	0.41
广　东	0.794	-16	0.664	-16.42	2.63
海　南	0.92	-12	0.808	-12.14	1.17

资料来源：根据《"十一五"各地区节能目标完成情况表》整理

2. 减排指标

东部地区主要污染物减排成效最为显著。2010 年东部地区化学需氧量排放量和二氧化硫排放量分别为 421.05 万吨和 671.41 万吨，分别比 2005 年下降 17.52% 和 20.81%，化学需氧量排放量和二氧化硫排放量降幅为各区域最大。该区域中，北京市化学需氧量排放量和二氧化硫排放量降幅最高，较 2005 年分别降低 20.69% 和 39.74%；化学需氧量排放量降幅最小的是海南省，降幅为 2.84%；二氧化硫排放量降幅最小的是福建省，较 2005 年降低 8.03%。

3. 节能减排综合指数

根据节能减排综合指数对各区域省份进行分类，可以把节能减排综合指数在［2.00－3.00）之间的为第一类地区，在［1.50－2.00）之间的为第二类地区，在［1.00－1.50）之间的为第三类地区，［0.00－1.00）在之间的为第四类地区。

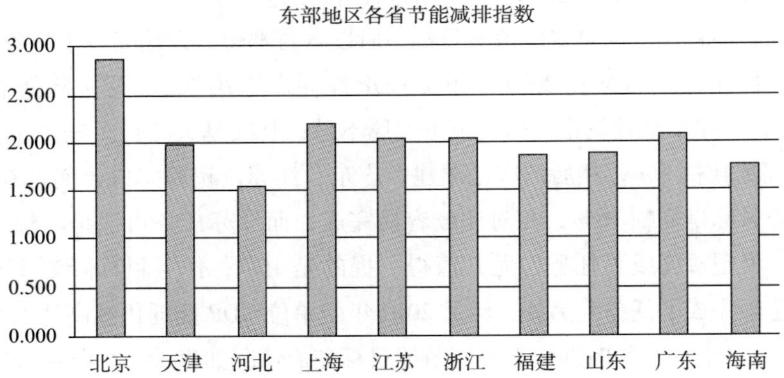

图 15－3　东部地区节能减排综合指数柱状图

Figure 15－3　Bargraph of Energy Conservation and Pollution Reduction Composite Index of Eastern Provinces

资料来源：根据图 14－9 整理

东部地区有五个省市分布在第一类地区，分别是北京市、上海市、江苏省、广东省、浙江省，节能减排综合指数得分分别为：2.867、2.194、2.076、2.031 和 2.027，东部地区在各区域中节能减排效果整体最好。从效果指数、努力程度指数和潜力实现三个二级指数来看，在全国 31 个省市排名中北京得分分别为 1.000、0.867 和 1.00，均为第一；上海两项二级指标第二，一项指标第四；广东两项二级指标第三，一项指标第七。图 15－3 为东部地区节能减排综合指数柱状图。

东部地区另五个省市分布在第二类地区，分别是天津、河北、福建、山东和海南，且从全国来看，在二类地区节能减排综合指数最高的是天津市为1.977。从效果指数、努力程度指数和潜力实现指数来看，天津市的效果指数和潜力实现指数最高，分别为0.925和0.925，在全国均排第六位；努力程度指数为0.127，在全国排第16。山东和福建的综合排名分别为全国的第九位和第11位，也是相对不错的位置。只有河北综合排名为全国的第21位，相对落后一些。

东部地区综合节能减排指数没有排在第三类和第四类的省市，这也说明东部地区的节能减排在全国是最好的。东部地区节能减排综合指数及总排名情况如表15-9所示。

表15-9 东部地区节能减排综合指数及总排名

Table 15-9 Energy Conservation and Pollution Reduction Composite Index of Eastern Provinces and the Ranking

省市	综合指数及排名		二级指数					
	节能减排综合指数	总排名	效果指数	排名	努力程度指数	排名	潜力实现指数	排名
北 京	2.867	1	1.000	1	0.867	1	1.000	1
上 海	2.194	2	0.951	2	0.293	4	0.950	2
广 东	2.076	3	0.948	3	0.181	7	0.948	3
江 苏	2.031	4	0.927	5	0.179	8	0.925	5
浙 江	2.027	5	0.930	4	0.167	9	0.930	4
天 津	1.977	6	0.925	6	0.127	16	0.925	6
山 东	1.877	9	0.871	8	0.133	14	0.873	8
福 建	1.857	11	0.906	7	0.044	25	0.908	7
海 南	1.761	15	0.871	9	0.017	27	0.873	9
河 北	1.536	21	0.722	21	0.095	21	0.720	21

资料来源：根据表14-9整理

15.1.4 节能减排的主要成就及展望

“十一五”时期，国家的节能减排战略得到东部地区各省政府空前的重视，尤其是2008年的北京奥运会、2010年的上海世博会两大国际盛会的举办，带动东部地区的节能减排工作登上了一个新的历史高度。除北京、上海外，其他各省也都是省市一把手亲自抓，进一步完善了监督机制，指标层层落实，取得了预期的成就。

1. 北京

“十一五”以来，北京市以筹办奥运会为契机，坚决退出高耗能、高污染产业。除保留满足城市功能需要的基础性产业外，基本退出高耗能、高污染、低附加值的产业，加快淘汰高污染、低效益的产品和落后的工艺、技术、设备。主要包括首钢分阶段关停钢铁生产能力800万吨，截至2010年底石景山地区冶炼、热轧项目全部停产；实施北京焦化厂、东方石化公司所属北京化二股份公司、有机化工厂停产搬迁；关停小火电机组，淘汰10万千瓦以下燃煤火电机组；淘汰水泥立波尔窑、中空窑生产线，关停或搬迁五环路以内混凝土搅拌站和水泥制品企业；按国家有关规定关闭16种小煤矿。

在减排方面重点控制煤烟型污染、防治机动车排气污染、深化工业污染治理、控制扬尘污染，全面削减大气污染物排放。推进实施高井、华能、京能、国华等燃煤电厂烟气脱硝工程；推进全市20蒸吨以上燃煤锅炉安装烟气脱硫除尘设施及自动监测设备；加快建设新城集中供热工程并配套建设烟气脱硫除尘设施；推进加油站、油罐车、部分油库油气回收治理工作；加强施工工地管理和裸露农田扬尘治理，市区城市车行道路机扫率和洒水率均达到95%，郊区城市车行道路机扫率达到60%；开展与周边省市的合作，控制大气污染物的产生和远距离传输。

2. 天津

天津市在“十一五”期间经济保持快速增长、能耗需求不断增加的情况下，保持了单位能耗持续下降。

一是建立完善节能体系。天津市政府相继出台了《天津市节能监督检测管理办法》、《固定资产投资项目合理用能评估和审查暂行管理办法》等一系列配套法规和规范性文件，使节能法律法规得到进一步细化、量化，法规体系更完善，操作性更强。

二是全面推进重点领域节能。重点抓好冶金、化工、石油石化、电力四大行业，占工业能耗65%的20户“千家企业”，以及94种重点耗能产品和年能耗500吨以上标准煤的设备，使工业能耗持续大幅下降，成为带动全市单位GDP能耗下降的主要动力。

三是加快实施重点节能工程。围绕国家十大重点节能工程建设，积极推动企业节能技术改造，全面推广工业窑炉改造、余热余压回收利用、电机系统改造、能量系统优化改造等重点工程，努力实现建设实施一批、组织报批一批、筹备策划一批，做到每年都有新的节能大项目、好项目，不断提升持续节能能力。

四是积极推广节能新技术。发挥节能专项资金“四两拨千斤”的作用，调动企业和社会资金投入，形成政府引导、企业主体、社会参与的多元投入机制，积极推广节能新技术，取得明显成效，规模以上万元工业增加值能耗五年累计下降38.1%。

“十一五”期间，天津市完成了全部火电机组脱硫治理工程，其中电力行业24台20万千瓦以上机组脱硫效率达到80%，电力行业和工业企业自备电厂43台20万千瓦以下机组脱硫效率达到37%；供热锅炉447台10吨/小时以上脱硫效率达到51%。化学需氧量方面，新（扩）建污水处理厂44座、升级改造污水处理厂16座，铺设排水主干管网2700公里；淘汰落后产能，关停小化工、小造纸等企业80余家；推动工业企业污染治理设施69台（套）。

3. 河北

河北将改善城市环境质量定为城镇面貌“三年大变样”的首要目标，为此发布了《河北省城镇面貌三年大变样环保行动计划》，并对城市建成区内124家不符合功能定位的重点污染企业进行搬迁改造；726家国控、省控重点企业实现达标排放，所有城镇污水处理厂出水达到国家《城镇污水处理厂污染物排放标准》一级标准的A标准或B标准并实现在线监控。

出台《关于进一步加强规划环境影响评价工作的通知》，明确了规划环评审查程序要求，进一步强化了重点区域、行业规划源头污染防治措施。共完成146个工业聚集区、开发区规划环评审查。其中，列入国家计划的5个工业园区，全部完成了规划环评审查；列入省计划的44个工业园区，32个完成了规划环评审查；列入省计划的100个产业聚集区，75个完成了规划环评审查；其他类别聚集区，39 个完成了规划环评审查。

按照“扶优汰劣”、“关小上大”的原则，不断加大淘汰落后产能工作力度。“十一五”期间，全省共关停小火电机组 77 座，总装机 306 万千瓦，关停落后钢铁产能 4300 万吨、水泥产能 4805 万吨、玻璃产能 4055 万重量箱、焦炭产能 508 万吨。

4. 上海

上海市“十一五”期间以全力保障世博会环境质量及为城市未来可持续发展奠定基础为目标，大幅提高了节能减排的标准和要求。

一是加大调整和淘汰落后产能力度；二是严格控制高耗能、高排放行业增长，严格项目审核管理，对未经审核通过的项目一律不准开工；对违规在建项目，责令停止建设；对违规建成的项目，依法停电停水；三是加快实施节能减排重点工程，“十一五”期间完成建筑实施节能改造 3000 万平方米；四是抓好重点领域节能减排，着力抓好工业、建筑、交通三大重点领域节能；五是加快推进合同能源管理和节能服务产业发展，推动公共机构、大型公共建筑及重点用能单位优先采用合同能源管理方式实施节能改造；六是完善节能减排经济政策，强化用能管理，加强预测预警，加大资金投入力度。

通过加快推进污染企业结构调整，先后实施了桃浦、吴淞、吴泾等传统工业区环境综合整治。环境综合整治成效显著，创建了 682 平方公里“基本无燃煤区”和 728 平方公里“扬尘控制区”；全面推进机动车污染控制，提前实施机动车“国Ⅳ”排放标准和成品油质量标准升级。2010 年城镇污水处理率达 81.9%，生活垃圾无害化处理率达 84.9%，基本实现电厂脱硫全覆盖，危险废物、医疗废物得到集中安全处置。

5. 江苏

江苏省“十一五”期间累计淘汰钢铁落后生产能力 1162 万吨、焦化产能 444 万吨、造纸产能 50.3 万吨、皮革产能 25 万标张、化纤产能 16.5 万吨，关停小火电机组 728.6 万千瓦，淘汰水泥落后产能 3320 万吨。连续五年开展化工污染专项整治，关闭化工生产企业 5034 家，占全省化工生产企业总数的 43.5%。

截至 2010 年底，全省城镇污水日处理能力达 1100 多万立方米，比“十五”末翻了一番。所有大型燃煤机组全部安装了脱硫设施，全省燃煤机组总装机 6384 万千瓦，脱硫率近 100%，新增脱硫能力是“十五”末的 3 倍。

加强太湖流域水环境综合整治方案，组织实施《江苏省太湖流域省界断面总体达标方案》和《太湖一级保护区整治方案》，已完成投资 416 亿元；全面开展治淮行动，出台《江苏省淮河流域水污染防治规划执行情况考核暂行办法》，开展通榆河环境治理，印发《关于加强通榆河水环境综合整治的意见》，完成投资 77.36 亿元。

6. 浙江

浙江“十一五”期间出台了《浙江省实施〈中华人民共和国节约能源法〉办法（修订草案）》，使 11 个市、60 余个县（市、区）的能源监察执法机构将真正有法可依。为确保节能减排长效机制，首先强化机制保障，将节能减排任务完成情况作为领导干部综合考评的重要指标，实行严格的问责制和“一票否决制”。相继出台了实施方案、监测、统计、考核和管理等一系列文件，将节能减排目标分解到各县（市、区）和企业，逐步形成全社会的节能减排责任网络。

加强电力、钢铁、有色、石油化工、化工、建材等重点工业行业节能降耗管理；在重视工业节能的同时，实现节能工作向建筑、交通运输、商用及民用节能等社会各领域深化推进；加强新建建筑节能监管，全省城镇新建建筑执行节能强制性标准的比例达到95%；为把环境监控工作落到实处，在2008年率先全面建成环境质量和污染源自动监控系统，并实现全省联网，使浙江省污染源监控系统的信息化指数和信息技术应用水平处于全国前列。

将经济手段引入环境治理，连续几年对电解铝、铁合金、水泥、钢铁等8个行业实行差别电价政策，运用价格杠杆引导高耗能企业加快节能降耗技术改造，加快淘汰落后产能。

7. 福建

结合国家“十一五”十大重点节能工程，福建省选取节能潜力较大的电力、石油、石化、钢铁等高耗能特种设备集中使用的行业，积极推动工业锅炉（窑炉）、余热利用、区域热电联产、能量系统优化等节能工程的推广应用；进一步推进建筑、交通、公共机构节能，加强建筑节能监管与车辆用油定额考核；开展“四个一”节能工程：即开展“1000台在用燃煤工业锅炉运行能效快速测试方法应用试点，100个安全与节能管理标杆锅炉房建设，10个锅炉设计文件节能审查试点，10项节能新技术应用示范”；采取奖励资金与节能量挂钩的办法，对年可形成1000吨标准煤以上节能能力的项目按每吨150元标准奖励。

强化重点耗能单位节能管理。组织开展重点企业节能监测、能源审计和节能诊断工作；深入开展行业性能效对标达标活动，率先在水泥、合成氨、炼钢等行业开展全行业对标。

切实加强用能管理，开展省控200家重点用能企业主要能源品种的消费计量数据在线同步监控和采集；加大节能发电调度力度，优先安排可再生能源发电机组和大容量、高效燃煤脱硫机组发电，限制未脱硫、能耗高的低效机组发电。

强化政策引导，加大实施差别电价政策力度。对超能耗（电耗）限额标准的企业，实行惩罚性电价，对超过限额标准1倍以上的，比照淘汰类电价加价标准执行。积极落实国家支持节能减排的税收优惠政策。

8. 山东

为了更加注重生态文明，大力发展绿色经济、循环经济和低碳经济，山东省“十一五”期间主动要求加压，把单位生产总值能耗由国家布置的五年降低20%提高到22%，把COD排放由五年累计减少14.9%提高到18%，并采取了一系列富有成效的措施：

一是加大淘汰落后产能力度。要求对未完成淘汰落后产能任务的地区，暂停项目环评、供地、核准和审批；对未完成任务的企业，不予审批和核准新的投资项目，不予批准新增用地，加大执法处罚力度。

二是加强高耗能、高排放行业源头控制。严格节能评估审查管理，落实新增能耗等量淘汰和“上大压小”政策，坚决遏制高耗能、高排放行业过快增长。

三是抓好重点领域节能。着重抓好省千户企业节能工作，确保实现年节能量400万吨标准煤以上。大力推进建筑节能，完成了既有居住建筑1900万平方米的改造任务。积极发展节能型交通运输体系，大力推广新能源汽车和不停车收费，淘汰老旧汽车，实现交通节能。

四是组织好重大节能减排项目建设。加大节能资金投入，积极推进200个重大节能技改项目建设。实施好“节能产品惠民工程”，全面推广高效节能空调、节能电机，淘汰低效照明产品，创建1000个“高效照明产品示范村”。

五是加快发展新能源产业和节能环保产业。把新能源作为战略性新兴产业重点培育，全年推广太阳能光热与建筑一体化面积1200万平方米，着力打造新能源产业基地，促进新能源产业规模化集约化发展。加大对节能环保产业扶持力度，建设12个省级节能环保产业基地。

六是强化节能减排监管。加强对节能减排工作的组织领导，严格实行问责制度，搞好督促检查。加快节能减排统计、监测和考核体系建设，建立全省节能降耗预警机制，对能耗出现异动的市、超过能耗限额和使用国家明令淘汰用能设备（工艺）的企业、未经节能评估审查的固定资产投资项目等进行预警和调控。

9. 广东

广东省成立了以省长挂帅、省直26个部门主要负责同志等为成员的省节能减排工作领导小组，建立了节能减排统筹协调机制，领导小组办公室设在省经济和信息化委员会。

省政府将“十一五”节能任务分解下达给各地级以上市政府，并与21个地级以上市政府、重点耗能企业签订节能目标责任书；狠抓重点领域节能减排，加强领导，落实责任。严格控制“两高”和产能过剩行业新上项目，不再审批、核准、备案“两高”和产能过剩行业扩大产能项目；对于那些违规在建项目，要求责令停止建设，金融机构一律不得发放贷款。

积极支持实施国家和省十大重点节能工程，从2008年起设立省节能专项资金，支持节能工作的开展，对年节能量1万吨标准煤以下、2000吨标准煤以上（后扩展到1000吨标准煤）的项目进行奖励，每节约1吨标准煤奖励200元。淘汰落后产能、推动产业转型升级，建立健全工作机制、加强财政资金引导、建立政策激励机制、严格执行差别电价和环保政策等措施。

10. 海南

“十一五”以来，海南按照建设生态省和国际旅游岛建设，以及构建具有海南特色的经济结构和产业结构的要求，将节能减排目标分解到市、县和重点耗能企业，同时为进一步明确责任和强化监管，省、市县政府成立了节能减排工作领导小组，并建立了节能减排统计、监测和考核体系，使全省形成了节能减排市县负责、部门推动的工作新格局。在加强节能减排的宏观管理上，省政府先后出台了《海南省建设节约型社会近期重点工作实施方案》、《海南省节能减排综合性工作方案》和《海南省落实“十一五”期间主要污染物排放总量控制指标工作方案》等一系列文件。

共建成海防林16万亩，使森林覆盖率每年提高近1个百分点，从2005年的55.5%提高到2010年的60.2%。建设完成25个污水处理厂、19个垃圾处理工程，初步形成覆盖全省的垃圾污水处理体系，到2010年底全省城镇生活垃圾无害化处理率、污水处理率分别达到86%和70%。

15.1.5 东部地区节能减排存在问题与不足

“十一五”时期，尽管东部地区的经济取得了快速发展，在节能减排方面总体上也取得了很大进步，全部完成了国家下达的任务，但由于节能减排是一个复杂的系统工程，以及东部地区内部发展也不平衡，仍然还存在很多问题。

1. 各省发展不均衡，差异较大

一些地方政府仍存在将经济增长与节能减排目标相互对立，片面追求经济增长的倾向。特别是在受金融危机影响、经济出现下滑的时候，一些地方认为经济增长必须靠投资拉动，又开始盲目投资高耗能产业，到期末完不成任务又开始“拉闸限电”，人为造成了经济的波动和节能减排的被动。

2. 工业节能减排形势比较严重

资源消耗支撑的现象仍较普遍，发展付出的资源环境代价过大；高耗能行业仍保持较快增长，工业发展面临的资源约束进一步加大；工业生产过程余热余压浪费巨大，能源转换利用技术水平仍然落后，节能与科学用能的研究和指导薄弱。

3. 能源消费结构不尽合理

煤炭在一次能源消费结构中比重依然过高，石油比重有上升趋势，天然气等清洁能源供应不足。

4. 环境压力依然存在

例如，2007年5月，由于连续高温高热，导致太湖蓝藻在短期内积聚爆发，水源水质恶化，最终使城区出现了大范围的自来水发臭现象。专家分析主要原因：一是气温高及水位低因素，由于高温且降水量偏少，日照强烈，加之水位为近50年最低，使藻类生长高峰期大大提前；二是水体因素，氮、磷含量相对较高，水体富营养比较严重，养分充足，适合藻类的生长需求。

15.2 “十二五”时期东部地区节能减排展望

15.2.1 节能减排的战略目标

（1）节能目标。根据国家发改委下发的“十二五”节能目标分解方案征求意见稿，第一类地区全部是东部地区省份，包括天津、上海、江苏、浙江和广东，其单位GDP能耗降低率为18%；第二类地区包括东部的北京、河北和山东，单位GDP能耗降低率为17%；福建属于第三类地区，其单位GDP能耗降低率为16%；海南为第五类地区，其单位GDP能耗降低率为10%。

（2）减排目标。根据国务院《“十二五”节能减排综合性工作方案》关于减排的要求，东部地区各省2015年的减排目标如下：

- 北京：化学需氧量排放总量控制在18.3万吨，比2010年减排8.7%（其中工业和生活控制在9.8万吨，减排9.8%）；氨氮排放总量控制在1.98万吨，比2010年减排10.1%（其中工业和生活控制在1.47万吨，减排10.2%）；二氧化硫排放总量控制在9.0万吨，比2010年减排13.4%；氮氧化物排放总量控制在17.4万吨，比2010年减排12.3%。
- 天津：化学需氧量排放总量控制在21.8万吨，比2010年减排8.6%（其中工业和生活控制在11.2万吨，减排9.2%）；氨氮排放总量控制在2.50万吨，比2010年减排10.5%（其中工业和生活控制在1.95万吨，减排10.4%）；二氧化硫排放总量控制在21.6万吨，比2010年减排9.4%；氮氧化物排放总量控制在28.8万吨，比2010年减排15.2%。
- 河北：化学需氧量排放总量控制在128.3万吨，比2010年减排9.8%（其中工业和生活控制在40.7万吨，减排10.8%）；氨氮排放总量控制在10.14万吨，比2010年减排12.7%（其中工业和生活控制在6.10万吨，减排12.6%）；二氧化硫排放总量控制在125.5万吨，比2010年减排12.7%；氮氧化物排放总量控制在147.5万吨，比2010年减排13.9%。
- 上海：化学需氧量排放总量控制在23.9万吨，比2010年减排10.0%（其中工业和生活控制在20.1万吨，减排10.5%）；氨氮排放总量控制在4.54万吨，比2010年减排12.9%（其中工业和生活控制在4.21万吨，减排12.9%）；二氧化硫排放总量控制在22.0万吨，比2010年减排13.7%；氮氧化物排放总量控制在36.5万吨，比2010年减排17.5%。

• 江苏：化学需氧量排放总量控制在112.8万吨，比2010年减排11.9%（其中工业和生活控制在75.3万吨，减排12.8%）；氨氮排放总量控制在14.04万吨，比2010年减排12.9%（其中工业和生活控制在10.40万吨，减排13.2%）；二氧化硫排放总量控制在92.5万吨，比2010年减排14.8%；氮氧化物排放总量控制在121.4万吨，比2010年减排17.5%。

• 浙江：化学需氧量排放总量控制在74.6万吨，比2010年减排11.4%（其中工业和生活控制在53.7万吨，减排12.5%）；氨氮排放总量控制在10.36万吨，比2010年减排12.5%（其中工业和生活控制在7.84万吨，减排12.5%）；二氧化硫排放总量控制在59.3万吨，比2010年减排13.3%；氮氧化物排放总量控制在69.9万吨，比2010年减排18.0%。

• 福建：化学需氧量排放总量控制在65.2万吨，比2010年减排6.3%（其中工业和生活控制在43.1万吨，减排6.0%）；氨氮排放总量控制在8.90万吨，比2010年减排8.4%（其中工业和生活控制在5.67万吨，减排8.0%）；二氧化硫排放总量控制在36.5万吨，比2010年减排7.0%；氮氧化物排放总量控制在40.9万吨，比2010年减排8.6%。

• 山东：化学需氧量排放总量控制在177.4万吨，比2010年减排12.0%（其中工业和生活控制在54.6万吨，减排12.9%）；氨氮排放总量控制在15.29万吨，比2010年减排13.3%（其中工业和生活控制在8.70万吨，减排13.5%）；二氧化硫排放总量控制在160.1万吨，比2010年减排14.9%；氮氧化物排放总量控制在146.0万吨，比2010年减排16.1%。

• 广东：化学需氧量排放总量控制在170.1万吨，比2010年减排12.0%（其中工业和生活控制在113.8万吨，减排12.9%）；氨氮排放总量控制在20.39万吨，比2010年减排13.3%（其中工业和生活控制在15.16万吨，减排13.5%）；二氧化硫排放总量控制在71.5万吨，比2010年减排14.8%；氮氧化物排放总量控制在109.9万吨，比2010年减排16.9%。

• 海南：化学需氧量排放总量控制在20.4万吨，其中工业和生活控制在9.2万吨，均与2010年持平；氨氮排放总量控制在2.299万吨，与2010年持平（其中工业和生活控制在1.37万吨，可增排1.0%）；二氧化硫排放总量控制在4.2万吨，可比2010年增排34.9%；氮氧化物排放总量控制在9.8万吨，可比2010年增排22.3%。

15.2.2 节能减排的战略选择

东部地区与全国其他地区相比，经济比较发达，节能减排技术相对先进，“十一五”时期大部分省市超额完成了目标，且单位GDP能耗的平均水平远低于其他地区，而且“十二五”时期大部分省的节能目标都排在第一、二类地区，进一步加大了完成节能减排任务的难度。

1. 北京

“十二五”时期，北京市将适应特大型城市能源运行特点和绿色发展要求，以确保能源安全为前提，加快能源结构调整，进一步加强节能减排工作。

更加注重制度建设，以建筑节能和管理节能为重点，依靠科技进步、标准带动、价格和利益机制引导，加快向“内涵促降”转变，下更大的力气把节能降耗提升到一个更高水平。

大幅度提高建筑节能水平。抓住城市化加快发展、建筑规模较大的关键时期，着眼于长远，大力推广绿色建筑，突出抓好建筑标准制定、新材料和新技术应用以及结构、设计等关键环节，大幅度提高建筑节能水平。提高建筑节能标准，新建居住建筑实施75%节能设计标准。推广高效照明产品，基本淘汰白炽灯。

深度推进工业节能降耗，强化产业退出标准和产品设备淘汰目录约束作用，强化企业节能管理，鼓励企业建立健全全流程绿色管理体系，深入实施清洁生产，加快实施节能改造，系统提升工业能源利用效率。

全面推进公共机构节能降耗，实行公共机构定额用能管理制度，实施办公建筑节能、公务车节油、空调和数据中心节电等关键环节节能措施。完成2000家公共机构节能改造，政府机构率先全部完成。

深入强化目标责任考核，增加对重点行业进行节能减排目标责任分解；健全能源统计计量体系，建成市、区、重点用能单位三级节能监测管理综合平台；实施能评全过程管理机制，推进重大用能企业技术改造；继续发布节能节水减排技术（产品）推荐目录，推进热计量改革，全面实施阶梯电价。

继续控制生产型污染。进一步优化能源结构，大幅增加天然气等清洁能源利用，减少煤炭使用，严格控制煤烟型污染，控制餐饮油烟等低矮面源污染；加大资源消耗型、污染型企业淘汰力度，坚决退出中小型水泥、建材、玻璃、化工等高排放企业；建立氮氧化物排放总量控制制度，推广低氮燃烧技术，水泥窑全部进行脱硝治理。完善挥发性有机物产品准入标准和监控体系，有效治理化工、涂料、家具制造、包装印刷等行业挥发性有机物污染。

加大治理机动车污染。实施国家第五阶段机动车污染物排放标准，加速淘汰老旧机动车，鼓励使用节能环保型汽车，在公交、环卫、出租等公共服务领域推广使用新能源汽车，支持物流企业建立“绿色车队”。

2. 天津

“十二五”期间，天津预计将新增加人口300万人，占现有常住人口总数的近1/4，经济发展和人口增加将带来能耗增长的刚性需求和排放的大量增加。“十二五”期间，天津的节能减排重点应注重以下方面：

一是加快金融、创意、科技服务等高端行业发展，使服务业占GDP的比重年均提高1个百分点，到“十二五”末达到50%。

二是大力发展节能环保、新能源、新材料、高端装备等战略新兴产业；同时加大对高耗能、高污染行业调整力度，加快淘汰落后产能。

三是大力发展新能源和可再生能源，提高燃气比重，严格控制煤炭增量，遏制主要污染物增量的过快增长，进一步推动风能、太阳能、生物质能等非化石能源的利用。

四是大力发展循环经济。以建设国家循环经济示范试点城市和国家低碳城市试点为契机，大力发展循环经济和低碳经济。将泰达、子牙、临港、北疆、华明的循环经济发展模式向全社会推广。发展资源循环利用产业，在化工、冶金、电子、医药、建材等领域，培育一批具有行业代表性的循环经济试点企业，建成20条具有较高水平、较大规模、行业特色突出的循环经济产业链条。

3. 河北

河北“十二五”时期将重点抓好保定国家级低碳城市发展试点和邯郸国家级循环经济试点；强力推进节能减排，积极推广应用清洁能源，深入谋划和实施一批重大节能减排工程，严控新建高耗能、高排放项目；综合运用法律、市场、经济和必要的行政手段，坚决淘汰钢铁、煤炭、水泥、玻璃、造纸、制革等行业落后产能。建立有利于促进绿色发展的长效机制，健全节能减排市场化机制，加快推行合同能源管理，逐步建立碳排放交易市场，强化统计监测和目标责任考核，全面推进节能、节地、节水、节材和资源综合利用。

4. 上海

“十二五”期间，上海将建立全市重点用能单位节能监控平台，深入推进工业节能，完善重点用能单位能效管理，推进能源审计、对标管理和清洁生产，实施工业用电设备节电、工业锅炉节能等技改专项工程；强化建筑节能，新建建筑全面实行65%的节能标准，加强既有建筑节能改造，提高建筑用能管理水平；加强交通节能，鼓励和引导市民绿色出行，有序发展公共租赁自行车，积极推广节能环保型交通运输工具和运输方式；对机关、学校、医院、商场、旅游宾馆等公共机构和大型公共建筑，实施能耗定额管理制度。

强化节能目标责任考核，完善节能法规和标准；健全节能市场化机制，推广合同能源管理，培育一批大型专业化节能服务公司；运用价格、财税等手段推进节能降耗，完善重要能源价格形成机制，进一步健全体现差别化、限制性、惩罚性的价格政策。

有效控制化学需氧量、氨氮、总磷等污染物排放总量，推进污水处理厂和收集管网建设，完成白龙港污染处理厂扩建二期、白龙港南线东段及郊区污水管网工程，实施初期雨水处理和雨污混接改造，基本完成全市建成区直排污染源截污纳管；加快竹园污泥处理工程、郊区污水厂污泥处理工程建设，在全国率先实现污水厂污泥基本有效处理；加强对重大环境风险源的动态监测和风险预警及控制；强化环保责任制和执法监督，健全重大环境事件和污染事故责任追究制度；加强流域、区域环境合作和污染联防联控。加强杭州湾北岸化工集中区、宝山南大地区、高桥石化地区、青东农场等重点地区的环境综合整治。

5. 江苏

“十二五”时期，江苏将继续推广合同能源管理，实施重点节能工程；健全完善倒逼机制、约束机制和激励机制，全力推进工业、建筑、交通等重点领域节能，继续组织开展重点耗能企业节能行动计划，积极推广节能技术和产品；加强资源综合利用，重点实施资源综合利用领域重点技术创新及重大示范工程项目；到2015年，一般工业固体废物综合利用率达到95%以上。

重点推进冶金、化工、纺织、造纸等重点行业和涉及重金属行业开展清洁生产审核，对超标、超总量排污和使用、排放有毒有害物质的重点企业依法实施强制性清洁生产审核；推广应用先进清洁生产工艺技术与设备，建成一批清洁生产先进企业。

严格控制化学需氧量、氨氮及太湖流域总氮、总磷污染排放总量；继续实施重点污染源入园进区、搬迁治理、提标改造，合理布局和建设污水处理厂及配套管网，提高污水管网配套率，逐步实现污水统一纳管处理，适当提高排污费征收标准；新建城镇污水处理厂必须配套建设除磷脱氮设施，执行国家城镇污水处理一级A排放标准，推进已建污水处理厂的除磷脱氮改造工程；建设中水回用工程，提高中水回用比例；解决徐州、宿迁等城市尾水出路问题；加强农村生活污水治理和畜禽养殖污染治理设施建设；到2015年，建制镇污水处理设施覆盖率达到90%，规模畜禽养殖场粪便综合利用率达到80%。

深入实施太湖流域水环境综合治理总体方案，全面实施小流域综合整治，实施结构调整、控源截污、调水引流、蓝藻治理、生态修复等措施，推动太湖流域水质持续改善，保护太湖周边饮用水安全；切实加强洪泽湖、高邮湖、骆马湖等重点湖泊的保护与治理，加快推进长江流域、淮河流域、京杭大运河及南水北调东线江苏段和通榆河水污染防治，建设清水走廊，限期治理入江、入湖河流污染。

6. 浙江

“十二五”时期，浙江将积极开展结构节能、技术节能、管理节能和全民节能，深入实施节能降耗“十大工程”；建立严格的耗能行业准入制度，严格控制新上高耗能项目。突出抓好工业、建筑、交通运输和公共机构等领域的节能，大力开展节能产品进机关、企业、学校、社区和家庭活动。

全面落实综合减排措施，主要污染物减排实现国家下达目标，严格控制重金属、持久性有机污染物等有毒有害污染物排放。强化源头控制，实行空间、总量、项目“三位一体”环境准入制度；加大重点区域、行业、企业污染整治力度，加快推进城镇污水处理厂和配套管网、垃圾收集和处理设施建设，有效提高城镇污水处理率、垃圾无害化处理率、再生水利用率和污泥无害化处置率。

进一步加大流域、平原河网、湖库水环境综合整治力度，建立重点流域和湖库蓝藻、海洋赤潮的预警应急机制，八大水系、主要湖库水质达到或优于Ⅲ类标准的比例超过75%；继续强化工业、城镇生活和农业等各类水污染源治理，县级以上集中式饮用水源地水质达标率超过90%；深入实施“清洁空气行动”，有效控制酸雨、灰霾和光化学烟雾等污染。加强危险废弃物、医疗废物全过程监管和处置设施建设。

7. 福建

“十二五”期间，福建将加大投入力度，加快建设覆盖生产、流通、消费等各环节的资源循环利用体系，形成有利于循环经济发展的税收、价格、信贷、政府采购政策体系和评估考核机制；构建循环型产业体系，鼓励企业建立循环经济联合体，实现内部生产工艺间节能和物料循环利用，重点在电力、交通、建筑、冶金、化工、石化等行业推行循环型产业链和共生产业模式；加快建设以城市社区和乡村分类回收站点为基地、集散市场为枢纽、分类加工利用相衔接的再生资源回收利用体系，推动垃圾源头减量和分类收集，加强废金属、废纸、废塑料、废旧轮胎、废弃电子电器产品、废旧机电产品、废弃包装物等的回收和循环利用，加强共伴生矿及尾矿综合利用，推进工业固废和建筑废弃物综合利用，工业固体废物综合利用率达75%；推进汽车零部件、工程机械等机电产品再制造试点；开展城市餐厨废弃物资源化利用。组织实施国家级和省级循环经济试点，强化清洁生产审核，推进循环经济示范城市、园区和企业建设。

加速淘汰能耗高、效率低、污染重的落后工艺、技术和设备，控制碳密集型生产项目的布局；严格执行商品能效标准，限制低能效产品消费；加强低碳技术研发和产业投入，大力发展低碳产业，加快建立以低碳排放为特征的产业体系，形成低碳产业新的增长点；强化循环式生活方式的宣传教育和政策引导，推动居民节水、节电，鼓励消费者购买使用节能节水产品、节能环保型汽车和绿色建筑。

8. 山东

“十二五”期间，山东省将在超额完成“十一五”的“上大压小”任务基础上，继续淘汰落后机组，大力发展新能源，力争全省燃煤火电所占比重由目前的92%下降到71%；新能源装机达到1340万千瓦，占电力总装机比重由5%提高到12%。

通过制定有针对性的产业、财税、价格、环保等政策，实施好重大节能减排项目建设，加大淘汰落后产能力度；努力推进再生资源回收利用，充分挖掘废弃物资源价值，构建再生资源回收网络体系，原材料企业废旧资源利用要达到规定标准；强化节能减排监管。严格问责制度，加快节能减排的统计、监测和考核体系建设，完善全省预警机制，切实完成节能减排任务。

加强环境保护，以保障南水北调沿线和重点区域水环境安全为重点，巩固提高省控59条主要污染河流治污成果；以治理工业废气、城市扬尘和机动车尾气为重点，突出抓好城市建成区大气污染防治；以重金属污染治理为重点，加快构建环境安全防控体系；以农村污水和垃圾处置为重点，开展农村环境综合整治；以海岸带、海域、海岛整治修复为重点，打造生态海岸、生态海洋。

9. 广东

“十二五”广东的节能重点在于提高能源利用效率，抓好工业、建筑、交通运输等重点领域节能，继续实施十大重点节能工程和节能产品惠民工程；提高产业招商引资项目的能耗准入门槛，推进高耗能企业节能技改或转移；大力发展绿色社区和绿色建筑，推进既有建筑节能改造；加强用电需求侧管理，建立节电管理长效机制；推广先进节能技术和产品，加快推进合同能源管理，促进节能服务产业发展；建立健全用能单位主体责任制，完善能耗统计、监测体系，加大节能执法力度。

强力推进淡水河、石马河、前山河、练江等流域水污染综合整治，强化重点污染源工业废水达标排放，推进工业废水集中处理，加快建设和完善城镇生活污水处理厂及配套管网建设；开展重点流域农村连片整治，加强畜禽养殖污染防治，逐步减少化肥农药使用量，降低农业面源污染；到2015年，全省城镇污水集中处理率达到75%以上，其中珠三角达到85%以上；国控、省控断面水质达标率达到75%以上，基本解决珠三角河涌污染问题。

实施大气污染联防联治，重点防治电力、钢铁、水泥等行业的污染排放，控制二氧化硫、氮氧化物、颗粒物和挥发性有机物的排放，重点解决珠三角大气灰霾问题；在电力行业全面推行低氮燃烧技术，推动火电机组配套建设烟气脱硫脱硝设施；加强机动车氮氧化物排放控制，提高新车准入门槛和车用成品油标准。实施生产企业的挥发性有机物排放控制；到2015年，全省地级以上市空气质量达到二级以上标准的天数占全年比例保持在95%以上。

10. 海南

“十二五”期间，海南将由外延粗放利用资源向内涵集约利用资源转变，以提高能源资源利用效率为核心，以节能、节水、节地、节材、能源资源综合利用和发展循环经济为重点，逐步形成节约能源资源和保护生态环境的产业结构、增长方式、消费模式，提高全省能源资源利用水平。

节能方面，完善节能法规和标准，提高节能环保市场准入门槛，彻底淘汰立窑水泥、黏土砖等落后产能；强化节能目标责任考核，加强重点用能单位节能管理，抓好建筑、交通、公共机构等重点领域节能减排。推进节能减排重点工程，大力推广蓄能、变频、高效节能空调、节能汽车、节能电机等节能新技术新产品。

从生产、流通、消费等环节入手，构建资源循环利用体系，大幅度提高资源产出效率；加快循环经济区建设，继续推进洋浦经济开发区、东方工业区、老城开发区、昌江工业区等循环经济示范园区建设，加强以余热利用、冶金废物利用、粉煤灰回收利用等行业为重点的循环经济建设；以沼气为纽带，推进畜禽养殖业废弃物、糖业蔗渣等综合利用，发展循环农业。以城镇和旅游区为重点，借鉴国际经验，建立全岛再生资源分类回收利用体系，发展再生资源产业，促进垃圾资源化利用。

编制和实施全省低碳技术应用与推广计划，优化能源结构，增加非化石能源占一次能源消费比重；积极推动低碳产业发展，建设以低碳排放为特征的工业、建筑和交通体系，推动经济社会发展向高能效、低能耗、低排放模式转型。加强蓄能、变频、建筑节能等低碳技术和洁净煤、新能源汽车、节能灯等产品推广应用；力争“十二五”时期海口市、三亚市公交车、出租车基本普及新能源汽车；推广使用更高排放标准的汽车清洁燃料，完善汽车加气、充电站点的布局和建设；积极开展低碳城镇、低碳工业园区和低碳旅游景区试点，加大低碳技术研发和应用。

16 中部地区的节能减排

16.1 “十一五”中部地区节能减排综述

16.1.1 中部地区的经济发展

随着经济全球化和区域经济一体化的进一步深化，我国经济发展从东到西、从沿海往内陆梯度推进力度不断加强，国内经济结构也在发生着深刻的变化，东部沿海地区向中部地区的产业转移进程明显加快，转移产业主要集中在劳动密集型等传统产业，转移产业的层次进一步提高，汽车及零部件、家电、高新技术、现代服务业等产业逐步成为转移的重点，这些都为中部地区加快发展创造了机遇。

1. 从地理位置来看

中部地区承东启西，位于东部沿海地区与西部地区之间，由于东部沿海地区生产要素成本和综合商务成本的逐渐提高，中部地区更容易接受到东部沿海地区的经济辐射和影响。随着中部崛起发展战略的全面实施，中部地区已经成为国家鼓励和支持发展的重要区域。中部地区凭借要素供给充裕、综合成本相对较低、产业配套能力强、区位条件优越等诸多有利条件，已经发展成为东部发达地区产业和资本转移的首选地区和重要基地。至“十一五”末，中部地区经济规模占全国19.7%的份额，比“十五”末提高了0.9个百分点。同时，作为国家综合运输网络的中心区域和重要的能源与原材料基地，中部地区的生产力布局与经济发展空间正在不断优化，湖南长株潭城市群“两型社会”、湖北武汉城市圈“两型”社会、河南中原城市群、安徽皖江城市带以及山西国家资源型经济转型综合配套改革试验区等国家级战略的相继出台，为中部地区崛起描绘了美好的发展前景。

2. 从经济发展规模来看

“十一五”时期，中部地区各省的GDP年均增长速度比“十五”时期明显加快。截至2010年，河南省的GDP超过2万亿元大关，达到22942.7亿元；GDP超过1.5万亿元的有湖南和湖北二省，分别为15902.1亿元和15806.1亿元；安徽省的GDP也超过了1万亿元，达到了12263.4亿元；只有山西和江西的GDP不到1万亿元，分别为9088.1亿元和9435亿元。可见，中部地区各省经济发展速度日趋加快，经济规模差异暂时还不像东部地区和中部地区等其他地区相差悬殊很大，但随着中部大开发战略的不断向纵深推进，各省的GDP曲线呈现加快发散的趋势，差异化有逐步扩大的势头，如图16－1所示。

3. 从产业结构上看

总体上，中部地区各省的产业结构大都表现为“二三一”结构。第一产业比重偏高，高于

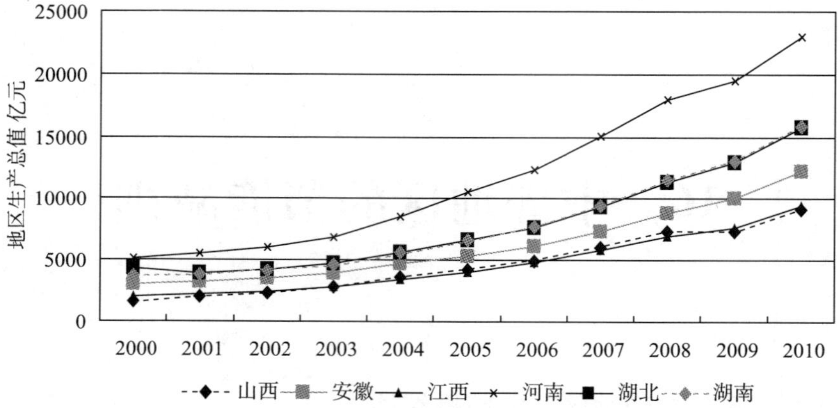

图 16－1　中部地区各省地区生产总值

Figure 16－1　Regional GDP of Central Provinces

资料来源：《中国统计年鉴》(历来)、《中国统计摘要 2011》

10%的省份有湖南、河南、安徽、湖北和江西五省，分别为 14.7%、14.2%、14.1%、13.6%和 12.8%，仅山西省第一产业比重低于 10%，为 6.2%；第二产业比重也偏高，第二产业比重超过 50%的省份有河南、山西、江西和安徽四省，分别为 57.7%、56.8%、55.1%、52.1%，湖北和湖南二省的第二产业比重也只是略低于 50%，分别为 49.1%、46%；第三产业比重偏低，发展滞后，第三产业比重最低的河南省，为 28.1%，其他五个省份的第三产业比重都超过了 30%，最高的湖南省为 39.3%，这说明中部地区各省产业结构还有待进一步优化，如图 16－2 所示。

从三次产业结构的变化来看，与“十五”末相比，在“十一五”末中部地区各省的三次产业结构表现出不断优化的趋势，第一产业所占比重有所下降，第二产业所占比重大幅上升，第三产业所占比重也略有下降。其中，第一产业比重下降幅度达到 5 个百分点左右的有江西和湖南二省，分别为 5.1% 和 4.9%，安徽、河南和湖北三省的第一产业比重分别下降了 3.9%、3.7%和 3%，而山西省则仅仅下降了 0.1 个百分点；第二产业所占比重升幅最大的是安徽，为 10.8%，升幅最小的是山西，为 0.5%，江西、湖南、湖北和河南四省第二产业所占比重升幅都在 5%之上，分别为 7.8%、6.1%、6% 和 5.6%；第三产业所占比重降幅最大的是安徽，为

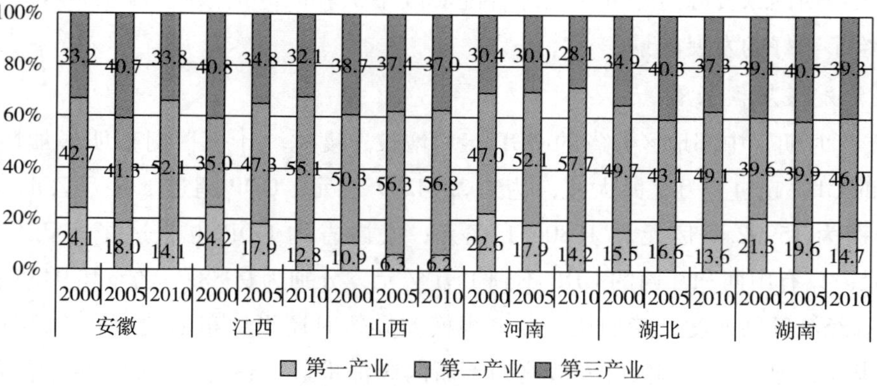

图 16－2　中部地区的产业结构变化

Figure 16－2　Industrial Restructuring of Central Provinces

资料来源：《中国统计年鉴》(2001，2006)、《中国统计摘要 2011》

6.9%，降幅最小的是山西，为0.4%，湖北、江西、河南和湖南四省第三产业所占比重降幅分别为3%、2.7%、1.9%和1.2%。比较分析三次产业结构的变化，江西、安徽和湖南三省的产业结构有较大改进，河南和湖北二省的产业结构改进程度不大，山西省的产业结构改进不大。

4. 从工业化发展程度来看

按照“工业化率达20%—40%为正在工业化初期；40%—60%为半工业化国家；60%以上为工业化国家”的标准。“十五”末，工业化率低于40%处于工业化初期的省份仅有湖南省，为39.46%，其他五省的工业化率均高于40%而不足60%，处于半工业化阶段，其中山西和河南二省的工业化率超过50%，江西、安徽和湖北三省的工业化率在50%之下，如图16-3所示。

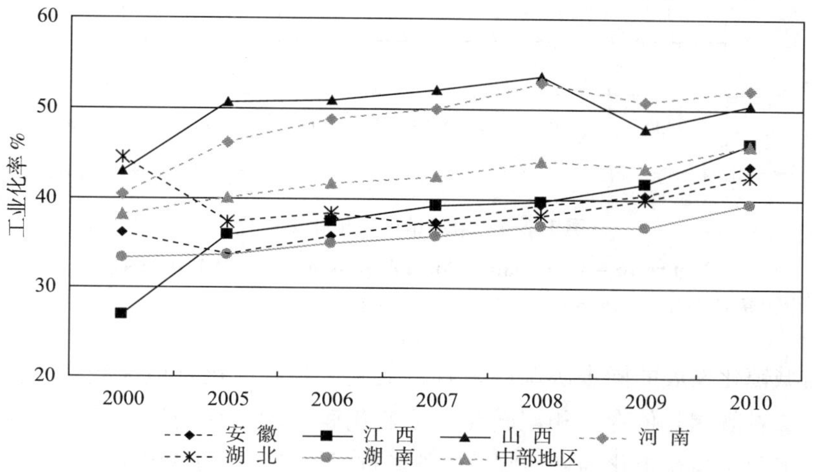

图16-3　中部地区工业化率变化

Figure 16-3　Fluctuation in Industrialization Rate of Central Provinces

资料来源：《中国统计年鉴》(历年)、《中国统计摘要2011》

从工业化的趋势来看，与“十五”末相比，“十一五”末，中部地区各省的工业化程度在稳步提高。除了由于2008年美国次贷危机引发的全球范围经济危机的影响，中部地区部分省份2009年的工业化水平出现了一定程度的滑坡外，在其他年份中部地区的其他省份的工业化水平都在稳步提高。相对于2005年，2010年中部地区工业化率的平均水平（45.95%）提高了5.89个百分点。其中，江西和安徽二省的工业化率提高幅度较大，分别为10.32%和9.91%；然后依次为河南、湖南和湖北，分别为5.85%、5.83%和5.19%；只有山西省的工业化率出现了微小的下降，降幅为0.2%。

5. 从城市化水平来看

当一国（地区）人均GDP超过3000美元时，其城市化率会超过50%，达到55%左右。2009年中国的人均GDP已达到3734.8美元，而城市化率为46.59%，远低于55%的水平，可以说，中国的城市化是落后于经济发展水平的。对于我国中部地区而言，“十一五”期间各省的城镇化进程有所加快，城市化率由“十五”末的36.54%上升到2009年的42.26%，年均增幅1.14个百分点，城镇化水平稳步提高；但是中部地区城市化水平42.26%总体落后于全国城市化率46.59%的4.33个百分点，各省城市化水平也均落后于全国城市化的平均水平。具体地，2009年中部地区城市化水平较高的省份有湖北和山西，分别为46%和45.99%，然后依次是湖南、江西和安徽，城市化率分别为43.2%、43.18%和42.1%，河南的城市化水平最低，仅为

37.7%。以上分析表明，中部地区各省城市化率都处在高于30%与低于60%的城市化急速发展阶段，可以判断，中部地区各省已经进入城市化快速发展阶段，图16－4为中部地区城镇化率变化情况。

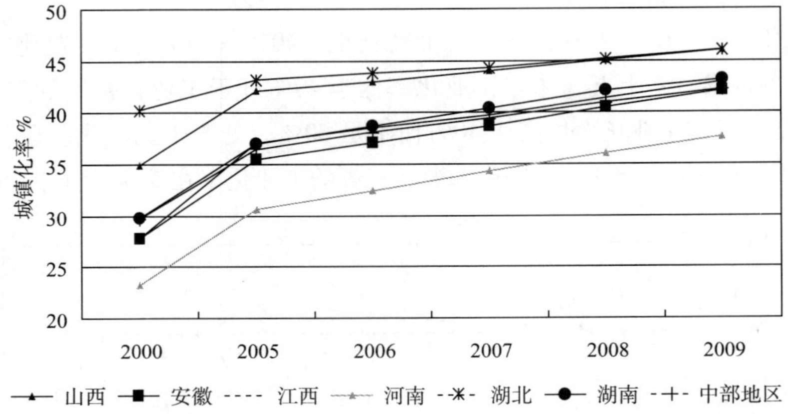

图16－4　中部地区城镇化率变化

Figure 16－4　Urbanization Progress of Central Provinces

资料来源：《中国统计年鉴》（历年）、《中国统计摘要2011》

从工业化与城镇化发展的同步协调性来看，工业化和城市化是地区社会经济发展现代化进程的两个轮子，二者既相互依存、相互转化，工业化要以城市化为发展动力，城市化则靠工业化为基本支撑。工业化与城市化的良性联动，能够促进经济社会协调发展。从图16－5中可以看出，中部地区城镇化发展滞后于工业化的发展水平，其城市化与工业化发展并没有同步发展。

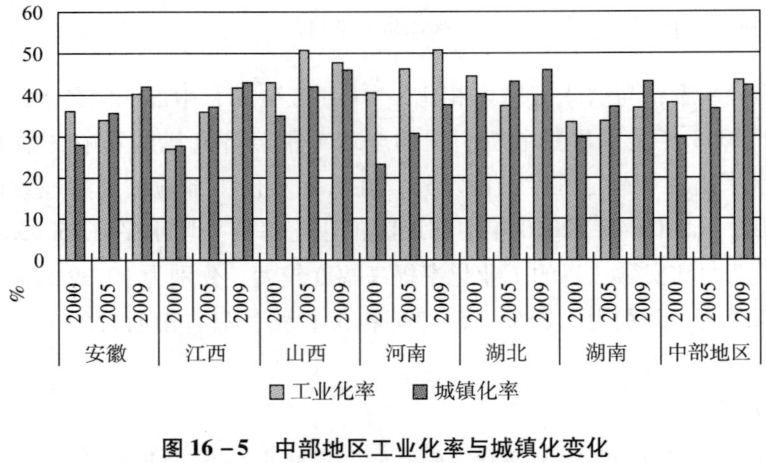

图16－5　中部地区工业化率与城镇化变化

Figure 16－5　Central Provinces' Changes in Industrialization and Urbanization Rates

资料来源：《中国统计年鉴》（历年）

16.1.2　中部地区节能减排

1. 能源消费

随着我国中部崛起发展战略不断深入推进，以及国家在中部地区实施的多个国家级发展战略的部署，中部地区经济规模迅速发展起来，同时也带动了中部地区能源消费的迅速增长，至“十一五”末，已经发展成为我国能源消费的第三大地区，2009 年中部地区能源消费量已经达到 77074 万吨标准煤，比 2005 年的能源消费增长了 32.98%，年均增长率为 7.39%。从终端能源消费情况来看，中部地区 2009 年的终端能源消费量达到 45073.07 万吨标准煤，比 2005 年的能源消费增长了 35.73%，年均增长率为 7.94%。其中，煤及煤制品消费量为 31414.32 万吨标准煤，石油及制品消费量为 8872.01 万吨标准煤，天然气消费量为 1079.4 万吨标准煤，电力及其他能源消费量 3707.34 万吨标准煤，在中部地区终端能源消费量所占比重分别为 67.9%、19.68%、2.39% 和 8.23%。

从各省的能源消费总量上看，截至 2009 年，能源消费量接近 20000 万吨标准煤的省份有河南省，为 19751 万吨标准煤，占中部地区能源消费量 25.63%；能源消费量超过了 10000 万吨标准煤的省份有山西、湖北和湖南三省，分别为 15576 万吨标准煤、13708 万吨标准煤和 13331 万吨标准煤，在中部地区能源消费中所占比重为 20.21%、17.79% 和 17.3%；能源消费量不超过 10000 万吨标准煤的省份有安徽和江西二省，分别为 8896 万吨标准煤和 5813 万吨标准煤，在中部地区能源消费中所占比重为 11.54% 和 7.54%。可见，中部地区各省能源消费总量大，能源消费增幅快，且各省之间的差异程度也比较大。图 16－6 为 2009 年中部地区各省能源、消费比例情况，图 16－7 为“十一五”期间中部地区各省能源消费量情况。

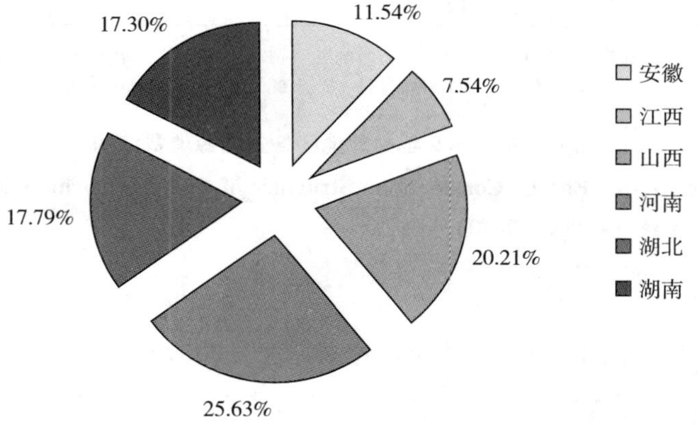

图 16－6 2009 年中部地区各省能源消费比例

Figure 16－6 Energy Consumption Proportions of Central Provinces in 2009

资料来源：《中国能源统计年鉴》(2010)

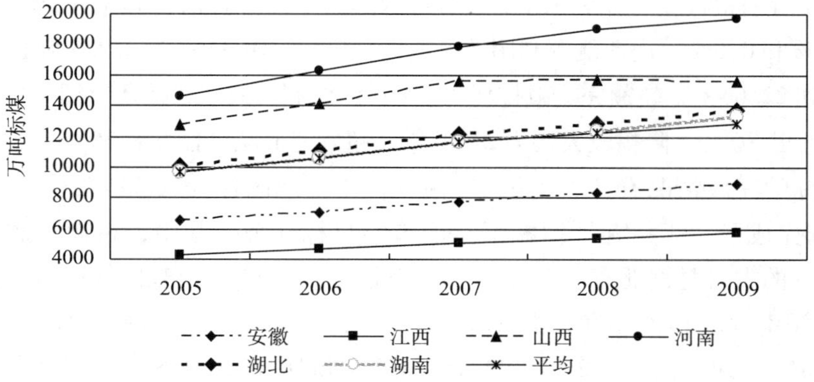

图 16－7 “十一五”期间中部地区各省能源消费量

Figure 16－7 Energy Consumption of Central Provinces during China's 11th Five－year Plan Period

资料来源：《中国能源统计年鉴》(历年)

2. 能源结构

2009年，中部地区煤炭消费比重均高于全国煤炭消费的平均水平。其中，消费比重超过75%的省份有山西和河南二省，分别为76.6%和75.36%；比重超过65%的省份有安徽、江西和湖南三省，分别为73.32%、67.81%和67.04%；仅湖北省的比重低于65%，为59.39%。油品消费比重高于30%仅有湖北省，其他省份的油品消费比重均低于全国油品消费的平均水平。中部地区各省天然气消费消费比重均低于全国天然气消费的平均水平。电力及其他能源消费方面，除了湖南的比重高于全国平均水平，为13.29%，其他五省均比全国平均水平低。2009年中部地区各省能源消费结构情况如图16-8所示。

通过以上分析可以发现，湖北省属于原煤与原油消费并重的能源消费结构，其他五省则均属于以原煤为主的能源消费结构。

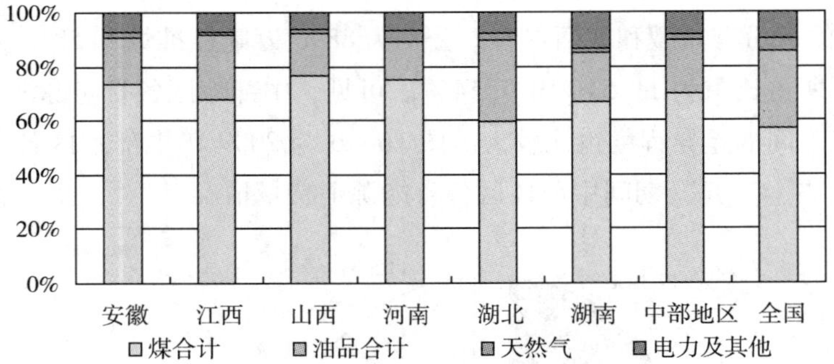

图16-8 2009年中部地区各省能源消费结构

Figure 16-8 Energy Consumption Structure of Central Provinces in 2009

资料来源：《中国能源统计年鉴》(2010)

3. 节能效率

从能源利用效率看，中部地区在国内处在中等水平。2009年中部地区平均单位GDP能耗为1.308吨标准煤/万元，比2005年下降12.53%；从各省的能源利用效率情况来看，江西能源利用效率最高，单位GDP能耗为0.845吨标准煤/万元，但降幅最小，为20.04%，仅超额完成减排任务0.2%；能源利用效率较高的还有安徽，单位GDP能耗为0.845吨标准煤/万元，降幅为20.36%，超额完成减排任务1.8%；湖南和河南能源利用效率在中部地区处于中等水平，单位GDP能耗为1.17吨标准煤/万元和1.115吨标准煤/万元，降幅为20.43%和20.12%，超额完成减排任务2.15%和0.6%；山西能源利用效率最低，单位GDP能耗为2.235吨标准煤/万元，但降幅最大，达到22.66%，超额完成减排任务3%；能源利用效率较差还有湖北，单位GDP能耗为1.183吨标准煤/万元，降幅较大，为21.67%，超额完成减排任务8.35%。

从节能目标的完成情况看，"十一五"时期，中部地区各省的节能工作取得了显著成效，至2010年全部完成了"十一五"国家下达的节能目标任务。其中，超额完成"十一五"节能目标任务最好的省份是湖北省(8.35%)，完成"十一五"节能目标任务超过1%的省份有山西(3%)、湖南(2.15%)和安徽(1.8%)，完成"十一五"节能目标任务低于1%的省份有河南(0.6%)和江西(0.2%)。"十一五"期间中部地区各省节能目标完成情况如表16-1所示。

表 16 – 1 "十一五"期间中部地区各省节能目标完成情况表

Table 16 – 1 Energy Conservation Performance of Central Provinces during China's 11th Five – year Plan Period

地 区	2005 年		2010 年		超额完成数（%）
	单位 GDP 能耗（吨标准煤/万元）	"十一五"时期计划降低（%）	单位 GDP 能耗（吨标准煤/万元）	"十一五"时期实际降低（%）	
山 西	2. 89	–22	2. 235	–22. 66	3. 00
安 徽	1. 216	–20	0. 969	–20. 36	1. 80
江 西	1. 057	–20	0. 845	–20. 04	0. 20
河 南	1. 396	–20	1. 115	–20. 12	0. 60
湖 北	1. 51	–20	1. 183	–21. 67	8. 35
湖 南	1. 472	–20	1. 17	–20. 43	2. 15

资料来源：《"十一五"各地区节能目标完成情况表》

4. 减排分析

从减排目标的完成情况来看，"十一五"时期，中部地区各省的减排工作取得了显著成效。截至 2010 年，中部地区化学需氧量排放量和二氧化硫排放量分别为 316. 63 万吨和 511. 18 万吨，分别比 2005 年下降 10. 05% 和 14. 25%，全面完成了"十一五"国家下达的减排目标任务。从各省减排情况来看，各省份都超额完成了"十一五"主要污染物减排目标任务，其中，化学需氧量减排完成情况为：超额完成目标 20% 左右的省份有湖北（42. 86%）和河南（29. 81%）二省；超过 10% 的省份有安徽（14. 00%）和江西（13. 35%），低于 10% 的省份有湖南（6. 14%）和山西（5. 51%）。二氧化硫减排完成情况为：陕西、湖北、湖南、江西、山西和河南六省分别超额完成二氧化硫排放量的 68%、51. 09%、42. 33%、30. 43%、25. 71% 和 25. 64%。中部地区各省主要污染物"十一五"期间减排目标完成情况如表 16 – 2 所示。

表 16 – 2 中部地区各省主要污染物"十一五"期间减排目标完成情况

Table 16 – 2 Emission Reduction of Major Pollutants in Central Provinces during China's 11th Five – year Plan Period

省份	化学需氧量					二氧化硫				
	2005 年排放量（万吨）	"十一五"降低目标（%）	2010 年排放量（万吨）	"十一五"实际降幅（%）	超额完成幅度（%）	2005 年排放量（万吨）	"十一五"降低目标（%）	2010 年排放量（万吨）	"十一五"实际降幅（%）	超额完成幅度（%）
山西	38. 7	–13. 20	33. 31	–13. 93	5. 51	151. 6	–14. 0	124. 92	–17. 60	25. 71
安徽	44. 4	–6. 50	41. 11	–7. 41	14. 00	57. 1	–4. 0	53. 26	–6. 72	68. 00
江西	45. 7	–5. 00	43. 11	–5. 67	13. 35	61. 3	–7. 0	55. 70	–9. 13	30. 43
河南	72. 1	–10. 80	61. 99	–14. 02	29. 81	162. 5	–14. 0	133. 92	–17. 59	25. 64
湖北	61. 6	–5. 00	57. 20	–7. 14	42. 86	71. 7	–7. 8	63. 25	–11. 79	51. 09
湖南	89. 5	–10. 10	79. 91	–10. 72	6. 14	91. 9	–9. 0	80. 13	–12. 81	42. 33

资料来源：根据《"十一五"期间全国化学需氧量排放总量控制计划表》、《"十一五"期间全国二氧化硫排放总量控制计划表》整理

5. 节能减排综合指数分析

根据节能减排综合指数的分类，即：第Ⅰ类地区节能减排综合指数的区间值为［2.00－3.00)，第Ⅱ类地区节能减排综合指数的区间值为［1.50－2.00）之间的为，第Ⅲ类地区节能减排综合指数的区间值为［1.00－1.50)，第Ⅳ类地区节能减排综合指数的区间值为［0.00－1.00)，中部地区各省节能减排综合指数排名如表16－3所示。

综合比较中部地区节能减排综合指数类型的分布情况，全国属于Ⅰ类地区的五个省份均在东部地区；属于Ⅱ类地区的17个省份中，中部地区占5席；属于Ⅲ类地区的四个省份中，中部地区仅占1席；属于Ⅳ类地区的三个省份全部都在西部地区。可见，从全国范围来中部地区节能减排排名处于中等水平。在中部地区的范围内，Ⅰ和Ⅳ类地区中都没有出现中部地区省份，中部地区各省主要分布于Ⅱ类地区。具体地，属于Ⅱ类地区有湖北、安徽、河南、江西和湖南五省，属于Ⅲ类地区仅有山西省。

从中部地区各省的节能减排综合指数来分析，湖北和安徽两省的节能减排综合指数排名较靠前，位居全国排名的前十之列，分别第七、第八位；河南、江西和湖南三省排名居于全国的中间位置，排名分别在第13、第14和第18位；山西省的节能减排综合指数排名靠后，位居第26位。从中部地区各省的节能减排综合指数的构成来分析，属于Ⅱ类地区的湖北、安徽、河南和湖南和的属于Ⅲ类地区的山西共五省的努力指数排名均高于其综合指数排名，说明尽管受到各种局限性的影响，以上各省区的综合指数排名还并不高，但它们客观上都能够努力去实现节能减排工作目标；而属于Ⅱ类地区的江西省的努力指数排名低于其综合指数排名，说明江西省的努力程度更滞后于其他方面，其节能减排工作还有待进一步努力提高；河南省的效果指数、努力指数和潜力指数排名都高于其综合指数排名，说明河南省在节能减排工作中的努力程度、完成效果和可挖掘潜力上都高于目前的综合排名；湖北省的努力指数排名很高，而效果指数与潜力指数排名较低，说明该省在节能减排的实际工作效果及深入挖潜方面还有待提高。

表16－3 中部地区各省节能减排综合指数

Table 16－3 Energy Conservation and Pollution Reduction Composite Index of Central Provinces

	综合指数及排名		二级指数					
	节能减排综合指数	排名	效果指数	排名	努力程度指数	排名	潜力实现指数	排名
山　西	1.116	26	0.495	26	0.129	15	0.493	26
安　徽	1.921	8	0.837	10	0.247	5	0.838	10
江　西	1.763	14	0.822	11	0.121	18	0.820	11
河　南	1.784	13	0.812	12	0.162	10	0.810	12
湖　北	1.959	7	0.788	15	0.384	2	0.788	15
湖　南	1.649	18	0.745	19	0.162	11	0.743	19

资料来源：根据表14－9整理

16.1.3 节能减排的主要经验及存在的问题

“十一五”以来，根据《国民经济和社会发展第十一个五年规划纲要》、《“十一五”期间各地区单位生产总值能源消耗降低指标计划》和《“十一五”期间全国主要污染物排放总量控制计划》提出的“十一五”期间单位国内生产总值能耗降低20%左右的目标，主要污染物排放总量降低10%的减排任务，在将节能减排约束指标具体落实到中部地区六省后（如表16－1、表16－2），再由各省分解落实到各市（地）、县及有关行业和重点企业，并制订有关具体实施方案，明确责任主体，层层抓好落实，严格执行，以确保总量控制目标的实现。截至2010年，中部地区各省的节能减排工作成绩显著，顺利完成了节能减排的目标任务。但是，中部地区各省的节能减排工作也面临着非常严峻的形势。2011年8月，发改委发布的本年度上半年全国节能减排完成情况晴雨表显示，有八个省（区）预警等级被列为最高级别一级预警①，其中属于中部地区的省份有江西、河南二省，高耗能行业的快速发展是节能形势严峻的重要原因。总结“十一五”以来节能减排工作的主要经验和存在的问题表现在以下几个方面：

1. 主要经验

第一，严格执行节能减排工作问责制。建立健全节能减排工作责任制和问责制，将节能减排指标完成情况作为政府领导干部综合考核评价和企业负责人业绩考核的重要内容、实行“一票否决”制，让政府主要领导作为节能减排工作的第一责任人。“十一五”期间中部地区各省节能减排实践经验表明，将节能减排指标完成情况纳入各级地方政府政绩考核体系中的“自上而下”的问责机制能够在短期快速扭转了节能减排工作的不利局面。

第二，强化节能减排工作领导力量。中部地区各省均成立了由政府主要领导任组长的节能减排工作领导小组和节能减排办公室，协调解决当地节能减排工作中的重大问题，并要求各县、市也建立相应领导小组。而节能减排办公室则在节能减排工作中履行牵头总抓的职责，综合协调节能减排日常工作，负责全社会节能减排的监管和考核评价，以及重点企业节能减排的监督检查等工作。强有力的节能减排工作领导机构为顺利开展节能减排工作提供了组织保障。

第三，着力优化产业结构。中部地区是我国重要的煤炭、有色金属、磷化矿生产基地，创建一批以中部地区能源、原材料工业等为基础的循环经济型企业，提高资源利用效率；加强对机械工业企业的技术改造，促进产业结构调整，强化制造装备升级，提高产业整体竞争力，促进新型工业化发展，实现节能降耗、减排治污；结合自然资源禀赋、产业基础状况及其潜能激发特征，充分利用优势性、特色性、专有性、制高性地产资源或功能要素，选择具有巨大后发优势或强劲竞争潜能的良好成长性产业，进行重点培育；在经济发展水平较高的城市和地区，积极发展知识密集型产业和新兴产业，实施产业创新，培育新的经济增长点。

第四，突出节能减排的重要抓手。按照以项目确定资源配置的原则，对重点行业、重大项目、重点工程、重点企业予以重点支持，严格环保准入标准，着力抓好节能重点工程、环境治理重点工程、循环经济重点工程。以燃煤电厂和冶金、有色、建材等重点行业脱硫、脱硝为重点，着力推进大气污染防治。重点抓好化工、纺织印染、造纸、食品加工等行业水污染治理。大力引导重点企业向园区集中，实行污染集中治理。

第五，健全节能减排法规和制度。不断完善节能减排工作评价、考核制度，提高节能减排信息通报制度的透明度，实行单位生产总值能耗和主要污染物排放总量公报制度，定期向社会公布辖区的节能减排主要指标，接受社会监督。加快重点产品的能耗限额标准、建筑业的能耗

---

① 按照发改委的解释，一级预警代表节能形势十分严峻，二级预警代表节能形势比较严峻，三级则表明节能进展顺利。

标准和节能设计规范等标准和规范的制定和完善工作，建立健全节能减排的评估审查制度，建立有利于节能减排的利益导向机制。

第六，完善节能减排财政政策。中部地区各省通过设立节能减排专项资金有力地支持了重点节能减排项目建设。例如，2007 年湖北省省财政共安排 5.2 亿元专项资金用于小造纸、小水泥、小火电及四湖流域综合治理，经过各地的共同努力，2007 年全省共关停小煤矿 200 个，关停并验收小造纸企业 185 家，关停小火电 20 台共 60.3 万千瓦机组，关停小水泥企业 91 家，均完成或超额完成年计划任务；筹措安排资金 8854 万元，支持主要污染物减排的监测、指标和考核体系建设，重点支持污染源普查、国控重点污染源监督性监测和自动监控中心运行，对“百家企业”开展节能检测，完成 65 家企业的主要耗能设备检测和测试等工作，为掌握生态环境有关数据提供了资金保障。湖北省省财政在 2008 年预算中安排 3 亿元用于“两型社会”建设激励性转移支付，其中 5000 万元用于重点企业节能技术改造。另外，省财政还安排环保专项资金 1 亿元，环保监测能力项目资金 4000 万元，支持生态环境保护和加强环保监测执法能力建设。

2. 存在的问题

第一，节能减排观念意识有待提高。在中部地区，一些地方政府往往为了追求经济增长而忽视节能降耗、削减排放，片面认为经济增长必须靠投资拉动，盲目投资高耗能产业。还有少数企业，为了企业的一己之私利，不惜以牺牲环境为代价，非法排放大量废水废气、有毒有害烟尘等，危害生态环境。

第二，节能减排的组织、运行、管理和监督体系亟待完善。在“十一五”的节能减排目标考核中，政府基本上采用“算总账”方法。容易导致节能减排中的“前松后紧，临时突击”问题的发生，许多地方在初始阶段人为压低节能减排效率、在考核后半阶段才加大节能减排力度，等到考核期末之时，尽管能源强度下降的指标达到了，但能源消费总量大幅上涨。建立一种长效激励与监管机制势在必行，充分发挥政府、企业和民众等利益相关者参与决策的能动性。

第三，节能减排的配套政策体系有待完善。“十一五”期间，虽然中部地区各省均出台了大量支持节能减排的相关配套政策，但是，能够得力到位、落到实处的并不多，而且与其他相关政策之间缺乏协调，这种情况已经在一定程度上影响着节能减排工作的进一步推进。

16.2 “十二五”中部地区节能减排展望

16.2.1 节能减排的战略背景

“十二五”时期面临的国内外环境。当今世界正处于大发展大调整大变革之中，世界多极化和经济全球化深入发展，科技创新孕育新突破，绿色发展蓬勃兴起；国内外产业转移深入推进，区域发展分工合作不断深化；同时，国际金融危机影响深远，世界经济增长速度减缓，全球需求结构出现明显变化；而且，气候变化以及能源资源安全、粮食安全等全球性问题更加突出。国内经济发展方式加快转变，发展环境和条件呈现许多新特点、新变化；国家坚持扩大内需战略，实施区域发展总体战略和主体功能区战略，推动经济布局逐步从沿海向内陆延伸；工业化城镇化加速推进，消费结构加快升级，持续投入效应不断释放，引发全国各省（市、区）都在抢抓机遇竞相发展，区域发展格局正在发生重大而深刻的变化。综合分析判断，国内外环境总体有利于中部地区六省的快速发展，特别是国际国内产业转移加速，国家促进中部崛起政策支持力度加大，国家鼓励和支持长株潭城市群“两型”社会、武汉城市圈“两型”社会、中原城市群、环鄱阳湖城市群、皖江城市带以及山西国家资源型经济转型综合配套改革试验区等建设，为中部地区各省加快发展提供了重大历史机遇，未来五年将是中部地区崛起战略的黄金发展期。

同时也必须清醒看到，未来五年面临的挑战前所未有。世界经济进入深度调整时期，我国经济发展的条件和动力也在发生深刻变化，中部地区发展不平衡、不协调、不可持续的题日渐显现。经济社会发展的基础仍然薄弱，综合竞争力不强；结构性矛盾突出，产业结构层次较低，第一产业比重偏高，第三产业比重过低，工业结构中能源原材料工业比重过大，处于产业链前端和价值链的低端，经济增长的质量和效益不高，科技创新能力不足；经济发展方式粗放、经济素质不高；中心城市带动力不强，城乡区域发展不协调；资源环境约束加大，面临加快发展与加快转型双重压力。此外，中部地区部分省份产业结构趋同及竞相升级也会产生相互挤压的负面效应。

1. 节能战略方面

在节能战略方面，中部地区着重抓好以下几个方面问题：

第一，坚持同步推进结构节能、技术节能和制度节能。严格控制高耗能、高排放行业低水平扩张和重复建设，依法淘汰落后产能，强化各行业用能管理，推进技术性节能；加强共性、关键和前沿节能降耗新技术、新工艺的引进、研发和应用，重点开展燃煤锅炉（窑炉）改造、工业余热余压利用、电机系统节能改造，突出抓好建筑、工业、交通、商业、公共机构等领域节能，积极实施可再生能源建筑应用和绿色照明工程，全面推进绿色建筑行动，推进技术性节能；实施强制性能耗物耗标准，强化节能降耗目标责任评价考核，积极推行合同能源管理、节能产品政府强制采购、建筑能耗定额和超定额加价制度，严格投资项目节能评估审查制度，提高准入门槛，推进制度性节能。

第二，突出工业节能在节能战略重要地位。“十二五”期间，中部地区各级政府和有关部门需要采取一系列强有力的政策措施，综合运用法律、经济、技术及必要的行政手段，加强工业节能管理，以工业生产源头、过程、产品三个重点，实施能效提升计划；鼓励企业建立节能计量、台账和统计制度，实施电力需求侧管理、能效标识、政府节能采购等管理方式；加强工业节能技术的研究开发和技术储备，加快资源综合利用与清洁生产关键技术产业化，推进先进高效技术、装备、产品的推广应用，支撑工业发展方式转变。

2. 减排战略方面

在减排战略方面需要努力做好以下几个方面的工作：

第一，加强结构减排，工程减排和管理减排并重。结构减排的重点坚持由淘汰高耗能低端产能减排向实现产业优化、减少排放强度转变；工程减排逐步启动县域减排工程建设，加强减排工程和设施运行的监督和检查，提高管理减排的效果。

第二，发展循环经济，重点推行循环型生产。从推行清洁生产、提高资源综合利用水平、建设改造各类产业园区等方面入手，实现土地集约利用、废物交换利用、能量梯级利用、废水循环利用和污染物集中处理，实现产业循环式组合。

第三，加大推广工业清洁生产。大力推行清洁生产，积极推行工业领域生态设计，加强规划环境影响评价，提高环保准入门槛，从全生命周期减少资源消耗和污染物排放，加大重点行业清洁生产审核和清洁生产改造力度，加强清洁生产重大技术的研发和推广应用，着力解决大气污染、水污染等危害人民身体健康的突出问题。

第四，改善能源消费结构。能源消费结构改善的重点是积极推进煤炭清洁高效化利用技术，降低煤炭在终端消费中的比例，提高煤炭消费用于发电的比重。

综合分析各方面因素，未来五年，中部地区即将进入快速发展黄金时期，同时也可能带来一系列的节能减排问题。因此，必须进一步增强机遇意识和忧患意识，科学把握发展规律，主动适应环境变化，充分利用各种有利条件，积极面对节能减排的严峻形势和挑战，着力解决突出矛盾和问题，努力实现经济社会又好又快发展。

16. 2. 2 节能减排的战略目标

1. 发展目标

中部地区是我国重要的煤炭、有色金属、磷化矿生产基地，创建一批以中部地区能源、原材料工业等为基础的循环经济型企业，提高资源利用效率；加强对机械工业企业的技术改造，促进产业结构调整，实现节能降耗、减排治污；在经济发展水平较高的城市和地区，积极发展知识密集型产业和新兴产业，实施产业创新，培育新的经济增长点；建立完善的节能减排法规体系、发展机制和框架，通过制度创新和观念创新，大力推进资源集约型、环境友好型社会建设，促进中部地区和谐可持续发展道路。促成中部地区各省单位生产总值能源消耗和二氧化碳排放大幅下降，主要污染物排放得到有效控制，使单位地区生产总值能耗，非化石能源占一次能源消费比重，单位地区生产总值二氧化碳排放量，化学需氧量、二氧化硫、氨氮、氮氧化物等主要污染物排放总量等达到或控制在国家下达的指标内。

2. 节能目标

根据国家发改委下发的《“十二五”节能目标分解方案（征求意见稿）》，中部地区六省的单位 GDP 能耗降低率的最高降幅均为 16%，属于第三类地区；再结合六省“十二五”规划纲要的主要目标内容，六省节能目标如下：

- 山西：全面完成国家下达的节能减排目标。
- 安徽：全面完成国家下达的节能减排目标。
- 江西：单位生产总值能耗累计下降 16%。
- 河南：非化石能源占一次能源消费比重占 5%，单位生产总值能耗控制在国家下达的指标内，单位 GDP 能耗降低率为 16%。
- 湖北：全省单位生产总值能源消耗比“十一五”末降低 16%。
- 湖南：非化石能源占一次能源消费比重达 11. 4%，单位地区生产总值能耗降低 16%。

3. 减排目标

根据 2011 年 9 月 7 国务院《关于印发“十二五”节能减排综合性工作方案的通知》（国发〔2011〕26 号），结合中部地区各省“十二五”规划纲要的主要目标，将各省减排目标内容总结如下：

- 山西：全面完成国家下达的减排目标，即与“十一五”末相比较，至“十二五”末化学需氧量排放总量下降 9. 6%、氨氮排放总量下降 12. 2%、二氧化硫排放总量下降 11. 3%、氮氧化物排放总量下降 13. 9%。

● 安徽：全面完成国家下达的节能减排目标，森林覆盖率提高到29%，生态环境质量明显改善。具体地，与“十一五”末相比较，至“十二五”末化学需氧量排放总量下降7.2%、氨氮排放总量下降9.9%、二氧化硫排放总量下降6.1%、氮氧化物排放总量下降9.8%。

● 江西：单位生产总值二氧化碳排放量、主要染污物排放量控制在国家下达的目标范围内，即与“十一五”末相比较，至“十二五”末化学需氧量排放总量下降5.8%、氨氮排放总量下降9.8%、二氧化硫排放总量下降7.5%、氮氧化物排放总量下降6.9%；城镇生活污水集中处理率、城镇生活垃圾无害化处理率分别提高到85%、80%以上。

● 河南：二氧化碳排放量、化学需氧量、二氧化硫、氨氮及氮氧化物等主要污染物排放量控制在国家下达的指标内。即与“十一五”末相比较，至“十二五”末化学需氧量排放总量下降9.9%、氨氮排放总量下降12.6%、二氧化硫排放总量下降11.9%、氮氧化物排放总量下降14.7%。

● 湖北：全省主要污染物排放总量比“十一五”期末明显减少，单位生产总值二氧化碳排放降低17%，重点水域水质进一步改善。具体地，与“十一五”末相比较，至“十二五”末化学需氧量排放总量下降7.4%、氨氮排放总量下降9.7%、二氧化硫排放总量下降8.3%、氮氧化物排放总量下降7.2%。

● 湖南：按照湖南省“十二五”规划纲要的主要目标，至“十二五”末二氧化碳排放量降低17%，二氧化硫、化学需氧量排放累计分别减少8%，氮氧化物、氨氮排放累计分别减少10%。根据国家下达的减排目标，与“十一五”末相比较，至“十二五”末化学需氧量排放总量下降7.2%、氨氮排放总量下降9.8%、二氧化硫排放总量下降8.3%、氮氧化物排放总量下降9%。“十二五”中部地区各省减排目标情况如表16－4所示。

表16－4 “十二五”中部地区各省减排目标

Table 16－4 Pollution Reduction Targets of Central Provinces for China's 12th Five－year Plan Period

地区	化学需氧量排放总量控制计划			氨氮排放总量控制计划			二氧化硫排放总量控制计划			氮氧化物排放总量控制计划		
	2010年排放量	2015年控制量	2015年比2010年下降	2010年排放量	2015年控制量	2015年比2010年下降	2010年排放量	2015年控制量	2015年比2010年下降	2010年排放量	2015年控制量	2015年比2010年下降
	（万吨）	（万吨）	（%）	（万吨）	（万吨）	（%）	（万吨）	（万吨）	（%）	（万吨）	（万吨）	（%）
全国	2551.7	2335.2	－8.5	264.4	236.6	－10.5	2267.8	2067.4	－8.8	2273.6	2021.6	－11.1
山西	50.7	45.8	－9.6	5.93	5.21	－12.2	143.8	127.6	－11.3	124.1	106.9	－13.9
安徽	97.3	90.3	－7.2	11.2	10.09	－9.9	53.8	50.5	－6.1	90.9	82	－9.8
江西	77.7	73.2	－5.8	9.45	8.52	－9.8	59.4	54.9	－7.5	58.2	54.2	－6.9
河南	148.2	133.5	－9.9	15.57	13.61	－12.6	144	126.9	－11.9	159	135.6	－14.7
湖北	112.4	104.1	－7.4	13.29	12	－9.7	69.5	63.7	－8.3	63.1	58.6	－7.2
湖南	134.1	124.4	－7.2	16.95	15.29	－9.8	71	65.1	－8.3	60.4	55	－9

资料来源：根据国务院《关于印发“十一五”节能减排综合性工作方案的通知》整理

16.2.3 节能减排的战略选择

中部地区位于东部沿海地区与西部地区之间，地理位置上承东启西，是国家综合运输网络的中心区域，“综合交通枢纽”地位非常突出，而且，也是我国重要的能源与原材料基地。为此，中部地区各省的节能减排战略选择应该按照区域自然条件、资源环境承载能力、经济社会发展基础，以及中部地区崛起规划和主体功能区战略，选择符合各省具体实际条件的节能减排战略。

1. 山西省节能减排战略

实现经济社会可持续发展，必须坚持“绿色、低碳、洁净、健康”的发展理念，加快构建资源节约型、环境友好型的生产方式和生活方式，大力推进国家资源型经济转型综合配套改革试验区建设，努力建设绿化山西、气化山西、净化山西、健康山西。

（1）加快发展循环经济。以循环经济试点省建设为契机，大力发展循环型农业、循环型工业、循环型服务业，建设循环型社会，使循环经济成为基本经济形态和产业发展模式。按照“多联产、全循环、抓高端”的思路，全面推进各类产业特别是传统产业循环化发展，延长煤焦化、煤电材、煤电铝等资源循环产业链，提高煤矸石、粉煤灰、工业废渣、矿井水、中水等废弃物综合利用水平。构建循环经济关键、共性技术研发体系，加快技术服务平台建设，引进、开发和应用源头减量、循环利用、再制造、零排放和产业链接技术。加强循环经济地方立法和执法工作，完善循环经济标准规范和政策体系，增强循环经济的科技支撑、政策激励和法律保障。

（2）强化能源节约与管理。推进产业结构转型，大力发展现代装备制造、现代煤化工、新能源、新材料、节能环保以及现代服务业等接续替代产业，实现支柱产业由单一向多元、产品由低端向高端转变。继续实施和完善煤炭工业可持续发展试点政策，进一步促进煤矿安全生产、生态环境治理、煤炭企业改革和转型转产。加快财税体制改革，搭建支持资源型经济转型的投融资平台，引导社会资金参与资源型经济转型发展，积极推进煤炭行业资源税从价征收试点工作，按照国家统一部署开征环境保护税。

（3）加大工业减排力度，严格控制污染物排放。实施环境容量总量控制，严格煤炭、焦炭、冶金、电力、化工、建材等“两高一资”产业准入门槛，加快淘汰落后产能，积极发展低碳、环保型产业。积极实施重点行业和重点领域节能减排工程，大幅降低能源消耗强度和二氧化碳排放，有效控制二氧化硫、化学需氧量、氮氧化物、氨氮以及烟尘、粉尘等主要污染物排放。

（4）加强环境保护监管。落实节能减排目标责任制，加快推行合同能源管理，加大环境保护和节能降耗地方立法力度，建立健全污染者付费、水电差别价格和排污权交易等制度。落实节约优先战略，全面实行资源利用总量控制、供需双向调节、差别化管理，加强资源节约和管理。严格污染物排放标准和环境影响评价，强化对排污企业和重点耗能企业的监测监管，加强执法监督，健全重大环境事件和污染事故责任追究制度。强化应对气候变化意识，形成人人关心节能减排、全社会推进节能减排的强大合力和良好氛围。

2. 安徽省节能减排战略

（1）大力进推节约能源。继续抓好工业、建筑、交通运输、公共机构和商业旅游等重点领域节能，切实加强重点耗能企业节能管理，实施重点行业能效水平对标，降低能源消耗强度。以节能增效为主线，发挥工程性节能的主导作用，重点实施节能改造优化工程、节能技术产业化示范工程、节能产品兴企惠民工程、绿色照明产品推广工程、合同能源管理推进工程、节能基础能力建设工程等六大节能工程。加大节能新技术、新工艺、新产品的开发、引进和推广应用，重点推广高效节能家电、汽车、电机、节能灯等节能产品，加大政府对环保产品认证宣传和资金扶持力度。提高产业准入门槛，严格固定资产投资项目节能评估审查，从源头抑制高耗能产业过快增长。调整能源消费结构，增加非化石能源比重。完善政策机制，形成反映资源稀缺程度、供求关系和环境成本的价格形成机制。加大财政、税收、金融对节能的支持力度。把大幅度降低能源消耗强度作为重要约束性指标，分解落实，完善目标责任制。强化依法节能，健全配套法规和标准，加强节能执法监督检查。加强宣传教育，弘扬节约环保的消费模式和生活习惯。

（2）大力发展循环经济。全面推进循环经济示范城市、园区和企业建设。合理规划园区布局，鼓励企业通过共享资源、废弃物利用等途径发展循环经济，构建循环型产业体系。提高共伴生矿产资源、工业废弃物、农林废弃物、建筑垃圾和秸秆转化等综合利用水平。推进废旧汽车零部件、工程机械、机床、废旧轮胎、办公设备等再制造产业发展。积极推进“城市矿产”示范基地建设，打造铜陵矿山、滁州报废汽车循环经济示范园等。推进废旧家电、废杂金属及电子产品等再生资源的回收利用。提高餐厨废弃物资源化和无害化水平。加快推进铜陵有色、淮南矿业、马钢等重点行业企业循环经济示范项目建设。

（3）加强企业污染防治，加强污染防治。严格环保准入标准，淘汰落后工艺和产能；加快城乡环境综合整治，强化汽车尾气、烟尘、噪声等环境综合治理，加快城市污水设施建设，扩大污水管网覆盖范围，加快规模乡镇污水处理设施建设，推广适用乡村污水处理技术；加强重点流域污染防治，加强重污染河流整治，加大淮河、巢湖、长江安徽段、新安江流域水污染治理和保护力度，加强重点排污企业和入河入湖排污口监控。

（4）完善环境保护机制。实行主要污染物排放总量控制制度，落实环境保护目标考核制、责任追究制和行政首长负责制。加强环境影响评价和“三同时”管理，推进空间准入、总量准入、项目准入和专家评价、公众评价的环境准入制度。实行主要污染物排放许可证管理，推行排污权有偿使用和交易。提高重点行业和重点区域污染物排放标准。完善重大环境安全隐患防范制度。强化环境保护执法，健全环境监管体制，实行环境质量公告和企业环保信息公开制度，鼓励社会公众参与环保监督。加强宣传教育，增强全民环保意识。鼓励创建国家环境保护模范城市。

（5）积极应对气候变化，控制温室气体排放。大力发展低碳经济，控制工业生产温室气体排放。加强国际合作。积极引进和推广国外的先进技术和管理经验，充分利用清洁发展机制，争取更多 CDM 项目进入国际市场。完善应对方案，不断提升全省应对气候变化和可持续发展的能力。

3. 江西省节能减排战略

牢固树立绿色、低碳发展理念，促进经济与生态相融合、人与自然相协调，加快建设资源节约型、环境友好型社会。

（1）大力发展循环经济。创建循环型生产方式，大力推进冶金、化工、建材、造纸、印染、制药等重点行业进行清洁生产，积极创建生态工业园区和循环经济工业园区，优化企业结构和布局，鼓励企业间通过资源共享、废弃物利用等途径发展循环经济；实施绿色标识认证制度，倡导绿色消费模式鼓励消费者购买使用节能节水产品、节能环保型汽车和节能省地型住宅，推进政府绿色采购，逐步提高节能节水产品和再生利用产品比重，加强宣传教育推动形成绿色生活方式和消费模式；完善法律法规，加强规划指导、财税金融等政策支持，落实生产者责任延伸制度和再生产品标识制度，建立完善全省循环经济统计评价制度，推进生产、流通、消费各环节循环经济发展。

（2）强化节能降耗。完善节能法规，制定并严格执行能耗限额和产品能效标准，完善固定资产投资项目节能评估和审查制度。控制高耗能、高排放产业过快增长，突出抓好工业、建筑、交通、公共机构等重点领域节能，加强重点用能单位节能管理。健全节能市场化机制，加快推行合同能源管理和电力需求侧管理，实行能效标识制度、节能产品认证制度、节能产品政府强制采购和财政补贴制度。开展企业节能低碳行动，全面推进绿色建筑行动，深入开展节能降耗全民行动，创建一批示范性节能公共机构。加强节能能力建设，力争“十二五”期间节能量达到1000万吨以上标准煤。加强节能目标责任考核，完善奖惩制度。

（3）加大环境保护力度，控制污染物排放。实施化学需氧量、氨氮、二氧化硫和氮氧化物总量控制。执行严格的饮用水源地保护制度，继续推进重点流域环境管理和水污染防治，加强对“五河一湖”和东江源头等主要江河湖泊的排污管制，防治地下水污染。大力推进火电、钢铁、有色、建材等行业二氧化硫和氮氧化物治理，强化脱硫脱硝设施稳定运行。推行燃煤电厂脱硝，开展非电行业脱硝示范，防治机动车尾气污染，加强颗粒物污染防治，有效控制城市噪声污染。加强重金属污染综合治理，以鄱阳湖流域为重点，开展重金属污染治理与修复试点示范。建立健全区域大气污染联防联控机制，控制区域复合型大气污染和酸雨。继续实施城镇生活污水处理设施和配套管网建设，分阶段集中建设全省工业园区、有条件的特色产业基地、重点企业污水处理设施，大力开展城乡垃圾无害化处理。

（4）强化环境保护监管。健全环境保护法律法规及标准体系，加强环境监测、预警和应急能力建设。完善环境保护科技和经济政策，建立健全污染者付费制度。加强环境监管能力建设，严格环保准入门槛，依法开展环境影响评价。严格落实环境保护目标责任制，健全重大环境事件和污染事故责任追究制度。加强执法监督，建立环保社会约束和监督机制。

（5）控制温室气体排放，积极应对气候变化。综合运用节约能源、提高能效、增加森林碳汇以及调整产业结构和能源结构等多种手段，大幅降低化石能源消耗强度和二氧化碳排放强度。加大低碳技术和低碳产品的研发、示范、推广，减少工业、交通、建筑等领域温室气体排放。探索建立低碳产品标准、标识和认证制度，逐步建立碳排放市场，开展低碳城市和低碳县（市、区）试点，积极推进清洁发展机制（CDM）项目建设。

4. 河南省节能减排战略

（1）大力发展循环经济。加快发展资源循环利用产业，壮大循环经济规模，建设全国工农业复合型循环经济示范省。抓好资源开发、资源消耗、废弃物产生、再生资源利用和社会消费等五个关键环节，实施八大循环经济重点工程，重点打造有色、煤炭、非金属矿、农业和再生资源等领域五大循环产业链。按照循环经济要求，规划、建设和改造各类产业园区，加强土地集约利用、能量梯级利用、资源综合利用、废水循环利用和污染物集中处理。建成100个循环经济示范企业、20个示范园区和10个示范城镇，建设安阳安西、鹤壁宝山、邓州等一批特色循环经济试验区。加强资源综合利用和再生利用，构建以城市社区和乡村站点回收为基础、集散市场为枢纽、分类加工利用为目的的再生资源循环利用体系。规范发展再制造产业，加强再制造产品流通和监管体系建设。完善循环经济法规和标准体系，健全循环经济财税、价格、投融资政策，创新循环经济发展模式。

（2）加强资源节约和管理。大力推进节能降耗，合理控制能源消费总量，实施工业节能综合改造、建筑节能、低碳交通、公共机构节能、重大节能技术装备产业化示范、节能产品惠民、节能服务体系建设、节能能力建设等八大重点节能工程。加快淘汰电力、煤炭、建材、钢铁、有色、化工、造纸、发酵等高耗能、高排放行业落后产能，对能耗高、污染重、技术落后的工艺、设备及产品实行强制性淘汰制度。推广先进节能技术、设备和产品。落实固定资产投资项目节能评估制度，加强节能监察执法能力建设，健全节能统计监测考核体系，强化节能目标责任考核。

（3）加强环境保护，实施工业污染全防全治。健全工业污染源头预防、过程控制、末端治理、回收利用的全防全治体系，推进重点排污企业深度治理，全面推行清洁生产。建立更加严格的环境准入制度和重污染企业退出机制，对造纸、皮革、发酵、化工等重点行业实行污染物排放总量控制。以区域大气污染防治和高污染行业为重点，推进大气污染联防联治。推进火电行业脱硫设施稳定运行，全面完成钢铁、有色、建材、化工等行业脱硫设施建设。加大火电、钢铁、化工、水泥等行业企业氮氧化物污染防治，完成30万千瓦以上火电机组脱硝改造。加强铅、汞、铬等重金属、危险废物、固体废物污染治理，加快处理设施建设，建成省危险废物处置中心和南阳危险废物处置中心。

（4）严格环境管理。落实环境目标责任，严格环境目标考核。健全环境法规和标准体系。严格环境执法，加大环境违法违规成本约束和处罚力度，严厉打击各类环境违法行为。完善全省环境自动监控系统，建立健全环境应急监管体系，完成环境监管标准化建设。全面推行排污许可证制度，开展排污权有偿使用和交易。健全水环境断面考核制度。加大污水处理费、垃圾处理费征收力度，提高城市污水垃圾处理设施运行效率。实施差别电价、差别水价，提高重污染落后产能企业信贷风险等级。加强环境宣传教育。

（5）积极应对气候变化。坚持减缓和适应并重，充分发挥技术进步的作用，完善体制机制和政策体系，提高应对气候变化能力。综合运用调整产业结构和能源结构、节约能源和提高能效、增加森林碳汇等多种手段，有效控制温室气体排放。加快节能环保、低碳能源、再制造等绿色产业发展，积极培育以低碳为特征的新经济增长点，构建低碳型工业、建筑、能源和交通体系。鼓励开发清洁发展机制项目，推行绿色税收、绿色信贷和绿色保险，扩大政府绿色采购，构建有利于绿色发展的政策体系。建立完善温室气体排放统计监测制度，完善温室气体排放管理体制和机制，加快碳排放权交易市场建设。

5. 湖北省节能减排战略

大力推进绿色发展，加快形成节约资源能源、保护生态环境的产业结构、生产方式和消费模式，扎实推进两型社会建设，建设生态湖北。

（1）着力淘汰落后产能。综合运用法律、经济、技术及必要的行政手段，加快淘汰落后产能，抑制过剩产能盲目扩张，限制高耗能、高污染产业发展，适度发展资源型产业。充分运用市场机制，提高落后产能企业和项目使用能源、资源、环境、土地的成本。严格市场准入，强化指标约束。对淘汰落后产能实行问责制，加强舆论和社会监督。加强环保监测和执法检查，加大执法处罚力度。

（2）大力发展循环经济和低碳产业。积极支持循环经济技术研发、示范推广和能力建设。推进矿产资源综合利用、工业产品再制造、工业废物回收利用、余热余压发电和生活垃圾资源化利用，加快发展循环农业。继续深化国家和省循环经济试点建设，并逐步推广。加快建设一批生态工业园区、再生资源基地。积极推进资源枯竭型城市转型发展。有效控制温室气体排放。积极推进国家低碳试点省建设，探索低碳发展模式和有效运行机制。大力开发低碳技术，壮大低碳装备制造业，推进低碳技术及产品产业化。

（3）抓好重点行业、重点企业和重点项目的节能减排。加强工业、交通、建筑、农业、公共机构等重点领域节能。大力推进电力、有色、建材、化工、钢铁等高耗能行业和企业节能。加快淘汰电力、钢铁、建材、造纸等行业的落后工艺、技术、设备，提高能源利用效率。扎实推进十大重点节能工程，鼓励推广节能工艺、技术、设备。建立节能减排的长效机制。全面推行清洁生产与节能技术，实行强制性淘汰消耗高、污染重、技术落后的工艺、设备及产品制度。建立排污权交易市场，强化排污总量控制和许可证制度，将环境成本纳入企业生产成本核算。加大节能减排新技术、新工艺、新设备、新产品和新材料应用推广力度。推进节水型社会建设。

（4）着力提高生态文明水平。以“两型”社会建设为主线，加快建设生态文明，实现绿色增长。积极探索能源资源合理开发、优化配置、高效利用和有效保护的新途径，制定实施生态补偿、排污权交易、绿色信贷和环境资源税制改革等政策。实施促进绿色消费政策，形成绿色消费方式。开展生态文明创建活动，提升全社会生态文明意识。

6. 湖南省节能减排战略

加强生态文明建设，把资源节约和环境保护作为生态文明建设的重要战略任务，以节能减排为重点，推进绿色发展，健全激励和约束机制，加快构建节约能源资源和保护生态环境的产业结构、增长方式和消费模式，增强可持续发展能力，建设绿色湖南。

（1）加快发展绿色经济。严格控制高耗能、高污染、高排放产业发展，利用绿色技术改造提升传统产业，培育和发展节能环保等绿色新兴产业，提高新能源和可再生能源比重。推广循环经济典型模式，加强资源节约和综合循环利用，全面推进节能、节水、节地、节材。加强低碳、清洁生产、高效节能、污染治理等绿色技术的研发、引进和应用，加快建立资源节约型、环境友好型技术体系和生产体系。大力推广绿色交通、节能建筑，积极倡导公众绿色消费，形成文明、节约、绿色、低碳的生活方式和消费模式。建立完善绿色科技经济政策体系，建立多元环保投融资机制，制定有利于环境保护的财税政策，建立绿色经济统计、跟踪和评价机制。

（2）全面推进节能减排。以推进“两型”社会建设为导向，增强资源环境危机意识，健全节能减排激励机制和约束机制，提升科技支撑水平，强化目标责任考核，大幅降低能源消耗强度和二氧化碳排放强度，明显减少主要污染物排放总量。

（3）大力推进结构性节能、技术性节能和制度性节能。综合运用经济、法律等手段，严格控制高耗能、高排放行业低水平扩张和重复建设，依法淘汰落后产能，强化各行业用能管理，抑制能耗不合理增长；加强共性、关键和前沿节能降耗新技术、新工艺的引进、研发和应用。深入实施“万家企业节能行动”，重点开展燃煤锅炉（窑炉）改造、工业余热余压利用、电机系统节能改造，突出抓好建筑、工业、交通、商业、公共机构等领域节能，积极实施可再生能源建筑应用和绿色照明工程，全面推进绿色建筑行动；实施强制性能耗物耗标准，强化节能降耗目标责任评价考核，积极推行合同能源管理、节能产品政府强制采购、建筑能耗定额和超定额加价制度，严格投资项目节能评估审查制度，提高准入门槛，健全能源统计制度和节能计量统计体系，完善省、市、县三级节能执法监察和评估体系，加快建立湖南国家节能计量中心。到

2015 年，能源消费总量控制在 2.3 亿吨标准煤以内，单位规模工业增加值能耗累计下降 18%。

（4）加强环境治理，减少污染物排放。加强重点地区、重点行业、重点企业污染治理，实现主要污染物排放总量减少和达标排放。严格控制水体污染，减少化学需氧量和氨氮排放总量，加强工业废水、城镇污水无害化处理，统筹协调湘、资、沅、澧四水流域及洞庭湖水资源管理和环境综合治理。严格控制大气污染，积极应对气候变化，有效控制温室气体排放，减少二氧化硫、氮氧化物排放总量，实施重污染产业退出计划，推进电力、钢铁、有色、化工等行业二氧化硫治理，推行燃煤电厂脱硝，开展非电行业脱硝示范，加强机动车尾气治理，强化颗粒物污染控制。严格执行污染物排放标准和环境影响评价制度，强化执法监督，健全重大环境污染事故责任追究制度。有序推进主要污染物排污权交易试点。

（5）加强环境监管。完善环境监测网络，加强环境监测、预警和应急能力建设，建立健全省市县三级环境污染事故应急预警系统，全面提高环境的检测、监控、预警水平和应急处置能力。加强环境执法，严格执行排放许可、限期治理、建设项目的环境影响评价制度，强化项目环境准入控制。严格落实环境保护目标责任制，强化总量控制指标考核，探索实行重点企业年度环境报告制度，健全重大环境事件和污染事故责任追究制度，建立环保社会约束和监督机制。

16.2.4 “十二五”中部地区节能减排的政策措施

1. 加大实施产业和产品结构调整的步伐

加快产业结构调整，提高传统产业的集中度、降低污染物排放，提升第三产业在国民经济中的比重。为此，在今后较长时期，应根据中部地区各省份的具体发展实际，加快长江中游沿岸地区的开发，使之成为推动我国经济布局由东向西逐步转移的重要纽带；制定钢铁、有色、水泥等高耗能行业发展规划、政策，建立产业退出机制，提高行业准入标准，加快能源、原材料工业建设，促进区域节能减排；运用高新技术和先进适用技术改造提升中部地区传统产业，促进传统产业结构优化和升级，推进节能减排和环境保护；搞好传统服务业发展的同时，着力发展现代服务业，做大第三产业经济总量、提升第三产业结构层次方面争取较大突破。

2. 加快科技进步和技术创新

结合中部地区各省份高新技术发展的具体实际，加快科技进步和技术创新。为此，一是积极推进中部地区高技术产业发展，开展区域特色高技术产业链试点，引导大项目布局和建设，推动中部地区高技术产业基地和信息化建设，充分发挥国家高新技术开发区的集聚、引领和辐射作用，降低资源消耗和环境污染，提高产品质量和档次，促进完成中部地区节能减排目标；二是实施电力电子、新型元器件、新型显示器、生物医药、生物育种、生物基材料、高性能纤维复合材料等高技术产业化，加快中部各省高新技术产业发展；三是从项目和政策上支持高压输变电设备、大功率交流传动机车和高速铁路动车组重要组件、先进农业机械、大型施工机械、数控机床等重点领域的发展；四是围绕大型清洁高效发电设备、高档数控机床和基础制造装备等关键领域，推动重大装备、关键零部件及元器件自主研发和国产化。

3. 大力发展循环经济

发展循环经济，促进资源的循环式使用、产业的循环式组合，是转变经济增长方式的重要举措。中部地区有许多资源型城市和工矿企业，它们都选择了发展循环经济。“铜都”铜陵被列为安徽省的清洁生产试点城市并成为国家级循环经济试点市；“江南煤都”萍乡也选择了循环型城市的发展道路，昔日污染严重的市区，出现了绿树成阴、碧水蓝天景象。三门峡、交城、湖南等省市循环经济都得到了长足的发展。为此，一是大力发展循环经济，发展循环型服务业和资源回收产业，实施产业创新，培育新的经济增长点；二是创建一大批循环经济型企业，提高资源利用效率；三是构建以产业体系、基础设施体系、生态保障体系为核心的循环型区域，建立完善的循环经济法规体系、发展机制和框架，通过制度创新和观念创新，引导全民建设资源循环型社会，形成新型的经济发展模式，使中部地区走上和谐可持续发展道路。

4. 建立政策导向和激励机制，利用经济杠杆促进节能减排

在经济手段方面，充分发挥价格杠杆促进资源节约与环境保护的作用，对具备竞争条件的产品，要引入竞争机制，在市场竞争中形成最优价格，以提高企业开展技术创新、节能降耗、环境保护的积极性；同时，采取适度的行政手段，建立节能减排的长效机制，通过宏观调控，促进节能减排、遏制环境恶化。

在财税政策方面，出台一系列有利于节能技术、节能产品推广应用的政府扶持政策，鼓励发展节能产业。建立节能专项基金，加大节能改造基金的投入和对节能工作引导的力度，支持节能宣传培训和节能奖励。设立节能投资的担保基金，引导社会资金投入，支持中小企业通过市场直接融资，进行节能技术改造，创新节能激励机制，借鉴发达国家及地区的经验，建立覆盖节能产品生产、销售、消费全过程的节能激励机制。实行现金返还、税收减免、政府投资、政策补贴等多形式节能资金投入机制和精神奖励机制。

5. 完善管理节能减排

完善节能法律法规和相关政策，加强节能减排监察队伍建设，强化节能监管。强调环保优先，强化舆论监督，通过广泛深入的宣传教育，在全社会营造节能减排的浓厚氛围。加大中部地区污染减排工作力度，有效防范和妥善处理突发环境事件，加强对各类开发建设活动的环境监管。

17 西部地区的节能减排

17.1 “十一五”期间西部地区节能减排综述

17.1.1 “十一五”西部地区经济发展

随着全球经济重心日渐东移，我国经济规模迅速扩张，而且，伴随着经济全球化和区域经济一体化的进一步深化，我国经济发展从东到西、从沿海往内陆梯度推进力度不断加强，国内经济结构也在发生着深刻的变化。西部大开发战略实施10年来，西部地区的经济增长年均增速达到11.9%，至“十一五”末，其经济规模占全国18.7%的份额，比“十五”末时提高了1.6个百分点。特别是在“十一五”期间，国家深入实施西部大开发战略并摆在优先位置，西部地区的GDP平均增长速度更比“十五”加快了2.4个百分点。同时，西部地区的生产力布局与经济发展空间不断优化。成渝、关中—天水、环北部湾经济区正成为西部地区三大“增长级”，重大基础设施、公共服务、生态保护、教育卫生、社会保障等的投入力度不断加大，基础设施建设也取得突破性的进展，青藏铁路、西气东输、西电东送等标志性的工程相继建成；已经集中了一批重点航天军工、电子信息、装备制造、新能源设备等具有雄厚实力的产业，优势能源资源产业发展势头良好，有力地推动了新形势下西部地区经济发展，积极扩大了拉动内需的现实要求，努力拓展了西部地区的经济发展空间。

1. 从经济发展规模来看

“十一五”时期，西部地区各省（区、市）的GDP年均增长速度比“十五”时期明显加快。截至2010年，GDP超过1万亿元大关的有四川、内蒙古和陕西三省（区），分别为16898.6亿元、11655亿元和10021.5亿元；而不足2000亿元的省（区）多达三个，分别为宁夏、青海和西藏，其GDP仅仅分别为1643.4亿元、1350.4亿元和507.5亿元。可以了解，西部地区内部各省（区、市）经济规模差异非常悬殊，并随着西部大开发战略的不断向纵深推进，各省（区、市）的GDP曲线呈现加快发散的趋势。西部地区各省地区生产总值情况如图17－1所示。

2. 从产业结构上看

总体上，西部地区各省（区、市）的产业结构大都表现为“二三一”结构，仅有西藏和贵州属于“三二一”结构，这说明西部地区的第三产业发展滞后，还处于较低水平，西藏和贵州则由于社会、经济和自然生态等因素导致的第二产业发展滞后第三产业虚高的现象。深入到三次产业内部，相比“十五”末期，在“十一五”末西部地区各省（区、市）的三次产业结构呈现快速优化的趋势，第一产业所占比重明显下降，第二产业所占比重快速上升。“十五”末期，

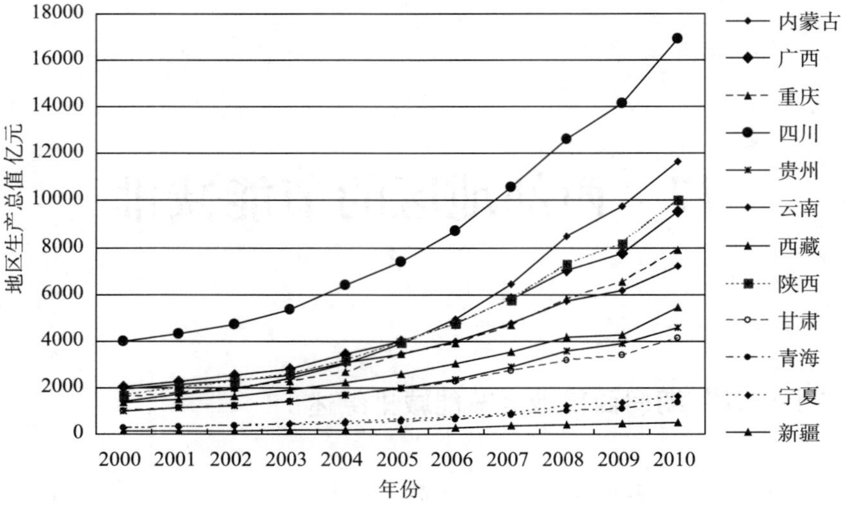

图 17－1 西部地区各省地区生产总值

Figure 17－1 Regional GDP of Western Provinces

资料来源：《中国统计年鉴》（历年）、《中国统计摘要2011》

所有省（区、市）的第一产业比重都高于10%，有些甚至高达20%以上，仅仅只有三个省（区）的第一产业比重低于15%，例如青海为12%、宁夏和西藏均为11.9%。只有的省（区、市）在有广西、新疆、云南、四川、甘肃和贵州等，在“十一五”末，大多数省（区、市）的第一产业比重均低于15%，且低于10%的就包括重庆、内蒙古、宁夏、陕西和青海等五省（区、市），而第一产业比重最高的是新疆，为19.9%，也略低于20%。第二产业所占比重为超过50%的省（区、市）在“十五”末期仅有陕西省，多数省（区、市）的第二产业比重为40%左右；到“十一五”末期，多数省（区、市）的第二产业比重都超过50%，有重庆、青海、内蒙古、陕西、宁夏和四川等六省（区、市）。西部地区的产业结构变化情况如图17－2所示。

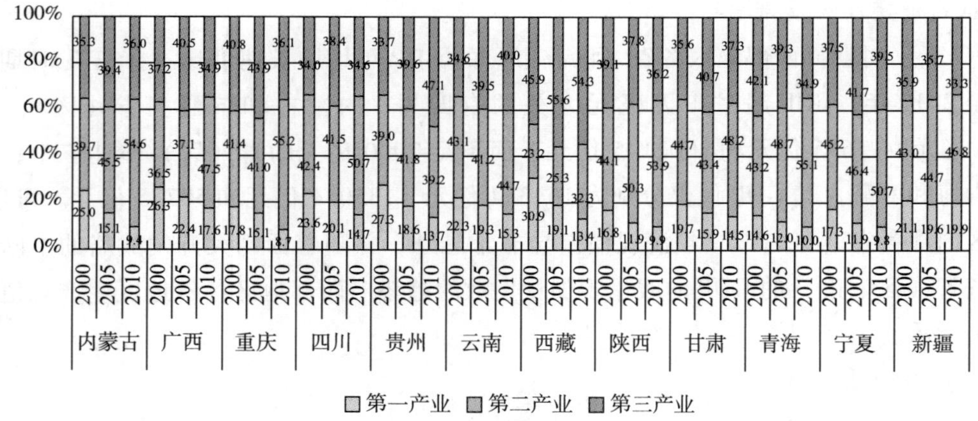

图 17－2 西部地区的产业结构变化

Figure 17－2 Industrial Restructuring of Western Provinces

资料来源：《中国统计年鉴》（2001，2006）、《中国统计摘要2011》

3. 从工业化发展程度来看

按照“工业化率达20%—40%为正在工业化初期；40%—60%为半工业化国家；60%以上为工业化国家”的标准，在“十五”末期西部地区各省（区、市）的工业化率均低于50%，高于40%的仅有陕西，而工业化率低于35%的有近一半的省（区、市），包括四川、云南、重庆、广西及西藏，可见，到“十五”末期，除陕西外，西部地区其他省（区、市）都还处在工业化初期阶段；至“十一五”末期，西部地区各省（区、市）的工业化率均有较大幅度提升，有一半的省份都高于40%，它们是内蒙古、重庆、青海、陕西、四川和广西，最高的内蒙古工业化率接近50%，为48.21%，这些省（区、市）达到了半工业化水准；而宁夏、甘肃和新疆的工业化率也接近40%，分别为39.46%、38.91%和38.85%，都接近半工业化水平。西部地区工业化率变化情况如图17－3所示。

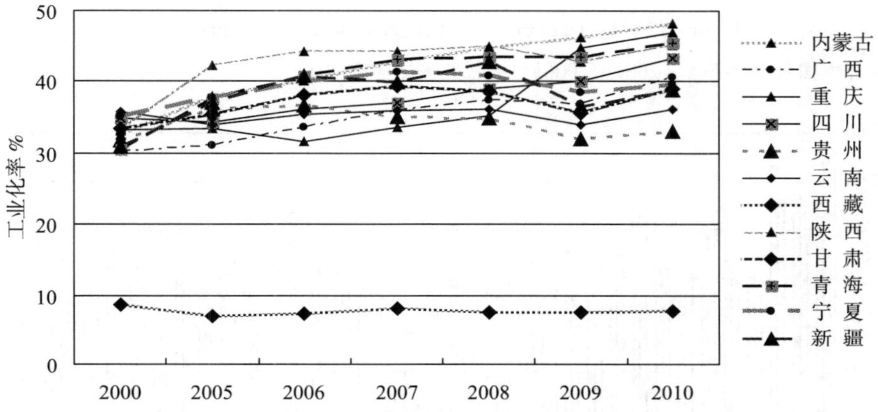

图17－3　西部地区工业化率变化

Figure 17－3　Fluctuation of Industrialization Rate in Western Provinces

资料来源：《中国统计年鉴》(2001，2006)、《中国统计摘要2011》

4. 从城市化水平看

我国西部地区“十一五”期间各省（区、市）的城镇化进程有所加快，城市化率由“十五”期末的34.6%上升到2010年的39.4%，年均增幅0.8个百分点；但是西部地区城市化水平总体落后于全国城市率49.68%的平均水平10.28个百分点，城镇化水平持续提高。就各个省（区、市）的具体情况来看，2009年西部地区各省（区、市）的城市化水平相差悬殊。其中，仅内蒙古和重庆的城市化率略高于全国的平均水平，分别为53.4%和51.59%；而云南、甘肃城市化率仅略高于30%，分别为34%、32.65%；贵州和西藏更不足30%，分别为29.89%和23.8%；其他各省中，宁夏为46.1%，陕西为43.5%，青海为41.9%，新疆为39.85%，广西为39.2%，四川为38.7%。由此可以判断，西部地区各省（区、市）将步入城市化发展的快车道。西部地区城镇化率变化情况如图17－4所示。

从工业化与城镇化发展的同步协调性来看，工业化和城市化是地区社会经济发展现代化进程的两个轮子，二者既相互依存又相互转化，工业化要以城市化为发展动力，城市化则靠工业化为基本支撑。工业化与城市化的良性联动，能够促进经济社会协调发展。从图17－5中可以看出，西部地区城镇化发展滞后于工业化的发展水平，其城市化与工业化发展还没有同步发展。

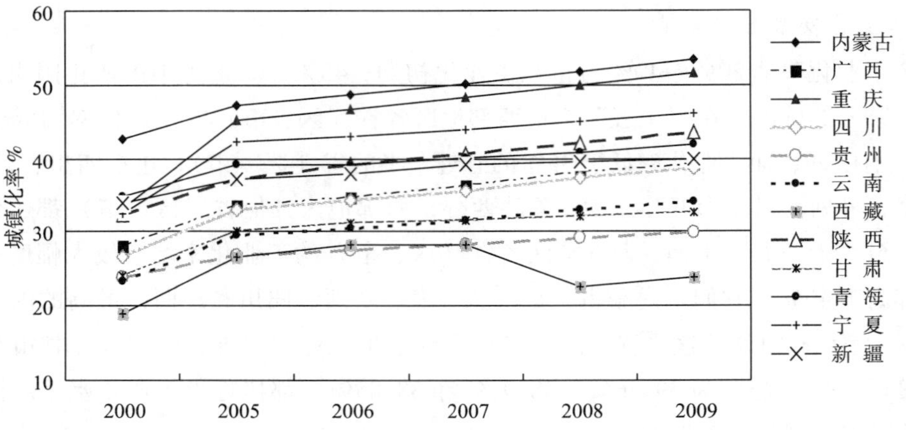

图 17－4　西部地区城镇化率变化

Figure 17－4　Urbanization Progress of Western Provinces

资料来源：《中国统计年鉴》(2001，2006)、《中国统计摘要2011》

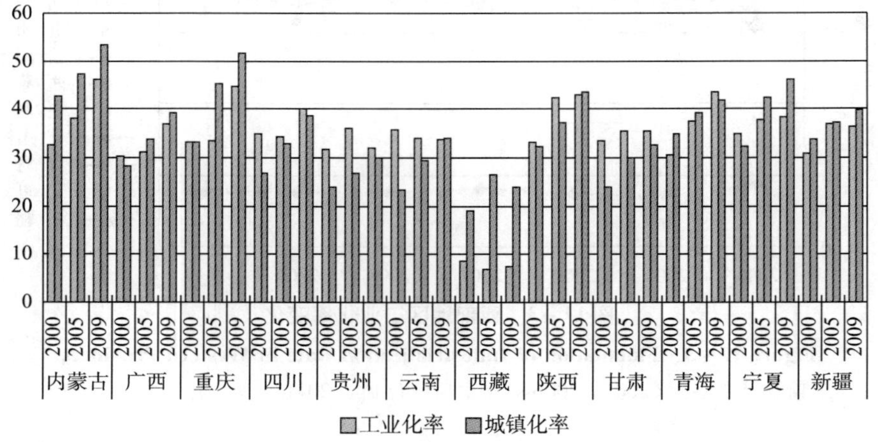

图 17－5　西部地区工业化率与城镇化变化

Figure 17－5　Western Provinces' Changes in Industrialization and Urbanization Rates

资料来源：《中国统计年鉴》(历年)

17. 1. 2　“十一五”西部地区节能减排

1. 能源消费

随着我国西部大开发战略不断深入推进，西部地区经济规模迅速发展壮大，促进了西部地区能源消费的迅速增长，至“十一五”期末，已经发展成为我国能源消费的第二大地区。2009年西部地区能源消费量达到88155万吨标准煤，比2005年的能源消费增长了40.8%，年均增长率为8.93%；从终端能源消费情况来看，西部地区2009年的终端能源消费量达到49179.11万吨标准煤，比2005年的能源消费增长了39.78%，年均增长率为8.73%。其中，煤及煤制品消费量为31406.02万吨标准煤，石油及制品消费量为10606.75万吨标准煤，天然气消费量为3879.15万吨标准煤，电力及其他能源消费量3593.7万吨标准煤，在西部地区终端能源消费量所占比重分别为63.24%、21.57%、7.89%和7.31%。

从各省的能源消费总量上看，截至2009年，能源消费量超过10000万吨标准煤的省份有四川、内蒙古，分别为16322万吨标准煤、15344万吨标准煤，分别占西部地区能源消费量18.51%、17.41%；能源消费量不超过5000万吨标准煤的省份有青海、宁夏，分别为2348万吨标准煤、3388万吨标准煤，在西部地区能源消费中所占比重为2.66%、3.84%；而其他省份的能源消费量都超过5000万吨标准煤。其中，能源消费量最高的四川是能源消费最低的青海的6.95倍之多。从各省的能源消费增长幅度来看，“十一五”期间，四川省能源消费增幅最大，为58.74%；能源消费增幅超过40%有广西、陕西、重庆和青海，分别为45.32%、44.37%、42.22%和40.58%；能源消费增幅超过30%有四川、新疆、贵州、宁夏和云南，分别为38.13%、36.67%、34.12%、33.58%和33.34%；能源消费增幅最低的是甘肃，为25.5%。可见，西部地区各省能源消费大，增幅快，且各省之间的差异程度也比较大。图17－6为2009年西部地区各省能源消费比例情况，图17－7为“十五”末以来西部地区各省能源消费量情况。

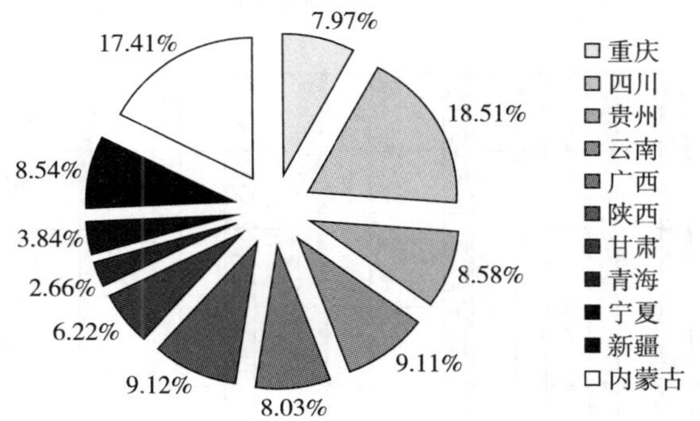

图17－6　2009年西部地区各省能源消费比例

Figure 17－6　Energy Consumption Proportions of Western Provinces in 2009

资料来源：《中国能源统计年鉴》(2010)

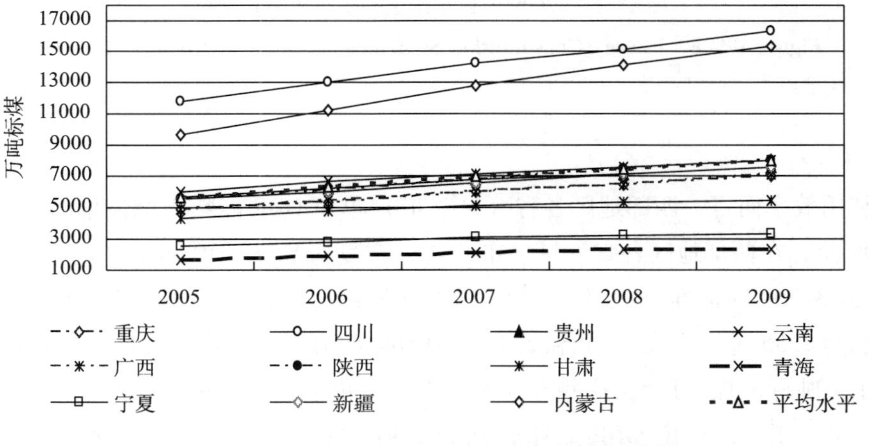

图17－7　“十五”末以来西部地区各省能源消费量

Figure 17－7　Energy Consumption of Western Provinces Since the End of China's 10th Five－year Plan Period

资料来源：《中国能源统计年鉴》(历年)

2. 能源消费结构

2009年，西部地区煤的消费比重超过75%的省份有重庆、贵州、云南等三个省份，分别为

82.05%、78.74%、77.86%，煤消费比重低于全国能源消费结构平均水平的省份有新疆、青海和甘肃，分别为48.21%、53.13%和55.87%；西部地区各省的油品消费比重均低于全国能源消费结构平均水平，其中，低于20%的省份有重庆、青海、宁夏和贵州，分别为17.95%、17.94%、16.61%和15.07%；天然气消费比重超过10%的省份有青海、新疆、四川和陕西，分别为23.17%、16.22%、15.07%和12%，而除四川外的重庆、云南、广西和贵州西南诸省天然气消费比重非常低；电力及其他类能源消费比重超过10%的省份有甘肃、新疆、内蒙古和宁夏，分别为15.34%、13.86%、11.89%和10.26%，而重庆、云南的电力及其他类能源消费比重非常低。可见，西北地区的内蒙古、宁夏以及西南地区的重庆、贵州、云南煤炭消费消费比例高；西部地区油品消费比重都不高，均低于全国平均水平；西南地区的四川以及西北地区的青海、新疆和陕西天然气消费比例较高；西北地区的甘肃、新疆和宁夏以及内蒙古的电力及其他类能源消费比重较高。图17-8为2009年西部地区各省能源消费结构图。

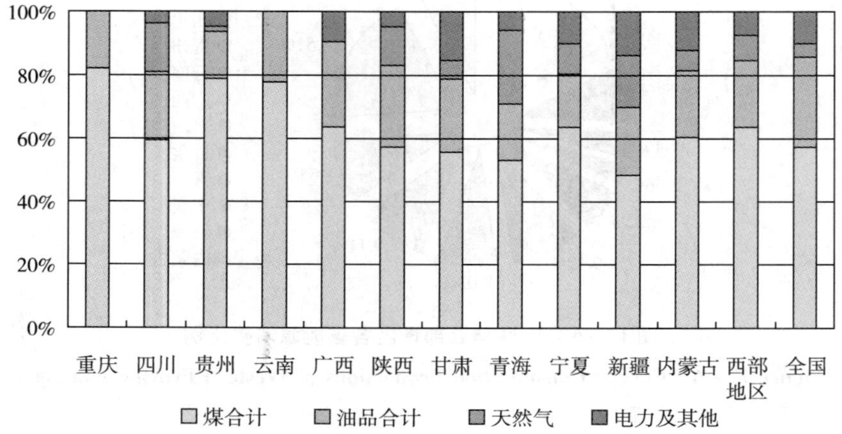

图17-8 2009年西部地区各省能源消费结构

Figure 17-8 Energy Consumption Structure of Western Provinces in 2009

资料来源：《中国能源统计年鉴》(2010)

3. 节能效率

就能源利用效率而言，西部地区各省（区、市）单位GDP能耗指标值偏高，落后于全国平均水平。2009年，在西部地区各省（区、市）中，宁夏的单位GDP能耗最高，超过了3吨标准煤/万元，达到3.308吨标准煤/万元；单位GDP能耗超过了2吨标准煤/万元的省份有青海和贵州二省，分别为2.55吨标准煤/万元和2.248吨标准煤/万元；其他省（区、市）的单位GDP能耗也均超过了1吨标准煤/万元，这些省（区、市）有内蒙古、甘肃、云南、四川、陕西、重庆和广西七省（区、市），单位GDP能耗分别为1.915吨标准煤/万元、1.801吨标准煤/万元、1.438吨标准煤/万元、1.275吨标准煤/万元、1.129吨标准煤/万元、1.127吨标准煤/万元和1.036吨标准煤/万元。从节能目标的完成情况来看，"十一五"时期，西部地区各省的节能工作取得了显著成效，至2010年全面完成了"十一五"国家下达的节能目标任务。其中，超额完成"十一五"节能目标任务较好的省份有重庆（4.75%）、内蒙古（2.82%）和云南（2.41%）三个省（区、市），完成"十一五"节能目标任务超过1%的省份有四川（1.55%）、广西（1.47%）、陕西（1.3%）和甘肃（1.25%），完成"十一五"节能目标任务低于1%的省份有宁夏（0.45%）、贵州（0.3%）和青海（0.24%）。（如表17-1）可见，就能源利用效率而言，西部地区大多数省（区、市）的能耗指标值偏高，能源利用效率落后于全国平均水平，但节能工作取得了较大进展，成效显著。

表17-1 “十一五”西部地区各省节能目标完成情况表

Table 17-1 Energy Conservation Performance of Western Provinces during China's 11th Five-year Plan Period

省份	2005年		2010年		
	单位GDP能耗（吨标准煤/万元）	“十一五”时期计划降低（%）	单位GDP能耗（吨标准煤/万元）	“十一五”时期实际降低（%）	超额完成数（%）
内蒙古	2.475	-22	1.915	-22.62	2.82
广西	1.222	-15	1.036	-15.22	1.47
重庆	1.425	-20	1.127	-20.95	4.75
四川	1.6	-20	1.275	-20.31	1.55
贵州	2.813	-20	2.248	-20.06	0.30
云南	1.74	-17	1.438	-17.41	2.41
陕西	1.416	-20	1.129	-20.25	1.25
甘肃	2.26	-20	1.801	-20.26	1.30
青海	3.074	-17	2.55	-17.04	0.24
宁夏	4.14	-20	3.308	-20.09	0.45

资料来源：根据《“十一五”各地区节能目标完成情况表》整理

4. 减排分析

从减排目标的完成情况来看，“十一五”时期，西部地区各省的减排工作取得了显著成效，截至2010年，西部地区化学需氧量排放量和二氧化硫排放量分别为334.32万吨和758.26万吨，分别比2005年下降9.08%和10.23%，全面完成了“十一五”国家下达的减排目标任务。从各省减排情况来看，各省份都超额完成了“十一五”主要污染物减排目标任务，其中，化学需氧量减排完成情况为：超额完成目标20%左右的省份有陕西（20.86%）和云南（19.58%），超过10%的省份有重庆（14.46%）、贵州（13.42%）和内蒙古（10.06%），低于10%的省份有四川（8.6%）、广西（2.8%）、甘肃（2.75%）和宁夏（1.33%），青海没有承担化学需氧量减排任务；二氧化硫减排完成情况为：超额完成目标20%左右的省份有陕西（29.58%），超过10%的省份有重庆（17.98%）、广西（17.7%）和内蒙古（11.88%），低于10%的省份有四川（8.68%）、贵州（2.65%）、云南（2.01%）和宁夏（0.94%），甘肃和青海没有承担二氧化硫减排任务。西部地区各省主要污染物“十一五”减排目标完成情况如表17-2所示。

表17-2 西部地区各省主要污染物“十一五”减排目标完成情况

Table 17-2 Emission Reduction of Major Pollutants in Western Provinces during China's 11th Five-year Plan Period

省份	化学需氧量					二氧化硫				
	2005年排放量（万吨）	“十一五”降低目标（%）	2010年排放量（万吨）	“十一五”实际降幅（%）	超额完成幅度(%)	2005年排放量（万吨）	“十一五”降低目标（%）	2010年排放量（万吨）	“十一五”实际降幅（%）	超额完成幅度(%)
内蒙古	29.7	-6.70	27.51	-7.37	10.06	145.6	-3.8	139.41	-4.25	11.88

续表

省份	化学需氧量					二氧化硫				
	2005年排放量（万吨）	“十一五”降低目标（%）	2010年排放量（万吨）	“十一五”实际降幅（%）	超额完成幅度(%)	2005年排放量（万吨）	“十一五”降低目标（%）	2010年排放量（万吨）	“十一五”实际降幅（%）	超额完成幅度(%)
广西	107	-12.10	93.69	-12.44	2.80	102.3	-9.9	90.38	-11.65	17.70
重庆	26.9	-11.20	23.45	-12.82	14.46	83.7	-11.9	71.95	-14.04	17.98
四川	78.3	-5.00	74.05	-5.43	8.60	129.9	-11.9	113.10	-12.93	8.68
贵州	22.6	-7.10	20.78	-8.05	13.42	135.8	-15.0	114.89	-15.40	2.65
云南	28.5	-4.90	26.83	-5.86	19.58	52.2	-4.0	50.07	-4.08	2.01
陕西	35	-10.00	30.77	-12.09	20.86	92.2	-12.0	77.86	-15.55	29.58
甘肃	18.2	-7.70	16.76	-7.91	2.75	56.3	0.0	55.18	-1.99	/
青海	7.2	0.00	8.31	15.42	/	12.4	0.0	14.34	15.65	/
宁夏	14.3	-14.70	12.17	-14.90	1.33	34.3	-9.3	31.08	-9.39	0.94

资料来源：根据《“十一五”期间全国化学需氧量排放总量控制计划表》、《“十一五”期间全国二氧化硫排放总量控制计划表》整理

5. 节能减排综合指数分析

根据节能减排综合指数的分类，即：第Ⅰ类地区节能减排综合指数的区间值为［2.00－3.00)，第Ⅱ类地区节能减排综合指数的区间值为［1.50－2.00）之间的为，第Ⅲ类地区节能减排综合指数的区间值为［1.00－1.50)，第Ⅳ类地区节能减排综合指数的区间值为［0.00－1.00)，西部地区各省节能减排综合指数排名（如表17－3)，Ⅰ类地区中没有出现西部地区省份，属于Ⅱ类地区有重庆、陕西、四川和云南四个省份，属于Ⅲ类地区有内蒙古、甘肃和广西三省区，属于Ⅳ类地区有青海、贵州和宁夏三省区。

比较其他地区节能减排综合指数类型的分布情况，全国属于Ⅰ类地区的五个省份均在东部地区；属于Ⅱ类地区的17个省份中，西部地区仅占4席；属于Ⅲ类地区的四个省份中，西部地区仅占3席；属于Ⅳ类地区的三个省份全部都在西部地区。可见，西部地区节能减排任重道远。

从西部地区各省的节能减排综合指数的构成来分析，重庆市的效果指数、努力指数和潜力指数排名都高于其综合指数排名，说明重庆市在节能减排工作中的努力程度、完成效果和可挖掘潜力上都高于目前的综合排名；属于Ⅱ类地区的陕西和云南、属于Ⅲ类地区的内蒙古和广西、属于Ⅳ类地区的贵州和宁夏的努力指数排名均高于其综合指数排名，说明尽管受到各种局限性的影响，以上各省区的综合指数排名还并不高，但它们客观上都能够更大的努力去实现节能减排工作目标；而属于Ⅱ类地区的四川、属于Ⅲ类地区的甘肃和属于Ⅳ类地区的青海的努力指数排名均低于其综合指数排名，说明它们的努力程度更滞后于其综合排名。

表17－3 西部地区各省节能减排综合指数

Table 17－3 Energy Conservation and Pollution Reduction Composite Index of Western Provinces

省份	节能减排指数及排名		二级指数					
	节能减排指数	排名	效果指数	排名	努力指数	排名	潜力指数	排名
内蒙古	1.328	23	0.613	23	0.102	19	0.613	23

续表

省份	节能减排指数及排名		二级指数					
	节能减排指数	排名	效果指数	排名	努力指数	排名	潜力指数	排名
广西	1.203	25	0.564	25	0.077	22	0.563	25
重庆	1.736	16	0.788	14	0.161	12	0.788	14
四川	1.554	20	0.741	20	0.071	23	0.743	20
贵州	0.744	28	0.347	28	0.049	24	0.348	28
云南	1.510	22	0.706	22	0.097	20	0.708	22
陕西	1.706	17	0.775	17	0.156	13	0.775	17
甘肃	1.204	24	0.589	24	0.027	26	0.588	24
青海	0.872	27	0.434	27	0.003	29	0.435	27
宁夏	0.393	29	0.190	29	0.013	28	0.190	29

资料来源：根据表14－9整理

17.1.3　节能减排的主要经验及存在的问题

“十一五”以来，根据《国民经济和社会发展第十一个五年规划纲要》、《“十一五”期间各地区单位生产总值能源消耗降低指标计划》和《“十一五”期间全国主要污染物排放总量控制计划》提出的“十一五”期间单位国内生产总值能耗降低20%左右的目标，主要污染物排放总量降低10%的减排任务，在将节能减排约束指标具体落实到西部地区十一省（区、市）后（如表17－1、表17－2），再由西部地区各省（区、市）分解落实到各市（地）、县及有关行业和重点企业，并制订有关具体实施方案，明确责任主体，层层抓好落实，严格执行，以确保总量控制目标的实现。截至2010年，西部地区各省（区、市）的节能减排工作成绩显著，顺利完成了节能减排的目标任务。但是，西部地区各省（区、市）的节能减排工作也面临着非常严峻的形势。2011年8月，发改委发布的本年度上半年全国节能减排完成情况晴雨表显示，有八个省（区）预警等级被列为最高级别一级预警，其中属于西部地区的省（区）有内蒙古、甘肃、青海、宁夏、新疆五个省（区）。据了解，内蒙古、宁夏、新疆等省份都是煤化工等高耗能项目集中的省份，高耗能行业上半年的快速发展是节能形势严峻的重要原因。总结“十一五”以来节能减排工作的主要经验和存在的问题表现在以下几个方面：

1. 主要经验

第一，“自上而下”的节能减排工作问责机制。自从中央将节能减排指标完成情况作为政府领导干部综合考核评价和企业负责人业绩考核的重要内容、实行“一票否决”制以来，各地都着手建立健全节能减排工作责任制和问责制，政府主要领导作为第一责任人。例如，甘肃省不仅对不达标县市实行一票否决制，还要对重点排污企业和恶意排污行为实行重罚，严重的甚至追究刑事责任。“十一五”期间西部地区各省（区、市）节能减排实践经验表明，将节能减排指标完成情况纳入各级地方政府政绩考核体系中的“自上而下”的节能减排工作问责机制能够在短期内迅速地扭转节能减排工作的不利局面，并取得显著成效的最直接原因。

第二，强有力的节能减排工作领导机构。西部地区各省（市、区）都成立了由政府主要领导任组长的节能减排工作领导小组和节能减排办公室，协调解决当地节能减排工作中的重大问题，并要求各县、市也建立相应领导小组。而节能减排办公室则在节能减排工作中履行牵头总抓的职责，综合协调节能减排日常工作，负责全社会节能减排的监管和考核评价，以及重点企业节能减排的监督检查等工作。“十一五”期间，西部地区各级政府和有关部门采取了一系列强有力的政策措施，综合运用法律、经济、技术及必要的行政手段，大力推动落后产能淘汰工作，上大压小、关停小火电机组，淘汰了大部分过剩、落后的炼铁、炼钢和水泥等的产能。强有力的节能减排工作领导机构成为了顺利开展节能减排工作的组织保障。

第三，调整优化中的产业结构升级。西部地区生态环境脆弱，开发过程中稍不注意就容易引起生态环境退化，优化经济结构，改善西部地区生态环境刻不容缓。为此，西部各省应着力发展能源、矿业、机械制造、农果业、中药材、加工制造和特色旅游等产业，严格控制高耗能、高污染行业过快增长，加快淘汰落后产能，以改善产业结构，从而实现提高能源利用效率、减少污染排放，挖掘西部地区各省节能减排的潜力。例如，青海省把结构调整作为转变经济发展方式的主攻方向，坚持传统产业的改造升级与发展新兴产业相结合，培育形成新能源、新材料、盐湖化工、有色金属、油气化工、煤化工、装备制造、特种钢、特色纺织和生物医药等十大特色优势产业，力争全省规模以上工业增长15%以上，园区工业增加值增长25%以上，循环经济增速高于整体工业。

第四，强化节能减排工作的重要抓手。按照以项目确定资源配置的原则，对重点项目、重点工程、重点企业予以重点支持，着力抓好节能重点工程、环境治理重点工程、循环经济重点工程，大力推动煤炭、电力、钢铁、有色、化工、建材等重点行业和耗能大户节能，关闭和淘汰污染严重的企业和生产工艺设备，推进重大项目、重点工程和重点企业的节能减排。例如，新疆把煤制气、煤制烯烃、煤制二甲醚、煤制油、煤基多联产等项目作为发展现代煤化工产业的制高点和节能减排的典型。宁夏自治区政府严格执行节能减排“十大铁律”，实施节能改造、节能产品惠民、合同能源管理等工程，全面开展建设项目能耗评估和清洁生产审核，突出抓好83户重点企业的节能降耗，认真抓好石嘴山市、宁东基地等国家级循环经济建设试点，毫不动摇地淘汰落后产能，严格项目准入，坚决杜绝能耗高、污染重的项目上马。陕西省切实抓好汉中循环经济产业聚集区、商丹循环经济工业园、铜川董家河煤电铝产业园等项目建设，大力支持各地建设以低碳和循环经济为特征的产业园区。

第五，不断加强的节能减排法规和制度保障。不断完善节能减排工作评价、考核制度，提高节能减排信息通报制度的透明度，实行单位生产总值能耗和主要污染物排放总量公报制度，定期向社会公布辖区的节能减排主要指标，接受社会监督。加快重点产品的能耗限额标准、建筑业的能耗标准和节能设计规范等标准和规范的制订和完善工作，建立健全节能减排的评估审查制度，建立有利于节能减排的利益导向机制。

第六，激励性的节能减排财政政策。西部地区各省（区、市）通过设立节能减排专项资金有力地支持了重点节能减排项目建设。例如，广西壮族自治区加大财政支持减排力度，2008年自治区本级财政新增安排2.1亿元支持节能减排；同时，积极创新支持方式，提出以奖代补，实行财政奖励资金量与节能减排量挂钩，进一步推动减排工作的开展。根据“以奖代补、重点倾斜、专户管理、专款专用”的方式，2.1亿元节能减排奖励资金与全区当年的节能减排目标挂钩，重点支持60个节能技术改造项目和317个主要污染物减排项目建设。“十一五”期间，新疆自治区节能减排专项资金重点支持节能减排项目、高效节能产品和节能减排新机制的推广、节能减排示范试点工作的开展、节能减排管理能力建设以及自治区财政厅、经贸委根据国家、自治区有关方针政策确定的其他节能减排项目及活动，对列入“国家千家企业节能行动名单”的企业、自治区重点用能企业、自治区工业经济领域循环经济试点单位实施的节能减排项目及活动给予优先支持。

2. 存在的问题

第一，节能减排观念意识有待提高。在西部地区，许多地方政府仍片面地将发展经济与节能减排目标相对立起来，认为经济增长必须靠投资拉动，盲目投资高耗能产业，往往为了追求经济增长而忽视节约能源、降低排放。也有一些企业，为了追逐高额利润，不惜以牺牲环境为代价，非法排放大量废水废气、有毒有害烟尘等。

第二，亟待健全节能减排的组织、运行、管理和监督体系。在“十一五”的节能减排目标考核中，政府基本上采用到最后一年整体考核是否完成节能减排目标，即采取“算总账”方法。这样一来，许多地方往往采取在初始阶段人为压低节能减排效率、在考核后半阶段才加大节能减排力度，等到考核期末之时，尽管能源强度下降的指标达到了，但能源消费总量大幅上涨。正是由于各个节能减排主体缺乏节能减排自觉能动性和利益驱动力，现行的“自上而下”的管理体制并不能解决长期存在的节能减排问题，如节能减排中的“前松后紧，临时突击”问题，“一刀切”问题，治标不治本问题等，建立一种长效激励与监管机制势在必行，充分发挥政府、企业和民众等利益相关者参与决策的能动性。

第三，节能减排指标分配方案有待进一步优化。西部地区各省基本属于欠发达省份，大多属于能耗强度正处在上升期的重化工业阶段，单位 GDP 能耗强度下降潜力较小，节能减排和发展经济之间的矛盾较为突出。中央在制定节能减排政策的时候应该充分考虑到西部地区的经济社会发展水平、资源禀赋、技术能力和管理水平等方面的差异，因地制宜的安排西部地区各省（市、区）的节能减排指标和目标。

第四，节能减排的配套政策体系亟待完善。“十一五”期间，西部地区节能减排的相关配套政策的制定严重滞后，已经在一定程度上影响着节能减排工作的进一步推进，并且已出台的配套政策也需要加强与相关政策的协调性。为此，为了更有效地推进西部地区节能减排工作，相关配套措施必须得力到位、落到实处。

17. 2 “十二五”时期西部地区节能减排展望

17. 2. 1 节能减排的战略背景

当前和今后一个时期，西部地区正处于机遇和挑战并存、机遇大于挑战、可以大有作为的重要战略机遇期。世界多极化、经济全球化深入发展，我国与世界各国的经济合作不断深化，为西部地区加快发展营造了良好的外部环境。同时，西部地区正处于工业化和城镇化快速推进阶段，经济社会发展的基本态势长期向好，国家扩大内需战略的实施，并且把西部大开发放在全国区域协调发展总体战略的优先地位，中央企业转型发展和向西部扩张加快，东部产业加快转移，以及扶持西部贫困地区加快发展和西部民族地区跨越发展，都将为西部地区的发展提供了巨大空间，都有利于西部地区各省通过争取中央支持和借助东部地区资源加速发展。而且，随着国际国内生产要素流动和产业转移步伐加快，东部沿海发达地区辐射带动能力的日益增强，以及国家更加重视西部大开发战略的深入实施与纵深推进，这些因素都为西部地区跨越式发展创造了新的机遇。后金融危机时期孕育着新的科技革命和产业革命，为西部地区各省发挥后发优势，实现技术跨越，抢占科技和产业制高点提供了良好契机。

同时，西部地区还属于欠发达地区，未来发展还面临不少挑战和困难。主要表现为：

一是产业发展不充分，农牧业基础薄弱，工业整体水平不高，服务业发展不足，产业结构有待优化、产业等级提升。

二是经济增长过度依赖于对资源的开发，非资源型产业发展严重滞后，科技创新能力不强，产业竞争力较弱。

三是生产力布局不尽合理，区域发展不平衡，大中城市发育不足、带动力不强。

四是生态环境脆弱，基础设施瓶颈制约问题突出。

五是制约科学发展的体制机制障碍依然较多，对外开放的力度需要进一步加大。

“十一五”期间，西部地区节能减排工作持续稳步推进，节能减排工作取得显著成就的同时，也为“十二五”节能减排工作的开展打下了良好的基础。

1. 节能战略方面

在节能战略方面，西部地区“十二五”期间需要侧重于以下几个方面：

第一，坚持以结构节能为主，加大对高耗能产业的技术改造和产品升级，严格控制高耗能产业在区域的新的布局，坚决淘汰技术落后、浪费资源、污染严重的产品和产业，按照科技含量高、经济效益好、资源消耗低、环境污染少的新型工业化发展要求，加快工业结构的调整，逐步形成有利于资源节约和环境保护的工业体系，降低单位产出的能源消耗和排放。

第二，加大工业节能力度，注重产品节能。加大重点行业包括钢铁、有色、石化等高耗能、高污染行业工业节能力度，重点发展余热余压利用、工业锅炉（窑炉）改造、热电联产等，提高工业能源利用率效率。提高财政补贴力度，加大家用电器、交通工具、照明产品等高效节能产品在西部区域的推广应用。

2. 减排战略方面

在减排战略方面，西部地区“十二五”期间需要努力做好以下几个方面的工作：

第一，加大工程减排在污染物的削减的作用，继续完善城市污水、工业废气的处理设施和废水深度治理工程；提高规模以上燃煤脱硫、脱硝机组、钢铁烧结机烟气脱硫设施的比重。

第二，推进资源综合利用，实现循环发展。围绕能源、大宗短缺、稀缺金属等重要矿产资源，推进综合开发及共伴生矿综合利用；以尾矿、赤泥等大宗工业固废为重点，开展固废综合利用示范基地建设。

第三，加快构建企业为主、政府指导、市场驱动、全民参与的清洁生产工作格局；通过抓好清洁生产审核、清洁生产评价，在重点区域、重点领域实施清洁生产重大工程，带动企业实施清洁生产改造。

第四，积极调整优化能源消费结构，降低煤炭在终端消费中的比重，继续推进“上大压小”，大力发展大容量、高参数机组火电，不断提高一次能源的加工转化比重，特别是提高煤炭消费用于发电的比重。

综合分析各方面因素，西部地区是有条件继续保持较长的快速发展期，同时也可能带来一系列的节能减排问题，而且，由于“十二五”时期的发展环境和面临形势将更加复杂，不确定因素增加，与“十一五”时期相比已经发生了重大变化，西部地区的节能减排形势非常严峻，所面临的问题和挑战依然还很多。

17.2.2 节能减排的战略目标

1. 发展目标

根据西部地区的资源禀赋和产业基础，充分利用西部地区的能源资源优势，在促进资源、环境与经济发展相协调的前提下，以市场为导向，调整优化能源结构，促进集中布局，提高优势资源加工增值比重，着力发展能源、矿业、机械制造、农果业、中药材、加工制造和特色旅游等产业，建设成为国家重要的能源供应基地、风情独特的旅游目的地和生态环境建设的示范基地。促成西部地区各省单位生产总值能源消耗和二氧化碳排放大幅下降，主要污染物排放得到有效控制，使单位地区生产总值能耗，非化石能源占一次能源消费比重，单位地区生产总值二氧化碳排放量，化学需氧量、二氧化硫、氨氮、氮氧化物等主要污染物排放总量等达到或控制在国家下达的指标内。

2. 节能目标

西部地区各省（区、市）的单位 GDP 能耗降低率的分类分布情况：单位 GDP 能耗降低率降幅为 18% 的第一类地区和单位 GDP 能耗降低率降幅为 17% 的第二类地区中均没有西部地区省份，单位 GDP 能耗降低率降幅为 16% 的第三类地区中有重庆、四川和陕西三省（市），单位 GDP 能耗降低率降幅为 15% 的第四类地区中有内蒙古、广西、贵州、云南、甘肃和宁夏六省（区），单位 GDP 能耗降低率降幅为 10% 的第五类地区中有西藏、青海和新疆三省（区）。各地区“十二五”节能目标分解方案如表 17－4 所示。

根据西部地区各省（区、市）“十二五”规划纲要的主要目标内容，11 省（区、市）节能目标如下：

- 内蒙古：单位地区生产总值能源消耗降低 15%，非化石能源占一次能源消费总量比重达到 5%。
- 广西：非化石能源占一次能源消费比重达到 20%，确保完成国家分解下达我区的单位地区生产总值能源消耗降低目标任务。
- 重庆：单位地区生产总值能耗减少 15%。
- 四川：单位生产总值能源消耗达到国家要求，非化石能源占一次能源消费量的比重明显提高。
- 贵州：单位生产总值能源消耗控制在国家下达的指标范围内。
- 云南：非化石能源占一次能源消费比重达到 30% 左右，单位生产总值能耗进一步降低，确保国家下达的“十二五”节能减排等约束性指标圆满完成。
- 陕西：单位生产总值能源消耗降低 16%。
- 甘肃：单位地区生产总值能耗达到或控制在国家下达的指标内。
- 青海：单位生产总值能耗控制在国家下达指标以内。
- 宁夏：节能指标控制在国家下达任务以内。
- 新疆：工业单位产品能耗达到国家限额标准或国内先进水平，有效控制万元生产总值能耗。

表 17-4 各地区"十二五"节能目标分解方案

Table 17-4 Classified Energy Conservation Targets of Different Regions for China's 12th Five-year Plan

类别	单位 GDP 能耗降低率最高降幅	包含的省份
第一类	18%	天津、上海、江苏、浙江和广东
第二类	17%	北京、河北、辽宁和山东
第三类	16%	山西、吉林、黑龙江、安徽、福建、江西、河南、湖北、湖南、重庆、四川和陕西
第四类	15%	内蒙古、广西、贵州、云南、甘肃和宁夏
第五类	10%	海南、西藏、青海和新疆

资料来源：根据《"十一五"节能目标分解方案初定》整理

3. 减排目标

根据2011年9月7国务院《关于印发"十二五"节能减排综合性工作方案的通知》（国发〔2011〕26号），结合西部地区各省（区、市）"十二五"规划纲要的主要目标内容，11省（区、市）减排目标（如表17-5）如下：

- 内蒙古：单位地区生产总值二氧化碳、化学需氧量、二氧化硫、氨氮、氮氧化物排放达到国家的要求。即与"十一五"末相比较，至"十二五"末化学需氧量排放总量下降6.7%、氨氮排放总量下降9.7%、二氧化硫排放总量下降3.8%、氮氧化物排放总量下降5.8%。
- 广西：确保完成国家分解下达我区的单位地区生产总值二氧化碳排放降低和化学需氧量、二氧化硫、氨氮、氮氧化物排放减少目标任务。即与"十一五"末相比较，至"十二五"末化学需氧量排放总量下降7.6%、氨氮排放总量下降8.7%、二氧化硫排放总量下降7.9%、氮氧化物排放总量下降8.8%。
- 重庆：主要污染物排放总量持续减少，具体地，与"十一五"末相比较，至"十二五"末化学需氧量排放总量下降7.2%、氨氮排放总量下降8.8%、二氧化硫排放总量下降7.1%、氮氧化物排放总量下降6.9%，单位地区生产总值二氧化碳排放减少17%。
- 四川：单位生产总值能源消耗和二氧化碳排放降低、主要污染物排放减少等指标达到国家要求。即与"十一五"末相比较，至"十二五"末化学需氧量排放总量下降7%、氨氮排放总量下降8.6%、二氧化硫排放总量下降9%、氮氧化物排放总量下降6.9%。
- 贵州：单位生产总值二氧化碳排放降低，主要污染物排放总量减少控制在国家下达的指标范围内。即与"十一五"末相比较，至"十二五"末化学需氧量排放总量下降6%、氨氮排放总量下降7.7%、二氧化硫排放总量下降8.6%、氮氧化物排放总量下降9.8%。
- 云南：单位生产总值二氧化碳排放量、主要污染物排放总量进一步降低，确保国家下达的"十二五"减排等约束性指标圆满完成。具体地，与"十一五"末相比较，至"十二五"末化学需氧量排放总量下降6.2%、氨氮排放总量下降8.1%、二氧化硫排放总量下降4%、氮氧化物排放总量下降5.8%。
- 陕西：单位生产总值二氧化碳排放降低15%，主要污染物排放总量减少10%。具体地，与"十一五"末相比较，至"十二五"末化学需氧量排放总量下降7.6%、氨氮排放总量下降9.8%、二氧化硫排放总量下降7.9%、氮氧化物排放总量下降9.9%。
- 甘肃：单位地区生产总值二氧化碳排放量，化学需氧量、二氧化硫、氨氮、氮氧化物等主要污染物排放总量达到或控制在国家下达的指标内。即与“十一五”末相比较，至“十二五”末化学需氧量排放总量下降6.4%、氨氮排放总量下降8.9%、二氧化硫排放总量不降反而增长2%、氮氧化物排放总量下降3.1%。到2015年，建成国家级循环经济示范区。

- 青海：循环经济增加值比重达到80%，可再生能源生产比重达到40%，单位生产总值二氧化碳排放量、主要污染物排放总量均控制在国家下达指标以内。即与“十一五”末相比较，至“十二五”末化学需氧量排放总量、氨氮排放总量、二氧化硫排放总量和氮氧化物排放总量均不降反而增长，分别增长18%、15%、16.7%和15.3%。

- 宁夏：减排指标控制在国家下达任务以内。即与“十一五”末相比较，至“十二五”末化学需氧量排放总量下降6%、氨氮排放总量下降8%、二氧化硫排放总量下降3.6%、氮氧化物排放总量下降4.9%。

- 新疆和西藏：与“十一五”末相比较，至“十二五”末新疆和西藏二区的主要污染物排放均控制指标均未变化。

表17-5 “十二五”西部地区各省（区、市）减排目标

Table 17-5 Pollution Reduction Targets of Western Provinces (Autonomous Regions and Municipalities) for China's 12th Five-year Plan

地区	化学需氧量排放总量控制计划			氨氮排放总量控制计划			二氧化硫排放总量控制计划			氮氧化物排放总量控制计划		
	2010年排放量	2015年控制量	2015年比2010年下降	2010年排放量	2015年控制量	2015年比2010年下降	2010年排放量	2015年控制量	2015年比2010年下降	2010年排放量	2015年控制量	2015年比2010年下降
	（万吨）	（万吨）	（%）	（万吨）	（万吨）	（%）	（万吨）	（万吨）	（%）	（万吨）	（万吨）	（%）
全国	2551.7	2335.2	-8.5	264.4	236.6	-10.5	2267.8	2067.4	-8.8	2273.6	2021.6	-11.1
内蒙古	92.1	85.9	-6.7	5.45	4.92	-9.7	139.7	134.4	-3.8	131.4	123.8	-5.8
广西	80.7	74.6	-7.6	8.45	7.71	-8.7	57.2	52.7	-7.9	45.1	41.1	-8.8
重庆	42.6	39.5	-7.2	5.59	5.1	-8.8	60.9	56.6	-7.1	38.2	35.6	-6.9
四川	132.4	123.1	-7	14.56	13.31	-8.6	92.7	84.4	-9	62	57.7	-6.9
贵州	34.8	32.7	-6	4.03	3.72	-7.7	116.2	106.2	-8.6	49.3	44.5	-9.8
云南	56.4	52.9	-6.2	6	5.51	-8.1	70.4	67.6	-4	52	49	-5.8
西藏	2.7	2.7	0	0.33	0.33	0	0.4	0.4	0	3.8	3.8	0
陕西	57	52.7	-7.6	6.44	5.81	-9.8	94.8	87.3	-7.9	76.6	69	-9.9
甘肃	40.2	37.6	-6.4	4.33	3.94	-8.9	62.2	63.4	2	42	40.7	-3.1
青海	10.4	12.3	18	0.96	1.1	15	15.7	18.3	16.7	11.6	13.4	15.3
宁夏	24	22.6	-6	1.82	1.67	-8	38.3	36.9	-3.6	41.8	39.8	-4.9
新疆	56.9	56.9	0	4.06	4.06	0	63.1	63.1	0	58.8	58.8	0

资料来源：根据《关于印发“十二五”节能减排综合性工作方案的通知》整理

17.2.3 节能减排的战略选择

西部地区幅员广阔，区域内差异较大，大多数地区生态环境脆弱。为此，西部地区的节能减排战略选择应充分考虑深入实施的西部大开发战略大背景并将其摆在优先位置，充分考虑西部地区生态环境建设与可持续发展战略的要求，选择符合各省（区、市）具体实际条件的节能减排战略。因此，西部地区各省的节能减排战略既有其共性的一面也有其独特性和个性的一面。

1. 内蒙古节能减排战略

以坚持绿色发展，建设环境友好型和资源节约型社会为发展目标，以保护优先和自然恢复为主，从源头上扭转生态环境恶化趋势，建设祖国北疆生态安全屏障。切实加强环境保护与污染治理，有效提高污染源综合治理水平和环境质量。强化资源节约集约利用，发展循环经济，逐步建立低投入、高产出、低消耗、少排放、能循环、可持续的绿色经济发展体系。

（1）加大环境保护力度。一是改善区域和城市大气环境质量。从注重重点行业削减向全社会减排转变，全区70%的主要城市空气质量好于二级标准的天数超过292天。在重点区域推行大气联防联控机制，协调解决大气污染防治重大问题。加强燃煤电厂脱硫脱硝，加快钢铁、焦化、烧结、有色行业等脱硫工程建设，强化脱硫设施稳定运行。推广脱硫石膏盐碱地改良技术。在大中城市及其近郊严格控制除热电联产外的新、扩建燃煤电厂，推进集中供热，推行清洁能源替代，控制工业烟尘、粉尘和城市扬尘的排放。

二是开展农村牧区环境综合整治。对土壤污染进行全方位评价，提出土壤分区控制、利用和保护对策，建立土壤污染防治和修复机制，开展典型区域、典型类型污染土壤修复示范。控制农药、化肥和农膜等面源污染，加大畜禽养殖污染防治力度。推广经济适用的污染防治和废弃物综合利用技术，提高秸秆综合利用率和畜禽粪便资源化水平。加快推动农村牧区垃圾集中处理，开展农村环境集中连片整治。严格禁止工业固废、危险废物、城镇生活垃圾及其他污染物向农村牧区转移。

三是加强环境监管。严格落实环境保护目标责任制，强化污染物总量控制目标考核，健全重大环境事件和污染事故责任追究制度，加大问责力度。强化环境执法监督，严格污染物排放标准、环境影响评价和污染物排放许可制度，进一步健全环境监管体制，提高环境监管能力。加快环境应急体系建设，提高突发事件应对能力，有效防范环境风险。

四是节约利用能源。推动煤炭、电力、钢铁、有色、化工、建材等重点行业和耗能大户节能，关闭和淘汰污染严重的企业和生产工艺设备。支持和鼓励建筑领域采用节能型建筑结构、材料和产品，逐步推行分户供热和分户计量。加强公共机构节能，党政机关率先垂范，重点实施建筑物及采暖、空调、照明系统节能改造。加强能源生产、运输、消费各环节的制度建设和监管，实现管理节能。推行汽车燃油经济性标准，加快淘汰老旧运输设备。对能耗大户进行能源审计，推广合同能源管理，加强节能监察。

（2）大力发展循环经济。一是构建循环型工业体系。以煤炭、电力、化工、冶金、建材等行业为重点，培育100户国家级循环经济骨干企业，大力开展以节能、降耗、减污、增效为目标的清洁生产，抓好煤炭、稀土、有色金属共伴生矿产资源综合利用，推进粉煤灰、煤矸石、冶金和化工废渣及尾矿等工业废物利用。以不同行业的骨干企业为龙头，围绕资源的循环利用、节能减排和产业链延伸，培育发展横向关联配套、纵向延伸拓展的产业网络，在主要工业行业重点形成一批循环经济产业链。加快重点开发区和工业园区改造提升，建设一批国家级和自治区级循环经济园区。

二是构建循环型农牧业体系。发挥农牧业生态系统的整体功能，提高集约化水平，大力推广保护性耕作、合理施肥（药）、节水灌溉、旱作农业、集约化生态养殖、沼气与秸秆综合利用等循环利用技术。以农村牧区沼气建设为依托，推动规模养殖、特色种植、庭院经济和无公害、绿色、有机农畜产品的发展，提升农村牧区经济发展水平。

三是构建循环型城市与社区。建设再生资源回收利用系统，推进全区各城市和社区废纸、废旧金属、废旧轮胎和废弃电子产品等回收利用，推广车载桶装密闭式垃圾收运模式和密闭式垃圾自动收集系统，社区生活垃圾全部分类收集。推进可持续消费，倡导节约和循环型消费观念，在商场、酒店、机场、车站、公园和旅游景点等场所杜绝使用不可降解、不可循环使用的产品。推行政府绿色采购，提高政府采购中可循环使用的产品、再生产品以及节能、节水、绿色有机产品比例。

（3）积极应对气候变化。一是加快培育低碳经济体系。加强战略规划和试点示范，落实《内蒙古自治区应对气候变化实施方案》，推进低碳经济合理有序发展。建设森林碳汇基地，增强草原碳汇功能，探索建立草原固碳标准体系，培育碳汇交易市场，推动开展碳汇交易。发挥科技对发展低碳经济的支撑作用，加快低碳技术的引进、研发、示范和产业化步伐。倡导循环使用、低碳消费、低碳经营的理念，推行低碳生活方式。

二是建立低碳经济政策体系与评估机制。全面提高全社会应对气候变化的意识，开展气候变化统计、监测、评估和科学研究，增强农牧业、生态、水利、交通、卫生等领域应对极端气候变化的能力。建立有利于低碳经济发展的投资、财税、价格、政府采购政策体系和评估考核机制。

2. 广西节能减排战略

广西以建设全国生态文明示范区为契机，坚持大力发展循环经济、加强节能管理、加强资源节约和管理、强化环境保护和加强生态建设和防灾减灾体系建设等举措。

（1）大力发展循环经济。以提高资源产出效率为目标，加强规划指导、财税金融等政策支持，推进生产、流通、消费各环节循环经济发展。鼓励企业建立循环经济联合体，在制糖、铝业、锰业、石化、建材等重点行业构建循环利用产业体系。加强共伴生矿及尾矿、废石综合利用。鼓励产业废物循环利用，完善再生资源回收体系和垃圾分类回收制度，推进资源再生利用产业化。鼓励河池发展有色金属循环经济，支持崇左、贵港发展糖业循环经济，建设梧州、玉林再生资源循环利用示范园区和华润（贺州）循环经济产业园区，推广循环经济典型模式。

（2）加强节能管理。把大幅降低能源消耗强度和二氧化碳排放强度作为约束性指标，综合运用节能减排、增加森林碳汇等手段，积极应对气候变化。调整能源消费结构，合理控制能源消费总量，抑制高耗能产业过快增长，提高能源利用效率。强化节能目标责任考核，完善节能法规和政策，健全节能市场化机制和对企业的激励与约束，实施重点节能工程，推广先进节能技术和产品，加快推行合同能源管理，重点抓好工业、建筑、交通运输等领域节能。

（3）强化环境保护。落实减排目标责任制，强化污染物减排和治理，加大西江流域重点江河及大中型水库水污染防治力度，有效控制城市大气、噪声污染，加强重金属、危险废物、土壤污染治理，强化核与辐射监管能力。严格污染物排放标准和环境影响评价，强化执法监督，健全重大环境事件和污染事故责任追究制度。完善环境保护、科技和经济政策，建立健全污染者付费制度。加强环境监测、预警和应急能力建设。建立多元环保投融资机制，大力发展环保产业。

3. 重庆节能减排战略

以构建资源节约型和环境友好型社会作为发展目标，树立绿色、低碳发展理念，加快资源节约型、环境友好型社会的建设，全面提升可持续发展能力，促进人与自然协调发展。

（1）加强环境保护。坚持预防为主、综合整治、强化监管，以创建国家环保模范城市、污染物总量减排、保障三峡库区环境安全为重点，着力解决危害群众健康和影响可持续发展的突出环境问题，不断改善城乡环境质量。

（2）大力推进污染物减排。完善污染物减排责任制，强化工程减排、结构减排和管理减排措施，推进主要污染物排放权有偿使用交易，大幅度减少污染物排放量。

（3）加强大气污染防治。实施二氧化硫和氮氧化物总量控制，继续加强燃煤电厂脱硫脱硝，推进燃煤锅炉废气污染治理，提升工业窑炉废气治理水平，严格控制机动车尾气排放，加快排放超标车淘汰更新。加强餐馆及摊贩油烟排放治理。严格控制建筑工地等城市扬尘污染。建立健全区域大气污染联防联控机制，控制复合型大气污染。

（4）构建环境保护预防体系。提高环境准入标准，严格控制高耗能、高污染、低产出项目，加快淘汰落后产能。严格执行主要污染物排放许可证制度。推行绿色信贷，稳步推进排污权有偿使用和交易。加强对重点环境风险源的动态监测与风险控制。健全环境风险预防和处置体系。全面推行城乡污水、垃圾收费制度。

（5）大力发展循环经济。建设中西部发展循环经济的示范区。推进工业园区上中下游产业链、水电气热联供、基础设施配套、物流配送服务和生产生活环保生态管理等“五个一体化”。推进企业生产内部工艺之间的能源梯级利用和物料循环利用，推动产业循环式组合。加快建设再生资源回收体系。开展大宗工业固体废物综合利用工程试点，工业固体废物综合利用率达到80%。

（6）建设国家低碳试点城市。大力发展低碳经济，降低温室气体排放强度，不断提高减缓和适应气候变化的能力。建立促进低碳发展的政策法规、标准和统计、监测及目标考核体系，制定财税、金融、土地、生产原料价格等方面支持政策，鼓励和引导社会资金投向低碳经济领域。鼓励低碳技术创新、引进、研发和应用，强化低碳发展的技术支撑。以调整产业结构、提高能源效率、优化能源结构、促进低碳技术研发应用为重点，打造以低碳排放为特征的产业体系，着重抓好工业、建筑和交通等重点领域碳减排，促进重点行业和支柱产业向低碳转型。提高低碳能源供给总量，到2015年非化石能源占一次能源消费比重达到13%。构建低碳标准及产品认证体系，开展低碳企业和产品认证、低碳园区试点和示范。增加森林蓄积量和碳汇储备，探索建立碳排放交易市场。倡导绿色低碳生活方式和消费模式，培育壮大低碳产品的消费市场。完善气候观测系统，加强三峡库区气候变化与局地气候效应研究，增强适应气候变化特别是应对极端天气气候事件能力。开展应对气候变化和低碳发展的国内外合作。

4. 四川节能减排战略

（1）加快构建长江上游生态屏障，加大环境保护力度。以解决突出环境问题为重点，加强综合治理和污染防控，改善环境质量，化解环境风险。加强岷江、沱江、嘉陵江等重点流域水污染防治，严格保护城乡集中式饮用水水源地。推进城市大气复合型污染防治和酸雨防治，改善大气环境质量。加强重金属污染治理。实施总量控制，减少主要污染物排放。完善环境保护政策，建立健全污染者付费制度，建立多元环保投融资机制，大力支持环保技术研发和环保产业发展。

（2）加强能源基础设施建设。推动能源生产和利用方式变革，优化能源结构，实现传统能源清洁高效利用，提升能源普遍化服务水平。在保护生态和切实做好移民安置工作的前提下优先发展水电，优化发展火电，积极发展核电，加快天然气开发建设，因地制宜有序开发生物质能、太阳能、风能、煤层气、沼气等新能源。加强智能电网、分布式能源、新能源汽车供能设施建设，加快煤电油气输送通道建设，优化能源区域布局，努力构筑安全、稳定、经济、清洁的现代能源体系。

（3）大力发展循环经济。按照减量化、再利用、资源化的原则，从生产、流通、消费等环节入手，推进资源循环利用和清洁生产。支持循环经济技术研发、示范推广和能力建设，加快循环经济园区建设，推进循环经济城市试点。加大低碳技术等先进节能减排技术的推广应用，实施重点节能工程，加快推进合同能源管理，突出抓好工业、交通、建筑、商贸流通等重点领域节能降耗，最大限度挖掘全社会节约资源能源潜力。全面实行严格的节能减排考核制度。

5. 贵州节能减排战略

树立节约资源、保护环境、建设良好生态的可持续发展理念，大力发展循环经济、绿色经济，加快建设资源节约型、环境友好型社会，走生产发展、生活富裕、生态良好的文明发展之路。

（1）大力发展循环经济，提高资源产出效率。一是推进循环型生产方式，加快推行清洁生产。大力推进粉煤灰、磷石膏、煤矸石、黄磷尾气和炉渣等大宗工业固体废物和建筑、道路废物以及农林废物资源化利用，工业固体废物综合利用率达到60%以上；按照“煤—电—化—建材”、“煤—磷—电—化—建材”、“煤—电—冶金—建材”一体化产业循环模式，推进产业链接技术的开发运用，加快建设一批省级铝工业、磷化工、煤化工、钢铁、建材等循环经济工业园区和示范企业；继续抓好国家级循环经济试点单位（城市）建设，积极支持六盘水市建设国家级循环经济示范城市。

二是建立健全资源循环利用回收体系，推进再生资源规模化利用。以再生金属、废旧轮胎、废旧家电、电子产品回收利用为重点，加快完善再制造旧件物流回收体系，优化布局，壮大规模，积极发展再制造业；建立健全垃圾分类回收、密闭运输、集中处理体系，推进餐厨废弃物等垃圾资源化利用和无害化处理。

三是建立循环经济促进机制。综合运用投资、产业、价格、财税和金融等政策措施，建立面向市场，有利于发展循环经济的政策支持体系；推动发展循环经济的地方法规体系建设。建立和完善循环经济技术研发和咨询服务体系；编制和实施循环经济发展规划，引导中心城市、重点开发区、重点产业园区、重化工业集中地区，按照发展循环经济的要求集聚发展；倡导文明、节约、绿色消费理念，逐步培育形成绿色生活方式和消费模式。

（2）大力推进能源节约，提高能源利用效率。控制低水平高耗能企业发展，确保全省单位生产总值能源消耗降低达到国家要求；积极淘汰落后生产能力，到2015年，全省淘汰落后产能1000万吨以上，淘汰单机20万千瓦以下火电机组100万千瓦；突出抓好钢铁、有色、电力、化工、建材、煤炭等重点耗能行业和重点耗能企业的节能管理，加强电力需求侧管理和节能发电调度；强化建筑、交通运输和公共机构等领域的节能技改和节能管理；全面实施重点节能工程，推广普及先进节能技术和产品，加快推行合同能源管理，促进节能服务产业发展；严格实施新建项目节能评估和审查制度，完善能效标准和节约产品认证制度；落实能源节约经济政策，健全节能法律法规和标准体系；加强节能监察监测能力建设，建立完善节能统计、监测制度，强化目标责任考核，严格节能执法监察；全面实行政府优先和强制采购节能产品制度，普及节能知识，加强消费引导，实施全民节能行动方案。

（3）加强重点行业污染控制和区域大气污染防治。大力推进火电、钢铁、有色、水泥等重点行业脱硫设施建设和技术改造，开展火电机组脱氮设施建设；加强氮氧化物污染防治，开展挥发性有机物及有毒有害物质控制，推进多污染综合控制；加强城市颗粒物、机动车尾气排放、噪声等污染治理。

（4）强化环境保护机制。健全考核评估制度，严格落实环保目标责任制，强化政府环境管理职责；严格执行环境影响评价制度，推进规划环评实施。严格排放标准，全面推进排污许可证制度；坚持谁污染谁治理原则，建立和完善企业环保自我约束机制；加强环境管理能力建设，完善监管体系，强化执法监督，健全重大环境事件和污染事故责任追究制度；建立社会化、多元化环保投融资机制，运用经济手段加快污染治理市场化进程；深入开展环保宣传教育，动员全社会积极参与环境保护；推进贵阳市国家级环境保护模范城市建设，积极创建国家级和省级生态市（地、州）、县、乡（镇）、村。

（5）积极应对气候变化。坚持减缓和适应并重，充分发挥技术进步的作用，加强政策支持，提高应对气候变化能力；积极推进低碳技术的引进和转化，控制工业、建筑、交通和农业等领域温室气体排放；开发能源高效利用技术，大力发展可再生能源，优化能源结构；逐步建立和完善温室气体排放统计监测和考核体系，开展区域气候变化影响综合评估；推进贵阳市国家级低碳城市试点，开展适应气候变化重点示范区建设；加强应对气候变化知识普及，提高公众气候变化意识。

6. 云南节能减排战略

（1）加大环境保护和污染治理力度。深入实施“七彩云南保护行动”。加大矿山生态环境保护和治理力度，强化污染物排放治理，重点加强水、大气、土壤等污染防治，做好重金属污染防治工作，着力解决危害群众健康的突出环境问题。全面推进城镇污水、生活垃圾处理设施建设，有效控制城市大气、噪声污染。探索排污权交易，建立和完善污染者付费制度。严格环境准入制度，强化执法监督，健全环境监管体系和环境重大事件应急处理机制及责任追究制度。

（2）深入推进节能减排。强化节能减排目标责任制，严格实行能源消费和污染物排放总量控制，建立以改善环境质量为根本目标的减排指标、监测、考核体系和工作机制。运用环保、技术标准、产业和金融政策措施，从源头上控制高污染、高排放，下决心淘汰落后产能。实施重点节能工程，抓好重点领域节能，大力推进建筑节能、全民节能、电厂脱硫等工程，积极引进、消化吸收和推广先进适用节能减排技术及产品。继续开展节能减排全民行动。大力推广清洁生产。

（3）积极促进低碳发展。加强低碳、循环技术推广和应用，积极构建低碳经济体系，打牢绿色经济强省发展基础。全面推进低碳试点省建设。在资源生产加工、流通消费、废物利用各领域大力发展循环经济，强化再生资源回收利用。发挥碳汇优势，积极开展国际互惠交易和清洁发展机制合作。实施政府绿色采购，推进绿色办公。大力弘扬优秀民族生态文化，促使公众自觉参与低碳环保活动，倡导低碳生活。

7. 陕西节能减排政策措施展望

（1）深入推进节能降耗。以提高能源利用效率为核心，合理控制能源消费总量，突出抓好工业、建筑、交通运输和公共机构等重点领域节能，所有建筑设施都要应用节能技术和产品。加强电力、钢铁、有色、建材、化工、石油石化等高耗能行业和重点用能单位管理，大力推广先进节能技术和产品，积极实施节能重点工程，不断降低单位产品能耗。严格项目节能评估审查，促进科学合理用能。健全激励约束机制，强化节能目标责任考核和执法监察。加强节能基础工作，完善市县（区）节能监管体系和标准体系，充实节能管理力量。加快推行合同能源管理，运用市场手段促进节能。深入开展节能宣传，使节能成为单位、家庭、社会成员的自觉行动。

（2）大力发展循环经济。按照“减量化、再利用、资源化”原则，在生产、流通、消费各环节和企业、园区、社会各层面推动循环发展。合理规划循环经济产业布局，重点建设关中循环产业带，汉中、安康、商洛循环经济产业核心聚集区和陕北能源化工循环经济基地等“一带三区一基地”。深化循环经济试点，扶持循环经济示范项目200个、示范企业100个、产业园区50个，提高资源产出效率。加快实施循环经济重点工程，构建煤油气盐能源开发转化、钢铁有色采选冶炼、设备再制造、再生资源回收利用等循环经济产业链。集中力量研发推广一批源头减量、循环利用、再制造、零排放等关键共性支撑技术。建立健全循环经济政策法规体系和技术标准体系，实行生产责任者延伸制度。

（3）加大环境保护力度。处理好环保与发展、稳定、民生的关系，兼顾环境效益和经济效益的内在统一。大力发展清洁生产，减少资源消耗和污染排放，初步构建重点领域、重点行业适应清洁生产发展的科技创新体系、标准体系和审核体系。落实减排目标责任制，加强环境综合治理，建立和完善监测预警、执法监督和环境管理三大支撑体系，着力推进工程、结构和管理三大减排措施。严格污染物排放标准和环境影响评价，减少化学需氧量、氨氮、二氧化硫和氮氧化物排放总量，完成国家下达减排任务。实现所有规模以上火电机组脱硫设施稳定运行，建立光化学烟雾污染预警系统，加快氮氧化物控制技术研发和产业化进程，全面完成设区市危险废物和医疗废物处置工程。鼓励使用生活垃圾发电及分类处理技术，启动西安等城市生活垃圾发电项目。着力解决重金属污染和城市汽车尾气、噪声污染等突出问题，加强核与辐射安全监管，逐步实现在公共场所禁烟。继续推进矿产资源开发中的环境综合整治。全面开展农村面源污染综合治理，推广垃圾集中运送处理，加快城乡垃圾、污水处理一体化。搞好排污权交易示范。

（4）创建国家低碳示范省。以大幅降低能源消耗强度和二氧化碳排放强度为目标，加快发展低碳产业和清洁能源，推广低碳产品和技术。通过技术、市场、政策等多种手段降低工业、交通、建筑等领域温室气体排放。大力开展低碳城市、园区、企业、社区、村镇建设活动，推动全民参与低碳社会建设。牢固树立低碳理念，广泛宣传和普及低碳生产、生活知识，引导低碳消费，倡导低碳生活。大力开展植树造林，增加森林蓄积量，增强固碳能力。二氧化碳排放强度下降15%，固碳能力年递增2%左右。

8. 甘肃节能减排战略

甘肃以建设资源节约型、环境友好型社会和生态文明省为目标，大力发展循环经济，切实加强生态建设和环境保护，全面推进节水型社会建设，提高可持续发展能力。

（1）大力发展循环经济。全面实施《甘肃省循环经济总体规划》，落实扶持政策，着力建设七大循环经济基地和培育16条产业链，实施72大类循环经济重点支撑项目，努力形成循环经济产业集群。以资源高效循环利用为重点，开展工业废弃物循环利用。抓好餐厨废弃物资源化利用和无害化处理城市试点及城市矿产示范基地建设，构建西北地区废旧汽车、家电、有色金属等资源再利用中心，扶持和推进再制造产业发展，促进再生资源规模化回收和加工利用水平。积极推行合同能源管理，促进节能服务产业发展。继续加大节能减排力度，鼓励企业生产使用节能产品、工艺、技术和设备，开展清洁生产。推动企业向产业园区集中，实现集聚生产、集中治污、集约发展，提高能源、水资源和废弃物的循环利用率。加快白银、玉门资源枯竭型城市转型，推进金昌、嘉峪关产业转型步伐，努力推进资源型工矿城市可持续发展。

（2）加强节能减排和环境保护。实施重点节能工程，调整能源消费结构，提高能源利用效率。抓好工业、建筑、交通运输、公共机构等重点领域节能，推广先进节能技术和产品。建立并完善节能减排统计监测制度，健全节能市场化机制和激励约束机制，强化节能目标责任考核。积极推进以污染减排为重点的重大环境保护项目建设，着力解决黄河流域以及渭河、泾河等支流水污染、矿区环境污染、兰州和金昌等城市大气污染等重点流域区域环境污染问题。加强饮用水源地保护，确保饮用水安全。严格执行环境影响评价制度，推进火电、钢铁、有色、化工等行业二氧化硫治理，加大机动车尾气治理力度。推进城镇生活污水、生活垃圾、危险废物、医疗废物处置等环保基础设施建设，提高污水处理率和垃圾无害化处理率。到2015年，设市城市污水处理率达到80%，生活垃圾无害化处理率达到90%。开展农村环境综合整治，推进农村垃圾集中收集处理，综合治理土壤污染，防治农药、化肥、农膜等面源污染和规模化养殖场污染。实施化学需氧量、氨氮、二氧化硫、氮氧化物等主要污染物总量控制。加强核设施及核辐射监测工作。

9. 青海节能减排战略

（1）环境污染防治。一是继续控制主要污染物排放总量，建立健全工业污染防控体系，加大水环境综合治理力度，重点推进湟水流域、青海湖流域水环境保护治理。加强西宁、格尔木、德令哈市和河湟沿岸城镇环境空气、噪声污染综合治理，基本完成县城生活污水处理以及重点小城镇、主要旅游景点垃圾处理设施建设。

二是大力发展环保型产业，加强环保技术的研发及产业化，加快电动汽车充电站建设，扩大环保专用清洁车生产规模。完善再生资源回收体系，强化综合利用及污染防治，推进资源再生利用产业化。加快处置历史堆存和遗留的危险废物。发展清洁生产，推进排污权有偿使用和市场交易试点工作。到2015年，集中式饮用水源水质达标，黄河干流水质达到Ⅲ类以上，湟水流域好于Ⅳ类以上水体达到60%，城市空气质量优良率达到82%以上，人均公共绿地面积达到7.5平方米；二氧化硫、氮氧化物、化学需氧量、氨氮等主要污染物排放减少率均控制在国家下达的指标之内。

（2）强化节能管理。实施节能科技示范工程，坚决淘汰浪费资源、污染环境的落后产能。突出抓好钢铁、有色、煤炭、电力、化工、建材等产业和重点用能企业的节能工作。组织实施好热电联产、省柴节煤灶、建筑节能、公共机构节能等节能重点工程。加强能源制度建设和合同能源管理，实施强制性标准，最大限度地降低单位产品能耗。到2015年，全面淘汰落后产能，低能耗产业形成规模，资源节约型、环境友好型社会建设实现新突破，循环经济增加值占全省工业增加值的80%以上，矿产资源综合利用率在2010年的基础上提高3—5个百分点。

（3）积极应对全球气候变化。合理控制能源消费总量，提高能源利用效率，有效控制温室气体排放。强化节能目标责任考核，完善节能法规和标准，健全节能市场化机制和对企业的激励与约束，实施重点节能工程，推广先进节能技术和产品，加快推行合同能源管理，抓好工业、建筑、交通运输等重点领域节能。调整能源消费结构，增加非化石能源比重。加强适应气候变化特别是应对极端气候事件能力建设。建立温室气体排放和节能减排统计监测制度，加强气候变化科学研究，加快低碳技术研发和应用。

10. 宁夏节能减排战略

宁夏以改善生态环境、构建“两型”社会为契机，树立生态、绿色、低碳发展理念，以节能减排和生态建设保护为重点，健全激励和约束机制，加快构建资源节约型、环境友好型的生产方式和消费模式，增强可持续发展能力。

（1）积极应对全球气候变化。一是采取得力措施，以降低能源消耗强度和二氧化碳排放强度为重要抓手和突破口，减缓气候变化。进一步强化节能目标责任制，加强重点企业节能管理，完善落后产能退出机制，限制和淘汰低效落后的工艺设备。严格执行新建项目节能评估审查，推进合同能源管理。到2015年，高耗能行业单位产品能耗指标达到或超过国内同行业先进水平。二是建设低碳产业园区，大力发展低碳产业，鼓励发展新能源及新能源设备制造产业，进一步延伸风力发电、太阳能光伏发电产业链。优化能源消费结构，提高非化石能源消费比重，广泛开展农村沼气综合利用，鼓励低碳产品消费。三是坚持用低碳理念指导城市规划编制，发展低碳城市。应用绿色节能建筑技术，发展节能节地建筑，执行新建住宅建筑和公共建筑推广节能65%的标准。优先发展城市公共交通，推广使用节能环保型汽车，建设绿色照明工程。

（2）大力发展循环经济。坚持减量化、再利用和资源化，全面推进循环经济发展，从生产、流通、消费各环节入手，构建资源循环利用体系，大幅度提高资源产出效率。以沿黄经济区为重点，全面推进清洁生产，打造煤炭、化工、冶金、造纸等高效循环生态产业链，着力发展循环型农业。继续推进宁东基地国家级循环经济示范区、石嘴山市国家级循环经济示范城市、永宁县农业循环经济示范县、灵武市可再生资源利用循环经济示范园、大地循环经济产业园等建设，为创建全国循环经济示范省区打下坚实基础。

（3）加大环境保护力度。一是围绕生态文明建设，加快建设环境友好型宁夏，促进经济发展方式转变，不断开创经济、社会和环境共赢的新局面。到2015年，黄河干流宁夏段水质全面达到Ⅲ类，饮用水水源达标率90%以上，县级以上城市空气质量Ⅱ级标准天数比例达到80%以上，确保二氧化硫、氮氧化物、化学需氧量、氨氮等主要污染物控制在国家下达的指标内。二是以环境保护优化经济发展，围绕创建宁东能源化工基地等生态工业园区、实施沿黄城市带区域环境全防全控、确保中部干旱带饮用水安全、加强南部山区生态保护为重点，严格执行环境影响评价、产业和项目准入制度，设置鼓励、限制和禁止建设的产业环保准入标准。改善全区环境质量，实施环保民生工程，切实解决老百姓关心的突出环境问题。

11. 新疆节能减排政策措施展望

坚持环保优先、生态立区，走资源开发可持续、生态环境可持续的道路，推动形成资源节约型、环境友好型的生产方式和消费模式，增强可持续发展能力，确保新疆山川秀美，绿洲常在。

（1）大力推行节能降耗。强化节能减排目标责任评价考核，加强项目节能评估，严格执行重点用能行业单位产品能耗限额标准，提高能源利用效率；健全节能市场化机制，加快推进合同能源管理、能效标识管理和节能产品认证管理。从严控制二氧化碳排放强度，积极应对全球气候变化；突出抓好工业、交通运输、建筑、公共机构等重点领域节能；加快钢铁、煤炭、电力、有色、化工、建材等重点用能企业的节能技术改造，淘汰落后产能，鼓励推广应用节能环保新技术、新工艺、新设备、新材料。调整优化交通运力结构，降低空载率；鼓励购买小排量汽车，加大淘汰旧汽车力度；积极推广节能环保型建筑和绿色建筑，大力实施新建建筑热计量收费管理；深入开展节能全民行动，推广高效节能产品，全面实行政府优先和强制采购节能产品制度，大力实施“节能惠民”工程。

（2）发展循环经济。按照循环经济的要求，规划、建设和改造各类产业园区，实现土地集约利用、废物交换利用、能量梯级利用、废水循环利用和污染物集中处理；实施循环经济和资源节约示范工程，构建循环经济产业链；加快构建全社会的资源循环利用体系，加快推行清洁生产，从源头上减少废弃物产生和排放，加强共伴生矿及尾矿综合利用，强化对工业、建筑业、农业等重点行业和城市生活垃圾的废物资源化利用，强化循环经济试点工作，形成具有行业特色、发挥示范和带动作用的循环经济模式。

（3）加强环境保护。加强城市及周边企业的污染治理，对重点排污企业实行挂牌整治，有计划地逐步搬迁城区内的重污染企业，力争到2015年，城镇污水集中处理率达到75%，城镇生活垃圾无害化处理率达到45%以上；加强环境监管基础能力和应急体系建设，提升监测预警水平；加强环境敏感区的风险防范；大力提高保护环境消费意识，鼓励低碳消费方式，实行环境标识、环境认证和政府绿色采购制度。

17.2.4　“十二五”西部地区节能减排的政策措施

西部地区幅员广阔，区域内差异较大，部分地区生态环境脆弱，因此该地区的节能减排政策应与生态环境建设相匹配。

1. 加强产业结构调整

产业结构调整是实现西部地区各省节能减排的关键，而产业结构调整又必须以西部地区各省的资源优势特色与发展阶段背景为基础，在发挥各自独特的资源优势的前提下，大力调整产业结构调整。一是资源利用深度加工化发展，由原来生产原材料等上游产品、初级产品、粗加工产品，更多地向生产下游产品、高级产品、精加工产品方向转变，拉长其产业链，增加资源附加值，优化产业和产品结构，有效降低单位产值的能源消耗和污染排放。二是加大优化工业结构力度，制定并严格执行能源利用效率和污染物排放等控制指标一票否决的准入制度，严格限制资源能源效率低、污染物排放强度高的产业发展，坚决淘汰技术落后、浪费资源、污染严重的产品和产业。三是发挥西部地区丰富的自然和文化旅游资源的优势，大力发展旅游业等第三产业，以推动产业结构的进一步优化升级。

2. 加快科技进步和技术创新

推进科技进步和技术创新是推进节能降耗减排的重要措施。一是根据国家和地方政府关于节能减排的总体部署，实施科技进步和技术创新工程，加快建立以企业为主体的节能技术创新体系，引导企业对市场效益明显的节能减排技术加大投资力度，大力开发并推广相关技术改造示范项目和相关新产品、新技术、新设备。二是积极采取集中规模化生产，调整过于分散的产业布局，促进产业联合重组和集中生产，促进集群化发展，以此推动节能降耗减排工作。

3. 优化能源结构，提高清洁能源消费比例

优化一次能源供应结构。大力发展坑口电厂，降低煤炭在一次能源供应总量中的比重；在妥善处理生态环境等保护和移民安置等基础上大力发展水电，积极发展风电，鼓励发展生物质能源、太阳能等可再生能源；同时，优化能源转换结构，发展火电，发展大容量、高参数机组，上大压小，不断提高一次能源的加工转化比重，特别是提高煤炭消费用于发电的比重。

4. 大力发展循环经济

发展循环经济，非常适合西部地区各省资源依赖型经济的发展，是西部地区各省切实可行的一种可持续发展模式。西部地区应该在三个层次上构筑循环圈：首先，构筑企业内部的循环和企业间的循环，实现生产资源在同一个企业和不同企业的循环利用，通过资源利用的最大化，实现经济效益的最大化；其次，构筑产业间循环，利用产业间上下游产品的特点，推动产业间"循环链"的对接，使各类资源在产业之间循环利用，实现资源在产业间的有效配置和合理利用；再次，树立全民的环保意识，建设循环型、和谐型的绿色社会。政府应该鼓励境内外各种资本进入资源循环产业，为循环经济的发展提供资金和技术支持；最后，要在全社会形成节约资源，保护环境、发展经济的社会氛围。

5. 加大国家支持力度，促进西部地区节能减排

一是增加环保投入。中央政府在安排环境建设时，尽可能把有限的资金补助给西部地区，用财政转移支付的手段，对西部地区进行环境保护补偿。例如，在青海、西藏等省区建立长江、雅鲁藏布江等江河源头保护基金，以及增加对西部地区重点工业污染源污染防治资金的投入；二是中央和地方政府应加大那些对于治理环境污染成绩显著的企业的资金转移力度，促使企业积极进行污染治理工作；三是加大国家支持力度，通过规划、政策、资金、人才等手段大力支持和促进西部地区发展资源深度加工、高端装备制造，支持西部地区的产业高度化，改变西部主要提供初级产品的产业结构格局。

6. 完善节能减排管理体系

完善节能减排法律法规和相关政策，强化节能减排监管。认真贯彻落实《中华人民共和国节约能源法》、《中华人民共和国环境保护法》、《清洁生产促进法》等法律和法规，依法组织对煤炭、电力、冶金、建材、化工、造纸等行业及企业和大型商场、宾馆等场所进行节能专项检查，对能效标准、高耗电产品限额标准、建筑节能标准及行业节能设计规范执行情况进行检查；加强资源环境管理能力建设，建立更加有效的节能环保监督管理体系，坚决依法惩处各种违法违规行为，严格实行建设项目审批环保一票否决制；加大节能减排的宣传力度，不断提高广大人民群众的节能意识、环保意识、责任意识和参与意识，积极支持各类环保民间组织和环保热心人士开展工作，形成全民参与监督的机制。

18 东北地区的节能减排

18.1 “十一五”期间东北地区节能减排综述

18.1.1 “十一五”时期东北地区经济发展

由于国家振兴东北战略的强大刺激，“十一五”期间东北地区经济发展速度加快，主要经济指标也保持良好增长势头。2005 年，东北地区国内生产总值为 17181.23 亿元，到 2010 年就已达到 37090.40 亿元，占全国总量的 8.60%，按不变价折算，五年增长了 88%，年均增长 17.6%。

“十一五”末，辽宁、吉林、黑龙江三省的国内生产总值分别是 18278.3 亿元、8577.1 亿元、10235.0 亿元，五年总增长速度为 92%、100%、76%，年平均增长速度分别为 13.9%、14.9%和 12.0%，其中吉林增长速度最高，黑龙江最低。但从经济总量看差别却较大，GDP 最高的辽宁是最低的吉林的 2.13 倍，较前几年差距有拉大趋势。“十一五”期间东北地区生产总值增长情况如图 18－1 所示。

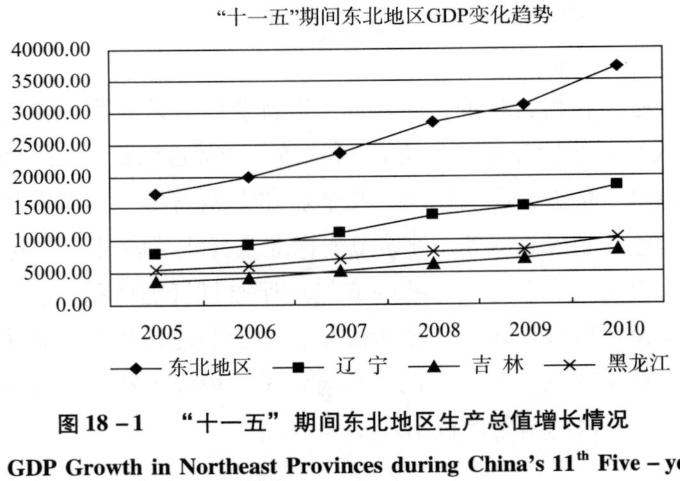

图 18－1 “十一五”期间东北地区生产总值增长情况

Figure 18－1 GDP Growth in Northeast Provinces during China's 11th Five－year Plan Period

资料来源：《中国统计年鉴》（历年）、《中国统计摘要》2011

从人均 GDP 水平来看，2010 年东北地区人均 GDP 为 33866 万元，在全国四个地区中处于第二位，仅次于东部。在区域内部，辽宁人均 GDP 为 35239 元，吉林为 26595 元，黑龙江为 22447 元。这表明东北地区内经济发展总体水平比较高，但各省之间经济发展也存在一定差距。

从三次产业结构看来，东北地区各省三次产业结构呈现出“二三一”结构态势，各省市三大产业增加值比例的分布：辽宁为 8.9∶54.0∶37.1、黑龙江为 12.7∶49.8∶37.5、吉林为 12.2

:51.5∶36.3。东北地区各省三次产业结构有相似之处：第一产业产值比重都不太低，除辽宁的8.9%，其他的都在10%以上，说明东北地区农业在经济结构中的比重较高；第三产业比重都不很高，均低于40%，说明东北地区第三产业有很大发展潜力；第二产业在地区经济发展中占主导地位，对各省GDP的贡献达到50%左右，最高的辽宁达到了54.0%，最低的黑龙江也达到49.8%。“十一五”期间东北地区三次产业比例情况如图18－2所示。

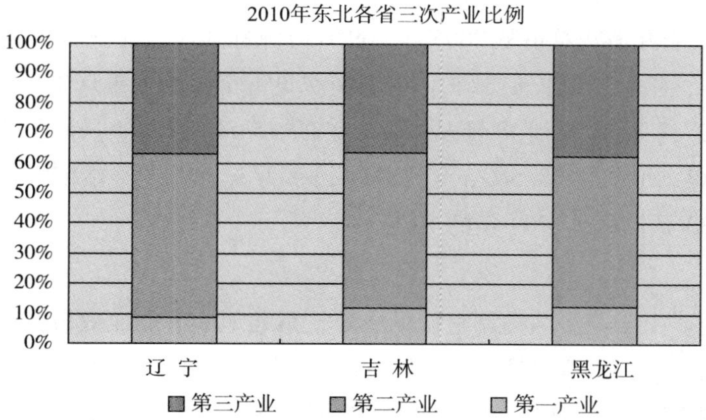

图18－2 “十一五”期间东北地区三次产业比例图

Figure 18－2 Industrial Structure of Northeast Provinces during China's 11th Five－year Plan Period

资料来源：《中国统计摘要2011》

“十一五”时期，随着我国城镇化进程的加快，东北地区也进一步加强了城市基础设施建设，提升了城市综合承载能力，致使城镇化水平得到进一步提高。从城镇化率的绝对水平来看，2010年东北地区的城镇化水平是全国各区域最高的地区，城镇人口占地区总人口的比重为56.9%，比2005年提高了1.7个百分点。辽宁、吉林和黑龙江三省的城市化率分别达到60.35%、53.32%和55.50%。

2010年，辽宁国内生产总值比2005年翻了一番多，升至全国第七位；地方财政一般预算收入年均增长24.3%，是2005年的3倍；全社会固定资产投资年均增长30.5%，累计达到5.1万亿元，是“十五”的4.2倍，名列全国前茅；实际利用外商直接投资年均增长42%，累计达到633亿美元，是“十五”的2.9倍，跃居全国第二位；城镇居民人均可支配收入由全国第15位上升到第九位，年均实际增长11.4%，农民人均纯收入年均实际增长9.3%，高出全国平均水平989元；城乡居民收入增幅均超过全国平均水平。

2010年，吉林国内生产总值比上年增长13.7%；工业总产值跃上万亿元台阶，规模以上工业增加值增长19.9%；社会消费品零售总额增长18.5%；民营经济主营业务收入增长30.7%；旅游总收入增长26.2%；进出口额增长43.5%；一般预算全口径财政收入1206亿元，地方级财政收入602.4亿元，分别增长23.4%、23.7%；规模以上工业企业实现利润794.7亿元，增长66.4%。

“十一五”时期，黑龙江全社会固定资产投资累计为2.1万亿元，年均增长31.5%，比“十五”增长2.2倍。社会消费品零售总额累计达到1.5万亿元，年均增长17.7%，比“十五”增长1倍。2010年，畜牧业产值达到965.8亿元，比2005年增长42.4%；农业机械化程度、绿色有机食品经济总量均列全国之首。规模以上工业增加值达到4003.5亿元，高新技术产业产值近

4000亿元，分别比2005年增长95%和200%。第三产业实现增加值3832.6亿元，比2005年增加1.1倍。外贸进出口总额达到255亿美元，比2005年增长1.7倍。利用外资累计达到113亿美元，比2005年增长1倍。

18.1.2 “十一五”时期东北地区节能减排

东北经济的快速发展，也带动了能源需求的增长，能源消费总量在全国的比重也有所上升。从总量看，东北地区能源消费总量从2005年26976万吨标准煤增加到2009年的37276万吨标准煤，增长了38.18%，年平均增长8.42%，能源消费量占全国的比重从2005年的10.24%上升到2009年的10.43%；终端能源消费量从2005年的15450.51万吨标准煤增长到2009年的21398.24万吨标准煤，增长了38.5%，年平均增长率为8.48%。终端能源消费量占全国的比重也从2005年的10.33%上升到2009年的10.46%。

1. 节能评述

“十一五”时期，由于经济结构调整初见成效，东北地区节能成效较为显著，各种能耗指标均出现较大下降。东北地区同属我国老工业基地，尽管各自产业结构、技术基础也有不同，但与其他地区相比，相同的比例要多一些，其2005年的单位GDP能耗也相差不多，“十一五”末各省都超额完成了节能目标任务。从超额幅度来看，完成目标最好的是黑龙江，任务目标是2010年比2005年单位GDP能耗降低20%，实际上降低了20.79%，超额完成比例3.95%；其次是吉林省，任务目标是降幅22%，实际完成降幅22.04%，超额完成比例0.18%，因为吉林的目标是东北地区最高的，仍然超额完成了任务，也非常不容易；值得一提的是辽宁，尽管也完成了节能任务，但2010年的单位GDP能耗仍然高达1.38吨标准煤/万元，节能潜力仍然很大，“十二五”时期需更加重视节能工作。“十一五”时期东北地区节能目标完成情况如表18-1所示。

表18-1 “十一五”时期东北地区节能目标完成情况表

Table 18-1 Energy Conservation Performance of Northeast Provinces during China's 11th Five-year Plan Period

省份	2005年单位GDP能耗（吨标准煤/万元）	“十一五”时期降低任务目标（%）	2010年单位GDP能耗（吨标准煤/万元）	2010年比2005年实际降低幅度（%）	超额完成比例（%）
辽宁	1.726	-20	1.38	-20.01	0.05
吉林	1.468	-22	1.145	-22.04	0.18
黑龙江	1.46	-20	1.156	-20.79	3.95

资料来源：《“十一五”各地区节能目标完成情况表》

2010年，辽宁、吉林和黑龙江三省的单位GDP能耗分别达到1.38吨标准煤/万元、1.145吨标准煤/万元和1.156吨标准煤/万元，比2005年各自下降了20.01%、22.04%和20.79%，均完成了“十一五”期间各自20%、22%和20%的目标任务。

“十一五”时期，吉林的单位GDP能耗最低，为1.209吨标准煤/万元；单位工业增加值能耗降幅最大的也是吉林，降幅为39.44%。黑龙江的单位工业增加值能耗最低，为1.382吨标准煤/万元；其单位GDP电耗也最低，为798.67千瓦小时/万元。而辽宁的单位GDP电耗降幅最大，达到17.57%。

总体分析，在节能方面吉林的效果最好，黑龙江处于中间位置，而辽宁还有一定差距。从全国范围看，东北地区在国内处比较领先的水平。单位 GDP 能耗为 1.0—2.0 吨标准煤/万元、单位 GDP 电耗小于1000 吨标准煤/万元的省份有黑龙江和吉林；单位 GDP 能耗为1.0—2.0 吨标准煤/万元、单位 GDP 电耗为 1000—2000 吨标准煤/万元的省份有辽宁。这说明黑龙江和吉林的能源利用效率比第一梯队地区略差而稍高于第二梯队，而辽宁则处于的第二梯队。

2. 减排评述

2010 年东北地区化学需氧量排放量为 133.85 万吨，二氧化硫排放量为 186.87 万吨，分别比 2005 年下降了 13.92% 和 10.46%，化学需氧量排放量降幅在各区域中仅次于东部地区，居于四个区域中的第二位。该区域中，辽宁省的化学需氧量排放量降幅和二氧化硫排放量的降幅最高，较 2005 年分别降低 15.90% 和 14.60%；黑龙江省的降幅最小，与 2005 年相比分别降低 11.77% 和 3.52%；但三个省都完成了责任目标。“十一五”期间东北地区各省减排目标完成情况如表 18－2 所示。

表 18－2　“十一五”期间东北地区各省减排目标完成情况

Table 18－2　Pollution Reduction Performance of Northeast Provinces during China's 11th Five－year Plan Period

省份	化学需氧量					二氧化硫				
	2005 年排放量（万吨）	“十一五”降低目标（%）	2010 年排放量（万吨）	“十一五”实际降幅（%）	超额完成幅度（%）	2005 年排放量（万吨）	“十一五”降低目标（%）	2010 年排放量（万吨）	“十一五”实际降幅（%）	超额完成幅度（%）
辽　宁	64.4	－12.90	54.16	－15.90	23.26	119.7	－12.0	102.22	－14.60	21.69
吉　林	40.7	－10.30	35.22	－13.46	30.68	38.2	－4.7	35.64	－6.71	42.77
黑龙江	50.4	－10.30	44.47	－11.77	14.27	50.8	－2.0	49.01	－3.52	76.00

资料来源：根据《“十一五”期间全国化学需氧量排放总量控制计划表》、《“十一五”期间全国二氧化硫排放总量控制计划表》整理

如果按单位 GDP 减排或人均减排考量，东北地区 2010 年与 2005 年相比，其化学需氧量排放量和二氧化硫排放量都有较大幅度的降低。按单位 GDP 测算，东北地区的化学需氧量减排幅度达到 52.01%，二氧化硫减排幅度达到 51.23%。其中，吉林的化学需氧量减排幅度达到 60.16%，辽宁的二氧化硫减排幅度达到 55.34%。按人均测算，东北地区的化学需氧量减排幅度为 11.12%，二氧化硫减排幅度达到 9.44%。其中，吉林的化学需氧量减排幅度达到 21.20%，辽宁的二氧化硫减排幅度达到 16.82%。三省当中，黑龙江的人均减排效果相对差一些。2010 年东北地区各省比 2005 年单位减排情况如表 18－3 所示。

表 18－3　2010 年东北地区各省比 2005 年单位减排情况

Table 18－3　Pollution Reduction in Northeast Provinces from 2005 to 2010

省份	按单位 GDP（吨/万元）				按人均（吨/万人）			
	化学需氧量排放量		二氧化硫排放量		化学需氧量排放量		二氧化硫排放量	
	2010 年	减排（%）	2010 年	减排（%）	2010 年	减排（%）	2010 年	减排（%）
辽　宁	37.98	－52.77	66.74	－55.34	134.27	－12.01	235.95	－16.82
吉　林	44.79	－60.16	47.78	－54.72	118.13	－21.20	125.99	－10.45

续表

省份	按单位 GDP（吨/万元）				按人均（吨/万人）			
	化学需氧量排放量		二氧化硫排放量		化学需氧量排放量		二氧化硫排放量	
	2010 年	减排（%）	2010 年	减排（%）	2010 年	减排（%）	2010 年	减排（%）
黑龙江	52.02	-43.11	51.96	-43.63	131.81	-0.15	131.66	-1.05
地区平均	43.71	-52.01	58.08	-51.23	129.36	-11.12	171.89	-9.44

资料来源：根据《“十一五”期间全国化学需氧量排放总量控制计划表》、《“十一五”期间全国二氧化硫排放总量控制计划表》整理

3. 节能减排综合指数分析

根据节能减排综合指数对各区域省份进行分类，可以把节能减排综合指数在［2.00－3.00）之间的为第一类地区，在［1.50－2.00）之间的为第二类地区，在［1.00－1.50）之间的为第三类地区，［0.00－1.00）在之间的为第四类地区。

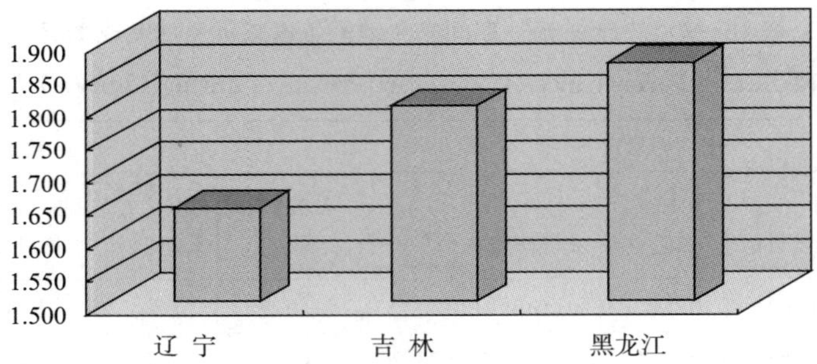

图 18－3　东北地区节能减排综合指数柱状图

Figure 18－3　Energy Conservation and Pollution Reduction Composite Index Bargraph of Northeast Provinces

资料来源：根据图 14－9 整理

东北的黑龙江、吉林和辽宁三省的节能减排综合指数均排在第二类地区，综合指数分别为 1.866、1.800 和 1.643，在全国分别排第 10、第 12 和第 19 位。从效果指数、努力程度指数和潜力实现指数三个二级指数来看，黑龙江在全国分别排第 16、第 3 和第 16 位；吉林在全国分别排第 13、第 6 和第 13 位，说明这两个省的节能减排效果在全国也处于较高的位置，只有辽宁稍差一些。总的来说，东北三省节能减排效果在全国处于中上水平。图 18－3 为东北地区节能减排综合指数柱状图，表 18－4 为东北地区节能减排综合指数及排名情况。

表 18－4　东北地区节能减排综合指数及排名

Table 18－4　Energy Conservation and Pollution Reduction Composite Index of Northeast Provinces and the Ranking

	综合指数及排名		二级指数					
	节能减排综合指数	排名	效果指数	排名	努力程度指数	排名	潜力实现指数	排名
黑龙江	1.866	10	0.782	16	0.301	3	0.783	16
吉　林	1.800	12	0.799	13	0.203	6	0.798	13
辽　宁	1.643	19	0.759	18	0.124	17	0.760	18

资料来源：根据表 14－9 整理

18.1.3 主要成就与做法

1. 辽宁

“十一五”期间，辽宁以辽河治理和污染减排两大攻坚战为着重点，全省环境质量不断改善。长期以来，由于过度开发，辽河资源破坏严重，生态用水缺乏，生态环境恶化，成为我国污染最重的河流之一。

从2008年开始，辽宁为加速治理辽河重点实施了以下“三大工程”：

一是以造纸企业整治为重点，实施工业点源治理工程。强令全省417家造纸企业停产治理，最终彻底关闭了294家，占全省造纸企业的70%。同时，按照“上大、压小、提标、进园”的思路，积极推进造纸企业的升级改造，鼓励造纸企业进入园区规范发展。“十一五”期间，辽宁累计关闭涉水企业或生产线1000余个，累计削减化学需氧量10万余吨。

二是以县级污水处理厂建设为核心，推进污水处理厂建设和运行工程。2008—2009年，辽宁投资约百亿元建设99座污水处理厂，新增日处理能力315万吨，是过去12年总和的两倍。2009年底，辽宁实现县县都有污水处理厂的目标，一举还清历史欠账。目前，辽宁共有136座污水处理厂，设计日处理能力为642.92万吨。

三是以支流河整治为抓手，开展河流生态治理工程。目前，辽河干流城市段已全面建成生态景观带，过去污染严重的河流城市段，如今已成当地著名景观和市民休息娱乐的好去处。

2. 吉林

“十一五”期间，吉林省累计关停小火电机组137.7万千瓦，淘汰落后浆纸产能46.9万吨、水泥产能420万吨、酒精产能4.9万吨，不断加大产业结构调整力度，累计形成化学需氧量、二氧化硫减排量4.08万吨和4.82万吨。严格环境准入，坚决不上“两高”项目，共拒批建设项目200余个，完成各类各级开发区、工业集中区规划环评66个。

“十一五”期间是吉林省污染防治设施建设最多、污染减排能力提升最快的时期。全省污水处理厂总数达到42座，形成污水日处理能力233万吨，形成化学需氧量实际减排能力10万吨。重金属污染排查整治、饮用水水源保护、化冰期风险隐患排查等一系列环保专项行动，集中解决群众反映强烈的环境违法问题，共限期整改企业1549家，停产治理616家，取缔关闭365家，挂牌督办202家，对613家企业实施了行政处罚，重点工业企业排污达标率由“十五”末的65%提高到90%以上。

3. 黑龙江

在推进企业节能技术进步方面，制订了全省工业节能产业发展行动计划，滚动推进100项重大节能、节水、综合利用技术改造项目和100项节能产业化项目，大力开展了节能新产品、新技术、新工艺、新设备在重点行业、领域的推广应用工作，加强了项目的培育和管理，并取得了成效。此外，还加大了淘汰落后产能力度，累计完成炼铁17万吨、炼钢20万吨、电解铝0.6万吨、铁合金0.95万吨、焦碳284.5万吨、水泥381.4万吨、玻璃168万重量箱、造纸行业39.2万吨、酒精行业7.5万吨、味精行业1.2万吨、柠檬酸行业0.5万吨的淘汰落后产能任务。

全省共实施重点减排项目310项。新建17座污水处理厂，强化已建成16座污水处理厂的运行管理，新建脱硫项目72个，关闭淘汰不符合国家产业政策的企业或生产线112家（条），淘汰落后产能焦炭111万吨、水泥121万吨、造纸44.6万吨；鹤岗市煤炭洗选企业洗煤废水全部实现闭路循环改造，年减排废水600余万吨，回收煤泥40万吨；对鹤岗鑫塔水泥等20余家企业进行了清洁生产强制审核。

2009年《黑龙江省松花江流域水污染防治条例》开始实施，明确了政府责任和考核目标，强化了流域水污染防治监管，建立了跨界协管机制，流域法制化监管走在全国前列。对环境保护部确定的环境风险源，邀请国内专家提出三级防控体系评估意见，隐患基本得到有效监控。

18.1.4 存在的问题

东北地区是我国的老工业基地，产业结构以重工业和矿业采掘为主，所以节能减排的基础相对薄弱，尽管有国家振兴东北老工业基地的政策支持，但产业转型也不是一朝一夕的事情。“十一五”期间，东北地区的节能减排主要存在以下问题：

一是东北地区总体上处于工业化中期阶段，对能源的总量需求上升较快，供需矛盾进一步加剧。

二是作为老工业基地，东北地区工业结构中重化工业比重较大，能源消耗高，而且能源结构中花狮能源比重较高，污染排放大。

三是东北地区环境污染问题还没有从根本上得到解决，主要江河的部分支流污染问题仍然较严重，尤其是松花江屡次发生污染事故，潜在环境风险隐患较多。

四是价格、财政、税收、金融等政策还不完善，节能减排工作长效机制尚未形成。

五是市场作用发挥不够，企业节能减排动力不足。同时，能源统计、计量和节能监察等基础有待进一步提高，监管能力有待进一步加强。

六是自主创新能力不强，资源环境矛盾突出、城乡区域发展不协调、城乡居民收入有待提高以及制约科学发展的体制机制性问题尚未根本解决。

18.2 “十二五”时期东北地区节能减排展望

18.2.1 节能减排的战略目标

1. 节能目标

“十二五”期间，东北地区的节能减排任务仍然比较艰巨，个别省份的节能减排任务可能高于全国平均水平。从节能任务目标来看，辽宁省属于第二类地区，“十二五”承担的节能目标是17%；吉林、黑龙江两省属于第三类地区，“十二五”承担的节能目标是均为16%。

2. 减排目标

从主要污染物减排任务来看，东北地区的“十二五”减排目标与全国平均水平相当。根据国务院《“十二五”节能减排综合性工作方案》关于减排的要求，东部地区各省2015年的减排目标如下：

- 辽宁：化学需氧量排放总量控制在124.7万吨，比2010年减排9.2%（其中工业和生活控制在42.1万吨，减排10.4%）；氨氮排放总量控制在10.01万吨，比2010年减排11.0%（其中工业和生活控制在6.69万吨，减排11.5%）；二氧化硫排放总量控制在104.7万吨，比2010

年减排10.7%；氮氧化物排放总量控制在88.0万吨，比2010年减排13.7%。

- 吉林：化学需氧量排放总量控制在76.1万吨，比2010年减排8.8%（其中工业和生活控制在26.1万吨，减排9.4%）；氨氮排放总量控制在5.25万吨，比2010年减排10.5%（其中工业和生活控制在3.49万吨，减排10.9%）；二氧化硫排放总量控制在40.6万吨，比2010年减排2.7%；氮氧化物排放总量控制在54.2万吨，比2010年减排6.9%。
- 黑龙江：化学需氧量排放总量控制在147.3万吨，比2010年减排8.6%（其中工业和生活控制在43.4万吨，减排9.3%）；氨氮排放总量控制在8.47万吨，比2010年减排10.4%（其中工业和生活控制在5.49万吨，减排10.6%）；二氧化硫排放总量控制在50.3万吨，比2010年减排2.0%；氮氧化物排放总量控制在73.0万吨，比2010年减排3.1%。

18.2.2 节能减排的战略选择

1. 辽宁节能减排展望

当前，辽宁的经济社会发展已经进入了快车道，在深入实施东北振兴战略基础上，辽宁沿海经济带开发开放上升为国家战略，沈阳经济区被确定为国家新型工业化综合配套改革试验区，为“十二五”的节能减排打下了较好的经济基础。

“十二五”时期，辽宁将继续依托重点节能、水污染治理、大气污染治理、资源再生利用等工程，重点发展节能、环保产业、循环经济等关键技术和装备及环保材料与药剂，加快建设大连国家生态工业示范园区（静脉产业类）、辽宁（铁岭）静脉产业、沈阳、大连、鞍山节能环保成套装备、沈阳静脉产业、大连绿色低碳环保、辽宁（丹东）环保、朝阳节能环保除尘设备、铁岭开原循环经济、盘锦市环保等产业基地。

全面开展辽河、大小凌河流域生态环境集中整治；禁止河道滥采滥挖、滥建滥排，全面恢复河滩地生态，实现水质状况进一步改善；完善大伙房、观音阁等水库上游生态补偿机制，加大辽河三角洲湿地保护和海洋污染防治力度。

大力推广地源、污水源、海水源等供暖新技术。积极推进清洁燃料开发使用。在建设城市、县城垃圾处理场和污水处理厂工程的同时，启动实施乡镇垃圾处理场和污水处理厂建设工程。

2. 吉林节能减排展望

“十二五”时期，吉林将着力优化能源结构，重点发展风电、核电、生物质能、太阳能等清洁能源，打造国家重要的新能源基地；积极促进风电、生物质能、光伏、核电设备制造产业发展，重点开发2.0兆瓦以上的大型风电机组和AP1000核电主管道，提升新能源装备制造能力；重点发展城市热电联产机组项目，在中小型城市积极推广背压机组热电联产供热项目建设；有序开发水电，开工丰满大坝重建、敦化抽水蓄能电站等工程项目；大力发展风电，建设国家千万千瓦级风电基地；加快千万吨生物质能综合利用基地建设；积极发展太阳能、地热能等其他新能源，促进分布式能源系统的推广应用。

淘汰煤耗高的火电机组、供热小型燃煤锅炉，削减污染物排放；严格控制高耗能、高排放行业过快增长，控制新增排放量。继续推进重点排污企业深度治理，降低污染物排放强度；全面推进燃煤机组脱硫工程，着力推进30万千瓦以上燃煤机组脱硝工程、规模化畜禽养殖污染治理工程，加大机动车尾气治理力度；强化污染减排目标责任考核，严格污染物排放标准；健全完善脱硫优惠电价、脱硫机组优先发电调度、污水管网建设以奖代补、政府绿色采购等环境经济政策，探索建立污染治理设施稳定运行的长效机制，积极开展松花江流域主要污染物排污权交易试点。

加大工业污染治理力度，着力解决工业结构性污染问题，严格环境准入，推行清洁生产。到2015年，全省工业企业污染物排放稳定达标率达到90%；开展国家环保模范城的创建活动，在“十二五”初期县建成污水处理厂基础上，加快建设重点建制镇污水处理厂，加强现有污水处理厂的升级改造和配套管网建设，城镇污水集中处理率平均达到70%；提高生活垃圾无害化处理水平，县以上城市全部建成垃圾处理厂，城市生活垃圾无害化处理率达到80%。

3. 黑龙江节能减排展望

“十二五”时期，黑龙江将积极探索合理控制能源消费总量的有效措施，严格固定资产投资项目节能评估和审查，完善能评监督和管理制度，严把投资项目合理用能关；实施节能减排重点工程。加大资金支持力度，加快实施重点节能改造工程、节能产品惠民工程、合同能源管理推广工程等，形成实现目标任务要求的节能减排技术能力。继续推进化学需氧量减排治理项目以及电力和非电力二氧化硫减排工程建设，开展氨氮和氮氧化物减排项目建设，组织实施好城镇污水垃圾处理设施建设，推动污泥和垃圾渗滤液的处理处置，不断改善环境质量；加强重点用能单位和重点领域节能减排。

组织357户年耗能万吨标准煤以上的企业开展“万家企业节能行动”，落实目标任务，加强节能管理。实施重点用能企业能源利用状况报告制度，开展能源管理负责人培训。启动绿色建筑评选工作，全面推行新建建筑65%节能标准，制定可再生能源与建筑一体化技术标准等地方标准。发展公共交通运输，推广替代燃料，推进智能交通运输技术。加强公共机构节能精细化管理。推进农村新能源项目建设，提高农机作业综合机械化水平；加强节能能力建设。建立健全省级、市（地）和区（县）级节能监察机构，确保法律法规的贯彻落实。

拓展农作物秸秆、畜禽粪便、林业三剩物利用途径，加快农林废弃物综合利用。以建设“城市矿产”示范基地为重点，加大对有色金属、废钢铁、废橡胶等回收利用力度，推进资源循环利用产业发展。抓好城市生活垃圾和餐厨废弃物、粉煤灰、煤矸石、城市建筑垃圾等大宗废弃物综合利用。强化资源综合利用认定管理，落实好国家鼓励资源综合利用的财税政策，力争资源综合利用认定企业数量达到100个。加大清洁生产关键、共性技术的研发和推广，建设一批清洁生产示范项目和示范企业。

抓好重点领域和重点企业节能减排，淘汰落后产能。推进松花江流域污染综合防治，加大对高耗能高污染行业治理力度，加快大中型污水处理设施建设。加强对森林、草原、湿地、水资源保护，科学有序开发林木和矿产资源，大力培育和发展以生态旅游、特色种植养殖、绿色食品加工、北药开发、清洁能源、林木和矿产资源开发及精深加工等为主导的生态型经济。到2015年，生态环境有新改善，森林覆盖率有新提高，接续和替代产业有新发展，建立较为完备的林业生态体系、发达的产业体系和繁荣的生态文化体系。

第4篇　环保篇

粗放型的经济增长方式为环境带来了很大的压力，环境问题日趋严重，我国政府和民众都认识到问题的严重性，政府也出台各种政策来限制工业污染物的排放，并在国家发展规划中将环境问题作为一个重要的经济问题解决，将建设资源节约型、环境友好型社会作为未来我国经济、社会的发展方向。

为此，我国要大力发展节能环保、新能源等战略性新兴产业，加大对该类产业的投入和政策支持。战略性新兴产业在国民经济具有战略地位，对经济社会发展和国家安全具有重大的和长远影响的行业，具有成为一个国家未来经济发展支柱产业可能性。

但是，我国节能环保产业整体处于国际分工中下游，在国际竞争中处于弱势。产业集中度低，没有形成分工合作、协同发展的产业链条。目前我国对节能环保产业的引导，往往通过上“大项目”和补贴末端产品的方式来进行示范和支持。要推动节能环保产业发展，不仅要促进其外延性增长，也要提高其内生性增长能力，二者缺一不可。

本篇我们重点分析以大气污染防治、污水处理、固体废物利用为主体的环保产业，侧重介绍其发展动态以及相关技术。

19　环保产业概述

我国对环保产业的界定，始见于1990年国务院《关于积极发展环保产业的若干意见》，其中对环境保护产业是这样定义的：环境保护产业是国民经济结构中以防治环境污染，改善生态环境、保护自然资源为目的所进行的技术开发、商业流通、资源利用、信息服务、工程承包、自然保护开发等活动的总称，主要包括环保机械设备制造、自然保护开发经营、环境工程建设、环境保护服务等方面。按照这个定义，环保产业的范围是相当广泛的，像生态农业工程、自然保护事业、盆景培育技术等都被纳入环保产业中。由于环保产业是新兴产业，人们对环保产业的认识有一个不断深入的过程，其内涵和范围也处于不断变动之中。现在大致可以确定环保产业范围包括污染防治、资源综合利用和自然生态保护三个方面。2001年，原国家经贸委等八部门联合发布的《关于加快发展环保产业的意见》明确指出，当前国家优先发展的环保产业重点领域：一是环保技术与装备、环保材料和环保药剂；二是资源综合利用；三是环境服务。

节能环保产业是一个跨产业、跨领域、跨地域，与其他经济部门相互交叉、相互渗透的综合性新兴产业，被喻为新世纪的朝阳产业，节能环保产业将面临巨大的发展机遇。

19.1　国际环保产业的新趋势

经过30多年的迅速发展，发达国家环保产业的产值已经占到了国内生产总值的10%—20%，已经次于风头正劲的信息业与制药业之间，并且开始高于计算机行业，而且还以高于GNP增长率一到两倍的速度发展着。从环保产业发展规模来看，它正朝着综合化、大型化、集团化的方向发展。日本现有环保企业1000余家，从业人员3万多人，年产值300亿美元；德国、法国、英国等国的环保产业实力雄厚，年产值分别达到270亿美元、120亿美元和90亿美元。发运国家的环保技术正向深度化、尖端化方向发展，产品不断向普及化、标准化、成套化、系列化等方向发展。目前，新能源技术、新材料技术、生物工程技术正逐渐被引进环保产业。

随着环境保护和可持续发展观念日益深入人心，环保业被视为耗费资金和人力的观点已发生变化，许多国家的企业家发现，环保产业将发展成为一个有利可图的大市场。保护生态平衡，不仅可以提供大批就业机会，而且可以节省大量资金。因此，各国环保投资不断增长，越来越多的国家力求在环保市场获取优势。国际经济界普遍认为，环保产业已成为国际经济发展中新的增长领域，是实施可持续发展战略的需要。

19.1.1　世界环保产业的新特征

1. 环保产业内涵日益扩展

关于环保产业，国际上至今没有一个统一的定义，例如美国称之为“环境产业”，日本称之为“生态产业”或“生态商务”。而且环保产业的分类，国际上也各不相同。美国环保产业分为环保服务、环保设备和环境资源三大类；国际经济合作发展组织（OECD）认为，环保产业是指在防治水、空气、土壤污染及噪声，缩减和处理废物及保护生态系统方面提供产品和服务的部门；日本将环境产业分为环境保护、环境恢复、能源供给、清洁生产、洁净产品、废弃物处置和利用等六个部分。不过，人们已经形成了一个共识：保护环境，不仅仅是支出，更重要的是，它可以带来经济效益，形成强有力的国际经济新的增长点，并成为许多国家革新和调整产业结构的重要目标和关键。

随着科技与经济的发展，环保产业的内涵越来越丰富，并且促进了其他行业的快速发展。过去人们所指的环保产业，仅限于治理空气污染、废水、垃圾、噪音、土壤及海洋污染和环境监测的产品、设备、服务及技术项目。而现代社会的环保产业，其内涵已经延伸到发展具有防止和减少污染、节省能源和减少资源投入等效应的新领域，由此促进了多种新的产业和服务的开发和成熟，例如促进新能源产业中的太阳能、风能、氢能、生物能等的快速市场化，仅电池一项未来几年就可形成近百亿美元的大市场。日本由环境产业推动的其他产业规模将由2000年的41万亿日元增加到2025年的103万亿日元，就业人数大大增长。

由于环保市场和环保内涵的扩大化趋势，越来越多的企业将在相关行业中扩大环保投资，这样将会刺激这些产业的进一步良性发展，从而更好地改善环境。

2. 鼓励通过创新占领环保技术制高点

随着世界范围的经济增长和对自然资源需求增长的压力，现有的技术不再能够保障实现可持续发展。一些以占领世界环保产品市场为目的，以争夺环保技术制高点为中心的国际竞争已拉开序幕。在回收资源上，不少国家推出新的标准。这不仅是为了制约生产厂商，更主要是为了改善本国环境，避免浪费环保投资，同时节省大量有用资源。1994年后，国际贸易开始与“环保标准”结合起来，环保标准已被视为国际贸易往来的一种筹码，在许多情况下它可变成新的贸易壁垒而加以利用。实际上，美国、欧洲和日本都扩大了以保护环境为由的禁止进口的商品品种，或者将进口产品的环保标准修订得更为严格。显然，取得环保技术优势的国家，正在迫使贸易伙伴国接受它们的新标准和采用由它们提供的环保技术。在这种情况下，工业化国家常常可以利用环保技术优势获得更多向外扩张的机会。

此外，发达国家也一直鼓励并不断地进行技术创新，以达到对环境的影响最小的目的。欧洲委员会2004年1月28日制订的环境技术行动计划（Environmental Technologies Action，ETAP）提出吸引更多的私有和公共投资用于环境技术的开发和示范应用，鼓励创新工艺，奖罚分明将创造从实验室带到市场。欧盟采取了两大创新行动：一是建立技术平台；二是建立测试网络。其目的是想了解如何建立私有/公共合作关系，使研究更贴近市场。技术平台就是将所有感兴趣的商人集中起来，建立长期合作关系，开发和促进特定技术或解决特定问题。

3. 环保投资增加迅速

世界各国对环保事业的投资迅速增加是推动国际环保业快速发展的重要原因之一。1995—2004年发达国家用于环保治理的费用年均递增14.5%，而此前的10年为10.3%。同期，发达国家私人企业用于发展环保产品（服务）的投资年均递增18.5%，而此前10年为11.8%。其中，德国增长最快，近10年工业企业的上述支出年均增长25%。该国2005年国内环保市场规模约600亿欧元，可为100万个劳动者提供就业机会。

经济发展速度居全球之首的亚洲，近年来在环保领域的投资也持续上升。1999—2004 年，东盟泰国、马来西亚、印尼和菲律宾四国的环保支出约 400 亿美元。

4. 政府与民间密切合作

美、欧、日各国政府与民间的合作非常密切，很多事业采用公私合作（Public – Private Participation）形式，这种组合运作的典范是美国 Rouge 河道走廊的治理。

5. 更加注重国际间的交流与合作

为了改善环境状况，创造清洁环境，同时扩大贸易和投资，亚洲国家尤其是中国、韩国和日本，已经在环境问题和政策措施的信息交流以及亚洲国家环境产业发展的促进方面，开展了许多有益的合作。

东盟作为一个地区性组织，在保护地区环境和生物多样性方面起到了积极的作用。欧洲正在兴起最佳信息交流技术（BAT）在各国之间担当了协调员的角色。联合国环境规划署在日本设立了国际环保技术中心 United Nations Environment Programme（UNEP）—Intenational Environmental Technology Center（IETC），它的主要任务是促进环保先进技术的采用和推广，协助进行人员培训，组织环保技术的交流，以解决各国城市等方面的环境问题。近五年来已在波兰、匈牙利、罗马尼亚、马来西亚、中国等太平洋地区和拉丁美洲牙买加等地先后开展多项活动，例如，开办各种讲习训练班、代培环保技术专业人员。

19.1.2 美国环保产业发展动态

从国际环保产业格局的变化来看，近 10 年里，美国环保产业或清洁技术实际上在走下坡路。在新能源领域，美国在风能、太阳能乃至新能源汽车等传统优势领域已被德国、日本等发达国家超越；在一些技术创新和专利申请方面，美国甚至不及中国；在新能源与清洁技术的投资方面，美国也落后于中国、土耳其、巴西、英国和意大利。这一趋势越来越引发美国各界的关注。奥巴马政府上台后积极采取一些措施，例如通过现金补贴等方式来刺激新能源、清洁技术和气候变化等新兴产业，以此来扭转其环保产业的被动局面，带动新一轮增长。

2009 年 2 月 17 日，奥巴马政府正式通过了《美国再生再投资法》，实施总额为 7872 亿美元的经济刺激政策。其中，大约 580 亿美元投入到环境与能源领域。绿色新政由绿色能源、能源效率、温室气体排放、向低碳经济转型共四个部份组成。主要做法包括：

（1）能源部分。该部分主要包括二氧化碳回收储藏、可再生能源、低碳交通、智能电网四个方面。在二氧化碳回收储藏方面推行碳回收储藏技术，把燃煤排放的二氧化碳存入地下。在可再生能源方面要求提高生物能、风能、地热、太阳能等可再生能源发电比例。在低碳交通方面制定低碳交通运输燃料标准，促进先进生物燃料及其他清洁能源在交通运输领域的运用。在智能电网方面推广智能电网，减少企业高峰用电，促进新型家电适应智能电网。

（2）能源效率部分。2009 年 5 月 19 日，美国公布关于汽车节能减排的新计划。该计划规定从 2012 年开始，美国汽车制造商必须逐年改善汽车的燃油效率；到 2016 年，美国小型汽车和轻型卡车每 100 公里所耗费的汽油量将不超过 6.6 升。

（3）为低碳经济发展提供财政支持。为了支持发展低碳能源，奥巴马上任之初即支持能源部 2010 年申请 263 亿美元资金的预算，侧重开发新一代低温室气体排放的可再生和替代能源，帮助美国转型为低碳能源经济。

19.1.3 日本环保产业发展动态

日本作为亚洲最发达国家已经面对本国经济环境问题，主要分为两个阶段。第一阶段是在快速发展期间进行了一套吸收外国的先进技术的政策。如此的外汇控制与技术转让战略在今天的世界是不可模仿的，大部分是由于世界经济布局的大量变化和世界贸易机制等国际组织的影响。第二阶段是先进企业和管理方式在一个资源有限和大众关心环境问题对生活的影响的情况下，逐步执行一套节能和监督性措施解决该问题。

日本是一个资源比较匮乏的国家，因此也是新能源开发最领先的国家，在太阳能发电、海洋能、地热、垃圾发电、燃料电池等新能源领域都走在世界的前列。2006 年，日本政府首次制定了国家能源战略——《新国家能源战略》。2009 年 4 月，日本政府公布了《绿色经济与社会变革》的政策草案，提出通过实行削减温室气体排放等措施大力推动了低碳经济发展。

（1）推进交通运输领域的低碳化。汽车产业是日本的核心产业，目前，日本加快低碳汽车的技术开发，促进传统汽车向低碳汽车的转换，在促进低碳汽车的技术开发方面采取领跑者计划的制度，对开发成功者提供补助金。

（2）建立绿色金融体系。日本国民的金融资产高达1500 万亿日元，日本社会整体的资产更加雄厚庞大，日本的当务之急是建立绿色金融体系，将这些雄厚而庞大的资金运用到低碳技术的创新投资中。

（3）促进可再生能源的开发与普及。资源缺乏的日本要结合温室气体的减排大力发展风力、水力、太阳能、潮汐、低热、废弃物等可再生能源的开发，以此取代大量排放二氧化碳的化石燃料，为此日本将进行相关制度的设计并且对现有的社会基础设施进行根本的改造。

19.1.4 英国环保产业发展动态

英国是全球低碳经济的积极倡导者和先行者，政府在政策法规建设方面的许多做法均具有开创性，创造了多个世界第一：它不仅是世界上第一个征收气候税的国家，还是第一个为温室气体减排目标立法的国家；同时，也是世界上第一个立法约束“碳预算”的国家。英国在过去10 年间，实现了200 年以来最长的经济增长周期，经济增长达到28%，温室气体排放减少量达到8%。英国10 年的低碳经济增长之路向我们证明，低碳排放和经济增长是可以实现共存的。2009 年7 月，英国政府颁布的《英国低碳转换计划》内容涉及能源、工业、交通和住房等多个方面，该文件标志着英国向低碳经济转型的帷幕正式拉开。

（1）提高能源使用效率，降低能源使用成本。采用先进技术提高能源使用效率，可以降低各项社会服务成本。例如，热电联产技术（CHP）的实施，此项技术成功使用了发电时一般会浪费掉的热能，可以将燃料利用的总效率增加到70%—90%，相对于传统发电方法的30%—50%还可节省40%的燃料费。

（2）采用新兴能源技术，引进低碳生产方式。重视可再生能源、核能、碳捕获和储存技术、输电网等能源基础设施建设，将其作为未来英国发展低碳产业的重要方向。例如，现阶段英国政府非常关注碳捕获与埋存技术（CCS），这项技术可以最大限度防止二氧化碳进入大气层，通过将二氧化碳气体注入地表以下数千米的地质层中，使其可以被应用在发电站和石油及天然气生产过程中的脱碳工艺中。

（3）发放财政补贴，生产低碳产品。为使英国成为全球低碳汽车开发和生产领先者，英国政府从2011年开始给予低碳汽车最高5000英镑的补贴，鼓励低碳汽车消费。凡是购买达到安全指标和低排放标准的电动车、插电式混合动力车或氢燃料电池车的车主均可获得车价25%的补贴。

从短期来看，推动“低碳经济”发展可从以下三个方面创造新的商机和就业机会，进而助力英国经济复苏。一是有效促进太阳能、风能等新能源产业的发展；二是对传统产业的低碳化升级改造可促进生产和消费；三是通过碳排放交易获利。根据联合国“清洁发展机制”等规定，没有减排指标的发展中国家或减排工作做得较好的机构，可以将碳排放配额拿到市场上交易。

从长远来看，一旦新能源技术获得突破，英国对石油的依赖程度会下降，其二氧化碳排放也将随之趋稳甚至下降，气候变暖问题将得到缓解。此外，英国的现实基础也为进一步推进“低碳经济”提供了有力支撑。通过之前的政策法规建设、低碳技术研发推广到国民认知等诸多方面的良好发展，英国已初步形成了以市场为基础、以政府为主导、以全体企业、公共部门和居民为主体的“低碳经济”互动体系。

19.1.5 发展中国家环保产业的发展

发展中国家，无论是从影响还是被影响的角度来看，都是未来气候变化国际对话的主要参与者，“G8 +5”就是未来国际上多边对话的雏形，参与其中的发展中国家，急需维护自己各方面权益。

发展中国家应对方式各有不同。以墨西哥为代表的发展中国家，因为国家规模大、气候和环境问题较为突出，所以国家较早重视气候和环境问题，在政府机构中设立专门机构，并且随着气候变化问题严重程度和国际社会重视程度的加剧，机构的规格不断提升；同时，这些机构通过制定相关法律、法规直至修正宪法，强化国家层面的决策。这种作法在很多方面已经领先于包括中国在内的许多更有影响的发展中大国，因此在未来国际社会对话中让墨西哥的意见更有说服力和示范意义（因为背后有国家法律的强力支持），也会使该国更快地适应未来更为严酷的国际环境变化，从而保证可持续发展免遭因政策巨变带来的停滞甚至倒退。

以印度为代表的发展中国家则通过培育国家研究机构使国家获得应对环境变化的主要依据，同时积极参与国际事务，通过强调“人均”和发展中国家生活方式更为节约等，维护发展中国家的利益和发展优先地位。其中，印方将涉及气候变化的自然和社会科学各个方面开展综合性的研究。此举值得我们借鉴，中国的相应研究机构，应该积极与印度的专门研究机构合作，使双方受益并在国际上占据更强更高的位置。这样可以改变中国以往再两大领域各自为战，无法合力提出更全面和有效的政策建议的局面。例如，中国气象局创办了支持气候变化的自然和社会科学研究的杂志《气候变化研究进展》，可以说走出了联合的第一步。

有一些国家，例如阿根廷等国，已经开始重视气候变化问题，但相应的政府咨询和决策机构还没有形成规模，推进较为困难。中国科学家如果关注这些国家环境状况及其与发展的关系，必要时在人力和财力上支持其开展相应的研究，就能够争取这些国家对中国的信任和支持。

此外，还有一些亚洲国家，例如韩国，在几次世界峰会后，也开始重视气候变化问题，采取了政府设立跨部门的专门机构，政府总理任主要领导的国家一体化应对体系，同时也加速了相应的国家行动计划的制定，体现了亚洲国家全民一致性的特征。中国在亚洲如何将经济特色十分不同的国家利益通盘进行考虑，研究它们与中国在应对气候变化和可持续发展问题上利益的异同，也是今后一段时间面临的挑战。

19.2 我国环保产业的发展

19.2.1 中国环境与经济发展的矛盾

要达成中国经济保持稳定高速增长的目标，普遍提高社会生活水平，大量能源消耗是不可回避的。尤其是在中国拥有庞大的13亿人口基数，且地域发展存在差异大部分地区技术、市场成熟度与发达国家相比仍有一定距离的情况下，要求中国在短期内达到发达国家的减排水平是不现实的，也显得不近人情。然而，尽管人均能耗水平较低，中国的能耗总量逐年攀升，并于去年在二氧化碳排放量世界排位上超过了传统能耗大国美国，这也是无可争辩的事实，国际社会的压力亦随之增大。

1. 经济增长方式与环境保护的矛盾

粗放型的经济增长方式为环境带来了很大的压力，环境问题日趋严重，我国政府和民众都认识到问题的严重性，政府也出台各种政策来限制工业污染物的排放，并在国家发展规划中将环境问题作为一个重要的经济问题解决，将建设资源节约型、环境友好型社会、建设新农村，构建和谐社会作为未来我国经济、社会的发展方向。

2006年国家“十一五”规划中提出每年能耗降低4%、污染物排放降低2%两项硬性指标，令环境治理、环境保护看到希望，然而当年这两项指标都没有完成，与此相对的我国的GDP增长速度却又登上了一个新的台阶，两者之间似乎只能选其一的矛盾越发的突出。

一方面治理环境污染，改善环境质量需要大量的资金，造成对环境治理的投资多而对固定资产的投资减少，这可能会阻碍经济的增长速度；另一方面由于我国是以能源、资源消耗为代价的经济增长，经济增长得越迅速，环境破坏得越严重，两者之间的矛盾越来越激化。

2. 国内需要与国际压力之间的矛盾

地球资源的稀缺性以及环境影响跨国界的特征使得国家之间必然会为能耗问题互相施压，具体到中国，体现为发达国家标准与发展中国家标准的冲突。如果中国要在不承担过大压力的情况下满足国内发展的需求，必须通过积极参加环境保护问题的国际讨论等方式，借由这些互动交流得到必要的发展空间、资源和国际支持。

《京都议定书》没有被美国政府接受主要是国会拒绝批准。主要冲突是针对比较大的发展中国家，特别是中国和印度，因为在协议里面没有严格规定减少发展中国家二氧化碳排放量。美国在国际论坛上一直把这点作为重点之一。此后，发展中国家在全球二氧化碳排放量的比例不断增加，而随着排放量的增加，发达国家对发展中国家的要求也同步增加。虽然现在从总数来看这些发展中国家占比较大的比例，总数字隐藏了人均与历史上的不公平。现在发达国家与发展中国家的总排放量比例是54:46，而中国占全球18.9%，但是如果按人均排放量计算，这个比例就少得多。

中国领导人在国际论坛、会议等场所不断面临其他国家对中国在全球变暖中所扮演角色的担忧。G82007年巴黎峰会和2008日本峰会，都有很多其他国家领导人对中国在这方面提出要求。国际层面的广泛关注使得外交部2007年专门在国际论坛上做了针对全球变暖问题的外交宣传。2010年哥本哈根会议将会讨论《京都议定书》的后续计划。在筹备过程中，中国一直强调以下几点：第一，国家不仅要对现在的排放量负责，而且由于二氧化碳在空气的保留时间比较长也要对历史上的排放量负责任；第二，人均排放量比总排放量更能有效反映真实的情况；第三，发达国家已经经过最高排放量的发展阶段，但发展中国家还在面临这个阶段，而国际协议应该反映并照顾到这个事实。这三点把全球变暖的主要责任归因在发达国家，而强调排放量限制与交易体系的作用，并强调发达国家应承担在资金供应与技术转让方面的责任。

19.2.2　我国环保产业的战略定位

在2010年的政府工作报告中，温家宝总理指出，战略性新兴产业在国民经济具有战略地位，对经济社会发展和国家安全具有重大的和长远影响的行业，具有成为一个国家未来经济发展支柱产业可能性。要大力发展新能源、新材料、节能环保、生物医药、信息网络和三网融合等战略性新兴产业，加大对战略性新兴产业的投入和政策支持。

在《让科技引领中国可持续发展》中，温总理明确指出："战略性新兴产业必须掌握关键核心技术，具有市场需求前景，具备资源能耗低、带动系数大、就业机会多、综合效益好的特征。"

为支持节能环保产业发展，2010年财政将从生产、加工等各个环节加以谋划，在科技研发、示范推广、资金投入、机制创新等各个方面，给予支持，同时加强宏观指导，统筹规划，以此促进新能源和节能环保产业又好又快发展。节能环保产业被列为战略性新兴产业重要的一项，体现发展节能环保产业的重要性和紧迫性，也体现出国家对发展节能环保产业的重视。

2011年3月16日，《中华人民共和国国民经济和社会发展第十二个五年规划纲要》（以下简称《纲要》）正式发布。《纲要》提出，今后五年，要确保科学发展取得新的显著进步，确保转变经济发展方式取得实质性进展。要坚持把建设资源节约型、环境友好型社会作为加快转变经济发展方式的重要着力点。深入贯彻节约资源和保护环境基本国策，节约能源，降低温室气体排放强度，发展循环经济，推广低碳技术，积极应对全球气候变化，促进经济社会发展与人口资源环境相协调，走可持续发展之路。

在"十二五"开局之年，节能环保更是放在七大产业之首。《纲要》第三篇第10章"培育发展战略性新兴产业"提出，要大力发展节能环保、新一代信息技术、生物、高端装备制造、新能源、新材料、新能源汽车等战略性新兴产业。节能环保产业重点发展高效节能、先进环保、资源循环利用关键技术装备、产品和服务。"十二五"将是节能减排产业的一个新起点。

全国31个省市自治区被分为五类地区，每类地区确定一个节能指标，其单位GDP能耗降低率分为10%—18%区间以下五类任务：

- 第一类地区包括天津、上海、江苏、浙江和广东，其单位GDP能耗降低率最高，为18%。
- 第二类地区包括北京、河北、辽宁和山东，其单位GDP能耗降低率为17%。
- 第三类地区包括山西、吉林、黑龙江、安徽、福建、江西、河南、湖北、湖南、重庆、四川和陕西，其单位GDP能耗降低率为16%。
- 第四类地区包括内蒙古、广西、贵州、云南、甘肃和宁夏，其单位GDP能耗降低率为15%。

- 第五类地区包括海南、西藏、青海和新疆，其单位GDP能耗降低率为10%。

与“十一五”规划相比，《“十二五”规划纲要》提出的经济社会发展主要指标中，资源环境类指标变化非常大。“非化石能源占一次能源消费比重达到11.4%”作为一项约束性指标被列入“十二五”时期经济社会发展主要指标，意味着未来几年我国将加大清洁能源产业建设力度，更加注重资源节约和环境保护。

除了非化石能源占一次能源消费比重这一约束性指标外，还新增单位国内生产总值二氧化碳排放降低17%的约束性指标。

“主要污染物排放总量减少”指标在原来的化学需氧量和二氧化硫两个子约束指标基础上，新增氨氮、氮氧化物各减排10%；将原来的森林覆盖率指标细化为森林覆盖率和森林蓄积量。这意味着减排工作将从单一污染物控制转向多元污染物控制，也意味着环保产业在未来五年间将迎来发展黄金期，其投资机会值得关注。

从近期来看，发展节能环保产业有利于拉动内需，扩大就业。节能环保产业潜在市场强，就业机会多，综合效益好，社会责任也大，具备广阔的发展空间，是促进消费、扩大投资，稳定出口的一个重要结构点，也是提高国际结构和国际竞争的结构点。

从长期来看，发展节能环保产业，有利于提升竞争力，形成战略性新兴产业。一方面节能环保产业可缓解国家保持经济平稳快速发展的压力；另一方面，关于我国发展节能环保产业的目标和任务。长远目标是提升竞争力，形成战略性的新兴产业，为建设资源节约型、环境友好型社会提供雄厚的物质基础和技术保障。

19.2.3 我国环保产业的发展空间

据相关统计数据表明，2007年我国环保产业产值为7025亿元，相当于GDP的2.6%，从业人数300多万。其中，环保装备制造、环保服务业产值分别为625亿元、500亿元，相当于美国环保装备制造、环保服务业产值的6.1%和11%；2008年节能环保产业总产值达1.55万亿元人民币，占国内生产总值（GDP）的5.17%。其中节能产业占2700亿元、资源循环利用产业占6600亿元、环保产业4800亿元，就业人数超过3000万人。

节能环保列为七大战略性新兴产业的首位，是国家未来的支柱性产业，未来市场空间广阔。作为国家重点发展的方向，节能环保将不缺乏政策刺激。数据显示，2001—2005年我国环保总投资为0.80万亿元；2006—2010年，我国环保总投资为1.5万亿元，增长近100%。我们预计在相关法规的支持下，未来五年环保总投资可望再次翻番，达3万亿元，节能环保产值可达4万亿元—5万亿元。

据悉，作为“十二五”期间的重点发展领域，《节能环保产业发展规划》目前已基本编制完成。规划将对节能产业、环保产业和循环利用产业提供技术、产品和服务等支持。到2015年，中国节能环保产业总产值将达到GDP的7%—8%，成为新的支柱产业。

环保行业属于公用事业，政府资金投入和政策支持是核心推动力。上海证券预计，节能环保产业规划将于第二季度出台。作为战略性新兴产业，“十二五”期间国家将在行业政策、税收优惠、资金投入上给予大力支持，行业产值将有更大提高。据国务院发展研究中心“重点产业调整转型升级”课题组测算，2015年环保产业产值可达2万亿元。“十二五”期间，我国环保投资额预计将达3.1万亿元，是“十一五”期间1.6万亿的2倍左右。

从规划细节来看，我国将把高效节能技术和装备、高效节能产品、节能服务产业、先进环保技术和装备、环保产品与环保服务六大领域列为重点支持对象，在财政、税收、金融等方面提供政策支持。

工业和建筑是我国目前能耗最为集中的两个领域，工业节能和建筑节能显得非常重要。据悉，国家发改委的《节能环保产业发展规划（征求意见稿）》提出：“十二五”期间，全国新增污水配套管网建设能力20万吨，新增污水处理规模9000万吨，升级改造污水处理规模5000万吨；新增污泥日处理能力4.7万吨；新增垃圾填埋日处理能力25万吨，新增垃圾焚烧日处理能力15万吨；新增燃煤电厂脱硝设施1.8亿千瓦；开展重金属污染治理示范试点；支持建设一批清洁生产的示范工程。

从主流观点来看，机构毫不掩饰对行业龙头企业的偏好，“强者恒强”的规律将在节能环保行业中持续发挥作用。碧水源（300070）、万邦达（300055）是水处理中的名企；固废处理龙头桑德环境（000826）在行业内实现了业务的全覆盖；电子垃圾处理龙头的格林美（002340）能实现钴镍二次循环利用；九龙电力（600292）的脱硝催化剂生产线和核废料处理工艺均是国内独有的，龙净环保（600388）、三聚环保（300072）是工业脱硫脱硝领域的重点企业；合康变频（300048）、英威腾（002334）分别是高压和中低压变频产品的领先制造商，泰豪科技（600590）是建筑节能产业龙头。

19.2.4　我国环保产业重点行业

中国的新一轮针对环境保护的中央政策中，政府指定的经济发展方向为高附加值高科技生产和运作。符合这个标准的环保技术行业不在少数，而我们可以通过产业结构调整、补贴、促进政策和指导意见了解为中央政府所看重的重点行业。其中，大部分举措与新能源有关，包括太阳能、风能和沼气等。另外，中央政府也鼓励新的交通工具和计划、推广以更节约资源的方式进行生产和利用能源的措施以及加强生态保护。

《国务院关于加快培育和发展战略性新兴产业的决定》提出：到2015年，战略性新兴产业增加值占国内生产总值比重力争达8%左右；到2020年，战略性新兴产业增加值占国内生产总值比重力争达15%左右；节能环保、新一代信息技术、生物、高端装备制造四大产业成为国民经济支柱产业，新能源、新材料、新能源汽车三大产业成为国民经济先导产业。

战略性新兴产业发展需要长期规划，需要相关目录加以指导。“对战略性新兴产业每一个细分领域，都有专门司局在做具体研究，从而最大程度地对这些产业加强指导，确保其有序开展、良性发展。”13个重点领域专项规划将非常细化。

在环保产业方面，一是发展先进环保技术和装备，包括污水、垃圾处理，脱硫脱硝，高浓度有机废水治理，土壤修复，监测设备等，重点攻克膜生物反应器、反硝化除磷、湖泊蓝藻治理和污泥无害化处理技术装备等。

二是发展环保产品，包括环保材料、环保药剂，重点研发和产业化示范膜材料、高性能防渗材料、脱硝催化剂、固废处理固化剂和稳定剂、持久性有机污染物替代产品等。

三是发展环保服务，建立以资金融通和投入、工程设计和建设、设施运营和维护、技术咨询和人才培训等为主要内容的环保产业服务体系，加大污染治理设施特许经营实施力度。

19.2.5 我国环保产业发展策略

我国节能环保产业整体处于国际分工中下游，在国际竞争中处于弱势。产业集中度低，没有形成分工合作、协同发展的产业链条。目前我国对节能环保产业的引导，往往通过上“大项目”和补贴末端产品的方式来进行示范和支持。要推动节能环保产业发展，不仅要促进其外延性增长，也要提高其内生性增长能力，二者缺一不可。

2011年9月27日，国务院召开会议，全面动员和部署“十二五”节能减排工作。当前，节能减排形势还相当严峻，国务院日前印发了《“十二五”节能减排综合性工作方案》（简称《方案》），推出多项促进节能环保产业发展的政策措施。

1. 节能环保产业布局应遵循“竞争力”原则，做到因地制宜

《方案》公布的节能减排目标显示，“十二五”期间，东部地区的节能减排目标大幅高于西部地区。例如广东、浙江等东部沿海省份，化学需氧量排放总量要分别减少12%、11.4%，单位GDP能耗降低率均为18%；而在西部地区，云南的化学需氧量排放总量减少6.2%，单位GDP能耗降低率15%，在新疆、西藏等环境容量较大的地区，化学需氧量排放总量减少为零，单位GDP能耗降低率为10%。

节能环保产业布局应遵循“竞争力”原则，做到因地制宜。只要企业附加值减去各种要素成本所获得的数值较高，无论建在任何区域都是合理的。此次《方案》还提出了全方位的节能减排管理，例如增加了农村、公共机构、商业和民用的节能减排要求，无论城乡都产生了巨大的市场需求。

2. 努力让“看不见的手”发挥作用

“十二五”期间，我国节能环保产业将由“行政手段驱动”转向“市场手段驱动”，这是决定节能环保产业能否由“会走”转向“会跑”的关键时期。《方案》提出了政府为主导、企业为主体、全社会参与、市场有效驱动的原则，其中，在培育市场方面还应加大力度。

“目前我国对节能环保产业的引导，往往通过上‘大项目’和补贴末端产品的方式来进行示范和支持。”中国社会科学院中国循环经济与环境评估预测研究中心副主任杜欢政表示，市场培育涉及政府、企业、消费者等，需在理念、政策、技术三个方面下功夫。只有营造“大市场氛围”，才能形成巨大的市场需求。

以节能荧光灯和LED节能灯推广为例，由于价格比白炽灯贵不少，于是政府通过补贴来鼓励消费者购买。政府用来补贴的资金，今后可以将一部分用来做宣传教育，培养消费者购买节能环保产品的责任意识；另一部分则用来支持研发和改进工艺，以降低生产成本和满足市场需求。这样的节能环保产品兼具生产成本低、市场价格低、使用成本低的“三低”优势，完全可以挤掉高能耗、高污染的低价产品。要推动节能环保产业发展，不仅要促进其外延性增长，也要提高其内生性增长能力，二者缺一不可。

3. 产业起步晚，技术研发投入少、强度低，需创新引领，集约发展

中国节能环保企业起步晚、底子薄，只好从产业链的下游做起，但技术研发投入少、强度低，这是一个“硬伤”。欧美节能环保企业的科技投入一般占销售收入的15%—20%，而国内高新技术企业平均为2%。没有形成创新驱动的发展模式，就不可能到达产业链“上游”。

我国节能环保产业发展潜力巨大，大量产业资本和金融资本正向这个领域聚集。“十一五”

期间，我国节能环保产业保持15%—20%的年复合增长率，大幅超过工业增速。到2015年，总产值将超过3万亿元，占GDP的8%左右。然而，我国节能环保产业整体处于国际分工中下游，在国际竞争中处于弱势。此外，产业集中度低，没有形成分工合作、协同发展的产业链条，在国内的发展也面临危机。

专家指出，《方案》在产业结构调整、节能减排管理，以及价格、财政、税收、金融等方面提出了有利于节能环保产业发展的经济政策，但节能环保产业更应重视解决自身存在的问题，这样才能做大做强。

"长期以来，传统的节能环保产业被当成简单的服务业，分散在各个耗能、排放部门，依附于耗能排放主体产业，这已不适应当前的节能减排形势需求。"中国工业节能与清洁生产协会会长王小康表示，现代节能环保产业集资本、技术、装备、工程、运营服务为一体，具有典型的生产性服务业特征，更应注重产业发展的集中度，认真研究其发展的规律。

19.2.6 我国环保产业的产业政策

在一系列政策即将密集出台的同时，环保部还在制定《国家环境保护标准"十二五"规划》，在污染物排放总量控制背景下，将在未来五年出台100项涵盖水、大气、固体废弃物和危险化学品污染物以及环境噪声等排放标准体系。

100项标准体系重点包括20项水污染物、30项大气污染物和20项固体废弃物排放标准。其中，水污染物排放标准涉及钢铁工业、有机化合物制造、无机化合物制造、合成氨、农药、畜禽养殖业、公共污水处理厂等领域，大气污染物涉及火电厂、钢铁、石油炼制、炼焦、水泥和锅炉等，重金属污染物控制涉及稀土工业、再生金属工业、电子工业、电池工业、火化厂等。

除此前已发布征求意见稿的火电厂、稀土、电池等污染物控制标准外，今年内还将制定出台钢铁、冶炼、水泥、电子等行业标准。在标准公布之前，环保部还将研究出台各行业污染物达标治理的工程技术规范，以引导产业环保技术升级。业内分析指出，政策上意图通过提高环保门槛促进产业升级换代的信号比较明确，这同时也加速了各产业领域的"优胜劣汰"进程。未来五年，类似于此前火电厂、电池等行业污染物排放标准公布给行业带来的冲击，其他一些重污染行业领域的环保冲击波预计也将此起彼伏。

由国家发改委环资司牵头制定的《"十二五"节能环保产业发展规划》，经过了三轮征求意见，即将上报国务院，预计最快2011年9月出台。其中，针对节能环保企业税收减免、特许经营权、提高产业集中度等内容成为该规划亮点。

该规划将按节能、环保、资源综合利用三类，对相关行业进行政策扶持，涉及的领域将包括高效节能技术和装备、高效节能产品、节能服务产业、先进环保技术和装备、环保产品与环保服务六大领域。这将是国家有关部门对节能环保产业支持力度最大的产业政策规划。该规划将在产业、金融、税收等多方面为节能环保产业发展给予政策支持，并提出一些方向性目标以指导产业发展。

19.2.7 我国环保产业的财税政策

《"十二五"节能减排综合性工作方案》披露了"十二五"期间节能减排主要目标，提出要通过调整优化产业结构、实施节能减排重点工程、完善节能减排经济政策等多种手段，确保

"十二五"节能减排目标实现。方案提出：到2015年，要实现全国万元国内生产总值能耗下降到0.869吨标准煤（按2005年价格计算）的目标，比2010年的1.034吨标准煤下降16%。"十二五"期间，实现节约能源6.7亿吨标准煤。到2015年，全国化学需氧量和二氧化硫排放总量分别控制在2347.6万吨、2086.4万吨，比2010年的2551.7万吨、2267.8万吨分别下降8%；全国氨氮和氮氧化物排放总量分别控制在238.0万吨、2046.2万吨，比2010年的264.4万吨、2273.6万吨分别下降10%。为实现"十二五"节能减排目标，我国将完善财政激励政策，健全税收支持政策、强化金融支持力度，进一步完善有利于节能减排的经济政策。

（1）财政方面。要加大中央预算内投资和中央财政节能减排专项资金的投入力度，加快节能减排重点工程实施和能力建设。深化"以奖代补"、"以奖促治"以及采用财政补贴方式推广高效节能家用电器、照明产品、节能汽车、高效电机产品等支持机制，强化财政资金的引导作用。

国有资本经营预算要继续支持企业实施节能减排项目。地方各级人民政府要加大对节能减排的投入。推行政府绿色采购，完善强制采购和优先采购制度，逐步提高节能环保产品比重，研究实行节能环保服务政府采购。

（2）税收方面。落实国家支持节能减排所得税、增值税等优惠政策。积极推进资源税费改革，将原油、天然气和煤炭资源税计征办法由从量征收改为从价征收并适当提高税负水平，依法清理取消涉及矿产资源的不合理收费基金项目。

积极推进环境税费改革，选择防治任务重、技术标准成熟的税目开征环境保护税，逐步扩大征收范围。完善和落实资源综合利用和可再生能源发展的税收优惠政策。调整进出口税收政策，遏制高耗能、高排放产品出口。对用于制造大型环保及资源综合利用设备确有必要进口的关键零部件及原材料，抓紧研究制定税收优惠政策。

（3）金融方面。加大各类金融机构对节能减排项目的信贷支持力度，鼓励金融机构创新适合节能减排项目特点的信贷管理模式。引导各类创业投资企业、股权投资企业、社会捐赠资金和国际援助资金增加对节能减排领域的投入。

提高高耗能、高排放行业贷款门槛，将企业环境违法信息纳入人民银行企业征信系统和银监会信息披露系统，与企业信用等级评定、贷款及证券融资联动。推行环境污染责任保险，重点区域涉重金属企业应当购买环境污染责任保险。

（4）进一步推进价格和环保收费改革。深化资源性产品价格改革，理顺煤、电、油、气、水、矿产等资源性产品价格关系。推行居民用电、用水阶梯价格。完善电力峰谷分时电价政策。深化供热体制改革，全面推行供热计量收费。对能源消耗超过国家和地区规定的单位产品能耗（电耗）限额标准的企业和产品，实行惩罚性电价。要严格落实脱硫电价，研究制定燃煤电厂烟气脱硝电价政策。进一步完善污水处理费政策，研究将污泥处理费用逐步纳入污水处理成本问题。改革垃圾处理收费方式，加大征收力度，降低征收成本。

此外，由国家发改委环资司牵头制定的《"十二五"节能环保产业发展规划》，在税收政策方面，该规划有可能在节能环保企业已获"三免三减半"税收优惠基础上，再比照国家需重点扶持的高新技术企业的标准，按15%征收节能环保企业所得税。

2008年1月1日颁布的《中华人民共和国企业所得税法实施条例》第二十七条规定，企业从事公共污水处理、公共垃圾处理、沼气综合开发利用、节能减排技术改造、海水淡化等项目的所得，自项目取得第一笔生产经营收入所属纳税年度起，给予“三免三减半”优惠。但是“三免三减半”优惠期结束后，对一次性投入高、回报周期长的节能环保企业而言，25%的常规企业所得税率给企业带来的负担比较大。例如，高新技术企业，今后可按15%的税率征收节能环保企业所得税，将极大降低节能环保企业税负，从而进一步提高企业发展能力。

此外，节能环保企业土地使用税与房产税，有可能获相应减免。特许经营权、收费质押贷款、节能环保产业基金等相关内容，也可能在该规划中有所体现。

目前国际上，工业企业污染治理一般采用专业化运营模式，即污染企业通过付费将本企业产生的污染交由专业环保公司治理，拥有特许经营权的节能环保企业的专业化运营除可降低污染治理成本外，还能刺激环保产业发展。

在我国污水、垃圾处理等收费制度尚未健全情况下，节能环保企业运营资金短缺问题突出，通过鼓励开展收费质押贷款与发展各类节能环保产业基金，可有效解决上述问题。

19.3 我国环保产业最新动态

19.3.1 国家统计局推出2011年绿色发展指数

北京师范大学、西南财经大学、国家统计局中国经济景气监测中心9月24日联合发布《2011中国绿色发展指数报告》。该报告测算了全国30个省（市、自治区）及34个重点城市的绿色发展指数，并进行排序。

绿色发展指数的测算从2010年开始，今年中国统计学会又推出了综合发展指数，与当前使用的GDP指标相比，这样的指标无疑在转变经济发展方式方面能起到更大作用。尽管类似的研究仍处在起步阶段，但是不难看出有关部门正在尝试设计新的统计指标体系以加快我国经济结构调整的步伐。

据了解，绿色发展指数的研究主要是“要解决资源、环境与经济发展的矛盾，实现经济可持续发展。”秉持这一原则，今年的报告得出的一个重要结论是：绿色与发展可以同时实现。

这是通过对全国30个省（市、自治区）、34个重点城市的绿色发展指数排序得出的。根据课题组测算，2011年公布的绿色发展指数排在前十位的省份依次是：北京、上海、青海、天津、海南、浙江、云南、福建、江苏和广东。从地理区位看，除青海和云南位于西部外，其他八个省份均处于东部地区。排在前十位的城市依次是：深圳、海口、昆明、北京、合肥、广州、大连、青岛、长沙、福州。有六个城市来自东部地区，数量最多。

19.3.2 北京CBD核心区将建设成为低碳环保示范区

中经网北京讯2011年9月15日，在2011北京CBD商务节暨世界CBD联盟年会开幕式上，中国国际期货有限公司、阳光保险（集团）股份有限公司、三星生命保险株式会社、中国国际金融有限公司、中国民生银行股份有限公司、中国中信集团公司等相关企业负责人共同签署了《绿色建筑先行，创建低碳商务区》的承诺书，承诺将CBD核心区建设成为北京市乃至全国领先的低碳、绿色、环保示范区。据了解，这些入驻CBD核心区的企业，将坚持低碳理念，把“建设低碳绿色建筑”理念始终贯穿于项目开发的过程中，在项目设计、建设、运营和管理等各个环节引进国际最先进的绿色设计理念和建筑标准。通过高标准设计、高质量建设和高水平管理，力争全面达到绿色建筑三星级标准或美国 LEED 认证系统金奖级标准，力争建筑节能率达到 65%以上。

此外，在节能减排方面，CBD 核心区建设过程中将强化绿色技术应用和绿色施工管理，更加注重环境友好材料的选择、可再生能源和再生水资源的利用，以及国内外节能新技术新产品的综合运用等。

19.3.3 广东拟在 2013 年建成地方性碳交易体系

据悉，中科院广州能源研究等机构向英国繁荣战略基金（SPF）申报的“广东省碳排放交易机制设计及重点行业优先探索项目”日前获中英双边主管机构批准，从 SPF 获得 29.8 万英镑的研究经费。研究项目将借鉴包括英国在内的其他国家碳排放交易体制机制建立的经验，提出适合广东省碳排放权交易体系的基本思路与可行方案，并选择广东省典型行业开展碳排放权交易的试点工作。广东省力争在 2013 年 3 月前建立地区性碳交易体系，成为国内首个区域性碳交易市场。

根据课题目标，到 2013 年，广东省要建立起碳排放权交易的运行机制、政策体系和方法学，为中国建立跨省际区域、跨行业领域、多交易品种的碳排放交易体系积累经验并发挥示范作用。在根据不同行业和企业制定碳排放配额前，本次研究项目首先在电力和建筑材料两个行业展开试点工作。电力行业排放高，但排放核算相较简单，而选择建筑材料行业主要是希望通过碳核算和总量控制帮助相关产业实现转移和升级。

2010 年，国家启动低碳试点工作，广东省成为“五省八市”试点之一。《广东省开展国家低碳省试点工作实施方案》中明确提出将开展碳排放交易试点工作，指出在 2011 年底将研究提出控制二氧化碳排放目标的分解方案及考核办法。

20 大气污染防治

由于人类活动或自然过程，向大气中排放一些有害物质（称污染物），当排入污染物浓度达一定限度，则使原来洁净空气的品质下降，若这种情况维持时间够长，便会对人类、动物、植物和大气中的物品产生危害和不良影响，这种大气状态称为空气污染。

大气有一定的自净能力，现代大工业发展以前，因自然过程等排入大气的污染物，与由大气自净过程而从大气移除的量基本平衡。但是20世纪五六十年代以后，现代大工业迅速发展，人类排入大气的污染物量大大超过了大气的自净能力，致使目前全球大气都遭到不同程度的污染。这引起了各国的关注，我国已把保护大气环境作为国家的基本国策之一。

20.1 大气污染物的危害

近年来，随着城市工业的发展，大气污染日益严重，空气质量进一步恶化，这不仅危害到人们的正常生活，而且威胁着人们的身心健康。大气污染的主要危害包括：

（1）气候变暖。温室气体具有这样的特性：让阳光通过，但强烈吸收地而发出的长波辐射，从而使气温上升，起到了温室的作用。21世纪以来全球平均气温升高了0.5℃，如果温室气体按目前的速度增加，到2030年，全球平均气温将再提高2℃—3℃，灾害性天气和异常天气将更加频繁。

（2）平流层臭氧层变薄。距地而20千米—30千米的平流层中存在一个臭氧层，它能强烈吸收阳光中的紫外线，从而保护地球生物免遭伤害。但是，工业生产中释放的氯氟烃气体强烈地破坏臭氧层，使之减少变薄，甚至出现南极臭氧空洞。

（3）陆地和海洋生物受到污染。工业生产中排放的含有氯、铅、汞、砷的有毒物质影响生物的生长发育，使生物组织中含有有毒物质，其中有些是致癌物质。

（4）产生酸雨。工业生产中释放的二氧化硫和氧化氮使降水中酸性成分增加，形成酸雨。酸雨可使湖泊酸化，造成鱼类死亡。酸雨还可遏制森林生长，甚至导致森林死亡。

20.2 大气污染防治的关键技术

人类的生活与生产活动，特别是其中与能源转化与利用有关的过程，常常伴随着有害物质的产生，而成为大气污染最主要的成因。通常将大气污染源分为固定源和移动源两大类，大气污染控制技术也可以分为固定源污染控制技术和移动源污染控制技术两大类。

20.2.1 固定源污染控制技术

固定源是指排放污染物的固定设施，例如排放二氧化硫、氮氧化物、颗粒物及其他有害物的锅炉、加热炉、工业窑炉等。固定源的单体排放量通常较大，运行条件相对稳定。

（1）颗粒物污染控制。从废气中除去颗粒物（包括烟尘与粉尘）的技术又称做除尘技术，常见的除尘技术可以分为以下几类。① 机械除尘：利用力学原理实现颗粒物与气体的分离；② 电除尘：将废气通过电场，使颗粒物带电，然后在静电力的作用下把它们沉积在集尘板上可以实现净化的目的；③ 湿式除尘：将废气与液体（通常是水）接触，利用水滴与颗粒物的碰撞及其他相互作用可以捕集颗粒物，净化废气，过滤除尘。过滤除尘技术是使废气通过过滤材料而将颗粒物从气流中分离去除的技术。

（2）二氧化硫污染控制。最主要的二氧化硫污染排放来自燃料的燃烧，因此可以对燃前、燃时和燃后三个阶段采用针对性措施降低二氧化硫的排放，其他污染源的二氧化硫排放也有相似的处理技术。① 燃料脱硫：我国主要的含硫燃料是煤和重油，脱除其中的硫元素是降低二氧化硫排放的有效途径。煤中的硫可以通过选煤工艺与灰分同时去除，我国成熟应用的选煤工艺包括重力分选、浮法分选、磁力分选等。② 废气脱硫：矿石冶炼和造纸等工艺过程产生的废气，由于二氧化硫的浓度较高，具有回收利用价值，可以将其用于硫酸等的生产。经过多级催化处理，可以将废气中99.7%的二氧化硫转化为硫酸。

（3）氮氧化物污染控制。大部分的氮氧化物排放污染是在燃料高温燃烧时生成的，其余的来自硝酸生产和硝化等工艺过程，其控制技术主要包括以下方面。① 低氮燃烧：燃烧过程中氮氧化物排放的条件包括温度、气体组分和停留时间等，可以通过适当降低锅炉内的氧气浓度减少氮氧化物排放，氧气浓度低限水平以不造成一氧化碳的排放增加为宜。② 废气脱氮：对排放废气进行处理，减少氮氧化物排放的技术又称为废气脱硝。最成熟的废气脱氮技术是选择性催化还原技术，它采用氨气作还原剂，通过催化剂作用，使其不与氧气而只与氮氧化物反应，将其还原为氮气和水，净化效率可以达到60%—90%。

（4）挥发性有机物（VOCs）污染控制。① 燃烧法：通过氧化反应将VOCs转化为二氧化碳和水而去除；② 吸收法：对于一些高浓度VOCs废气，如果其温度较低且压力较高，可采用某些溶剂吸收其中的VOCs，使其得到净化；③ 冷凝法：对VOCs废气进行降温处理，达到一定的冷凝温度时，VOCs会以液态从废气中冷凝析出，实现净化目的。

20.2.2 移动源污染控制技术

移动源主要指机动车、移动式工程机械、船舶和飞机等在移动过程中排放废气的污染源。目前，飞机的污染控制技术还不够成熟，基本处于研究阶段。

机动车排放的主要污染物包括一氧化碳、氮氧化物、碳氢化合物、颗粒物和其他污染物等，他们来自能量转化过程，即发动机燃烧燃油（或其他燃料），利用化学能产生机械能的过程。从全过程控制的角度出发，可以将机动车污染控制技术划分为改善燃油质量、机内净化和排气后处理等几个主要方面：

（1）改善燃料质量。车用燃料是机动车排放污染的源头，控制机动车排放污染必须从源头抓起。

（2）提高燃油的清净性。在发动机的工作过程中，燃油可能在进气阀、燃烧室和喷嘴等部位产生结焦的现象，燃油成分中的烯烃、芳烃和某些含氮化合物都易于产生结焦。作为减少排放污染的有效措施，燃油清净剂已经在我国得到广泛应用。

（3）控制燃油中有害物质的含量。受原油成分、炼制和储运过程等多方面因素的影响，燃油中常常含有一些杂质元素，例如硫、铅、锰、铜、磷、铝等。它们的存在可能对发动机的正常工作和排气后处理技术的某些环节产生严重的危害，使污染物排放增加。目前，我国正在研究制定更加严格的标准，对此类有害物质的含量进行控制。

（4）机内净化。机内净化技术是指从各种有害污染物的生成机理及影响因素出发，通过对发动机进行改进，实现减少污染物排放量的各种技术。

（5）废气再循环。在保证发动机性能稳定的前提下可以将适量的排气导入进气管，并进入气缸参与燃烧。废气再循化还原技术和颗粒物过滤器技术等。

20.2.3 其他控制技术

（1）光触媒。在市场上各种治理室内空气的办法中，光触媒技术是新近从国外引入，应用较多的一种，对重度污染具有治理见效快的显著特点，但价格也最高。而且，据资料，光触媒在进行光合作用，发生化学反应过程中，有可能产生少量二次污染，对壁纸、木制家具的油漆表面等会有所影响。

（2）臭氧。这是目前国际公认的室内空气治理的一种常用、安全的物理方法，适用于中度、轻度污染。该技术主要是利用臭氧强氧化性，净化空气，杀除空气中的有害成分。这一技术已经被医院等公共场所广泛采用。其最大的特点是不会生成任何残留物及二次污染，但采用这种技术对居室进行治理时，人要暂时离开房间，避免臭氧中毒。

（3）高压电负离子。该技术利用高压电离分解苯、甲醛等有害气体的办法，主要是用一种产生高压电的仪器，使苯、甲醛等有害气体经高压电离，快速氧化成负离子，与空气结合后，还原成氧气、水和二氧化碳。这种办法具有见效快、无污染，不留死角的特点，可以定期采用，作为定期集中治理室内空气超标问题的选择。

（4）炭。竹炭、活性炭等都是利用炭的吸收异味、吸附有害气体的原理，来治理室内空气污染。该方法成本低廉，无毒无副作用，但是见效较慢，可以作为室内空气轻微超标的房间长期治理采用。

20.3 大气污染防治的国际经验

20.3.1 日本大气污染防治

日本1968年制定《大气污染防治法》，1970年对该法进行大规模修订，以后又多次修订，1990年的修订中扩充了环境影响评价和大气污染物排放总量控制的内容，目前实行的是1996年修订的32号法。为控制机动车污染，1992年制定《指定区域机动车排放氮氧化物总量控制特别措施法》。2000年4月29日颁布的《大气污染防治法》，以法律的形式规定了大气污染总量控制制度，并陆续在地方性环境法规中得到体现。2003年7月1日起施行的《排污费征收标准管理办法》更加强化了对这一制度的实施力度。此外，在《电气事业法》、《煤气事业法》、《道路交通法》和《道路运输车辆法》等法律中亦有相关的大气污染防治内容。

日本总量控制区针对的是“遭受大气污染危险”的地区，即酸雨、二氧化硫污染严重的地区。日本总量控制区适用的大气污染因素除含硫氧化物外，还有烟尘、特定有害物质及机动车废气。1992 年制定的《指定区域机动车排放氮氧化物总量控制特别措施法》规定了削减汽车排放氮氧化物，其原因是日本民用汽车的快速发展使机动车排放的废气污染加重。

除机动车废气及时控制制度外，日本在大气污染总量控制立法中明确规定了无过错责任原则，并将刑事立法的罚役、罚金及“双罚制”引入《大气污染防治法》处罚，且所针对的违法行为主要是排放超标、违反燃料标准和设备标准等。处罚程度上，罚役最轻被关押服劳役不超过 3 个月，最重不超过 1 年；最轻罚款不超过 200, 000 日元（折合人民币约 12500 元），最重罚款不超过 1, 000, 000 日元（折合人民币约 62500 元）。另外，日本环境刑法引入了“处罚危险犯”。危险犯不同于结果犯，它是指行为人实施的足以造成某种危害危险的犯罪行为，虽然其危害结果尚未发生，但危险状态已造成即构成犯罪既遂，这种犯罪就是危险犯。

日本在制定大气污染总量控制标准和拟定总量降低计划时，规定了公众参与，其目的是便于公众了解和理解标准，使标准更切实际和完善，更便于实施。为有效实施环境影响事前评价制度和“公开性”原则，保障公众参与、保证公众意见和意志能反映到污染控制对策和落实到环境监督管理，日本的《大气污染防治法》中规定了居民有要求环境保护部门对污染源进行调查并公布调查结果的权利，有要求政府对可能造成环境破坏的开发计划和建设项目以及污染源实行限制的权利，有要求设立由居民代表组成的公害监督委员会的权利。日本《大气污染防治立法》中的“征询意见”、“公布周知”的规定，体现了立法中的民主原则和参与意识。立法的民主原则是现代社会应体现和遵循的基本原则。

在对法定职责的规定中，体现了中央政府、地方政府和环保行政部门的分工与合作。日本是以内阁政令的形式规定含硫氧化物的最高排放容量，所适用的环境标准则以公害对策基本法的形式规定。日本在中央与地方的关系协调中体现出了中央对地方的意见征询，这样可使总量控制区的规定更具可操作性、便于地方执行。

分析可见，日本大气污染控制法规中的总量控制制度、监督管理处罚制度和公众参与制度在可操作性执行力度等方面都比较完善。

20. 3. 2　英国大气污染防治

立法是英国治理大气污染的主要措施，其中以国家立法为主。近 100 多年来，英国相继颁布了一系列空气污染控制法规和标准，主要有：1926 年颁布《公共卫生（烟害防治）法》、1930 年制定《道路交通法》、1848 年的《公众健康卫生法》、1863 年的《工业发展环境法》、1874 年的《第二个工业发展环境法》、1906 年的《制碱及化学工厂法》、1956 年的《空气洁净法》、1968 年的《修订空气洁净法》、1974 年的《烟气排放法》、1990 年的《环境保护条例》，并制定有 78 个行业标准，均是由政府和行业共同商量制定，进行成本和污染平衡，选定最适用技术在各行业推行。目前，英国为进一步改善其空气质量采取了一系列的措施，主要包括：

- 积极制定新的更严的空气质量标准，其中包括空气污染物的控制目标。
- 积极发展利用最新技术，保证空气目标的实现。
- 英国建立了完善的空气监测体系，超过 1600 个包括自动和半自动监测站目前已经遍布整个英格兰。自动监测系统为英国政府随时监控大气环境质量状况提供了强有力的支持。其中有 120 个站自动向公众提供每小时的监测数据。目前，英国的环境监测站仍在不断增加。
- 加强国际合作以改善国外污染物向本国的迁移。

● 实行交通管制，加大污染控制力度。在大气污染严重的城市，限制某些种类和型号的车辆上路或限制行驶速度；推广柴油机替代燃料，在工业和汽车用户中推广使用清洁燃料；推广大气污染防治技术。

● 广泛进行经济分析。经济分析指的是应对事物进行全方位系统的评估，要求充分考虑环境的价值。

英国的大气污染控制是在排放物容量不明确的基础上执行的，其特色体现在“切实可行的措施”（the Best Practicable Means，简称 BPM）之中，BPM 的基本要素有：

● 不能容忍造成某种公认的或长或短的健康公害的排放物。

● 在考虑地方条件与环境、控制技术知识现状、所排放物质的后果、财政状况和所使用的措施等基础上，必须根据排放物浓度和质量，将其减至最低的适当的水平。

● 为确保排放物达到最低的适当的水平，必须规定气体排放的高度，以致通过稀释和消散使残留的排放物无害而不令人生厌。

英国治理污染的一个特点是“污染者支付原则”（the Polluter Pays Principle）是大气污染控制的关键因素。它意味着防治污染的费用，或者减轻因污染造成的环境损害的费用应由对此负责的方面负担，是依照法令要求个人或组织使用切实可行的措施，来减少有害或有毒气体的排放。

英国治理污染的另一个特点则是科研力量的参与。英国的不少研究机构以及高等学府在相当广泛的基础上参与了环境监测和防止污染的研究工作。自 1960 年起，以华伦·斯普林实验室为重心，根据遍布全英的 1600 多个监测站的测定结果，对空气中烟尘和二氧化硫的含量进行估计，据此提出不同地区控制污染的不同措施，并研究新发现的污染物对环境的影响。里丁大学、阿斯顿大学、帝国理工学院、威尔士大学、谢菲尔德大学和利兹大学等高等学府，也分别对车辆排放污染物、空气质量标准、控制污染物的排放、空气污染对农作物和土壤的影响、测定灰尘及其他污染物的仪器的改进，以及烟囱的设计与安装位置等问题进行了研究。科研力量的广泛参与，为大气污染的防治以及其他环境问题的解决，提供了有力的理论支撑。

20.3.3　美国船舶大气污染防治

在美国，尽管汽车、货车等机动车大量增加，但机动车造成的污染却得到了很好的控制，美国的空气质量有了显著的提高。尤其是从 1970 年批准《清洁空气法》开始，轿车和卡车的废气排放减少了 90% 以上，6 个最主要的空气污染物（“年度排放标准”）下降 160 万吨。然而，船舶造成的污染却是这种趋势的一个例外。船舶和港口设施造成的空气污染越来越严重，并随着对外贸易的急剧增长而不断加重。在很多城市，船舶运输业已成为造成空气恶化的最大污染源。

洛杉矶问题尤为严重。据当地空气质量管理区（AQMD）的报告：该地区排放的二氧化硫有 70% 来自船舶和港口设施。远洋船只排放的氮氧化物，超过所有在南海岸盆地的发电厂及炼油厂排放的氮氧化物之和。问题并不仅限于洛杉矶地区，据美国环境保护署（EPA）统计，美国全境有 40 多个处于“超标”的地区。

针对船舶产生的大气污染问题，美国采取了一下措施进行整治：

（1）1997 年，国际海事组织（IMO）通过了 MAIU，MARPOL 附则 VI《防止船舶造成大气污染规则》，这项规则在2005年已经得到约占世界74.7%航运吨位的国家批准，对于船舶废气中的硫氧化物和氮氧化物的排放含量，做出了严格的限制规定：远洋船舶燃料的含硫量限制在4.5%以下。（相比之下，美国在路机动车柴油燃料的硫含量是15%）；限制在波罗的海、北海和英吉利海峡行使的船舶燃料油的含硫量在1.5%以下；限制新引擎和发动机的氮氧化物排放量在9.8克/千瓦时—17.0克/千瓦时之间（相比之下，美国东部电厂的氮氧化物排放量限制在0.45克/千瓦时—0.73克/千瓦时以下）；限制在港口排放具有挥发性的有机化合物；禁止使用消耗臭氧的燃料；禁止船舶焚烧多氧联苯，禁止焚烧含有卤家化合物的精炼石油产品。2008年，国际海事组织海洋环境保护委员会（MEPC）对《防止船舶造成大气污染规则》进行修正，新的修正案从2010年开始生效。本次修正案最大的变化是逐步减少船舶硫氧化物的排放，从当前的4.50%减少到3.50%（2012年1月1日生效），然后减少至0.50%（2020年1月1日生效）。硫排放控制区的适用标准将从当前的1.50%减少到1.00%（2010年1月1日生效），最后减少到0.10%（2015年1月1日生效）。除此之外，MEPC还对氮氧化物技术规则、废气净化系统、挥发性有机物等方面作了一些新的规定。

（2）其他法律法规的制定与执行。1999年10月，美国环保局就设立了一个自愿认证序，以便使发动机制造商能证明他们的发动机均符合防止船造成大气污染规则。该规则适用于任何在美国注册的船只，任何在美国水城内的港口、码头登记的船舶，任何在美国航水域或指定排放控制区航行的船舶，任何通过美国水域船舶（即使他们目的地不是美国）。该规则也适用于在美国专属济区的任何其他船舶。

参议院和其他委员会也针对船舶造成的污染制定了相关法案，例如，美国环境和公共工程委员会在2008年修正《清空气法案》，要求任何进入或离开美国港口和码头的远洋船只使的燃料含硫量不超过千分之一。并从2010年12月31日开始，将限制扩大到距离美国海岸或大海岸线2加英里以内的海域。除燃料标准，该法案还要求美国环保局针对进入或离开美国港口远洋船舶的发动机制定排放控制标准，以减少氮氧化物、颗粒物、氮氢化合物和一氧化碳的排放。考虑到技术成本和具体实行的行性，该标准将于2012年1月1日起施行。

此外，污染地区（如加利福尼亚）自行制定相关法律法规确保船舶大气污染得到有效控制。

20.3.4 国际机动车碳税经验

自芬兰于1990年首次实施碳税以来，碳税已有20年实践经验。实施碳税的初衷是为了降低高能耗行业的化石能源消耗。根据OECD（2010）统计，OECD国家交通领域的环境税收入占总环境税收入的90%，包括交通燃料（64%）和机动车（26%）环境税，而供暖（5%）、电力（3%）、废弃物（1%）和其他污染税仅占总税收的10%。在实施碳税的国家，交通成为了碳税的主体税源。

OECD国家道路运输二氧化碳排放占交通总排放二氧化碳的88.57%。航空和水运交通燃料在大多数发达国家基本是减免碳税，轨道交通以用电为主，因而交通碳税主要是指道路交通碳税。交通领域的碳税可以分为两大部分：第一部分是针对交通燃料，即针对机动车用汽油、柴油等燃料的课税；第二部分是直接针对机动车所征碳税。

机动车二氧化碳排放是发达国家重点排放领域之一，OECD国家2008年道路交通二氧化碳排放占能源燃烧排放的23.75%，仅次于电力和热力生产（39.53%），并且增长很快。法国对于机动车二氧化碳排放课以重税，其2008年排放水平也仅达到2000年的水平。机动车二氧化碳排放是发达国家减排的核心领域之一，并且对其征收碳税不存在影响企业竞争力和碳泄漏问题，同时有利于减少机动车尾气污染物排放，环境协同效应明显，因而尽管存在许多批评声音，机动车碳税却在发达国家越来越受欢迎。

20.4　我国大气污染物防治现状与对策

20.4.1　我国大气污染防治现状

我国大气污染主要呈现为煤烟型污染特征。城市大气环境中总悬浮颗粒物浓度普遍超标；二氧化硫污染保持在较高水平；机动车尾气污染物排放总量迅速增加；氮氧化物污染呈加重趋势；全国形成华中、西南、华东、华南多个酸雨区，以华中酸雨区为重。

在1995年我国由于酸雨和二氧化硫污染造成农作物、森林和人体健康等方面的经济损失就约为1100多亿元。2010年，全国地表水污染依然较重，长江、黄河、珠江、松花江、淮河、海河和辽河等七大水系均受到污染；全国近岸海域水质总体为轻度污染；受调查的494个城市（县）中，超过一半出现酸雨，达249个，其中酸雨程度严重或较重的城市有107个，占21.6%。

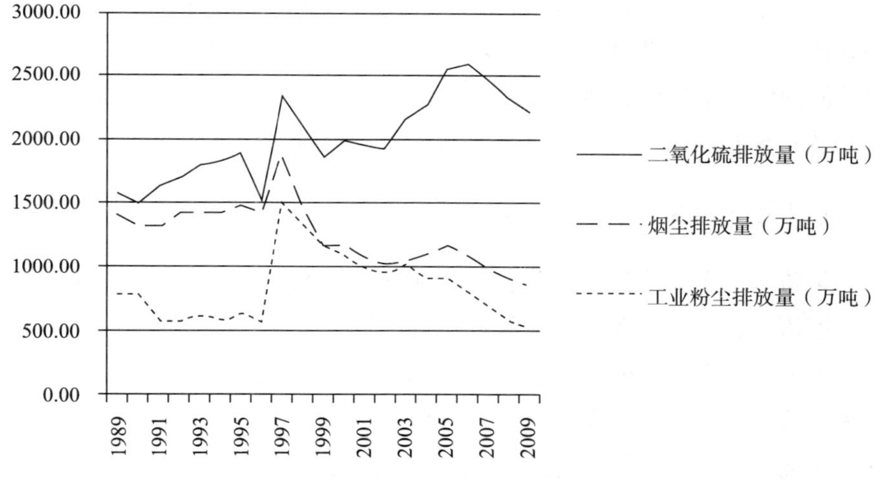

图20-1　我国近20年废气排放走势

Figure 20-1　Fluctuation of China's Waste Gas Emission in the Last 2 Decades

资料来源：中国国家环境保护总局

由图20-1可以看出，从1989—1998年，废气排放量呈上升趋势，到1997年、1998年时达到高，这阶段正值“八五”期间我国经济增长是通过大量消耗能源、资源来实现的。随着经济发展方式的转变，节能减排逐渐被政府和广大群众重视。从1999年开始，废气排放量迅速降低。但是由于城市的快速发展，特别是机动车的急速增加，一些被淘汰的落后产能死灰复燃，使得最近几年我国废气排放量，特别是二氧化硫排放量降速减缓并呈缓慢上升的趋势。

我国大气污染防治取得的成效主要表现在以下几个方面：

一是主要大气污染物排放总量得到有效控制。到2008年底，全国已建成脱硫设施的火电装机容量累计3.63亿千瓦，形成年脱硫能力约1000万吨。与2005年相比，二氧化硫排放总量减少了8.95%。

二是城市大气环境综合整治不断加强。各地区积极推动燃煤锅炉清洁能源改造，鼓励发展热电联产、集中供热，努力控制面源污染问题。与2000年相比，2008年全国城市空气中二氧化硫、可吸入颗粒物和二氧化氮年均浓度分别下降28.5%、33.3%和31.5%。

三是积极探索大气污染区域联防新机制。为确保北京奥运会空气质量达标，环境保护部与北京、天津、河北、山西、内蒙古、山东六省（区、市）以及各协办城市建立了大气污染区域联防联控机制，完全兑现了奥运环境承诺，并创造了近10年来北京市和华北地区空气质量最好水平。

四是产业结构调整力度进一步加大。有关部门先后出台《促进产业结构调整暂行规定》、《产业结构调整指导目录》，限制高排放、高耗能行业盲目扩张。建立更加严格的环境准入制度，提高了铁合金、焦化、电石等10多个行业的环境准入条件。

五是清洁能源利用和节能工作扎实推进。国家制定了可再生能源中长期发展规划，出台了可再生能源电价补贴和配额交易方案，先后实施了“西气东输”、“西电东送”等清洁能源重点工程，积极鼓励核电发展。

六是大气污染防治法制建设和执法工作不断深入。有关部门已连续五年开展了整治违法排污企业保障群众健康环保专项行动，查处环境违法企业12万多家（次），取缔关闭违法排污企业两万多家。严格实施“区域限批”政策，集中解决了一批突出环境问题。

七是污染防治基础能力建设力度不断加大。近两年，环保能力建设投资超过150亿元，全国环境空气质量监测网络基本形成，地级以上城市共建成911套空气质量自动监测系统，配备主要环境监测仪器设备4.5万台（套），648个环境监察机构通过了标准化建设验收。

20.4.2 我国大气污染防治现存问题

当前大气污染防治工作仍然存在一些问题，随着国民经济的持续快速发展，能源消费的不断攀升，发达国家历经近百年出现的环境问题在我国近二三十年集中出现，呈现区域性和复合型特征，存在发生大气严重污染事件的隐忧，大气环境形势非常严峻。

一是以煤为主的能源结构导致大气污染物排放总量居高不下，2006—2008年，我国煤炭消费量增加了6亿多吨，其中火电行业增加了4亿多吨，预计到2010年二氧化硫排放总量仍将达2200万吨左右。

二是城市大气环境形势依然严峻。2008年，全国23.2%的城市空气质量未达到国家二级标准，113个重点城市中，有48个城市空气质量达不到二级标准，城市空气中的可吸入颗粒物、二氧化硫浓度依然维持在较高水平。

三是区域性大气污染问题日趋明显，我国长三角、珠三角和京津冀地区等城市群大气污染呈现明显的区域性特征，而目前我国城市大气污染治理“各自为战”，尚未建立有效的区域空气联防联控机制，难以从根本上解决区域和城市的大气环境问题。

四是机动车污染问题更加突出，2008年，我国汽车保有量超过6400万辆，汽车尾气排放成为大中城市空气污染的重要来源，加剧了大气污染治理的难度。

五是环境法规和保障体系有待进一步加强，现行的大气污染防治法规在排污许可证、机动车污染防治、区域性大气污染防治等方面尚不能满足工作需要；环境违法成本低、守法成本高的问题仍然存在；相关经济激励政策体系不完善。

20.4.3 “十二五”大气污染防治目标

针对下一步大气污染防治工作，以污染减排为中心，加大污染治理力度，着力解决危害群众健康的大气环境问题，努力改善大气环境质量，推动经济社会又好又快发展。

大气污染防治“十二五”规划正在制定，政策调控目标是：到2015年主要大气污染物总量显著下降，重点企业全面达标排放，重点区域主要大气污染物排放达到国家二级标准。火电、钢铁、有色、石化、水泥、化工被确定为防控的重点行业。为保障政策落实，有关部门将加大资金投入，进一步严格火电、钢铁、水泥、电解铝等行业上市公司的环保核查。

进一步完善大气污染防治监管和保障体系，积极配合做好《大气污染防治法》修订工作，严厉打击各类环境违法行为，健全环境经济政策体系。强化大气污染防治的科技支撑。加强“十二五”期间大气污染防治规划框架体系、指标体系、环保投入体系及政策法规体系研究，努力提高规划的指导性、针对性、可操作性和执行力①。

“三区六群”被确定为大气污染防控重点区域。“三区”为京津冀地区、长三角地区、珠三角地区；“六群”为辽宁中部城市群、山东半岛城市群、武汉城市群、长株潭城市群、成渝城市群、海峡西岸城市群。其中，“三区”是防控的重中之重。新疆乌昌地区、陕西关中地区、山西太原和大同沿线地区也可能成为大气污染防控重点区域。

二氧化硫、氮氧化物、颗粒物、VOC（挥发性有机物）被确定为重点防控的大气污染物。其中，前两项污染物是全国约束性控制污染物，“十二五”期间，二氧化硫减排8%，氮氧化物减排10%。后两项污染物是大气污染防控重点区域的减排指标，具体指标正在制定。我国将从以下四个方面减排氮氧化物：

第一，严格控制“两高一资”的项目，对千万人口以上的城市，实行机动车总量控制，对电力和水泥行业实行氮氧化物的总量控制，对长三角、珠三角、京津冀地区，建议探索实行煤炭总量控制。加快实施机动车“国四”排放标准，尽快落实在全国范围内实行“国四油”的供应。

第二，大规模地推行现役电厂和干发水泥生产线的脱硝改造工程。环保部将于近期颁布实施修订后的火电厂大气污染排放标准，将成为世界上最严格的排放标准。

第三，加大淘汰落后产能的力度。加大黄标车的淘汰力度。

第四，加强环保监管，继续安装自动在线监测装置，同时加大检查的力度，确保企业治污设施能够稳定达标运行。

“十二五”期间，电力行业的减排任务依然沉重。据介绍，“十二五”期间二氧化硫减排贡献率约为24%，而氮氧化物则高达73%左右。“十二五”期间，环保部明确新建火电项目要实现80%以上的脱硝效率。环保部预测，新标准实施后，到2015年，全国需要新增烟气脱硝容量8.17亿千瓦，共需脱硝投资1950亿元，2015年需运行费用612亿元/年。到2020年，需要新增烟气脱硝容量10.66亿千瓦，共需脱硝投资2328亿元，2020年需运行费用800亿元/年。

大力推进机动车污染防治工作，进一步加快机动车排放控制标准的实施进程，力争到2020

---

① 资料来源：国家环保局网站，“周生贤向全国人大常委会报告大气污染防治工作时指出：加大污染治理力度努力改善大气环境质量”，http://www.zhb.gov.cn/zhxx/hjyw/200904/t20090423_150765.htm。

年，新车排放控制水平与发达国家接轨；加大新车环境保护监管力度，确保新车出厂稳定达标；加强在用车环保年度检测，在全国实施统一的环保合格标志。“十二五”期间，我国将逐步健全包括汽车、摩托车、低速货车等的排放标准体系。目前，欧盟已将车辆碳排放纳入标准体系中，我国也可能会将这一项列入国Ⅴ标准中。在汽车排放标准的实施方面，轻型汽车目前采用的是国Ⅲ标准，2011年将在全国范围内启动国Ⅳ标准。在监管方面，环境保护部也将对排放不达标的车辆下重手。未来五年将从三方面加强监管：一是加强新生产车辆的排放控制，包括制定更严格的排放标准，完善达标车辆的申报制度，加强生产的一致性检验；二是强化在用车辆排放监督，继续实施标志制度，强制维修排放不达标的车辆，开展排放超标车辆的召回工作，加速淘汰高排放车辆等；三是严格控制车用油品有害物质，包括制定标准、要求备案等。

20.4.4 我国二氧化硫防治现状

在我国电源结构中，火电设备容量占总装机的75%以上，在2010年全国总装机容量达到540GW，火电装机容量约为400GW以上，需要电煤约10亿吨。到2020年预计火电装机在600GW左右，约需煤炭14亿吨。根据国家环境保护部《火电厂烟气脱硫关键技术与设备国产化规划要点（2000—2010年）》的介绍，如果不加处理，到2010年我国仅火力发电就会造成二氧化硫排放约10000万吨。而早在1998年，我国二氧化硫的排放已经达到2090万吨，成为世界上二氧化硫排放最多的国家。

“十一五”期间，全国单位GDP能耗下降19.1%，全国二氧化硫排放量减少14.29%，全国化学需氧量排放量减少12.45%，基本完成了“十一五”规划《纲要》确定的目标任务。节能减排的成效主要体现在为保持经济平稳较快发展提供了有力支撑。“十一五”期间，我国以能源消费年均6.6%的增速支撑了国民经济年均11.2%的增速，能源消费弹性系数由“十五”时期的1.04下降到0.59，缓解了能源供需矛盾，扭转了我国工业化、城镇化加快发展阶段能源消耗强度和主要污染物排放量上升的趋势。“十五”后三年全国单位GDP能耗上升了9.8%，全国二氧化硫和化学需氧量排放总量分别上升了32.3%和3.5%；“十一五”期间，全国单位GDP能耗下降了19.1%，全国二氧化硫和化学需氧量排放总量分别下降了14.29%和12.45%。

2010年与2005年相比，环保重点城市二氧化硫年均浓度下降26.3%，地表水国控断面劣五类水质比例由27%下降到16.4%，七大水系国控断面好于三类比例由41%上升到59.9%，为应对全球气候变化作出了重要贡献。“十一五”期间通过节能降耗减少二氧化碳排放14.6亿吨，得到国际社会的广泛赞誉，也体现了负责任的大国形象。

在“十一五”期间，31个省、自治区、直辖市和新疆生产建设兵团以及国家电网公司和华能、大唐、华电、国电、中电投五大电力集团公司都较好地完成了《“十一五”期间全国主要污染物排放总量控制计划》下达的总量控制任务。

2011年上半年部分污染物减排效果未达预期，化学需氧量、二氧化硫排放量继续下降，但氨氮排放量仅下降0.73%，氮氧化物排放量反而上升6.17%，氨氮和氮氧化物排放量与年度计划减排目标和“十二五”规划目标差距很大。

20.4.5 机动车排气污染防治现状

国家环境保护部日前发布《中国机动车污染防治年报（2010年度）》（以下简称《年报》），

首次公布我国机动车污染物排放情况。结果显示，2009 年我国首次成为世界汽车产销第一大国，机动车污染日益严重，机动车尾气排放成为我国大中城市空气污染的主要来源。

据《年报》统计，我国机动车保有量呈快速增长态势。2009 年，全国汽车产、销量分别达到 1379. 1 万辆和 1364. 5 万辆，同比增长 48. 3% 和 46. 2%；机动车保有量接近 1. 7 亿辆，同比增长 9. 3%；与 1980 年相比，全国机动车保有量增加了 25 倍。其中，汽车 6209. 4 万辆，摩托车 9453. 1 万辆。按汽车排放控制水平分类，达到国 III 及以上排放标准的汽车占汽车总保有量的 25. 4%，国Ⅱ汽车占 31. 8%，国 I 汽车占 25. 7%，其余 17. 1% 的汽车还达不到国 I 排放标准。随着机动车保有量的增加，机动车排放污染物对环境的影响日趋严重，给城市和区域空气质量带来巨大压力。2009 年环境监测显示，全国 113 个环保重点城市中 1/3 的城市空气质量不达标，很多城市尤其是大中城市空气污染已经呈现出煤烟型和汽车尾气复合型污染的特点，加剧了大气污染治理的难度。同时，我国一些地区酸雨、灰霾和光化学烟雾等区域性大气污染问题频繁发生，部分地区甚至出现了每年 200 多天的灰霾天气，这些问题的产生都与机动车排放的氮氧化物、细颗粒物等污染物直接相关。2009 年，全国机动车排放污染物 5143. 3 万吨，其中一氧化碳（CO）4018. 8 万吨，碳氢化合物（HC）482. 2 万吨，氮氧化物（NOx）583. 3 万吨，颗粒物（PM）59. 0 万吨。汽车是机动车污染物总量的主要贡献者，其排放的 CO 和 HC 超过 70%，NOx 和 PM 超过 90%。按车型分类，全国载客汽车 CO 和 HC 排放量明显高于载货汽车，其中轻型载客汽车贡献率最大；载货汽车排放的 NOx 和 PM 明显高于载客汽车，其中重型载货汽车是主要贡献者。按燃料分类，全国汽油汽车 CO 和 HC 排放量明显高于柴油汽车，超过排放总量的七成以上；柴油汽车排放的 NOx 接近总量的六成，PM 超过九成以上。

实施严格机动车排放标准的减排效果显著，1980—2009 年，全国汽车污染物排放量呈逐年上升趋势；1980—2000 年期间污染物排放量与汽车保有量呈线性关系增长；2000 年后，污染物排放量增速有所减缓，这与不断实施严格机动车排放标准和淘汰高排放的“黄标车”有关。2009 年，占汽车保有量 17. 1% 的国 I 前标准汽车，其排放的四种主要污染物占总排放量的 50% 以上；占保有量 25. 4% 的国Ⅲ及以上排放标准汽车，排放量不足总排放量的 5%。

经过近 30 年的发展，我国机动车环保管理工作取得了较大进展，具体体现在以下三个方面：

一是机动车环保法规、标准体系不断完善。30 年间先后颁布了 90 多项机动车环保标准，七年内实现了机动车国 I 标准到国 III 标准的升级，缩小了与发达国家排放控制水平的差距，2009 年，新生产轻型汽车的单车污染物排放量比 2000 年下降了 90% 以上。

二是机动车环保监管制度已基本建立。各级环保部门积极推进机动车环保定期检验工作，在全国范围内实行统一的机动车黄色和绿色环保标志管理，提高机动车尾气排放合格率；有关部门和地方政府先后出台老旧机动车淘汰补贴经济政策，颁布对大排量乘用车提高消费税税率、对购置节能与新能源汽车给予一定补贴等政策，全面加快了“黄标车”淘汰进程。

三是车用油品环保管理和油气污染治理不断强化。京津冀三省市先后完成了 1976 座加油站、48 座储油库和 1265 辆油罐车的油气回收改造，珠三角、长三角和全国设区城市油气污染治理工作陆续启动。

鉴于目前机动车尾气排放已成为我国大气污染的主要来源之一，环境保护部今后将进一步加大工作力度，不断完善新生产机动车、在用机动车环境监管，全力削减机动车污染物排放总量；同时与有关部门密切协助，从行业发展规划、城市公共交通、清洁燃油供应等方面采取综合措施，缓解机动车尾气排放对大气环境的影响。

曾为我国淘汰含铅汽油、机动车排放标准制定等工作提供咨询的国际清洁交通委员会主席沃尔什近日表示："经过十几年的不懈努力，中国建立了完备的机动车环境管理体系，并成功实施了一系列严格的污染防治措施，取得了令人瞩目的成绩。"同时他也指出，中国机动车污染防治仍存在诸多必须克服的问题和不足。一是低硫燃油供应滞后是影响中国机动车环保技术发展的最大障碍和瓶颈，车油不配套大大削弱了污染物排放标准的实施效果，同时也严重制约了机动车污染控制技术水平的提升；二是机动车环保监督管理能力严重不足，人员设备配置和资金支持与机动车污染防治形势不相适应；三是缺乏激励与惩罚并行的环保政策措施，没能有效利用税收、价格等手段的杠杆作用，油品、车辆生产企业及消费者的环保积极性没有得到有效调动。

为了应对未来中国机动车保有量仍将保持快速增长所带来的污染物排放问题，报告提出了相关建议：

一是继续严格实施机动车污染物排放标准，缩小与发达国家环保技术水平的差距。

二是继续降低车用燃油含硫量，统筹推进机动车污染物排放标准和车用燃油标准。

三是强化用车检验、维修制度，建立机动车环保检验数据库。

四是大力推动高排放黄标车淘汰，政府可以考虑创建基金，为黄标车维修或报废提供补贴。

五是强化在用车环保标志管理，考虑延伸标志用途，将标志改为销售前核发，引导环保消费。

由于城市经济的迅猛发展及城市化进程的加快，城市交通需求量急剧上升，机动车排放造成的空气污染已经越来越引起人们的关注。有研究表明，我国大城市60%的CO，50%的NO_x，30%的HC污染，是机动车排放尾气造成的。交通系统产生的大气污染物包括CO、NO_x、非甲烷碳氢化物THC及其他有害物质（如铅氟氯化烃排放物的二次衍生物光化学烟雾等）。其中CO、NO_x是主要的大气污染物，对城市大气环境和人体健康都带来诸多不利影响，已成为导致心血系统、神经系统、呼吸系统以及人体各种组织发生病变的重要因素。

根据我国现阶段经济发展的形势和现有的技术水平，在短时期内，我国的汽车尾气污染还会有进一步加剧的趋势。

2011年上半年全国113个城市中约有2/3的二氧化氮排放，呈现上升的态势。其中，锦州的排放量同比增加了100%。环境保护部部长周生贤2011年9月27日透露，"十二五"期间预定氮氧化合物要下降10%，但是实际2011上半年排放量反而上升了6.17%，与年度计划减排目标和"十二五"规划确定的目标差距很大。目前，氮氧化合物的排放主要由机动车、电厂和工业企业排放，约各占1/3左右。其中，在大城市中机动车的排放占据了重要的比例。

面对严峻的汽车尾气污染形势，世界各国都在积极防治汽车尾气污染。美、日、欧洲的发达国家已经取得了巨大的进步，他们已经拥有了先进的低污染汽车制造技术、高效的尾气控制技术和完善的尾气污染控制法律法规体系。我国在这方面起步较晚，还存在以下几个方面问题。

（1）对尾气进行有效净化的技术基础落后。我国到1997年才在北京、上海和广州率先使用无铅汽油，到2000年才在全国全面实现汽油无铅化。因此，我国各种汽车催化净化技术、装置的运用受到限制，造成我国机动车单车尾气排放量大。据报道，我国国产车的碳氢化合物（HC）、一氧化碳（CO）、氧化氮（NO_X）分别是1992年美国新车排放量的14.5倍、11.8倍、3.3倍。

（2）法律法规体系零散，未形成系统。我国于1987年制定的《大气污染防治法》中虽有关于机动车尾气污染控制的原则性规定，但由于当时有关的国际公约和一些国家的法律没有规定关于机动车的监督管理问题，同时，我国缺乏相关的经验和配套法规，也未有关于汽车尾气排放的相关排放标准和监测手段，因此，我国的汽车尾气排放实际上处于无控制状态。直到2000年修订的《大气污染防治法》才比较多地体现有关机动车尾气污染控制的规定，该次修正案设立专章对机动车进行管理，体现了全程控制的思想，从机动车制造、在用车使用和维修、燃油质量、监督检查等几个方面分别作了规定。就机动车排放标准而言，我国现行的排放标准基本上是参考欧洲、美国和日本的排放法规的基础上制定的，控制水平较低。我国在2000年1月1日实施（欧1号）的汽车尾气排放标准，欧洲在1992年左右就实施了。

（3）机动车尾气及排放的管理体制存在较多问题。机动车排放管理体制存在很多问题，例如未明确各有关单位的职责，尤其是环保部门的职责范围，使对不合格车辆的抽检、治理、处罚难以实现，除汽车尾气污染控制比较发达的少数几个城市外，多数城市的环保部门没有对汽车维修行业进行统一的监督管理。体现为多部门管理，各个部门沟通不足，管理力度不够。承担主要治理任务的环保部门基本没有参与汽车维修行业的监督管理；在管理中又存在很多漏洞，例如检测设备不符合要求，检测质量没有保证，数据不可靠等；对在用车的年检、路检和抽检中只有少数城市是公安环保联合实施的，大部分的城市环保部门没有参与。这些都需要在相关制度中明确给予规定。

我国虽然在汽车尾气污染控制方面存在一些不足，但也具有很好的控制治理条件。自从2000年《大气污染防治法》修改以来，我国的汽车尾气污染控制观念逐渐被广泛接受，污染控制净化技术也取得了较大的发展。一方面，在经济全球化的今天，我国完全可以借鉴利用发达国家日趋成熟的汽车排放控制技术及其法律制度；另一方面，在借鉴吸收的同时应加强自身特色的法律法规的建设，因为制度规范对技术进步起着引导作用，有怎样的排放法规就要求有怎样的汽车排放控制技术和相应的燃油标准，二者缺一不可。所以，我国在进行汽车尾气污染控制的过程中要侧重加强控制的对策，尤其是法律控制对策的研究。

20.5　我国大气污染治理动态

20.5.1　《火电厂大气污染物排放标准》正式出台

火力发电是重要排污行业，2010年火电装机容量达7.07亿千瓦，占全国总装机容量的73%，火电发电量占全国发电量的80%以上。虽然我国的电力装机容量已经跃居世界第二，但是人均水平仅为0.74千瓦，而美国为3.5千瓦，日本为2千瓦，欧盟为1.4千瓦，因此电力行业的污染排放在我国尚有很大上升空间，而降低排放强度的主要措施就是制定和实施更加严格的排放标准。

所谓标准减排，就是以制定严格的排放标准为核心，开展环境监测规范、分析方法标准、监控技术规范等配套标准制定、修订工作，形成若干由相关标准组合而成的减排标准体系，并通过严格实施这些标准提高排放控制水平，实现压增量、削存量、减总量的目标。随着国家排放标准体系建设的推进，标准减排的作用将更加凸显。

2011年9月21日，备受瞩目的《火电厂大气污染物排放标准》正式出台。在环保部公布的正式标准中，氮氧化物污染物排放限值为：从2012年1月1日开始，所有新建火电机组氮氧化物污染物排放标准为100毫克/立方米；从2014年7月1日开始，现有火电机组氮氧化物污染物排放标准为100毫克/立方米（特殊规定的执行200毫克/立方米的标准）。而重点区域火电机组的氮氧化物污染物排放标准则统一为100毫克/立方米。

标准在具体适应范围方面做了一定程度的细化。主要有以下四点：

● 现有火电机组执行新标准的时间由二次征求意见稿的2014年1月1日，推迟到了2014年7月1日，时间宽限了半年。

● 对于汞及其化合物污染物排放标准，规定2015年1月1日起开始执行，执行对象为燃煤机组。与二次征求意见稿规定的执行对象为“新建火电机组”相比，在正式标准中扩大了执行对象的范围，且把执行时间推迟了一年。

● 对于脱硫标准进一步细化。对二次征求意见稿中的高硫煤地区做了详细规定：把广西、四川、重庆、贵州四省市（自治区）规定为高硫煤地区。

● 在氮氧化物排放标准方面，对范围进行了微调。执行200毫克/立方米的燃煤锅炉范围除了二次意见稿中的“2003年12月31日前建成投产或者通过建设项目环境影响报告书审批的火电发电锅炉”外，还特别增加了“采用W型火焰炉膛的火电发力锅炉和现有循环流化床火电锅炉。”

此外，本次标准与2003年的旧标准相比，在总体上收紧了污染物排放限值，提高了新建机组和现有机组排放控制要求；适当地合并了不同时期建设的现有机组的排放控制要求；进一步强化了标准的适用性，删除了在实际工作中难以发挥作用的内容；还进一步完善了污染物指标体系，增设了汞的排放限值和燃气锅炉排放限值；最后还增设了适用于环境敏感地区的大气污染物特别排放限值。

新标准实施后到2015年，电力行业的氮氧化物与二氧化硫排放量可分别减少580万吨和618万吨，而且电力行业颗粒物和汞等污染物也有明显的减排效果。

国家环保局测算，根据新排放标准，到2015年时，全国需要进行去除氮氧化物改造的现有机组和新增的机组容量约为8.17亿千瓦，估计需投资1950亿元，年运行费用约612亿元；除尘和脱硫设施的改造和新建投资费用约需650亿元。

尽管标准之争尘埃落定，但是其具体执行的道路仍然存在众多问题。“鉴于2003年的‘旧火电标准’从出台到真正有效控制二氧化硫之间一共用了超过五年的时间，‘新标准’的有效实施还需要环保部制定并实施进一步的行动方案。”

20.5.2 “十一五”减排提前实现五大电企脱硫成绩不俗

环境保护部部长周生贤在日前举行的“2010年全国环境保护工作会议”上提出，2010年的环保工作目标为：二氧化硫排放量力争比2009年再削减40万吨；分别淘汰电力、炼铁、水泥、焦化和造纸落后生产能力1000万千瓦、2000万吨、5000万吨、2000万吨和52万吨；确保新增燃煤电厂脱硫装机容量5000万千瓦等。二氧化硫“十一五”减排目标已经提前一年实现。据初步测算，与2008年相比，2009年全国二氧化硫约下降了4.57%。

减排目标的实现得益于三个因素：一是减排目标责任考核力度不断加大；二是减排工程建设和落后产能淘汰取得新进展；三是治理设施运行监管不断强化。去年新增燃煤脱硫机组总装机容量 1.02 亿千瓦，2006 年以来累计新增燃煤脱硫机组装机容量 4.11 亿千瓦，其中新增现役燃煤脱硫机组装机容量 1.81 亿千瓦。

2010 年全国环境保护工作报告中指出，重点城市二氧化硫年均浓度为 0.043 毫克/立方米，较 2008 年下降 10.4%，较 2005 年下降了 24.6%。密切关注中国减排问题的美国，通过其全球卫星观测系统对我国减排情况进行观测和分析后认为，2007 年下半年，中国二氧化硫排放量的确出现拐点，随后二氧化硫浓度呈现较大幅度下降。自 2000 年以来，政府出台了一系列的政策，促进烟气脱硫产业化的快速发展，燃煤电厂的二氧化硫排放控制能力得到明显提高。

"十一五"以来，电力二氧化硫排放量逐年降低，已由 2006 年最高的 1350 万吨，下降至 2008 年底的 1050 万吨，二氧化硫排放绩效值由 2006 年的 5.7 克/千瓦时下降到 3.8 克/千瓦时。大唐集团 2009 年新增脱硫机组 2195.5 万千瓦，除列入关停计划的机组外在同行业中率先实现了脱硫装备率 100%。华能集团 2009 年新增脱硫机组 2181 万千瓦，到 2009 年底，公司安装脱硫设施的机组容量达到 8148 万千瓦，规划脱硫机组全部实现脱硫，完成"十一五"环保责任书目标的 165%，预计到 2010 年底，将实现燃煤机组全部脱硫，所有燃煤电厂建成节约环保型燃煤发电厂。中电投集团开展烟气脱硫无旁路试点，完成了 22 台机组脱硫改造，新增二氧化硫减排能力 24 万吨/年。华电集团单位二氧化硫排放绩效 3.15 克/千瓦时，同比下降 8%。国电集团 2009 年二氧化硫排放量控制在 90 万吨以内，提前一年完成了"十一五"节能减排目标。

各大电企也纷纷提出 2010 年的火电脱硫目标：华电集团提出 2010 年单位二氧化硫排放绩效 3.05 克/千瓦时，比上年下降 3 个百分点。国电集团 2010 年将实现二氧化硫排放量控制在 75.5 万吨的目标。中电投集团提出 2010 年减排二氧化硫 9.96 万吨的目标。五大发电集团二氧化硫减排量占到了发电企业总减排量的 50% 以上。我国尚有 40% 左右的火电机组未装备脱硫设施，要进一步加大脱硫改造力度，同时积极推广应用脱硝、脱氮、二氧化碳捕捉封存技术，推广合同能源管理模式，集中利用综合节能技术，大幅提高发电系统运行效率，同时大力发展新型清洁煤发电技术。

20.5.3　山西钢铁行业两年内烧结机烟气脱硫

目前，山西省有规模以上烧结机 75 台，钢铁行业烧结工序排放的二氧化硫占钢铁企业排放总量的七成以上。作为山西省"十二五"期间二氧化硫减排的重点，钢铁企业必须同步配套建设烧结机烟气脱硫装置。2011 年全省 11 个市烧结机总面积的 60% 要完成烟气脱硫建设。2012 年，山西省钢铁行业所有运行烧结机烟气全部实现脱硫，吨钢二氧化硫排放量不超过 1.8Kg。到 2013 年底，山西省将逐步淘汰钢铁企业 90 平方米以下的的烧结机，剩余烧结机必须分期分批完成脱硫治理。据测算，由此山西省可减排二氧化硫 20 万吨，相当于该省年排放量的约 1/6。

山西省环保厅表示，要将钢铁行业烧结机脱硫设施工程纳入对各地政府节能减排目标责任制的考核范围，实施严格的奖惩制度。对未在限期内完成治理任务的钢铁企业，将依法进行关停，并停止其二氧化硫排放建设项目环评文件及新增排放总量指标的审批，取消对该企业的政策资金支持；对脱硝设施通过环保部门验收并正常运行的燃煤机组，山西省物价局将积极争取国家脱硝电价政策支持。对进展缓慢或未能按期完成任务的燃煤电厂，环保部门将给予处罚，直至报省政府决定关停。

21　污水处理

我国是一个贫水国家，水资源总量2.8万亿立方米，居世界第六位，人均水资源占有量不足2400立方米，仅为世界人均的1/4，列世界第88位。我国国民生产总值虽为美国的1/14.1和日本的1/5.6，但实际取水量已达5000亿立方米，超过了美国且为日本的5.6倍。且近年来美、日两国的取水量很少增长，甚至有所下降，而我国仍保持持续增长趋势，并导致废水排放量的急剧增长，引起水环境污染日益加剧，造成优质水资源锐减，长此发展下去前景堪忧。

造成我国水环境持续恶化的原因有两点：一是用水效率低。在工业方面，单位产品用水定额往往比发达国家高几十倍以上。二是废水处理率低。一方面城市缺水十分严重；另一方面大量的城市污水白白流失，既浪费了资源，又污染了环境。

21.1　污水源

按照污水来源分类，污水处理一般分为生产污水处理和生活污水处理。生产污水包括工业污水、农业污水以及医疗污水等；生活污水就是日常生活产生的污水，是指各种形式的无机物和有机物的复杂混合物，包括：①漂浮和悬浮的大小固体颗粒；②胶状和凝胶状扩散物；③纯溶液。

按照污水的性质来分，水的污染有两类：一类是自然污染；另一类是人为污染。当前对水体危害较大的是人为污染。

水污染可根据污染杂质的不同而主要分为化学性污染、物理性污染和生物性污染三大类。污染物主要有：

- 未经处理而排放的工业废水；
- 未经处理而排放的生活污水；
- 大量使用化肥、农药、除草剂的农田污水；
- 堆放在河边的工业废弃物和生活垃圾；
- 水土流失；
- 矿山污水。

21.1.1　城市污水

城市污水按来源可分为生活污水、工业废水和径流污水。生活污水主要来自家庭、机关、商业和城市公用设施。其中主要是粪便和洗涤污水，集中排入城市下水道管网系统，输送至污水处理厂进行处理后排放。其水量水质明显具有昼夜周期性和季节周期变化的特点。

工业废水在城市污水中的比重，因城市工业生产规模和水平而不同，可从百分之几到百分之几十。其中往往含有腐蚀性、有毒、有害、难以生物降解的污染物。因此，工业废水必须进行处理，达到一定标准后方能排入生活污水系统。生活污水和工业废水的水量以及两者的比例决定着城市污水处理的方法、技术和处理程度。

城市径流污水是雨雪淋洗城市大气污染物和冲洗建筑物、地面、废渣、垃圾而形成的。这种污水具有季节变化和成分复杂的特点，在降雨初期所含污染物甚至会高出生活污水多倍。

21.1.2　农村污水

农村生活污水是农村居民在生活和家庭养殖等过程中产生的污水的总称。农村生活污水包括厨房炊事废水（淘米洗菜用水、泔水等）、洗衣、洗浴和厕所污水、家养畜禽粪便、废水等。与城镇生活污水相比，农村生活污水无论是排放点分布还是排放量大小，无论是污染物浓度还是资源化需求都具有明显特征。

生活污水污染是当前农村主要环境问题之一。据测算，全国农村每年产生生活污水 80 多亿吨。农村生活污水污染不仅是农村水源地潜在的安全隐患，而且会引发农村水质性缺水，对农业生产和农产品品质构成不利影响，进而直接影响广大村民的生存环境与身体健康。重视农村地区生活污水污染控制，加强农村生活污水的收集、处理和资源化利用，积极探索并全面实施适应农村实际的污水处理方案是当前仍至整个“十二五”期间农村环境保护的重点和难点。

21.2　污水的危害

21.2.1　城市污水危害

城市污水中 90% 以上是水，其余是固体物质。水中普遍含有以下各种污染物：

（1）悬浮物。一般为 200 毫克/升—500 毫克/升，有时候可超过 1000 毫克/升。其中，无机和胶体颗粒容易吸附有机毒物、重金属、农药、病原菌等，形成危害大的复合污染物。悬浮物可经过混凝、沉淀、过滤等方法与水分离，形成污泥而去除。

（2）病原体。包括病菌、寄生虫、病毒三类。常见的病菌是肠道传染病菌，每升污水可达几百万个，可传播霍乱、伤寒、肠胃炎、婴儿腹泻、痢疾等疾病。常见的寄生虫有阿米巴、麦地那丝虫、蛔虫、鞭虫、血吸虫、肝吸虫等，可造成各种寄生虫病。病毒种类很多，仅人粪尿中就有百余种，常见的是肠道病毒、腺病毒、呼吸道病毒、传染性肝炎病毒等。每升生活污水中病毒可达 50 万个—7000 万个。

（3）需氧有机物。包括碳水化合物、蛋白质、油脂、氨基酸、脂肪酸、酯类等。其浓度常用五日生化需氧量（BOD5）来表示。也可用总需氧量（TOD）、总有机碳（TOC）、化学需氧量（COD）等指标结合起来评价。常用 BOD5/COD 指标来反映污水的可生化降解性，用微生物呼吸氧量随时间变化曲线来反映生化降解的快慢，据此选择处理方案。城市污水 BOD5 一般为每升 300 毫克—500 毫克，造纸、食品、纤维等工业废水可高达每升数千毫克。

（4）植物营养素。生活污水、食品工业废水、城市地面径流污水中都含有植物的营养物质——氮和磷。城市污水中磷的含量原先每人每年不到 1 千克，近年来由于大量使用含磷洗涤剂，含量显著增加。来自洗涤剂的磷占生活污水中磷含量的 30%—75%，占地面径流污水中磷含量的 17% 左右。氮素的主要来源是食品、化肥、焦化等工业的废水，以及城市地面径流和粪便。硝酸盐、亚硝酸盐、铵盐、磷酸盐和一些有机磷化合物都是植物营养素，能造成地面水体富营养化、海水赤潮和地下肥水。硝酸盐含量过高的饮水有一定的毒性，能在肠胃中还原成亚硝酸盐而引起肠原性青紫症。亚硝酸盐在人体内与仲胺合成亚硝胺类物质可能有致畸作用、致癌作用。

城市污水中除含以上四类普遍存在的污染物外，随污染源的不同还可能含有多种无机污染物和有机污染物，如氟、砷、重金属、酚、氰、有机氯农药、多氯联苯、多环芳烃等。如果城市污水不经处理就排入地面水体，会使河流、湖泊受到污染。但城市污水水量非常大，例如，全部进行污水二级处理，投资极大。因此，结合具体情况研究经济有效的处理措施，是环境保护的重大课题之一。

21.2.2 农村污水危害

过去，我们一直把环保工作重点放在大中城市，而忽视了占全国总面积近90%的广大农村，从而致使农村环境问题日益恶化，其中水污染环境问题尤为突出，呈现出迅速恶化趋势。例如，地表水体富营养化，地下水质逐渐恶化等。据统计：目前，全国有80%以上的河流受到不同程度的污染，我国农村有近7亿人的饮用水中大肠杆菌超标，1.7亿人的饮用水受到有机污染，而且，由于农药等化学物质的广泛使用，致使许多地方的地下水已经不适于饮用，严重影响了人民群众的身体健康和农村经济的健康发展。

我国人多地少土地资源的开发已接近极限，近年来随着科技的进步我国农业生产方式也发生了重大变化，往昔的农家肥等有机肥料被农药、化肥的广泛使用所取代。因此，化肥和农药是主要的农业污染源。由于化肥、农药使用量大的蔬菜生产发展迅猛，使得我国已成为世界上使用化肥、农药数量最大的国家。但是，在化肥施用中还存在各种肥之间结构不合理等现象，化肥利用率低、流失率高，不仅导致农田土壤板结，耕地质量差，土壤和肥料养分易流失，还通过农田径流造成了地表水、地下水的污染、湖泊富营养化。

随着镇、村工业及工业园区的建设和发展，一些污染严重的冶金、建材、化工工业等污染严重的企业落户农村，大量污水直接排入，给农村河流造成严重污染。这些企业中相当一部分属于效益较差、能耗较大、环境污染严重的企业，并且企业技术含量低，尤以造纸、煤炭、非金属矿制品、化工工业为主。其中，造纸业的废水排放量占总排放量的44.9%，其次，乡镇企业领导和职工的环境意识淡薄，整体环境管理水平落后，工业布局不合理，使其环境呈现出脏、乱、差的局面，流经乡镇企业区域的河流水质往往较差。工业污水灌溉农田，可使土壤受到污染，土壤中重金属逐渐积累，致使粮食、蔬菜、水果品质下降，所含重金属超标。含酸、碱、盐的污水灌溉往往使作物减产或枯死，或者使土壤盐碱化。

生活污水是人们日常生活中产生的各种污水混合物，例如各种洗涤水和人畜粪便等是农村水体污染的主要污染源之一。由于农村无地下排水管道，生活污水处理系统建设滞后，大部分生活污水都直接进入河流、湖泊，直接造成水体污染。一是洗涤污染。新农村建设初期，由于水源利用方式比较原始，大都在河塘中洗涤，污垢直接溶解在水体之中，造成水体中氮、硫、磷的含量较高，在厌氧微生物的作用下，易产生硫化氢、硫酸等具有恶臭气味的物质。村庄周边池塘及小沟经常出现水体混浊，呈黄绿色以及黑色的现象。二是人畜粪便污染。农村由于缺乏完善的排水系统及对粪便废水的处理设施，都带来了人粪尿对大气、地表水、地下水的污染，以及粪便中的病原菌造成流行病的传播。通过调查发现，集镇、村庄等人口集中的河段，生活污水对水体污染影响十分严重。水中病原微生物对人体健康形成的危害历史长久，受病原微生物污染的水体中会有大量的致病菌、病毒等，可引起伤寒、痢疾、霍乱和腹泻等肠道疾病。大肠杆菌是肠道传染病菌，属于人类肠道中的正常菌群，但现已发现其中某些致病菌株可引起腹泻，严重者可致死。

近几年，城乡畜牧业规模发展迅速，城镇郊区建立起大批养殖场，畜禽养殖户绝大多数养殖场采用圈养，由于缺乏对畜禽养殖及其污染物排放的有效管理，没有排水设施，更没有污水处理设施，粪便连同冲洗水在圈内堆积、漫流，清理出的粪便随意堆积，相关的屠宰场、孵化场往往直接将动物血、废水、粪便、蛋壳等倾倒入附近的水体，导致大量的氮、磷流失和河道的水体变黑，富营养化严重，由此带来大量畜禽粪便的排放和污染问题。粪尿中的大量氮磷渗人地下，使地下水的硝态氮、硬度和细菌总数超标，尤其到了汛期，雨水把粪便从高处带到低处，直接污染地表水。

21.3 污水处理典型技术

21.3.1 城市污水处理技术

1. 生物接触氧化法

生物接触氧化法的技术特点与工艺生物接触氧化法是一种介于活性污泥法和生物膜法的生活污水生物处理技术，兼备两者的优点。其主要构筑物为生物接触氧化池，生活污水与生物膜广泛接触。在微生物的作用下，有机污染物得到去除，生活污水得到净化。由于池内具备适于微生物栖息增殖的良好条件，因此，生物膜上物相相当丰富。食物链长，微生物浓度高，活性强，不产生污泥膨胀，污泥生成量少，且易于沉淀。生物接触氧化法具有多种净化功能，除有效去除有机物外，如运行得当，还能够脱氧和除磷，因此生物接触氧化法是住宅小区生活污水处理较早采用的技术之一。对于中小水量，生物接触氧化法是目前用得最多、最普通的方法。SBR 法具有流程简单、投资少、操作简便等多种优点。

2. 活性污泥法的技术特点与工艺

（1）AB 法。该工艺对曝气池按高、低负荷分为二级供氧。A 级负荷高，曝气时间短，产生污泥量大，其负荷较高，通常为普通活性污泥法的 50—100 倍；B 级负荷低，污泥龄较长。A 级和 B 级亦可分期建设，A 级与 B 级间设中间沉淀池。两级池子的 F/M（污染物量与微生物量之比）不同，形成不同的微生物群体。AB 法尽管有节能的优点，但不适合低浓度水质。

（2）SBR 法。此法进水、曝气、沉淀、出水在同一座池子中完成，常由 3—4 个池子构成一组，轮流运转，一池一池地间歇运行，故称序批式活性污泥法。这种一体化工艺的特点是工艺简单，由于只有一个反应池，不需二沉池、回流污泥及相关的设备，一般情况下不设调节池，多数情况下可省去初沉池，故节省了占地和投资，耐冲击负荷且运行方式灵活，可以从时间上安排曝气、缺氧和厌氧的不同状态，实现除磷脱氮的目的。

（3）A/A/O 法。城市污水处理对出水有去除氮和磷的要求，此种由厌氧—缺氧—好氧组成的工艺，利用了生物处理法进行脱氮除磷，可获得优质的出水，是一种深度二级处理工艺。A/A/O 法的可同步除磷脱氮机制由两部分组成。一是除磷：生活污水中的磷在厌氧状态下（DO <

0.3mg/L)，释放出聚磷菌，在好氧状况下又将其更多地吸收，以剩余污泥的形式排出系统；二是脱氮：缺氧阶段要控制 DO < 0.5mg/L，由于兼氧脱氮菌的作用，利用水中 BOD 作为氢供给体（有机碳源），将来自好氧池混合液中的硝酸盐及亚硝酸盐还原成氮气逸入大气，达到脱氮的目的。

（4）普通曝气法。普通曝气法出现得最早，其实际处理效果好，可处理大的生活污水量，对于大工厂可集中建设污泥消化池，所产生的沼气可作能源利用传统普曝法的不足之处是只能作为常规二级处理，不具备脱氮除磷功能。近几年，在工程实践中，通过降低普通曝气池的容积负荷，可以达到脱氮的目的；在普通曝气池前设置厌氧区，可以除磷，亦可用化学法除磷。采用普通曝气法去除 BOD，在池型上有多种形式，例如氧化沟，工程上称为普通曝气法的变型工艺，亦可统称为普通曝气法。

3. 城市住宅小区生活污水处理技术方案

城市住宅小区生活污水就其处理技术而言，可以采用目前城市生活污水处理的成熟技术和工艺。

住宅小区生活污水流量小，可生化性好。宜优先采用生物膜法处理技术生物膜法具有生物相应丰富、微生物浓度高、食物链长不会发生污泥膨胀、污泥沉降性能好等优点。

住宅小区用地紧张时，应优先考虑占地省的生活污水处理工艺并在设计中采取一定措施，一般设计成地下式或半地下式，形成地下为生活污水处理站，地面为绿地花坛的格局，同时，美化环境设计时应注意通风要求和臭气处理等问题。

由于受小区管理人员人数和专业素质的限制，应优先选用运行维护管理较方便的工艺，并努力提高运行管理自动化程度。

住宅小区建设工程工期要求生活污水处理设备由构筑物向设备的转化，似是一种必然趋势。采用装配式生活污水处理设备安装简捷、工期短、便于维护适应管理。

随着对出水水质要求的提高，单一工艺难以满足需要。组合式生活污水处理技术和设备得以发展。目前组合式主要有多级好氧处理：厌氧 + 好氧处理、厌氧 + 缺氧处理等。从降低能耗、回收生物能方面来看，厌氧生物处理有着广阔的前景，生活污水中的有机物质本身都具有一定的潜在能量。厌氧处理时，一方面勿需曝气充氧，可降低能耗；另一方面，其生物沼气可回收利用。供小区生活采暖和供热，形成小区生态平衡系统。

4. 生活污水处理技术与工艺的创新应用

（1）膜分离技术。用膜分离代替沉淀进行泥水分离，可带来活性污泥工艺的以下变化：一是不再存在污泥膨胀问题。在调控活性污泥系统时，不必再考虑污泥的沉降眭能，从而使工艺控制大大简化；二是曝气池的污泥浓度将大大提高，从而使系统可在超大泥龄、超低负荷状态下运行，充分满足去除各种污染物质的需要；三是在同样的处理要求下，可使曝气池容积大大减小，节省了处理厂的占地面积；四是污泥浓度的提高，要求较高的曝气速率，因而纯氧曝气将随着膜的分离而被大量采用。

（2）分子生物技术。目前，分子生物技术已开始应用于生活污水处理领域。为探索聚磷菌除磷的生化机理，已开始用分子诊断技术获取聚磷菌的遗传信息。在活性污泥中发现 30 多种丝状菌，其中只有四种被准确命名及生物分类学定位，因为这些丝状菌大部分无法进行分离和纯培养。目前，用分子诊断技术将进行这些丝状菌进行生物学定位，以进一步准确了解其特性。

随着分子诊断技术的大量应用和活性污泥微生物基因库的建立，在此基础上，可采用基因技术培育出具有高效活性的污泥菌种，进一步提高处理效果将是未来的发展方向。

21.3.2 农村污水处理技术

在新农村建设中，生活污水处理应根据各地自身特点找到合适的处理方法。我国现有污水处理工艺中80%以上采用的是活性污泥法。AB法、氧化沟工艺、A/O工艺、A2/O工艺、SBR等方法在我国城市污水处理厂也已经得到了应用。污水处理工艺也由过去只注重去除有机物发展为具有脱氮除磷功能。但是受经济发展水平制约，近年来，污水处理厂因为运行成本等问题全部或部分停运现象时有发生。

1. 生态处理技术

污水生态处理技术以土地处理方法为基础，是污水土地处理系统的进一步演化和发展。它以自然净化功能为核心，强调在污染物降解过程中植物—微生物体系或动物—处理环境（如土壤）的相互关系，特别注意对外部条件的优化与调控。

污水土地处理技术是利用土壤—微生物—植物系统的自我调控机制和对污染物的综合净化功能进行污水处理的一种工艺，具有造价低、能耗低、结构简单、运行维护方便、处理效果好的优点，其净化水质可达二级处理，甚至三级处理水平。

（1）慢速渗滤技术。慢速渗滤是将污水投配到土壤的表面，污水在流经系统时通过土壤—微生物—作物对污水进行净化的处理技术，它适用于渗水性良好的土壤、砂质土壤及蒸发量小、气候润湿的地区。该技术优点：可种植经济作物、污水净化效率高和出水水质优良。其缺点是：处理效果受季节和植物营养需求的影响很大、水力负荷小和占用土地面积大。

（2）快速渗滤技术。快速渗滤是使污水在重力作用下快速流过土壤，依靠土壤微生物将被土壤截留的溶解性和悬浮有机物进行分解，使污水得以净化的工艺，是一种高效、低耗、经济的污水处理与再生方法。适用于渗透性非常良好的土壤，如砂土、砾石性砂土、砂质砂土等。该技术优点：管理方便、投资省、土地面积需求量少。其缺点：对水文地质条件的要求较其他土地处理系统更为严格；对TN（总氮，Total Nitrogen）的去除率不高，处理出水中硝态氮可能导致地下污染。

（3）地表漫流技术。地表漫流是以表面布水或低压、高压喷洒形式将污水有控制地投配到生长多年生牧草、坡度和缓、土地渗透性能低的坡面上，使污水在地表沿坡面缓慢流动过程中被植物摄取、蒸发和渗入地下而得以充分净化的污水处理技术。它适用于渗透性低的黏土或亚黏土，地面最佳坡度为2%—8%。深圳某工业区采用人工地表漫流田处理园区污水站二级处理出水，实践表明，培植美人蕉的地表漫流系统，对去除废水中的BOD5、COD和色度都有显著的效果。该技术优点：投资少、管理简单、处理出水可达到二级或高于二级处理的出水水质。其缺点：系统受气候、作物需水量、地表坡度的影响大。

（4）人工湿地技术。人工湿地是由水体、基质、水生植被和微生物构成，利用物理、化学和生物协同作用来净化污水的生态处理系统，既能净化污水，又能美化景观，具有投资少，运行、维护简单等特点。按污水流经的方式可以分为：表面流人工湿地、潜流人工湿地、垂直流人工湿地。垂直流人工湿地构造要求高，卫生条件差，容易滋生蚊蝇，但潜流人工湿地占地面积小，无臭味，卫生条件好，处理效果好并且不受季节影响，相对于表面流人工湿地的造价稍高。表面流人工湿地最接近自然湿地，它比另外两种人工湿地的造价低，但其占地面积大，处理效果差，夏季会散发出臭味，冬天会结冰，影响处理效果，不适合特别寒冷的地区。广州市南沙区红湖村建有两座人工湿地处理系统，处理全村2/3生活污水，监测结果表明：该人工湿地处理系统对COD的处理效率可以达到47%以上，其处理出水水质达到广东省《水污染物排放限值》第二时段二级标准的要求。

（5）地下渗滤技术。地下渗滤处理技术是将污水投配到具有一定构造和良好扩散性能的地下土层中，污水在经毛管浸润和土壤渗滤作用向周围和向下运动过程中达到处理、利用要求的污水处理工艺类型。上海市金山区陆家浜自然村采用此技术处理全村的生活污水，监测结果显示：该系统对COD、BOD5、NH3－N·TP的处理，都取得了立竿见影的效果，处理后的水质达到了国家一级B排放标准。该技术的优点：无损地面景观、不受外界气温影响、运行与维护简单和运行费用低、对进水负荷的变化适应性强与耐冲击、出水可回用。其缺点：受场地和土壤条件的影响较大、投资相比要高、且在生物脱氮时硝化反应易受到土壤内部还原性质的影响。

（6）稳定塘技术。稳定塘是利用水生植物（主要是藻类）以及水中的微生物共同作用用降解有机物净化污水的自然净化技术。我国于20世纪50年代就开始了对稳定塘污水处理技术的研究，建成的规模较大的稳定塘有：日处理20万立方米城市污水的齐齐哈尔稳定塘、日处理17万立方米城市污水的西安漕运河稳定塘、日处理3万立方米城市污水的山东胶州氧化塘和日处理8万立方米化工废水的湖北鸭儿湖氧化塘等。该技术优点：稳定塘系统能充分利用地形，基建投资少，运行费用低，无需污泥处理，易于管理和维护，对有机物和病原体有很好的去除效果。其缺点：处理系统占地面积大；容易出现短流现象，导致实际水力停留时间小于设计水力停留时间；在温度较高的时候容易臭气四散、滋生蚊虫；对氮磷的去除效果不稳定。

（7）蚯蚓生态滤池处理技术。蚯蚓生态滤池是在普通生物滤池基础上进行改进的一种新型生物处理技术。根据蚯蚓具有吞食有机物质、提高土壤通透性能和促进有机物的分解转化等特点而设计。江苏宜兴某镇建造了16个科研性塔式蚯蚓生态滤池示范工程，处理农户生活污水及能用明沟收集到的地表径流，经过一年多的运行，滤池出水各项指标可达到《城镇污水处理厂污染物排放标准》（GB 18918－2002）的一级B标准。该技术优点：处理效果好，出水水质优于传统二级活性污泥法；工艺流程十分简单，物耗、能耗低；剩余污泥零排放；系统繁殖的蚯蚓可作为畜禽和鱼类的高蛋白饲料，蚓粪可作为农肥或园艺养花肥料。其缺点：水力负荷低于传统的二级处理工艺；土建工程量较土地处理技术大；对温度、湿度的要求较高。

2. 生物处理技术

（1）厌氧生物技术。厌氧生物处理是在没有游离氧存在的条件下，兼性细菌与厌氧细菌降解和稳定有机物的生物处理方法。在厌氧生物处理过程中，复杂的有机化合物被转化为简单的化合物，同时释放能量。无动力、地埋分散式厌氧系统是以厌氧为主、兼性生物滤池为辅的生活污水处理系统。浙江省在2007年就将该技术在850个村进行了推广应用，共建池容为25×104立方米，处理生活污水量为5615×104立方米，受益农户达21.3万户，出水基本可达到《污水综合排放标准》（GB 8978－1996）一级标准。该技术优点：占地少、无能耗、管理简单、污泥量少。其缺点：处理能力有限氮磷去除效果不足，缺乏技术规范。厌氧生物滤池是一种淹没式的固定填料生物膜法，即采用填充材料作为微生物载体的高效厌氧反应器。北京市新农村建设试点村污水处理工程中胡家垡村采用了低能耗的厌氧生物滤池工艺，对该系统在进水污染物浓度偏低的情况下，出水水质可达到北京市《水污染物排放标准》（DB 11/307－2005）的二级标准。该技术优点：生物浓度高、抗冲击负荷能力强、启动容易、运行稳定和管理简便。其缺点：占地面积大；滤料费用贵，滤料易堵塞。生活污水净化沼气池是一种新型的农村生活污水处理技术，是在我国农村沼气池和传统的三格式化粪池技术基础上，借鉴国外及台湾等地的农村生活污水处理技术基础上发展而来的。优点：运行无能耗；可回收能源；运行稳定；维护管理简便。缺点：处理负荷低；出水水质不稳定，氮磷去除效果差。截至2005年，该系统在四川有5.28万处，水质生化指标达到《四川省环境污染物排放标准》中三类污染物排入一类水域的甲级标准。

（2）好氧生物技术。好氧生物处理是在有游离氧（分子氧）存在的条件下，好氧微生物降解有机物，使其稳定、无害化的处理方法。传统的污水处理工艺包括：A/O、A2/O、氧化沟、生物转盘、生物接触氧化和SBR法等。该技术优点：处理效果好，工艺成熟。其缺点：能耗大，管理维护复杂，适用于管网建设较完备、运行维护资金充足的地区。2006年北京市新农村建设试点村污水处理工程共涉及13个区县的78个行政村，共建设污水处理站130座，污水处理工艺包括：A/O、A2/O、生物接触氧化、生物转盘等工艺及组合。

3. 生物——生态组合技术

为得到较好的出水水质，特别是对氮、磷的去除效果，常采用生态及生物—生态组合技术以强化系统对有机物及脱氮除磷的能力。目前，“生物＋生态”组合模式在江苏农村应用较多，人工湿地是目前应用最普遍的一种后续生态处理技术。主要的组合工艺有“厌氧—跌水充氧接触氧化—人工湿地”、“厌氧—滴滤—人工湿地”、“厌氧滤池—氧化塘—生态渠”和“水解—脉冲滴滤池—人工湿地”等。浙江淳安县某村采用人工湿地/稳定塘工艺对污水进行生态处理，出水水质达到GB8978－1996的一级标准。桂双林等人开发出的“厌氧池—射流充氧生物滴滤塔—人工湿地”组合工艺，是一种比较适合湖泊流域农村分散式生活污水处理的技术，出水水质达到GB 8978－1996的一级B标准。

4. 其他技术

（1）膜生物技术。膜生物反应器（简称MBR）是将生物降解作用与膜的高效分离技术结合而成的一种新型高效的污水处理与回用工艺。目前，北京市新农村治污工程建设中MBR/CWT工艺的应用越来越普遍，且水质净化效果显著。该技术的优点：省掉二沉池、活性污泥浓度高、水力停留时间（HRT）和污泥停留时间（SRT）可以分别控制。其缺点：管理与操作要求程度高、存在膜易污染定期清洗困难、能耗较大与运行费用高（约为1.3元/立方米－2.0元/立方米）等。

（2）净化槽技术。净化槽是由日本研发的一种处理分散式污水的一体化装置，通过功能分区，使COD、氮磷等污染物分别在不同处理单元内得到削减。净化原理包括生物降解、物理沉淀、物理吸附和化学絮凝反应等。太湖地区通过引进日本生活污水处理设施——净化槽，开发适合于太湖流域的分散型生活污水治理示范技术。研究结果表明，净化槽同时具备去除有机物、氮和磷的能力。在正常运行状况下，能够达到预定的出水水质目标，即BOD≤10mg/L、TN≤10mg/L和TP≤1mg/L。该技术优点：节省动力、管理操作简便、实现了污水和污泥一体化、常年不排泥、运行期间较稳定且无需专人管理和出水水质优良。其缺点：净化槽在国内的应用的条件还不成熟，技术设计参数还需进行大量研究，配套设备及填料问题还有待解决。

（3）微生物固定化技术。微生物固定化技术是利用化学或物理方法将游离的微生物（主要是细菌和真菌）限定于一定的空间区域内，使其保持生物活性，且可反复循环使用的一种基本技术。目前，国内对微生物固定化技术的研究集中在工业废水处理上，包括含酚废水、重金属废水、含氮废水及高浓度有机废水处理等。利用从农村生活污水中筛选、驯化的高效菌 SHJ－1，以沸石为载体进行固定化，研究了固定化细胞的降解特性。实验表明：最优条件为温度为 30℃，pH 值为 8，接触时间 60 分钟，生物沸石投加量 15g/L，氨氮去除率为 99.72%。该技术优点：效率高、稳定性强、污泥产量少、能纯化并保持高效菌种，固液分离效果好。其缺点：载体价格偏高，且使用寿命短。

21.4 污水处理国际经验

21.4.1 美国

1. 美国水污染治理的法律控制机制

美国联邦政府对全国水域进行规范开始于 19 世纪，其目标是改善水路运输与促进贸易往来，并于 1899 年制定了《河流与港口占有法》。作为通航水路控制污染的一种手段，该法被认为是美国最早的有关水污染控制方面的立法。

自 1940 年以来，一切发生了变化，随着工业、农业的现代化和城市的扩大，美国的河流、湖泊、海湾和地下水等水体受到严重的污染。主要有三个方面的因素加剧了这种污染形势：第一，都市化与工业化的发展空前高涨；第二，在 1940—1947 年间实际上没有修建任何市政或工业废水处理工厂；第三，巨大的技术进步带来了一系列新型的污染，例如合成化学品与放射性废物。

伴随着水污染形势的加剧，20 世纪中叶，美国联邦政府的水管理目标从专注于保护水体的适航性转向避免环境退化的关怀。现代的水污染控制立法是 1948 年的《联邦水污染控制法》。它是现行《水污染控制法》的最初形式，是资助州和地方政府解决水污染问题的第一个法令。它把联邦在水质管理方面置于一个很次要的位置，联邦的主要责任是对地方的水污染控制规划给予技术上的服务和财力上的支持，它明确承认水污染控制主要是州和地方政府的责任。由于该法把消除污染仅仅看成是一个地方性问题，不能很好地控制污染形势的发展，于是国会先后于 1956 年、1961 年、1965 年、1966 年对其进行了几次重大修改。

进入 70 年代，美国当时面临酸雨性矿物的排放、核电站的污染、工程队的疏浚物、船舶倒出的垃圾以及公共工程和石油的污染，特别是油轮事故造成严重污染，尼克松签署了《水质改善法》。该法是因为两起石油泄漏事件而做出的修改，即 1967 年位于英国海岸的托雷凯伦号（the Torrey Canyon）失事导致的油污泄漏和 1969 年在加利福尼亚州圣·芭芭拉海峡附近海域一座海上石油钻井平台发生井喷而引起的油污泄漏事件。该法并没有改变水污染控制法的基本结构，针对油污泄漏，建立了严格责任制度。

1969 年圣·芭芭拉海峡石油污染事故把环境问题带入人们的视线，环境问题终于成为人们普遍关注的焦点和公众舆论的中心话题；1970 年 4 月 22 日的首次地球日纪念活动把这种关注推向了高潮，从而也影响了美国国会在 1972 年对《联邦水污染控制法》的修订，即《清洁水法》（Clean Water Act），它是所有联邦水法中最严格的一部法律。随后，国会分别于 1977 年、1981

年、1987年、2002年，2005年就联邦水污染控制法进行了重新修订。如上所述，美国的水污染控制立法在其近60年的历史发展过程中，既取得了辉煌的成绩，又收获了成功的经验与失败的教训。

美国水污染控制的成功得益于成功的法律控制机制：

第一，命令控制机制。美国是命令控制型的环境管理模式的发源地。这种模式所推崇的强制性的行政管理逐渐发展成为世界各国水污染防治和水环境管理的主要方式。该模式在美国主要表现为由联邦机构制定水污染控制的基本政策和污水排放标准，由各州实施的强制性管理制度。这种管理制度以基于污染控制技术的排放标准管理为主，以水质标准管理为补充，以总量控制和排污许可证为主要内容。《联邦水污染控制法》半个多世纪的变迁可以清晰地体现出命令控制机制的形成和发展。

第二，利益协调机制。资源分配本身就是一种利益的分配。在美国的水污染控制历史上，没有联邦政府的财政支持，州和地方政府不大可能实行充分的污染控制。在美国，绝大多数的公有处理厂使用二级处理仅仅是因为联邦政府为新处理设施的建设承担最大份额的费用，而地方政府对暴雨排放的处理并不积极，这是因为它们要自己承担全部费用。很难想象没有联邦政府在财政上的大力支持，美国的水污染控制现在会是一个什么状况。

第三，经济刺激机制。在上世纪70年代，为了使环境法不以公共利益为名过多地损害市场主体的根本利益，调动市场主体参与水污染控制工作的积极性，运用市场手段、遵循市场规律，使公共利益和个体利益能达到双赢，美国在其环境法和环境管理活动中较好地运用了经济学的理论和思想，在环境法中确立了水污染控制的经济刺激机制。

经济刺激措施一般有肯定的刺激措施，如环境补贴和课税扣除，以及否定的刺激措施，如环境税收和环境税费。环境补贴和课税扣除一般由政府对治理污染的企业给予无偿财政援助或扣除其上缴的税收，而环境税与环境费是由工业企业或其他排污单位向国家上缴一定的税和一定的资源费。

另外，低息或优惠贷款与价格机制也是政府的选择。为弥补企业兴建防污设施资金的不足，美国政府往往提供比市场利率低的贷款以及通过使企业将外部不经济行为的污染防治，视为商业活动的一部分，将污染防治边际成本内化为必要的生产成本的办法来刺激企业的水污染控制。

排污权交易或许可证交易也常常被用来减少水污染。例如，影响布什政府环境决策的保守派就极力主张通过许可证交易来控制水污染，他们认为许可证交易有如下优势：一是许可证交易能够减少获得环境目标的总成本；二是许可证交易能够减少水污染控制规则的成本，并以此为基础制定更严厉的环境规则；三是许可证交易能够鼓励新的环境技术的开发，并能带来更低成本；四是许可证交易能够给非赢利的环境组织提供从事许可证买卖的机会，也因此能直接导致环境的改善。因此，它不但能够消除企业之间过高的污染处理成本，充分利用低成本的污染处理，而且还提供了在时间上的合理调节以降低污染处理成本。另外，该交易政策规定：企业自己减少的排污量如果不出售，可以存在指定的信托机构里，以备日后再用，或者以后再出售给其他企业，这就给了企业充分的回旋余地。企业利用设备更新或技术改造的机会减少污染物的排放，可以节约经费。如果没有排污权的信托贮存，企业可能没有积极性来利用设备更新的机会增加一些投资以减少排污量。实践证明，美国政府采取的各种经济刺激手段既激发了企业的积极性，同时，也很好地控制了水污染形势，真正实现了双赢。

第四，公众参与机制。公众参与是符合环境管理特点的富有成效的制度，在美国环境法理论与实践中具有重要的地位和作用。公众是推动美国的环境保护、环保立法最为关键的因素。在《联邦水污染控制法》开篇的国家目标和政策宣言中也明确规定：各州及联邦环保局长应当明确规定、鼓励、资助公众参与联邦环保局或任何州政府依照本法建立的项目、计划、排污限制、标准、规定的制定、修改和执行。该条明确了公众参与机制在水污染控制中的地位，明确了国家对公众参与水污染控制的积极态度，为公众参与机制发挥作用奠定了法律基础。

公众参与机制是命令控制、利益协调机制和经济刺激机制的有效补充，公众是自己利益的最好评价者，当他们的利益受损，他们会主动地维护自己的利益，具有积极性和参与的广泛性。而且，公众参与机制还有利于提高民众的环境保护意识，有利于保证政府环境保护部门的廉洁性。因此，公共参与机制在美国的水污染控制中取得了巨大的成功。

第五，法律执行机制。有效的法律执行机制是法律控制能够成功的关键。在20世纪六七十年代以前的美国在水污染控制方面的效果很小，主要是因为没有一个有效的执行机制。这种情况在1972年修订案中有了改善，该法案规定了行政、刑事与民事方面的惩罚措施。更重要的是，1972年立法授权环境组织在环保署没有履行责任的时候可以执行法案。1987年修订案规定行政机关可以强制执行民事罚款使行政执行措施更有效。

2. 美国水污染治理的资金运作机制

除了完善的法律法规作为保障，兴建城市生活污水处理厂以及相关的环境基础设施，无论对于发达国家还是发展中国家所面临的首要问题都会是项目的资金问题。在缺少资金或者没有有效的资金运作机制时，这些重要的环保项目都无法进行。

美国在1987年开始实施《清洁水法》（Clean Water Act）。该法律授权联邦政府为各州设立一个滚动基金来资助它们实施污水处理以及相关的环保项目。经过多年的努力，到目前为止，各州全都已经有了比较完善的滚动基金计划。

在滚动基金中，资金来自联邦政府和州政府。这些资金作为低息或者无息贷款提供给那些重要的污水处理以及相关的环保项目。贷款的偿还期一般不超过20年。所偿还的贷款以及利息再次进入滚动基金用于支持新的项目。

因此，滚动基金计划是被美国环保局所认定的最成功的一项环保项目资助计划。联邦政府的投入在目前的全部250亿美元的基金中占171亿，各州的投入共占36亿。各州依靠滚动基金发行的债券达88亿美元。另外，统计数字还显示1993年的偿还贷款加利息仅为5亿美元。而到2000年，偿还贷款加利息已经接近22亿美元。仅仅对于污水处理厂这类项目的支持，自从1988年以来其资金已经达到289亿美元。其中，对于二级处理项目的资助为132亿，三级处理项目为157亿。这些不仅说明滚动基金的运作良好和收益显著，而且正是由于这些项目的完成，美国全国水环境状况才得到了根本的改善。

3. 纽约污水治理的实践经验

美国纽约市现有14家污水处理厂，污水管线长达6000英里，5万个污水垃圾截流池，93个污水泵站，494个合法污水放流口，日均处理工业和生活污水14亿加仑，并设有37个水质监测站。长期以来，纽约市政府十分重视城市污水处理，采取了一系列措施改善城市水质，在创建绿色城市方面发挥了重要作用。

第一，企业排污前须对污水进行预先处理。纽约市规定，企业排放污水前，必须对污水预作处理，清除其中有毒物质。经过长年努力，工业排污重金属含量大为减少，已从1987年的日均2000磅降至目前的37磅。如此显著成效，与纽约市采取的两项措施密不可分：一是污染物追查机制。如果在港口水域发现多氯联苯、芳烃碳氢化合物、汞和其他有机化学成分，纽约市环保局即会追查污染源头，帮助问题企业安装或改进污水预处理设备。二是四氯乙烯减排规划。为了控制大量四氯乙烯排入城市污水管网，纽约市制定了特别排污规则，向干洗店之类的重点监管企业发放排污许可证，明确标明排污指标，并定期检查污水预处理设备的运行情况，一旦发现问题，即发出违规通知书或局长命令，要求当事单位限期整改。

第二，化废为宝，变污泥为沃土。污泥是污水处理过程中的自然产物，纽约市日产下水道污泥1200吨。纽约市将污泥加工成一种富含养分和有机成分的、可回收再利用的生物固体，并作为肥料和土壤改良剂用于公园、田地、草坪、高尔夫球场等地方。其主要做法如下：一是建造8个污泥脱水设备；二是与3家专业化企业签订五年合同，年投入1亿美元，由这些企业承包将下水道污泥转化成生物固体，并加工成土地增肥剂。

2011年，美国纽约市计划将污水处理厂，以及大量的淤泥、甲烷与废水“副产品”作为一项资产——潜在的可再生能源。纽约居民每天会产生13亿加仑污水，污水处理厂处理后会残余1200吨淤泥，这些废物最后送往垃圾填埋场填埋。目前环保局正在为这些“垃圾”寻找买家以便发挥新的效用（1加仑约合3.79公升）。最大的潜在能源是污水处理厂脱硫罐里的甲烷气体。该市污水处理厂产生甲烷气的一半已经用于提供14家污水处理厂所需20%的能量，而这14家厂每年的电费高达5000万美元。现在纽约市想将剩余的另一半甲烷推入市场，避免浪费。目前该市已经与美国国家电网建立合作，纽约布鲁克林区的Newtown Creek废水处理场明年有望给该市的天然气网络增加甲烷气体，为2500户家庭供热。

第三，重点监管油脂排入下水道。油脂排入下水道是引起污水管网阻塞的重要原因。纽约市立法规定，纽约市餐饮业等产生油脂的企业必须配备、使用型号合适的油脂截留器，并进行设备保养清洗。为了确保监管成效，纽约市重点抓四个方面的工作：一是对餐饮业进行定期检查，确保油脂截留器设备到位，型号合适，运转正常，截油有效；二是组织宣传教育活动，配发文字材料，对全市餐馆、洗衣店、水果蔬菜摊、养老院等进行油脂排污教育；三是教育普通市民不要向厨房和厕所倾倒油脂，不要用开水或洗涤剂清洗油脂，否则，油脂进入下水道冷却积累后同样会引起管道阻塞；四是对违法者予以惩处，每天、每次最高罚款1万美元，有时当事人还必须到庭受罚。

第四，研发合流制污水溢流控制技术。与其他老城市一样，纽约市采用的是合流制污水管网，生活污水和雨水共用一条排水沟管。每当雨季来临时，由于污水处理厂不胜负荷，一些排污管道水满为患，常常出现污水溢流现象，并将城市垃圾带入水中，增加了河水中的细菌和有机物，危及水中动物和植物的生存。为此，纽约市环保局投入18亿美元采取了一系列创新解决方案，改善污水溢流现象。一是在受污水溢流影响严重的海湾和支流建造大蓄水箱，储蓄雨水；二是在没有排污管的地方单独安装排污管；三是改进污水处理厂处污能力，雨季对排污系统采取截流措施，控制污染。此外，纽约市还重点监管旱季将原污水排入水域的现象，通过新设污水管，严厉惩治将排污口接入雨水管的非法行为，彻底杜绝旱季非法排污现象。

第五，倡导节约用水，缓解污水处理压力。据统计，纽约市人均每天用水100加仑，是大多数欧洲城市和美国其他许多城市的两倍。在一个拥有900万人口的国际大都市，市民生活用水越多，废水处理厂压力就越大。为此，纽约市从1989年开始发起了一项综合性节水计划：一是安装电子漏水检测设备；二是限用自来水浇灌草坪和冲洗人行道；三是给消防栓上锁，防止人们自行取水；四是改变水费每年只计费，不计量的做法，新增63万个家用水表；五是对市民和中小学生进行节水教育。此外，纽约市在收取水费的同时，还要收取排污费，计费标准目前为用水量的159%。以上措施成效显著，过去10年，纽约市日均减少用水约2亿加仑。

第六，纽约市环保局加强监测不放松。一是长年不断监测水情。每年的5—9月每周采样一次，10月至次年4月每月采样一次，检测内容包括细菌、混浊度、温度和溶解氧水平等多种关键指标；环保局据此评估工作成效并监测水质趋势；二是公布违规排污机构名单；三是根据联邦和地方法规，对过去12个月违反排污规定的机构名单予以曝光；四是处罚严厉。美国《联邦水污染控制法》规定，凡是违法将污水排入江河者，每天、每次罚款可高达15万美元。据统计，2002年，整个纽约地区处罚违规排污2688起，罚款总额达到1070万美元。

纽约市环保局港口年度调查报告显示，纽约市污水处理工作卓有成效，水质不断改善，水中溶解氧含量增加，甚至超过了20世纪初第一次记录的水平，同时，粪生大肠杆菌群明显减少。1974年以前，纽约市大片水域被定级为最差等级，“仅供鱼类生存”，能够游泳的水域屈指可数。1985年，这种局面得到了很大改善，“仅供鱼类存活”的最差水域面积急剧下降，到1988年仅剩下两处，而适合游泳的水域越来越多。2001年以后，情况彻底改观，纽约市再没有发生过海滩因水污染而关闭的事件。

21.4.2 日本（以福冈为例）

福冈市现有人口约140万人，每日自来水用量50万吨—60万吨，最高日可达到70万吨。福冈市污水处理设施作为市政公用设施，由政府负责规划、建设和管理。福冈市下水道局是污水处理行业的行政主管部门，下设管理部、建设部、总务部、河川部四个业务管理部门，主要负责雨水、污水收集系统和处理设施的建设维护和运行管理等职能。下水管道和污水处理厂的建设费用主要是靠政府投资和银行贷款来实现的。

1. 排水管网情况

日本原有的排水系统是雨、污合流的排水体制，经过近20年的完善和改建，目前排水系统已基本实现雨、污分流。福冈市的下水道普及率达到98.8%，已经在130多万市民的生活中普及了下水道。

（1）污水的收集处理。排放的污水有两种收集处理方式：一是部分污水由单独的污水管道收集后，输送到污水处理厂进行集中处理；二是部分大型建筑物内排放的污水就近经深度处理后直接回用。

（2）雨水的收集处理。降雨期间，大部分雨水通过路面的透水沥青直接渗入地下或由道路两侧的雨水边沟收集后渗入地下，作为地下水的补充水源，过剩雨水通过雨水管道排至受纳水体。

2. 污水处理厂情况

福冈市污水处理厂的运行经费主要靠收缴的下水道使用费维持。在福冈市，排放1吨污水需要交纳200日元（折合人民币13元）的下水道使用费，污水处理厂90%的运行经费来自下水道使用费，不足部分由财政收入填补。福冈市共有5座城市集中污水处理厂，设计处理能力90万吨/天，实际处理量67万吨/天，污水处理率达到98%以上。

3. 污泥及臭气处理情况

全日本共有2000余座污水处理厂，处理后的污泥利用率可达60%以上。福冈市的污泥处理主要采用焚烧和堆肥的处理方式，处理后的污泥97%用于制作肥料和建筑材料。

污水处理厂的处理构筑物全部加盖封闭，产生的臭气通过管道收集后压入土壤，利用土壤中的微生物处理除臭后排入大气。

4. 再生水利用

为了创建循环型都市和节水型都市，福冈市政府相继出台了《福冈市节水推进条例》和《福冈市再生水利用下水道事业相关条例》。条例规定：建筑面积在5000平方米以上的新建大型建筑物必须配套再生水回用设施；在再生水供给范围内建筑面积达到3000平方米的大型建筑物必须使用再生水。福冈市目前再生水设计供给能力为9600立方米/天，实际使用量为8800立方米/天，再生水的供给范围达到1165平方公里，再生水主要用于冲厕、公园和道路的冲洗以及树木的绿化等方面，产生的污水经过处理后形成再生水继续使用，从而达到节约水源，循环利用的目的。

21.5　中国污水处理现状及展望

21.5.1　我国污水排放现状

2010年，全国废水排放总量为617.3亿吨，比上年增加4.7%；化学需氧量排放量为1238.1万吨，比上年下降3.1%；氨氮排放量为120.3万吨，比上年下降1.9%。

1996年后，我国污水处理行业进入快速增长期，到2006年中国建成污水处理厂为815座，年增长率约为11%（2001—2006），2006年中国污水处理能力为242.2亿立方米，年增长率约为14%（2001—2006），2006年污水处理率为55.7%。我国污水处理能力具体数据如表21－1所示。

表21－1　我国污水处理能力

Table 21－1　China's Capacity in Sewage Treatment

年份	污水处理厂数量（座）	污水年处理能力（亿 m^3）	污水年排放量（亿 m^3）
1978	37	2.3	149.4
1980	35	2.6	195.1
1985	51	5.6	231.8
1990	80	10.1	293.9
1995	141	26.1	350.3
2000	427	78.8	331.8
2001	452	113.4	328.6
2002	537	130.6	337.6
2003	612	155.3	349.2
2004	708	179.3	356.5
2005	792	209.0	359.5
2006	815	242.2	362.5
2011			384.5

资料来源：国家环保总局网站

截至2010年底，中国已建成投运城镇污水处理厂2832座，处理能力1.25亿立方米/日，分别比2005年增加了210%和108%。全国城市污水处理率达到77.4%，比2005年提高25个百分点。

国家统计局2011年发布的《2010年国民经济和社会发展统计公报》，公报显示，在监测的331个城市中，年末城市污水处理厂日处理能力达10262万立方米，比上年末增长13.4%；城市污水处理率达到76.9%，提高1.6个百分点。

21.5.2 “十一五”期间我国污水处理现状

“十一五”时期我国城镇污水处理处于快速发展阶段，污水处理量实现了大幅度增加。2008年较2000年，我国污水处理量增加了140%，由此，中国也成为世界上唯一实现10年内污水处理量翻番的国家。其次，我国污水处理厂建设也正突飞猛进。截至2009年10月，我国污水处理厂数量已达1817座，总处理能力为9958万立方米/日，分别是2000年的3.8倍和4.5倍。另外，城镇污水处理率也实现了突破，2008年城镇污水处理率已达65%，较2000年翻了一番。此外，在管网建设、污水处理厂处理量、设施利用效率、污染物削减等方面也均取得了较大进展。到2009年年底，我国城镇污水处理厂能力已超过1亿立方米/日，相当于2005年的1.7倍；目前我国正在建设的1986个污水处理项目的污水处理能力达5530万立方米/日。我国已建和在建污水处理项目的总处理能力将接近1.6亿立方米/日，这将与美国整个国家的污水处理能力相当。我国污水处理率目前已完成“十一五”规划目标，污水处理新增COD削减能力大大增加，已超过了“十一五”规划预计。

21.5.3 “十二五”期间我国污水处理目标

我国城镇污水处理行业在经历快速发展的同时也面临着发展不平衡、部分设施能力闲置、污染物削减效率亟待提高、污泥二次污染问题突出等问题。我国城镇污水处理的未来仍旧任重道远。在2010年和“十二五”期间，国家将进一步加大对城镇污水处理的支持力度。对于技术工作者和工程实践者来说，应在“十二五”期间注重将污水集中与分散处理相结合；污染企业零排放与强化雨污分流收集相结合；雨水收集与再生水就地循环利用相结合；传统工艺优化与技术创新相结合，力争于“十二五”期间实现：城市和县城基本实现污水处理的全覆盖；重点建制镇全面推进污水处理设施建设；管网配套和污泥无害化处置进一步加强；再生水利用和城市水系修复将重点推进。在此基础上，尽早建立相对完善的城镇污水处理基础设施，从根本上解决城镇水污染加剧的问题，确保水安全。2020年前，我国城镇化率的不断提高、消费转型、农业和农村现代化的发展、贸易与国际履约带来的环境问题和压力，将使得我国环境问题变得更为复杂和不确定。因此，“十二五”期间应以提升环境质量为切入点，以削减总量为重要抓手，在实现环境形势趋好的前提下，严格防范环境风险，保障环境安全。国家“十二五”水污染防治规划范围包括淮河、海河、辽河、松花江、黄河中上游、三峡库区极其上游、太湖、巢湖、滇池、南水北调东线、南水北调中线水源地11个流域，涉及23个省、市、自治区。重点流域总面积约275万平方公里，水资源总量8128.7亿立方米，总人口约7.88亿。重点流域水污染得到有效控制，全国的水环境质量将得到明显改善。

国家发改委、住建部、环保部联合制定的《全国城镇污水处理及再生利用设施建设规划

(2011—2015 年)》和《全国城镇生活垃圾无害化处理设施建设规划 (2011—2015 年)》的编制工作于 2011 年 8 月正式启动。面对占环保产业半壁江山的污水处理和垃圾处理两大领域，上述规划将列明“十二五”期间各地污水、垃圾处理设施建设任务、投资规模等。

根据上述规划测算，“十二五”期间，包括中央政府、地方政府和业主个人投资在内，城市污水处理总投资达到 4500 亿元，城市生活垃圾处理总投资达到 2800 亿元，两者共计达 7300 亿元。环保产业投资对产值的拉动效应大概是 1∶1.2，到“十二五”末，城市污水处理和生活垃圾处理的产值就将达到大约 8760 亿元，是环保产业的两大支柱。

在生活污水处理领域，污水处理项目 50% 以上属于土建投资，因此水处理产业依次利好于土建工程企业、市政工程企业，以及环保企业。据测算：“十二五”期间，光是按照环保企业的窄口径统计，污水处理产业年均增速至少在 20%。“十二五”末将较“十一五”末翻一番，产值将达到 3000 亿元。如果算上土建和市政，产值将达到 1 万亿。在环保产业中，设备制造商、工程安装商、项目运营商将分享上述利润。

21.5.4 我国污水治理主要动态

1. 国家力保地下水五年污水治理投入万亿

地下水污染防治工作十分迫切。目前，我国地下水开采总量已占总供水量的 18%，北方地区 65% 的生活用水、50% 的工业用水和 33% 的农业灌溉用水均来自地下水。全国 657 个城市中，有 400 多个以地下水为饮用水源。环保部数据显示，我国 90% 城市地下水不同程度地遭受有机和无机有毒有害污染物的污染。

出于地下水污染防治等方面需要，污水处理将继续得到政策的支持。环保部环境规划院日前发布的一份报告预测，未来五年，我国城镇生活污水、工业污水的治理投资将大幅增加。预计“十二五”期间，我国污水治理累计投入将达到 1.06 万亿元。

2011 年 8 月 24 日召开的国务院常务会议上讨论通过了《全国地下水污染防治规划 (2011—2020 年)》。会议指出，到 2015 年，要初步控制地下水污染源，初步遏制地下水水质恶化趋势，全面建立地下水环境监管体系；到 2020 年，对典型地下水污染源实现全面监控，重要地下水饮水水源水质安全得到基本保障，重点地区地下水水质明显改善，地下水环境监管能力全面提高，建成地下水污染防治体系。

为达到上述目标，会议提出以下措施：抓紧开展地下水污染状况调查和评估，划定地下水污染治理区、防控区和一般保护区；严格地下水饮用水水源保护与环境执法，依法取缔饮用水水源保护区内的违法建设项目和排污口，限期治理地下水污染隐患；严格控制影响地下水的城镇污染，推进管网系统改造，提高城镇生活污水处理率和回用率，加强垃圾填埋场建设和治理；加强重点工业行业地下水环境监管，防范石油化工行业和地下工程设施、地下勘探、采矿活动污染地下水，控制工业危险废物对地下水的影响；分类控制农业面源对地下水的污染，逐步减少使用化肥和农药，在水源保护区内实施退耕还林还草；有计划地加快推进地下水污染修复，在地下水污染突出区域进行修复试点，开展海水入侵综合防治示范，切断废弃钻井、矿井等污染途径等。

2. 中央将投入 460 亿元治理长江中下游污染

2011 年 9 月，五部委联合印发《长江中下游领域水污染规划 (2011—2015 年)》(以下简称

《规划》)。该《规划》由环境保护部、发改委、财政部、住建部、水利部联合制定，并延续以往工程治理为主的思路，在未来五年确定了837个骨干工程项目，共需投资459.81亿元。在工程治理之外，《规划》也提出要建立问责制，对严重干扰正常环境监管执法及因决策失误造成重特大环境事故的领导干部和公职人员实行责任追究。

资料显示，长江中下游流域包括长江流域自三峡库区以下至长江口的广大区域，涉及广西、湖南、湖北、河南、江西、安徽、江苏、浙江、陕西和上海等10省（区、市），共66个市（州）505个县（市、区）。《规划》指出，与全国相比，长流中下游流域内的污染排放严重，其废水排放量占全国排放总量的17.8%，化学需氧量、氮氧化物排放量分别占全国排放总量的17.9%和19.2%，镉、砷等有毒有害污染物排放量分别占全国排放量的56.6%、63.5%。规划区内的重金属污染等有毒有害物质主要集中在湖南，汞、镉、铬、铅、砷等排放量分别占区域排放总量的55%、81.1%、46.8%、77.6%和70.6%。

长江中下游流域污染问题与其工业布局有关，主要是工业结构性污染，主要排污企业集中在化学原料及化学制品制造业、造纸及纸质品业、黑色金属冶炼及压延加工业、有色金属冶炼及压延加工业，而且沿江分布的工业园区加剧了流域结构性污染的特征。为此，《规划》提出了规划区污染治理的目标，即在48个考核断面中，15个水质达到II类，20个水质达到III类，7个水质达到IV类，6个水质达到V类，海洋功能区和近岸海域功能区水质的达标率达到40%以上。

在此基础上，《规划》确定了873个骨干工程项目，投资约459.81亿元。其中，工业污染防治项目343个，投资约92.88亿元；城镇污水处理设施建设优选项目（含管网完善及污泥处置）284个，新增污水日处理能力897.35万吨，投资207.46亿元（其中在建项目175个，新增污水日处理能力628.65万吨，投资148.52亿元），另有78个城镇污水处理设施设备项目，累积新增污水日处理能力347.67万吨，投资96.68亿元（未纳入投资统计范围）；重点区域污染防治项目210个，投资159.47亿元。

3. 2013年江西鄱阳湖生态经济区工业园区污水处理率将达80%①

“十二五”时期，江西省在推进鄱阳湖生态经济区建设中，将集中力量推进十二大生态经济工程建设，工业园区污水处理设施建设是其中的一项重要工程。按照统一规划、分期实施的原则，我省优先在鄱阳湖生态经济区范围内选择了10个入园企业多、污水排放量大、条件比较成熟的园区进行示范建设。10个园区污水处理厂采取BOT（建设—经营—转让）模式建设，总投资5亿元，明年年底前全部建成投运，并形成日处理污水25万吨的能力。到2013年底，我省将实现工业园区污水处理设施全覆盖，工业园区污水日处理能力达到200万吨，污水处理率达到80%，削减化学需氧量14.6万吨。

整个工业园区污水处理工程总投资100亿元，主要在江西省94个工业园区全面建成污水处理设施，并建立较为完善的园区工业废水管网收集、雨污分流系统和在线监测系统。

按照统一规划、分步实施的要求，第一步，优先在鄱阳湖生态经济区范围内选取了10个入园企业多、污水排放量大、条件比较成熟的园区，作为示范工程，率先组织建设。南昌经济技术开发区、共青城全国青年创业基地等10个园区已于2010年11月30日集中开工；第二步，在

① 资料来源：江西省信息中心，2011-01-17。

总结示范建设经验的基础上，2011年新开工建设30个园区污水处理项目；第三步，2012年6月底前，再开工建设33个园区污水处理设施项目。加上已经建成和正在建设的21个园区污水处理设施项目，到2013年底，实现工业园区污水处理设施全覆盖，全省工业园区污水日处理能力可达到200万吨，污水处理率达到80%，削减化学需氧量14.6万吨。

江西省委关于制定全省国民经济和社会发展第十二个五年规划的建议中，明确提出将实施工业园区污水处理设施建设工程，推进城镇生活污水处理设施二期工程建设，完善污水收集管网，初步建立城镇排水管网雨污分流模式，确保城镇污水集中处理率达到80%以上。

4. 北京沿五环建6座污水处理厂①

“十二五”期间，朝阳区将在垡头、东坝、孙河等基础设施相对薄弱地区新建6座污水处理厂，6座处理厂全部位于五环周边地区。

目前，朝阳区已经建成4座污水处理厂，分别为高碑店、酒仙桥、北小河及北苑污水处理厂，合计日处理能力达134万立方米。“4座污水处理厂位于五环内地区，主要用于处理中心城区的生活排放污水。”朝阳区水务局相关负责人介绍，与城区相比，五环外的农村地区配套污水管线系统相对较弱，大量未经处理的污水通过明沟或雨水管直接排放入河道，造成污染。

将在五环周边新建的6座污水处理厂，分别位于垡头、定福庄、东坝、楼梓庄、苇沟以及孙河等地。其中已经正式开工建设的垡头污水厂位于东五环化工桥外，规划总流域面积44.6平方公里，排放水质达到一级B标类标准，其规划污水流域范围主要为垡头集团的规划建设用地，南起东五环路，北至大郊亭路，西起京津塘高速路，东至通惠河，明年完工后的日处理污水能力2万立方米，远期建设规模为日处理污水10.5万立方米。投入使用后，附近的萧太后河将得以还清，从而提高当地及下游水环境质量。

6座污水处理厂全部建成后，朝阳区五环外及昌平、顺义、通州等周边区县的污水处理能力将得到增强，从而降低河道污染。另据了解，经过处理厂处理后的污水将全部达到排放水标准要求，一方面用于补充河道用水；另一方面也将作为景观灌溉、洗车以及冲洗厕所用水。

5. 抚矿集团公司200万吨造纸项目实现污水零排放②

造纸工业在生产中产生的废水、废气、废渣、毒性物及噪声等能对环境造成严重污染，对其进行控制、防治或消除处理，使周围环境不受或少受影响，已成为造纸工业一项重要内容。抚矿集团公司200万吨造纸项目将通过水、固体废弃物等零排放将这一高污染行业与环保理念、循环经济联系在一起，实现污水零排放。

琥珀纸业有限责任公司生活用纸生产线目前正在进行设备安装与调试工作，2011年9月末将投入生产。届时，日产180吨的质量上乘的高档卫生卷纸、纸抽纸、手帕纸、餐巾纸等三大项纸产品将走进抚顺市民的生活。据了解，抚矿集团琥珀纸业有限责任公司200万吨造纸厂分二期建设。一期110万吨工程主要包括两条5万吨生活用纸生产线、30万吨箱纸板生产线、30万吨瓦楞纸生产线等。

这个项目一期工程的两条生活用纸生产线采用从国外进口纯木浆进行生产，从而避免了在造纸制浆过程中工业污水的产生。另外，正在建设中的箱板纸和瓦楞纸生产线95%的原料则是来源于废纸再利用，这种再生造纸也是我国鼓励的再生资源回收利用项目。对于生产过程中产

① 资料来源：北京晚报，2011-01-06。

② 资料来源：国家信息中心，中经网数据库。

生的少量污水，将通过自建的污水处理厂处理，处理后的污水化学需氧量将低于50毫克/升，再排放到望花污水处理厂进行再处理，实现造纸过程中污水的零排放。另外，这一项目还与煤矸石热电厂有机结合，建设循环经济产业园，相互之间利用废弃物、排放物，能够实现园区内的零排放，是一个经济效益、环保效益非常好的循环。

22 固体废物利用

固体废物是指人类在生产、加工、流通、消费和生活等过程中所产生的、并被抛弃的所有固体和泥状物质的总称。随着人类社会文明的发展，人们在索取和利用自然资源从事生产和生活活动时，由于受到客观条件的限制，总要把其中的一部分作为废物丢弃。另外，由于各种产品本身有其使用寿命，超过了寿命期限，也会成为废物。但"废物"具有相对性，一个过程的废物，往往可以成为另一过程的原料，所以有人说固体废物是"被错待了的原料"，应该加以利用。如果弃置不用，则成为土地、环境卫生和环境美观的污染源。

22.1 基于循环经济理念的固体废物利用

循环经济（Cyclic Economy）即物质闭环流动型经济，是指在人、自然资源和科学技术的大系统内，在资源投入、企业生产、产品消费及其废弃的全过程中，把传统的依赖资源消耗的线型增长的经济，转变为依靠生态型资源循环来发展的经济。循环经济是针对传统的经济发展模式提出的新理念。以资源的循环利用为特征，将废旧物生产成再生资源进行再利用的一种循环模式。

循环经济是一种将经济体系与环境资源紧密结合的生态经济模式，它建立在物质不断循环利用的基础上，要求经济运行遵循"资源—产品—再生资源"的物质反复循环流动的环境友好型的经济发展模式，循环经济主张"减量化、再使用、再循环"的经济活动原则。固体废物循环利用是循环经济的一种具体体现形式，本质上也是一种生态经济，是按照生态规律利用自然资源和环境容量，实现固体废物的生态化转向，可以有效地减少固体废物的产生量、排放量，使其成为一种原料资源从而创造新的经济价值。对于固体废物，可通过回收可再生资源、各行业自行处理及垃圾综合利用来实现部分固体废弃物向有用资源的转化。

22.1.1 我国固体废物在循环利用中的问题

数据显示，工业是中国资源消耗和污染物排放的主要行业，能源消耗占全国70%以上，二氧化硫和COD排放分别占全国排放总量的86%和37%。2009年万元工业增加值用水量为116立方米，高于发达国家的平均水平，工业废水排放占全国总量40%以上，仍有8%左右的废水未达标排放；目前废旧手机及相关充电器电池每年产生的电子垃圾量高达两万吨，可以说中国工业目前是典型高消耗、高排放、综合利用率低的粗放型发展模式。

1. 资源回收率低

在固废的循环利用中，不少问题仍然十分突出，例如固体废物的分类回收体系和分类回收机制尚未建立和完善，很多可以循环利用的固体废物没有得到充分回收和利用，一些不易回收利用的再生资源丢弃现象严重，资源的回收多为低级的材料回收，例如废纸、废玻璃、废钢铁等。我国的循环经济实践用来解决电子垃圾问题还处于起步阶段，但是由于目前的电子垃圾污染严重，建立社会化的回收处理服务体系，实现电子垃圾的减量化、资源化和无害化势在必行。

2. 经济上受到制约

虽然实施循环经济在理论上会带来长远的和宏观上的经济效益增长，但在近期必然要求环境保护方面增大投入。以垃圾处理为例，非循环的填埋处理成本大大低于分类处理和购入焚烧设备进行处理的成本。成本的增大，对于我国经济发达地区一般是可以承受的，但对于欠发达地区却是难以承受的。我国最需要实施循环经济的地区，恰恰是那些经济欠发达、环保成本高、环保水平很低的地区，所以，国家在制定相关政策法规时，应在财政上适度倾斜。

3. 技术工艺相对落后

对于一些行业，产品回收利用技术开发投入严重不足，技术与装备水平不高，严重影响回收及利用效果。

22.1.2 我国循环经济理念下固体废物利用的发展方向

我国自2009年开始实施循环经济促进法后，在行业、企业及地方等各个层面都涌现了许多典型的发展案例，起到了示范和带动作用。但从总体看来，还面临一些突出的问题，从工业领域来讲，还没有在全行业树立循环发展的理念，缺少适用可行的循环经济技术，国家支持循环经济发展的相关政策还没有完全落实。

我国可以从以下五个方面推动工业的循环经济发展：

一是把发展循环经济作为“十二五”工业发展的一项重要任务，纳入规划体系。

二是突出重点领域，重点行业，加快提升循环经济水平。“十二五”期间，要抓住钢铁、有色、建材、化工等高耗能行业全面提高工业节能水平，全面实施工业尾矿综合利用规划和再生有色金属规划，制定并实施大宗工业固体废物综合利用规划。

三是选择产业链长，废物产生量大的行业开展试点，探索循环经济发展模式。

四是加大循环经济技术开发和应用推广力度，“十二五”期间，要分类分行业提出技术需求，力争突破一批重大的节能、节水、节材技术以及固体废物再利用技术、再制造技术等，支持循环经济发展。

五是积极研究支持循环经济发展的有关政策。

权威部门统计显示，“十一五”期间，中国工业固体废物综合利用量从2005年7.7亿吨增加到2010年的15.2亿吨，综合利用率由55.8%提升至69%。截至2010年，中国煤矸石、粉煤灰、钢铁渣、尾矿、工业副产石膏的综合利用量分别达到4亿吨、3亿吨、1.8亿吨、1.7亿吨和0.5亿吨，再生资源的回收利用量达到1.4亿吨。

“十二五”期间国家把加快培育和发展战略性新兴产业提升到产业结构战略调整的重要位置，其中节能环保资源循环利用产业被列为战略性新兴产业之一。据预测，到2015年，节能环保产业产值将达到4.5万亿元以上，约占当年GDP的8%，其中资源循环利用产业规模将快速增长，产值将达到1.5万亿元。

面对日益严峻的资源和环境压力，近年来国家大力发展循环经济，资源综合利用产业也实现快速发展，为中国工业经济加快产业结构调整奠定基础。

22.2 工业固体废物利用

工业固体废物指再生产、经营活动中产生的所有固态的、半固态和除废水以外的高浓度液态废物，产品的生产过程就是废物的产生过程。工业固体废物按危害状况可分为一般工业固体废物和危险废物，一般工业固体废物包括粉煤灰、冶炼废渣、炉渣、尾矿、工业水处理污泥、煤矸石及工业粉尘等；危险废物指易燃、易爆，具腐蚀性、传染性、放射性有毒有害废物，除固态废物外，半固态、液态危险废物在环境管理中通常也划入危险废物一类进行管理。固体废弃物的种类繁多，成分繁杂，数量巨大，是环境的主要污染源之一，其危害程度不亚于水污染和大气污染造成的危害。

22.2.1 工业固体废物利用的关键技术

与城市生活垃圾相比，工业固体废物成分相对单一，排放相对稳定，综合利用相对方便。它们主要被用于生产水泥、制砖、混凝土、筑路、充填材料、骨料等。对于不能利用的部分，一般采取填埋的方式进行处置。归纳起来有如下几种途径：

1. 提取各种金属

首先提出最有价值的各种金属是工业固体废弃物再资源化的重要途径。但应该说我国目前资源综合利用水平与国际水平比较还很低。由于综合利用水平还不高，造成资源消耗高、浪费大、利用率低，这不仅使企业成本上升、经济效益低下，而且由于废弃物排放量大，造成对环境的污染。

2. 生产建筑材料

利用固体废弃物生产建筑材料是我国处理工业固体废弃物的重要途径。由于用工业固体废弃物生产建筑材料一般不会产生二次污染，因而也是消除污染，化害为利的较好途径。主要有以下几个方面的用途：

- 生产碎石：一些冶金矿渣，例如高炉渣、铁合金渣、钢渣以及矿山废石用作混凝土骨料，道路材料、铁路道砖等。
- 生产水泥：有些工业废渣的化学成分与水泥接近，具有水硬性，可作为水泥工业原料。
- 生产建筑制品：用粉煤灰、尾矿、赤泥、煤矸石、电石渣等生产砖、砌块、大型墙体材料。
- 生产铸石和微晶玻璃：铸石是钢材，有色金属的良好待用材料，微晶玻璃在工业和建筑中有广泛的用途。用某些工业固体废弃物可生产铸石和微晶玻璃。中国地质科学院研制出用尾矿石生产微晶玻璃，并进入工业试验阶段。
- 生产矿渣棉和轻骨料：用高炉矿渣、煤矸石、粉煤灰等生产矿棉，用粉煤灰或煤矸石生产陶粒，用高炉渣生产膨胀矿渣等。轻骨料和矿渣棉在工业和民用建筑中具有越来越广泛的用途

22.2.2 工业固体废物利用的国际经验

1. 俄罗斯

俄罗斯联邦的耕地，有4300万公顷的酸性土壤。非黑土区有77%的耕地是酸性土壤，其中

46%是严重酸土和中等酸土。热电站每年约产生1亿吨粉尘和炉渣，其中只有1000多万吨作为石灰施用于酸性土壤，油页岩灰利用最为广泛，工业废物是补充石灰材料的重要来源。有些废物如爱沙尼亚的油页岩灰、捷利亚宾斯克电冶金联合工厂等企业的熔渣，作为钙肥已使用了许多年。苏联从1986—2000年期间，经济和社会发展的主要方向是在农业化学化和扩大有机肥利用的基础上，有计划地提高土壤肥力。

俄罗斯下塔格尔斯克、诺沃列别茨克等冶金联合企业加工的冶金炉渣加工成施用于酸性土壤所必需的石灰产品，这在很大程度上能缓和非黑土区石灰质缺乏的程度，而且要广泛利用包括生产性废物在内的农家肥资源增加化学化资源，必须努力合理利用工业废物，减少环境污染，制止歼石和垃圾侵占农田。根据俄罗斯农业化学组织的创议，农业化学化设计勘测站测定了对提高土壤肥力有希望的废物现存量。许多废物都可用来补充土壤有机质的含量，例如制材厂、纸浆厂、制糖厂等部门的废物都可利用。对于利用这些废物，俄罗斯已经积累了许多经验。

2. 法国

近年来，法国在城市固体废物管理和综合治理上，坚持以降低污染和提升循环再生水平为主导方向，逐步完善废物管理立法，推进垃圾废物的综合治理进程。坚持避免、减量、循环再生和处理处置的序列，坚持分类收集、分类清运和分类处置，突出废物资源循环再生，促进了垃圾废物的再循环。例如，巴黎市由政府职能部门负责城市固体垃圾的综合治理规划，筹集治理投资，推广垃圾处理技术，推进公共意识教育等，而具体的生活垃圾收集、清运和处理处置业务则交由各类专门的处理公司负责。这些处理公司在进行垃圾处理经营活动的同时要接受政府职能部门的监督，包括对资金管理及环境治理的监控，以确保其市场经济符合国家的有关法律、法规，合理使用政府提供的财政资金。

此外，巴黎没有设立城市固体垃圾管理的专门税种（垃圾税或垃圾费），其在城市生活垃圾综合治理过程中产生的各种费用由市政府从统一收缴的税费中，按审计预算结果依法划拨。企业在取得政府资金补贴的同时，也接受政府职能部门的监管。政府直接安排公务人员管理企业资金，准确评估垃圾清运、处理和处置所涉及的成本和收入，认定成本回收事项，为确定补贴提供标准。同时政府公务人员通过监督企业的生产经营，掌握各种有利于降低成本、1992年的一项法律规定了一种有益于国家环境和能源控制的生活垃圾和普通工业垃圾的税，这种税由垃圾存放设施经营企业上交，从而为环境部所属的垃圾现代化基金会提供资金。该基金会致力于城市垃圾处理技术和发展，以及处理生活垃圾的设施建设，它给那些新市镇间的生活垃圾或类似垃圾的处理设施或是扩建处理设施的市镇提供资金帮助。

3. 荷兰

荷兰的《废物焚烧法令》是禁止对生活垃圾和商业垃圾进行直接焚烧处理的。另外，荷兰也对危险废物的焚烧处理制定了特别的法规。在荷兰制定的废物处理法律法规中，垃圾焚烧设施的排放标准在全世界也是最严格的。目前，荷兰最大的焚烧厂是AVR，其处理量占整个垃圾焚烧市场的33%。其次是Essent，占24%。荷兰三家主要的大垃圾焚烧处理厂的处理量几乎占到总处理量的75%。目前，荷兰有34家垃圾填埋场，其中27家由州级或市级政府部门拥有。公营的垃圾填埋场占76%，其余由7家私营垃圾填埋场处理。2000年荷兰垃圾填埋场总收入为3.09亿欧元，盈利5100万欧元。此外，垃圾分选在垃圾处理系统中也占很重要的位置。2000年荷兰共收集260万吨纸，其中100万吨来自生活垃圾，约35万吨废玻璃由市政当局收集，6万吨由企业收集。

从2003年3月开始，荷兰实施了国家废物管理计划。这个废物管理计划提出了到2010年全国废物的83%要得到有效利用的战略目标。实现这一战略目标主要依靠三项措施：一是通过推进废物回收利用，使废物资源尽量得到循环再生；二是把无法直接回收利用的废物作为燃料实现废物的能源化；三是依靠其他的废物再利用方式来实现废物的有效利用。这个废物管理计划的核心就是最大限度地合理利用废物资源。荷兰政府还大力支持用废物制燃料事业（垃圾焚烧和生物制能等），即利用废物制取能源，并给予财政上的支持。

4. 英国

英国制定了一系列废物管理法规，包括1974年的《污染控制法》、1989年的《污染控制法（修正案）》、1990年的《环境保护法》、1991年的《可控废物管理规定》、1994年的《废物许可证管理规定（修正）》、1996年的《特别废物管理规定》、1997年的《生产者责任义务（包装废物）管理规定》、1998年的《包装废物管理规定》、1998年的《废物减量法》、1999年的《污染防治法》。这些法律法规的颁布实施使英国废物管理和处置状况得到改善。废物管理法规和许可证制度最初集中在废物处置上，自从欧盟颁布了废物管理框架法令后，许可证制度也包括了废物的贮存、处理和回收利用。1994年英国通过了废物管理许可证法规，开始实行“废物管理许可证制度规定”。

英国从2003年6月开始出台新法令禁止填埋橡胶轮胎（轮胎作为填埋场工程用料除外）。英国环境部还公布了《填埋渗滤液监测及地表水和地下水污染防治指导方针》，规定了渗滤液防治污染过程中的实际监测方法和控制方法。

22.2.3　我国工业固体废物利用

1. 我国工业固体废物利用现状

随着我国经济和工业化进程的迅猛发展，工业固体废物造成的污染日益突出，严重影响了生态环境，威胁着人们的身体健康。我国工业固体废物主要是粉煤灰、尾矿、炉渣、冶炼废渣和煤矸石。据有关文献统计，我国每年工业固体废弃物的产生量已经达到近10亿吨，累计堆存量超过67亿吨，占用土地达到65412万平方米，由于综合利用水平不高，造成资源消耗高、浪费大、利用率低，不仅使企业成本上升、经济效益低下，而且由于废弃物排放量大，对环境造成严重污染。1981年中国工业固体废弃物总产量为3.37亿吨，1995年增长到6.45亿吨，1996年为6.59亿吨。2009年中国环境统计年报数据显示，全国工业固体废物产生量203943万吨，比上年增加7.3%；工业固体废物排放量710万吨，比上年减少9.2%。工业固体废物综合利用量138186万吨，比上年增加11.9%；工业固体废物贮存量20929万吨，比上年减少4.4%。工业固体废物处置量47488万吨，比上年减少1.7%。产生量大的几种工业固体废弃物是尾矿2.47亿吨，煤矸石1.87亿吨，粉煤灰1.15亿吨，炉渣0.90亿吨，冶炼废渣0.8亿吨。

据估算以每吨工业固体废物堆存的经济损失14.25元计，每年造成经济损失203.93亿元。如果改变其资源—产品—废物的线性物质流程，转化为可以继续利用的资源，形成“资源—产品—再生资源”闭合流动的经济系统，不仅能降低工业固体废物对自然环境的影响，而且以平均综合利用每吨工业固体废物的经济效益为9.98元计算，每年可以获得经济效益142.82亿元。可见，资源转化能使我国的工业固体废物在综合利用方面每年增效346.75亿元。2010 年全国工业固体废物产生及处理状况如表 22－1 所示。

表 22－1 2010 年全国工业固体废物产生及处理状况

Table 22－1 National Status of Generation and Treatment of Industrial Solid Wastes in 2010

单位：万吨

产生量		综合利用量		储存量		处置量	
合计	危险废物	合计	危险废物	合计	危险废物	合计	危险废物
240943.5	1586.8	161772.0	976.8	23918.3	166.3	57263.8	512.7

资料来源：《中国环境状况公报》2010 年数据

目前，我国有色金属矿山和冶炼企业综合回收的伴生黄金占全国黄金产量的 10% 以上，伴生白银占白银产量的 90%，伴生硫产量的 47%，铂族金属全部是冶炼厂回收的。钢材产量的 1/3 是以废钢铁的原料生产的。应该说我国目前资源综合利用水平与国际水平比较还很低。目前，我国矿产资源总回采率仅为 30%，比世界平均水平低 10—20 个百分点，对共生、伴生矿进行综合开发的只占 1/3。如果我国矿产资源的总回采率提高 10 个百分点，则至少可少开发、开采 1/4 的矿产资源。我国主要工业产品的能源、原材料消耗比国外先进水平高 30%—90%。大量可利用的资源作为“三废”白白浪费掉，得不到充分利用，每年生产的可利用而未利用的工业固体废弃物的资源价值已超过 250 亿元，每年因再生资源的流失而造成的经济损失已达 250 亿元—300 亿元。

利用固体废弃物生产建筑材料是我国处理工业固体废弃物的重要途径。由于用工业固体废弃物生产建筑材料一般不会产生二次污染，因而也是消除污染、化害为利的较好途径。一些工业固体废弃物具有潜在能源可以利用，如粉煤灰中含煤达 10% 以上，可以通过分选把煤选出然后利用；煤矸石发热量为 0.8MJ/Kg—8MJ/Kg，可利用煤矸石发展坑口电站。目前我国煤矸石发电利用率已达 40%。

2. 分类工业固体废物排放与处理情况

（1）煤矸石。全国国有煤矿现有矸石山 1500 余座，堆积量在 20 亿以上（占中国工业固体废物排放总量的 40% 以上）。目前，主要综合利用的途径是煤矸石发电、生产水泥制砖、井下充填、复垦造田、筑路等。2010 年煤矸石的综合利用率达 70%。

（2）粉煤灰。我国粉煤灰综合利用技术发展很快，已由最初的只是工程回填方面的利用技术发展到建材制品、建筑工程方面的应用。目前我国非常重视高附加值方面的应用，例如水处理，提取微珠、碳铁、铝、洗煤重介质、冶炼三元合金、高强轻质耐火材料和耐火泥浆。

（3）赤泥。我国氧化铝产量达到 346 万吨，每年产赤泥量也与此相当。但由于赤泥中的碱含量高，限制了它水泥生料的配比用量。因此目前大部分仍是堆存，少量的用于生产水泥。我国应用赤泥最好的山东铝业公司也只利用了产量的 1/3。

（4）钢渣。钢渣的利用途径，一是作为烧结矿原料和炼钢的返回料；二是在建筑材料行业应用，例如用作路基材料、回填材料和水泥原料等。但由于钢渣的不稳定性，利用也受到了一定制约。

（5）污泥。现阶段污泥处置仍是以填埋为主，利用为辅。填埋处理由于占地大、渗滤液会污染地下水、后续处理管理费用高等问题，应用受到限制。

3. 我国工业固体废物利用存在的问题

原规划到2005年，全国危险废物、医疗废物和放射性废物基本实现安全贮存和处置，但由于危险废物和医疗废物集中处置项目实施缓慢，从2003—2010年已历经七年的时间，项目建设严重滞后，到目前为止，建成并投入运行的危险废物处置设施仅有94项，占规划总数的28%。已建成的项目在运行过程中也在一些问题。我国危险废物种类繁多、产生量大，但目前仍然存在着危险废物产生底数不清、废物收集难度较大的问题。危险废物集中焚烧处置设施普遍存在实际处置量达不到设计规模的现象，大量焚烧设施没有得到充分利用，造成设备浪费。原因之一是设施在设计时对本地区危险废物产生、收集和处置现状不清楚，盲目加大处置规模的建设；另外也有收费标准不落实或收费过高、过低的情况均影响了项目的实施。全国工业固体废物产业排放及利用情况如表22－2所示。

表22－2 全国工业固体废物产生、排放及利用情况表

Table 22－2 National Status of Generation, Discharge and Utilization of Industrial Solid Wastes

单位：万吨

年度（年）	产生量	排放量	综合利用量	储存量	处置量
2005	134449	1655	76993	27876	31259
2006	151541	1302	92601	22398	42883
2007	175632	1197	110311	24119	41350
2008	190127	782	123482	21883	48291
2009	203943	710	138186	20926	48438
09年比08年增长率（%）	7.3	－9.2	11.9	－4.4	－1.7

资料来源：中国环境保护产业协会固体废物处理利用委员会，《我国工业固体废物处理利用行业2010年发展综述》，2011.

22.2.4 我国工业固体废物利用发展建议

1. 加强宣传引导

利用电视、互联网、报刊杂志、文艺会演等多种形式大力宣传《循环经济促进法》，讲述循环经济的内容，提高全民发展循环经济的意识和节约资源、废物资源化回收利用、保护环境的观念，使循环经济理念深入人心。

2. 提高技术能力

依靠技术进步，为固废综合利用提供有力的技术支撑。加快运用高新技术和先进适用技术。选择具有标志性目标和有广泛推广前景的先进适用技术，在重点行业、重点企业组织实施一批重大示范工程；加快先进适用技术的推广，开发适合于我国的工业固体废弃物处理、处置技术和装备，推进工业固体废弃物处理产业化。

3. 加强法律规范

加快地方性相关法律、法规的配套建设，加快固体废弃物的法制建设，尽快完善固体废弃物污染防治的法律、法规和标准，为企业提供固废综合利用的政策保障体系。坚持鼓励与限制相结合，形成资源综合利用的激励机制。

4. 制定污染控制规划

引导企业尽量减少废弃物的产生，努力实现固体废弃物的“减量化、资源化和无害化”。运用经济手段，按照污染者负担的原则，合理征收工业固体废弃物排污费。推行清洁生产，把固体废弃物尽可能消灭在生产过程中。

我国应积极发展固体废物污染治理设备运营服务业，深化经济体制改革和政策机构与运行机制改革，依靠技术进步和先进的管理来获得污染治理设施运营的高效率、高水平、高效益和低成本，并采取措施，强化市场准入和市场秩序，以确保行业朝着健康和规范的方向发展。

22.3 生活固体废物利用

生活固体废物是指在日常生活中或者为日常生活提供服务的活动中产生的固体废物，以及法律、行政法规规定视为生活垃圾的固体废物。生活垃圾一般可分为四大类：可回收垃圾、厨余垃圾、有害垃圾和其他垃圾。目前常用的垃圾处理方法主要有综合利用、卫生填埋、焚烧和堆肥。可回收垃圾包括纸类、金属、塑料、玻璃等，通过综合处理回收利用，可以减少污染，节省资源。

生活固体废物如果处理不当也会对环境造成危害。塑料类垃圾难以分解，会破坏土质，焚烧会产生有害气体；电池含有有毒重金属汞；剩餐剩饭则促使垃圾中细菌大量繁殖，产生有毒气体和沼气，引起垃圾爆炸；油漆和颜料等生活垃圾是危险的易燃品，并含有重金属，对健康不利。

22.3.1 生活固体废物利用的关键技术

目前，生活固体废物利用的技术解决生活固体垃圾问题的目标是：将垃圾进行减容、减量、资源化、能源化及无害化处理。主要有以下六种技术。

（1）压实技术。压实是一种通过对废物实行减容化、降低运输成本、延长填埋寿命的预处理技术，压实是一种普遍采用的固体废弃物的预处理方法，例如汽车、易拉罐、塑料瓶等通常首先采用压实处理，适于压实减少体积处理的固体废弃物，不宜采用压实处理，某些可能引起操作问题的废弃物，如焦油、污泥或液体物料，一般也不宜作压实处理。

（2）破碎技术。为了使进入焚烧炉、填埋场、堆肥系统等废弃物的外形减小，必须预先对固体废弃物进行破碎处理。固体废弃物的破碎方法很多，主要有冲击破碎、剪切破碎、挤压破碎、摩擦破碎等，此外还有专有的低温破碎和混式破碎等。

（3）分选技术。固体废物分选是实现固体废物资源化、减量化的重要手段，通过分选将有用的充分选出来加以利用，将有害的充分分离出来；另一种是将不同粒度级别的废弃物加以分离，分选的基本原理是利用物料的某些特性方面的差异，将其分离开。

（4）固化处理技术。固化技术是通向废弃物中添加固化基材，使有害固体废物固定或包容在惰性固化基材中的一种无害化处理过程，经过处理的固化产物应具有良好的抗渗透性、良好的机械性以及抗浸出性、抗干湿、抗冻融特性，固化处理根据固化基材的不同可分为沉固化、沥青固化、玻璃固化及胶质固化等。

（5）焚烧和热解技术。焚烧是固体废物高温分解和深度氧化的综合处理过程，好处是大量有害的废料分解而变成无害的物质。由于固体废弃物中可燃物的比例逐渐增加，采用焚烧方法处理固体的废弃物，利用其热能已成为必须的发展趋势。热解是将有机物在无氧或缺氧条件下高温（500℃—1000℃）加热，使之分解为气、液、固三类产物，与焚烧相比，热解则是更有前途的处理方法，它最显著的优点是基建投资少。

（6）生物处理技术。生物处理技术是利用微生物对有机固体废物的分解作用使其无害化可以使有机固体废物转化为能源、食品、饲料和肥料，还可以用来从废品和废渣中提取金属，是固化废物资源化的有效的技术方法，目前应用比较广泛的有：堆肥化、沼气化、废纤维素糖化、废纤维饲料化、生物浸出等。

上述生活固体废物利用的技术往往会对环境造成污染，目前我国在结合国内外城市生活垃圾的组成特点和处理现状的基础上研制开发了旋风分选新技术。该工艺的主要分选方法是采用旋风分离器。一级旋风分离的作用是将塑料、纸等轻物料从重物料中分离出来。一级旋风分离后的垃圾被分成两个组，即重物料组（厨余、无机物、玻璃等）和轻物料组（塑料、纸等）。其中，重物料可根据不同情况进行填埋等处理；轻物料内由于塑料、纸等一些轻物质的物理特性非常接近，分离不明显，需进行二级气动剥离，以提高回收和分选率。二级分选时，为了将塑料、纸等轻物质分离，首先要对其进行特殊工艺变重处理，再进行旋风分选，将轻物质有效分离出来，然后根据不同用途进行利用。采用旋风分离的城市垃圾回收系统如图 22 - 1 所示。

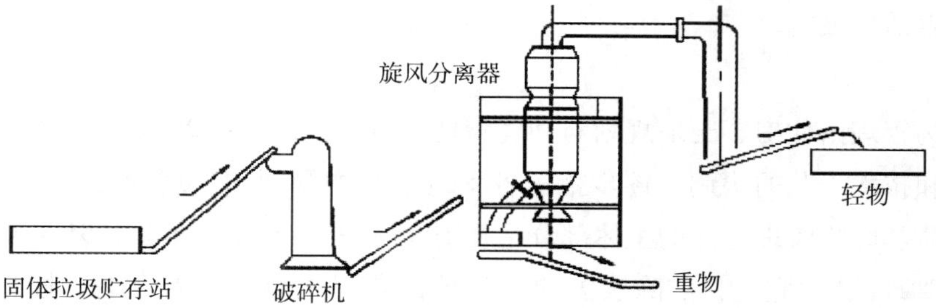

图 22 - 1 采用旋风分离的城市垃圾回收系统

Figure 22 - 1 Urban Waste Recovery System Adopting Cyclone Separation

22.3.2 生活固体废物利用的国际经验

1. 日本

日本是个资源贫乏的国家，人口、环境、资源压力大。为了合理、充分利用资源，保护环境，日本制定了一整套完备的构建循环型社会的法律体系，其完备的法规体系为垃圾分类回收提供了依据和指南。2000 年 12 月公布实施的《促进建立循环社会基本法》。在垃圾分类方面，该法从宏观层面对政府、事业者、地方团体、公民的责任义务做出了规定。此外，日本根据各种产品的性质制定的具体法律法规，例如《促进容器与包装分类回收法》、《家用电器回收法》、《建筑及材料回收法》、《食品回收法》、《汽车再循环法》以及《绿色采购法》。

日本日野市垃圾改革成功，是因为改革的推进者日野市当局较好地运用了“全民动员”的手法。首先，在行政机关内部建立了改革推进体制。1998 年 4 月，该市将“再生利用课”改组为“再生利用推进课”，由其负责垃圾改革方案的设计。其次，事先向市民提起问题，宣讲改革的宗旨、目的。这当然是为促使广大市民认识到改革的必要性，以顺利推进改革。

1998年12月，日野市当局向“垃圾减量等推进审议会”提出了题为“请求居民负担废弃物处理费”的咨询。随后，1999年3月，日野市市长在市议会上发表演讲，说明了废除垃圾箱的政策构想。这一系列活动可以理解为一种“舆论铺垫”，使得改革真正推行起来，市民已经有足够的心理准备。

2. 法国

近年来，法国在城市固体废物管理和综合治理上，坚持以降低污染和提升循环再生水平为主导方向，逐步完善废物管理立法，推进垃圾废物的综合治理进程。为方便清除垃圾，环境部提供给市镇和管理企业一定的经济和技术上的帮助，1992年的一项法律规定了一种有益于国家环境和能源控制的生活垃圾和普通工业垃圾的税，这种税由垃圾存放设施经营企业上交，从而为环境部所属的垃圾现代化基金会提供资金。

在废物管理政策方面，法国提出了资源化和能源化技术互补原则。法国政府既不强制执行家庭生活垃圾管理方式，也不指定结合何种基本技术。技术互补原则可以使废物处理费用在给定资源化率的条件下实现最小化，特别是当把以实现能量回收为目的的垃圾焚烧技术运用到垃圾废物处理计划中时，根据废物管理政策的规定，将废物分拣、分选与以再生为目的的废物回收和垃圾堆肥等技术有机结合起来实施综合治理，这种管理方式在废物治理实践中取得了很明显的治理效果。几年来，法国所有销纳公共垃圾的填埋场都只接收经过再循环处理并已没有任何使用价值的生活垃圾。

3. 荷兰

荷兰的废物管理由住房、土地规划与环境管理部负责。荷兰从20世纪80年代开始注重废物管理立法程序和技术工艺的协调，逐步推进各项垃圾废物管理和处理的法律、法规和技术标准的和谐统一。荷兰废物管理立法的主体部分是在1988—1991年制定的。1995年又颁布了一部综合性的《环境管理法》，它是目前除法国的《环境法典》之外，世界上综合性最强的一部环境法。当前，荷兰的环境立法与废物管理立法体系已经很完整，它为荷兰的环境治理与废物综合治理提供了强有力的法律保障和支撑。

垃圾分类收集和垃圾分流是荷兰积极推进的生活垃圾与废物的收运方式。在荷兰，住别墅的居民家庭每家都配有带轮子的垃圾桶，桶上有门牌号码。不同颜色的垃圾桶分装不同成分的垃圾，例如，绿色的盛装有机垃圾，灰色的盛装无机垃圾，蓝色的盛装废报纸和旧杂志。住公寓的居民都必须将分类后的垃圾送到楼下的垃圾房或整齐的摆放在街道的路边，由清洁公司统一收运。在荷兰，尽管采取了避免和减量措施，抑制了废物的增长速度，但是垃圾废物的增长还是比预期快得多。实践证明，单纯依靠废物的回收利用和循环再生已经不能满足废物处理处置的需求，必须全方位的安排垃圾废物的综合治理问题。

4. 英国

据统计，英国居民家庭每年平均产生2820万吨生活垃圾，其中庭院垃圾、废纸和厨房垃圾占一半以上。2005年英国回收利用了33%的废旧包装物，有35000亿只玻璃瓶、10000亿只塑料瓶、45000亿只铝罐被回收利用。目前英国所使用的包装物中有将近一半被回收利用。从2001年英国城市生活垃圾处理状况分析，其生活垃圾处理方式已经出现了显著变化。2001年英国城市生活垃圾收集量为2820万吨，比上一年增长了2.7%；生活垃圾填埋量2210万吨，比上一年下降了2%；家庭垃圾再循环率为1.1%，比1997年增长了4%。家庭垃圾路边收集为51%，比上一年增长3%。

英国制定了一系列废物管理法规，包括1974年的《污染控制法》、1989年的《污染控制法（修正案）》、1990年的《环境保护法》、1991年的《可控废物管理规定》、1994年的《废物许可证管理规定（修正）》、1996年的《特别废物管理规定》、1997年的《生产者责任义务（包装废物）管理规定》、1998年的《包装废物管理规定》、1998年的《废物减量法》、1999年的《污染防治法》、2003年的《家庭生活垃圾再循环法令》等。这些法律法规的颁布实施使英国废物管理和处置状况得到改善。

按照欧盟填埋导则的要求，2005年英国还出台了《生物降解垃圾的填埋限令》。该法令明令限制填埋可生物降解垃圾，并制定了分阶段降低可生物降解垃圾填埋的计划。按照欧盟填埋导则的要求，2010年英国运往填埋场可生物降解的垃圾必须降到1995年运往垃圾填埋场的可生物降解垃圾总量的75%，2013年要降到50%，2020年降到35%。这不仅体现了英国政府正在进行从垃圾处置到垃圾资源管理的战略转移，也是英国政府执行欧盟有关法规规定的具体举措，达到实现可生物降解垃圾的治理目标。

22.3.3 我国生活固体废物利用

1. 我国城市生活固体废物利用现状

我国城市垃圾的产生和处理现状，是随着我国城市化进展的加快和人民生活水平的不断提高，城市生活垃圾产生量平均以每年4.8%的速度持续增长。据统计，2003年，全国生活垃圾清运量为14857万吨，比上年增加了88%；其中生活垃圾无害化处理量人7550万吨，比上年增加了20%，生活垃圾无害化处理率为50.8%。大量城市生活垃圾露天堆放或简易填埋，对环境造成巨大危害。城市生活垃圾处理已成为影响人们生存环境和可持续发展的重要因素。

2003年发布的《北京市生活垃圾治理白皮书》显示：北京市当年年产垃圾421万吨；垃圾处理的主要方式是填埋、堆肥和焚烧，所占比重分别为89.6%、5.8%和4.6%。而经多年努力，在2009年，北京市垃圾处理的主要方式仍然是填埋（在各种处理方式中所占比重为90%），每年约占用500亩土地，垃圾填埋场使用寿命仅剩4年左右。

统计显示，目前中国90%以上的城市垃圾无法处理，只能堆放和填埋，每年因垃圾造成的资源损失价值在250亿元人民币—300亿元人民币，如果采用分类收集处理，实现垃圾资源化，每年可以创造的产值至少在3000亿元左右。可事实上，除废钢铁的回收率较高外，废塑料的回收率为25%，废橡胶的回收率为32%，废纸的回收率为35%，废玻璃的回收率只有13%。目前我国垃圾的收集、管理和处理尚处于粗放阶段。混合垃圾人工分拣困难，全国每天的45.4万吨城市垃圾，至少需要增加4.54万人去分拣，需1.5亿平方米的场地及与其配套的车辆、设施等，从经济、管理的角度都不可行。据《中国城市建设统计年鉴》，卫生填埋是我国处理MSW的主要手段，2008年全国卫生填埋量占总处理量的62.9%，其中北京市的卫生填埋量比重为93.34%，吉林最小，仅34.17%。

通过调查发现，无锡市新区每天产生约415吨生活固废，人均每日生活垃圾产生量约为0.98Kg，略低于全国平均水平（全国平均水平为1.0Kg/人·天）。就季节分布而言，5月—9月期间由于天气原因，饮料、瓜果等消耗量大，这个时段产生的生活固废也略高其他季节。通过进一步分析发现，新区每天集中处置约306吨生活固废，集中处置率约为74%，其中约186吨运往垃圾发电厂进行焚烧处置，120吨运往填埋场进行卫生填埋。上海市生活垃圾主要采取填埋、焚烧、生化处理、临时堆放等方式进行处理。上海统计年鉴数据显示：2007年，上海市城市生活垃圾中填埋量为377万吨/年，约占53.7%；焚烧量为110万吨/年，约占15.7%；生化处理量为70万吨/年，约占10.0%；临时堆放量为145万吨/年，约占20.6%。目前上海市共有6座大型生活垃圾处置设施，其中填埋场3座（处理能力6400吨/天），焚烧厂2座（处理能力2000吨/天），生化处理厂1座（处理能力1000吨/天）。

2. 我国农村生活固体废物利用现状

随着我国农村经济与农民收入水平的快速提高，农村生活固体垃圾产生与排放的数量快速增加，已经严重影响了农村环境、农民健康和农业可持续发展，成为我国建设社会主义新农村必须面对和尽快解决的问题。一些研究表明，我国农村2000年的垃圾产量达到1.4亿吨，而且继续以每年10%的速度增加。

在甘肃省及河北省的调查数据显示，大多数村没有处理生活固体垃圾的设备，而且地区差别非常明显。从两个地区的总体情况来看，只有30%的村有处理生活固体垃圾的设备，而没有处理设备的村高达70%（如表22－3、表22－4）。与河北省相比，甘肃省村级拥有的生活固体垃圾的处理设备状况要好一些。甘肃接近一半的村有处理设备，达到43%；而在河北仅有5%的村有处理生活固体垃圾的设备。

表22－3　村级生活固体垃圾的处理现状和计划加强固体垃圾管理的村的比例

Table 22－3　Treatment Status of Domestic Solid Waste at the Village Level, Share of Villages Planning to Strengthen Management of Solid Waste

项目	现状		计划管理	
	有设备	无设备	有计划	无计划
总体（%）	30	70	38	62
甘肃（%）	43	57	47	53
河北（%）	5	95	32	68

资料来源：中科院农业政策研究中心，2008年数据

表22－4　农户生活固体垃圾的处理现状

Table 22－4　Treatment Status of Domestic Solid Waste at the Household Level

项目	农户比例（%）		
	总计	甘肃	河北
投资至村里公用设备	30	43	4
其中分类垃圾桶	1	2	0
随便扔	48	36	74
出售	3	3	1
焚烧	12	17	3
填埋	6	1	18

资料来源：中科院农业政策研究中心，2008年数据

此外，中国科学院农业政策研究中心2008年对全国5个省（江苏、四川、陕西、吉林和河

北省）2020个农户的资料，刘莹分析了我国农村生活固体垃圾的处理现状。调查数据显示，目前我国农村生活垃圾的处理方式以丢弃为主，只有19.4%的垃圾被扔到规定地点进行统一清运处理；49.1%的垃圾处于无人管理的裸露状态；垃圾还田和燃烧的比率分别为15.3%和12.2%，另外还有其他少量垃圾被掩埋或丢弃到水体，占4%。就不同省份而言，在江苏垃圾得到统一处理的比率远高于其他省份，而四川、陕西、河北的处理比率则偏低。江苏生活垃圾的主要处理方式是丢弃到规定地点进行统一清运、处理，占38.6%，这一比率高于随意丢弃的比率（占30.3%）。而在其他省份则主要以随意丢弃为主，其中河北的随意丢弃比率最高，占64.6%。相对而言，在这五个省份中，陕西的垃圾还田率最高，占20.6%，而四川的垃圾燃烧比率最高，占23.8%。2008年农村生活垃圾不同处理方式的比例情况如表22－5所示。

表22－5 2008年农村生活垃圾不同处理方式的比例

Table 22－5 Proportions of Different Treatment Patterns for Rural Domestic Waste in 2008

单位:%

省份	非规定地点	规定地点	还田	燃烧	其他	合计
江苏	30.3	38.6	8.2	16.4	6.5	100
四川	48.3	7.1	19.0	23.8	1.8	100
陕西	53.9	11.4	20.6	8.8	5.3	100
吉林	48.5	25.4	12.7	7.6	5.8	100
河北	64.6	13.1	16.2	5.8	0.3	100
平均	29.1	19.4	15.3	12.2	4.0	100

资料来源：《农村生活固体垃圾排放特征、处理现状与管理》．农业环境与发展，2011，(2).

3. 垃圾利用中存在的问题

国内城市生活垃圾，目前采用单一处理方法居多，每一种处理方法只是对垃圾中一种或几种组分有效，而对其他组分常常不起作用。为了发挥某一种方法的优点，避开缺点，采用多种方法联合处理，是实现垃圾处理资源化、减量化、无害化的根本保障，而垃圾分选是实现垃圾综合处理的关键。

我国目前在MSW（城市生活垃圾）处理方面存在垃圾日处理量小，设备容量小，垃圾资源利用率低，垃圾无害化处理率低等问题，尤其是自然填埋的处理方式对土壤、地下水、大气等都会造成现实的影响和潜在的危害；国产焚烧炉装置配套不完善，缺乏可靠、经济的烟气净化设备，导致烟尘及有毒气体，形成二次污染；堆肥肥效不高，客观上造成再次堆积，由MSW一次污染转为二次污染。

22.3.4 我国生活固体废物利用发展建议

1. 加强立法、执法建设

发展循环经济，推进垃圾处理产业化，是实现经济可持续发展的客观要求，是社会参与、政府扶持与垃圾处理产业化运作的有机结合，是实现城市垃圾资源循环再利用的要求。为了发展循环经济，推进垃圾处理产业化，实现城市生活垃圾综合管理，促进废弃物的资源化利用，必须有国家法律、法规保障和政策的支持，对垃圾的收运和处理等各环节都给予明确的规范。

2. 建立健全有效监督机制

垃圾处理产业是以为公众提供"公共产品"为特点的朝阳产业，必须把社会效益和环境效益放在重要位置，实行产业化发展后，政府和企业追求的目标不完全一致。目前，农村的环保意识不强，则环境保护的效应对于村民的总效来说就微不足道，所以短期内使村民转变环保态度，首先必须靠政府的引导和强制迫使村民的环保效应内化。从长期考虑，我们需要建立社区家庭等外部环境的道德约束、舆论和监督机制等。

3. 加强宣传教育提高垃圾分类收集意识

现阶段我国关于垃圾分类回收处理的宣传和教育非常少，大多数人不了解生活垃圾处理的相关知识，"从娃娃抓起"的基础教育也十分不足。因此，我国需积极营造浓厚氛围，通过各种媒介不断地利用环境观、垃圾资源观、资源危机观等来教育市民，让国民了解保护环境的重大意义。

4. 建立生活垃圾收费模式

计量收费和定额收费是垃圾收费的两种基本模式，国外经济发达地区许多城市的生活垃圾收费采用计量收费制，有效地促进了垃圾减量化。计量收费的方式主要有三种：垃圾桶（箱）方式、专用垃圾袋方式、垃圾袋标签方式。国内城市大多采用定额收费制。定额收费制操作简便，对住宅楼式居民占人口多数的城市比较适用，因为可以节约管理成本和监督成本，但不利于垃圾减量化。因此，国内城市应借鉴国外经验，并参考本地区的实际情况，慎重选择垃圾处理费的收费方式。

22.4 电子垃圾的利用

电子电器废弃物俗称电子垃圾，含有金、银、钯、铜、铝、锡等宝贵资源，有很高的利用价值。但是如果处理不当，其拆解过程产生的"三废"不仅直接危害人体健康，而且会对土地、水源造成严重污染。随着市场经济的迅猛发展，电子信息产业也随之取代传统工业成为主导产业。由于电子电器产品和普通人生活密切相关，而且技术更新换代很快，所以社会需求量极大，并仍在继续增长。与此同时，电子电器产品除了给人类的生活带来了前所未有的方便和舒适外，也带来了一系列问题。目前全球电子垃圾排放情况不容忽视。

22.4.1 电子垃圾利用的关键技术

电子废物的再生回收工艺主要是：拆解、分类、分离、再生（再冶炼）等。其中拆解和分离技术是电子废物资源再生的关键技术。电子废物的拆解分为机械拆解和人工拆解。由于电子废物种类、型号繁多，难以采用机械方式进行分部件拆解（即电子电气设备组装的逆向操作），所以通常进行机械式破碎，然后再进行分类分离。

机械破碎可以最大程度地减少对人力的依赖，改善工人操作环境，大大提高处理效率和污染控制水平，但是由于不能进行精细拆解，资源利用率不会太高；而人工拆解则可以解决这一问题。人工拆解可以利用组装的逆操作结合必要的剪切、破碎、热熔等拆解工艺，最大程度地分离电子废物中的各种材料，并进行再生利用。然而，人工拆解的最大问题是人工效率较低，工作环境恶劣和污染控制困难。特别是采用热熔拆解过程可以使废物中的铅、镉等低挥发点中金属物质挥发到大气环境中去，然后冷凝降落到土壤和地表水中，造成拆解工作场所周围环境的重金属污染。最佳拆解工艺应该是人工与机械相结合的组合拆解工艺，即通过人工方式进行单一物质材料部件的拆解，最大程度地进行物质分类回收；对于复合材料和需要用化学分离或热分离的组件，则采用机械分离或破碎方式进行处理，结合后续分离再生技术回收其中的有用物质。

分离技术是采用机械或化学方法将复合材料或者难分离部件组合材料中有用物质分离出来，进行再生处理的过程。机械分离通常是将所分离材料破碎，然后根据物质特性分别采用风力分选、电磁分选和密度分选等方式将其按类分离，分别进行物质再生。化学分离可分为火法和湿法两大类。火法即在高温环境下，将材料中有机物氧化分解，从而使金属材料从有机材料中分离出来；或者利用金属材料不同的熔点和挥发点，将不同的金属分离出来。湿法通常采用酸溶的方法将材料中的金属物质溶解到酸溶液中，然后通过选择性化学沉淀、萃取等分离方式将各种金属分离出来。化学分离最容易产生二次污染，无论是火法处理过程中的烟气排放，还是湿法处理过程中废酸的处置，都非常容易产生重金属和酸性物质对环境的污染，而这些污染也正是通常所说的电子废物污染的主要形式。因此，选择适宜的分离工艺及其污染控制技术，是控制电子废物污染的关键环节。电子废物拆解、回收和再生过程中产生的残渣也是造成电子废物污染的重要环节。由于这些残渣中含有大量的重金属物质和酸性物质，如果不进行无害化处置而随意堆置，将会严重污染堆置所在地土壤和地下水，甚至污染地表水。而这些残渣的处置方式依据其性质不同可以分别采用焚烧、填埋处置，或者在水泥窑中进行共处置，同时按照危险废物污染控制要求进行管理和处置。

电子废物中的污染物质包括电子产品中含有的有害物质，如铅、镉、汞、铬、锌等重金属物质，以及电路板和其他组件中含有的溴代阻燃剂等有机有害物质。这些污染物质的释放过程主要是前述的电子废物拆解、分离和再生等过程，因此应该根据其污染产生过程和产生机理采取相应的污染控制措施，其主要治理技术包括焚烧（火法处理）烟气的处理、废酸的处理和残渣的处置等。

22.4.2 电子垃圾利用的国际经验

1. 德国

德国是欧洲最早关注电子垃圾管理的国家，于 1991 年 7 月颁布了《电子废弃物法规》，1992 年起草了《关于防治电子电器产品废弃物产生和再利用法草案》，1996 年公布了更为系统的《循环经济和废弃物管理法》。

德国利用各市区直属的市政企业对电子垃圾进行回收，提供网络化服务，上门收集废旧电器。在柏林等大都市，旧电器不允许随便丢弃，只要给制定机构打个电话、发个文件或电子邮件，就会有人上门服务，但居民需要根据电器的体积大小缴纳 10 欧元—30 欧元的运输费用。德国环境部发布的一份有关废旧电器免费回收的工作草案中，规定电器制造商必须承担回收利用旧电器的义务，并于 2005 年 8 月其开始实施消费者免费送还废旧电器的规定；草案再次明确电器制造商必须在生产时就考虑废旧产品的再利用问题，以尽量减少环境污染；草案同时还规定，待有关规定开始生效后，消费者可免费将诸如废旧电脑、电视机等家用电器送交社区回收点，由电器商统一处理。此外，德国环保政策中最重要的“谁污染、谁负责”也是治理电子垃圾的重要原则，根据这一原则要求，制造商负有主要责任，进口商、消费者也承担相应责任。

2. 日本

日本同样通过法律法规应对电子垃圾处理问题。1996年日本制定了《家电回收利用法》（2001年开始实施），该法贯彻“谁扔垃圾谁付钱”的原则，规定市民应负担废弃家电的回收处理费用。2001年4月开始实施的《资源有效利用促进法》中规定生产厂家有义务回收废旧电脑并将其进行再商品化或再生资源化处理。

日本每年都会产生大量的废旧电脑，尤其是2005年以后，每年约有500万台废弃电脑，废旧电脑的回收处理已经成为令地方政府十分棘手的问题，但日本高度重视电子垃圾的处理问题。2001年3月，日本电子工业振兴协会就颁布了2005年度关于生产厂家处理废旧电脑和废旧显示器应达到的平均再利用目标。该目标规定，再利用的零件类和二次资源化的材料应达到回收比重的60%。目前，日本的制造商已经开始研究回收方案，其中大部分将采用通过快递公司回收的方式，回收及再利用所需的费用有用户负担。这些再生企业将把从家电中拆解出来的废旧零件全部转化为再生资源，即转化为玻璃、塑料、铜、铝、铁等，总回收利用率达57%以上。

3. 美国

加州参议院于2005年6月4日批准的《电子废物再生法》要求电子制造商制定回收计划并且提供回收资金，同时规定了视频显示设备回收再生和限制物要求，该法案与2007年1月1日起强制实施。

美国是世界上最大的电子产品生产和消费国，同时也是电子垃圾的最大产生国。据分析，2000年美国约有2000万台电脑被废弃，其中仅有10%的废旧电脑被回收利用，其余大多数则被堆放在家庭或公司的仓库里。《美国新闻周刊》也报道，目前世界上各地废弃的电脑软盘加在一起，每隔20分钟就可形成一座100层高的“摩天大厦”。美国仅在加利福尼亚居民家中的废旧电脑就有600万台，每天有6000台电脑被淘汰。美国对电子垃圾的早期处理方法是掩埋和焚烧。大量的电子垃圾被填埋，成为潜在的环境隐患。近年来，由于公众反对的呼声日益增高，美国开始禁止填埋电子垃圾。2003年，IBM、惠普、柯达等电脑及电子产品生产商开始回收其电子产品。美国电脑生产商、地方政府和环保人士已经达成一项协议，准备建立覆盖全美的废旧电脑回收网络。在已经建立的几个废旧电脑回收中心，处理过程首先是整机或部件的再利用，然后是对不可用的部件进行分拆，塑料、金属、玻璃分门别类，最后进行熔炼。美国加州的立法机构通过了一项提案，要求顾客在购买新的电脑或电视机时，必须缴纳每件10美元的“电子垃圾回收费”，旨在为环保提供额外资金。

4. 欧盟

2004年8月13日，欧盟各成员国（包括新入盟的15国）完成了关于电子垃圾的立法工作，欧盟《电子垃圾处理法》正式出台。2005年8月13日，该法规正式开始实施。《电子垃圾处理法》的出台是依据2003年欧盟的两个指令完成的，这两个指令分别是《关于电子电气设备废弃物的指令》（WEEE）和《关于电子电气设备中限制使用某些有害物质的指令》（ROHS）。

20世纪90年代末，欧共体国家每年的废旧家用电子电器为540万吨—600万吨，每个公民每年平均产生17公斤—20公斤的电子或电器垃圾，占城市固体废物的4%。目前每年以3%—5%的速度增长，其增长速度是城市固体废物增长速度的3倍。以英国为例，每年更新手机1500万部，废弃手机形成1500吨固体垃圾。西欧曾对电子产品进行的一项市场销量调查表明，2002年，各种电子产品的总销量约为700万吨，电子废弃物总量约为400万吨，占整个欧洲废物流的

2%—3%。为降低电子垃圾对环境造成的严重污染，欧盟有关部门在2002年底决定，要求欧盟境内的电器制造商回收给累废旧电器，并在电器生产过程中更严格地限制使用特定的毒原材料。此外欧洲部分国家也有自己的相关条令规定，例如瑞典的法律规定废旧电子产品的处理费用由制造商和政府承担，挪威、荷兰、瑞士和意大利等国政府已要求电子设备制造商负责回收电子垃圾。法国更强调全社会要共同尽责，规定每人每年要回收4千克电子垃圾。电子垃圾回收遵循“谁生产、谁销售、谁使用、谁就负担相关环保费用”的权利和义务对等原则。根据该法令，自2005年8月13日起，从电脑、电视、电冰箱、洗衣机到电话和电吹风机，所有新出厂的电器都须印有小垃圾桶标志，表示其生命完结之后可以回收再利用。电子产品商作为回收主力，承担其产品未来的回收及循环再利用费用。欧洲主要国家废旧家用电子电器回收体系如表22-6所示。

表22-6 欧洲主要国家废旧家用电子电器回收体系

Table 22-6 Electronic Apparatus Recovery Systems of Major European Nations

国家	生产商的角色	废旧家用电子电器收集点	收集的家用电子电器种类	所需费用来源
奥地利	建立专门的回收系统	零售店、地方相关部门	冰箱、冷冻机、空调以及可以再利用的各种家用电子电器设备	消费者在购买时支付处理费用
德国	生态设计，提供有关的资料和信息	地方政府	大型家用电子电器、视机、办公电器等	消费者在购买时支付处理费用；地方政府从税收中拿出部分用于小型家用电子电器回收
法国	产品设计，提供相应的资料和信息	地方政府	大型家用电子电器、电视机、办公电器等	消费者在购买时支付处理费用
瑞士	建立并资助有组织的自愿者回收	商业网点、供货商、车站商用和家用大型耐用电子电器	消费者在购买时支付处理费用	
荷兰		零售网点或地方当局	大型家用电子电器、电视机、办公电器等	消费者在购买时支付处理费用

资料来源：《国外废旧家用电子电器回收再利用研究》. 中国人口资源与环境，2003

通过对各国做法的分析可以发现，对电子垃圾处理的宏观把握主要从以下几个方面着手：

（1）立法是解决电子垃圾污染问题的根本保障。许多国家都通过法律手段处理电子垃圾污染问题。

（2）市场环节是控制电子垃圾污染的重要一环。欧盟的WEEE和ROHS两项指令规定含有铅、汞等共六种有害物质的电子类产品禁止流通入欧盟市场。2001年4月开始实施《家用电器回收法》，强迫制造商家用电器的同时，也要求消费者承担大部分回收循环。发达国家大多通过价格调节，引导生产者、销售者和消费者绿色生产、销售和消费活动，保护生态环境。

（3）“生产责任制”是控制电子垃圾污染的基本原则。生产责任延伸制度（EPR）是欧美等国家针对电子实施的一项主要制度。生产者在产品的生命周期内承担责任，包括废旧产品的回收、处置等。此举有利于促进电回收和循环利用，同时也可刺激生产者采用易循环无害材质的使用。

22.4.3 我国电子垃圾利用

据统计，全球每年产生上千万吨电脑、手机、打印机等电子垃圾，其中70%以上通过各种“灰色”渠道进入我国。与此同时，国内每年产生的电子垃圾如今已达230万吨，仅次于美国的300万吨。电子垃圾中含有铅、镉、锂、聚溴二苯醚和聚溴联苯等几百种物质，其中50%对人体有害，在回收过程中如果处理不当，将严重污染环境，危害人身健康。令人担忧的是，目前国内数量庞大的电子垃圾进入循环利用体系的还不足10%。

我国是家用电子电器生产和消费大国，据统计，2008年我国电视机产量为9015万台、冰箱4600万台、洗衣机3900万台、空调6850万台、电脑13800万台、打印机6167万台、移动电话6亿部。由于电视机、冰箱等电器都在20世纪80年代中后期进入家庭，按照10年—15年的使用寿命计算，现在已经步入报废高峰期。据统计，2003年，我国每年至少有500万台电视机、400万台冰箱、500万台洗衣机要报废，而截至到2007年起，我国每年至少有1000多万台电视机、800多万台冰箱、1200多万台洗衣机、1000多万台电脑以及上亿部手机要报废。我国电子垃圾污染形式势严峻，究其原因，一方面是国内消费产生的电子垃圾总量大，目前的电子废弃物年产量约为111吨；另一方面是国外进入的，随着欧洲环保法令的日益严厉，中国正迅速成为电子垃圾的主要出口国和避风港——世界上有80%的电子垃圾被运往亚洲，而中国就接纳了其中的90%，中国进口的电子垃圾增长速度高达100%，1990年进口量为99万吨，进口额为2.6亿美元，占当年进口总额的0.49%，到2000年增加到1750万吨。我国家用电器年废弃量及预测值如表22-7所示。

表22-7 我国家用电器年废弃量及预测值

Table 22-7 Estimated Annual Quantity of Discarded Household Appliances of China

年废弃量（单位：万台）

年份	电视机	家用冰箱	家用洗衣机	室内空调
2012	3917.88	1086.99	2530.44	2524.40
2013	4041.73	2094.18	1374.37	3875.04
2014	4251.48	1242.00	1673.12	2992.61
2015	4449.13	1714.78	1519.46	3250.11

资料来源：《我国废旧家电产品回收现状与发展对策》. 长春大学学报（社会科学版），2008，(6).

电子废弃物行业在国内门槛过低，再加上不正当竞争，导致这一行业的发展面临尴尬。据中国再生资源回收利用协会统计，目前仅华北地区每年产生电子废物超过15万吨，到2015年我国电子废物年产生量将达到500万吨。但进入循环体系的电子废弃物不足10%，根本原因在于巨大的成本差价，使得正规的企业竞争不过小商贩、小作坊，比如手工作坊处理1斤电子垃圾成本只要6毛—7毛，而正规企业则要翻倍。为了追求经济效益，这些小商贩、小作坊大多采用就地焚烧、硫酸浸泡等原始落后方式提取贵金属，随意排放废气、废渣，对水汽和土壤造成严重污染。许多人为了从废旧电脑芯片中提取金属，经常使用王水，在这个过程中会导致大量酸性气体挥发。

2011年1月1日，《废弃电器电子产品回收处理管理条例》将正式实施，为保障和支持条例顺利实施，环境保护部等部门近期出台了相关配套政策。我国废弃电视、冰箱、洗衣机、空调和电脑等年产生量约2500万台，合50多万吨。《废弃电器电子产品回收处理管理条例》规定了废弃电器电子产品处理目录、处理发展规划、基金、处理资格许可、集中处理、信息报送等一系列制度。

国家鼓励制造企业参与电子垃圾处理，提供自主产品运作的回收标准，与科研单位合作开发废旧材料的再循环技术，积极借鉴国际先进电子产品垃圾处置和循环再利用经验，依靠技术引进、建立示范处理处置基地和建立循环经济型工业小区等措施，减少电子垃圾的污染。例如，无锡新区与新加坡伟成工业有限公司合作，由伟成工业有限公司投资6500万美元建设电子垃圾处理工厂，建成后将实现电子垃圾社会化回收服务体系。越来越多的科研院所也为电子垃圾处理提供了强有力的技术支持。

22.4.4　我国电子垃圾利用发展建议

让电子垃圾回收既节约资源又减少排放，规范产业发展是必然选择。对于资源短缺的中国来说，处理电子垃圾不仅是一个环保问题，也是一个资源战略问题。

实际上，对电子废弃物进行资源化利用，可以大大减少对环境的污染。例如，与生产新钢材相比，利用废弃电子电器产品提炼再生钢材，可以减少97%的矿废物、80%左右的污染，节约90%的原材料，而且两种钢材的性能相同。对于电子废弃物处理中产生的二次污染问题，可以通过建立循环经济园区，利用新工艺、采用集中产业和集中治污的办法来解决，简单取缔等于因噎废食，不能解决现实问题。

一是提高公众环保意识。政府和各级环保部门应加大环保宣传力度，使消费者提高认识，鼓励消费者购买环保产品以及正确处置废旧家电产品，主动配合做好废旧家电回收工作。同时应鼓励专家参与其中，制定高科技产品制造和处理的相关计划，避免涌现大量电子垃圾，争取在电子垃圾处理上走可持续发展道路。

二是完善相关法律体系。制定相应法律、法规，鼓励电子产品生产厂家使用无污染原料，或对使用含毒性强的生产原料的厂家征收高额的环境保护税。我国已于2009年3月4日发布了《废弃电器电子产品回收处理管理条列》，并于2011年1月1日起实施。在此基础上，政府有关部门应制定出相应的法规，使废旧家电的回收再利用步入“从商品到商品”的循环经济轨道。

三是要求企业负责垃圾回收。电子产品企业对自己的产品承担全部责任直至产品的废弃阶段，是电子产业的大势所趋，也是从源头上预防电子垃圾污染。

四是建立健全废旧家电回收体系。我国可以构筑以生产厂商为主体的定点或上门回收服务、以零售商为中心的回收服务和个体家电回收服务的综合回收网络体系。

22.5　我国固体废物利用动态

22.5.1　工业固废综合利用率剑指50%相关规划即将出台①

从中国资源综合利用协会获悉，由该协会牵头起草的大宗工业固体废物综合利用“十二五”专项规划即将正式出台。专项规划提出，到2015年，国内大宗工业固废平均综合利用率将达

① 资料来源：中国证券报，2011－09－28。

50%，比2010年提高10个百分点；同时，将出台示范项目政策，推动产业规模化发展。

在即将出台的《节能环保产业"十二五"发展规划》中，资源综合利用被明确为除节能和环保产业之外的三大重点支持板块之一。根据专项规划，到2015年，我国大宗工业固废总生产量将达30亿吨，重点涉及尾矿、赤泥、钢铁渣、煤矸石、粉煤灰、工业副产石膏和氨碱白泥等七大领域。其中利用量将达15.5亿吨，综合利用率达50%，2010年全国大宗工业固废平均综合利用率为40%。

专项规划指出，"十二五"期间，上述七大工业固废综合利用领域要攻克一批自主重大共性关键技术，建成一批示范项目，引导大宗工业固废综合利用产业向规模化、集约化发展，形成多途径、大规模、高附加值综合利用产业格局。

据了解，目前大宗工业固废处理及利用成效不一，例如，煤矸石和粉煤灰综合利用率已达65%和68%，而尾矿和赤泥利用率仅为14%和4%。规划提出的尾矿和赤泥利用率到"十二五"均达20%以上，提升空间巨大。

在工业固废处理利用领域，目前国内已有包括首钢股份、中国铝业、中国中冶等涉足尾矿、赤泥和钢铁废渣处理利用，大同煤业大股东同煤集团、兖州煤业大股东兖矿集团等已开始规模化加工利用煤矸石和粉煤灰。

22.5.2 垃圾发电或成新的能源增长点①

环境问题是制约当代经济社会可持续发展的重大主题。据统计，全球每年排放各类垃圾近5亿吨，中国主要城市年产生活垃圾1.5亿吨，并且还在以每年8%—10%的速度攀升。中国城市每年因垃圾造成的损失高达250亿元—300亿元。中国97%的城市垃圾只能采用堆放或填埋的方式简单处理。垃圾困局，已经衍变成一场席卷全社会的生态危机。

受垃圾处理技术的制约影响，目前中国城市垃圾处理主要是卫生填埋和焚烧、堆肥三种方式。卫生填埋是最传统的处理方式，需要占用大量的土地面积，无法实现垃圾的减量化，还会污染地表水、地下水，由于资金、技术的局限，相当多的"卫生填埋"基本上无法达到"卫生"的要求。堆肥方式同样无法实现垃圾的减量化，而且堆肥产品的市场接受率低，无法与化肥形成有效竞争，在发达国家基本上是不被采用的。而传统的原生垃圾焚烧方式由于自身技术的不完善，燃烧不充分，容易释放二恶英类污染物质，目前遭遇到了民众的强烈抵制。他认为，从长远来看，燃烧无疑是对垃圾减量化处理效果最好的一种方式，而利用燃烧产生的热能发电更加符合目前全球范围内环保低碳、资源的循环永续利用这一主题。

为了弥补目前国内城市居民垃圾分类意识的不足，中能佳程采用高度机械化的前分选设备将原生垃圾依据再生利用的渠道不同分为四大类，并提取其中可回收的金属、塑料等资源进行再利用。业内专家认为，该技术以良性方式弥补了城市发展的后顾之忧，为大中型城市的发展带来了全新的理念，有望成为突破"垃圾围城"的首要选择。在煤电资源日渐紧张的今天，生活垃圾综合处理技术带来的电能资源无疑将会是新的能源增长点。

22.5.3 河北省公布废弃电器电子产品处理发展规划

2011年8月22日，《河北省废弃电器电子产品处理发展规划（2011—2015）（征求意见

① 资料来源：凤凰网，2011-09-08。

稿)》出炉。其中规划，到2015年，全省拟建电子产品拆解企业13家，建成覆盖全省的废弃电器电子产品回收网络和安全处理体系，废弃电器电子产品处理能力达到每年430万台以上。

1. 全省年产电子垃圾超410万台

据统计，截至到2009年，河北电视机保有量2540万台、冰箱1470万台、洗衣机1950万台、空调950万台、电脑640万台。这些电器多数是20世纪80年代中后期进入家庭的，按10年—15年使用寿命算，从2003年起每年至少有410万台电器被淘汰，成为电子垃圾。

而《河北经济年鉴》中收集的数据显示，2011—2015年家庭用五类（电视机、电冰箱、洗衣机、空调器、微型计算机）废弃电器电子的产生总量约为2133.49万台。这其中，包括废旧电视机约373.23万台，洗衣机431.6万台，电冰箱400.79万台，微型计算机434.95万台，空调器492.91万台，电子垃圾总重量超62万吨，而这些数据还不包括机关、企事业单位产生的大量废弃电器。

2. 拆解能力每年仅220万台

2010年6月1日，河北省被列为“家电以旧换新”全国试点省市，随后我省53家企业被确定为以旧换新实施企业，全省建立销售、回收网点1万多个，遍及全省各市、县（市、区）、乡镇。

据统计，家电以旧换新实施半年时间后，累计回收旧家电375996台，其中电视机占到75.76%。但回收率仅占一年产生量的18.3%，回收率较低，电子垃圾未能得到全面控制。

另外，河北省电子垃圾拆解能力不足，截至2010年底，我省共有3家电子废弃物拆解单位列入电子废物拆解利用处置单位名录，合计拆解处理能力仅约220万台/年。由于正规拆解企业少，拆解能力不足，导致一些电子废物在小作坊进行简单拆解，资源利用率低，并对土壤、地下水源等造成严重的污染。

3. 五年内建成废弃电器回收网络

《规划》要求，到2015年，要建成覆盖全省的废弃电器电子产品回收网络，处理能力达到430万台/年以上。同时，要行业实现专业化、规模化、产业化发展；除经维修进入二手市场外，其余100%进入正规拆解企业；有利用价值的拆解产物再利用率100%；拆解产生的危险废物无害化处理率达到100%。此外，还将继续增加家电以旧换新回收网点，并拟到2015年，全省确定新增电子产品拆解企业10家。

4. 电子垃圾处理走规模化道路

目前，居民处理废旧家电一般有以下三种渠道：商家以旧换新、供销合作社下属的废品收购站，以及收购废旧家电的小商贩。然而，由于个体经营者环境意识差、监管难度大，回收后的废旧电器无法得到规范处理，甚至可能导致氟利昂、铅、汞等对环境的污染。处理电子垃圾应走规模化道路，这样才能专业化，污染也会相应降低。

第5篇　新兴能源篇

“十一五”期间，我国风能、太阳能、核能、生物质能等新能源产业发展迅猛，产业规模不断扩大，产业层次快速提升，产业政策体系逐步完善。

2011年作为“十二五”规划的开局之年，其重要性不言而喻。各项促进新兴能源产业发展的规划政策纷纷出台，备受关注的《新兴能源产业发展规划》业已上报国务院待批。

可以预见，未来五年新兴能源将进入新一轮的快速发展期，其发展将不仅限于产业规模的迅速扩展，更多地将体现在技术水平的提高，产业结构的优化和产业层次的进一步提升。

本篇在延续《2010中国节能减排产业发展报告》结构的同时，对篇章内容进行了调整。整体回顾了新兴能源产业在“十一五”期间的发展状况，同时，对“十二五”期间的发展前景进行了展望。在此基础上，本篇重点关注了太阳能、风能、生物质能与核能四大产业，进一步集中分析探讨了这四大产业的发展现状、产业政策、重点项目及发展趋势。

23　新能源与可再生能源概述

BP 能源预测显示，到 2030 年，世界能源需求在现有的基础上，仍将上升 45%，相当于在现在世界能源消费水平上增加两个美国的消费量。但依据目前化石能源的开采水平和新能源的发展速度，未来 20 年的能源消耗不仅对环境造成巨大的破坏，而且能源供应能否具有可持续性，能源价格能否在人们的承受范围之内都存在着重大的不确定性。

从全球范围来看，目前化石能源仍在能源结构中占有绝对优势，如果未出现重大能源创新，不改变现有的能源消费结构，那么未来几十年，无论是能源供应水平，还是碳排放压力，都可能超出人们的承受底线。

新能源和可再生能源开发利用日益受到国际社会的重视，许多国家提出了明确的发展目标，制定了支持可再生能源发展的法规和政策，使可再生能源技术水平不断提高，产业规模逐渐扩大，成为促进能源多样化和实现可持续发展的重要能源。欧洲“2020”战略中将在未来 10 年着重发展“低碳经济”；美国将“智能电网”作为其国家战略中的核心部分。欧盟不仅提出在 2014 年之前建成欧洲单一的能源市场，而且将 2020 年可再生能源比例调整为 20%。美国总统奥巴马在近期的国情咨文中表示，2015 年要见到逾百万辆电动汽车行驶在公路上，到 2035 年，美国的电力将有 80% 由可再生能源提供。

IEA（国际能源署）在《世界能源展望 2010》（World Energy Outlook 2010）中给出的“新政策前景”预测：全球能源需求将持续增长，但增长步伐将比近几十年放缓，化石燃料的份额将从 2008 年的 81% 降至 2035 年的 74%。这一预期是根据世界各国政府所宣布的范围广泛的政策承诺作出的。中国已宣布其在《哥本哈根协议》下的承诺，至 2020 年全国单位国内生产总值二氧化碳排放比 2005 年下降 40%—45%，非化石能源占一次能源消费的比重提高至 15% 左右。

考虑到中国当前人均能源使用量较低，持续的经济增长趋势显著，IEA 预计在今后的几十年内，中国的能源需求仍会保持强势增长趋势。根据 IEA 的预测，2035 年中国能源消费量将达到世界能源消费量的 22.3%，如图 23－1 所示。

要在保持经济快速增长的同时应对能源需求和气候变化的挑战，新能源与可再生能源的开发利用将起到极为重要的作用。在过去的几十年中，中国已经采取了多项重要的政策措施，旨在推动非化石能源的发展。据国家统计局公布的数据，2010 年中国单位 GDP 能耗比 2009 年减少了 4.01%。“十一五”期间，中国单位 GDP 能耗下降了 19.1%，电力是全球能源总消耗的关键组成部分。根据中国国家能源局公布的数据，截至 2010 年底，火力发电在总装机量中仍占主导地位，非化石能源装机比重合计占 26.6%（如图 23－2、图 23－3），比上年提高 1.1 个百分点，累计发电量 7862 亿千瓦时，按发电煤耗折算约合 2.63 亿吨标准煤。2010 年，全国电源工程建设完成投资3641亿元，非化石能源建设投资占电源建设总投资的比重达到63.5%。

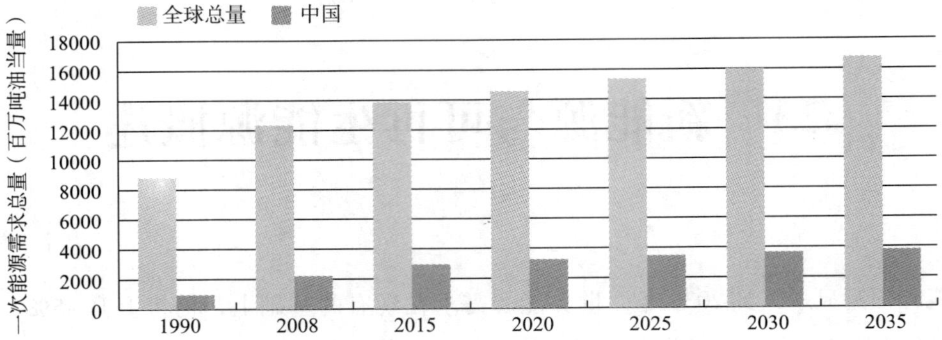

图23－1　能源需求量增长预计

Figure 23－1　Estimation for Increase in Energy Demands

资料来源：IEA World Energy Outlook 2010－New Policies Scenario，SEMI

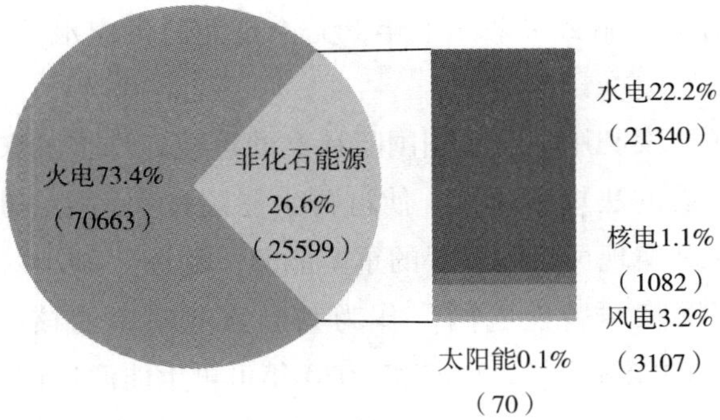

图23－2　2010年中国发电装机总容量（万千瓦）

Figure 23－2　China's Total Installed Capacity of Power Generation in 2010（10000 KW）

资料来源：中国国家能源局

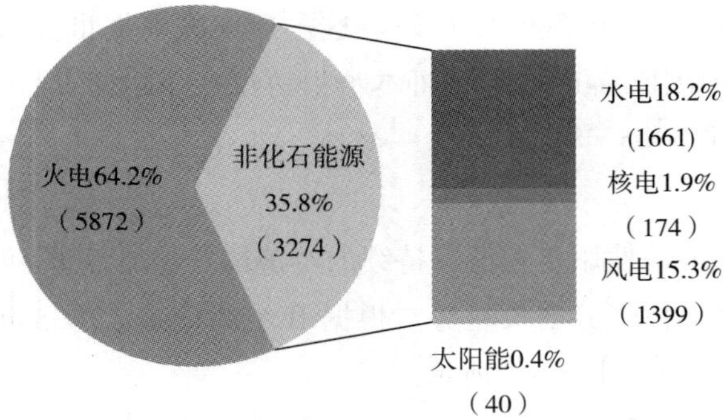

图23－3　2010年中国新增发电设备容量（万千瓦）

Figure 23－3　Newly Installed Power Generation Capacity of China in 2010（10000 KW）

资料来源：中国国家能源局

我国“十二五”规划中特别提出“加强现代能源产业，推动能源生产和利用方式变革，构建安全、稳定、经济、清洁的现代能源产业体系。”

为实现到2020年的提高碳减排和非化石能源比重“两个目标”，在未来10年内，中国政府将进一步扩大和推动新能源领域的投资。已上报国务院审批的《新兴能源产业振兴规划》为2011—2020年间的中国新能源开发利用和传统能源的升级变革作出了具体部署规划，拟在2011—2020年的10年内，以开放市场吸引内外资的方式，累计增加直接投资5万亿元人民币，发展中国新兴能源产业。

23.1 新兴能源产业“十一五”回顾

“十一五”期间，我国太阳能、风能、核能、生物质能等新能源产业发展迅猛，产业规模不断扩大，产业层次快速提升，产业政策体系逐步完善。核电等重要领域进入世界先进水平，在未来能源竞争中占据领先地位。

“十一五”期间，我国水电建设新增机组接近我国有水电以来前95年的总和，水电装机突破2亿千瓦。龙滩、景洪、构皮滩、拉西瓦等大型水电站先后建成，累计投产9000万千瓦。三峡左右两岸26台机组，1820万千瓦全部并网发电，累计发电量达到4500亿千瓦时。

我国核电利用达到世界先进水平，连续两年成为全球核电在建规模最大的国家。2005年以来，国家先后核准了13个核电项目，共34台机组，3702万千瓦。目前，在建机组28台，3097万千瓦，在建规模占世界的40%以上。设备技术与运行稳定性均处于世界领先水平。

太阳能产业快速发展。目前，我国已成为世界光伏第一产能大国。2010年中国的光伏产品制造产量占全球产量的50%以上，产品占据世界主要市场的主要份额。全国光伏发电装机规模达到60万千瓦，年产量达到800万千瓦，主要企业已形成完整产业链，从多晶硅原料到光伏系统安装，以及玻璃板、封膜等配件，中国都已占据世界优势，“两头在外”的状况已得到初步改观。行业建立起两个国家重点实验室，逐步迈向高端研发。国内光伏发电市场有序启动，2009年敦煌1万千瓦光伏电站项目实施招标，现已建成，全部并网发电。2010年又在西部六省区组织28万千瓦光伏发电项目招标，带动了一批光伏电站项目建设。

“十一五期间”以太阳能热水器为代表的太阳能热利用行业成就显著：2010年真空管热水器生产4600万平方米，增长了15%，平板热水器300万平方米，增长了50%，总计热水器比上一年增长了16.7%，而这一增长量是“十一五”期间增速最慢的一次。2010年保有量为1亿6800万平方米，增长了15.9%，表明产量在五年中翻了6.7倍，保有量翻了6.5倍。2010年末产量接近世界的60%，保有量达到80%，显示出中国是太阳能热水器生产及应用大国。2010年产值为735亿，比2009年增长了22.5%，230多个企业的产品出口世界154个国家，已有五六个企业开始在海外建厂。

风电产业迅猛发展。全国风电吊装容量累计达到3100万千瓦，连续五年翻番增长。从2005年开始，中国的风电总装机连续五年实现翻番。2010年中国风电总装机比上年增长约62%，位列世界第一。24个省市建设了风电场，河北、内蒙古、甘肃等地的国家级风电基地，进入快速成长期。2010年有4家中国企业进入了世界风电装备制造业10强。2010年12月，中国首座大型海上风电场全部风机并网发电，它采用华锐风电制造的中国第一批具有完全自主知识产权的3MW海上风机，填补了大型海上风电机组的空白，标志着中国大功率风电机组装备制造业跻身世界先进行列。

坚强智能电网建设起步。为支持风电发展和大范围消纳，国家电网公司加大电网建设力度，截至2010年底，共投资418亿元，建成投运风电并网线路2.32万公里，国家电网经营区域内风电并网装机达2826万千瓦，占全国最大用电负荷的4.2%，连续五年实现翻番。为最大限度接纳风电、太阳能等清洁能源，国家电网加快技术标准制定和关键设备的研发应用，提高电网优化配置能力。目前已制定15项智能变电站标准，形成了世界首套智能变电站系列技术标准，申请专利126项，整体技术水平国际领先。截至目前，已建成陕西、延安、江苏等8个智能变电站，成为引领世界变电站技术发展的中坚力量；已在北京、重庆和河北建成6个智能小区，在上海建成1座智能楼宇，构建了便捷的智能家居生活。

23.2 新兴能源产业“十二五”展望

“十二五”期间，中国将新能源产业列入国家重点支持的七大领域之一。日前，备受关注的新能源规划最终定名为《新兴能源产业发展规划》。业内人士认为，预计新能源发展规划出台后，未来10年我国新能源投资将达5万亿元。

这一规划重点支持的领域集中在风能、太阳能、核能、生物质能、水能、煤炭的清洁利用、智能电网等七大方面。在具体实施路径、发展规模以及重大政策举措方面，对新能源的开发利用和传统能源的升级变革进行了部署。未来五年，在国家政策推动和新能源技术驱动下，我国新兴能源产业将继续保持良好的发展态势。

预计到2020年，中国新能源发电装机2.9亿千瓦，约占总装机的17%。其中，核电装机将达到7000万千瓦，风电装机接近1.5亿千瓦，太阳能发电装机将达到2000万千瓦，生物质能发电装机将达到3000万千瓦。据规划预计，新兴能源产业规划实施后，到2020年将大大减缓对煤炭需求的过度依赖，能使当年的二氧化硫排放减少约780万吨，当年的二氧化碳排放减少约12亿吨。规划期累计直接增加投资5万亿元，每年增加产值1.5万亿元。

23.2.1 大力开拓国内光伏市场

2010年我国光伏产量占到全世界的49.4%，大陆和台湾产量占全世界的64.8%，中国是名副其实的第一太阳能大国。但是从装机量看，2009年世界光伏市场仍主要在欧洲，占世界总安装量的78%以上，德国占据世界最大市场，占世界总安装量的52%。2010年德国安装量为7GW，约占世界总安装量的55.2%；而我国装机量只为400MW，仅占世界总安装量的2.76%，太阳能电池连续五年96%以上出口，所以中国的光伏产业完全依赖国际市场，急需开拓国内市场。

为了大力开拓国内光伏市场，“十二五”期间，中国将建设100座新能源示范城市，并进一步建设并网光伏发电站，在中西部等太阳能资源丰富的地区发展微电网示范区，通过多种手段推动光伏发电在国内的应用。2011年，国家将继续在西部地区开展光伏电站项目特许权招标，总规模在50万千瓦左右（500MW）。国家能源局正在编制《光伏发电十二五规划》，据透露，到2015年，中国国内的光伏装机容量目标将达到10GW；到2020年，目标至少要到50GW。这一目标大大超出了光伏企业的预期。业内人士纷纷表示，未来几年中国市场将会呈现爆发式增长。

23.2.2 发展太阳能热利用产业

“十二五”期间，国家将进一步扶持太阳能热利用产业，规定在内蒙、甘肃、青海、新疆、西藏等适宜地区开展热发电试点，还将2015年太阳能热水器定下4亿平方米的发展目标，同时计划在农村推广太阳能热水器、太阳灶等热利用装置，实现村镇的阳光浴室工程。相比较于光伏发电，光热发电在世界范围内都是刚刚起步。热发电被列入了我国战略性新兴产业规划、可再生能源中长期发展规划。据估测，在未来的五年，太阳能热发电的发电量会激增18倍，从现在的年生产5亿8千万瓦增长到大约1万兆瓦。“十二五”末期，我国太阳能热发电将开始初步的商业化运营。

“十二五”期间，要以科技进步、扩大利用、产业升级、扩展市场为主要指导思想，全面提升太阳能热利用在可再生能源利用的比例，提高节能减排的贡献力。以大力推广中温系统为主要方向（参见专栏23-1），培养有国际竞争力的大型骨干企业，装备向自动化方向迈进。工业应用也是发展重点，要向大型供热系统迈进。热发电商业化还需一段时间，集热管的实践还有待于时间的考验。预测到2015年，太阳能热水器年产量将达到12000万平方米，保有量达到4亿平方米左右。

专栏23-1

全球首个CPC中温太阳能工业热力系统落成

5月10日，全球第一个CPC中高温系统在工业热能领域的应用项目——力诺瑞特CPC中高温太阳能锅炉落成暨“工业绿动力”计划在山东济南启动，标志着中国太阳能光热产业正式进入工业热能应用阶段。据了解，力诺瑞特太阳能锅炉项目是目前规模最大、系统最完善、水温最高、太阳能与工业结合最好的系统，它依托于其耗时五年、投资1.2亿元研发的太阳能热能应用系统技术，意在为燃煤、燃气、燃油锅炉作高效热力补充，实现对全国工业热能市场的开发。

我国工业能耗自1996年以来一直占据全社会总能耗的70%以上，工业与太阳能的结合始终是难以攻克的课题。对此，业内专家称，工业用热温度较高，以造纸、食品、烟草、木材、化工、医药、纺织、塑料等八大工业行业为例，其用热温度大部分在80℃—250℃之间，要使太阳能在这些产业中顺利应用，必须在核心太阳能功能部件上达到较高温度。力诺瑞特太阳能锅炉项目正是在CPC中温集热器与锅炉集成热力系统之上，采用企业独有的全玻璃真空中温镀膜管、CPC反光板等技术，通过聚焦吸收更多的太阳热能，将15℃左右的冷水加热至95℃，再将95℃热水加热成150℃蒸汽，突破了国内原有集热器的低温限制，解决了中国太阳能中高温工业热利用技术的难题。

资料来源：解放日报，2011-05-14

23.2.3　海陆风电双重发展

“十二五”期间，国家将有序推进大型风电基地建设，重点发展内蒙古、甘肃、新疆及东北地区等千万千瓦级风电基地，同时也将加快海上风电开发。我国风电开发已经逐渐由陆上风电转向海陆风电双重发展，东部沿海的风电发展将成为今年风电发展的新增长点。据“十二五”规划，我国7个千万千瓦级风电基地中的甘肃、河北、吉林、蒙东、蒙西、江苏沿海要基本建成。2011年，我国将陆续开工建设甘肃酒泉二期500万千瓦、新疆哈密200万千瓦、内蒙古开鲁200万千瓦、吉林通榆150万千瓦风电项目；同时启动江苏新的100万千瓦海上风电项目，大力推动河北、山东、浙江、福建等地海上风电项目，争取并网风机累计达到5500万千瓦，并抓紧建设上海东海大桥海上风电项目二期工程（参见专栏23-2）。

专栏 23 -2

海上风电成为角逐的重点战场

在陆上风电增长渐趋平稳的背景下，海上风电成为角逐的重点战场。

近日，国家能源局为中国海上风电定下的目标为：2015 年建成海上风电 500 万 KW，2020 年达到 3000 万 KW。为此能源局决定于今年下半年启动第二批海上风电特许权项目的招标准备工作，预计明年上半年完成招标，总建设规模在 150 万 KW—200 万 KW。跟第一批特许权项目相比，这次招标不仅规模翻番，而且覆盖范围更广，将从江苏延伸至河北、山东、浙江甚至广东等地区。

2010 年全球陆上风电增速放缓，海上风电则延续 2008 年和 2009 年持续增长势头，新增装机出现翻倍增长。据 BTM 咨询公司统计，截至 2010 年底，全球海上风电新增 1444MW，累计装机量达 3554MW。陶刚表示，和陆上风电不同，海上风电的入网条件比较好，海上风电开发集中在东部沿海经济发达的区域，用电需求大。近日有消息称，风电并网技术国家标准已通过终审有望出台，如果并网标准确立则将为海上风电的发展提供更好的环境。

资料来源：海上风电第二轮招标在即价格竞争或免恶战覆辙．南方报业网

http：//nf. nfdaily. cn/nfdsb/content/2011 -07/13/

23. 2. 4　从快速增长向提升质量转变

未来我国风电装机容量增速将相对放缓，但绝对数量仍相当可观，风电产业发展将从数量的快速增长转向研发能力和产品质量的不断提升。

2010 年，几起恶性的风电事故给近年来突飞猛进的风电发展敲响了警钟，风电设备的质量问题凸显出来（参见专栏 23 -2）。我国风电机组关键设计技术依然依托欧美，风电机组的检测认证体系还不健全，在风电场缺乏长期运营维护的经验之下，中国风电质量大考已经开始。

目前我国风电产业存在以下几大问题：

第一，研发与自主创新能力较弱，多数企业尚未解决自主知识产权问题，核心技术、核心部件研发对外依赖性强。

第二，质量与可靠性是市场竞争的首要关键因素，然而中国整机企业运营历史普遍较短，缺少经验和技能的积累，过去几年的高速增长掩盖了一些发展中的问题。

第三，欧美同业已建立了相对完善的运营体系，而中国企业应如何充分利用全球资源，在世界范围实现本土化的资源配置是一大挑战。

第四，学习如何建立现代化的经营管理体系，培养和引进全球一流的风电管理与技术人才是迫在眉睫的需求。

“十二五”期间，中国风电企业应着力打造技术先进、质量可靠的产品，努力提升综合服务能力，实现客户价值最大化。积极开发风光互补、智能电网、风能海水淡化、储能技术、大型非并网发电系统等风能应用的高端产业链，多参与国际竞争与合作，进一步做大做强。

专栏 23－3

甘肃“二·二四”酒泉风电事故

2月24日，西北电网甘肃酒泉风电基地因桥西第一风电场35千伏电缆馈线电缆头三相短路故障，导致近600台风电机组脱网，损失电力84万千瓦。酒泉风电是国内最大的陆上风力发电场之一，电监会称该起事故是近几年我国风力发电发生的对电网影响最大的一起事故。

国家电网有关人士表示，此次事故暴露出部分并网运行风电机组实际不具备低电压穿越能力。据记者了解，具备低电压穿越能力的风机，当电网发生故障时，风力发电机组能够不间断并网，从而提高电网的安全稳定性。但目前国内已投运的风电机组多数不具备此技术。国家电网人士介绍，目前具有低电压穿越能力的风机成本较高。

资料来源：电监会通报“二·二四”酒泉风电事故．新京报

23.2.5　生物燃料为发展重点

生物质能不同于风能、太阳能。生物燃料是唯一可以作为气态、固态、液态三种形态存在的可再生能源，因此是其他可再生能源无法相比的。我国目前生物质发电全国装机达到50MW以上，今后发展将从科技到产业重点发展气体燃料、固体燃料和液体燃料。生物质能的资源来源是多元化的，包括传统可再生能源、新型可再生能源、将来潜在的利用新技术开发新的资源品种。

生物质能更强调综合性，生物质能燃料是“十二五”发展的重点。“十二五”期间要由生物质能技术示范转变为规模化应用、产业化发展，提出建设新能源示范城市，并加大推动绿色能源县建设的力度，因地制宜，加强规划协调，多能互补，加快发展。

传统生物质资源的问题是资源量不能满足燃料工业生产队原料的需求，农林生物质资源，效率低下，生长周期过长，并且资源分散，收集半径太大，生产生物燃料成本过高。这些因素导致了对新型生物质资源（例如，培育藻类等非食用能源作物专用新品种）的迫切需求。目前生物质能技术发展的总趋势是：（1）原料供应从以传统废弃物为主向新型资源选育和规模化培育发展；（2）高效、低成本转化技术与生物燃料产品高值利用始终是未来技术发展核心；（3）生物质全链条综合利用是实现绿色、高效利用的有效方式。

未来生物质能发展的思路是以发展优势、重点突破、多元互补为原则，以高新技术与核心技术研究开发为手段，逐步向资源供给多元化、转化技术高效化；设备装备成套化、终端产品高值化；多元产品联产化、产业布局区域化方向发展。

23.2.6　可再生能源科技发展目标

科技发展重点支持大规模利用风能、太阳能和生物质能等可再生能源技术与系统，突破了一批关键技术；重点发展数兆瓦级风力发电、高效太阳电池、兆瓦级太阳能热发电技术装置，建设了一批产业化示范装置与系统。

“十二五”期间科技发展的总体目标为：（1）加大可再生能源的规模化和产业化技术开发；（2）在可再生能源利用系统和关键部件的设计、制造、安装和运行方面攻克一批制约产业发展的瓶颈问题，掌握核心技术；（3）培育一批具有参与国际竞争能力的可再生能源相关企业，建

立有效的产学研用结合机制；（4）培养造就一支高水平、高素质的科技创新队伍；（5）创建可再生能源国家技术创新体系。低成本薄膜和新型太阳电池产业化关键技术、百兆瓦级大型并网光伏电站关键技术与装备、太阳能热发电站关键技术、陆上风电场及电机组关键技术、近海风电技术、生物质液体燃料制备关键技术、海洋能发电技术和地热能低温利用和中高温发电技术作为可再生能源主要技术方向。

23.2.7 建成坚强智能电网，解决电网消纳问题

如果要完成中央提出的2020年非化石能源占一次能源消费15%的目标，那么2020年水电要达到4亿吨标准煤，比现在翻一倍，要在保证生态及安全的基础上加快水电发展。保持风电产业平稳较快发展最大的制约因素不是经济问题，也不是技术和制造，而是电力系统对于风电的消纳能力。同时也要研究风电消纳低谷期的用电消耗方法，比如农业灌溉使用低谷电等。

国家电网宣布，到2015年，将基本建成坚强智能电网，智能化程度达到国际先进水平，实现接入风电1亿千瓦和光伏发电500万千瓦的目标；到2020年，坚强智能电网可实现二氧化碳减排量约16.5亿吨，二氧化碳排放强度比2005年下降8.8%，对实现我国2020年碳减排目标的贡献率可达20%—23%。“十二五”期间，国家电网公司计划建设的特高压交流骨干网架将连接大媒电、大水电、大核电及大型可再生能源基地与主要负荷中心，促进能源资源更大范围的优化配置，推动清洁能源大规模开发和利用。

23.3 小结

新能源和可再生能源，对优化能源结构，保障能源安全，拉动经济增长，改善生态环境，应对气候变化，促进经济社会可持续发展均具有重要意义。世界能源委员会的报告中称，到2020年全球新能源市场的商机将达2万亿美元，很多国际机构预计，未来的3年—5年，绿色能源产业每年将有20%的持续增长力。预计在2010—2020年间，大多数可再生能源技术可具有市场竞争力，在2020年以后将会有更快的发展，并逐步成为主导能源。

目前，中国已跻身于世界主要的能源消费国和温室气体排放国家之列，在全球能源市场上的重要性日益凸显。为实现到2020年的提高碳排放和非化石能源比重“两个目标”，在未来10年内，中国政府将进一步扩大和推动新能源领域的投资。不难预见，在国家政策和新能源技术进步的双重驱动下，我国的新兴能源产业将继续保持良好的发展态势。

24　太阳能

随着全球经济的快速发展，煤炭、石油等不可再生能源供应日趋紧张，开发使用新能源已成当务之急。太阳能作为一种丰富、洁净和可再生的新能源，它的开发利用对缓解能源危机、保护生态环境和保证经济的可持续发展意义重大。

当前太阳能发展的主要方向是光伏发电和热利用。我国在光伏产品制造、太阳能光伏发电和太阳能热利用方面均取得了较大发展。目前，我国已成为光伏第一产能大国，产业链日趋成熟。同时，在政策的推动下，国内市场有序启动。在太阳能热利用方面，太阳能热水器行业成就显著，产量和保有量均居全球第一，与建筑一体化的结合更是有望给该行业带来可观的市场机会。光热发电产业随着技术的成熟，在国家政策的大力扶持下，亦将得到快速的发展。

“十二五”期间，国家将进一步加快发展太阳能利用产业，做大经济总量，调优产业结构，推进经济转型。

24.1　现状分析

近年来，全球光伏发电成长速度惊人。从1998年开始，世界光伏市场出现了供不应求的局面，年增长率保持在30%—40%，2004年更是创造历史新高，达到67%。2010年国际光伏产业在经历了2009年的短暂低迷期后，又迈入了新一轮的增长期。

而在太阳能热利用方面（参见专栏24－1），除了常规的太阳能热水器产业外，光热发电产业也渐渐升温，开始进入投资者和战略决策者的视野。光热发电较之光伏发电和风电，更有利于电网的稳定，并且没有了光伏发电中昂贵的硅晶光电转换，节约了成本，避免了污染。欧美各国已经开始将光热发电作为未来替代传统能源的主导选择。而随着位于内蒙古鄂尔多斯的50MW光热发电项目特许权招标的开始，我国光热发电产业也正式启动。

专栏24－1

太阳能利用方式

太阳能利用方式可简单分为光利用和热利用。前者通过转换装置把太阳辐射能转换成电能进一步利用，光电转换装置通常是利用半导体器件的光伏效应原理进行光电转换的，因此又称太阳能光伏技术；后者通过转换装置把太阳辐射能转换成热能进一步利用，属于太阳能热利用技术。

基于利用方式的不同，太阳能产业划分成太阳能光伏产业和太阳能热利用产业。以硅材料的应用开发形成的产业链条称之为光伏产业，包括高纯多晶硅原材料生产、太阳能电池生产、太阳能电池组件生产、相关生产设备的制造等。太阳能热利用产业主要用于热水器，因此往往又被称为太阳能热水器产业。

24.1.1 太阳能光伏产业现状

1. 国际光伏产业现状

2009年世界太阳能光伏产业受全球金融危机冲击，进入了一轮阶段性低迷期，而2010年随着全球宏观经济环境的逐步好转，全球光伏产业又迈入了一轮新的增长期。市场研究机构Solarbuzz 3月16日发布的最新年度太阳能市场报告指出，2010年全球太阳能市场安装量达18.2GW，相较于前年增长139%，其中，欧洲地区太阳能市场规模为14.7GW，占全球81%的需求。而欧洲区域前三大太阳能市场国家依序为德国、意大利与捷克，三个国家市场规模共计12.9GW。2010年日本与美国市场的成长非常迅速，年成长率分别达到101%与96%。总体来说，2010年有超过100个国家为太阳能产业需求的爆发式成长或多或少多地作出了贡献。

2010年，美国、印度等光伏产业新兴国家加大了对光伏产业的政策扶持力度，促进了光伏市场的快速发展；而作为世界上最大光伏市场的欧洲开始盛行削减光伏上网电价补贴，这一举措引发了一轮抢装潮，在一定程度上刺激了2010年光伏产业的爆发式成长。由于市场普遍预期欧洲主要国家的光伏发电上网电价补贴会持续下调，并且这种下调是梯次性，因此许多项目业主都想赶在光伏上网电价下调之前建成相关项目，获取相对较高的上网电价补贴，这就引发了一波抢装潮。在世界上光伏装机容量最大的德国，这种现象尤为明显。早在2009年德国政府就表示，要在2010年下调光伏发电上网电价的补贴，最终确定的方案是2010年分两次下调：即7月1日开始下调13%；10月1日再在7月1日补贴额度的基础上下调3%。由于补贴削减的幅度较大，所以许多项目业主都想抢在补贴下调之前安装光伏发电系统，这实际上是在透支2011年德国新增光伏装机容量。

在光伏市场带动下，光伏电池产量也在持续增大，2010年全球光伏电池产量达到21GW，与2009年的10.5GW相比，翻了一番。其中尚德公司和中国晶澳太阳能公司分别以1.572GW和1.46GW的产量超越美国First Solar公司，位居全球第一、第二位。接下来则分别是美国Frist Solar公司和日本夏普公司，其产量分别为1.41GW和1.174GW。全球前10大电池片企业的总产量达到11GW，占据今年全球光伏新增装机量近64.7%。在这10家电池企业中，中国大陆有4家，而台湾地区则有2家。全球主要光伏企业电池产能/产量情况如表24-1所示。

表24-1 全球主要光伏企业电池产能/产量情况

Table 24-1 Production Capacity/Volume of Major Photovoltaic Cell Manufacturers in the World

序号	企业	产量（MW）	产能（MW）	2011年产能预计（MW）
1	尚德	1570	1800	2400
2	晶澳	1460	1900	3000
3	First Solar	1410	1502	2250e
4	夏普	1174	1300	1680e

续表

序号	企业	产量（MW）	产能（MW）	2011年产能预计（MW）
5	英利	1061	1000	1700e
6	天合	1060	1100	1900e
7	Q－Cells	1014	1235	1300e
8	茂迪	850	1150	1750e
9	昱晶	800e	930	1500e
10	京瓷	600e	600	1000e

资料来源：CPIA，SEMI

在全球前十大光伏组件企业中，以生产晶体硅太阳能电池为主的中国无锡尚德以1.572GW的产量超越美国First Solar公司位居全球首位；美国First Solar则以1.41GW的产能位居次席，其电池产品主要是碲化镉薄膜电池，同时该公司也是全球最大的薄膜电池生产企业；位居全球第三位的是日本夏普公司，产量为1174GW，其产品结构中除了晶体硅电池外，也有硅基薄膜电池。在全球前七大组件企业中，我国占据有5席，而这七家组件企业总产量也达到7.87GW，约占据全球光伏装机量的45%。全球主要光伏企业组件产能情况如表24－2所示。

表24－2 全球主要光伏企业组件产能情况

Table 24－2 Spare Part Production Capacity of Major Photovoltaic Cell Manufacturers in the World

序号	企业	产量（MW）	产能（MW）	2011年产能预测（MW）	2011年出货量预测（MW）
1	尚德	1570	1800	2400	2200
2	First Solar	1410	1502	2378	2200
3	夏普	1174	1300	1680	1680
4	英利	1061	1000	1700	1700
5	天合	1060	1100	1900	1700
6	阿特斯	803	980	1850	1250
7	韩华（林洋）	797	900	1500	1150

资料来源：CPIA，SEMI

专栏24－2

晶澳太阳能：布局全球向千亿迈进

晶澳成立于2005年，两年之内在美国完成上市。在过去的五年，晶澳一直以平均100%的速度成长，去年增长率更是超过了180%。在当前光伏行业发展低迷期，企业却呈现出超强的发展力，并持续保持着高速发展态势。根据国际权威统计机构的统计，去年第三季度，晶澳太阳能电池发货量超过美国第一太阳能，成为全球最大的专业制造太阳能电池片制造商。

上海晶澳公司CEO方朋在多种场合中表示，目前光伏行业调整已经进入第二阶段，晶澳正在全力布局全球市场，向千亿规模目标迈进。

目前，太阳能行业已经发展到了需市场规范、企业整合的阶段，太阳能行业已经经过第一个周期，刚刚进入第二个周期。太阳能行业发展到目前规模后，行业整合需求明显，通过行业周期性的调整，优秀的企业会继续壮大。

从产能、规模来说，晶澳是国内、国际上最大的光伏企业之一，其目标是要达到全球15%—20%的市场占有率。晶澳对企业现行布局始终掌握节奏，在不影响公司财务的结构下稳步推动企业发展。现在晶澳产能已达到2GW，加上现在的产能规划，未来会达到6GW—8GW的产能，晶澳正在向千亿规模集团的目标迈进。

资料来源：晶澳太阳能：布局全球向千亿迈进．中国能源报

同时，在产能方面，受到全球光伏市场强劲需求的带动，这七大光伏企业也纷纷扩产，全球产能将再攀新高。在光伏组件方面，预计2011年七家企业总产能将达到13.48GW，增长率达到54.7%。其中，中国无锡尚德2010年电池组件已达到1.8GW，而在2011年将继续扩产，预计将达到2.4GW；美国Frist Solar公司则继续在越南、美国、马拉西亚等地继续新增生产线，预计到2011年，其电池生产线将达到38条，以目前每条生产线产能62.6MW算，其产能将由目前的1.5GW增长至2.3GW。日本夏普公司预计将产能从1.3GW增至1.68GW；而中国英利、天合光能和阿特斯也将分别由目前的1GW、1.2GW和980MW扩展至2011年的1.7GW、1.9GW和1.85GW。

在电池片方面，除了尚德、天合和英利等几家企业的垂直一体化发展之外，中国晶澳太阳能公司（参见专栏24-2）的电池片产能将会有一个质的飞跃，将由2010年的1.9GW扩产至2011年底的3GW，中国阿特斯的电池片产能将在2011年中旬达到1.3GW，而台湾地区的茂迪公司也将其产能从2010年的1150MW扩产至1750MW。届时，这10家电池片企业的产能将由2010年的12GW增长至2011年底的18.48GW，增长率达到54%，与组件的增长率有着惊人的相似。

2010年是光伏市场异常火爆的一年（参见专栏24-3），随着光伏组件售价的不断下降，使得在德国、意大利等国家投资光伏系统变得非常有利可图，内部收益率甚至可达到15%以上。特别是众多系统集成商为了赶在德国等光伏补贴下调前安装，以获得高额的投资回报，使得对光伏组件的需求发生了跨越式的增长。电池组件制造工厂里的机器没日没夜的咆哮可却依然不能满足源源不断的订单需求，一些大型企业甚至不得不寻求代工，巨大的出货量也使得这些企业在2010年赚得盆满钵溢，例如美国First Solar公司的营业额达到25.6亿美元，第一季度的毛率甚至达到49.7%，实行垂直一体化的企业由于在生产成本方面的优势，其毛利率普遍比单经营某一环节的企业为高，例如中国英利和常州天合公司的营收也在18亿美元以上，毛利率也在30%以上，而主要以电池组件为主要业务的中国尚德和晶澳公司，收入虽然分别达到29亿美元和17.8亿美元，但其毛利率分别为17.4%和21.7%，2010年走垂直一体化之路的企业获益颇丰。

专栏24-3

欧洲再现光伏抢装潮国内光伏行业酝酿反转

行业内“马太效应”凸显

经历欧洲削减补贴、产品大幅降价的光伏市场下半年有望重新提振。目前看来，这一趋势

正越来越明朗化。和去年相似，由于削减补贴的“靴子”已经落地，而安装截止日又将临近，欧洲正掀起新一轮光伏“抢装潮”。但对国内大大小小的光伏企业来说，随着竞争更趋激烈，强者愈强的“马太效应”也开始显现。

“目前国内光伏市场两极分化情况特别严重。”中国最大的多晶硅及硅片生产商之一赛维LDK一位管理人士告诉本报记者，一些光伏组件或电池小厂即使再降价也拿不到订单。相比之下，大厂出货却一直非常顺畅，如赛维LDK目前都不敢随便接订单，因为怕产能不够，生产不过来。

他指出，由于去年下半年到今年上半年有大量新产能投产，目前全球光伏总产能仍大于需求。在此背景下，一线大厂与小厂的销售情况就出现明显分化，意味着“马太效应”正在光伏行业出现。

据他介绍，这一点在国内市场更加明显。一方面，不少项目投资方怕出问题，都会刻意选择一些大厂的产品。另一方面，大厂在国内的项目公关能力也相对更强。

欧洲市场“利空”出尽

来自行业机构的信息显示，欧洲主要国家太阳能市场需求在6月出现一个久违的小高潮。尽管意大利政策停滞一度导致下游系统厂商观望，但随着政策在5月后明朗，光伏电站的建设需求重新开启。此前，为了保证现有的安装系统能够在补贴下调之前顺利并网，意大利政府已将现有的三号法案延期执行，这极有可能刺激部分新增安装量的需求提前到最近几个月开工建设。

而德国则一直是全球光伏安装量最大的国家，其去年总装机容量达到了7.25GW，同比增长了约90%，也带动了全球太阳能市场的景气度上升。与意大利相似，德国政府也不断下调政府补贴以促进光伏发电成本下降。而从去年该国月安装量来看，每逢补贴下调之前德国的月安装容量都会出现明显上涨，这主要是因为系统厂商希望通过“抢装潮”以搭乘补助下调前的末班车。

“根据德国政府原先的一个设想，如果今年3—5月份的装机超过一定的数量则会进一步下调补贴，这一消息会在6月底之前宣布。而目前看来，装机量并没有超出德国政府的预设限额，而时间已经到了6月底，因此进一步削减补贴已几无可能。对光伏行业来说，至少下半年不会有更大的利空消息，是否要进一步削减补贴须等到明年。”一位行业人士对本报记者说。

事实上，为了使下半年的政策不会过分严苛，德国系统厂商在3—5月份的安装量一直处在低潮，以使量化到全年的安装量处于一个较低的区间。而随着进一步削减补贴的威胁远去，系统厂商为了赶上政策末班车，其安装需求在6月集中释放，令去年的“抢装潮”重现。

值得注意的是，9月30日与10月31日分别是意大利与德国的光伏安装截止日。“如果不能在这两个时间之前上报项目的话今年就赶不上了，所以大部分企业肯定会在这两个时间点之前抓紧申报。”上述人士说。

他表示，接下来将进入夏季这一光伏安装的传统旺季，行业有望重新趋热。

资料来源：陈其珏．上海证券报

2. 国内光伏产业现状

得益于欧洲光伏市场的拉动，中国的光伏产业在2004年之后经历了快速发展的过程，连续五年的年增长率超过100%。从2007年至今，中国已经连续四年光伏电池产量居世界首位。

2010 年中国光伏电池产量已超过全球总产量的 50%，目前已有数十家光伏公司分别在海内外上市，据估算，行业年产值超过 3000 亿元人民币，直接从业人数超过 30 万人。

2009 年中国年度光伏新增装机量达到 160MW，超过了截至 2008 年底的累计安装总量；而 2010 年国内光伏发电新增装机容量有望达到 500MW（如表 24－3），大大的超过了 2009 年同期水平，2010 年实际新增装机量超过 500MW。

中国光伏产业走上了快速发展之路，已经掌握了包括太阳能电池再制造、多晶硅生产等关键工艺技术，设备及主要原材料逐步实现国产化，产业规模快速扩张，产业链不断完善，产业链垂直整合趋势明显，制造成本持续下降，具备较强的国际竞争能力。以晶体硅电池制造企业为例，包括尚德、晶澳、英利、天合在内的多家企业不仅制订了宏伟的电池生产扩产计划，还向产业链上下游拓展，业务延伸至多晶硅材料、硅片、组件及系统安装等。中国 2010 年光伏行业主要环节产能及产量统计如表 24－3 所示。

表 24－3 中国 2010 年光伏行业主要环节产能及产量统计

Table 24－3 Production Capacity/Volume of Major Links in China's Photovoltaic Industry in 2010

主要环节	2010 年产能	2010 年产量
多晶硅	85000 吨	45000 吨
硅锭/硅片	23GW	11GW
晶体硅电池	21GW	8.5GW
薄膜电池	2.5GW	0.5GW

资料来源：CPIA，SEMI

中国光伏市场近几年的增长速率令人印象深刻，但中国的光伏装机量从全球角度看仍然相当小，2009 年中国光伏安装量占全球总安装量的份额约为 2%，2010 年上升约 1 个百分点，达 3.1%，如表 24－4 所示。

但是这并未改变总体市场规模偏小的现状，目前国内光伏企业 90% 的光伏组件产品仍然需要出口，国内光伏产业市场在外的现状没有得到实质性改观。面对迟迟未能启动的中国光伏市场，相关部门正在大力推进"金太阳"工程，试图利用这一杠杆来撬动国内光伏市场。

表 24－4 全球及中国光伏年度装机量

Table 24－4 Annual Installed Capacity of Photovoltaic Industry in China and the World

	2006	2007	2008	2009	2010
全球总量	1603	2932	5950	7380	16000
中国	10	20	40	160	500
中国占世界的百分比	0.6%	0.7%	0.7%	2.2%	3.1%

资料来源：CPIA，SEMI

2010 年 9 月，第二轮光伏电站特许权招标结果出炉，13 个项目所报出的中标电价均低于 1 元/千瓦时，其中最低价甚至只有 0.7288 元/千瓦时，这标志着国内光伏上网电价正式迈入1 元时代。

现阶段中国光伏产业的发展充满了机遇与挑战。国际、国内竞争进一步加剧，新技术和新产品不断涌现。同时，国外大型电子制造企业和大型财团都以不同方式进入光伏产业，其资本优势和技术优势为中国光伏产业的继续发展带来了极大的未知因素。为了打破光伏市场技术差异性相对较小的特点，行业领军企业都极为重视技术的积累进步和新产品的开发。多家设备企业和原材料供应商也在致力于技术的更新换代，加大了在新兴技术方面的投资。可以预见到，若干关键技术的突破将会改变整个光伏产业的发展进程和竞争格局。

24.1.2 太阳能热水器产业现状

太阳能光热利用，除太阳能热水器外，还有太阳房、太阳灶、太阳能温室、太阳能干燥系统、太阳能土壤消毒杀菌技术等。太阳能热水器产业是中国发展最早和规模较大的一个可再生能源产业，太阳能热水器技术日趋成熟，应用领域已从生活热水扩大到泵水、采暖、制冷空调、海水淡化、工业加热、热发电等。相对于太阳能光伏产业而言，中国的太阳能热利用方面既有产业，又有市场。

目前，我国已成为世界上最大的太阳能光热应用市场，也是世界上最大的太阳能集热器制造中心。太阳能光热技术不仅在民用领域得到应用，而且在造纸、饮料、机械、纺织、食品、养殖等工农业生产方面得到了广泛应用。

中国是世界上最大的太阳能热水器生产和应用国。仅太阳能热水器行业就有3000多家企业，2010年迎来了井喷，产值达735亿元，年生产能力4900万平方米，全国累计装备太阳能热水器接近1.5亿平米。2009年太阳能热水器的产量和保有量分别占到了世界的80%和76%以上。2007年、2008年、2009年中国太阳能热水器产量年增长率均保持在30%以上，分别为30%、32.5%和35.5%，如表24-5所示。

表24-5 1998—2009年中国太阳能热水器年生产量和保有量

Table 24-5 Production Volume and Inventory of Solar Water Heaters in China during 1998—2009

年份	总产量		比上年增长（%）	保有量		比上年增长（%）	能源替代标准煤（万吨）
	万平方米	兆瓦热输入量		万平方米	兆瓦热输入量		
1998	350	2450		1500	10500		225
1999	500	3500	43	2000	14000	33	300
2000	640	4480	28	2600	18200	30	390
2001	820	5740	28	3200	22400	23	430
2002	1000	7000	22	4000	28000	25	600
2003	1200	8400	20	5000	35000	25	750
2004	1350	9450	12.5	6200	43400	24	930
2005	1500	10500	11.1	7500	52500	21	1130
2006	1800	12600	20	9000	63000	20	1350
2007	2300	16100	30	10800	75600	19.4—20	1620
2008	3100	21700	32.5	12500	87500	15.7	1880
2009	4200	29400	35.5	14500	101500	16	2180

资料来源：《中国太阳能热水器产业报告》

2009年，中国太阳能热水器热利用面积已达1.45亿平方米，中国太阳能产业规模已居世界第一，而且太阳能产业的产值也在逼近千亿元大关。2010年太阳能热水器市场持续增长，年产量约为4900平方米，占世界年产量80%左右，销售额约为735亿元，出口额约为2.5亿美元，同比增长15.7%，与2009年相比，增速放缓。

太阳能热利用专委会新近完成的企业调查，在一类年产值过亿的12家大型太阳能骨干企业中，2010年上半年销售的平均增幅为34.2%，其中12家企业中只有北京的一家企业与去年持平，1家企业增幅为10%；在年产值5000万元—6000万元的二类企业中，北京的3家太阳能企业中只有1家是-20%的下滑，山东的8家企业最低都保持了43%的增长，江苏7家平均是30%的增长，二类企业的平均增幅为37.4%。上半年品牌企业还是保持了30%的增长，只是速度相比往年有所放缓。2000多家太阳能中小企业，上半年放假、停工和关闭的各占1/3，地区小企业停产和倒闭的则各占一半。

太阳雨2010年上半年实现倍速增长，年产有望突破200万台。力诺瑞特上半年产量、销量、销售收入等主要经营指标实现了近40%的增长。桑乐上半年销售收入增长速度超过60%，重点市场甘肃、广西销售收入均超过去年全年。日利达太阳能上半年同比增幅达30%。

专栏24-4

江苏太阳雨太阳能有限公司

历经10年持续发展，江苏太阳雨太阳能有限公司目前已是全球最大的太阳能热水器及热水系统供应商，中国太阳能光热产业领导者。公司经营范围涉及全玻璃真空集热管、太阳能热水器、光电等产品，产品覆盖全球100多个国家和地区。

- 下辖7大基地，连云港基地、山东基地、洛阳基地、华北基地、东北基地、西南基地、华南基地，同时正在考察海外基地。
- 2010年1—6月份，江苏太阳雨太阳能有限公司取得98.74%的同比增长，半年销售即突破2009年全年，2010年底该公司有望突破200万台的业绩。
- 2009年末以来，江苏太阳雨太阳能有限公司以系统内最高销量，市场最高占有率12.54%，持续领跑中国太阳能光热产业。
- 公司的千亩工业园、两园一区一镇规划业已进入项目实施阶段。从国内到国际，从光热到光伏再到风光互补等新能源的综合利用，江苏太阳雨太阳能有限公司已成为中国新能源领域的整合创新者。
- 未来五年里，江苏太阳雨太阳能有限公司发展战略是产值过100亿元，完成国内五大产业链基地的建设并形成国际化产业布局。

有分析认为，太阳能热水器行业在经历2009年的高涨和2010年的萧条之后，2011年将开始转入沉寂，整个行业将面临洗牌的巨大阵痛。而综合2011年各方数据来看，似乎也证实了这一观点。以近年来的增长率一直居于行业前列的太阳雨为例，根据1—5月份的统计数据，太阳雨的销售规模对比去年同期增长率为58.87%，而去年上半年，太阳雨的增长率为120%左右。撇开基数增大带来的速度不谈，太阳雨增长率的变化直接可以感受到行业的冷暖（参见专栏24-4）。事实上，2011年1—5月份，太阳能热水器行业的整体增长率不足10%，与2010年全年的增长率15.7%相比，行业的增长速度进一步下滑，而且幅度不小。

2011年以来，太阳能热水器行业发展遭遇诸多阻碍：下乡政策效应减弱、市场需求下滑、农村竞争白热化、面向城市转型不力、原材料价格居高不下等。而在前几年的快速增长中，行业自身暴露出来的诸多问题，也成为阻碍行业进一步增长的最大阻力。频频暴露出的质量问题、安装问题和售后服务问题，对太阳能热水器行业无疑是雪上加霜。

目前，中国太阳能热水器行业共有23项国家标准，不过都是推荐性标准，企业可以参照，但无需强制执行，在一定程度上造成国内太阳能热水器产业“鱼龙混杂”，市场准入门槛不高。有统计数据显示，国内贴牌热水器占到整个太阳能热水器市场的20%以上，这种所谓贴牌产品对整个太阳能热水器市场的负面影响不可低估。

然而，造成当前太阳能热水器市场增长乏力的最根本原因并不是产业政策或者产业标准的缺失，而是整个行业缺乏行之有效的战略创新与需求创新。虽然经历了10多年的发展，但国内太阳能热水器市场却仍然停留在原始“作坊式”的产业发展初级阶段，没有找到一种能够迅速放大市场优势的战略路径，仍处于一种被五金等原材料价格左右的低层级粗放式发展阶段。

有鉴于太阳能热水器行业的诸多问题，从2010年开始，一批政策性标准和行业自律文件密集出台。2010年，由国家16个部委共同发文的《家用太阳能热水系统能效限定值及效率等级》启动。这是太阳能热水器行业第一个国家强制性执行标准，意味着今后凡是进入太阳能热水器行业的企业都要贯彻执行这个标准。

另外，太阳能热水器的能效标准也已经在2010年10月底完成全部制定工作。近日，中国资源综合利用协会可再生能源专业委员会主任朱俊生表示，该标准已上报国家标准委员会，半年后有望出台。该标准将有关能效标准修订为强制性标准。太阳能热水器能效标准将太阳能热水器按能效分为三级，一级能效最高，三级能效最低，不能达到三级标准的企业将被淘汰。

这些强制性标准的制定实施，将抬高太阳能热水器的市场准入门槛，规范行业发展。虽然短期内未必能起到立竿见影的效果，但从长远看，必然会推动行业向更高层次迈进。

同时，能否积极拓展海外市场、实现技术创新和产业创新、搭上太阳能与建筑一体化的快车，将成为太阳能热水器产业实现持续增长的关键。

24.1.3　太阳能光热发电产业现状

24.1.3.1　产业链概述

完整的光热发电产业链包括基本材料、主机设备的供应、系统的设计集成、电站的管理运营等部分。其中，原材料主要包括钢铁、玻璃、导热物质和储热材料，这些材料需要满足一定指标要求，例如玻璃的透光率、储热材料的工作温度范围等；设备方面，太阳能光热电站的最重要设备包括反光镜、接收器、涡轮机以及储热储能设备等。

光热发电的关键技术（参见专栏24－5）也主要集中在上述环节，包括高反射率、高精度反射镜，高温吸热器技术、高温储热技术、高热流密度下的传热以及各种相应设备的制造工艺和电站系统的设计、集成和控制技术等。

太阳能光热发电站一般建在日照时间较长的地方，要求直接辐射在每年1900千瓦小时/平方米以上，这个资源条件全世界只有美国、欧洲的西班牙、非洲和中国具备。美国和西班牙已捷足先登，建立了各自的光热发电项目，而在中国，内蒙、新疆、青海和西藏等广袤地区，尚有大片的光热资源等待开发。

专栏 24－5

太阳能光热发电技术概述

聚光太阳能发电（CSP）技术通过聚光器将阳光聚起来发电，其技术路线主要包括聚光光伏发电（CPV）和聚光光热发电，前者在聚光处进行光伏转化来实现发电，后者利用集热器产生热空气或热蒸汽，用传统的电力循环发电。聚光光热发电技术根据聚光方式又主要分为四种：槽式聚焦、塔式聚焦、碟式聚焦以及反射菲涅尔聚焦。

槽式光热发电是利用槽式聚光镜，直接将太阳光反射到位于镜面焦点处的集热管，将内部传热物质转化为蒸汽从而驱动涡轮发电；塔式光热发电是将吸收到的太阳能射线集中到塔中，对传热工作物质加热进而发电的一种聚光太阳能发电技术；碟式光热发电是利用旋转抛物面反射镜，将入射太阳光聚焦到焦点上，焦点处放置的斯特林发电装置进行发电；菲涅尔式光热发电的工作原理类似于槽式光热发电，只是采用菲涅尔结构的聚光镜来替代抛面镜，这使得发电成本相对较低，但效率也相应降低。

槽式系统是目前技术发展最为成熟的系统。据统计，目前在全世界运行的槽式太阳能热发电系统占整个 CSP 装置的 88%。但这一比例在各国情况不尽相同，例如，在美国使用槽式技术的 CSP 项目只占了 40% 左右。碟式技术近年来也得到了迅速发展，尤其是在美国加州得到了大量的应用。

2. 国际光热发电产业现状

世界上最早的太阳能光热发电站是美国加州的 Solar One，建成于 1982 年，但运行后由于电站集成技术方面存在阻碍，投资巨大且降低造价十分困难，所以各国在光热发电领域有较长时间的停滞。

之后美国和以色列联合组成路兹公司从 1985—1991 年在美国加州沙漠建成 9 座槽式太阳能热电站 SEGE Ⅰ－Ⅸ，从最初的 30MW 装机容量发展到 9 号的 80MW，并成功进入了商品化运行阶段。从此，世界各国相继建立起不同形式的示范装置（如表 24－6），促进太阳能光热发电技术的发展，并将太阳能热发电作为未来替代能源的关键候选之一。

表 24－6　全球早期光热电站资料

Table 24－6　Early Statistics of Photo－thermal Power Stations in the World

项目名称	地点	建成时间	容量（MW）
Solar One	美国 California	1982	10
Themis	法国 Targasonne	1982	2.5
CESA－1	西班牙 Almeria	1983	1
MSEE	美国 Albuquerque	1984	0.75
SEGS－1	美国 California	1984	14
Vanguard 1	美国	1984	0.025
MDA	美国	1984	0.025
C3C－5	俄罗斯 Crimea	1985	5

资料来源：国金证券研究所

如今，太阳能光热发电正成为世界范围内可再生能源领域的发展热点，太阳能光热电站遍布美国、西班牙、德国、法国、阿联酋、印度、埃及、摩洛哥、阿尔及利亚、澳大利亚等国家。

截至2009年底，全球已并网运行的光热电站装机容量约600兆瓦，全部集中在美国和西班牙。在建项目近1300兆瓦，已经宣布建设的装机容量则为1.26万兆瓦。这两个国家拥有丰富的太阳能资源以及相对成熟的光热电站管理经验，并且构建了比较完整、系统的产业政策支撑体系，促进光热发电产业的发展。世界各国太阳能光热发电规模比较如图24-1所示。

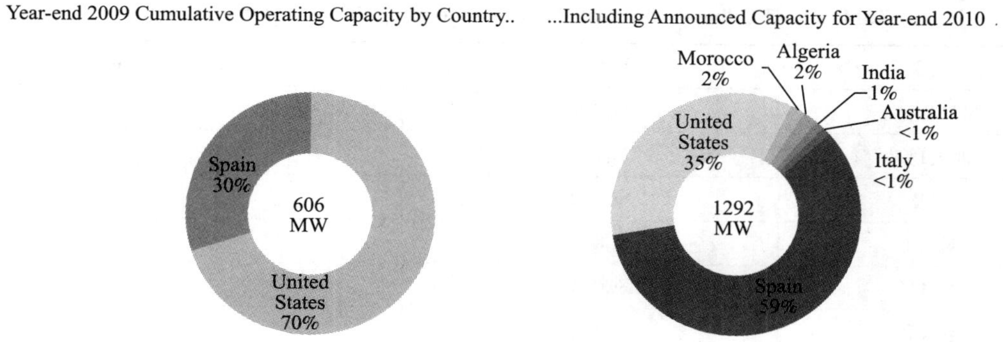

图24-1 世界各国太阳能光热发电规模

Figure 24-1 Photo-thermal Power-generating Capacity of Different Countries

资料来源：国金证券研究所

美国目前投入商业运行中的太阳能光热电站总装机容量达432兆瓦，位居世界首位，并且有大量电站在建和拟建中。2009年有3座总计12兆瓦的太阳能光热发电站投入运行，2010年还会再新增3座：位于Phoenix，Ariz的2兆瓦碟式电站、位于Grand Junction，Colo的4兆瓦槽式电站以及Martin County，Fla的75兆瓦中心混合槽式电站。这几座电站如果能如期完成，将更加巩固其在世界聚光光热发电领域的领先地位。预期在2020年，美国太阳能光热发电规模将达到5万兆瓦级别如图24-2所示。

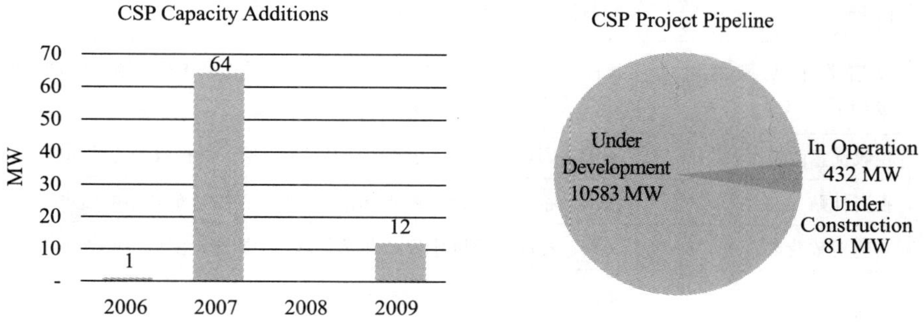

图24-2 美国太阳能光热发电规模

Figure 24-2 Photo-thermal Power-generating Capacity of US

资料来源：国金证券研究所

西班牙目前的总装机容量达330兆瓦，仅次于美国。2009年11月举行的西班牙部长会议提出2010—2013年太阳能热发电装机容量的目标为2440兆瓦，而其远期的规划则是在2020年达到2万兆瓦的装机容量。

截至2009年底，西班牙预登记的太阳能光热发电项目已达57个，预计在2013年投入运营。照此趋势，到2013年其光热发电规模将达到2.3GW。

除美国和西班牙之外，中东及北非地区将建立一系列联网的太阳能光热发电站，即“沙漠行动计划”，以满足欧洲15%的电力需求以及电站所在国家的部分电力需求，亚洲国家也在逐步建设示范性项目如表24-7所示。

表24-7 世界其他各国光热发电建设状况

Table 24-7 Photo-thermal Power-generating Status of Other Countries

国家	技术类型	装机容量（MW）
阿尔及利亚	ISCC槽式	43
澳大利亚	菲涅尔式	1
埃及	ISCC塔式	31
墨西哥	ISCC槽式	29
摩洛哥	ISCC槽式	30
意大利	槽式	40
印度	ISCC槽式	30
希腊	蒸汽循环槽式	50

资料来源：国金证券研究所

太阳能光热发电在世界范围内迅速发展，一定程度上得益于各国政府对太阳能产业的支持政策。世界各国对太阳能发电的支持政策情况如表24-8所示。

表24-8 世界各国对太阳能发电的支持政策

Table 24-8 Supporting Policies for Solar Power Generation of Different Nations

政策类型	具体政策	推行状况
公共政策	可再生能源组合标准（RPS）	美国、英国、荷兰、丹麦、澳大利亚实施较为典型，包括规定可再生能源发电配额量、用户购买最低限量绿色电力义务等
	政府收购太阳能电力（FLT）	德国、西班牙以0.32欧元—0.43欧元的上网电价收购；欧洲多国、澳大利亚、韩国及美国几个州亦推行中
财税政策	税收优惠政策（ITC）	美国法国均有ITC，美国对太阳能企业给予的税收减免达30%
	投资补贴	法国、希腊对部分太阳能项目给予财政补贴，最高额度达40%
	其他财政政策	美国、法国、西班牙等对投资人制定了加速折旧成本回收制度
政府直接投入		美国2010年投资6200万美元 日本2009年拨款90亿日元

资料来源：国金证券研究所

3. 国内光热发电产业现状

从自然环境来看，我国的太阳日照时数在2200小时以上地区约占国土面积的2/3以上，年接收太阳辐射的能量相当于上万个三峡工程的发电量，相当于17000亿吨标准煤的热量，太阳光资源非常充足。

目前，我国光热发电还处在技术研发阶段，尚没有一座真正投入商业应用的太阳能热发电厂。太阳能光热发电项目由科技部组织，进行自主技术的研发和示范电站建设，逐步从科研院所向企业转移。经过几年的研发，我国已实现技术上的突破，尤其是在最难的储热技术方面，独立建造大规模太阳能光热电站成为可能。

国内最早的太阳能热发电示范电站是南京玻纤院春辉公司与以色列魏兹曼研究院合作的南京江宁区 70KW 塔式太阳能热发电试验工程，于 2005 年建成并发电成功。

“十一五”期间，在国家 863 计划支持下，中国科学院电工研究所等 10 家单位于北京八达岭开始建设 1MW 的太阳能塔式热发电实验项目，这是亚洲首座太阳能塔式热发电技术应用项目。该电站将于 2010 年底实现并网发电，每年的发电量将达到 270 万千瓦时。

然而，光热发电要实现产业化的高速发展，道路并不平坦。相较于蓬勃发展的光伏发电产业，光热发电显然才刚刚起步，国家和各地方政府对光热发电产业的政策支持体系还未形成。但是，光热发电的优势和发展前景依然受到了各方的关注与重视。2009 年 10 月，中科院电工研究所、华电集团、保定天威、金晶集团、皇明太阳能、西安航空动力集团等企业和研究所联合成立了“中国太阳能光热产业技术创新战略联盟”，计划“十二五”期间在中国西部建设 1000 兆瓦级规模的太阳能热发电站。

甘肃、山东、江苏等在内的各地方政府都开始光热发电项目的建设准备工作并制定相应规划。目前，国内规模最大的项目是内蒙古 50 兆瓦槽式太阳能热发电项目，该项目落户内蒙古鄂尔多斯市巴拉贡创业区，于 2010 年 10 月进行招标。

各大电力集团也开始积极布局，在全国各地开展光热发电前期工作，并预留了大量土地以备将来的大规模扩张，如表 24－9 所示。

表 24－9 五大电力集团光热发电项目

Table 24－9 Photo－thermal Power Generation Projects of China's Five Power Giants

	光热发电项目状况
华电集团	参与八达岭 1MW 太阳能塔式热发电实验项目建设； 2009 年与澳大利亚雄狮国际签署合作协议，准备合作在格尔木开发 100 万千瓦光热发电项目； 2010 年与嘉峪关市签署包括发展光热发电在内的战略合作协议
国电集团	与青松建化合作在新疆吐鲁番开发太阳能热发电 30 万千瓦，目前示范项目仅 100 千瓦
华能集团 大唐集团 中电投集团	暂无相关公开资料

资料来源：国金证券研究所

目前，光热发电已成为全世界在能源领域推进商业化应用的重点，预计到 2015 年，全球总装机容量将达到 17GW 以上。由于技术的优势和雄厚的实力，美国和西班牙仍将继续占据领先地位。

中国拥有丰富的太阳能资源，光热发电的规模发展对于实现我国 15% 的非化石能源的结构调整目标具有重要的战略意义。2007 年《可再生能源中长期发展规划》中提到，“十一五”时期，在内蒙古、甘肃、新疆等地建设太阳能热发电示范项目；到 2010 年太阳能光热发电总容量 50 兆瓦；到 2020 年，全国太阳能光热发电总容量达到 200 兆瓦。太阳能光热发电也已经被纳入了国家重点发展的战略性新兴产业之一，在国家电力“十二五”规划中，明确指出，将在甘肃、宁夏、新疆、内蒙古选择条件适合地点建设太阳能热发电示范电站。据相关研究机构统计，如果所有已经公布的项目均能实施，在2015年前，我国的太阳能光热发电装机容量将达到300万千瓦左右规模，按照光伏系统可比成本计算，市场总量达450亿元人民币。

2010年10月，我国第一个光热特许权招标项目的启动，标志着我国光热行业的全面启动。新行业将在全国范围内进行规模扩张，相关技术、市场也将逐渐迈向成熟。

24.2 行业政策

相较于太阳能热利用行业，光伏行业无疑更受市场及政府的关注，政策扶持的力度也远大于其他行业。本章节将重点分析光伏行业的政策现状。

光伏行业作为朝阳行业，其前期发展主要依赖国家的政策扶持，引导装机需求。而随着行业的逐渐成熟以及成本的下降，其发展将实现从完全的政策驱动向市场化过渡。当前国际国内的政策趋势正是如此。作为光伏行业的主要市场，欧洲逐渐实现从完全政策驱动向市场化过渡，而一些新兴国家则通过加大政策扶持力度，逐步培养太阳能行业。世界主要光伏国家的补贴政策情况如表24－10所示。

表24－10 世界主要光伏国家的补贴政策

Table 24－10 Photovoltaic Subsidy Policies of Different Countries

传统市场政策收紧		新兴市场政策扩张	
德国	自2010年7月1日，德国对屋顶光伏系统和移除耕地农场设施的补贴额将减少13%，对转换地区补贴额将减少8%，其他地区将减少12%； 自2010年10月1日，总补贴额再减少3%； 2011年德国可能出台补助安装上限规定	美国	2010年7月通过“千万屋顶计划”，2012—2021年，总安装容量将达30GW—50GW，补贴不超过净安装成本50%； 洛杉矶贸易委员会提出CLEAN LA议案，计划允许商业和住宅太阳能用户将所发电量出售给电力公司，总量150MW，预计于2011年7月开始实施
西班牙	自2010年8月，削减上网电价幅度达45%； 对大型屋顶太阳能光伏装置的上网电价将下降25%，小型屋顶下降5%； 2011年可能限制得到补贴的发电小时数	日本	电价补贴政策原本计划在2010年4月开始实行，提前到2009年11月开始实施； 道、府、县、村等地方政府开展的居民住宅安装光伏系统支持项目高达400个； 展开由公共绿色能源基金对于公共设施的光伏发电进行电力买回计划与补贴借贷
意大利	自2010年12月31日，在2011年前四个月里，规模5MW以上的项目补贴将平均减少9.3%，之后每四个月调整一次，逐步削减补贴	澳大利亚	2010年对光伏上网电价补贴法案提出修订计划，南澳地区的太阳能用户额外发出的电，将获10澳分的新增补贴
捷克	2010年10月底，提出泡沫式光伏电站投资的惩罚方案，对光伏电站销售电力的所得征收26%的“太阳能税”； 自2011年3月，农业用地上的太阳能发电厂将不再获得政府补贴，仅屋顶太阳能系统和建筑光伏一体化（BIPV）系统将获得入网电价补贴	中国	2009年，太阳能屋顶计划与金太阳计划出台； 2011年即将出台《新兴能源产业发展规划》，在2011—2020年，对新能源预计累计增加投资5万亿元，每年可增加产值1.5万亿； 将2020年光伏装机规划从1.8GWp增加到20GWp
法国	自2010年9月1日，上网电价削减12%； 2011年3月10日起，补贴再将约20%。向9KW以下的光伏建筑一体化项目发放0.464欧元的补贴，向12MW以下的“其他类”项目发放每度0.12欧元的补贴	印度	2010年，新能源和可再生能源部门（MNRE）发布了“阳光印度”计划，制定了到2022年太阳能发电量达到20GW—22GW的目标； 2011年，印度再次将太阳能组件和太阳能日光灯重新划分为可再生能源项目。太阳能日光灯的关税将从10%下调到5%，而太阳能组件则完全免税

资料来源：国海证券研究所

24.2.1　国际政策

1. 德国最新光伏政策

德国在2011年初的时候限定降幅，若3—5月间的年均新增安装量超过6.5GW，届时7月份的补贴下调幅度将被限制在12%以下，最高可达15%。其中，对于屋顶安装的削减将从7月份开始，对地面电池组件的削减将从九月份开始。

装机小于2.5GW的补贴额度上调2.5%；装机在2.5GW—3.5GW不进行任何调整；装机量在3.5GW—4.5GW补贴下调3%；装机量在4.5GW—5.5GW补贴下调6%；装机量在5.5GW—6.5GW补贴下调9%；装机量在6.5GW—7.5GW补贴下调12%；装机量在7.5GW以上的补贴下调15%。

从德国政府提出的政策分析，德国不希望光伏装机量过快增长，但又想维持光伏产业稳定发展。

2. 意大利最新光伏政策

当地时间2011年5月5日，意大利工业部长和环境部长共同签署并批准了新的太阳能补贴法案。意大利政府对太阳能发电的补贴将持续到2013年，但补贴比例将逐渐降低；其后，补贴将与一定规模的装机量挂钩。意大利政府准备将每年用于太阳能发电补贴的资金限制在60亿欧元—70亿欧元之间，并一直保持到2016年。

新政策对补贴上限及金额提出了具体规划。新政策显示，2011年6月—2016年的光伏补贴上限为23GW，年均4.18GW。另外大型项目2012年前有3.1GW的上限要求，2013—2016年有9.77GW的上限。

而补贴金额下降为每年60亿欧元—70亿欧元，具体补贴下调如下：

- 2011年6月份，原有上网电价补贴将进一步削减4%—11%：小型系统下跌约4%，大型屋顶系统下降约11%。
- 6月份开始，上网电价将以月为单位进行调整，根据系统规模的大小，7月份将在6月份基础上下调2%—5%，预计到2011年底，FIT将在现在的基础上下降26%—42%。
- 2012年上半年和下半年将分两次进行8%—12%的进一步下调，预计到2012年底，FIT将再削减20%左右。
- 从2013—2016年，按照每个季度进行4%的下调。

意大利新政于6月1日生效。补贴额度上限的设置估计会让未来几年意大利市场的需求稳定在2.5GW—3.5GW的水平，而近两年对小型光伏系统不设限以及额外的补贴带来的需求值得期待。

3. 法国

法国的新补贴政策已于3月10日正式实施，新补贴政策向9KW以下的光伏建筑一体化项目发放每度0.464欧元的补贴，向12MW以下的“其他类”项目发放每度0.12欧元的补贴。以上政策将其上网电价下调20%。此外，新政策对100KW以上项目实施招标机制。此外，法国设置每年50MW的装机上限。

4. 印度

印度再次将太阳能组件和太阳能日光灯重新划分为可再生能源项目。自此，太阳能日光灯的关税将从10%下调到5%，而太阳能组件则完全免税。同时，产业链相关的光伏玻璃和银浆都将完全豁免基本关税。印度政府削减光伏产品关税将对其光伏市场的发展构成利好。

5. 北美

● 洛杉矶：洛杉矶贸易委员会提出 CLEAN LA 议案（美国最大的上网电价补贴方案），计划允许商业和住宅太阳能用户将所发电量出售给电力公司。此次提出的上网电价补贴计划为期五年，总量 150MW，预计于 2011 年 7 月开始实施。

● 加拿大安大略省：能源局授予 35 座大型光伏项目上网电价补贴许可。2011 年 3 月，安省能源局发放的第二轮许可，授予 35 家新建大型项目上网电价补贴许可。此次共有 872MW 项目递交了可再生能源补贴申请，仅有 257MW 获得补贴。后续安省能源局计划将为 500KW 以下的小型项目发放上网电价补贴许可。

24.2.2 国内政策

过去的两年内，随着光伏发电成本急速下降，中国国内光伏市场的政策也取得了一些重要进展，如图 24－3 所示。

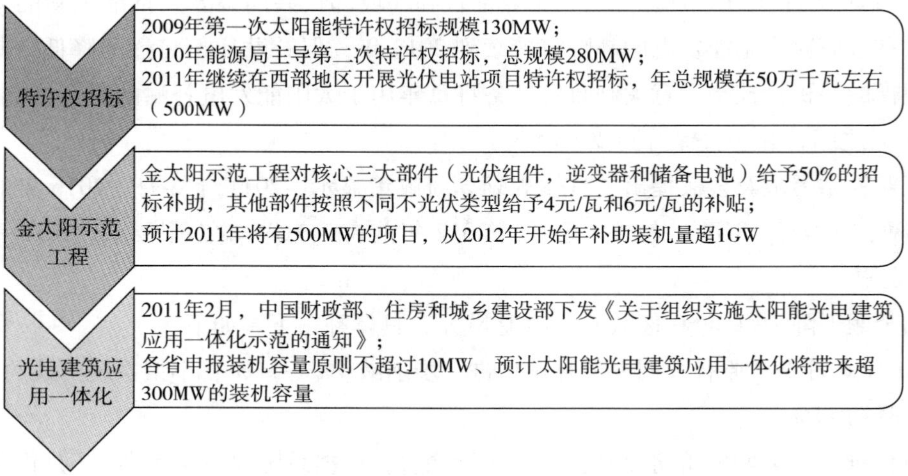

图 24－3 中国主要的光伏补贴政策

Figure 24－3 Major Photovoltaic Subsidy Policies of China

资料来源：国海证券研究所

2009 年 3 月，财政部会同住房和城乡建设部推出了促进 BIPV 和光伏屋顶应用的国家光伏补贴计划，该计划被视为中国光伏市场的转折点。2009 年 7 月，财政部会同科技部和国家能源局发布了第二个国家光伏补贴计划，即“金太阳示范工程”。当年批准了 201MW 的项目（2010 年取消了其中中标后未能实施的 54MW）。2009 年底，国家能源局举行了甘肃敦煌 10MW 并网光伏发电项目的特许权招标，最终中广核能源开发有限责任公司、江苏百世德太阳能高科技有限公司和比利时 Enfinity 公司组建的联合体以 1.0928 元/度的价格竞标成功。

2010 年开始，由财政部、科技部、住建部、国家能源局联合发布文件，对“金太阳示范工程和太阳能光电建筑应用示范工程”的有关政策进行了大幅调整，涉及设备招标、项目调整、补贴标准、项目并网等多个关键环节。2010 年新增了 272MW 的项目。此外，宣布在全国建立 13 个光伏发电集中应用示范园区，以此为依托推动中国光伏产业的的应用。并公开表示力争 2012 年以后每年国内应用规模不低于 1000MW。2010 年 8 月，国家能源局举行了 280MW 并网光伏发电项目特许权招标，最终中标价格分布在 0.7288 元/度—0.9907 元/度，远低于业内预期。

2011 年 1 月，全国能源工作会议上，能源局首次提出了把光伏产业培养成为中国先进的装备制造产业和新兴能源支柱产业。

能源工作会议称，在太阳能利用方面，2011 年将继续在西部地区开展光伏电站项目特许权招标，总规模在 50 万千瓦左右（500MW）。此前 2009 年的第一次招标共 130MW，2010 年的第二次招标是 280MW。

此外，能源局还提到建设青海格尔木太阳能发电基地，推进青海、内蒙古太阳能热发电等示范项目建设。同时，加快实施“金太阳工程”，支持建设若干个新能源示范城市，支持吐鲁番新城太阳能建设。

2011 年 2 月，中国财政部、住房和城乡建设部下发《关于组织实施太阳能光电建筑应用一体化示范的通知》在建筑一体化、并网技术、关键设备和建设周期，以及资金的申请、审核和拨付等方面进行了详细阐述。此次太阳能光电建筑应用项目各省申报总装机容量原则上不超过 10MW，预计太阳能光电建筑应用一体化将带来超过 300MW 的装机容量。

示范项目优先考虑采用用户侧并网的方式，实现自发自用。对于关键设备的质量要求，与国家统一招标入围企业名单相一致。非晶硅组件则可由项目业务自主采购，组件转换效率不低于 6%，输出功率衰减 2 年内不高于 4%、10 年内不高于 10%、25 年内不高于 20%。新建太阳能光电建筑示范项目的周期不超过 2 年，既有建筑的太阳能光电应用示范项目的建设周期不超过 1 年。

24.3 发展趋势

24.3.1 光伏产业趋势

1. 世界光伏市场增速放缓，国内光伏企业产能过剩的压力增大

2010 年世界光伏电站装机容量的增长额中，欧洲仍然占据了绝大多数的比重，这一地区的增长额中很大一部分是由德国、捷克等国下调光伏上网电价补贴引起的抢装效应带来的，这实际上是对 2011 年欧洲光伏市场扩大的一种透支。一方面，2011 年欧洲主要光伏大国会继续下调光伏上网电价的补贴，这就使得 2011 年欧洲光伏市场规模增速将明显放缓，加上 2010 年抢装潮带来透支效应，2011 年欧洲光伏市场将出现萎缩基本上是可以确定的；另一方面，美国、日本、中国等新兴光伏市场规模的增长速度相对有限。受以上这些因素的影响，2011 年将会是全球太阳能光伏市场的一个小年。有分析机构预计，要到 2013 年欧洲光伏市场的需求才能实现恢复性增长。另外，美国、印度等新兴市场的新增装机容量将难以弥补欧洲光伏市场需求大幅下滑造成的缺口，2011 年世界光伏市场的需求将不会超过 30%，如表 24 - 11 所示。

表 24 - 11 2011 年欧洲主要国家光伏政策及装机容量预测

Table 24 - 11 Photovoltaic Policies of Major European Nations in 2011 and Estimated Installed Capacity

相关预测 / 国家	政策调整预测	2009	2010	2011（E）
德国	2010 年 7 月和 10 月份两次下调了光伏发电上网补贴，累计下调补贴额度达到 16%，抑制国内光伏装机容量过快增长的目的基本上达到，所以德国 2011 年继续下调光伏发电上网补贴的可能性不大，但是有可能会实施规定新增装机容量上限的方案	3800	7000	4000

续表

相关预测 / 国家	政策调整预测	2009	2010	2011（E）
法国	法国政府已于2010年9月1日开始将光伏上网电价的补贴削减了12%，小于3KW的私人主宅电价维持不变；从近期法国政府采取的措施来看，2011年法国政府肯定会下调光伏发电上网电价补贴，下调幅度应该在20%之内	185	700	860
捷克	2010年10月底，捷克政府出台了一系列针对国内泡沫式光伏电站投资的惩罚式方案，该方案提出要对太阳能电站销售电力所得征收26%的“太阳能税”，甚至还要追溯性征收之前免去的税收，2011年不大可能再次下调补贴	410	1000	425
西班牙	2009年以来西班牙将国内光伏上网电价的补贴下调了4.5%左右后，西班牙光伏市场迅速萎缩，所以2011年西班牙政府没有必要再进一步下调补贴	70	650	750
意大利	意大利政府已经明确表示，2011年1—4月，规模5MW以上的光伏电站项目的上网电价补贴将下调9.3%；2011年4月后，再下调一次	710	1500	1250

资料来源：中投顾问

2. 国内市场将启动

国家发改委能源研究所副所长李俊峰在一个行业会议上表示，到2015年，中国国内的光伏装机容量目标将达到10GW；到2020年，目标至少要到50GW。这一目标大大超出了光伏企业的预期。全球最大的太阳能电池制造商尚德电力董事长施正荣表示，今后两年中国市场将会呈现爆发式增长。另一家制造商英利绿色能源董事长苗连生最近也做出了类似判断，“中国将迅速成为世界最重要的光伏市场之一。”

一个未经官方证实的消息也能证明国内市场开启已近在咫尺。据悉，国家发改委正计划给予过去两年已经并网但没有上网电价的光伏电站项目统一的上网电价补贴，价格可能为1.15元/度，涉及电站项目约100MW左右，这亦得到了几家主要光伏制造商的证实，政策可能在近期出台。

为了应对国内市场的启动，光伏制造商和投资商已经低调地开展布局行动。尚德电力低调成立了中国事业部；同时，赛维LDK也在加大对中国的投入，赛维公共事务总监姚峰在不久前被委任中国区营销总监，负责国内市场的开发；英利则通过参与金太阳工程等由政府发起的光伏项目，谋求在国内市场的竞争地位；不仅如此，包括天合光能、阿特斯等在内的光伏制造企业都在今年加大了国内的公关传播力度，提高其国内知名度。

不仅是产业链上游的制造商，长期在海外从事下游太阳能电站开发的投资商和服务商也开始不约而同地将目光投向了国内市场。光伏系统开发商和电力运营商天华阳光宣布成立“天华阳光中国”，开始将其在海外电站的经验复制到中国，针对中国市场，设计了“投资商+服务商”的特殊商业模式；在德国有29年光伏安装经验的德国IBC太阳能也宣布正式进军中国市场，定位于服务商，目标是为中国未来几年的光伏电站工程提供光伏发电系统。

3. 产业转移速度加快

光伏产业中游的电池片/组件工艺逐渐趋向成熟，该环节已逐渐发展成为劳动密集型产业。面对日益增加的组件成本下降诉求，全球晶硅电池组件制造企业加速向资源丰富、要素成本低、市场潜力大的地区转移。

- 美国 First Solar 公司（参见专栏 24 - 6）将除了继续在马来西亚扩产外，还在越南开设新的生产基地，美国的 SunPower 公司 2010 年除了增强其在菲律宾的工厂产能外，也与台湾友达公司在马来西亚建设新的工厂。
- 德国 Q - Cell 公司已在马来西亚地区建立了电池片制造工厂；日本京瓷公司在中国和墨西哥建设的工厂也已于 2010 年投产。
- 而我国台湾地区的光伏企业纷纷将电池组件制造业务往大陆地区转移。

与此同时，为了避免国际贸易纠纷所带来的影响，中国的一些光伏企业也纷纷向欧美国家布局。

- 无锡尚德在美国亚利桑那州开设的 30MW 工厂已经投产，预计今年产能将达到 50MW，明年年中将达到 120MW。
- 苏州阿特斯也在加拿大投资建立了一个产能为 300MW 的工厂。全球光伏产业的转移也使得资源得到合理的配置，将更有效地降低生产成本。

专栏 24 - 6

First Solar 在中国

太阳能产业全球最大公司 First Solar 的集团 CEO Rob Gillette 预计德国市场将在 2011 年大幅萎缩，所以该公司正将其业务重心从德国转移到亚洲和北美。

世界上最大的太阳能集团 First Solar，力图巩固其市场地位并要将产能几近翻番。这家美国公司计划到 2012 年，将产能从 1.4GW 增加到 2.7GW，届时将相当于全球近 1/5 的市场份额，远超任何竞争对手。First Solar 计划在亚洲、法国和德国建设新工厂和生产线，该公司的 CEO Rob Gillette 告诉德国商报《Handelsblatt》。

虽然 First Solar 扩充了其在法兰克福（奥德河畔）的生产基地，Gillette 先生打算减少该公司对德国市场的依赖。“北美将是我们明年最大的市场。但到目前为止，德国一直是我们的最大客户，”他说。

目前，First Solar 收入的一半来自德国。一年前，该比例甚至高达 70%。“到明年底，德国市场将仅占公司总收入的 25%—30%”，Gillette 表示，First Solar 在 2009 年公布的营业收入是 21 亿美元。

与 First Solar 一样，越来越多的德国企业也将创收重点转向海外。Solarworld、Aleo Solar 和 Phoenix Solar 最近都报道说，各自的海外市场都有大幅度增长。

但一些公司也在一些市场争夺中不敌中国对手尚德和英利，低价目前是中国公司的杀手锏。First Solar 也面临竞争的压迫，它为此也在不停地降价。但这家美国公司有成本优势，因为它采用的是不需要昂贵原料硅的更为便宜的薄膜技术，而 Solarworld 和中国的公司则生产硅模块。

向德国以外市场倾斜的原因有二：一是德国本土对太阳能补贴的减少；二是其他市场更有利的条件。在今年底明年初，从德国南部的帕绍（Passau）到北端的弗伦斯堡（Flensburg），德国太阳能发电的补贴将减少13%。因此，First Solar 的 CEO Gillette 预计，德国市场在经历“今年将达6500MW—7000MW 的惊人增长”后，在2011年将大幅度萎缩。

专家并不排除德国政府可能会进一步削减补贴的可能。德国政府正在讨论一个太阳能发电总输出容量的补贴上限，达到该容量后，新的太阳能项目将不会获得任何补助。由于这种风险，Gillette 先生没对2011年做出预测，因他认为情况太不明朗。“这可能使模块提价面临压力，”他说，“中国已被我们的低价打动”。

为应对德国市场的萎靡，Gillette 先生有意在全球采取广种薄收的战略。对 First Solar 来说，美国市场将变得举足轻重，并将贡献该公司明年1/3的总收入。美国各州的太阳能补贴政策不同，但基本上 Gillette 先生预期都会更优厚。First Solar 正在开发自己的太阳能电站（solar park）以期从这些优惠条件中获益。为此目的，该公司买下了3个项目策划公司。“他们提供了总计2.2GW 的项目容量，这些项目主要在加州。”

除了在美国的布局外，对于 First Solar 来说，亚洲和欧洲其他国家正变得越来越重要。First Solar 拟拓展在法国、英国、意大利和希腊的业务。“我们期望在法国和意大利，都有50%—60%的增长” Gillette 先生说。

Gillette 预期，First Solar 会在中国有所收获。First Solar 是第一批向中国出口模块的外国公司之一。此前，只有中国本地公司能在中国相关的招标中中标。“中国看来已被我们的低价打动”，Gillette 先生解释了为什么 First Solar 能在中国中标。First Solar 最新报价是0.77美元/瓦，是最便宜的公司。

Gillette 对 First Solar 的产品中含有的镉碲化合物会带来环境风险的指控进行了辩驳。“我们的产品经过了测试并为客户所接受”，Gillette 先生说。他补充道，到目前为止，已有的任何其他技术都无法以一种更好的方式把经济和生态优势结合起来。Solarworld 率先提出对碲镉的禁用，但没能得到支持。

资料来源：维库电子市场网 http：//www.dzsc.com/news/，

本文地址：http：//www.dzsc.com/news/html/2010-12-20/123490.html

24.3.2 光热发电产业趋势

1. 政策有望更加积极给力光热

美国已确立太阳能光热发电基础负荷地位，加州新批太阳能项目几乎全为光热。而在中国，今年5月发改委《产业结构调整指导目录》鼓励类新能源门类中，太阳能光热发电居于首位。在淡季电荒与核电可能放缓的背景下，发展可再生能源的必要性和紧迫性提升，光热以其并网优势和规模效应，在我国能源结构中的战略地位将显著提升，有望获得更多的政策倾斜。

2. 技术进步空间大，成本下降可期

太阳能光热发电原理简单，但系统复杂，各环节均有不同的技术路线，技术进步将促进系统效率提升。据 IEA 预测，在技术进步、工艺改进、规模化应用和市场竞争下，2020年单位投资成本将下降50%，电价有望降至10美分/KWh 以下；中性假设下到2015年全球 CSP 累计装机将达24.5GW，五年复合增速90%；到2020年 CSP 在全球能源供应份额中将占1%—1.2%，到

2050年占8.5%—11.8%，乐观假设下到2050年可满足全球电力需求的25%。

3. 光热发电国内市场起步

中国光伏产业迅速崛起，但“两头在外”并非可持续的清洁能源产业发展模式。光热在起步时就推出特许权招标，国内应用市场和制造环节同步启动，且有低温利用的产业基础，有望在国内形成从材料到制造再到系统集成以致最终应用的完整的清洁能源产业链。

有分析机构预测，中性假设下太阳能光热发电产业“十二五”时期直接市场容量将达500亿，发电和EPC、核心组件、基础零部件和材料以及电网建设相关公司都将受益。根据能源局官员最新表态，2015年末我国太阳能大型并网电站累计装机目标6.5GW，鉴于CSP在大规模并网发电领域的显著优势，预计光热将在新增大型并网电站中占半壁江山，即到2015年末，累计装机3GW，假设80%的国产化率，直接拉动的国内市场将接近500亿。

24.3.3 太阳能热水器产业趋势

1. 在太阳能与建筑一体化中寻找商机

与建筑结合，是太阳能企业近年来寻找到的一条新的发展之路，在老旧小区“上屋顶难”的时候，新建小区的一体化建设要现实得多。业内已经形成一个共识，在城市市场，太阳能行业正在从零售向房地产“打包销售”、新建小区整体安装的方式转变。各大品牌太阳能热水器企业也陆续推出了壁挂太阳能等适合于城市建筑的太阳能热水器产品，来迎合城市市场。同时，企业也专门下大力气来进行工程市场的拓展，寻找新的市场机会的突破。

“十二五”期间，我国绿色建筑将迈入快速发展阶段，节能减排与低碳成为建筑中的热点。我国“绿色建筑评价标识”工作已准备就绪，明确要求由国家财政投入的建筑无论是廉租房、经济适用房，还是限价房，首先“建筑节能”要达标。这对于太阳能行业来说，是很大的鼓舞，同时也带来了可观的市场机会。

如果说2010年的保障房是1000万套的话，那么对于太阳能热水器而言是至少1000万台的市场空间，针对这一市场机会，大企业已经开始布局。四季沐歌力推出的“幸福保障房阳光热水工程”正是意图在保障房市场有所建树。力诺瑞特也推出了“阳光保障房计划”，希望通过与各地政府积极合作，共同推进保障房建筑的低碳化。整个“十二五”期间，保障房的建设总数为3600万套，这对于太阳能行业来说，是一块巨大的蛋糕。

我国多个省市已经颁布实施了进一步推广利用太阳能热水器的政策，要求对太阳能热水器实行强制安装：其中大多数要求在12层（或9层）以下的民用建筑上必须安装太阳能热水系统，适用范围包括住宅建筑和公共建筑。这与保障房建设同时作用，为太阳能热水器在城市的发展铺路奠基。

2. 拓展海外市场

从未来产业持续健康发展的角度看，太阳能热水器企业一定要走出国门，树立国际品牌，推动中国从太阳能大国走向太阳能强国。

目前，世界各国政府都鼓励和支持新能源产业的发展，由此催生了惊人的市场份额。据相关资料显示，太阳能热利用市场正在全球快速增长。以美国为例，2009年美国的太阳能需求共增长了36%，而同期对石油的需求只增长了2%。太阳能热利用行业国际市场前景非常看好，有越来越多的厂商打破本土企业形象，朝着国际化发展，增加出口，扩张海外经营版图。同时，由于国内厂家的技术研发投入不足，多年来中国太阳能热水器主要停留在低端水平的利用领域，在核心技术上与国际领先水平有较大差距，而这背后也意味着庞大的国外高端市场尚待开启。在2010年举行的太阳能热利用产业高峰论坛上，国家能源局有关专家明确表示，现在全世界太阳能热利用行业在快速发展，欧洲、非洲、南美都有非常大的市场空间，中国企业应放眼世界，走出国门，建成全球化的体系。

24.4 典型企业动态

24.4.1 天威保变

天威保变电气股份有限公司秉承并发展了原保定变压器厂主要优良资产和大型变压器科研成果及产品品牌，已实现国有股份制改革，并与2001年2月登陆A股市场。公司立足于河北“打造沿海经济强省”和保定市“打造保定·中国电谷”的战略规划，大力发展变压器、太阳能光伏发电、风力发电设备以及其他输变电产业，建设中国新能源生产基地。目前，公司已成为1000KV级及以下各类变压器、互感器、电抗器、太阳能电池、风力发电设备、高压套管、变压器专用设备以及IT技术等多产业的综合经济实体。公司大力实施“科技兴企”战略，努力提高企业核心竞争力，相继研发出多台具有国际先进水平、在中国变压器发展史上名列“第一”的变压器产品，成为变压器单厂产量世界第一、拥有变压器行业核心技术最齐全的企业，被列入国家1000MW及以下火电机组、水电机组和750KV及以下变电站（所）主要设备重点生产厂家。公司通过了南非国家标准局生产能力认证，拥有进出口贸易经营权，产品已打入美国、加拿大、巴基斯坦、伊朗、马来西亚和中国香港等30多个国家和地区，公司先后通过了安全管理体系GB/T28001认证和环境管理体系GB/T24001认证，被评为“管理示范企业”及“中国企业信息化500强”。

2010年公司实现营业收入76.3亿元人民币，同比增长27%。其中，变压器产量8451万KVA、风电整机产量114台、风电叶片产量521片、薄膜电池产量30兆瓦、多晶硅实现年产1268吨，2010年公司实现利润总额7.22亿元人民币，同比增长6.5%。

2010年，面对复杂多变的经济形势、日趋激烈的市场竞争，公司坚持“双主业、双支撑”的发展战略，外拓市场，内抓管理，不断推进技术创新和管理创新，有效地提升了公司的综合实力。

在变压器产业方面，2010年公司全力开拓国内外市场，奠定变压器产业发展的龙头地位。2010年整个电力设备行业呈现供大于求的态势。为此，公司积极开展市场营销工作，全力开拓国内外市场，充分利用公司高端产品竞争优势，在国内市场，相继中标国网1000KV特高压、南网直流等重大项目；并进一步完善公司牵引变压器等特型产品市场结构，先后中标国内高铁和客运专线等重大项目。在海外市场，先后中标印度国网750KV级变压器，伊拉克华事德项目及出口美国调相变压器。

在新能源产业方面，公司全面加快新能源项目建设，进一步完善新能源产业布局。针对新能源产业，公司着眼于长远发展，加快新能源产业建设步伐，进一步巩固和完善了产业布局。2010年公司圆满完成了新津多晶硅项目的建设并进行全面调试生产、风电整机二期项目建设、风电叶片的一期建设、薄膜项目的一期技术升级的前期准备工作及长春新能源公司的股权收购项目。以上项目的建设完成，一方面，进一步完善了公司新能源产业布局；另一方面，随着新能源产业的逐渐复苏，也为公司新一轮跨越式发展提供强有力的保证。

公司遵循“科技兴企”的发展理念，不断进行技术创新，完善产品结构，优化产品设计，加大科研与新产品研发工作的管理和投入力度，在全面推行标准化工作的基础上，不断完善新工艺、新方法，增强企业的核心竞争力。

（1）在输变电产业方面。顺利试制成功向家坝－上海±800KV直流输电工程用换流变，完成葛—沪±500KV直流输电工程用换流变的制造，完成±800KV/7200MW直流输电线路用干式平波电抗器工程图纸开发设计与前期技术准备工作，稳固了天威保变在换流变产品的技术领先地位；完成了DFP－40万KVA/1000KV特高压交流升压变压器的研发，本产品的开发设计对加快特高压输变电设备的国产化进程具有重大意义；在大容量分裂式辅助变压器的研发方面取得新突破；开发了500KV、800KV现场组装变压器技术，确保了交流变压器的核心技术领先地位；着力开发了新型平衡牵引变压器，设计开发了三相一体全系列V V接线牵引变压器、研制成功了220KV新型平衡牵引变压器。

（2）在新能源产业方面。薄膜太阳能电池：公司迅速完成了生产线的产能爬坡并在原有技术基础上，通过工艺革新使转换效率提升了17%。通过原材料、备品备件等国产化工作，使成本下降了40%；同时，在加快项目一期技术升级、提高转化效率的同时，注重增强系统集成能力，提升纵向化服务能力，在做好现有传统光伏应用市场的基础上，力争走出一条差异化竞争之路，开发独具特色的BIPV、光伏农业大棚等薄膜光伏产品。

风电整机：整机方面逐步完善技术体系，规范研发设计流程制定了严密的图纸管理、工艺评价制度；公司已形成1.5MW 70米/77米/82米，2.0MW86米/93米，适应多种环境的产品谱系，产品得到市场认可；联合开发了拥有自主知识产权的SCADA系统；2.0MW产品成功下线，标志着公司进一步提升了制造能力；全力推进具有完全自主知识产权的3.0MW海上风力发电机组研发设计。

风电叶片：公司叶片产品全部取得了GL认证，并实现规模化生产，目前正在积极研制3MW海上风机叶片产品。

多晶硅：公司控股的天威硅业公司，其多晶硅生产线采用了目前国际先进的多晶硅生产工艺技术，其精馏系统、尾气回收系统、还原氢化系统等关键技术和配套工艺技术具有世界先进水平，已进行全面调试生产。

以上科研成果的取得，进一步加强了公司的设计研发基础，提升了公司的综合技术实力，为公司的全面可持续健康发展提供了有利保障。

在国内，当前节能减排和低碳经济是我国宏观经济发展的长期战略。在“十二五”期间，预计电源发展上，坚持优化发展煤电、大力发展核电、积极发展风电等可再生能源发电、适度发展天然气集中发电、因地制宜发展分布式发电、优先开发水电；在电网发展上，建设大型电源基地外送通道，构建坚强网架建设大型水电基地外送通道、加强省级500（750/330）千伏电网建设、促进城乡电网协调发展、推进电网智能化；着力转变经济发展方式，优化电源结构和布局，提高电网资源优化配置能力，积极推进水电、核电、风电、太阳能等非化石能源开发，提高非化石能源比重，并将水电和核电作为重点。电网实现更大范围的资源优化配置，将增强对风电、太阳能等新能源的消纳能力。

天威保变将继续实施“双主业、双支撑”的发展战略，通过技术改造、合资合作等方式，大力推进、完善产业园区和产业基地建设，在公司内形成以“输变电”和“新能源”两大产业为主的产业格局，支撑并促进公司的可持续发展。

输变电产业方面以“做精、做强、做大输变电产业”为目的，以建立海外销售网络为手段，充分利用天威自身的品牌和优势加大在国际市场的竞争实力。始终不移的坚持自主创新、不断提升产品的科技含量和质量；以保定为核心，充分发挥秦皇岛、合肥工厂的产能和地域优势，坚持技术统一、市场统一、采购统一的原则，进一步优化资源配置，充分发挥规模经济优势；突出公司优势产品地位，集中全力发展1000千伏交流变压器、高压直流变压器、核电变压器等优势产品，进一步稳定市场占有率；积极开拓国际市场，参与国际市场竞争，提高产品在国际市场知名度和美誉度。

新能源产业方面不断通过产品升级、管理升级、技术升级，提高公司新能源产品在国际、国内两个市场的品牌认知度和产品销售业绩。以完善产业链建设、形成系统集成能力为重点，进一步完善光伏产业制造业建设；尽快实现光伏系统集成能力，带动整个产业的发展；根据产业发展状况及市场时机，开展光伏建筑、光伏电站的建设；建立光伏技术研发中心，进行光伏产品、制造工艺和装备的研发，提升产业核心竞争力，形成产业的可持续发展能力，形成以整机及关键零部件制造为核心，以风电场开发、风电系统集成和技术服务为延伸的风电产业链。通过风电技术研究中心建设，研究关键技术、完善产品系列，打造天威风电品牌，提高市场竞争力，形成1兆瓦—1.5兆瓦、2兆瓦—2.5兆瓦、3兆瓦，双馈、直驱、永磁直驱等结构形式的陆地和海上风电机组系列，成为国内、国际领先的风电企业。

24.4.2 精功科技

浙江精功科技股份有限公司始创于1992年，2000年9月改制成为规范化的股份公司，2004年6月25日在深圳证券交易所中小企业板成功上市。精功科技拥有员工1300多人，资产近12亿，主要从事机电一体化的太阳能光伏专用装备、建筑建材专用设备、轻纺专用设备、专用改装汽车等高新技术产品的科研开发、生产制造、经营销售和技术服务。精功科技现拥有浙江精功新能源有限公司等4家控股子公司及2家分公司，主导产品连续多年保持生产规模和市场占有率国内第一，使用的“精大”牌注册商标在国内外具有相当的知名度，产品远销欧美等发达国家和地区。

公司系国家重点高新技术企业，已实施了国家级火炬项目14项，开发国家级重点新产品14项，32次获省、市、县科技进步奖，组织实施3项国家、省创新基金，完成15项国家级、省级技改、科技计划项目和28项省级新产品开发试制的实施，现已拥有68项具有自主知识产权的专利技术。

2010年公司完成营业收入974,686,637.80元，同比增长52.21%；实现营业利润106,660,887.81元，同比增长2870.21%；实现净利润108,048,961.746元，同比增长946.75%。

2010年是精功科技加快产业转型、升级，创建科学管理体系取得重要成就的一年，也是精功科技发展史上具有重要转折意义的一年。第一，公司成功实现了产业转型和升级。以公司太阳能多晶硅铸锭炉为代表的光伏专用装备和太阳能多晶硅片研发生产为代表的新兴产业迅速崛起，经济效益大幅增长，其中光伏设备订单总额184461.02万元，比2009年增长885.845万元；多晶硅订单额16121.35万元，比2009年增长153.515万元，这两项业务成为名副其实的公司核心产业。第二，以公司聚氨酯生产线为代表的建筑节能装备、以包覆丝机和气流纺纱机为代表的纺织专用设备及专用汽车等传统产业继续保持行业领先地位，销售实现了翻番目标。第三，公司圆满完成年初提出的经济目标并创历史新高。第四，公司顺利启动了再融资工作。上述各项工作的顺利实施，为公司2011年跨越发展奠定了坚实基础。

2011年是国家“十二五”规划开局之年，也是精功科技产业转型、升级后实现跨越发展的关键之年。公司核心产业均已进入良性循环轨道，尤其是光伏新能源和专用装备业务，经过前几年布局和产品结构调整，已经进入厚积薄发的黄金时期，必将面临良好的市场机遇和巨大的发展空间。

“十二五”期间精功科技的整体发展思路和发展目标：立足“新能源和建筑节能专用装备”两大主导产业，以提高经济效益和跨越发展为落脚点，以“创新、提产、降本、增效”为主线，通过产业经营与资本运行的良性互动、研发机制优化和科技人才集聚，集公司之力进一步打造和持续提高自身的自主创新能力、规模化生产能力、核心竞争能力，抢抓机遇，实现企业跨越发展。力争到2015年，公司成为管理规范、核心产业竞争优势明显、行业领先、国际公认的一流的光伏专用装备及建筑节能专用设备供应商。

2011年，公司光伏装备制造产业将在统筹考虑未来成长性的基础上，以战略的发展眼光，做好产品布局。对内进一步加大资源整合，提升光伏装备分公司、精功新能源公司及精功研究所等部门协同作战效应，在铸锭炉产品技术升级平台、工艺技术提升平台建设、铸锭炉创新与产品标准化、稳定性方案上展开各项研究；对外主要通过与国内外主流光伏企业进行技术和贸易合作的模式，在提升产品转换效率、行业前沿技术等方面展开专项研究。第一，继续按照2010年提出的“产业化、系列化、成套化”的发展思路，加快太阳能铸锭炉产品的更新换代速度、线剖锭机的产业化进程及相关延伸装备的开发，使企业具备太阳能光伏专用装备的成套供应和整套解决方案的能力；第二，攻坚并加快高转换效率太阳能多晶硅片新工艺研发进度；第三，重点对行业发展的共性技术和重要核心前沿技术进行研究与开发，进一步增加战略技术储备，确保公司行业领先地位。

2011年，公司新能源产业将通过资本市场融资等方式进行扩产、产业链延伸和投资布局。公司将充分发挥全资子公司浙江功新能源有限公司这一产业发展和对外投资平台，切实从光伏产业上下游协同发展等角度统筹考虑，对内进一步做好太阳能多晶硅铸锭工艺、硅片切片技术的研究，为公司光伏装备发展提供工艺技术支撑；对外适时以自身掌握的太阳能多晶硅铸锭优势装备、先进工艺、切片技术积极对外投资。第一，加快精功新能源公司搬迁暨太阳能多晶硅切片生产线技改项目的实施进度，力争在2011年2月底前完成整体搬迁并投入正常运行，使之成为公司2011年业绩增长的新亮点；第二，根据全球光伏市场多元化发展趋势，延伸产业链，为新能源公司产业链配套和业绩提增作储备；第三，完成总投资3000万元的1000KWp光电建筑并网发电金太阳示范工程项目竣工验收工作。

24.5　小结

“十二五”期间，太阳能产业无疑将得到快速发展。在光伏产品制造方面，我国将继续保持全球太阳能电池生产大国的地位。与此同时，国内光伏应用市场在政策的支持下将迅速成长，“两头在外”的局面将得到改善。光热发电将取得重大进展，在国家政策和研发的支持下，太阳能热发电站关键技术将取得突破。而随着行业标准及规范的出台，太阳能热水器产业也必将迎来新的发展机遇。

25 风能[①]

25.1 风电产业回顾与展望

2010 年，低碳环保仍然是全球倡导的主旋律，绿色能源与新兴能源持续被人们关注，作为其中重要成员的中国，其风电产业也借势继续行驶在高速发展的快车道上。纵观全年，我国风电产业的发展，喜忧参半可谓是最贴切的总结。喜的是我国风电产业继续保持增长势头，同时我国迈出了大力推进海上风电的发展步伐；忧的是风电产业规模继续扩大、市场竞争愈发惨烈，同时在风机不断求大追新的快速发展背后，风机质量安全成为隐患。

25.1.1 2010 年我国风电产业发展回顾

1. 装机容量继续增长，但增速明显放缓

2010 年，风电产业依旧是最受人们关注的新能源产业。在全球大力发展清洁能源的大好时机下，我国风电装机容量继续保持增长。全年新增装机 1892.7 万千瓦，累计装机 4473.3 万千瓦（如图 25 - 1）。与 2009 年新增装机 1380 万千瓦、累计装机 2580 万千瓦相比，2010 年风电新增装机增长率仅为 37%，累计装机增长率为 73%，相比过去四年里风电装机容量连年翻番增长的

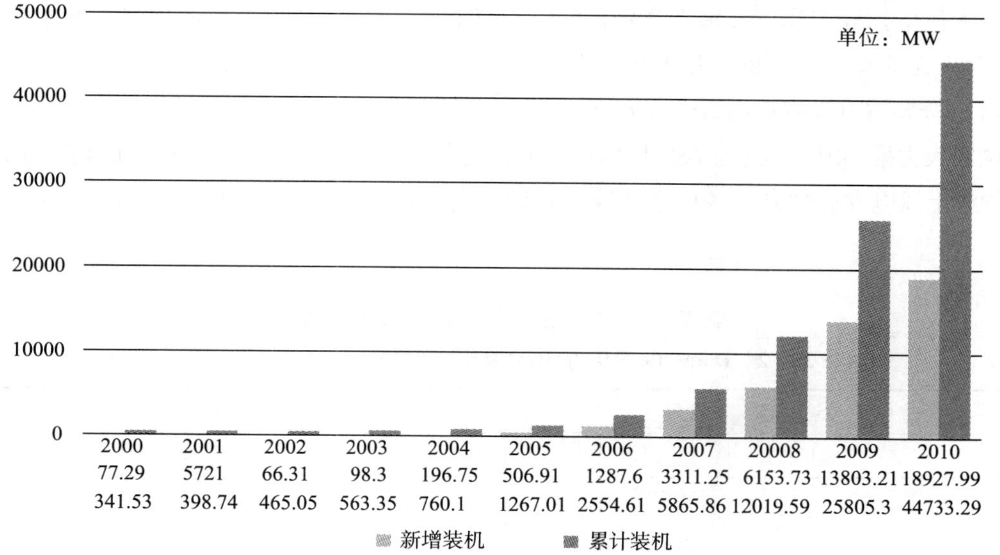

图 25 - 1　2000—2010 年我国风电装机容量

Figure 25 - 1　China's Installed Capacity of Wind Power during 2000—2010

资料来源：中国可再生能源学会风能专业委员会

① 本章由中国船舶重工集团公司经济研究中心胡颖撰写。

态势，2010 年我国风电装机容量增速明显放缓，风电产业正逐渐步入平稳增长。

2. 我国重视风电产业，海上风电成为新宠

为达到我国在哥本哈根会议上提出的“到 2020 年非化石能源占一次能源消费的比重达到 15%、单位 GDP 二氧化碳排放强度比 2005 年下降 40%—45%”的目标，2010 年国家陆续出台了一系列促进风电产业发展的法律、法规和产业政策（如表 25－1），发展目标更加明确，思路更加清晰，前景非常广阔。此外，根据国家最新的能源规划，2020 年前国家将在新能源领域增加 5 万亿元投资，其中可再生能源投资约 3 万亿元，风电占 1.5 万亿元。

表 25－1　国家促进海上风电发展的相关政策

Table 25－1　China's Policies for the Promotion of Offshore Wind Power Development

出台年份	政策名称	制定部门	主要内容
2010.1	《2010 年能源工作总体要求和任务》	国家能源局	2010 年，要继续推进大型风电基地建设，特别是海上风电
2010.1	《海上风电开发建设管理暂行办法》	国家能源局、国家海洋局	规范海上风电建设
2010.3	《风电设备制造行业准入标准》（征求意见稿）	工业和信息化部	明确表示有限发展海上风电机组产业化
2010.10	《国务院关于加快培育和发展战略性新兴产业的决定》	国务院	确定了七个产业发展的重点方向、主要任务和扶持政策，标志着新兴战略产业框架已成定局

资料来源：中国船舶重工集团公司经济研究中心

2010 年是我国海上风电加速发展的元年。我国能源局 5 月 18 日正式启动了总计 100 万千瓦的首轮海上风电招标工作，分别为滨海和射阳的两个 30 万千瓦的近海风电项目；大丰和东台的两个 20 万千瓦的潮间带项目，并于 9 月 10 日在北京开标。同年 6 月，我国首个海上风电项目——上海东海大桥 102 兆瓦项目全部并网发电。

在国家大力推动海上风电加快发展的形势下，上海、江苏、浙江、山东和福建等省市纷纷提交了各自的海上风电发展规划，各风电企业更是前赴后继进行海上风机的研制与生产，如表 25－2 所示。

表 25－2　内资企业海上风电机组研发情况

Table 25－2　Domestic Enterprises' R&D of Offshore Wind Power Units

公司名称	研发机型
金风科技	6.0MW
华锐风电	5.0MW
	6.0MW
上海电气	3.6MW
	5.0MW
明阳电气	3.0MW
	6.0MW

续表

公司名称	研发机型
东方电气	3.6MW
	4.5MW
湘电股份	5.0MW
保定天威	3.0MW
国电联合动力	3.0MW
海装风电	5.0MW
华仪电气	3.0MW
	5.0MW
北重	3.0MW
	5.0MW

资料来源：中国船舶重工集团公司经济研究中心

3. 风机大型化日益凸显，直驱双馈同台竞技

我国风电机组从2005年开始步入“兆瓦级时代”，当年仅占新增装机容量的21.5%；2009年，兆瓦级风机成为市场主流，占到当年新增装机容量的86.86%；2010年2.5兆瓦级以上风机成为行业重点发展产品，同年8月上海电气3.6兆瓦海上风机并网发电、10月华锐风电和湘电股份分别宣布5兆瓦海上风机下线。据悉，“十二五”期间将重点发展3—5兆瓦级陆地风机和5—10兆瓦级海上风机。

在技术路径的选择上，直驱与双馈各有利弊。随着对风力发电机组可靠性要求的不断提高，直驱式风机设计正在不断增加。由于风机单机容量提高，为应对极限载荷和疲劳载荷的挑战，新的直驱变速变桨技术正在逐步成为兆瓦级风机的主流技术，直驱风电机组的市场份额也在迅速扩大。尽管直驱风机可靠性高，运行稳定，但也存在单机结构体积和重量大、技术要求高、运输和安装较困难等问题。目前，已经有企业开始研发介于直驱式风机和双馈式风机之间的半直驱风机。在未来一段时间内，风机还将处于直驱式风机与双馈式风机同台竞技的格局。

4. 产业规模迅速扩张，市场竞争更加激烈

2004年，我国风电整机制造厂商只有6家。截至2010年底，作为全球最大的风电整机生产国，我国已有100多家整机商，有样机下线的就有近90家。此外，还有风电设备零部件制造企业上百家，叶片生产厂50余家，塔架制造厂100余家。越来越多的企业进入风电产业，风电设备市场的竞争也越发激烈，风电设备生产企业正面临新的行业洗牌。

5. 风电整机价格持续下降，企业利润严重缩水

面对激烈的市场竞争，企业之间展开了残酷的价格战。自2008年开始，我国风电机组的市场售价开始持续下滑。最典型的例子是1.5兆瓦风机的单价，已从2008年初的6500元/千瓦下降到了目前3800元/千瓦的水平。

整机价格的快速下降严重削弱了风电设备制造企业的利润，前两年高歌猛进的风电设备企业发展速度在2010年忽然放缓。以湘电股份为例，其2009年上半年财报业绩增长达300%，而2010年上半年归属于公司股东的净利润较去年同期仅增长50%以上。

6. 零部件制造业已基本形成体系，部分关键配套仍需增强实力

目前，我国风电零部件制造业逐步成熟，已形成涵盖叶片、齿轮箱、发电机、变桨偏航系统、轮毂、塔架等主要零部件的生产体系（如表25－3）。但部分关键配套实力与国际先进水平仍存在差距，例如控制系统、风电轴承产品等，相关研发和制造实力有待于进一步增强。

表25－3　国内主要配套设备研发生产情况

Table 25－3　Domestic Enterprises' R&D and Production of Major Corollary Equipment

配套设备	功率范围（千瓦）	生产能力及发展特点	核心企业或主要研究单位
叶片	750—3000	批量生产、基本满足需要	中航惠腾、中复联众、中材科技等
齿轮箱	750—3000	基本满足需要、受轴承制约	南高齿、重齿、大重减速机等
发电机	750—3000	基础较好，制造企业多	永济电机、南车电机、中船重工电机等
轴承		部分关键配件仍依赖国外	瓦轴、洛轴、天马、京冶轧机等
控制系统		产业化程度较低	金风、上海交大、上海自动化研究所
变流器		尚未形成大批量配套能力	禾望电气

资料来源：中国可再生能源学会风能专业委员会

7. 安全事故频发，质量成为隐患

2010年，几起恶性的风电事故给近年突飞猛进的风电发展敲响了警钟，风电设备的质量问题显现出来。在我国风电机组关键设计技术仍然依托欧美，风电机组的检测认证体系还不健全，在风电场缺乏长期运营维护的经验之下，中国风机质量大考已经开始。

2010年1月24日东汽风机倒塌，被视为2010年第一起风机事故。随后，华锐等品牌风机倒塌事故不断出现，国内前列的整机制造企业几乎无一幸免。风机质量出现的问题还包括风机叶片、主轴断裂，电机着火，齿轮箱损坏，控制失灵以及飞车等。

一般而言，风机在投入运行五年以后才进入对整机和零部件质量的真正考验期，而我国大部分风机都是最近两三年安装的。因此，未来几年将成为考验中国风机装备质量的关键时期。

25.1.2　我国风电行业未来发展展望

在刚刚过去的2010年，随着风电产业的快速发展，我国风电企业已由先前的盲目追“风”变得更加理性。展望未来，在新兴能源盛行全球、我国大力推进新能源产业发展的大环境下，我国风电产业将由快速发展转为健康发展。

1. 装机容量保持平稳增长

按照我国风电产业目前的发展速度，在继2009年我国风电新增装机跃升全球第一之后我国风电累计装机即将超过美国，占据世界首位，成为世界上最大的风电市场和风能设备制造中心。预计未来10年，我国每年新增风电装机容量仍可保持在1500万千瓦—2000万千瓦。

2. 海上风电仍将受宠

尽管海上风电受到开发技术难度大、投资成本高等因素限制，但是凭借其海上风能资源丰富、不受土地限制、年利用小时长、风速高、风资源持续稳定、机组发电量高、视觉及噪音影响少等优势仍将继续受到人们的关注，但其再技术、材料等方面势必面临很多挑战。

3. 风机大型化趋势不改

随着风电技术日趋成熟以及对降低风电发电成本的要求越来越高，风力发电机组正不断向

大型化发展，风电机组单机容量不断增加。2 兆瓦以上陆上风机和 3 兆瓦以上海上风机将成为未来市场的主力机型。

4. 直驱式风机市场份额将增大

直驱式风机由于省掉了齿轮箱，可靠性提高，运行更加稳定，国内更多的风电企业，以金风科技、湘电股份为代表，正在积极研究开发直驱式机组，预计直驱式风机的市场份额将进一步扩大。

5. 风机价格小幅下跌

未来我国风电市场竞争仍将十分激烈，为了能争得一块蛋糕，风电企业之间的价格战仍将继续。但是，由于目前的机组价格已经触碰了风电企业的底线，到达了亏本的边缘，很难继续大幅下跌。因此，预计未来风机价格将出现小幅走低。

6. 风机质量将放在首位

2010 年的几起风机事故已经给人们重重地敲响了警钟。作为战略新兴产业的风电产业已经开始转入健康稳健发展的轨道，风机质量必将成为未来风电设备市场角逐的重要因素。只有质量过硬、安全可靠的国产风电设备作支撑，中国风电产业才能健康长远地发展；只有品质优良、用户信赖的风机产品才能在激烈的市场竞争中屹立不倒。严把质量关是每一个风电企业必须高度重视的，也将是未来我国风电产业发展的重中之重。

7. WTO 规则将影响行业发展

2010 年 9 月 9 日，作为美国最大的工会组织之一，美国钢铁工人联合会（USW）按照《1974 年美国贸易法案》第 301 条款的规定，向美国贸易代表（USTR）办公室提交了长达 5800 页的诉状，称中国政府给予新能源企业高额补贴以提高中国风电、太阳能的价格优势，影响美国就业；同时指责中国对外国企业参与招标采取歧视性限制。10 月 15 日，美国贸易代表处决定对中国新能源进行调查。按照惯例，美国贸易代表办公室计划在 90 天时间内完成相关调查，该项调查将对我国风能、太阳能、高效电池和新能源汽车行业造成较大影响，涉及 154 家中国新能源企业。

在应对美国贸易代表处开展调查和批驳美国此举毫无根据且不负责任的同时，我国政府、行业、企业等各界人士都需要认真了解 WTO 规则，深入思考并积极备战。主动研究国际相关法律、法规，在未来制定政策时，更加谨慎小心，注意避免违背 WTO、GPA 等国际规则，最大限度保护中国风电行业利益。

25.2 海上风电的现状与趋势

哥本哈根会议将大力发展清洁能源推向了一个历史高潮。近年来，全球风力发电行业发展迅速，但陆上风电由于占用土地，资源紧俏，一些国家的陆上风电开发已趋饱和。海上风电凭借其海上风能资源丰富、年利用小时长、风速高、风资源持续稳定、机组发电量高、不受土地限制、视觉及噪音影响少等优势一跃成为“后哥本哈根”时代的新宠儿。因此，尽管目前海上风电开发技术难度较大，投资成本偏高，但是海上风电已成为许多沿海国家清洁能源的新发展方向。

25.2.1 世界海上风电发展现状

1. 世界海上风电发展迅猛

2009 年海上风电装机容量继续增长。截至 2009 年底，全球共有 12 个国家建立了海上风电场，其中10个位于欧洲，中国和日本有小规模的安装。

截至2009年底，世界海上风电累计装机容量达2110MW，较2008年增长48.5%，占到全球风电总装机量的1.2%。2009年世界海上风电新增装机容量达689MW，同比增幅超过100%，新增装机容量最大的前五个国家分别为英国、丹麦、中国、德国和瑞典，如表25－4所示。

表25－4　2009年世界海上风电安装情况

Table 25－4　Installed Capacity of Worldwide Offshore Wind Power in 2009

排序	国家	2009年底海上风电累计装机容量MW	2009年新增海上风电装机容量MW	2008年底海上风电累计装机容量MW	2008年新增海上风电装机容量MW
1	英国	894	306	588	194
2	丹麦	625.9	228	397.9	0
3	荷兰	246.8	0	246.8	120
4	瑞典	163.3	30	133.3	0
5	中国	63	63	0	0
6	德国	60	60	0	0
7	比利时	30	0	30	30
8	爱尔兰	25	0	25	0
9	挪威	2.3	2.3	0	0
	世界	2110	689	1421	344

资料来源：丹麦BTM咨询

2. 欧洲海上风电发展令世人瞩目

欧洲是海上风电发展最快的地区。根据欧洲风能协会（EWEA）的最新统计，2009年欧洲水域的八个海上风电场总计安装199台海上风力涡轮机并实现了并网，总容量为577MW，较2008年增幅超过50%。其中，最小装机容量为2.3MW（挪威的Hywind），最大装机容量为209MW（Horns Rev2）。另外，欧盟15个成员国和其他欧洲国家，有超过100GW的海上风力发电项目正在规划中。2009年欧洲各国海上风电装机情况如表25－5所示。

表25－5　2009年欧洲各国海上风电装机情况

Table 25－5　Offshore Wind Power Installed Capacity of European Nations in 2009

	英国	丹麦	瑞典	德国	挪威
风机数量（台）	84	98	10	6	1
装机容量（兆瓦）	284.4	230	30	30	2.3
风场数量（个）	3	2	1	1	1

资料来源：欧洲风能协会

在2009年装机并网的199台风机中，西门子风机（2.3MW和3.6MW两种机型）146台，维斯塔斯风机（3MW）37台，WinWind风机（3MW）10台，Multibrid风机（5MW）6台。除此之外，Repower风机（5MW）6台但尚未并网。2009年欧洲海上风电生产商市场份额如图25－2所示。

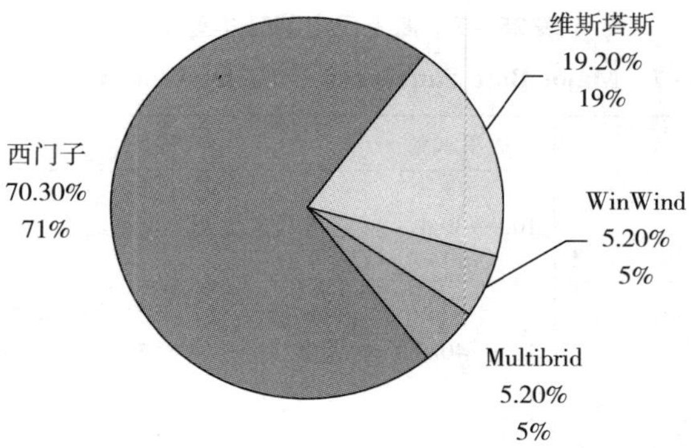

图 25 －2 2009 年欧洲海上风电生产商市场份额（按兆瓦计算）

Figure 25 －2 Market Share of European Offshore Wind Power Equipment Manufacturers in 2009 (by Megawatt)

资料来源：欧洲风能协会

3. 海上风电机组技术特点

目前，海上风电机组基本上是根据海上风况和运行工况，对陆地机型进行改造，其结构也是由叶片、机舱、塔架和基础组成。海上风电机组的设计强调可靠性，注重提高风机的利用率、降低维修率。

当今，海上风电机组呈现大型化的趋势，国外主要风机制造商生产的海上风电机组主要集中在2MW—5MW，风叶直径在 72m—126m。国外主要海上风电机组成技术情况如表 25 －6 所示。

表 25 －6 国外主要海上风电机组技术

Table 25 －6 Technological Parameters of Major Offshore Wind Power Equipment

生产商	风电机组型号	额定功率/万千瓦	叶轮直径/m	叶轮转速/ r · min －1	叶片数	功率调节
Vestas（丹麦）	V9022. 0MWOff shore	0. 2	80	可变速 9—19	3	变桨距
Bonus（丹麦）	Bonus3. 6MW	0. 36	107	可变速 5—13	3	变桨距
NEG－Micon（丹麦）	NEG－2Micon 72	0. 2	72	同步转速 12/18	3	主动失速
GE Wind Energy（美国）	GE3. 6MW	0. 36	104	可变速 8. 5—15. 3	3	变桨距
Nordex（德国）	N80 －22. 5MW Off shore	0. 2	80	可变速 10. 9—19. 1	3	变桨距
Enercon（德国）	E－2112	0. 45	114	可变速 8—13	3	变桨距
REpower（德国）	REpower 5M	0. 5	126	可变速 6. 9—12. 1	3	电动变桨距

资料来源：欧洲风能协会

海上和陆上风电机组的主要差别在于基础。为了承受海上的强风载荷、海水腐蚀和波浪冲击等，海上风电机组的基础远比陆上的结构复杂、技术难度高、建设成本高。目前，世界主要海上风电机组基础类型、适用水深及其优缺点如表 25 －7 所示。

表 25－7 海上风电基础类型汇总

Table 25－7 Major Base Patterns of Offshore Wind Power Equipment

基础类型	简要物理特征	适用水深	优点	局限性
单桩钢结构基础	单个支撑柱	10m—30m	生产工艺简单，安装经验丰富	施工噪音大，受海床和风机重量影响大
钻孔安装的单桩混凝土基础	单个支撑柱	10m—40m	有效施工方法的组合，成本低，环境（噪声）影响小，能实现工业化生产	重量大，运输困难
重力基础	混凝土结构，在Thornton bank 海上风电场使用过	>40m	没有打桩施工噪音，安装成本低	风机重量大，运输困难；需要对海床进行预处理；需要重型设备移动基础
沉箱基础	形似钢制桶，倒置于海底，借助海水压力压入海床	暂无数据	无需打桩，安装简单，移动方便	对海床要求严格
三角架或四角架	3 或 4 脚结构	>30m	强度高，适用于大型风机	结构复杂，重量大，运输困难
桁架式	网格结构	>40m	噪音小，适用于大型风机	目前造价昂贵；受海浪作用，容易疲劳失效；大型基础海上安装受天气影响较大
浮式基础	安装不受海床影响	>50m	适用于深水海域，该水域海上风力发电潜力大	海上风电缺乏重量、成本、稳定性等方面经验
柱型浮标式基础	安装于海床的浮式钢质圆筒	120m—700m	适用于超深水，钢材使用量较少	造价昂贵
半潜式基础	安装于海床的浮式钢质圆筒	blue H 的样机已在 113m 水深进行测试	适用于深水，钢材使用量较少	造价昂贵

资料来源：欧州风能协会

25.2.2 我国海上风电发展现状

根据2010 年1 月初中国气象局公布的我国风能资源详查数据：中国5m—25m 水深线以内近海区域、海平面以上 50m 高度可装机容量约2 亿 KW。从 2010 年起，我国海上风电产业开始加速发展。目前已有上海、江苏、浙江、山东和福建等多个省市区提出了各自的海上风电发展规划。根据这些规划，预计到2015 年我国海上风电装机容量将达到500 万千瓦；到2020 年增长至3000 万千瓦。

1. 我国主要风电整机企业争先“下海”

在中国陆上风机日趋饱和的情况下，进军海上风电市场成为中国主要整机企业的共同选择。各整机企业积极筹划、先入为主抢占市场。内资企业海上风电机组研发情况如表 25－8 所示。

- 华锐风电成功摘得中国第一个海上风电示范项目——上海东海大桥项目。2010 年 2 月 27 日，34 台 3MW 机组（共计 102MW）海上风电项目全部整体安装成功，并于6 月 8 日调试完毕，全部并网投入运行。

- 中国第二大风电整机企业金风科技已于2007年在渤海湾中海油的钻井平台试水了海上风机的所有工序。截至2009年6月，该海上风机已累计发电500万度。2009年11月18日，金风科技投资30亿元在江苏大丰经济开发区建设海上风电产业基地项目，并计划将其建设成为国内最大、世界领先的海上风电装备制造基地。

- 华仪电气宣布再融资11.46亿元，其中超过4.6亿元将投向3MW风力发电机组高技术产业化项目，用于备战海上风电；上海电气2009年5月宣布规模达50亿元的再融资计划，其中包括3.6MW海上风机的研发；湘电风能有限公司通过竞拍，以1000万欧元收购荷兰达尔文公司，并获得了该公司研究的DD115－5MW海上风机的知识产权，为进军海上风电奠定了基础；东方电气3.6MW海上机型正在研制中。

- 中船重工（重庆）海装风电充分依托集团公司在海洋工程领域的基础研究和试验基地等优势，整合风电整机和配套设备的研发实力，形成全产业链。现已组织实施了2MW近海潮间带批量装机工程，正致力于研发近海5MW风电机组，国家科技部授牌成立了“海上风力发电工程技术研发中心”。

表25－8　内资企业海上风电机组研发情况

Table 25－8　Domestic Enterprises' R&D of Offshore Wind Power Units

公司名称	研发机型	样机产出时间
金风科技	1.5MW	2007年
	3.0MW	2009年
	5.0MW	
华锐风电	3.0MW	2008年
	5.0MW	2010年
上海电气	2.0MW	2009年
	3.6MW	2010年
明阳电气	3.0MW	2009年
	6.0MW	2012年
东方电气	3.6MW	2010年
	4.5MW	2012年
湘电股份	5.0MW	2010年
保定天威	3.0MW	2010年
国电联合动力	3.0MW	2010年
海装风电	5.0MW	2011年
华仪电气	3.0MW	
	5.0MW	

资料来源：中国船舶重工集团公司经济研究中心

2. 国家政策推动海上风电发展

2008年已完成并发布了《近海风电场工程规划报告编制办法》和《近海风电场工程预可行性研究报告编制办法》，2009年完成并发布了《海上风电场工程可研报告编制办法》和《海上风电场工程施工组织设计编制规定》，印发了《海上风电场工程规划工作大纲》。

2010年1月，国家能源局在《2010年能源工作总体要求和任务》中称“2010年，要继续推进大型风电基地建设，特别是海上风电要开展起来”。2010年1月22日，国家能源局、国家海洋局联合下发《海上风电开发建设管理暂行办法》，规范海上风电建设。3月25日，工业和信息化部发布《风电设备制造行业准入标准》（征求意见稿），其中明确表示，“优先发展海上风电机组产业化”。近日，国家能源局启动了中国首轮海上风电首批特许权招标，并已经向辽宁、河北、天津、上海、山东、江苏、浙江、福建、广东、广西、海南等11个省份有关部门下发通知，要求各地申报海上风电特许权招标项目。可见，国家开发海上风电的步伐正在加快。

3. 政府大力推进海上风电项目

我国在哥本哈根全球气候变化会议做出了两项承诺：到2020年非化石能源在能源消费中的比例提高达到15%，单位GDP二氧化碳排放量比2005年减少40%—45%。经国内外多个部门和机构分析预测，为实现非化石能源达到15%的目标，我国风电装机容量应达到1.5亿千瓦。中国正希望从海上获得更多的风能，以完成这一目标。

东部沿海地区的滩涂及近海具有开发风电的良好条件，其中江苏、浙江两省将成为我国海上风电的重点省份，两省近海风资源到2020年规划开发容量分别为700万千瓦和270万千瓦。此外，全国各地酝酿及在建的海上风电场还包括广东湛江、广东南澳、福建宁德、浙江岱山、浙江慈溪、浙江临海、山东长岛等。目前，江苏、上海的海上风电规划已经完成，待其他地方海上风电规划全部完成之后，将汇总形成全国性海上风电规划。

我国将充分利用其丰富的近海风能资源优势，国家能源局5月18日正式启动了总计100万千瓦的首轮海上风电招标工作，包括两个30万千瓦的近岸风电项目和两个20万千瓦的滩涂风电项目。

25.2.3 我国海上风电发展建议

1. 加强国际技术合作，抢占海上风电市场

手握订单、拥有市场，企业才能生存发展，不断壮大。目前，国内海上风电还处在起步阶段，竞争格局尚未形成，及早抢占市场，打响品牌，对我国风电企业海上风电的长远发展有深远影响。

目前，我国海上风电产品集中在3MW，风电企业应不断提高技术水平，打造中高端产品，引领行业发展。加强国际技术交流与合作，发挥科研院所的研发优势，厂所合作，突破海上风电机组技术，在高起点上提高海上风电产品开发步伐和总体水平。

世界海上风电在未来的20年里将会迅速发展，尤其是欧洲，海上风电已获得欧盟及欧洲各国政府的大力支持，因此，海上风电的国际市场空间十分广阔。风电企业应加大开拓国际市场的步伐，组建高水平专业营销团队，先入为主，占领先机，并不断提高风电整机及配套设备在国际市场上的知名度和认可度。通过联合研发生产，收购并购等方式，主动进军国际市场。

2. 寻求电力企业合作，解决风电并网问题。

海上风电的良好发展前景促使“圈海”运动愈演愈烈，在机遇与挑战并存的情况下，风电企业应充分利用自身优势，积极寻求与国内电力企业合作，形成战略联盟，共同开发海上风电项目。由于海上发电同样面临着最终并网发电的问题，加强与电力集团的合作，解决风电入网

的各种技术问题，消除后顾之忧。

25.3 典型企业动态

25.3.1 华锐风电

华锐风电主营业务为大型风力发电机组的开发研制、生产及销售，当前产品主要为1.5MW系列和3MW系列双馈型风力发电机组。华锐风电发展速度极为迅猛，这家2005年底方成立的企业，仅用五年的时间就坐上了国内第一、全球第三的宝座。

2010年华锐风电主要业绩数据如下：

- 公司新增装机容量438.6万千瓦，继续排名中国第一，并且已跻身世界第二。
- 实现营业收入203亿元，同比增长48.03%；实现营业利润31.33亿元，同比增长49.09%；实现净利润28.56亿元，同比增长50.87%；1.5MW机组的销售较上年增长了42.42%，主营业务毛利率较上年降低了0.30%。
- 3MW机组的销售较上年增长了1777.39%，主营业务毛利率较上年增长了9.29%，主要系零部件采购价格下降，同时3MW风电机组大批量生产，实现了规模效益。
- 1.5MW机组、3MW风电机组主营业务成本分别较上年增长了42.96%、1572.85%。

华锐风电具体经营成果主要表现在：

1. 坚持科技创新，产品研发能力保持领先

完成了1.5MW、3MW风电机组向“技术创新型、电网友好型”的全面升级；完成SL1500/89低风速风电机组的研制并成功投运；自主研发并成功出产中国第一台5MW海上风电机组和中国第一台5MW陆地风电机组，填补了国内空白；完成6MW风电机组的技术开发；完成国家能源海上风电技术装备研发中心试验台群的研制；持续提升产品可利用率，陆地风电机组平均可利用率达到98.3%，海上风电机组平均可利用率达到98%以上。

2. 国内市场继续保持领先优势，国际市场取得突破

国内市场方面，在竞争日趋激烈的形势下，公司在国家风电特许权项目招标结果中排名第一：国家首轮100万千瓦海上风电特许权项目中标全部60万千瓦海上项目（全部采用3MW风电机组），新疆哈密及张家口地区330万千瓦特许权项目中标135万千瓦（其中采用3MW风电机组项目90万千瓦）。根据国家发改委、国家能源局批复，公司还中标江苏大丰沿海潮间带30万千瓦示范项目；另外，公司还中标384MW（采用3MW陆地风电机组）和45MW（采用5MW陆地风电机组）的甘肃酒泉千万千瓦级风电基地示范项目，并获得黑龙江大庆200MW的3MW风电机组订单。

国际市场方面，公司积极筹建覆盖美国、西班牙、加拿大、澳大利亚和巴西等地的全球化网络，并与上述国家的重点客户建立了战略合作伙伴关系，同时，加大金融支持力度，与国家开发银行签署了65亿美元的《开发性金融规划发展与战略合作协议》。截至2010年底，公司尚未履行完毕的在手订单及中标未签约项目合计14404.5MW。其中，尚未履行完毕的在手订单共3619.5MW，其中1.5MW风电机组3229.5MW，3MW风电机组390MW；中标未签约项目10785MW，其中1.5MW风电机组4182MW，3MW风电机组5610MW，5MW风电机组945MW，6MW风电机组48MW。

3. 全面推进3MW大功率风电机组的生产及规模化装机

公司自主研发的34台3MW海上风电机组在全球除欧洲以外的第一个海上风电场、中国第一个国家海上风电示范项目——上海东海大桥海上风电场全部并网发电，并顺利通过240小时预验收，打破了国外企业对高端风电机组制造技术的垄断；公司自主研发的潮间带3MW风电机组（SL3000/113）在江苏如东潮间带风电场成功投入运行；公司自主研发的3MW陆地风电机组已经在河北张家口、通辽北清河实现装机和并网发电。3MW系列化风电机组的全面推广，推动了中国风电技术向大功率风电机组的不断升级。

4. 全面实施全球产业布局

公司紧跟国家建设千万千瓦级风电基地的战略规划，结合自身发展战略，在大连、江苏盐城、内蒙古包头、甘肃酒泉基地实现规模化生产的基础上，建设了山东东营、吉林白城及内蒙古兴安盟等风电产业基地，在美国建立了子公司，并积极拓展上海、内蒙古乌拉特和赤峰、新疆哈密、大连长兴岛、天津、江苏射阳以及海外风电产业基地的建设工作。公司已经初步形成以北京为管理中心、产业基地遍布全球的战略布局。

5. 科学组织生产，持续提升产品质量

公司在建立和逐步完善适合全球化战略布局的生产和配套产业体系的基础上，进一步加强对产业链的控制和管理，提高产品的可靠性、稳定性；同时，公司结合市场动态，开拓多元供货渠道，并严格执行供应商评价和管理体系。公司选择的零部件配套厂商均为行业内领先的生产企业，公司已经与其建立长期了的战略合作关系，双方风险共担，这种行业协作、优势互补的模式有效保障了公司供应体系的稳定、可靠，获得的服务也更为完善，成本竞争优势显著提高。另外，2010年公司参与设立变频器制造企业大连国通电气有限公司并实现批量生产，进一步提高核心部件供给能力并降低成本。公司还成立了产品质量可靠性保障小组，建立以检测、预防、纠正、控制为核心的质量管理规范，深入开展质量控制，强化产品检验管理，全面实现质量管理过程的标准化，持续提升产品质量。

25.3.2 金风科技

金风科技作为国内风电行业的领先企业，已成功由单一的风电机组制造销售业务转向风电机组制造销售，进入零部件开发与生产制造、风电服务和风电场开发销售等环节；风电技术开发转让多种盈利模式相结合，成为风电整体解决方案提供商。公司是国家大型风力发电机组“863”项目及国家“九五”、“十五”、“十一五”科技攻关项目的重点承担单位，目前正在承担“十一五”国家科技支撑计划“大功率风电机组研制与示范”重大项目的三项课题。

公司共有控股子公司60家，其中直接控股子公司22家，间接控股子公司38家；另外，公司还有6家共同控制公司、3家联营公司及7家参股公司。

2010年公司完成营业收入1,759,552.06万元，较上年增加685716.54万元，同比增长63.86%；实现营业利润269059.29万元，较上年增加74367.27万元，同比增长38.20%；实现净利润238383.76万元，较上年增加59323.51万元，同比增长33.13%；实现归属母公司净利润228952.03万元，较上年增加54394.08万元，同比增长31.16%。

2010年公司产品成功转型，永磁直驱系列化风电机组得到市场全面认可，直驱永磁技术通过市场检验，与之相配套的技术研发、供应链、产品生产、质量控制、服务体系日趋完善，为适应市场需要，公司加快了系列化机型研发、生产及技术优化的速度，针对低风速、沿海、高海拔、高低温、极大风速等不同运行环境的系列化机组全部投入市场，全年销售 1.5MW 机组 2567 台，同时，2.5MW 永磁直驱风电机组已经进入批量化生产阶段。

公司的西安基地于 2010 年正式投产，成为西北片区重要的兆瓦机组生产基地。江苏大丰海上风电产业基地也已正式投产，这是公司海上风电战略的一个重要里程碑。公司将借助国家开发海洋能源的有利时机，携“海上风电整体解决方案”全面参与海上风电项目建设。至此，公司在国内的整机制造产能布局基本完成，共有 9 个生产基地，分别是：北京亦庄、新疆乌鲁木齐、内蒙古包头、甘肃酒泉、陕西西安、江苏大丰及辽宁阜新、河北承德及宁夏银川。随着技术、工艺的不断优化，1.5MW 机组的产能已达到 4000 台，这些基地同时能满足 2.5MW 机组的生产需求。

公司各项研发工作进展顺利，1.5MW 机组通过规模化工况条件下的低电压穿越的测试与认证，成为首批得到此项认证的永磁直驱机组。1.5MW 机组的可利用率已经稳定在 98% 以上，直驱永磁技术的优势通过市场的验证，客户满意度不断提高。公司研制的 2.5MW 机组已经分别在新疆、北京、江苏及德国等地完成 5 台样机的安装，实现并网运行，运行效果良好，上述样机分别适用于陆地及潮间带以适应不同市场需求。基于 2.5MW 直驱永磁机组良好的运行及检测结果，公司已开始进行该机型的系列化设计乃至容量升级工作。为适应市场竞争，抢占先机，公司加快了大型机组研发进度，2010 年启动了 6 兆瓦直驱永磁风电机组的开发工作。自 2008 年开始，公司积极推进战略转型，实施关键零部件的纵向一体化。2010 年公司与全球领先的半导体解决方案供应商英飞凌（Infineon Technologies AG）签署了核心模块技术引进协议，公司获权在国内生产兆瓦级风力发电机组变流器所需的英飞凌模块，在享有批量自产所带来的显著成本效益的同时，核心部件自给能力也得以持续提升。继电控系统的自主生产后，公司开始了叶片的自主设计与研发，2010 年公司收购了国内成熟的叶片供应商协鑫风电（江苏）有限公司及协鑫风电（锡林浩特）有限公司，意在整合双方的叶片设计与制造能力，快速实现叶片的自主生产，既能掌握核心技术，又能降低成本。随着公司直驱永磁机组逐步得到市场及客户的认可，公司订单持续增长，截至 2010 年底，公司持有待执行订单 2764.5MW，其中 1.5MW 机组 2716.5MW，2.5MW 机组 45MW，3.0MW 机组 3MW；另外，尚有已中标未签合同订单 3815.5MW，分别是 1.5MW 机组 2112MW，2.5MW 机组 1697.5MW，3MW 机组 6MW，其中包括 20 万千瓦海上项目，为后续的持续发展奠定了良好的市场基础。

公司国际化战略稳步推进，完成了海外目标市场的确定及业务布局，海外市场拓展初见成效。公司在美国的全资子公司 TianRun Shady Oaks, LLC 在美国市场成功中标电价并签署长期电价购买协议，为位于美国伊利诺伊州的 Shady Oaks 风电场自 2012 年起供应 20 年的电力，本次中标在公司国际市场战略进程上具有重要意义。此外，公司与中水顾问国际工程公司签署了风电设备供货合同，为其总承包的埃塞俄比亚 - Adama 风电场项目提供 34 台 1.5MW 直驱永磁风力发电机组，金风科技风力发电机组将首度落户非洲市场；此外，公司还将为塞浦路斯 Alexigros 项目提供 21 台 1.5MW 机组；公司在澳洲的业务拓展也取得突破，将为澳大利亚 Gullen Range 项目提供超过 70 台 MW 级机组。

2010 年 10 月，公司首次公开发行境外上市外资股（H 股）并在香港成功上市，成为深圳中小板第一家 A + H 股两地上市公司，募集资金 81.73 亿港元，增强了公司资本实力，为公司战略发展提供资金支持，通过国际路演及香港上市，将金风科技成功推介给全球资本市场，搭建了国际资本市场平台，推进了公司国际化进程。公司盈利模式日趋丰富，已形成多个盈利点，风电服务、风电场开发销售为公司盈利增长作出了很大贡献，风电场开发销售模式不仅带动了公司的整机销售，还开创了中国风电发展的新模式，在短短四年的时间里，累计开发风电项目容量超过100万千瓦；2010年公司控股子公司北京天源科创风电技术有限责任公司积极创新，推进“一站式服务体系”，服务内容涵盖风电项目开发建设的各环节，在项目咨询、工程建设、技术服务、信息技术业务等方面取得可喜成绩。

2010年国内风电行业保持了较好的增长，但市场竞争日益激烈，各大风机制造商均加快了产业布局的步伐，努力抢占市场，争取更大成长空间；随着相关政策的出台，优胜劣汰的行业整合趋势初现。

面对风电行业的发展趋势和竞争格局给公司未来发展带来的机遇和挑战，公司将持续不断地巩固及进一步提高在风力发电整机研发、制造及销售领域的市场地位，同时积极开拓风电技术服务、风电场开发与销售等多种盈利模式，将风电技术、制造、服务、项目开发等要素整合成金风科技的核心能力——成为为客户提供风电整体解决方案的风机制造商，为客户创造最大价值，做国内风电行业领先企业。坚持走国际化的发展方向，参与国际市场竞争，逐步实现技术、市场、人才、资本的国际化。凭借在中国风电行业多年成功的经验，持续缩小与国际一流风电设备制造企业之间的差距，力争在下一个五年内稳步进入全球风电设备制造前三位。

加大研发投入，加快新产品的研发速度；继续对现有产品进行技术优化改进，提高产品的综合竞争力。实现2.5MW直驱永磁风电机组的批量化生产，推出适应不同运行环境的2.5MW系列化风电机组；完成6.0MW直驱永磁风电机组的整体设计。

26　生物质能

生物质的资源量十分巨大，是排在化石能源煤、油、气之后的第四位能源。全球能源消费中，生物质能约占14%，仅次于煤炭、石油和天然气，在不发达地区可达到60%以上。而且，生物质是生物体经光合作用合成的有机物，是唯一可以直接生产气体、液体、固体等能源的可再生资源。人类数千年来依赖生物质能而生存发展，将来还会依赖其发展。

我国生物质产业起步较晚，目前还处于完全依靠政府补助促发展的阶段。根据国家能源局、农业部等多个部委的公开数据显示，“十一五”期间，生物质能发展规划多项指标未能完成，产业发展遭遇技术水平低、资源禀赋弱、市场规模小、配套政策缺失等问题。

国家能源局最新透露的“十二五”生物质发电装机为1300万千瓦，较过去五年约550万千瓦的数字实现翻番增长，这让市场仿佛看到了生物质发电产业的春天。然而，能否真正实现生物质能产业的较快发展，还有赖于技术的进步及国家的产业政策扶持力度。

26.1　现状分析

26.1.1　资源现状

1. 生物质能储量丰富

中国具有丰富的生物质能资源。每年可作为能源使用的农作物秸秆资源量约为1.5亿吨标准煤，林业剩余物资源量约2亿吨标准煤，小桐子（麻疯树）、油菜籽、蓖麻、漆树、黄连木和甜高粱等油料植物和能源作物潜在种植面积可满足年产5000万吨生物液体燃料的原料需求。工业有机废水和畜禽养殖场废水资源量，理论上每年可生产沼气近800亿立方米，相当于约6000万吨标准煤。现在，从技术角度看，每年可作为能源利用的生物质资源量相当于约5亿吨标准煤，实际利用量只有约2.5亿吨标准煤，而且主要是低效的传统方式燃烧利用。

我国生物质分布十分广泛，从全国范围来看，各省分布不平衡，1/2以上的生物质资源集中在四川、河南、山东、安徽、河北、江苏、湖南、湖北、浙江等9个省，广大的西北地区和其他省区相对较少。目前，我国年可获得生物质资源量达到3.14亿吨标准煤，其中秸秆和薪材分别占54%和36%，工业有机废水废渣占3%，禽畜粪便占3%，城市生活垃圾占3%，能源植物占1%。另外，我国还有5700万公顷宜林地和荒沙荒地，还有1亿公顷不适宜发展农业的边际土地资源，充分开发利用我国的土地资源，在不与农林作物（粮油棉）等争土地的条件下，发展林木生物质能源潜力巨大。

我国拥有丰富的生物质能资源，每年理论有生物质能资源50亿吨左右，其中农作物残留物占一半多。据初步估算，在我国，仅农作物秸秆技术可开发量就有6亿吨，其中除部分用于农村炊事取暖等生活用能、满足养殖业、秸秆还田和造纸需要之外，我国每年废弃的农作物秸秆约有1亿吨，折合标准煤5000万吨。照此计算，预计到2020年，全国每年秸秆废弃量将达2亿吨以上，折合标准煤1亿吨相当于煤炭大省河南1年的产煤量。

2. 生物质能应用广泛

生物质能是一种高效的清洁能源，生物质能利用主要包括生物质能发电和生物燃料。

生物质能发电主要是利用农业、林业、工业废弃物、城市垃圾等生物质能为原料，采取直接燃烧或气化的发电方式。按照生产技术的不同，生物质能发电主要包括生物质能直接燃烧发电、气化发电、沼气发电以及煤混合燃烧发电等技术。生物质能发电的主要利用形式有：

- 大型火电系统，例如直接燃烧生物质或与煤混燃产生蒸汽发电和供热。
- 大型生物质气化发电系统（10MW以上）。
- 每年集中处理农场和工业有机废水1万吨以上的厌氧发酵供热发电系统，垃圾填埋气回收供热和发电系统。

生物燃料是指通过生物资源生产的石油替代能源，包括生物乙醇、生物柴油、ETBE（乙基叔丁基醚）、生物气体、生物甲醇与生物二甲醚。

- 生物柴油是典型的“绿色能源”，是以大豆和油菜籽等油料作物、油棕和黄连木等油料林木果实、工程微藻等油料水生植物以及动物油脂、废餐饮油等为原料制成的液体燃料，是优质的石油代用品。

生物质能的主要利用方式如表26－1所示。

表26－1　生物质能的主要利用方式

Table 26－1　Major Utilization Patterns of Biomass Energy

利用形式	作用
燃料乙醇和生物柴油	车用燃料
生物质发电	直燃和气化发电
沼气和垃圾利用	居民燃料和发电
生物质颗粒成型燃料	发电和居民燃料

资料来源：《2010中国节能减排产业发展报告》

生物质能源是我国仅次于煤与石油的第三大能源，在全部能源消耗中约占15%，是唯一可运输并储存的可再生能源，既可作为燃料，又可发电。生物质能源目前主要用于发电和转化为“绿色汽油”与“生物石油”。

根据我国经济社会发展需要和生物质能利用技术状况，我国重点发展生物质发电、沼气、生物质固体成型燃料和生物液体燃料，生物质能将成为应用最广泛的可再生能源。

26.1.2　产业现状

英国知名新能源研究机构New Energy Finance预测表明，未来10年投资增长速度最快的能源依次为：燃料电池、太阳能发电、生物燃料、风电，如图26－1所示。

目前，国外的生物质能技术和装置大多数已实现了规模化产业经营。美国、瑞典和奥地利生物质转化为高品位能源利用方面已具有相当可观的规模，分别占该国一次能源消耗量的4%、

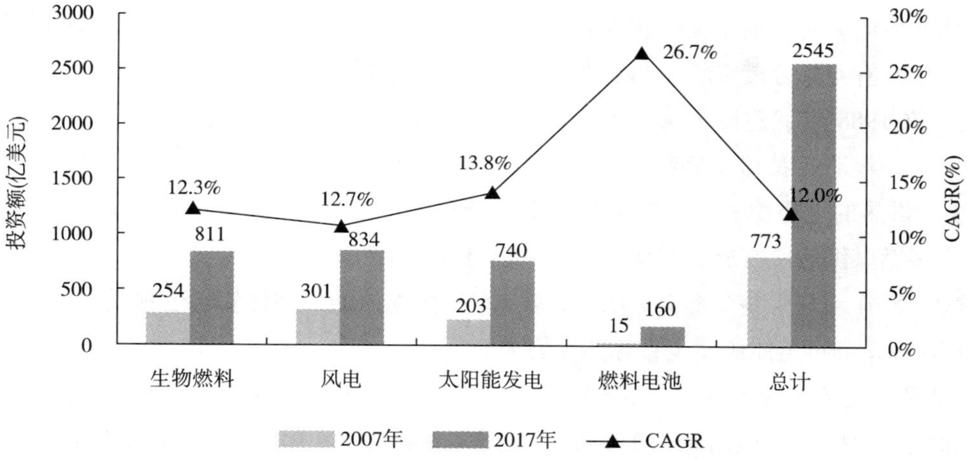

图 26－1 2007—2017 年全球可再生能源投资额增长预测

Figure 26－1 Estimated Increase in Global Investment for Renewable Energy during 2007—2017

资料来源：New Energy Finance

16% 和 10% 。

许多国家提出了生物燃料发展目标：

- 欧盟提出，到 2010 年生物燃料占交通燃料的份额达到 6% 。
- 美国提出，到 2020 年生物燃料在交通燃料中的比例达到 20% 。
- 瑞典提出，2020 年之后利用纤维素生产的燃料乙醇全部替代石油燃料，彻底摆脱对石油的依赖。

中国工程院院士、国家生化工程技术研究中心主任欧阳平凯预测，到 2020 年，全球将有 50% 有机化学品和材料来自生物质原料。有关资料显示，到 2020 年，西方工业国家 15% 的电力将来自生物质发电，而目前生物质发电只占整个电力生产的 1% 。届时，西方将有 1 亿个家庭使用的电力来自生物质发电，生物质发电产业还将为社会提供 40 万个就业机会。

在各国政策驱动，特别是通过补贴、减税、关税以及强制性将生物燃料与传统燃料混合等措施的综合推动下，生物燃料得以快速发展。2000—2007 年，全球生物燃料产量涨了 3 倍多，2007 年全球燃料乙醇产量 2597 万吨标准油，占全球汽车燃料消费的 2% ，近 11 年产量 CAGR 达 10% ，美国燃料乙醇产量 CAGR 达 17. 6% 。

1. 生物质能发电

自 1990 年以来，生物质发电在欧美许多国家开始大发展。2002 年约翰内斯堡可持续发展世界峰会以后，全球加快推进了生物质能的开发利用。截至 2004 年，世界生物质发电装机已达 3900 万千瓦，年发电量约 2000 亿千瓦时，可替代 7000 万吨标准煤。到 2005 年底，全球生物质发电总装机量增加到 5000 万千瓦。目前，国外的生物质能发电技术和装置大多数已实现了规模化产业经营。预计到 2020 年，发达国家 15% 的电力将来自生物质发电，而目前生物质发电只占整个电力生产的 1% ，生物质能发电在未来 10 年内将获得快速发展。

- 美国利用生物质发电处于世界领先地位，已经成为大量工业生产用电的选择。据美国能源信息署（EI）的统计数字，生物质能发电的总装机容量已超过 10000 兆瓦，单机容量达 10 兆瓦—25兆瓦，占美国可再生能源发电装机的40%以上。预计到2010年，美国将新增约1100万千瓦的生物质发电装机。根据美国政府制定的生物质能发展规划，生物质发电将达到13000MW装机容量，届时有400万英亩的能源农作物和生物质剩余物用作气化发电的原料。

• 丹麦在1988年就建设了第一座秸秆生物质发电厂。目前，丹麦已建立了15家大型生物质直燃发电厂，年消耗农林废弃物约150万吨，提供丹麦全国5%电力供应。同时，丹麦还有100多台用于供热的生物质锅炉。近十几年来，丹麦新建的热电联产项目都是以生物质为燃料，还将过去许多燃煤供热厂改成了燃烧生物质的热电联产项目。

• 欧盟15国，2000年全部电力的1.5%来自于生物质能，并计划将生物质能的开发作为其实现2010年22%可再生能源发电目标的主要内容。

• 奥地利成功地推行了建立燃烧木材剩余物的区域供电站的计划，生物质能在总能耗中的比例由原来大约2%—3%激增到约25%。印度研究用流化床气化农业剩余物如稻壳、甘蔗渣等，建立流化床系统，气体用于柴油发电机发电，可产生5000兆瓦的电力。

• 其他国家也大力发展生物质发电。例如，芬兰生物质发电量占本国发电量的11%；德国拥有140多个区域热电联产的生物质电厂，同时有近80个此类电厂在规划设计或建设阶段，德国在利用厌氧发酵处理废弃物发电技术方面，走在了世界的前列，目前已有1900个厌氧发酵厂，2004年装机27万千瓦。

目前，能源效率更高的BIGCC技术正处于工业示范阶段。预计2010年前，生物质发电技术将趋于成熟和完善。2010年以后，逐步具备市场竞争力，进入商业应用。

我国生物质能发电还处于导入期，2005年底以前，我国生物质能发电规模较小，发展速度较慢，生物质能发电总装机容量约200万千瓦，主要是农业加工项目产生的现有集中废弃物的资源利用项目，其中以甘蔗渣发电为主，总装机量约为170万千瓦，其余是碾米厂稻壳气化发电等。

随着《可再生能源法》和相关可再生能源电价补贴政策的出台和实施，我国生物质发电投资热情迅速高涨，启动建设了各类农林废弃物发电项目。我国生物质能发电技术产业呈现出全面加速的发展态势。2006—2009年，生物质发电的投资总额由168亿元增加到452亿元，年均增长率在30%以上；已经投产的总装机规模由2006年的140万千瓦增加到2009年的430万千瓦，年均增长率在30%以上（如表26-2）。从数据来看，生物质发电的增长逐年下滑，但仍然处于非常高的水平。

表26-2　2006—2010年生物质能发电行业规模统计

Table 26-2　Status of Biomass Energy Power Generation Industry during 2006—2010

	总装机规模（万千瓦）	增长率	投资总额（亿元）	增长率
2006年	140		168	
2007年	220	57.14%	256	52.38%
2008年	315	43.18%	347	35.55%
2009年	430	36.51%	452	30.26%
2010年	550	27.91%	586	29.65%

资料来源：世经未来整理

我国生物质能发电技术主要以秸秆直燃和垃圾焚烧发电为主。截至2009年底，我国秸秆直燃发电总装机容量为265万千瓦，占所有生物质能发电的62%；垃圾焚烧发电总装机容量为125万千瓦，占所有生物质能发电的29%；其他气化发电、沼气发电、混燃发电等所占比例很小，总共占有不到10%的比例，如图26－2所示。

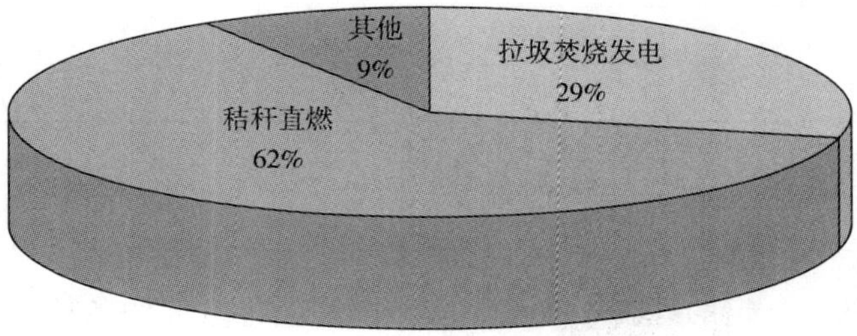

图26－2　2009年我国生物质能发电结构分析图

Figure 26－2　Structure of China's Biomass Energy Power Generation in 2009

资料来源：世经未来整理

我国生物质能发电行业的区域分布特征比较明显，从已经投产的生物质发电的地区分布来看，截至2009年12月底，生物质发电厂主要集中在华东地区，尤其是江苏和山东二省，投产生物质电厂数量分别占全国的19.5%和22.0%。

我国生物质发电产业规模还很小，在整个可再生能源发电中的比例只有0.5%左右，受到其他新能源发电的冲击较大；大多数的企业还处于亏损状态，少数效益较好的企业利润也不大，很多企业需要时不时处于停产状态，整个行业的生产状况不稳定。生物质发电的技术还不够成熟。

根据《可再生能源发展"十一五"规划》和《可再生能源中长期发展规划》，2010年中国生物质发电将达到550万千瓦，到2020年将上涨到3000万千瓦。截至2008年底，国家和各省发改委已核准项目100多个，总装机规模315万千瓦。2010年，国家对生物质能发电的各项政策仍在完善，各主要企业的在建项目也顺利进行，建成的生物质能发电总装机规模将超过550万千瓦。

除了国能生物质发电有限公司和中国节能投资公司两家国内最早介入秸秆发电项目的企业外，五大发电集团、粤电集团、皖能电力等诸多具备行业基础和资金、技术优势的大型国有企业，以及民营、外资等纷纷投资参与建设生物质发电的运营。截至2009年底，五大电力集团的总装机容量为109.6万千瓦，占所有生物质能发电装机容量的25%。五大电力集团秸秆直燃发电的总装机容量为90.7万千瓦，占全国秸秆直燃发电的34%。

从全国来看，目前生物质能发电规模最大的是国电集团和凯迪控股。2009年这两大集团的总装机容量达到了114.7万千瓦，占全国的26%，如表26－3所示。

表26－3　2009年主要企业生物质能发电份额

Table 26－3　Major Chinese Enterprises' Share in Biomass Energy Power Generation in 2009

企业	总装机容量	份额
国电	73.3万千瓦	17%
凯迪	41.4万千瓦	10%
华电	14.5万千瓦	3%

续表

企业	总装机容量	份额
江苏国信	11.5 万千瓦	3%
大唐	10.5 万千瓦	2%
中电投	6.3 万千瓦	1%
中节能	4.8 万千瓦	1%
华能	3 万千瓦	1%

资料来源：世经未来整理

2. 沼气

沼气是多种有机物在一定的湿度、浓度、酸碱度和隔绝空气的条件下，经过各类厌氧微生物的分解代谢而产生的一种可燃性气体。沼气是可再生的清洁能源，既可替代秸秆、薪柴等传统生物质能源，也可替代煤炭等商品能源，而且能源效率明显高于秸秆、薪柴、煤炭等。发展农村沼气，优化广大农村地区能源消费结构，是我国能源战略的重要组成部分，对增加优质能源供应、缓解国家能源压力具有重大的现实意义。沼气利用的意义可参见专栏 26－1。

专栏 26－1

沼气利用的意义

建设一个 8 立方米的户用沼气池，年均产沼气 385 立方米，相当于替代 605 公斤标准煤，可解决 3—5 口之家一年 80% 的生活燃料。一个年存栏 1 万头育肥猪场大中型沼气工程，每年可处理鲜粪 7200 吨左右，生产沼气约 55 万立方米，给居民供气相当于每年可替代 850 吨标准煤。如果按照规划到 2010 年建设 4000 万户沼气池和 4700 处大中型沼气工程，相当于年替代 2820 万吨左右标准煤。

沼气产业不仅是避免污染、提供能源和有机肥的产业，而且是有机体和有机物回归自然的途径，随着原料从畜禽粪便、生活垃圾到所有生物粪便、果实和尸体及其衍生物，用途从生活供气、发电、制肥到汽车燃料、制氢等，其内涵、用途和空间不断拓宽。

经过 40 年的持续发展，沼气产业已经成为社会各界的焦点。2009 年以来，各国政府纷纷出台扶持政策，发展沼气工程。

- 德国建造了世界上第一个供应 10 多万户的沼气市政管网供气系统，世界上最大日发电 20MW 的沼气热电联产工程。
- 瑞典加速了沼气火车、出租车和公交汽车的生产运营。
- 美国发明了一种分离沼气中的有毒物质硫化氢从而获得“绿色沼气”的新办法。
- 挪威大力推行以人类排泄物为原料产生沼气，用于新型生物燃料的城市公交车系统。
- 英国以牛粪、玉米和青贮为原料建设了第一个对外承包实施的沼气厂。

中国成功引进了车用沼气、秸秆沼气、餐厨垃圾沼气和干式发酵沼气生产技术，首次完成了户用沼气 CDM 交易。中国沼气产业扶持政策参见专栏 26－2。

中国于 2003 年开始实施农村沼气建设国债项目，近 10 年间，中国政府投入超过 260 亿元。

2006 年底全国已建成农村户用沼气池 1870 万口，生活污水净化沼气池 14 万处，畜禽养殖场和工业废水沼气工程 2000 多处，年产沼气约 90 亿立方米，为近 8000 万农村人口提供了优质的生活燃料。2008 年，中国户用沼气池达到 3000 多万口，大中型沼气设施达到了 1600 多座，沼气年利用量达到了约 140 亿立方米。

自 2000 年底—2009 年底，全国农村户用沼气池由 273 万户增至 3507 万户，年均增长 36.0%，农村户用沼气池的建设规模和使用量居世界之首。各类沼气工程也迅猛发展，从 2000 年的 1042 处猛增至 2009 年的 5.65 万处，总池容达 715 万立方米，年产沼气 7.65 亿立方米。其中，大中型工程 3717 处，小型工程 52817 处。技术标准体系也已初步形成。截至 2010 年底，已经颁布实施的各类沼气国家标准和农业行业标准项目共计 33 项。

专栏 26-2

沼气产业列入国家产业结构调整鼓励类产业

沼气产业化装备全面开发列入《产业结构调整指导目录（2011 年本）》，自 2011 年 6 月 1 日起施行。

为加快转变经济发展方式，推动产业结构调整和优化升级，完善和发展现代产业体系，前不久，国家发展和改革委员会主任张平发布了第 9 号令《产业结构调整指导目录（2011 年本）》。沼气产业化的主要内容都列入了该目录。

在该目录第一类鼓励类中，有 7 处直接提到了沼气和多处涉及沼气的条目。在农林业中，有秸秆沼气和农村可再生能源综合利用开发工程（沼气工程、“三沼”综合利用、沼气灌装提纯等）。在新能源业中，有以畜禽养殖场废弃物、城市填埋垃圾、工业有限废水等为原料的大型沼气生产成套设备，沼气发电机组、沼气净化设备、沼气管道供气、装罐成套设备制造。在机械业中，有沼气发生设备、沼气发酵及储气一体化（储气容积 300 立方米—2000 立方米系列产品）、沼气抽渣设备（抽吸量 1 立方米/分钟以上）等。在环境保护与资源节约综合利用中，有涉及沼气的城镇垃圾与其他固体废弃物减量化、资源化、无害化处理和综合利用工程，餐厨废弃物资源化利用技术开发及设施建设等。

该目录的发布施行，充分表明了国家更加重视沼气产业化发展工作，由此可以预见我国沼气产业化将会呈现出更加光辉的发展前景。

资料来源：中国沼气网，http://www.biogas.cn/

3. 生物乙醇

目前，燃料乙醇产量最大的依次为美国、巴西和中国（如图 26-3）。其中以中国增长最快，四年增长近 26 倍，达到 98.5 万吨标准油。美国其次，复合年均增长率（CAGR）为 18%，达到 1196 万吨标准油；巴西复合年均增长率为 4%，达到 1126 万吨标准油。其中，美国以玉米为燃料，巴西主要以甘蔗为原料，中国以玉米、小麦和木薯等为主要原料。

（1）美国。处于世界领先地位的美国开发出利用纤维素废料生产乙醇的技术，建立了 1 兆瓦的稻壳发电示范工程，年产乙醇 2500 吨，到 2012 年纤维乙醇至少要占到美国乙醇总产量的 3%，到 2022 年增至 44%，到 2022 年美国汽油燃料的替代性燃料（比如乙醇）掺混量达到 360 亿加仑，是目前水平的 5 倍。美国还利用其国内丰富的大豆资源，以豆油为原料炼制生物柴油。

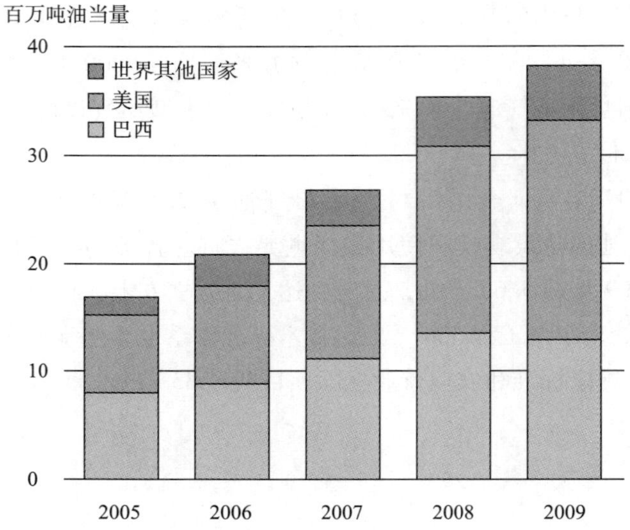

图 26－3　全球燃料乙醇产量

Figure 26－3　Global Production of Ethanol Fuel

资料来源：BP 世界能源统计，2010 年

（2）巴西。巴西政府于 1975 年大力推行以甘蔗为主要原料的“乙醇计划”，同时发展大型水电。尤其值得一提的是，经过多年努力，石油资源贫乏的巴西在 2006 年首次实现能源自给，其中甘蔗产业提供了全国能源需求的 15%，成为世界上唯一一个不使用纯石化汽油的国家。2007 年巴西可再生能源利用占能源总需求的 44.6%（如图 26－4），其中大水电约占 40%，甘蔗渣燃烧发电占 3%，而全球可再生能源平均利用率仅为 13.5%。可再生能源电力提供了巴西电力消费的 84%。

巴西还研究开发了乙醇和汽油混用的汽车发动机，目前巴西销售的新车一半以上是这种“灵活燃料”汽车。巴西政府计划在未来七年内，甘蔗产量将从目前的 4.27 亿吨增加到 6.27 亿吨，新建 89 家乙醇燃料生产企业。

巴西乙醇成本相当于石油 40 美元/桶，虽然 2001 年已完全取消对燃料乙醇的补贴，但目前每年乙醇消费量仍超千万吨，目前乙醇已经替代国内 40% 汽油消费。根据巴西甘蔗工业协会（UNICA）统计，2007 年甘蔗种植面积约 780 万公顷，占全国耕地面积 2.3%，燃料乙醇替代国内 50% 的汽油消费。巴西利用充足的耕地，大量生产甘蔗乙醇，不仅能够完全满足国内需要，每年还有大量乙醇出口，实现了贫油国家的能源出口，逐步摆脱油价巨幅波动带来的不利影响。

（3）中国。我国最先起步的生物质转化替代石油，即乙醇汽油。为缓解能源压力，我国政府未雨绸缪，有关生物能源和生物材料产业研究已有数十年历史，在生物质能加工转化及相关环保技术方面有了一定的积累。

全国主要有 5 家燃料乙醇生产企业，这 5 家生产厂分别是河南天冠燃料乙醇有限公司、安徽丰原生化股份有限公司、吉林燃料乙醇、黑龙江华润酒精有限公司、山东九九集团。其中，山东九九集团已建成年产 50 万吨燃料乙醇的绿色能源工程，其建设规模即生产能力跃居欧、亚、非三大洲之首，单机生产能力为世界第一，总投资 21 亿元。

另外，我国还在九省区（黑、吉、辽、冀、鲁、豫、苏、鄂、皖）试点乙醇汽油。截至 2006 年 12 月底，全国 4 个生产企业共生产销售燃料乙醇 243 万吨，2006 年全国共销售乙醇汽油

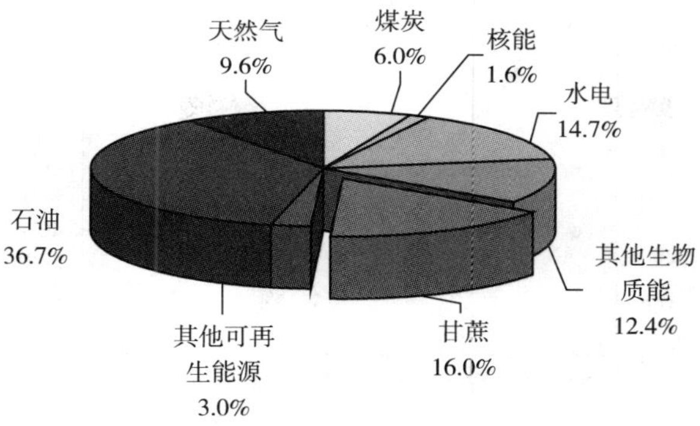

图 26－4　巴西能源结构

Figure 26－4　Energy Structure of Brazil

资料来源：Ministry of Mines and Energy Ben (Brazil，2008)

1300 万吨，占全国汽油消费量的 23.3%。截至 2008 年底，全国共生产销售燃料乙醇 165 万吨。我国已经成为全球第三大燃料乙醇生产国，仅次于巴西和美国。

随着陈化粮食逐步消耗殆尽和玉米价格节节攀升，考虑到玉米乙醇的发展对国家的粮食安全的威胁，国家 2006 年起停止新批玉米燃料乙醇企业，并大力鼓励发展非粮食作物为原料开发燃料乙醇。

“十五”期间，我国建成了年产 5000 吨规模的甜高粱秸秆制取乙醇燃料工业示范工程，及年产 600 吨规模的纤维素废弃物制取乙醇燃料的中试设施。用秸秆生产燃料乙醇，具有很大的发展前景，但最大的技术难点就在发酵环节。纤维素酶的成本能否下降，是纤维素乙醇能否实现工业化生产的关键因素。目前全世界只有两家企业能够生产用于纤维素工业化水解的酶，一家是丹麦的诺维信，另一家是美国的杰能科。目前这两家公司垄断着全球酶制剂市场份额的 65% 以上，拥有近 4000 项专利，而且新型酶制剂开发路径也被诺维信公司注册专利加以保护。目前，诺维信与中粮集团已在纤维素酶领域签署了战略合作协议，人们期待中国能够尽早在纤维素乙醇领域实现突破。

4. 生物柴油

近年来，为应对能源紧缺和油价震荡，生物燃料在全球范围内发展迅猛。以美国、欧盟和巴西等为代表的一些国家和地区把发展生物柴油作为解决能源问题的重要途径，制定具体的发展目标，并采取了相应的政策措施，积极推进生物柴油的产业化进程。

欧盟各国化石能源较为紧缺，生物质能利用比例较高。欧洲各国对生物柴油都实行了零税率，生物柴油的大型生产厂主要集中在欧洲。目前全球生物柴油产量最大的为德国，德国主要通过鼓励农民大量种植油菜，以菜籽油为原料，支撑国内生物柴油产业的发展。德国生物柴油产量如图 26－5 所示。

2007 年全球生物柴油生产能力达到 2600 万吨，生物柴油产量达到 900 万吨；到 2008 年底，全球生物柴油生产能力达到 4000 万吨，生物柴油产量达到 1200 万吨。全球生物柴油生产量从 2001 年的 9.59 亿升增长到 2009 年 157.6 亿升，年均增长率为 41.9%。有分析机构认为，在各国政府提高能源独立性和满足日益增长的能源需求的支持下，生物柴油市场可望在 2020 年生产 452.91 亿升，即在 2009—2020 年间年均增长率达到 10.1%。

欧洲是 2009 年生物柴油领先的市场，占生产份额的 49.8%，其次是美国，占生产份额的

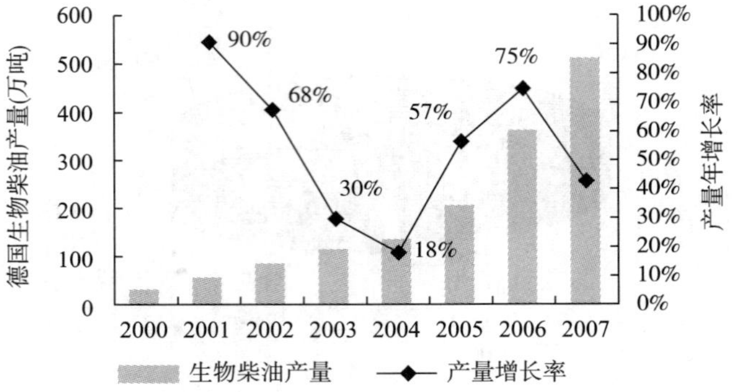

图 26－5 德国生物柴油产量

Figure 26－5 Germany's Production Volume of Biodiesel

资料来源：德国生物柴油协会（UFOR）2007

32.8%，亚太地区占4.4%。欧洲生产生物柴油的份额自2001年以来处于下降态势，而美国和亚太地区所占比例在上升。世界五大生物柴油生产国分别为德国、美国、法国、阿根廷和巴西，其生产总量占世界生物柴油总量的68.4%。澳大利亚是亚太地区最大的生物柴油生产国，其次是中国和印度。

近年来，中国的生物柴油行业也迎来了一个投资高潮，我国生物柴油行业年产能超过300万吨。据不完全统计，现有产能1万吨及以上的生物柴油企业有26家，其中，产能小于5万吨的有13家，5万吨—10万吨之间的有7家，产能达到和超过10万吨的有6家。以每吨生物柴油7000元计算，产值在3亿元以下的有13家，3亿元—10亿元之间的有12家，10亿元以上的只有1家。除了现有产能外，我国还有多项大规模的生物柴油项目正在建设中，累计约300万吨。

从2005—2007年，由于技术门槛比较低，曾经有大量企业涌入生物柴油行业，但是进入2008年后，国内生物柴油行业由于原料价格的保障失效。2007年中国生物柴油产量仅有约30万吨，产能利用率只有10%。2008年产量为28万吨，2009年约为33万吨。

2009年我国生物柴油产业逐渐进入推广阶段，并且在原料供应和技术应用方面已经逐渐形成自己的特色。海南正和公司在河北已开发了11万亩黄连木种植基地，每年可产果实20千吨—30千吨，可获得生物柴油原料8000吨—10000吨，该公司计划在此基础上建立年产生物柴油50千吨—200千吨的炼油化工厂。目前该公司在河北邯郸建成年产10千吨的生物柴油工厂。四川古杉集团建成年产30千吨生物柴油工厂。北京等省市也已经建立一定规模的生产线。上述生产线目前均是利用垃圾油或植物油脚、餐饮废油等为原料生产生物柴油。在今后五年内，我国将建成年产20千吨—50千吨规模的生物柴油产业化示范工程。

目前我国生物质能源的发展面临一些瓶颈问题，包括生物质资源不足、品质不佳、收集困难、难于转化；生物质催化与转化效率低下，过程能耗和水耗高；生物转化工艺难以低成本规模化放大以及生物能源终端产品品质不佳、产品标准欠缺等。

2010年4—6月，致公党中央调研组深入生物质资源丰富的山东、广东、广西、东北、河北以及北京周边的部分科研和生产单位，就生物质能源利用现状和产业发展进行了专题调研。调研中发现，尽管理论上生物质能是可行的替代能源，但实际应用并不尽如人意，还面临着以下挑战：

一是生物质能产业化发展受原料获得成本高的影响，如目前中试研究的每吨纤维素乙醇的原料消耗都在6吨以上，成本估算都在6500元/吨乙醇以上。大部分生产企业经营困难，需要获得额外的补贴、税收优惠才能赢利或生存。

二是生物质能扶持政策不完善，缺乏系统性和配套性，在多种能源产品和规模上未能给予明确的支持和指引，例如生物质发电的上网与电价补贴问题。

三是一些适应中国农业发展模式，适宜当地生物质资源特点的技术的生物质能产业发展模式缺乏市场竞争力，如木薯生物质燃料产业。

四是生物质能利用方式众多，例如秸秆汽化、燃烧发电等，各种能源开发技术发展水平参差不齐，转化成本高、效率低，运行可靠的国产化生物质能源开发装备、生物炼制关键技术、高蓄能抗逆能源植物开发、纤维素降解酶、微藻资源培育等技术急需突破。

因此，尽管有专家发出这样的感叹："中国生物质资源浪费严重，每年至少有2亿吨生物质资源在田间焚烧或丢弃，按电厂目前每吨300元的收购价格，就是600个亿！"但是由于上述的种种原因，也使一些人做出这样的判断，生物质能产业无疑还是一个幼稚产业，大的突破还要假以时日。

26.2　重点项目

近年来，我国生物质发电产业发展迅速，中国节能投资公司、国家电网公司、五大发电集团等大型国有企业，民营以及外资企业纷纷投资参与建设运营。截至2009年底，通过审批的生物质发电项目有70多个，在建的有30多个，投产电厂超过30座，发电总装机已超过300万千瓦。而根据国家能源局2010年一季度能源经济形势发布会公布的数据显示，目前全国投产、在建和开展前期工作的生物质发电项目有170多个，装机容量460多万千瓦，其中已投产50多个，装机容量100多万千瓦。

26.2.1　广东开建亚洲最大生物质能发电项目

2010年3月26日，亚洲最大的生物质能发电项目在广东湛江遂溪县开工建设，项目总装机容量为4＊50MW机组，规划总投资达25亿元人民币。该项目由广东粤电集团投资建设，主要开发利用湛江周边地区的秸秆、叶片、皮壳、树枝、树皮、林业加工边角料等农林废弃物进行直接燃烧，通过生物质能转换技术实现发电。

该生物质能发电项目计划在2011年初投产，届时将成为亚洲生物质能发电领域中单机容量及总装机容量最大的生物质发电企业。

该项目建设投产后，可直接反哺农业、带动农村就业，预计可以年创利税7000万元人民币，同时为当地农民增加收入两亿多元，每年可减少二氧化碳的排放量达到40万吨。

26.2.2　国内最大垃圾填埋气发电项目

中国石油济柴动力总厂发电机组中标国内最大的垃圾填埋气发电项目——高安屯卫生填埋场填埋气利用工程项目，打破了国外大功率机组在垃圾填埋气发电领域的垄断。

北京市朝阳区高安屯卫生天买成隶属于北京市朝阳区无害化处理中心，是全国首批循环经济产业示范园区，其垃圾填埋气发电醒目总装机容量3兆瓦，是国内最大的垃圾填埋气发电项目。济柴的中标产品为3台1000GF－NK型机组。此类型机组采用先进的闭环电控系统和稀薄燃烧控制系统，发电效率高，稳定性和可靠性好，各项性能指标均达到了国际先进水平。按照合同内容，济柴将负责该项目的设备供货、安装、调试总包工程，预计2010年7月份完成供货，2010年9月份正式并网发电。

26.2.3 大兴安岭将建10MW生物质发电项目

大兴安岭地区将利用林区三剩物（树皮、锯末、木耳栽培废料等）在塔河县建立总规模为10兆瓦的生物质发电项目。预计总投资11400万元，年消耗林区剩余物86100吨，仅原料收购一项每年可为当地群众增加4000万元收入。同时，与传统火电项目相比，每年可节省标准煤2.38万吨，并无烟尘、二氧化硫等大气污染物生成。该项目建成将缓解当地经济发展与能源利用之间的矛盾，是大兴安岭林业产业结构调整新的尝试。

26.2.4 大庆将建3.6亿元生物质发电项目

大庆肇州县政府与广东长青集团签署了36兆瓦生物质发电项目，该项目总投资3.6亿元，主要利用肇州县丰富的秸秆资源进行发电。项目将采取一些列系技术和新工艺拉长产业链，秸秆残留物草木灰还可制成有机肥，实现循环经济快速发展。项目占地14万平方米，已于2010年开工建设。建成后，每年可消耗秸秆22万吨，年可发电2.16亿度，可实现销售收入1.3亿元，利税2950万元。项目建成投产后，将在发展新能源、促进发展循环经济等方面发挥积极作用。

26.2.5 宁夏首个生物质发电项目启动

宁夏回族自治区首个生物质发电项目——中卫美利生物质（沼气）发电项目正式启动，这项利用污水处理厂厌氧沼气实现发电回用的循环经济工程，总投资4000万元，一期工程已于2010年7月并网发电，整个项目建成后可实现年发电量3500万千瓦时。

未来三年，宁夏节能投资有限公司将投资10亿元，在宁夏建设尾气、余热、新能源等节能发电项目，设计煤气发电、市政节能建设等多个领域，试图为宁夏节能和资源综合利用开辟一个新的途径。这些项目建成后，还将为宁夏年节约标准煤50万吨，可实现碳减排120万吨。

26.3 发展趋势

国家发改委就我国生物燃料产业发展作出三个阶段的统筹安排："十一五"实现技术产业化，"十二五"实现产业规模化，2015年以后大发展。预计到2020年，我国生物燃料消费量将占到全部交通燃料的15%左右，建立起具有国际竞争力的生物燃料产业。此前，国家发改委在可再生能源中长期发展规划中已提出：2010年生物燃料年替代石油200万吨，2020年生物燃料年替代石油1000万吨。

26.3.1 "十二五"或迎发展良机

国家《可再生能源中长期发展规划》和《可再生能源发展"十一五"规划》提出，到2010年，生物质发电总装机量要达到550万千瓦，生物质固体成型燃料年利用量达到100万吨，沼气年利用量达到190亿立方米，增加非粮原料燃料乙醇年利用量200万吨，生物柴油年利用量达到

20万吨。

然而，根据国家能源局、农业部等多个部委的公开数据，各项指标具体完成情况如下：截至2010年底，生物质发电装机约为550万千瓦，沼气年利用量约130亿立方米，生物质固体成型燃料年利用量为50万吨左右，非粮原料燃料乙醇年产量为20万吨，生物柴油年产量为50万吨左右。除生物质发电和生物柴油外，另外三项完成情况并不理想，而非粮燃料乙醇更是仅完成了既定目标的10%左右。

造成“十一五”期间生物质能发展规划指标未能完成的原因主要在于我国生物质能刚刚起步，尚无完整清晰的发展规划和相关的法律、法规，这就造成各地在发展生物质能时盲目投资，发展混乱，产业定位不清晰，甚至不符合本地的实际情况，发生与民争地、与民争粮等现象。

在“十二五”规划的开局之年，生物质能再次被列入新的五年规划。业内专家认为，“十二五”期间，国家应该会切实加大对生物质能的扶持力度，这一产业也将较快发展。国家能源局新能源和可再生能源司副司长史立山认为，生物质能要满足现代能源需求，必须能大规模生产质量合格、价格合理的产品，应该按照工业化、规模化和专业化的思路来发展。工业化是指要转变思路，按工业化的方式发展生物质能；规模化是指应积极探索推进大型沼气利用，通过提纯压缩，利用管道送到更广范围的农村市场；专业化是指要培养专业化的投资、服务公司开拓市场。

26.3.2 非粮食生物质能源将成为未来发展方向

以粮食为原料的第一代生物能源如果大规模扩张，最终要面临与人争粮，与粮争地的境地，发展非粮生物能源更有发展空间。第二代生物主要以非粮作物为生产原料，例如甜高粱、秸秆、柳枝稷和麻疯树等（如表26-4）。这些原料来源广泛，生长迅速，能耐贫瘠土壤，可以充分的利用边角地等，避免占用耕地。

从2000—2003年，国土资源部对中国耕地后备资源进行了详细调查，评估了31个省区（不包括港澳台地区）的耕地后备资源。调查结果显示，适合能源作物生长的可用耕地为653.7万公顷。

表26-4 中国主要能源作物特性

Table 26-4 Features of China's Major Energy Crops

作物	转化为生物燃料的潜力	特性
用于生产乙醇		
甜高粱	优良的能源作物，生物质转化事高，产量大； 茎杆汗液富含糖分，可转化为燃料乙醇，其纤维素也可通过酸法、酶法转化为乙醇； 每亩（1/15公顷）甜高粱每天经光合作用合成的碳水化含物可产3.2升酒精，而同样情况下玉米只能产1升，小麦0.5升，食用高粱为0.6升	杂交甜高粱是高产作物，其光合作用能力强，生物学产量高； 其生物质产出率最高可达3千吨/亩，是王米和甘蔗的1—2倍； 耐旱、耐涝、抗盐碱； 每生产1公斤干生物质，甜高粱需要25C公斤水，而小麦和豆类则需要500公斤水—700公斤水，林木则需要1000公斤水； 在PH值为5.0—8.5的土壤中，甜高粱仍然生长良好
甘薯	生产酒精的主要原料，生产工艺成熟，技术完善。甘薯的淀粉含量高、纤维少，并有适量的蛋白质、加工比较容易，淀粉的利用率较高	甘薯是高产作物，高产田亩产可达5000千克； 适应性和抗逆性强，再生力强； 耐旱、耐瘠薄、耐酸碱； 能抗风、雹、虫等自然灾害； 甘薯是喜温作物； 甘薯安全贮藏很重要，其收获质量直接影响贮藏效果

续表

作物	转化为生物燃料的潜力	特性
用于生产乙醇		
木薯	酒精工业的主要生产原料，生产工艺成熟，技术完善（加工工艺与甘薯基本相同）； 鲜木薯块根可制淀粉； 淀粉渣可做牛饲料或酿酒，综合利用价值很高（例如，其回收利用过程中能产生有价值的副产品，从而降低成本，并减少不必要的自然资源消耗）； 木薯的淀粉产量远远高于谷物，其加工性能优良	木薯生产不与粮食争地； 在热带或亚热带地区，四季都可种植，有利于全年原料供应（在温带为一年生作物）； 其生长适应性强、耐旱、耐瘠薄，在各种类型的土壤上都能生长； 现已广泛分布于华南地区，以广东、广西和海南栽培最多
用于生产生物柴油		
油菜籽	是中国最重要的油料作物； 种子含油量为33%—50%	油菜在中国分布极广，按生态划分，可分为冬油菜和春油菜两个大区； 长江流域是中国冬油菜的主产区，其产量占全国油菜产量的83.5%，是世界上最大的油菜生产带

资料来源：《中国农村生物质2020》

在生物能源、粮食安全与能源安全之间能找到一个均衡点，即生物能源发展的合理产量区域。根据我国目前生物质能发展现状及国内掌握的生物质能技术，我国完全有条件在“十二五”期间进行生物能源和生物材料规模工业化和产业化，可以在2020年形成产值规模达万亿元，在“石油枯竭拐点”形成部分替代能力。2020—2050年，生物质将逐渐成为主要能源之一。

26.3.3 技术水平不断提高

随着生物质能开发利用产业的发展，相关技术将得到长足的发展，主要有：纤维素分解生产乙醇技术，低成本、高效水解酶技术，工业规模甜高粱生产乙醇的成套设备和工艺技术，高产、优质含油能源作物育种、育苗和大面积栽培技术，高产木质能源林的育种、育苗和大面积栽培技术，生物质发电锅炉，小型生物质发电成套设备，大型沼气工程成套设备，生物基化工产品等。

有研究机构认为，在制定发展规划时，应因地制宜，选择合适的生物质能源技术，并根据各地区间的土地利用方式、农产品种类、农业废弃物利用方式和经济发展水平的差异，列出了建议中国各地区重点发展的生物质能源技术，如表26－5所示。

表26－5 各地区重点发展的生物质能源技术

Table 26－5 Prioritized Biomass Energy Technologies of Different Regions

地区	总人口（亿）	农村人口数量（亿、总量百分比）	农村家庭收入（元）	生物质资源	推荐技术
东北区	1.0753	0.4823（45%）	3416	人均资源量丰富； 冬季取暖需求高，但由于沼气池发酵时间较短，该地区发展户用沼气技术可能有一些不利因素	可利用生物质固体成型技术满足农户炊事和取暖的需求； 由于人均秸秆丰富，建议在该地区同时发展直燃发电技术
华北区	2.8033	1.6355（58%）	4642	主要的农业生产区，农业人口密度大； 农民人均秸秆资源丰富	户用沼气； 生物质固体成型燃料； 秸秆沼气； 畜禽场大中型沼气工程

续表

地区	总人口（亿）	农村人口数量（亿、总量百分比）	农村家庭收入（元）	生物质资源	推荐技术
黄土高原区	0.9662	0.6088（63%）	2308	农民人均秸秆资源量低； 经济发展落后； 秸秆资源有限，畜牧部门饲料需求量较大	高效炉灶
长江中下游区	3.6588	1.9101（52%）	4.596	农民人均秸秆资源量低； 气候条件适合发展户用沼气池和大中型沼气工程	如果技术和体制问题得到了妥善解决，可发展生物质气化技术
华南区	1.8198	0.9016（50%）	3.660	农民人均秸秆资源量低	根据气候条件，可发展沼气技术，包括户用沼气与中大型沼气工程
西南区	1.9172	1.2888（67%）	2.383	农民人均秸秆资源量低； 农民大多居住在山区，交通运输不便，经济落后，农村能源短缺	户用沼气 生物质固体成型燃料 生物质气化技术（如果技术和机制问题得到解决）
青藏高原区	0.0819	0.0532（64%）	2.115	中国的主要牧区，农民人均秸秆资源量全国最低； 农业和畜牧业之间的生态平衡非常重要	高效炉灶
蒙新区	0.4989	0.2865（57%）	2.660	中国主要的牧区，农民人均秸秆资源量最高，农村人品密度低； 根据政府的退耕还草政策，作物秸秆将主要被用做饲料	可结合户用太阳能技术，发展生物质固体成型燃料技术； 生物质气化技术（如果技术和机制问题得到解决）

资料来源：《中国农村生物质2020》

26.4　小结

现代生物质能发展的方向是高效清洁利用，将生物质能转化为优质能源。我国在生物质发电、燃气和液体燃料等方面均取得了一定成就。预计到2015年，我国生物质发电装机容量将达到720万千瓦，生物质液体燃料达到700万吨，沼气年利用量达到240亿立方米，生物质固体燃料达120万吨。

产业的长足稳定发展需要国家相关政策的支撑。生物柴油调和燃料B5标准名列的出台以及国家税务总局宣布的对纯生物柴油免征消费税的规定，表明了未来针对生物质产业的政策和标准将陆续出台，相关政策缺失的问题将在“十二五”期间得以解决。

27 核能

核电被认为是一种清洁、技术成熟、供应能力强、能大规模应用的发电方式，国际核能的应用经历了对核电机组的从第一代到第三代不断改进的过程。目前，国际第四代核能利用系统研究提出了反应堆设计和核燃料循环方案的新概念，我国核电已由起步进入发展阶段，具有自主设计建造第二代核电的能力，我国已做出积极推进核电发展的重大决定，加快我国核电建设，提高核电在电力供给中的比重，这将有助于缓解电力增长与交通运输、环境保护的矛盾，核能利用的发展前景将越来越广阔。

但是，2011 年由地震引发的日本福岛核电站核泄漏事故再一次将人们的目光聚焦到了核电的安全问题上。一时间，各国纷纷出台停核、缓核政策，对本国核电站的建设进程进行重新审视。核电的存废之争甚嚣尘上。在本章，我们将进一步分析探讨当前核电的发展现状，以及日本福岛核泄漏事故对全球核电发展产生的重大影响及其为其他新能源带来的发展机遇。

27.1 现状分析

27.1.1 全球核电稳步增长

国际原子能机构的数据显示，到 2007 年 1 月，世界各地有 435 座动力堆在运行。在这些在役的反应堆中美国有 103 个，提供电力份额 19%、法国 59 个，日本 55 个、俄罗斯 31 个、英国 19 个，仅这五个国家就占全球在役反应堆的 60% 以上。

另外，截至 2007 年 3 月，全球在建的核反应堆有 30 个，发电容量将达到 24.251 亿瓦，其中大多数在亚洲，共有 16 个（印度 7 个，中国 5 个，朝鲜、伊朗、日本和巴基斯坦各 1 个）。欧洲总计有 166 座核反应堆在运转中，有 6 座在建设中。芬兰、法国、保加利亚和乌克兰也有核能扩能计划。芬兰于 2005 年开始建设 Olkiluoto－3 核反应堆，该核反应堆是西欧自 1991 年以来第一套新建的核反应堆，如表 27－1 所示。

表 27－1 截至 2007 年 3 月全球运行中和在建的核反应堆一览表

Table 27－1 Nuclear Reactors in Operation and under Construction in the World by Mar 2007

国家	运行中的反应堆		在建的反应堆		2005 年核发电量[1]	
	数量	MW（e）	数量	MW（e）	TW（e）h	占总电量的比例（%）
比利时	7	5801	—	—	45.34	55.63
巴西	2	1901	—	—	9.85	2.48

续表

国家	运行中的反应堆		在建的反应堆		2005 年核发电量[1]	
	数量	MW（e）	数量	MW（e）	TW（e）h	占总电量的比例（%）
加拿大	18	12584	—	—	86.83	14.63
中国	10	7587	5	4220	50.33	2.03
法国	59	63363	—	—	430.90	78.46
德国	17	20339	—	—	154.61	30.98
印度	16	3483	7	3112	15.73	2.83
日本	55	47593	1	866	280.67	29.33
韩国	20	16810	—	—	139.29	44.67
俄罗斯联邦	31	21743	5	4525	137.27	15.78
乌克兰	15	13107	2	1900	83.29	48.48
英国	19	10822	—	—	75.17	19.86
美国	103	98145	—	—	780.47	19.33
其他	93	44923	10	9628	536.60	—
总计	435	368201	30	24251	2826.35	19.28

注：1. 该栏数据截至日期为 2005 年 12 月 31 日。
资料来源：根据 IAEA（国际原子能机构）数据汇总

根据国际原子能机构的预测，到 2030 年全球核电装机容量将达到 473GW—748GW，比 2007 年增长 27.15%—101%，占总装机容量的比重为 8%—9% 左右；届时，全球核电发电量将达到 3522TWh—5551TWh，比 2007 年增长 35.05%—112.85%，占全球总发电量的 12%—14% 左右。

现在核能发电站的扩建集中在亚洲：至 2006 年底建设中的 29 座就有 15 座在亚洲。最近建设的 36 座核反应堆已与电网联网的有 26 座在亚洲。

• 印度核能发电所占比例现小于 3%，但至 2006 年底，拥有建设中核电站的 1/4。在建设中 29 座核电站中拥有 7 座，计划到 2022 年将增长 8 倍，达到电力供应的 10%；到 2052 年将增长 75 倍，达到电力供应的 26%。75 倍的增长意味着年均增长 9.4%，与全球 1970—2004 年的平均增长率相同。

• 日本 55 座核反应堆在运转，一座在建设中，并计划使核能发电占电力份额从 2006 年 30% 提高到后 10 年内的超过 40%。

• 韩国于 2006 年投运第 20 座核反应堆，另一座在建设中，并已动工开始再建设 2 座。核能发电已供应其电力的 39%。

截至 2009 年 12 月底，全球共有 436 台现役核电机组，总装机容量约为 373GW；有 5 台核电机组处于长期停堆状态；另有 53 台核电机组正处于建设阶段。

2009 年，有 2 座新的核电机组并网发电；有 3 座老化的机组被最终关闭；11 台核电机组相继开工建设；有 4 台机组恢复建设；对现有核电机组进行功率提升使总装机容量增加 860MW。

2009 年，开工建设的核电机组共有 11 台，集中在亚洲和俄罗斯地区。其中，中国 9 台，韩国 1 台，俄罗斯 1 台，使开工建设的机组总数激增至 53 台。

根据国际原子能机构 2011 年 1 月公布的最新数据，目前全球正在运行的核电机组已经达到 442 个，还有正在建设的核电机组 65 个，核电发电量约占全球发电量的 16%。而至 2010 年底，全球已经有60多个国家提出了发展核电的计划，包括阿联酋等富油国家。来自国际原子能机构的预测，全球核能发电量在今后20年将会提高一倍。

5.1.2 中国核能发展现状

我国是继美、英、法、前苏联、加拿大和瑞典之后，世界上第七个能自主设计和建造核电站的国家，核电的发展状况与我国核大国的地位极不相称。2007年我国核电发电量和装机容量的比重分别达到1.92%和1.24%（如图27－1），在拥有核电的国家中是最低的。2008年我国核电总装机容量和发电量分别是885万千瓦和684亿度，仅仅占全国总装机容量的1.1%和总发电量的2%（如图27－2）。与世界水平17%相比，我国国内核电仍有很大的发展空间。

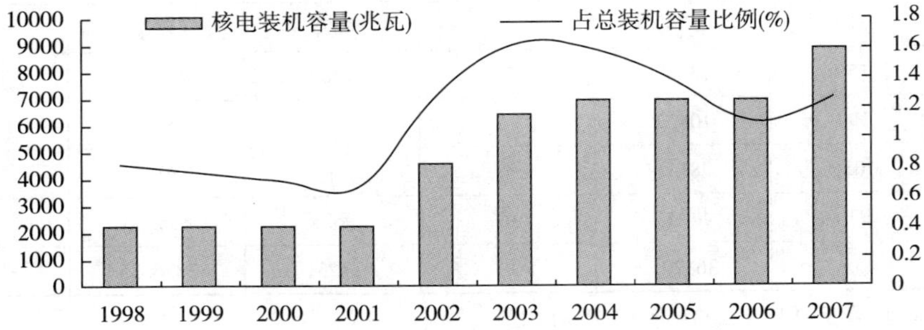

图27－1　我国核电装机容量及占总装机容量比重

Figure 27－1　China's Installed Capacity of Nuclear Power Units and Its Proportion in Total Installed Capacity

资料来源：东方证券研究所

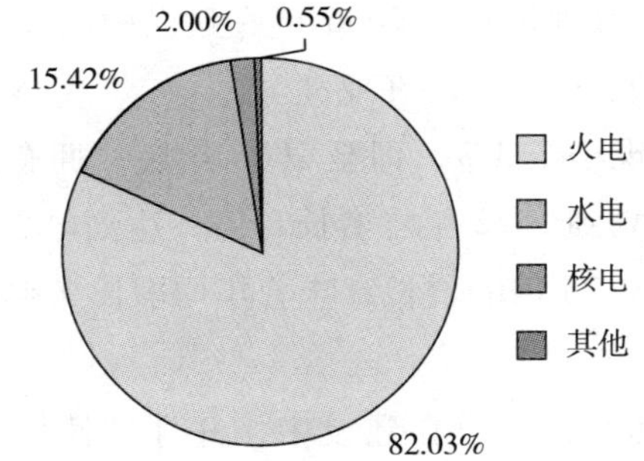

图27－2　2008年国内火电、水电、核电发电量占比结构

Figure 27－2　Shares of China's Thermal Power, Hydropower and Nuclear Power in 2008

资料来源：东方证券研究所

从装机容量来看，2001年以后核电装机份额基本上是在逐年上升，份额上升较快的年份通常是有核电机组投产的年份。例如，2007年田湾2台机组投产，导致当年核电装机份额从上一年的1.1%上升到1.24%。2008年、2009年没有新的机组投产，因此增长缓慢，如表27－2所示。

表 27－2 2001—2009 年核电发电装机容量分析（全口径）

Table 27－2 Installed Capacity of Nuclear Power Units during 2001—2009 (Inclusive)

单位：万千瓦

时间	发电装机容量	核电装机容量	核电装机份额
2001	33849	210	0.62%
2002	35657	447	1.25%
2003	39141	619	1.58%
2004	44239	684	1.55%
2005	51718	684	1.32%
2006	62200	685	1.10%
2007	71329	885	1.24%
2008	79253	885	1.12%
2009	87407	908	1.04%

资料来源：中国电力企业联合会

2009 年我国核电建设步伐加快，先后核准浙江三门 2 台 125 万千瓦、山东海阳 2 台 125 万千瓦、广东台山 2 台 175 万千瓦核电机组，并陆续开工建设。截至 2009 年底，中国在建核电机组 20 台，在建规模 2192 万千瓦。

我国核电利用达到世界先进水平，连续两年成为全球核电在建规模最大的国家。2005 年以来，国家先后核准了 13 个核电项目，共 34 台机组，3702 万千瓦。目前，在建机组 28 台，3097 万千瓦，在建规模占世界的 40% 以上。设备技术与运行稳定性均处于世界领先水平。

经过这些年的探索发展，我国核电在技术研发、建设管理水平等各个方面都取得了长足的进步，基本具备了第二代百万千瓦级核电站设计能力，在核电设备制造方面，60 万千瓦和 100 万千瓦核电站国产化率可达 70% 以上。核电设备国产化率的提高，不但使我国摆脱在关键领域受制于人的情况，而且大大降低了核电站的建设费用。

近年来我国不断调高核电发展的目标，根据最新目标，2020 年核电占电力总装机容量的比例将达到 8% 以上，照此计算，2020 年我国核电装机容量至少可以达到 7000 万千瓦，相当于未来 12 年我国平均每年有 5—6 台百万千瓦核电机组投产。根据中国核电产业发展规划，从沿海的广东、浙江、福建到内陆的湖北、湖南、江西将建设数十座核电站。到 2020 年，中国将建成 13 座核电站，拥有 58 台百万千瓦级核电机组，核电总装机容量达 4000 万千瓦，核电年发电量将超过 2600 亿千瓦时，核电占中国全部发电装机容量的比重 4% 左右，发电量比重占全国发电量的 6% 以上。

国家核电自主化工作领导小组提出我国核电发展最新目标是：到 2020 年，在运行核电装机容量 86GW；在建核电装机容量 32GW。

2008 年，国家发改委和能源局预计 2020 年核电装机目标有可能上调至 7000 万千瓦，在建 3000 万千瓦。我国核电中长期发展规划如图 27－3 所示。

27.2 日本核事故对核能产业的影响

2011 年 3 月 11 日发生在日本东北部地区的大地震，带来了巨大的人员伤亡和财产损失，地

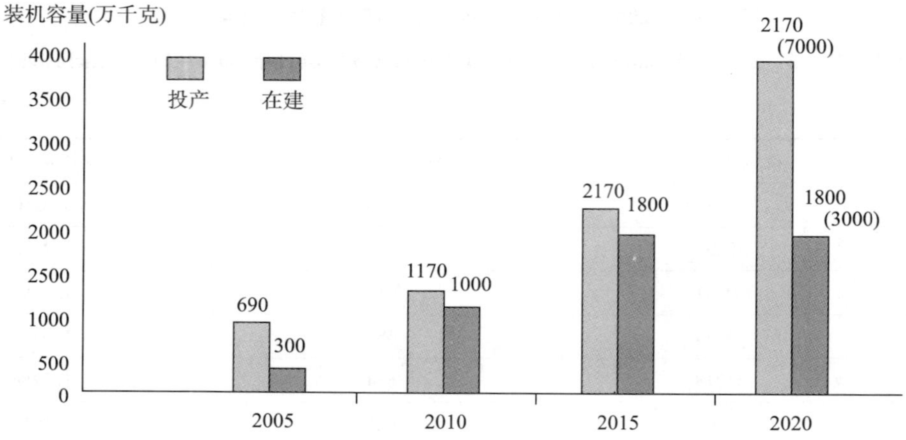

图 27－3 中国核电中长期发展规划（括号内为预期调整的目标）

Figure 27－3 Medium－and－long－term Plans for China's Nuclear Power Development (Figures in Brackets are Expected Revisions)

资料来源：国家发改委能源局

震引发的核泄漏也造成了重大影响，这不禁让人回忆起此前的切尔诺贝利核电站爆炸和美国三里岛核泄漏事故带来的巨大生态灾难，一时间世界上许多国家都将目光投向了核电的安全问题，重新审视本国核电站的建设进程。

- 3月14日，德国宣布暂缓延长核电站使用期后，3月15日又宣布关闭7座1980年之前建成使用的核电站，对143座核电站进行安全检修。德国已宣布在2022年前关闭其境内所有的核反应堆。
- 奥地利等本来就反对核电的国家更坚定了其反核主张。
- 瑞士也宣布将在2034年以前停止使用核能。
- 意大利政府关于重启核电的方案也在公民公投中被否决。

而一直准备在核电建设上实现跨越式发展的中国，也收紧了新建核电站的审批闸门，还有一些国家也采取了类似的举措（如表27－3）。全球核电的发展前景蒙上了阴影。

表 27－3 日本核危机引发世界各国的缓核、停核政策

Table 27－3 Worldwide Reactions to Japan's Nuclear Crisis

国家	对日本核危机的反应情况
中国	国务院总理温家宝提出，全面审查在建核电站、严格审批新上核电项目，在核安全规划批准前，暂停审批核电项目包括开展前期工作的项目
德国	3月14日，宣布暂缓延长核电站使用期后；3月15日，德国宣布关闭7座1980年之前建成使用的核电站，德国民间反核人士仍不断抗议核电运行
意大利	绿党主席波内利要求政府全面关闭意大利境内所有核能电厂，并冻结计划在2年后兴建新的核电厂
法国	法国核能部门评估本次核泄漏事故等级为6级；法国政府已经下令检查所有核电站安全性；此外，法国国内绿色组织呼吁法国逐渐改变核电的依赖
瑞士	瑞士联邦委员会宣布停止修建核电设施申请的审核；同时，联邦核安全监察部门将对所有的核电厂提前进行安全检查
美国	3月15日，奥巴马指示美国核管制委员会（NRC）要从福岛事故中吸取教训，确保美国的核安全，能源部部长称美国要推行能源多样化

续表

国家	对日本核危机的反应情况
韩国	韩国原计划建造 14 座新核反应堆，此次事故后可能重新审视核计划
印度	印度政府表示将立即检查核电厂安全性，确保核电厂能够经受环境波动考验

资料来源：国海证券研究所

显然，长期以来被视为清洁能源的核电，未来一段时间的全球范围内的装机容量增速将受到一定影响，全球核能发展总体步伐势必将放缓。然而，从长远来看，此次日本大地震引发的核泄漏事件对核电发展不会产生根本性的影响。国际原子能机构为总结日本福岛核事故教训专门召开机构成员国部长级核安全大会，多数国家，尤其是一些核能大国依然表示，要在“安全第一”的原则基础上坚持发展核电。核能的开发利用虽然因为这次核危机而暂时遭遇挫折，但如在建造和运行的过程中坚强风险控制和安全管理，在当今世界石油天然气资源面临枯竭的境况下，核电依然具有着无法忽视的潜力。

而从短期来看，此次核事故会促使各国更加重视风电、太阳能光伏产业等新能源产业的发展。益普索/路透 6 月 23 日公布的调查结果显示，如今核能受欢迎程度逊于太阳能、风能、水电、天然气和煤炭。此次调查涵盖全球 24 个国家近 1.9 万人，有近 3/4 的受访者认为，核能是一种有局限性、很快会被淘汰的能源。

据了解，到 2020 年全球核能总装机规划为 8.7 亿千瓦（870GW），而截至 2010 年底全球已有约 3.7 亿千瓦核电装机投入运行。如果此次日本核泄漏危机使全球未来新增装机容量下降约 20%，则会出现约 1 亿千瓦的电力缺口。若将这一缺口分摊给其他新能源，无疑将极大促进这些新能源产业的发展。而在本次日本核事故爆发之后，法国、英国等全球各国在检讨核电发展规划的同时，也加强了对太阳能、风能产业的补贴政策扶持力度。

27.3 发展趋势

新中国成立 60 年来，我国的核电产业从无到有、由弱到强。如今，全国核电装机容量 907.82 万千瓦，运行机组 11 台。新能源振兴规划将于年内出台，预计到 2020 年核电占电力总装机容量的比例将调高到 8% 以上，据此推算，预计将带动核电装备制造业 7000 亿元—7500 亿元的市场需求。

27.3.1 中国核电新规划触发投资市场

国家统计局此前发布的新中国成立 60 周年能源发展报告指出，根据中国核电产业发展规划，从沿海的广东、浙江、福建到内陆的湖北、湖南、江西将建设数十座核电站。到 2020 年，中国将建成 13 座核电站，拥有 58 台百万千瓦级核电机组；核电总装机容量达 4000 万千瓦，核电年发电量将超过 2600 亿千瓦时，核电占中国全部发电装机容量的比重 4% 左右，发电量比重占全国发电量的 6% 以上。

未来 10 年核电建设的投资规模将达到 1 万亿。未来十年新增核电装机 7000 万千瓦，按 1.5 万元/千瓦的造价计算，核电建设投资将高达 1 万亿，其中，用于设备和燃料采购的投资占到 50%，相关设备和材料制造商将分享巨大的核电蛋糕。预计 2020 年核岛、常规岛、辅助设备的国产化程度分别达 70%、80%、90%，相应的市场份额分别为 1700 亿元、1200 亿元、1100

亿元。

无疑，中国将迎来核电发展的高峰期。除了在投资建设上给予支持外，财政部和国家税务总局还下发了《关于核电行业税收政策有关问题的通知》，同意明确了鼓励核电发展的税收政策，这意味着中国将对核电实行长达15年的税收优惠政策。

日本核泄漏事故后，未来一段时间国内的核电项目的审批速度可能会放缓，核电建设也会有微调，但从国家大力发展新能源大方向来看，我国发展核电的长期战略不会受到太大影响。

27.3.2 短期核电发展速度放缓

在日本核泄漏事故后，中国宣布停止审批新核电项目，据推测，该政策可能延续到2012年初，凸显中国核安全审查的分量。在完成新安全规范和原子能法的制定工作后，我国最早可能于2012年恢复对核电项目的审批。2011年待批的9个核电项目将受到暂停审批影响。从中长期看，中国发展核电的方针不会变，但是，中国核电开发在两三年内将会放缓节奏。

据了解，《原子能法》的起草已于2011年10月前完成，《核安全新规》也将于年底前完成。新法规将覆盖从赔偿责任到安全及透明标准等各个方面，是对现行法规的补充。

中国核电行业的发展步伐远快于计划，迫使政府将2020年的核电发展目标提高了一倍。但是，中国目前没有完备的核安全规划，没有建立有效的应急应对体制。合格的核安全监察人员严重不足，核电的快速发展导致合格而又有经验的核电工程监督和运行检查人员无法覆盖电站设计、制造、建设、安装和安全运行全过程工作。中国目前已经运行、在建和筹备建设的第二代和二代半技术的动力堆实际多达70座，这些反应堆将在50年后才能退役，将长期存在危险因素，而真正采用第三代压水堆技术的仅有4座。这些问题都有可能成为未来核安全的致命隐患。因而，暂停审批核电项目，全面检查在建核电站，虽在短期内延缓了核电的发展，但无疑能够更好促进核电的长远稳定安全发展。

27.3.3 核电发展和核电安全同步

日本大地震造成日本福岛县第一核电站发生爆炸，我国核电站开发建设也引起了热议。国务院总理温家宝召开国务院常务会议，会议决定，立即组织对我国核设施进行全面安全检查，切实加强正在运行核设施的安全管理，同时要全面审查在建核电站，用“最先进的标准”进行安全评估，不符合安全标准的立即停止建设；在核安全规划批准前，暂停审批核电项目包括开展前期工作的项目。

历经30余年发展，我国是全球核电在建规模最大的国家，核电发展可以用飞速来形容。中国核电发展将安全放在了第一位，但是这也许并不影响中国核电的快速发展。来自电力协会的消息，核电也已列入电力“十二五”专项规划“加快发展”序列。规划成2011年开工建设首个内陆核电，力争2015年投产首台内陆机组。到2015年，我国核电装机4294万千瓦，2020年核电规划装机容量9000万千瓦。

核电站的选址正从沿海向内陆渗透，“十二五”规划纲要提出，“加快沿海省份核电发展，稳步推进中部省份核电建设”。

数据显示，我国投入运营的核电站有6个，12个正在建设，25个正在筹建中。从布局上看，我国将形成“东中部核电带”，即在辽宁、山东、江苏、浙江、福建、广东、广西、海南等沿海省区加快发展核电；稳步推进江西、湖南、湖北、安徽、吉林等中部省份内陆核电项目。“十二五”期间，我国还将核准10个核电项目。

中国将高效发展核电，前提是“确保安全”，目前核电发展规划和核安全规划正在编制论证。我国会吸取日本的教训，充分论证核电发展规划和核安全规划，将核电发展和核电安全规划同步实施。现在中国的核电装机容量只占全国装机容量的1.1%，相比日本、法国、俄罗斯和美国，比重非常小，还有发展空间。

27.4　小结

虽然，各国关于核电的存废之争尚无定论，但其对缓解全球能源危机，应对全球变暖的作用却是不言而喻。联合国原子能机构亦表示，全球核电利用规模预计将继续增长，只是增速将放缓。核电在我国的能源结构中占据着不可替代的重要作用。近年来，我国核电更是发展迅速，“十二五”规划已将其纳入“加快发展”的序列。为了更好地促进核电的长远稳定发展，核电的安全问题必将成为未来。

第6篇　技术篇

“十一五”期间，我国积极部署和加快推进节能减排科技创新，先后实施了节能减排科技专项行动和节能减排全民科技行动。通过将新技术应用到不同的节能环保领域，科技创新节能减排起到了支撑和引领的作用。例如，新一代可循环钢铁流程工艺技术取得突破并在国内多家钢铁企业实现工业应用，依托曹妃甸工程建立了洁净钢生产平台，实现了转炉功能的优化组合，每条生产线平均节约50万吨标准煤。超临界发电技术燃煤电厂的装机容量达到0.8亿千瓦，每年减少标准耗煤294万吨，水泥窑纯低温余热发电集成技术实现了工业应用，年发电量达2736万度，减少标准煤耗1.2万吨，减少二氧化碳排放1.6万吨。铬盐清洁生产工艺实现万吨级工程示范，推广后30%以上将实现清洁生产。大田膜下滴灌系统，比国外滴灌减少投入50%，比常规滴灌省水50%、省肥20%、增产20%，应用面积已达500多万亩。此外，我国成功研制出符合欧III标准的柴油车/摩托车尾气净化器并实现了产业化，生产的金属载体催化剂占国内摩托车净化市场60%以上。研制出湿法脱硫系统技术和设备，大幅度降低了工程造价，已投运项目58个，总容量34666兆瓦，累计减排二氧化硫255万吨，为企业减少排污费16.1亿元。

“十二五”期间，我国将加大节能减排技术产业化示范。实施节能减排重大技术与装备产业化工程，重点支持稀土永磁无铁芯电机、半导体照明、低品位余热利用、地热和浅层地温能应用、生物脱氮除磷、烧结机烟气脱硫脱硝一体化、高浓度有机废水处理、污泥和垃圾渗滤液处理处置、废弃电器电子产品资源化、金属无害化处理等关键技术与设备产业化，加快产业化基地建设。我国将重点推广能量梯级利用、低温余热发电、先进煤气化、高压变频调速、干熄焦、蓄热式加热炉、吸收式热泵供暖、冰蓄冷、高效换热器，以及干法和半干法烟气脱硫、膜生物反应器、选择性催化还原氮氧化物控制等节能减排技术。

在本篇我们将集中对上述产业化示范技术与重点推广技术从基本内涵、应用现状两个角度进行简要介绍。

28 节能减排“十二五”产业化示范技术

28.1 稀土永磁无铁芯电机[①]

28.1.1 技术内涵

电机系统用电量约占全国用电量的60%，其中风机、泵类、压缩机和空调制冷机的用电量分别占全国用电量的10.4%、20.9%、9.4%和6%。电机系统量大面广，节电潜力巨大。据行业协会统计，全国现有各类电机系统总装机容量约7亿KW，运行效率普遍比国外先进水平低10—20个百分点，相当于每年浪费电能约1500亿KWh。

为此，国家发改委早在“十大重点节能工程实施意见”中即提出：要推广高效节能电动机、稀土永磁电动机；同时推广变频调速、永磁电动机调速等先进电机调速技术，改善风机、泵类电机系统调节方式，逐步淘汰闸板、阀门等机械节流调节方式，并建议在以下领域推广应用稀土永磁电动机和调速系统：

- 电力：用变频、永磁电动机改造风机、水泵系统，重点是20万KW以上火力发电机组。
- 冶金：鼓风机、除尘风机、冷却水泵；加热炉风机、铸造除鳞水泵等设备的变频、永磁电动机调速。
- 机电：研发制造节能型电机、电机系统及配套设备。
- 轻工：注塑机、液压油泵的变频、永磁调速。
- 其他：企业空调和通风、楼宇集中空调的永磁电机系统改造等。

稀土永磁电机最显著的性能特点是轻型化、高性能化、高效节能。高性能稀土永磁电机是许多新技术、高技术产业的基础。稀土永磁电机的主要结构特点如下：

（1）高效节能。稀土永磁电机是一种高效节能产品，平均节电率高达10%以上，专用稀土永磁电机的节电率可高达15%—20%。美国GM公司研制的钕铁硼永磁起动电机与老式串激直流起动电机相比，效率提高了45%。在水泵、风机、压缩机采用永磁电机及变频调速技术后可节电率30%以上。

（2）轻型化。采用稀土永磁体可以明显减轻电机重量，缩小体积。例如，10KW发电机，常规发电机重量为220公斤；而永磁发电机重量仅为92公斤，相当于常规发电机重量的45.8%。

（3）高性能化。高性能化也是稀土永磁电机的突出优点，有些性能是传统标准电机所不能

---

① 资料来源：黄建民，贾廷纲．稀土永磁电机研究现状与发展．

及的。例如，数控机床用稀土永磁伺服电机，调速比高达1：10000。稀土永磁电机可以实现精密控制驱动，转速控制精度可达到0.1‰。

（4）结构简单。永磁电机与传统的电机相比，结构简单。采用稀土永磁铁后还可以增大气隙磁密，并把电机转速调整到最佳值，提高功率质量比。现代航空、航天用发电机几乎全部采用稀土永磁发电机。

28.1.2 应用现状

（1）航天航空领域。美国已研制成驱动航天飞机升降副翼用的12.6KW，9000r/min稀土永磁无刷直流电动机，效率为95%，仅重7.65公斤。

（2）船舶电力推进领域。高性能舰船一般均采用全电力推进系统。国外在电力推进系统中普遍采用永磁电机作为系统的推进电机，如西门子、ABB和Alstom等都已形成产业。德国西门子已制成了由6相变频电源供电的1095KW、230r/min的稀土永磁同步电动机；俄罗斯最近也试制成功3500KW的大型稀土永磁同步电机。

（3）风机水泵行业。主要有油田电力驱动系统、发电厂和钢铁生产企业中，水泵和通风机等机械负载的应用数量最大，这也是永磁稀土电机推广应用先进煤气化主要场合。我国在油井抽油机推广使用稀土永磁同步电机，已形成年产1万台以上的生产能力，在油田运行情况良好，取得了明显的技术经济效益。

（4）电梯行业。国外主要电梯制造商大都采用了永磁同步电动机驱动无齿轮曳引机，其效率可高达0.9以上，比交流异步电机+减速箱结构系统节能30%以上，国内三菱电梯、日立电梯和迅达电梯等均采用了该技术。

（5）空调行业。永磁同步电机及直流变频驱动技术是未来空调发展的主流，通过改变频率来调节压缩机功率，且电机在低频运转时的效率极高，是实现家电节能的最佳技术途径之一。目前，日本已掌握较先进的180°正弦波控制及驱动技术，上海海立与中央研究院合作也已掌握此技术。

（6）风电行业。国内湘潭电机厂，已研制成功2台2MW直驱永磁同步风力发电机，已应用于内蒙古卓姿风电场。力德风电与国家稀土永磁电机工程技术研究中心已经设计完成2.0MW直驱永磁风力发电机。

28.2 低品位余热利用①

28.2.1 技术内涵

目前，中国能源利用率仅为约30%，大量余热以各种形式被排放到大气中，可再生能源在能源结构中所占比例不足8%。因此，回收利用余热在提高中国一次能源利用率方面具有举足轻重的作用。目前，中国回收利用的余热主要来自高温烟气的显热和生产过程中排放的可燃气，中低温余热（即低品位余热）基本上还没有回收。相对于煤、石油、天然气等高品位能源而言，低品位余热在相同单位内包含的能量很低，利用难度大。但从能源利用的格局来看，低品位余

① 资料来源：李海燕，刘静．低品位余热利用技术的研究现状、困境和新策略．

热将作为产能和用能的关键环节，对节能减排的战略起到重要作用。

28.2.2 应用现状

工业余热主要指工矿企业热能转换设备及用能设备在生产过程中排放的废热、废水、废气等低品位能源，利用余热回收技术将这些低品位能源加以回收利用，提供工业、生活热水或者为建筑供热，不仅可以减少工业企业的污染排放，还可以大幅度降低工业企业原有的能源消耗。

工业余热资源十分丰富且广泛存在于各种生产过程中，特别在煤炭、石油、钢铁、化工、建材、机械和轻工等行业更是如此，被视为继煤、石油、天然气、水力之后的第五大常规能源。在中国，各主要工业部门的余热资源率平均达7.3%，而余热资源回收率仅为34.9%，回收潜力巨大。因此，充分利用余热资源是实现工业节能减排战略目标的主要手段之一。

（1）石油行业。目前，中国油田采出水总量在几亿立方米以上，常规油田采出水温度为38℃—43℃，稠油油田采出水温度为60℃—65℃，蕴藏着大量的热能资源。面对余热回收利用的巨大潜力，许多油田进行了积极的探索，例如大庆、辽河等油田将热泵回收采出水余热技术用于站内生活采暖，并提取采出水中的余热用于油水分离及原油输送过程的加热，收到明显的经济和社会效益。随着原油价格上涨及能量优化研究的深入，炼油工业中低温余热的回收利用也越来越受到重视，正成为炼厂节能的重点方向之一。

（2）钢铁行业。钢铁企业余热的主要来源有焦炭及烧结、转炉、加热炉的烟气，采用余热锅炉对这些烟气进行冷却，可以回收大量的蒸汽，这些蒸汽可用于发电、采暖等生产、生活领域。钢铁企业的各工序的余热回收利用潜力都在30%以上，而余热资源总量相当大，可见实现余热回收具有十分广阔的前景。

（3）水泥行业。水泥窑存在大量的余热，窑尾一级预热器排出的废气带走的热损失和窑头冷却机废气带走的热损失占总热量的近50%。水泥窑的余热除了工艺自身利用外，还有很大一部分热量，一般将其用于余热发电，目前广泛采用的是纯低温余热发电技术。另外，将余热转换后供给吸收式制冷机，用于企业或居民区的中央空调，也是一种有效的余热利用方式。

（4）其他行业。除以上各主要高能耗行业外，余热利用在汽车、船舶、造纸、纺织、酿酒、橡胶、冶金、铸造、建材、玻璃窑、陶瓷窑、油脂生产、海水净化、冷藏冷冻等行业和领域中也已渐得到重视，出现了许多与生产相结合的余热利用方法。

28.3 地热和浅层地温能应用

28.3.1 技术内涵

浅层地温能，又称浅层地热能，是指蕴藏在地表以下一定深度（一般小于200米）范围内岩土体、地下水和地表水中具有开发利用价值的、一般低于25℃的热能，是深层地热能与太阳能共同作用的产物。

浅层地温能的利用价值在于地层中比较恒定的温度与冬、夏两季外界空气存在较大的反向温差，这种温差即为潜在的可被利用的热能。浅层地温更接近于使人舒适的室温，当用热泵为建筑物供暖时，用热泵提温的幅度较小，使热泵机组运行工况更稳定、合理，所以当提供同样的热量时需要的电能较少；夏季制冷时，浅层地温能可提供比空气低得多的温度，使制冷机组的效率提高，因此节约电能。浅层地温能所贮存的空间有较大的热容量，且有较强的恢复能力，浅层地温能可在一个使用季节中通过暂时降低温度持续不断地输出热能，而在非使用期得到恢复，这种年内能够恢复的热输出量即为浅层地温能的合理开采量，这种多年可持续利用的热能就是浅层地温能的利用价值。

毋庸置疑，浅层地温能可以在一定的开采强度、合理的系统分布情况下，在冬、夏两季数月内连续为水源热泵机组提供稳定的低温热源（或热汇），例如冬天供暖时，浅层地温能为水源热泵机组提供的热能占到了热输出的70%—80%，在当前经济技术条件下具有很大的利用价值。所以根据以上定义，浅层地温能无疑是一种无形自然资源中的新能源。

浅层地温能具有如下特点：

（1）一种新型战略资源。浅层地温能资源既是清洁能源，又是可再生能源，所以开发利用浅层地温能资源是保障国家能源安全的必然选择，是确保我国能源结构调整的需要，是实现节能、减排战略目标的重要手段。

（2）分布广泛、能量巨大。浅层地温能资源广泛分布于除了赤道附近和高纬度地区之外的广大区域，尤其在中纬度、四季气温变化明显的地区有较高的利用价值。据专家测算，我国每百平方公里（相当于一个中等城市的面积）近百米深度内的土壤每年可采集的低温能量是1560×104 千瓦，而百米内地下水每年可采集的低温能量也有 21×104 千瓦。

（3）可循环利用。浅层地温能经过一个冬季的连续开采利用后，在换热区域内造成了局部地温场异常，存在一定的热亏损。由于异常区与周围正常地温存在一定的地温梯度，因而产生恢复性热流，这种热流可在其他三个非使用季节使换热区的热亏损得到自然补充，逐步恢复到平衡状态；有制冷需求的工程则可在夏季制冷时的反向利用（放热）中得到人工补给，使开发利用工程年内总的热开采量和补充量基本达到平衡。所以，浅层地温能在季节性利用后可通过自然和人工补给或冬、夏两个季节的反向温差利用后基本保持地温场的动态平衡，从而可长期循环利用。

（4）可就近利用。由于资源无处不在，利用工程可以就地取（排）热、就地取材，与地源热泵系统紧密结合为建筑物供暖和制冷。与传统能源相比可节省大量运输、传输和存放成本。如果大量采用则可提高资源的本地化水平，在一定程度上改善能源结构。

（5）可实现节能减排。浅层地温能资源的开发利用只通过消耗少量的电能，能够从浅层岩土体、地下水中提取大量的热量或冷量，而地下水的水质、水量不发生任何变化，在城区不向大气排放二氧化碳等气体及灰尘，对外界环境影响极小。与传统的锅炉供暖相比，地源热泵系统对一次能源利用率可以达到90%，而中型锅炉房则仅为65%左右，相对节能约38%。可以说浅层地温能资源的开发利用既达到了节能又起到了减排的作用，一举两得。因而，浅层地温能资源的开发利用为解决建筑节能减排提供了一条重要途径。

28.3.2 应用现状

我国浅层地温能的开发始于上世纪 90 年代，在前期的十几年内，虽然 31 个省（区、市）都不同程度地做了一定的工作，但由于没有进行系统的调查评价，资源分布不清，地质条件适宜区不明，开发工作进展缓慢。

2008 年，国土资源部在全国启动了浅层地温能调查评价试点工作，并于当年印发了《关于大力推进浅层地温能开发利用的通知》，发布了《浅层地温能勘查技术规范》、成立了浅层地温能研究推广中心、举办培训班培训了大批调查评价人才，为浅层地温能调查评价工作的顺利开展打下了良好的基础。国土资源部在2009年制定的《国土资源系统应对全球气候变化工作方案》和2010年印发的《关于做好“应对全球气候变化地质响应与对策”有关工作的通知》中，又对浅层地温能调查评价与开发利用工作进行了再部署。

目前，各省（区、市）国土资源部门已作出详细的工作方案，个别省份已开始实施，天津等试点地区已完成调查评价工作。截至2010年底，全国应用浅层地温能供暖/制冷的单位（住宅小区、学校、工厂等）约3400个，建筑面积近1.4亿平方米。项目比较集中的地区有北京、辽宁、河北、河南、山东、天津，80%集中在我国华北和东北南部地区。据估算，2010年浅层地温能的开发利用，使我国二氧化碳减排约2200万吨。随着我国能源结构的调整和地源热泵技术的逐步改善，浅层地温能必将成为我国重要的新型能源，具有广阔的开发前景。

28.4　生物脱氮除磷

28.4.1　技术内涵

近年来，随着社会的发展，人民生活水平的提高，大量含氮和磷的生活污水、印染废水、化肥和农药排入水体，使水环境污染和水体富营养化问题日益严重，人们对于污水处理厂出水中氮和磷排放量也越来越关注，国家对污水处理厂出水中剩余氮和磷的量也要求越来越严格，因此研究和开发高效、经济的生物除磷脱氮技术已经成为现今研究的一大热点。

生物脱氮是通过氨化、硝化、反硝化三个步骤完成的。生物除磷是通过聚磷菌在厌氧情况下释磷、好氧情况下吸磷这两个步骤完成的，并且好氧吸磷远大于厌氧释磷。

28.4.2　应用现状

“十一五”期间，COD的控制按预期进行，总体进展较大，氮磷污染则有加重趋势。总的判断是，我国水环境存在营养物质污染和重金属污染同时存在的情况，有55%的国控断面总氮超标，河流水系中氨氮污染尤其严重。氮磷污染关系着饮水安全，“十一五”期间我国在重点地区开展了除磷脱氮的试点，“十二五”期间准备全面开展。

目前初步判断，氮磷污染物质，氨氮已经超过总磷和高锰酸盐指数，成为主要污染因素。对河流污染情况分析，各污染指标排序为：氨氮、总磷、高锰酸盐指数、粪大肠菌。湖泊各污染指标排序则是：总氮、总磷、高锰酸盐指数。我国已经颁布一系列标准控制水体污染，在太湖流域发布了13项标准控制氮磷，在重点工业行业，氨氮标准已有100多项，标准体系还在不断完善。

我国脱氮除磷发展重点应该是：低投资、低能耗及低运行费的成套脱氮除磷污水处理设备和药剂；工业高浓度氮磷废水处理的成套技术、设备和自控系统；新型脱氮除磷工艺的工程化应月；农村和小城镇因地制宜的分散式处理系统；设备的集成化、系列化和自动化。

28.5　烧结机烟气脱硫、脱硝一体化

28.5.1　技术内涵

燃煤电厂烟气脱硫、脱硝一体化技术是目前控制SOx，和NOx排放的最有效手段之一。近

年来，烟气脱硫、脱硝一体化技术因其在同一套系统内实现同时脱硫与脱硝，具有设备精简、占地少、投资少和运行管理方便等优点，已成为大气污染控制领域中前沿性的研究方向。一般来说，同时脱硫、脱硝技术按照脱除剂及反应产物的状态可分为湿法、干法和半干法三大类。湿法工艺成熟、效率高、应用广泛，但存在成本高、占地面积与耗水量大、易产生二次污染、氨泄漏和设备腐蚀等问题；而干法、半干法虽然仍存在一些技术和经济等方面的缺陷，但由于具有耗水量少、运行成本低、设备简单、占地面积小等优点，因而成为极具发展前景的烟气净化技术。

28.5.2 应用现状

目前，我国钢铁企业有烧结机烟气脱硫设备和脱硫措施的还很少，新建、在建或计划建设烧结烟气脱硫的企业约有20多家，真正投入运行的只有几家，运行效果评价不一。脱硫方法有石灰—石膏法、氨法、循环流化床法、密相干塔高浓度有机废水处理法以及有机胺法。烧结脱硫工程的投资金额差距也很大，从2500万元—8000万元不等，与烧结机的大小和脱硫技术的选择有较大关系。同时，由于国内烧结烟气脱硫刚刚起步，行业竞争混乱，市场还有待规范。

国家环保局出台的《国家酸雨和二氧化硫污染防治“十一五”规划》中指出，落实14个烧结机烟气脱硫示范工程，在总结示范经验的基础上，制订钢铁行业二氧化硫减排规划，重点推进钢铁行业烧结机烟气脱硫工程。“十一五”期间，14个烧结烟气脱硫工程必须在2010年前投产，年减排$SO_2$10万吨。每套脱硫项目按5000万元计算，在此期间有近7亿元的脱硫市场。随着这14个烧结烟气脱硫装置的投产，必将推动全国近400台烧结机烟气脱硫项目的有效实施。考虑到小型烧结机上脱硫的可能性较小，按200台大中型烧结机计算，在未来5年—10年内预计将有100亿元的市场空间。目前，缺乏适合烧结烟气脱硫的成熟技术，是我国烧结烟气脱硫发展的主要障碍。我们要在已有的火电烟气脱硫技术的基础上，将脱硫与烧结工艺有机结合，选择有针对性的方法，力争投资少，运行成本低，有效减少SO_2排放。钢铁行业脱硫市场的参与者可通过合作研究、低价竞标、购买业绩、承接政府课题等途径寻求发展，钢铁厂自身也应主动担负起企业应有的社会责任，承担一些技术开发的费用和风险。

随着钢铁工业的飞速发展和环保政策的逐步落实，必将促进烧结烟气脱硫市场日益繁荣，开创我国烟气脱硫的新局面。尽管当前国内所上的烧结烟气脱硫装置运行稳定性和适宜性有待进一步验证，但这些事实表明，我国钢铁企业烧结烟气脱硫在稳步向前推进。随着环保产业的不断发展，脱硫技术也将不断进步，在减少投资费用、提高设备国产化率和谋求副产物资源化方面也将取得明显成效。

28.6 污泥和垃圾渗滤液处理处置

28.6.1 技术内涵

城市垃圾填埋场渗滤液的处理一直是填埋场设计、运行和管理中非常棘手的问题。渗滤液主要来源于降水和垃圾本身的内含水。由于液体在流动过程中有许多因素可能影响到渗滤液的性质，所以渗滤液水质变动较大。一般来说，其PH值在4—9之间，COD为2000mg/L—62000mg/L，BOD，为60mg/L—45000mg/L，重金属浓度和市政污水中重金属的浓度基本一致。

城市垃圾填埋场渗滤液是一种成分复杂的高浓度有机废水，若不加处理而直接排入环境，会造成严重的环境污染。

28.6.2　常用方法

目前，国内外对渗滤液的常用处理方法有好氧处理、厌氧生物处理、物化法及土地法四大类。

（1）好氧处理。可有效地降低 BOD5、COD 和氨氮，还可以去除另一些污染物质如铁、锰等金属。用活性污泥法、氧化沟、好氧稳定塘、生物转盘等好氧法处理渗滤液都有成功的经验。

（2）厌氧生物处理。厌氧生物处理能耗少，操作简单，因此投资及运行费用低廉，而且由于产生的剩余污泥量少，所需的营养物质也少。近年来，开发的厌氧生物处理方法有：厌氧生物滤池、厌氧接触池、上流式厌氧污泥反应器及分段厌氧硝化等。

（3）物化法。物化法是渗滤液后处理工艺中最常用的方法之一，包括絮凝沉淀、活性炭吸附、膜分离和化学氧化法等，但物化法处理成本较高，不适于大量渗滤液的处理。

（4）土地法。亦即土壤灌溉法，是人类最早采用的污水处理法，占地面积较大，多用于城市生活污水处理。

28.7　废弃电器电子产品资源化

28.7.1　技术内涵

废弃电器电子产品本身具有资源性和危害性兼有的双重属性，如能做到科学合理地回收，就可带来显著的经济效益和社会效益。但是，废弃电器电子产品中含有铅、镉、汞等重金属和有毒有害的物质，随意丢弃或进行不当的回收处理，就会造成严重的二次污染。

28.7.2　应用现状

电子废弃物是世界上数量增长最快的固体废物。根据 2010 年联合国环境署数据，全球电子废物产生量约 4000 万吨/年，中国已成为世界第二大电子废弃物产生国，仅次于美国。根据工信部数据：2009 年我国手机产量 6.2 亿部、电脑 1.82 亿台、彩电 9966 万台、家用空调 8153.27 万台、家用电冰箱 6063.58 万台、家用洗衣机 4935.84 万台，总产量已达 11.8 亿台之多，增长幅度明显。

目前，我国废弃电器电子产品的回收，虽然取得了一定的进展，但仍存在着回收体系不完善，回收处理方式相对落后等问题。在实行家电以旧换新的过程中，全国多数地区已初步形成了较为完善的废旧家电回收处理机制，但一方面流动的小商贩回收倒卖旧家电的情况仍然存在，另一方面部分目前在国内的电子废弃物处理领域，大多是采取焚烧和水洗等手工拆解的方式，处理技术落后，回收率低，对环境造成了极大的污染。

29　节能减排“十二五”重点推广技术

29.1　能量梯级利用

29.1.1　技术内涵

能量的形式多种多样，能量不仅在数量上具有守恒性，而且在质量上具有品位性，在其转换与传递过程中，存在品位的差异。热能是能量转换与传递的主要形式之一。电能、机械能可全部转变为热能，而热能却只能部分转变为机械能或电能，即电能、机械能的品位比热能高。在热能转换成机械能过程中，高温高压热源比低温低压热源转换潜力大，这说明温度高或压力高的热能品位高。分析能源结构并根据能量品位的不同合理分配能量，优化能源利用方式，可减少能量品位的降低，达到节约能源，提高能源效率的目的。

1988 年，吴仲华教授从能量转化的基本定律出发，阐述热能的梯级利用与品位概念和基于能的梯级利用的总能系统，提出了著名的“温度对口、梯级利用”原则，主要包括：（1）通过热机把能源最有效地转化成机械能时，基于热源品位概念的“温度对口、梯级利用”原则；（2）把热机发电和余热利用或供热联合时，大幅度提高能源利用率的“功热并供的梯级利用”原则；（3）把高温下使用的热机与中低温下工作的热机有机联合时，“联合循环的梯级利用”原则等。

29.1.2　应用现状

有分析指出，我国现行供热机组一般都是在原凝汽式机组基础上进行技术改造实现的，没有专门针对热电联产设计生产的产品，因此机组在运行时不可避免地存在着安全隐患及经济效益上的弊端。同时，能量梯级利用原理从能量品位合理匹配的角度出发，有效降低了不可逆过程的能量贬值。进一步而言，冷源损失的利用一直是热电联产事业亟待解决的问题，能量梯级利用原理应用于供热机组，本着温度对口，减小换热温差的原则，合理利用了供热抽汽，使机组在供热能力不变的前提下多发电，在环保和经济上的意义都是巨大的。

29.2　低温余热发电

29.2.1　技术内涵

目前，中国能源利用率仅为约 30%，大量余热以各种形式被排放到大气中，可再生能源在能源结构中所占比例不足 8%。因此，回收利用余热在提高中国一次能源利用率方面具有举足轻重的作用。目前，中国回收利用的余热主要来自高温烟气的显热和生产过程中排放的可燃气，中低温余热（即低品位余热）基本上还没有回收。相对于煤、石油、天然气等高品位能源而言，低品位余热在相同单位内包含的能量很低，利用难度大。但从能源利用的格局来看，低品位余热将作为产能和用能的关键环节，对节能减排的战略起到重要作用。现有的低品位余热的回收利用中普遍采用水冷介质，受到水资源、运输、地域等多方面的限制，一定程度上阻碍了余热的大规模应用。2010 年初，中国南方多省市及东南亚多个国家和地区遭受严重旱灾，使这一问题显得尤为突出。因此，寻找可重复利用的新型无水高效传热介质是推进低品位余热利用的关键所在。

29.2.2 应用现状

溴化锂吸收式热泵在工厂余热回收中的应用已经越来越广泛，它可以回收利用工厂低温余热热源的热量，将低品位热转变为高品位热，广泛应用于有余热资源或有低温热源的石油、石化、制药、酿造和钢铁等行业。另外，在炼油工业、天然气和热电冷三联供系统中可以配置吸收式系统实现对各种低温余热的高效利用。

吸附式制冷技术作为一种余热利用的新技术，其研究开发正日益成熟，它可以将太阳能或余热等低品位热源作为驱动热源，采用对环境友好的工质对，设备结构简单，一次性投资少，运行费用低，使用寿命长，无运动部件，无噪音，无环境污染，特别适用于有大量低品位余热排放的工业过程及有频繁震动的移动机械上。

针对冶金、玻璃、水泥、陶瓷等行业中高能耗的窑炉，回收烟气余热的传统做法是利用耐火材料的显热熔变化来储热，这种储热设备的体积大、储热效果不明显。改用相变储热材料，是研究较为广泛的一种方法，这样储热设备体积可减小 30%—50%，还可起到稳定运行的作用。目前正在研究的新型相变材料有潜热型功能热流体、纳米复合相变储能材料、定型相变蓄能材料和无机盐/陶瓷基复合相变蓄能材料等。

采用热电材料进行温差发电，也是工业余热的一个应用领域，如利用炼钢高炉等工业废热和利用汽车发动机的余热进行温差发电。另外，金属氢化物在空调、供热方面的应用也逐渐引起人们的重视。

热管技术。热管是一种由管壳和工质组成的高效导热元件，以相变（蒸发与凝结）换热作为传热的主要方式，具有传热能力大、温度控制能力强、传热效率高等特点。现在热管换热器已在电力、冶金、石化、玻陶、电子、轻工等行业的余热回收领域获得了广泛的应用。在钢铁企业中，有别于以往烟气通过各种换热器和余热锅炉转化为蒸汽或者热水进行热利用的方式，热管废热发生器能够直接利用烟气余热。

热声技术。热声热机是通过热声效应实现热能与声能的相互转化的装置。从声学角度来说，热声效应是由于处于声场中的固体介质与振荡的流体之间相互作用，使得距固体壁面一定范围内沿着（或逆着）声传播方向产生一个时均热流，并在这个区域内产生或者吸收声功的现象。热声热机可利用低品位热源，将难以利用的余热（如铝电解槽中槽壳侧部温度为 300℃的余热）转化成便于利用的电能，这也是它应用到工业中的立足点。

29.3 先进煤气化

29.3.1 技术内涵

煤炭气化是指煤在特定的设备内，在一定温度及压力下使煤中有机物质与气化剂（如蒸汽/

空气或氧气等）发生一系列化学反应，将固体煤转化为含有 CO、H_2、CH_4 等可燃气体的过程。煤炭气化时，必须具备三个条件，即：气化炉、气化剂、供给热量，三者缺一不可。

原料煤的性质，例如，煤中水分、灰分和挥发分的含量，粘结性，化学活性，灰熔点，成渣特性，机械强度和热稳定性，以及煤的粒度和粒度分布等，都对气化过程有不同程度的影响。因此，必须根据煤的性质和对气体产物的要求选用合适的气化方法。按煤在气化炉内的运动方式，气化方法可划分为三类，即固定床气化法、流化床气化法和气流床气化法。

（1）固定床气化。在气化过程中，煤由气化炉顶部加入，气化剂从气化炉底部加入，煤与气化剂逆流接触，相对于气体的上升速度而言，煤下降速度很慢，甚至可视为固定不动，因此称之为固定床气化；而实际上，煤在气化过程中是以很慢的速度向下移动的，比较准确的应称其为移动床气化。

（2）流化床气化。它是以粒度为 0—10mm 的小颗粒煤为气化原料，在气化炉内使其悬浮分散在垂直上升的气流中，煤粒在沸腾状态进行气化反应，从而使得煤层内温度均一，易于控制，提高气化效率。

（3）气流床气化。它是一种并流气化，用气化剂将粒度为 100um 以下的煤粉带入气化炉内，也可将煤粉先制成水煤浆，然后用泵打入气化炉内。煤在高于其灰熔点的温度下与气化剂发生燃烧反应和气化反应，灰渣以液态形式排出气化炉。

煤气化技术包括：备煤技术、气化炉技术、气化后工艺技术三部分，核心是气化炉。

29. 3. 2　应用现状

我国自煤的商业化和社会化以来迄今已 100 余年，但没有形成能与国际抗衡的商业化自主产权煤气化技术。上世纪 70 年代起 4 家厂引进德士古水煤浆气化技术；“九五”期间就“一体化煤气化联合循环（IGCC）关键技术（含高温净化）”立项，有十余个单位参加攻关；2004 年科技部对大规模高效气流床煤气化技术进行了基础研究；30 多年来我国共引进 10 余台德士古气化炉，国内配套完成了部分设计、安装与操作，积累了丰富的经验。

2007 年初，我国第一台日处理煤 1150 吨的新型气化炉，在山东兖矿集团国泰化工有限公司实现了连续稳定运行 33 天的最长周期。这项由兖矿集团拥有 7 项自主知识产权的新的煤炭气化技术，在我国首次进行大工业生产三个月后，即被熟练掌控，标志着 30 多年来世界发达国家对我国水煤浆气化技术的垄断从此打破，并同时创造了世界煤气化技术发展史上的新纪录。以这台气化炉为核心的新型水煤浆气化技术，是由兖矿集团承担、联合华东理工大学和中国天辰化学工程公司共同研发的国家“九五”攻关重点项目，研发当年即被列入国家科技部“863”攻关课题。2005 年 11 月 16 日，该技术在取得国家 7 项自主知识产权后，在兖矿集团国泰化工公司首次应用于大工业生产。试运行三个多月以来，国泰公司技术人员成立了 7 个攻关小组，设立了 120 多个攻关课题，克服了 60 余项原创性技术难题。2008 年 1 月 8 日，兖矿新型气化炉通过科技部验收，综合性能被评定为“国际领先”。目前，国际上水煤浆气化技术唯一推广应用的是美国的德士古气化炉。据测试，兖矿新型气化炉与德士古相比，消耗降低 7%，有效气成分高出 2%—3%，具有“转化充分，环保高效”的明显优点。同时，兖矿新型气化炉完全使用高硫煤为原料，每年在消耗 70 万吨高硫煤的基础上，还可回收生产硫磺 2 万多吨，对我国的能源政策具有重要意义。国泰公司是兖矿集团确立煤化工产业为支柱产业后投产的第一个煤化工企业，总投资27亿元人民币，生产规模为年产20万吨醋酸、24万吨甲醇、联产80000KW电力，年均销售收入约18亿元，利税6亿元，拥有3个“863”科技攻关项目。仅2005年试生产半个月的时间，国泰公司就实现产值2亿元，创利5千万元。

上世纪80年代末以前，我国煤气化完全依赖常压固定床技术，有常压固定床气化炉4000多台，配套小型合成氨生产装置及少量甲醇和联醇装置，这些气化装置中一部分至今仍在运转。上世纪80年代初，我国开始引进第二代煤气化技术，1家引进加压鲁奇技术，在山西潞城建厂，气化炉3开1备；共有5家引进德士古水煤浆气化装置，分别建于兖矿鲁南化肥厂、上海焦化总厂、陕西渭河化肥厂、安徽淮南化工厂、黑龙江浩良河化肥厂。这5套装置均用于生产合成气，制氨或甲醇。目前投产的煤气化技术有多套，分别是德士古气化炉（金陵、榆林、南京等）和壳牌气化炉（应城、岳阳、柳州、枝江、安庆、大连、安庆、曲靖等）。

目前，用于煤化工的气化炉主要为壳牌（SHELL）气化炉、德士古（TEXACO）气化炉和GSP气化炉。

29.4 高压变频调速

29.4.1 技术内涵

高压变频器是指输入电源电压在3KV以上的大功率变频器，主要电压等级有3000V、3300V、6000V、6600V、10000V等电压等级的高压大功率变频器。

我国的电动机用电量占全国发电量的60%—70%，风机、水泵设备年耗电量占全国电力消耗的1/3。造成这种状况的主要原因是传统的方法是通过挡板、阀门地截流消耗的功率大，而且通常在设计中，电机设计的容量比实际需要高出很多，存在“大马拉小车”的现象，效率低下，造成电能的大量浪费，因此推广交流变频调速装置效益显著。

29.4.2 应用现状

目前，在国内有大量的低压变频器厂商，其大部分为AC380V的中小功率产品，而在高压大功率变频器方面，却为数不多。能够研制、生产、并提供服务的高压变频器厂商，仅有少数的具备科研能力或资金实力的个别企业。

国内仅有少部分的中、高压电机进行了变频调速改造，且普通采用V/f控制方式。高压变频器的品种和性能还处于发展阶段，每年市场仍需大量进口。这些状况主要表现在如下几个方面：

- 国外各大品牌的产品正加紧占领国内市场，并加快了本地化的步伐。
- 具有研发能力和产业化规模的企业少。
- 国产高压变频器的功率等级较低，目前不超过3500KW。
- 国内高压变频器的技术标准还有待规范。
- 与高压变频器相配套的产业很不发达。
- 生产工艺落后，勉强满足变频器产品的技术要求，但是价格低廉。
- 变频器中使用的功率半导体，驱动电路，电解电容等关键器件完全依赖进口。
- 与发达国家的技术差距在缩小，具有自主知识产权的产品正应用在国民经济中。
- 已经研制出具有瞬时掉电再恢复、故障再恢复等功能的变频器。

• 能够进行四象限运行的高压变频器尚在研究与开发中。

有关资料显示，我国变频器总的潜在市场应为1200亿元—1800亿元，其中常压变频器约占市场份额的60%左右，中、高压变频器由于单台变频器功率大、售价高，应占市场总额40%左右。在国内高压变频器应用领域，活跃在我国市场上的国产品牌占80%左右，但市场份额仅占40%，尤其5000KVA以上超大功率产品市场基本上被ABB、西门子、东芝—三菱等跨国公司所垄断。

29.5 干熄焦

29.5.1 技术内涵

干法熄焦是目前国外较广泛应用的一项节能技术，其英文名称为Coke Dry Quenching，简称CDQ。干熄焦技术是利用冷的惰性气体（燃烧后的废气），在干熄炉中与赤热红焦换热从而冷却红焦，吸收了红焦热量的惰性气体将热量传给干熄焦锅炉产生蒸汽，被冷却的惰性气体再由循环风机鼓入干熄炉冷却红焦。干熄焦锅炉产生的蒸汽或并入厂内蒸汽管网或送去发电。

干熄焦法的技术特点如下：

（1）回收红焦显热。出炉的红焦显热约占焦炉能耗的35%—40%，这部分能量相当于炼焦煤能量的5%，如将其回收和利用，可大大降低冶金产品成本，起到节能降耗的作用。采用干熄焦可回收80%的红焦显热，平均每熄1吨焦炭可回收3.9MPA 450℃的蒸汽0.45吨—0.6吨。

（2）减少环境污染。由于干熄焦能够产生蒸汽（5吨—6吨蒸汽需要1吨动力煤），并可用于发电，可以避免生产相同数量蒸汽的锅炉燃煤对大气的污染，尤其减少了二氧化硫和二氧化碳向大气的排放。对规模为年产100万吨焦炭的焦化厂而言，采用干熄焦每年可以减少8万吨—10万吨动力煤燃烧向大气排放的各种污染物。

（3）可改善焦炭质量。国际上公认，大型高炉采用干熄焦焦炭可使其焦比降低2%，使高炉生产能力提高1%。在保持原焦炭质量不变的条件下，采用干熄焦可以降低强粘结性的焦、肥煤配入量10%—20%，有利于保护资源和降低焦炭成本。

29.5.2 应用现状

目前国内使用干熄焦技术的仅有3家企业，即宝山钢铁集团总公司（引进日本技术和设备）、上海浦东煤气焦化厂（全套引进乌克兰的技术和设备）和济南钢铁集团总公司（改造乌克兰技术，工艺和设备国产化率达90%）。据不完全统计，中国现有大小焦化厂140余家，400余座焦炉，其中大、中型以上焦炉120余座，年生产焦炭5000多万吨，按产汽率0.45计算，可回收蒸汽2250万吨，若用于发电，年发电量可达58亿KWh，价值26亿元，其他潜在效益则更大。可见，干熄焦技术在国内现有焦化厂的改造中将有很大的市场潜力。

29.6 蓄热式加热炉

29.6.1 技术内涵

蓄热式加热炉实质上是高效蓄热式换热器与常规加热炉的结合体，主要由加热炉炉体、蓄热室、换向系统以及燃料、供风和排烟系统构成。蓄热室是蓄热式加热炉烟气余热回收的主体，它是填满蓄热体的室状空间，是烟气和空气流动通道的一部分。在加热炉中，蓄热室总是成对使用，一台炉子可以用一对，也可以用几对，甚至几十对。在国内的一些大型加热炉上，最多用到四十几对。蓄热式余热回收的优点炉温更加均匀，由于炉温分布均匀，加热质量大大改善，产品合格率大幅度提高。燃料选择范围更大适合轻油、重油、天然气、液化石油气等各种燃料，尤其是对低热值的高炉煤气、发生炉煤气具有很好的预热助燃作用，扩展了燃料的应用范围。因此，炉子燃料消耗量大幅度降低。对于一般大型加热炉，可节能25%—30%；对于热处理炉，可节能30%—65%。如果采用传统的节能技术，助燃空气预热温度越高，烟气中 NO_X 含量越大；而采用蓄热式高温空气燃烧技术，在助燃空气预热温度高达800℃的情况下，炉内 NO_X 生成量反而大大减少。由于蓄热式燃烧是在相对的低氧状态下弥散燃烧，没有火焰中心，因此，不存在大量生成 NOx 的条件。烟气中 NOx 含量低，有利于保护环境。金属氧化烧损低氧燃烧的另一个好处是可降低被加热金属的氧化烧损。此外，蓄热式燃烧还可以提高火焰辐射强度，强化辐射传热，提高炉子产量。

29.6.2　应用现状

从1997年7月这项技术应用成功至今，蓄热式加热炉在钢铁企业建设不少于几百座，蓄热式换热技术在加热炉领域的应用逐步得到完善和成熟。钢铁企业轧钢领域蓄热式换热技术的应用：

- 从炉型上分：有推钢式加热炉；步进梁式加热炉；车底式炉；均热炉；罩式炉；带材连续式热处理炉及冶炼连铸领域的烘烤设备。
- 从匹配的轧机分：棒材、高速线材、中厚板、热带、H型钢、及型钢等加热炉，其中以步进梁炉为主。
- 从加热的钢种分，普碳、低合金和特殊钢加热炉（暂无加热硅钢坯的业绩）。
- 从炉子产量分：小时产量几十吨至几百吨。
- 从加热钢坯规格分：方坯最长达16m，断面最大为300mm×400mm；板坯最长达15.6m；板坯厚：250mm—300mm。

由上述情况可见，蓄热式加热炉的应用范围已经涵盖轧钢生产的各个领域。

采用这项技术的企业主要还是集中在地方中小型和民营钢铁企业，并且大部分都是烧纯高炉煤气的蓄热式加热炉，因此这些企业加热钢坯的燃烧成本低，钢铁产品的价格竞争性强，利润率高。最早利用蓄热式技术，加热炉烧高炉煤气的当属萍乡钢铁公司，他们先后有五座蓄热式加热炉小时产量60吨—150吨投入使用，加热炉的热耗指标都达到“特等炉”，其燃料费不到烧重油的一半。民营或合资企业中，当以唐山国丰，迁西津西、山西海鑫、天津荣程、山东日照及发展较快的建龙系等为代表。这些钢铁企业的轧钢厂都是烧高炉煤气的蓄热式加热炉。特钢企业以江阴兴澄特钢为代表，一期、二期轧钢步进梁式炉都是高炉煤气的蓄热式炉。还有一如张家港市沙钢集团，前期为“短流程”钢铁厂，建成十几座轧钢厂的步进梁炉和推钢炉都是烧重油的。后期建起了高炉，有了高炉煤气后，陆续将烧油加热炉改为烧高炉煤气的蓄热式加热炉。据沙钢主管加热炉改造的负责任称，就此一项技术改革，可为企业节省燃料费用以亿元计。再如，张家港市永联钢铁集团十几座加热炉也都是烧高炉煤气的蓄热式加热炉。

29.7 冰蓄冷

29.7.1 技术内涵

20世纪70年代，由于全球性能源危机，加之美国、日本和欧洲一些工业发达国家夏季电负荷增长和峰谷差拉大的速度惊人，致使发电站夜间处于低负荷下运转，由此冰蓄冷技术开始试验性地被引入集中式空调系统。此后在工程项目中采用冰蓄冷空调系统的愈来愈多。

冰蓄冷空调是利用夜间低谷负荷电力制冰储存在蓄冰装置中，白天融冰将所储存冷量释放出来，减少电网高峰时段空调用电负荷及空调系统装机容量，它代表着当今世界中央空调的发展方向。具体而言，具有如下优势：

- 削峰填谷、平衡电力负荷。
- 改善发电机组效率、减少环境污染。
- 减小机组装机容量、节省空调用户的电力花费。
- 改善制冷机组运行效率。
- 蓄冷空调系统特别适合负荷比较集中、变化较大的场合，如体育馆、影剧院、音乐厅等。
- 应用蓄冷空调技术，可扩大空调区域使用面积。
- 适合于应急设备所处的环境，计算机房、军事设施、电话机房和易燃易爆物品仓库等。

29.7.2 应用现状

在我国兴起已有近20年历史的冰蓄冷系统，其推广应用在暖通空调业界被普遍认作是利国利民的共赢之举，但多年来其市场状况却令人担忧。根据统计数据，我国2009年冰蓄冷市场总额只有15亿元，相比较于整体数千亿元的空调市场，规模十分之小。同样，相比于暖通空调行业，冰蓄冷企业数量也属凤毛麟角。目前，全国具有一定规模的只有华电华源、高灵等十几家主要厂商，且其中绝大部分是美国品牌。

业内人士看来，冰蓄冷市场发展缓慢与其市场关注程度和认知方面的被动处境不无关系。不过，“十二五”期间，冰蓄冷市场可能产生变革。2011年初出台的《电力需求侧管理办法》特别明确了相关主管部门的责任，鼓励低谷蓄能，在具备条件的地区实行季节电价、高可靠性电价、可中断负荷电价等制度，支持实施电力需求侧管理。同时，政策还特别提到，鼓励用户采用符合国家有关要求的高效用电设备和热泵、电蓄冷、电蓄热等技术，合理配置无功补偿装置，加强无功管理，优化用电方式，配合政府主管部门和电网企业开展电力需求侧管理。

蓄冷行业专家、天津大学教授张永铨分析认为，随着国家逐步重视建设能源节约型社会、强调突出抓好节能用电，电力及相关政府部门将进一步运用价格工具和市场杠杆，鼓励用户采用蓄冷设备。

与上述观点相印证的是，2011年7月中旬，位于天津开发区核心区域的泰达现代服务产业区（MSD）冰蓄冷区域供冷系统项目正式启动，这是我国目前最大的蓄冰桶蓄冷区域供冷项目。该项目总建筑面积51万平方米，采用高灵ICEBANK桶式蓄冰设备258个，总蓄冷量约为4.5万冷吨时，工程完成后将实现天津泰达现代服务产业区5A级高档办公楼及商业配套服务场所的集中供冷。

29.8　高效换热器

29.8.1　技术内涵

换热器是化工、石油、制药及能源等行业中应用相当广泛的单元设备之一。据统计，在现代化学工业中所用换热器的投资大约占设备总投资的30%，在炼油厂中换热器占全部工艺设备的40%左右，海水淡化工艺装置则几乎全部是由换热器组成的。上个世纪70年代初发生的世界性能源危机，有力地促进了传热强化技术的发展。为了节能降耗，提高工业生产的经济效益，要求开发适用不同工业过程要求的高效能换热设备。因此，几十年来，高效换热器的开发与研究始终是人们关注的课题，国内外先后推出了一系列新型高效换热器。

29.8.2　应用现状

2008年，中国换热器产业市场规模在360亿元左右，主要集中在石油、化工、冶金、电力、船舶、集中供热、制冷空调、机械、食品、制药等领域。其中，石油化工领域仍然是换热器产业最大的市场，其市场规模在100亿元以上；电力冶金领域换热器市场规模在60亿元左右；船舶工业换热器市场规模在30亿元以上；机械工业换热器市场规模约为30亿元；集中供热行业换热器市场规模超过25亿元。

按照产品类型的不同，我国换热器产业市场竞争主要集中在以下四大产品领域：

（1）板式换热器领域。国内外企业竞争激烈，大量外资企业已经完成在中国的布局。其中，四平巨元瀚洋、兰石换热设备公司、四平维克斯是我国板式换热器领域内资企业中的龙头企业，其板式换热器年产值都在2亿元以上。外资企业主要包括阿法拉伐（江阴）、舒瑞普（北京、苏州）、APV（上海、北京）、丹佛斯（天津）、传特（北京）、桑德克斯（上海、宁波）、风凯（常州）等企业，世界著名的板式换热器企业大都已经进入中国市场。此外，沈阳太宇、蓝科高新（原兰石所）、上海艾克森、湖北登峰、山东北辰、佛山渊石、上海南华等企业也是我国重要的板式换热器企业。

（2）管壳式换热器领域。我国生产企业众多，且规模都较小。其中，抚顺机械设备制造有限公司、兰石集团炼化设备公司、中石化南京化工机械是我国内资管壳式换热器的龙头企业，其管壳式换热器年产值都在2亿元以上；江苏中圣集团、无锡化工装备总厂、宝钛集［碉南京宝色股份、西安核设备制造厂（原国营五二四厂）、合肥通用特种材料设备有限公司是我国特种材料换热器领域的重要企业，其特种材料管壳式换热器年产值都在1.5亿元以上；中石化镇海石化建安工程有限公司、中石化北京燕化、中石化茂名重力石化机械制造有限公司等企业依托母公司中石化的市场优势，也形成了一定的换热器生产规模，年产值在1亿元—2亿元左右；此外，张家港化工机械、大连金重公司、湖北长江石化设备公司、大连东方亿鹏、合肥通用特种材料设备有限公司、西安大秦化工机械（原西安化工机械厂）、林德工程（大连）、天津国际机械（原天津市换热装备总厂）、大连东方亿鹏等企业也是国内管壳式换热器的主要生产企业，管壳式换热器的年产值都在1亿元以上。相对而言，管壳式换热器外资企业在华布点不多，比较知名的有日本森松（上海）、林德工程（大连）、美国艾普尔（苏州）、德国风凯（常州），这主要缘于我国石油化工领域换热器企业众多，生产能力较强，国外企业进入中国市场较为困难。

（3）空冷式换热器领域。哈空调是我国最大的空冷式换热器生产企业。此外，江苏双良股份、国电集团北京龙源冷却技术有限公司、四川简阳空冷器、蓝科高新（原兰石所）、兰州兰石集团长征机械、西安大秦化工机械（原西安化工机械厂）、湖北长江石化设备、江阴电力设备冷却器公司等企业也具有一定的竞争力。外资企业中，基伊埃（芜湖、廊坊）、斯必克（张家口）在空冷式换热器领域具有较强的竞争力。

（4）板翅式换热器领域。杭州杭氧股份和开封空分集团是我国石油化工领域著名的板翅式换热器企业，浙江银轮股份、贵州永红航空机械、无锡马山永红换热器则是我国汽车、工程机械、农业机械行业油冷器、中冷器领域的著名换热器企业。

29.9 膜生物反应器

29.9.1 技术内涵

膜—生物反应器（MembraneBio－Reactor，MBR）为膜分离技术与生物处理技术有机结合之新型态废水处理系统。以膜组件取代传统生物处理技术末端二沉池，在生物反应器中保持高活性污泥浓度，提高生物处理有机负荷，从而减少污水处理设施占地面积，并通过保持低污泥负荷减少剩余污泥量。主要利用沉浸于好氧生物池内之膜分离设备截留槽内的活性污泥与大分子有机物。膜生物反应器系统内活性污泥（MLSS）浓度可提升至8000mg/L—10000mg/L，甚至更高；污泥龄（SRT）可延长至30天以上。膜生物反应器因其有效的截留作用，可保留世代周期较长的微生物，可实现对污水深度净化，同时硝化菌在系统内能充分繁殖，其硝化效果明显，对深度除磷脱氮提供可能。

29.9.2 应用现状

我国对MBR的研究还不到10年，但进展十分迅速。国内对MBR的研究大致可分为几个方面：

- 探索不同生物处理工艺与膜分离单元的组合形式，生物反应处理工艺从活性污泥法扩展到接触氧化法、生物膜法、活性污泥与生物膜相结合的复合式工艺、两相厌氧工艺。
- 影响处理效果与膜污染的因素、机理及数学模型的研究，探求合适的操作条件与工艺参数，尽可能减轻膜污染，提高膜组件的处理能力和运行稳定性。
- 扩大MBR的应用范围，MBR的研究对象从生活污水扩展到高浓度有机废水（食品废水、啤酒废水）与难降解工业废水（石化污水、印染废水等），但以生活污水的处理为主。

在我国，MBR同时应用于生活污水与工业废水处理的研究。这些研究结果都表明：MBR对各种高浓度有机废水与难降解废水的COD等都达到良好的去除效果。

第7篇　数据篇

30 国际数据①

表 30－1 世界一次能源消费量

Table 30－1 World Consumption of Primary Energy

［时间］／［指标］	一次能源消费量（百万吨油当量）
1999 年	9151.4
2000 年	9382.4
2001 年	9465.6
2002 年	9651.8
2003 年	9997.8
2004 年	10482
2005 年	10800.9
2006 年	11087.8
2007 年	11398.4
2008 年	11535.8
2009 年	11363.2
2010 年	12002.4

表 30－2 世界煤炭产量和消费量

Table 30－2 World Production and Consumption of Coal

［时间］／［指标］	煤炭产量（百万吨）	煤炭（折油当量）产量（百万吨油当量）	煤炭（折油当量）消费量（百万吨油当量）
1999 年	4627.6	2329.3	2316
2000 年	4693.3	2352.5	2399.7
2001 年	4912.1	2458.9	2412.4
2002 年	4948.6	2477.7	2476.7
2003 年	5301.3	2666.1	2677.3
2004 年	5716	2893	2858.4
2005 年	6035.3	3064.4	3012.9
2006 年	6342	3237.1	3164.5
2007 年	6573.3	3362.4	3305.6
2008 年	6795	3470.3	3341.7

① 本章资料来源于国研网数据库《BP 能源统计》。

续表

[时间] / [指标]	煤炭产量（百万吨）	煤炭（折油当量）产量（百万吨油当量）	煤炭（折油当量）消费量（百万吨油当量）
2009年	6880.8	3511.8	3305.6
2010年	7273.3	3731.4	3555.8

表30－3　世界原油产量和消费量

Table 30－3　World Production and Consumption of Crude Oil

[时间] / [指标]	原油产量（百万吨）	原油消费量（百万吨）
1999年	3482.1	3526
2000年	3611.8	3571.6
2001年	3601.6	3597.2
2002年	3584.2	3632.3
2003年	3701.1	3707.4
2004年	3877	3858.7
2005年	3906.6	3908.5
2006年	3916.2	3945.3
2007年	3904.3	4007.3
2008年	3933.7	3996.5
2009年	3831	3908.7
2010年	3913.7	4028.1

表30－4　世界天然气产量和消费量

Table 30－4　World Production and Consumption of Natural Gas

[时间] / [指标]	天然气产量（10亿立方米）	天然气消费量（10亿立方米）
1999年	2333.2	2324.6
2000年	2413.4	2411.7
2001年	2478	2455
2002年	2519.4	2520.3
2003年	2616.5	2606.1
2004年	2694	2694.5
2005年	2778	2781.8
2006年	2880.7	2842.4
2007年	2950.5	2947.4
2008年	3062.1	3026.4
2009年	2975.9	2950.2
2010年	3193.3	3169

表 30－5 世界太阳能光伏发电累计安装容量

Table 30－5 World Accumulative Installed Capacity of Photovoltaic Power

[时间]／[指标]	太阳能光伏发电累计安装容量（兆瓦）
1999 年	979
2000 年	1455.3
2001 年	1779.2
2002 年	2248.5
2003 年	2839.8
2004 年	3989.1
2005 年	5426
2006 年	7013.5
2007 年	9571.5
2008 年	15900.6
2009 年	23042.5
2010 年	39777.8

表 30－6 世界风力发电累计安装容量

Table 30－6 World Accumulative Installed Capacity of Wind Power

[时间]／[指标]	风力发电累计安装容量（兆瓦）
1999 年	13932
2000 年	18450
2001 年	24927
2002 年	32037
2003 年	40301
2004 年	47912
2005 年	59397.7
2006 年	74306
2007 年	94005
2008 年	122158
2009 年	160086.6
2010 年	199522.8

表 30－7 世界水电消费量

Table 30－7 World Consumption of Hydropower

[时间]／[指标]	水电消费量（太瓦小时）	水电（折油当量）消费量（百万吨油当量）
1999 年	2618.2	592.4
2000 年	2648.9	599.4
2001 年	2585.6	585.1
2002 年	2634.6	596.2
2003 年	2633.7	595.9

续表

[时间] / [指标]	水电消费量（太瓦小时）	水电（折油当量）消费量（百万吨油当量）
2004年	2799	633.3
2005年	2910.6	658.6
2006年	3024.6	684.4
2007年	3078.1	696.5
2008年	3202.8	724.7
2009年	3253.9	736.3
2010年	3427.7	775.6

表30-8 世界核电消费量

Table 30-8 World Consumption of Nuclear Power

[时间] / [指标]	核电消费量（太瓦小时）	核电（折油当量）消费量（百万吨油当量）
1999年	2524	571.1
2000年	2582.4	584.3
2001年	2654.9	600.7
2002年	2699.2	610.8
2003年	2645.2	598.5
2004年	2763.1	625.2
2005年	2770.1	626.8
2006年	2807.9	635.4
2007年	2749.2	622.1
2008年	2736.5	619.2
2009年	2713.4	614
2010年	2767.2	626.2

表30-9 世界生物能源生产量

Table 30-9 World Production of Bioenergy

[时间] / [指标]	生物能源生产量（千桶/日）	生物能源（折油当量）生产量（千吨油当量）
1999年		
2000年	315.1	9176.3
2001年	344.3	10083.8
2002年	405.9	11929.6
2003年	501.5	14767.4
2004年	556.3	16452.2
2005年	661.5	19943.7
2006年	839.3	25742.5
2007年	1114.5	34512.1
2008年	1488.2	46293.8
2009年	1657.9	52097.6
2010年	1894.3	59261.2

表 30－10 世界地热发电累计安装容量

Table 30－10 World Accumulative Installed Capacity of Geothermal Power

[时间] / [指标]	地热发电累计安装容量（兆瓦）
1999 年	
2000 年	8165.5
2001 年	8275.7
2002 年	8386
2003 年	8386
2004 年	8912.4
2005 年	9127.9
2006 年	9492.3
2007 年	9968.7
2008 年	10364.1
2009 年	10716.2
2010 年	10906.2

表 30－11 世界其他可再生能源生产量

Table 30－11 World Production of Other Renewable Energy Sources

[时间] / [指标]	其他可再生能源消费量（太瓦小时）	其他可再生能源（折油当量）消费量（百万吨油当量）
1999 年	209.9	47.5
2000 年	226.3	51.2
2001 年	236.8	53.6
2002 年	267	60.4
2003 年	289.7	65.5
2004 年	329.2	74.5
2005 年	367.1	83.1
2006 年	409.3	92.6
2007 年	467.2	105.7
2008 年	540.6	122.3
2009 年	607.1	137.4
2010 年	701	158.6

表 30－12 世界二氧化碳排放量

Table 30－12 World Emission of CO_2

[时间] / [指标]	二氧化碳排放量（百万吨二氧化碳）
1999 年	24922.7
2000 年	25576.9
2001 年	25800.8
2002 年	26301.3
2003 年	27508.7

续表

［时间］／［指标］	二氧化碳排放量（百万吨二氧化碳）
2004 年	28875.2
2005 年	29826.1
2006 年	30667.6
2007 年	31641.2
2008 年	31915.9
2009 年	31338.8
2010 年	33158.4

31 国内能源数据[①]

表 31－1 能源生产总量及构成

Table 31－1 Total Production Volume and Structure of Energy

年 份	能源生产总量（万吨标准煤）	占能源生产总量的比重（%）		
		原 煤	原 油	天然气
2010	299000	76.8	9.6	4.3
2009	274618	77.3	9.9	4.1
2008	260552	76.8	10.5	4.1
2007	247279	77.7	10.8	3.7
2006	232167	77.8	11.3	3.4
2005	216219	77.6	12	3
2004	196648	77.11	2.8	2.8
2003	171906	76.21	4.1	2.7
2002	150656	73.51	5.8	2.9
2001	143875	73	16.3	2.8
2000	135048	73.2	17.2	2.7

表 31－2 能源消费总量及构成

Table 31－2 Total Consumption Volume and Structure of Energy

年 份	能源消费总量（万吨标准煤）	占能源消费总量的比重（%）		
		煤炭	石油	天然气
2010	325000	70.9	16.5	4.3
2009	306647	70.4	17.9	3.9
2008	291448	70.3	18.3	3.7
2007	280508	71.1	18.8	3.3
2006	258676	71.1	19.3	2.9
2005	235997	70.8	19.8	2.6
2004	213456	69.5	21.3	2.5
2003	183792	69.8	21.2	2.5
2002	159431	68	22.3	2.4

① 本章资料来源于《中国统计年鉴 2010》以及中经网数据库。

续表

年　份	能源消费总量（万吨标准煤）	占能源消费总量的比重（%）		
		煤炭	石油	天然气
2001	150406	68.3	21.8	2.4
2000	145531	69.2	22.2	2.2

表 31－3　能源生产增长与 GDP 增长比较

Table 31－3　Increase in Energy Production vs. GDP Growth

单位:%

年　份	能源生产比上年增长	电力生产比上年增长	国内生产总值比上年增长
2010	8.7	13.3	110.3
2009	5.4	7.1	109.2
2008	5.4	5.6	109.6
2007	6.5	14.5	114.2
2006	7.4	14.6	112.7
2005	10	13.5	111.3
2004	14.4	15.3	110.1
2003	14.1	15.5	110
2002	4.7	11.7	109.1
2001	6.5	9.2	108.3
2000	2.4	9.4	108.4
1999	1.6	6.3	107.6
1998	-2.7	2.7	107.8
1997	0.3	5.1	109.3
1996	3.1	7.2	110
1995	8.7	8.6	110.9
1994	6.9	10.7	113.1

注：国内生产总值增长速度按可比价格计算。

表 31－4　能源消费增长与 GDP 增长比较

Table 31－4　Increase in Energy Consumption vs. GDP Growth

单位:%

年　份	能源消费比上年增长	电力消费比上年增长	国内生产总值比上年增长
2010	5.9	13.1	110.3
2009	5.2	7.2	109.2
2008	3.9	5.6	109.6
2007	8.4	14.4	114.2
2006	9.6	14.6	112.7
2005	10.6	13.5	111.3
2004	16.1	15.4	110.1

续表

年 份	能源消费比上年增长	电力消费比上年增长	国内生产总值比上年增长
2003	15. 3	15. 6	110
2002	6	11. 8	109. 1
2001	3. 3	9. 3	108. 3
2000	3. 5	9. 5	108. 4
1999	3. 2	6. 1	107. 6
1998	0. 2	2. 8	107. 8
1997	0. 5	4. 8	109. 3
1996	3. 1	7. 4	110
1995	6. 9	8. 2	110. 9
1994	5. 8	9. 9	113. 1

注：国内生产总值增长速度按可比价格计算，2001 年为估算数。

表 31 –5 能源生产与消费弹性系数

Table 31 –5 Elasticity Coefficient of Energy Production and Consumption

单位：%

年份	能源生产弹性系数	电力生产弹性系数	能源消费弹性系数	电力消费弹性系数
2010	0. 84	1. 29	0. 57	1. 27
2009	0. 59	0. 78	0. 57	0. 79
2008	0. 56	0. 58	0. 41	0. 58
2007	0. 46	1. 02	0. 59	1. 01
2006	0. 58	1. 15	0. 76	1. 15
2005	0. 88	1. 19	0. 93	1. 19
2004	1. 43	1. 51	1. 6	1. 52
2003	1. 41	1. 55	1. 53	1. 56
2002	0. 52	1. 29	0. 66	1. 3
2001	0. 79	1. 11	0. 4	1. 12
2000	0. 28	1. 12	0. 42	1. 13
1999	0. 21	0. 83	0. 42	0. 8
1998	—	0. 35	—	0. 36
1997	—	0. 55	—	0. 52
1996	0. 31	0. 72	0. 31	0. 74
1995	0. 8	0. 79	0. 63	0. 75
1994	0. 53	0. 82	0. 44	0. 76

注：最新年份为估算数。

表 31－6 平均每万元国内生产总值能源消费量

Table 31－6 Energy Intensity By GDP

年 份 Year	能源消费总量（吨标准煤/万元）	煤炭（吨/万元）	焦炭（吨/万元）	石油（吨/万元）	原油（吨/万元）	燃料油（吨/万元）	电力（万千瓦小时/万元）
国内生产总值按 1980 年可比价格计算							
1980	13. 20	13. 36	0. 94	1. 92	2. 02	0. 67	0. 66
1981	12. 37	12. 60	0. 82	1. 94	1. 82	0. 59	0. 64
1982	11. 84	12. 23	0. 76	1. 57	1. 66	0. 54	0. 63
1983	11. 36	11. 82	0. 71	1. 44	1. 56	0. 49	0. 61
1984	10. 59	11. 20	0. 66	1. 29	1. 38	0. 43	0. 56
1985	10. 10	10. 74	0. 62	1. 21	1. 25	0. 37	0. 54
1986	9. 78	10. 40	0. 63	1. 18	1. 24	0. 36	0. 55
1987	9. 39	10. 06	0. 62	1. 12	1. 16	0. 34	0. 54
1988	9. 06	9. 68	0. 59	1. 08	1. 09	0. 32	0. 53
1989	9. 07	9. 68	0. 60	1. 08	1. 09	0. 32	0. 55
1990	8. 90	9. 51	0. 62	1. 04	1. 06	0. 30	0. 56
国内生产总值按 1990 年可比价格计算							
1990	5. 32	5. 69	0. 37	0. 62	0. 63	0. 18	0. 34
1991	5. 12	5. 45	0. 35	0. 61	0. 61	0. 17	0. 34
1992	4. 72	4. 93	0. 34	0. 58	0. 57	0. 15	0. 33
1993	4. 40	4. 59	0. 33	0. 56	0. 52	0. 14	0. 32
1994	4. 12	4. 31	0. 31	0. 50	0. 47	0. 12	0. 31
1995	3. 97	4. 16	0. 32	0. 49	0. 45	0. 11	0. 30
1996	3. 69	3. 83	0. 32	0. 48	0. 43	0. 10	0. 29
1997	3. 40	3. 44	0. 27	0. 48	0. 43	0. 09	0. 28
1998	3. 16	3. 13	0. 27	0. 46	0. 40	0. 09	0. 27
1999	3. 03	3. 00	0. 23	0. 45	0. 41	0. 08	0. 26
2000	2. 89	2. 80	0. 22	0. 45	0. 42	0. 08	0. 27
国内生产总值按 2000 年可比价格计算							
2000	1. 47	1. 42	0. 11	0. 23	0. 21	0. 04	0. 14
2001	1. 40	1. 35	0. 11	0. 21	0. 20	0. 04	0. 14
2002	1. 36	1. 30	0. 11	0. 21	0. 19	0. 03	0. 14
2003	1. 43	1. 40	0. 12	0. 21	0. 19	0. 03	0. 15
2004	1. 50	1. 46	0. 13	0. 22	0. 20	0. 03	0. 15
2005	1. 49	1. 47	0. 16	0. 21	0. 19	0. 03	0. 16
国内生产总值按 2005 年可比价格计算							
2005	1. 28	1. 25	0. 14	0. 18	0. 16	0. 02	0. 13
2006	1. 24	1. 22	0. 13	0. 17	0. 15	0. 02	0. 14
2007	1. 18	1. 15	0. 12	0. 15	0. 14	0. 02	0. 14
2008	1. 12	1. 08	0. 11	0. 14	0. 14	0. 01	0. 13
2009	1. 08	1. 04	0. 11	0. 13	0. 13	0. 01	0. 13

表31－7 能源加工转换效率

Table 31－7 Energy Transfer Efficiency

单位:%

年 份 Year	总效率	发电及电站供热	炼焦	炼油
1980	69.54	36.02	88.68	99.00
1981	69.28	36.68	90.89	99.06
1982	69.20	36.78	90.51	99.13
1983	69.93	36.94	91.18	99.16
1984	69.16	36.95	90.08	99.17
1985	68.29	36.85	90.79	99.10
1986	68.32	36.69	90.63	99.04
1987	67.48	36.75	90.46	98.81
1988	66.54	36.34	90.77	98.76
1989	66.51	36.74	90.30	98.57
1990	66.48	37.34	91.28	90.19
1991	65.90	37.60	89.90	98.10
1992	66.00	37.80	92.70	96.80
1993	67.32	39.90	98.05	98.49
1994	65.20	39.35	89.62	97.48
1995	71.05	37.31	91.99	97.67
1996	70.19	36.63	94.07	97.46
1997	69.76	35.89	94.01	97.37
1998	69.28	37.09	94.97	96.41
1999	69.25	37.04	96.13	97.51
2000	69.04	37.36	96.21	97.32
2001	69.34	37.63	96.48	97.92
2002	69.04	38.73	96.63	96.71
2003	69.40	38.83	96.13	96.80
2004	70.91	39.46	97.55	96.43
2005	71.55	39.87	97.57	96.86
2006	71.24	39.87	97.77	96.86
2007	70.77	40.24	97.56	97.17
2008	71.55	41.04	97.75	97.17
2009	72.01	41.73	97.38	96.63

表 31－8　主要高耗能产品的进、出口量

Table 31－8　Import and Export Volumes of Major Energy－intensive Products

指标	2000	2001	2002	2003	2004	2005	2006	2007	2008	2009
进口量										
钢材（万吨）	1596	1722	2449	3717	2930	2582	1851	1687	1543	1763
钢铁丝（吨）	336300	353771	427554	465540						
铜及铜合金（万吨）	81	95	133	156	138	142	97	173	170	347
铝及铝合金（万吨）	91	53	58	88	103	64	51	28	26	174
锌及锌合金（吨）	129974	141159	211722	310221						
烧碱（吨）	46458	27357	114834	104686						
纯碱（万吨）	13	7	29	30	20	7	14	4		
肥料（万吨）	1189	1092	1682	1213	1240	1397	1129	1169	622	411
纸浆（万吨）	335	490	526	603	732	759	796	847	952	1368
纺织用合成纤维（万吨）	100	92	104	106	99	84	62	51	32	35
出口量										
水泥（万吨）	605	621	518	533	704	2216	3613	3301	2604	1561
平板玻璃（万平方米）	5592	6123	11359	12427	14464	19925	26433	30917	27762	16643
钢材（万吨）	621	474	545	696	1423	2052	4301	6265	5923	2460
钢铁丝（吨）	190122	224484	310992	401235						
铜材（吨）	144484	123790	171710	232879	390023	463560	559122	499678	517522	455136
铝材（吨）	130052	135630	188744	273874	430988	711484	1240157	1853413	1900000	1390000
锌及锌合金（吨）	593336	562021	495987	484231	263149	146845	341465	276714	71320	29287
纸及 纸板（万吨）	65	68	74	114	101	167	305	422	361	362

表 31－9　高耗能产品产量

Table 31－9　Production Volume of Energy－intensive Products

年份	1995	2004	2005	2006	2007	2008	2009
钢材（万吨）	9536	31976	37771	46893	56561	58488	69626
水泥（万吨）	47561	96682	106885	123676	136117	140000	165000
平板玻璃（万重量箱）	15732	37026	40210	46575	53918	55185	56073
合成氨（万吨）	2766	4222	4596	4937	5171	4995	5136
机制纸及纸板（万吨）	2812	5413	6205	6863	7792	8391	——
烧碱（万吨）	532	1041	1240	1512	1759	1852	1832
纯碱（万吨）	598	1335	1421	1560	1765	1881	1938
乙烯（万吨）	240	630	756	941	1028	998	1066
焦炭（万吨）	13510	17749	25412	29768	33553	32359	34502
汽油（万吨）	3052	5250	5409	5595	5989	6348	7195
柴油（万吨）	3973	10162	11079	11656	12359	13324	14127
化肥（万吨）	2548	4805	5178	5345	5825	6013	6600
氧化铝（万吨）	——	699	859	1326	1947	2278	2379
精炼铜（万吨）	——	217	261	300	344	379	414

表31-10 工业分行业终端能源消费量(实物量)-2009

Table 35-10 Final Energy Consumption By Industrial Sector-2009 (Physical Quantity)

行业	煤合计	原煤	洗精煤	其他洗煤	焦炭	焦炉煤气	其他煤气	其他焦化产品	油品合计	原油	汽油	煤油	柴油	燃料油	液化石油气	炼厂干气	其他石油制品	天然气	热力	电力
(亿千瓦小时)		(万吨)	(万吨)	(万吨)	(万吨)	(万吨)	(亿立方米)	(亿立方米)	(万吨)	(万吨)	(万吨)	(万吨)	(万吨)	(万吨)	(万吨)	(万吨)	(万吨)	(万吨)	(亿立方米)	(万百万千焦)
	(104tn)	(104tn)	(104tn)	(104tn)	(104tn)	(108cu. m)	(108cu. m)	(104tn)	(104tn)	(104tn)	(104tn)	(104tn)	(104tn)	(104tn)	(104tn)	(104tn)	(104tn)	(108cu. m)	(1010KJ)	(108KWh)
工业	67755.90	58440.87	2916.87	5985.66	31583.02	467.24	2114.20	807.82	13153.74	826.16	670.98	32.04	2196.99	1227.48	477.41	932.69	6790.00	402.71	183874.13	24596.27
(一)采掘业	7299.19	6673.06	256.93	360.70	138.61	1.84	10.00	2.87	1158.88	486.91	65.44	4.64	498.17	32.47	0.29	33.60	37.37	91.18	4128.99	1761.77
煤炭开采和洗选业	6443.46	5854.58	233.35	351.02	27.83	1.07	8.46	2.87	139.70		21.19	2.90	108.40	4.83	0.05		2.34	1.86	374.91	691.13
石油和天然气开采业	141.45	141.45					0.01		837.26	486.91	25.04	0.05	230.21	27.02	0.20	33.60	34.23	89.07	1933.74	333.34
黑色金属矿采选业	166.01	151.96	6.42	3.63	83.82	0.77	1.54		59.26		6.46	0.45	51.92	0.32			0.10	0.03	0.20	289.73
有色金属矿采选业	94.26	83.85	9.16	1.25	14.89				27.58		7.64	0.87	18.85	0.04	0.02		0.16	0.04	1.33	215.72
非金属矿采选业	451.75	438.95	8.00	4.80	12.07				93.91		5.08	0.37	87.64	0.26	0.02		0.54	0.18	1818.81	155.26
其他采矿业	2.26	2.26							1.17		0.03		1.14					0.01		76.59
(二)制造业	57268.68	48676.66	2651.08	5542.18	31433.91	457.39	2096.44	790.81	11800.99	338.54	571.16	27.28	1590.50	1179.51	447.63	897.24	6749.13	308.71	173909.46	19685.98
农副食品加工业	1364.57	1271.23	7.16	81.59	10.56	0.03			108.74	0.10	32.98	0.22	51.83	12.68	2.73	7.24	0.96	0.65	3503.40	389.32
食品制造业	848.49	834.44	8.41	5.64	7.13	0.02	0.45		61.45		14.16	0.11	30.22	13.76	3.05		0.15	2.31	4851.24	166.54
饮料制造业	748.33	735.29	7.00	6.05	0.94	0.00			38.71	0.20	10.39	0.29	15.73	11.06	0.94		0.10	1.43	3159.97	117.03
烟草制品业	83.94	79.56	1.26	3.12		0.03	0.12		6.37		0.75		4.71	0.90	0.01		0.00	0.50	349.06	41.70

续表

行业	煤合计	原煤	洗精煤	其他洗煤	焦炭	焦炉煤气	其他煤气	其他焦化产品	油品合计	原油	汽油	煤油	柴油	燃料油	液化石油气	炼厂干气	其他石油制品	天然气	热力	电力
	(万吨)	(万吨)	(万吨)	(万吨)	(万吨)	(亿立方米)	(亿立方米)	(万吨)	(万吨)	(万吨)	(万吨)	(万吨)	(万吨)	(万吨)	(万吨)	(万吨)	(万吨)	(亿立方米)	(万百万千焦)	(亿千瓦小时)
	(104tn)	(104tn)	(104tn)	(104tn)	(104tn)	(108cu. m)	(108cu. m)	(104tn)	(104tn)	(104tn)	(104tn)	(104tn)	(104tn)	(104tn)	(104tn)	(104tn)	(104tn)	(108cu. m)	(1010KJ)	(108KWh)
纺织业	1842. 81	1808. 42	12. 73	14. 16	4. 10	1. 27	0. 01		99. 64	0. 20	26. 25	0. 43	43. 86	23. 96	3. 37	0. 25	1. 32	1. 35	18957. 39	1147. 50
纺织服装、鞋、帽制造业	204. 97	200. 90	2. 38	1. 69	1. 40	0. 01	0. 07		59. 23	0. 33	17. 18	0. 28	33. 62	6. 63	1. 12		0. 06	0. 24	625. 08	132. 54
皮革、毛皮、羽毛(绒)及其制品业	85. 73	84. 82	0. 36	0. 55	0. 31				34. 39	0. 03	9. 22	0. 29	16. 05	8. 43	0. 25		0. 12	0. 07	222. 13	78. 45
木材加工及木、竹、藤、棕、草制品业	424. 30	414. 67	6. 97	2. 66	2. 46				24. 50	0. 28	7. 94	0. 20	15. 23	0. 50	0. 29		0. 06	0. 34	520. 91	191. 53
家具制造业	29. 74	29. 35	0. 08	0. 30	1. 01		0. 01		21. 86		7. 84	0. 13	12. 62	0. 26	1. 00		0. 01	0. 44	118. 03	35. 59
造纸及纸制品业	2300. 21	2212. 88	25. 83	57. 00	4. 35	0. 12		1. 26	66. 82	0. 36	12. 83	0. 41	30. 54	18. 75	2. 58		1. 35	1. 06	12750. 56	482. 73
印刷业和记录媒介的复制	39. 19	38. 95	0. 08	0. 16	0. 56	0. 01	0. 01		25. 81	0. 00	8. 77	0. 23	13. 74	1. 44	1. 56		0. 07	0. 52	228. 60	83. 01
文教体育用品制造业	17. 63	17. 39	0. 03	0. 21	5. 06				24. 54	0. 04	4. 43	0. 12	15. 90	2. 14	1. 65		0. 26	0. 11	34. 92	48. 09
石油加工、炼焦及核燃料加工业	1421. 51	903. 92	441. 70	75. 90	97. 16	45. 92	38. 94	106. 14	5810. 25	142. 73	39. 81	1. 24	68. 59	201. 96	140. 08	816. 38	4399. 46	20. 57	26650. 20	475. 07
化学原料及化学制品制造业	10197. 35	9342. 07	433. 16	207. 71	2103. 77	14. 93	17. 91	299. 78	2849. 70	173. 47	49. 80	3. 69	200. 48	195. 33	75. 89	69. 86	2081. 18	170. 84	52480. 44	2907. 12

续表

行业	煤合计	原煤	洗精煤	其他洗煤	焦炭	焦炉煤气	其他煤气	其他焦化产品	油品合计	原油	汽油	煤油	柴油	燃料油	液化石油气	炼厂干气	其他石油制品	天然气	热力	电力
	（万吨）	（万吨）	（万吨）	（万吨）	（万吨）	（亿立方米）	（亿立方米）	（万吨）	（万吨）	（万吨）	（万吨）	（万吨）	（万吨）	（万吨）	（万吨）	（万吨）	（万吨）	（亿立方米）	（万百万千焦）	（亿千瓦小时）
	(104 tn)	(104 tn)	(104 tn)	(104 tn)	(104 tn)	(108 cu. m)	(108 cu. m)	(104 tn)	(104 tn)	(104 tn)	(104 tn)	(104 tn)	(104 tn)	(104 tn)	(104 tn)	(104 tn)	(104 tn)	(108 cu. m)	(1010KJ)	(108kW・h)
医药制造业	546.22	526.41	15.17	4.64	1.24	0.31	0.05		37.11	0.03	12.07	0.07	17.15	5.39	1.82	0.12	0.46	2.35	4707.71	188.99
化学纤维制造业	262.51	259.95	2.56		5.54			2.73	56.16	10.01	1.61	0.11	9.56	18.42	0.30	0.71	15.44	0.29	5496.44	269.93
橡胶制品业	408.30	399.37	3.44	5.50	3.25	0.28	0.03		32.96	0.40	11.06	0.08	7.72	11.90	0.46		1.34	0.69	1079.67	280.28
塑料制品业	292.10	284.49	3.92	3.69	3.07				85.07	0.02	18.24	0.23	46.65	14.51	2.05		3.37	1.36	465.12	460.29
非金属矿物制品业	23504.71	18175.08	492.71	4786.38	384.65	10.82	27.37	98.43	954.09	8.89	36.95	1.23	317.07	411.12	95.30	2.46	81.07	44.62	1072.30	2126.16
黑色金属冶炼及压延加工业	8747.65	7436.51	1041.13	193.44	27083.17	371.58	1881.71	273.41	224.06	0.04	15.52	1.39	111.72	58.61	28.58		8.20	18.76	20536.69	4020.52
有色金属冶炼及压延加工业	1432.15	1282.01	60.03	64.00	648.57	8.85	86.57	6.18	303.61	0.61	10.14	1.65	64.53	77.31	12.41		136.96	6.69	7767.77	2576.15
金属制品业	327.06	312.57	7.64	6.85	98.22	0.17	0.26		137.57	0.18	31.13	1.64	73.24	14.67	15.20	0.01	1.50	2.47	480.20	744.40
通用设备制造业	395.49	372.06	9.03	5.33	663.64	1.17	16.91	2.48	147.37	0.06	50.48	3.88	68.25	11.50	9.38	0.01	3.81	6.06	786.79	502.04
专用设备制造业	488.76	446.33	39.89	2.54	100.74	1.21	23.14	0.40	83.52	0.05	28.30	0.53	42.81	5.62	4.51		1.70	4.56	1780.65	256.94
交通运输设备制造业	605.44	590.99	10.30	4.14	163.67	0.32	1.00		186.00	0.10	42.74	7.16	105.60	12.75	14.43	0.02	3.20	12.12	2464.29	565.40
电气机械及器材制造业	169.75	161.44	7.12	1.19	23.21	0.11	0.39		139.66	0.12	35.60	0.50	68.55	9.63	19.88	0.18	5.20	3.03	1461.99	423.49
通信设备、计算机及其他电子设备制造业	139.96	137.96	1.68	0.32	1.74	0.06	0.14		121.32	0.25	19.25	0.22	69.74	27.34	4.40		0.12	4.88	865.36	558.96

续表

行业	煤合计	原煤	洗精煤	其他洗煤	焦炭	焦炉煤气	其他煤气	其他焦化产品	油品合计	原油	汽油	煤油	柴油	燃料油	液化石油气	炼厂干气	其他石油制品	天然气	热力	电力
	（万吨）	（万吨）	（万吨）	（万吨）	（万吨）	（亿立方米）	（亿立方米）	（万吨）	（万吨）	（万吨）	（万吨）	（万吨）	（万吨）	（万吨）	（万吨）	（万吨）	（万吨）	（亿立方米）	（万百万千焦）	（亿千瓦小时）
	(104 tn)	(104 tn)	(104 tn)	(104 tn)	(104 tn)	(108 cu. m)	(108 cu. m)	(104 tn)	(104 tn)	(104 tn)	(104 tn)	(104 tn)	(104 tn)	(104 tn)	(104 tn)	(104 tn)	(104 tn)	(108 cu. m)	(1010KJ)	(108kW · h)
仪器仪表及文化、办公用机械制造业	25.16	23.79		1.37	8.32		0.01		20.89	0.01	6.85	0.69	12.56	0.18	0.26		0.34	0.36	123.63	68.54
工艺品及其他制造业	304.28	286.01	9.29	3.51	1.81		1.18		34.04	0.02	8.18	0.23	18.15	2.28	4.01		1.17	0.06	363.60	335.49
废弃资源和废旧材料回收加工业	10.36	7.82		2.54	4.25	0.18	0.16		5.55		0.69	0.03	4.08	0.48	0.12		0.15		5.33	12.15
（三）电力、煤气及水生产和供应业	3188.03	3091.16	8.86	82.78	10.49	8.01	7.75	14.14	193.87	0.71	34.38	0.12	108.33	15.50	29.49	1.84	3.50	2.81	5835.68	3148.52
电力、热力的生产和供应业	3080.74	2984.31	8.60	82.60	8.13	0.04	3.13	12.50	139.04	0.50	26.46	0.12	93.92	14.49	0.03	0.20	3.32	0.47	5514.06	2821.76
燃气生产和供应业	80.07	79.66	0.25	0.16	2.27	7.97	4.62	1.64	45.07	0.21	2.88		10.21	0.51	29.44	1.64	0.18	2.22	283.94	70.23
水的生产和供应业	27.22	27.19	0.01	0.02	0.09				9.76	0.00	5.04		4.20	0.50	0.02			0.12	37.68	256.53

表31－11　分行业能源消费总量

Table 31－11　Consumption of Total Energy and Its Main Varieties By Sector

单位：万吨标准煤

行业	1995年	2000年	2005年	2006年	2007年	2008年	2009年
消费总量	131175.50	145530.86	235996.65	258676.30	280507.94	291448.29	306647.15
农、林、牧、渔、水利业	5505.20	3913.77	6071.06	6330.71	6228.40	6013.13	6251.18
工业	96191.30	103773.85	168723.53	184945.45	200531.38	209302.15	219197.16
采掘业	9941.00	10968.06	13914.67	14247.13	15240.74	17050.44	17585.17
煤炭开采和洗选业	5499.80	5286.08	7522.45	7672.90	8269.88	9356.17	10206.55
石油和天然气开采业	2812.60	3963.58	3710.92	3586.79	3651.26	4210.04	3945.90
黑色金属矿采选业	268.20	374.66	971.13	1138.05	1330.32	1408.03	1250.90
有色金属矿采选业	557.20	425.26	680.67	738.02	828.89	863.14	832.89
非金属矿采选业	553.40	690.46	927.84	978.21	1035.12	1028.27	1095.77
其他采矿业	249.80	228.02	101.68	133.17	125.27	184.79	253.16
制造业	78368.20	80771.97	137140.40	151274.67	164951.10	172106.52	180595.97
农副食品加工业	1972.50	1717.36	2207.28	2360.77	2537.65	2731.34	2795.37
食品制造业	1208.00	1055.55	1281.74	1389.67	1448.62	1544.66	1563.34
饮料制造业	1000.30	800.45	992.06	1081.60	1098.77	1161.85	1191.44
烟草制品业	223.80	303.98	257.08	255.38	249.03	232.60	233.80
纺织业	3531.30	3014.37	5281.32	6108.59	6528.30	6396.38	6251.01
纺织服装、鞋、帽制造业	329.20	348.43	577.42	661.68	711.44	725.34	713.08
皮革、毛皮、羽毛（绒）及其制品业	289.90	203.86	323.71	364.15	387.67	388.73	384.48
木材加工及木、竹、藤、棕、草制品业	380.00	360.66	754.77	843.10	893.67	981.91	1049.09
家具制造业	105.80	101.14	133.12	149.32	152.86	181.80	183.81
造纸及纸制品业	2138.40	2269.30	3574.91	3791.56	3643.36	3998.65	4101.00

续表

行业	1995 年	2000 年	2005 年	2006 年	2007 年	2008 年	2009 年
印刷业和记录媒介的复制	203.40	204.55	280.70	303.62	327.08	349.81	357.48
文教体育用品制造业	62.00	122.75	198.37	202.85	210.09	219.76	214.62
石油加工、炼焦及核燃料加工业	5567.30	7220.12	11923.57	12498.72	13445.30	13747.01	15328.29
化学原料及化学制品制造业	15821.60	14326.13	23848.69	25995.36	28621.16	28961.13	28946.07
医药制造业	1201.30	941.62	1203.58	1255.98	1261.41	1360.49	1354.58
化学纤维制造业	1278.00	1883.87	1382.82	1465.05	1575.02	1448.58	1436.85
橡胶制品业	644.10	687.98	1137.05	1246.95	1305.16	1335.83	1344.72
塑料制品业	541.90	696.39	1486.34	1604.91	1652.49	1852.37	1894.96
非金属矿物制品业	13058.00	13768.08	21310.46	22637.64	23111.66	25460.52	26882.28
黑色金属冶炼及压延加工业	18532.80	18962.27	39544.25	44729.92	50186.53	51862.92	56404.37
有色金属冶炼及压延加工业	2841.70	4079.30	7403.80	8861.83	10867.61	11287.99	11401.37
金属制品业	993.90	1210.42	2271.08	2632.41	2852.64	3023.79	3037.78
通用设备制造业	1650.50	1251.01	2149.71	2406.23	2649.23	2758.11	2985.24
专用设备制造业	1089.30	866.58	1314.71	1438.19	1579.44	1630.28	1671.52
交通运输设备制造业	1376.30	1502.38	2043.13	2247.85	2467.64	2732.58	3031.90
电气机械及器材制造业	629.20	648.35	1213.20	1353.04	1557.01	1791.10	1854.48
通信设备、计算机及其他电子设备制造业	321.40	692.89	1482.64	1754.05	1994.99	2197.44	2216.28
仪器仪表及文化、办公用机械制造业	142.60	156.70	197.45	232.82	260.90	284.98	291.92

续表

行业	1995 年	2000 年	2005 年	2006 年	2007 年	2008 年	2009 年
工艺品及其他制造业	1233.70	1375.45	1330.45	1349.93	1323.74	1401.72	1413.73
废弃资源和废旧材料回收加工业			34.98	51.49	50.64	56.84	61.11
电力、煤气及水生产和供应业	7882.70	12033.83	17668.45	19423.65	20339.53	20145.19	21016.02
电力、热力的生产和供应业	7052.70	10812.32	16326.50	18004.12	18892.27	18676.48	19574.86
燃气生产和供应业	341.30	605.09	643.19	663.11	650.79	634.60	565.25
水的生产和供应业	488.70	616.42	698.76	756.42	796.48	834.11	874.91
建筑业	1334.50	2178.53	3403.31	3760.73	4127.52	3812.53	4562.02
交通运输、仓储和邮政业	5862.90	11241.59	18391.01	20284.23	21959.18	22917.25	23691.84
批发、零售业和住宿、餐饮业	2017.80	3047.63	4847.75	5314.05	5689.38	5733.58	6412.26
其他行业	4519.00	5761.58	9254.56	10275.98	11158.19	11771.34	12689.81
生活消费	15744.80	15613.91	25305.43	27765.16	30813.90	31898.32	33842.88

表 31-12　主要高耗能产品单位能耗中外比较

Table 31-12　Comparison between China and Other Countries in Unit Energy Consumption of Major Energy-intensive Products

火电厂发电煤耗

Gross Coal Consumption Rate for Fossil-fired Power Plant

单位：克/千瓦小时

国家	Country	1990 年	1995 年	2000 年	2005 年	2006 年	2007 年	2008 年
中国①	China	392	379	363	343	342	332	322
日本②	Japan	317	315	303	301	299	300	292
德国	Germany	309	322	309	301	306		
韩国	Korea	332	322	311	302	300		

注：① 6MW 以上机组；

② 九大电力公司平均；

③ 2006 年发电能源结构。

国家	Country	电源结构（Power Generation by Source）				
		煤（Coal）	油（Oil）	气（Gas）	核（Nuclear）	水力（Hydro）
中国	China	81. 33	1. 46	4. 51	1. 92	14. 55
日本	Japan	26. 03	10. 58	24. 1	27. 76	8. 01
德国	Germany	48. 03	1. 52	12. 09	26. 98	3. 17
韩国	Korea	38. 01	5. 91	18. 06	36. 98	0. 86

资料来源：1. 国家统计局；

2. 中国电力企业联合会；

3. The Insititute of Energy Economics, Japan, Handbook of Energy and Economic Statistics in Japan, 2010, Edition;

4. International Energy Agency, Energy Statistics of OECD Countries.

火电厂供电煤耗

Net Coal Consumption Rate for Fossil – fired Power Plant

单位：克/千瓦小时

国家	Country	1990 年	1995 年	2000 年	2005 年	2006 年	2007 年	2008 年
中国	China	427	412	392	370	367	356	345
日本	Japan	332	331	316	314	312	312	310

发电厂自用率

Power Consumption Rate For Plant Use

单位：%

国家	Country	1990 年	1995 年	2000 年	2005 年	2006 年	2007 年
中国	China	6. 90	6. 78	6. 28	5. 87	5. 93	5. 83
日本	Japan	5. 82	5. 77	3. 71	3. 81	3. 86	3. 80
加拿大	Canada	2. 99	3. 08	3. 14	3. 19	3. 10	3. 31
韩国	Korea	4. 84	4. 08	4. 56	4. 26	3. 91	3. 96

注：同表 1。

资料来源：同表 1。

钢可比能耗

Comparable Energy Consumption for Steel

单位：千克标准煤/吨

国家	Country	1990 年	1995 年	2000 年	2005 年	2006 年	2007 年	2008 年
中国①	China	997	976	784	732	729	718	709
日本	Japan	629	656	646	640②	627②	610②	626②

注：①大中型钢铁企业平均值；

②根据钢产量和能源消费量估算。

资料来源：1. 中国钢铁工业协会；

2. 日本能源学会志。

水泥综合能耗

Fully Energy Consumption for Cement

单位：千克标准煤/吨

国家	Country	1990年	1995年	2000年	2005年	2006年	2007年	2008年
中国	China	201	199	181	167	161	158	151
日本	Japan	123	124	126	127		118	

注：综合能耗中的电耗，均按发电煤耗折算标准煤，日本为320gce/KWh。

资料来源：1. 国家统计局；

2. 工业和信息化部；

3. 日本能源学会志；

4. 日本水泥协会。

乙烯综合能耗

Fully Energy Consumption for Ethylene

单位：千克标准煤/吨

国家	Country	1990年	2000年	2005年	2006年	2007年	2008年
中国	China①	1580	1125	1073	1013	1026	1003
国外	Overseas	897①	714①	629②	629②	629②	629②

注：①主要用石油脑油作原料；

②中东地区平均值，主要用乙烷作原料；

③综合能耗中，电耗按发电煤耗折算标准煤。

资料来源：1. 国家统计局；

2. 中国化工节能技术协会。

合成氨综合能耗（大型装置）

Fully Energy Consumption for Sythetic Ammonia（Large Sized Plant）

单位：千克标准煤/吨

国家	Country	1990	1995	2000	2005
中国	China	1343	1347	1327	1340
美国	United States	1000	1000	970	970

资料来源：同表6。

纸和纸板综合能耗

Full Energy Consumption for Paper and Paperboard

单位：千克标准煤/吨

国家	Country	1990年	2000年	2005年	2006年	2007年	2008年
中国①	China	1550	1540	1380	1290	1255	1153
日本	Japan	744	678	640	627②	610②	626②

注：①自制浆企业平均；

②根据纸和纸板产量与能源消费量估算。

资料来源：1. 国家统计局；

2. 中国造纸协会；

3. 日本能源学会志；

4. The Institute of Energy Economics，Japan，Handbook of Energy and Economic Statistics in Japan，2010 Edition。

铁路货运综合能耗

Fully Energy Consumption for Railway Freight Traffic

单位：千克标准煤/万吨公里

国家	Country	1990 年	1995 年	2000 年	2005 年	2006 年	2007 年
中国	China	175. 4	134. 4	114. 5	97. 3	67. 5	64. 6
日本	Japan	84. 3	87. 1	85. 7	85. 7	82. 9	80. 0

资料来源：1. 国家发展和改革委员会综合运输研究所；

2. The Institute of Energy Economics, Japan, Handbook of Energy and Economic Statistics in Japan, 2010 Edition。

32　国内环境数据①

表 32－1　工业废水排放总量年度统计

Table 32－1　Yearly Statistics of Total Industrial Waste Water Discharge

单位：万吨

	2005 年	2006 年	2007 年	2008 年	2009 年
全　国	2431121	2401946	2466493. 307	2416511	2343856. 657
北京市	12813	10170	9134. 2507	8367	8712. 5326
天津市	30081	22978	21444. 3499	20433	19441. 0525
河北省	124533	130340	123537. 2629	121172	110057. 9976
山西省	32099	44091	41140. 3094	41150	39720. 212
内蒙古自治区	24967	27823	25020. 8419	29167	28616. 2246
辽宁省	105072	94724	95196. 7085	83073	75158. 5888
吉林省	41189	39321	39666. 3339	38353	37563. 4834
黑龙江省	45158	44801	38388. 2992	38910	34188. 2224
上海市	51097	48336	47569. 5309	41871	41192. 0286
江苏省	296318	287181	268761. 9951	259999	256159. 9698
浙江省	192426	199593	201210. 5518	200488	203441. 7094
安徽省	63487	70119	73555. 5526	67007	73441. 473
福建省	130939	127583	136407. 7725	139997	142746. 9854
江西省	53972	64074	71410. 3125	68681	67192. 4452
山东省	139071	144365	166573. 5535	176977	182672. 6358
河南省	123476	130158	134344. 317	133144	140324. 5647
湖北省	92432	91146	91001. 1289	93687	91324. 0928
湖南省	122440	100024	100112. 7053	92340	96395. 6901
广东省	231568	234713	246331. 0362	213314	188843. 8868
广西壮族自治区	145609	128932	183980. 8581	205745	161596. 4313
海南省	7428	7351	5960. 123	5991	7031. 2967
重庆市	84885	86496	69002. 8275	67027	65683. 845
四川省	122590	115348	114687. 1879	108700	105909. 5519
贵州省	14850	13928	12101. 0845	11695	13477. 6223
云南省	32928	34286	35352. 2824	32996	32375. 2054
西藏自治区	991	790	856. 4773	924	941. 5523
陕西省	42819	40479	48522. 7576	48477	49136. 7627
甘肃省	16798	16570	15856. 3141	16405	16363. 6072

① 本章资料来源于国家统计局数据库。

续表

	2005 年	2006 年	2007 年	2008 年	2009 年
青海省	7619	7168	7318. 0215	7098	8403. 9096
宁夏回族自治区	21411	18500	21089. 0412	20448	21542. 3672
新疆维吾尔自治区	20052	20558	20959. 5192	22875	24200. 7094

表 32 – 2　生活污水排放量年度统计

Table 32 – 2　Yearly Statistics of Domestic Sewage Discharge

单位：万吨

	2005 年	2006 年	2007 年	2008 年	2009 年
全　国	2813968	2966341	3102000. 851	3300290	3547020. 593
北京市	88196	94824	98682. 4439	104892	132100. 3511
天津市	30280	35909	35483. 8064	40796	40206. 3231
河北省	83991	91922	99377. 0907	113525	134930. 5165
山西省	62997	58765	63454. 0832	65761	66154. 837
内蒙古自治区	31274	33686	35384. 3846	41254	44538. 5134
辽宁省	113633	118229	125800. 0221	128948	141996. 089
吉林省	56816	57845	58191. 4933	69428	72151. 2578
黑龙江省	68883	70857	70584. 2865	72086	76319. 9385
上海市	148613	175419	179044. 545	181880	189325. 5
江苏省	223107	228368	236836. 3312	249702	266169. 3175
浙江省	120770	131101	136890. 3698	149889	161575. 4186
安徽省	93104	96352	101771. 5388	101663	106259. 7374
福建省	81453	88441	90590. 0946	96272	103265. 6439
江西省	69348	70444	69856. 38	70228	79888. 0938
山东省	141306	158272	167681. 181	181934	204058. 4411
河南省	139088	147864	162122. 9767	176049	193655. 5934
湖北省	144936	148524	155581. 4349	165187	174432. 655
湖南省	133198	144110	151960. 2771	157991	163882. 7152
广东省	406835	419706	444556. 0055	464038	498585. 2991
广西壮族自治区	125248	130789	135827. 4161	139610	143910. 661
海南省	27846	28006	29198. 5089	30197	30486. 4425
重庆市	60336	64117	65237. 5638	78086	81384. 9087
四川省	139061	137027	138274. 5969	153643	156799. 0988
贵州省	40818	41454	43011. 3394	44171	45681. 5508
云南省	42274	46192	48406. 6621	50869	55215. 4302
西藏自治区	3564	2462	2479. 08	2496	2513. 68
陕西省	40549	46086	50825. 3981	56406	62081. 98
甘肃省	26930	29151	28478. 951	31065	32906. 5562
青海省	11741	12239	12630. 47	12899	13767. 0241
宁夏回族自治区	14406	13296	16124. 3349	17500	19794. 0962
新疆维吾尔自治区	43367	44884	47657. 7841	51825	52982. 9228

表 32－3　2009 年各地区废水排放及处理情况

Table 32－3　Waste Water Discharge and Treatment Status of Different Regions in 2009

指标	废水治理设施数(套)	工业废水排放总量(万吨)	工业废水排放达标量(万吨)	工业废水中化学需氧量排放量(万吨)	工业废水中氨氮排放量(万吨)	生活污水排放量(万吨)	生活污水中化学需氧量排放量(万吨)	生活污水中氨氮排放量(万吨)	直接排入海的工业废水量(万吨)
全　国	77018	2343856. 6565	134695. 2439	2208743. 3185	439. 678047	27. 354439	3547020. 5927	837. 861328	95. 255193
北　京	524	8712. 5326	0	8574. 1519	0. 489811	0. 045279	132100. 3511	9. 394812	1. 257705
天　津	848	19441. 0525	583. 6323	19440. 139	2. 346936	0. 291549	40206. 3231	10. 953064	0. 906484
河　北	3869	110057. 9976	1162. 1422	108165. 824	23. 0355	1. 72428	134930. 5165	33. 973578	3. 787466
山　西	2548	39720. 212	0	32693. 932	14. 194064	1. 148631	66154. 837	20. 24741	2. 924928
内蒙古	889	28616. 2246	0	24366. 4156	12. 006058	0. 465193	44538. 5134	15. 845009	2. 92102
辽　宁	1798	75158. 5888	24434. 5172	64592. 957	21. 626013	0. 961086	141996. 089	34. 636928	5. 287914
吉　林	638	37563. 4834	0	30620. 7432	14. 716753	0. 262042	72151. 2578	21. 363259	2. 599295
黑龙江	1465	34188. 2224	0	31378. 5485	11. 189145	0. 675634	76319. 9385	35. 00807	4. 078184
上　海	1730	41192. 0286	2535. 8839	40686. 6005	2. 903062	0. 198284	189325. 5	21. 439146	2. 781132
江　苏	6877	256159. 9698	1622. 7281	251289. 552	25. 125266	1. 376535	266169. 3175	57. 04358	5. 156565
浙　江	8202	203441. 7094	11641. 0364	193847. 2486	24. 045192	1. 517144	161575. 4186	27. 334978	2. 581535
安　徽	1987	73441. 473	0. 001	70657. 3487	12. 880583	1. 442998	106259. 7374	29. 529998	3. 236641
福　建	3949	142746. 9854	73810. 8576	141032. 4967	7. 542507	0. 622823	103265. 6439	30. 028126	2. 385174
江　西	1826	67192. 4452	0	63047. 2611	10. 353645	0. 729454	79888. 0938	33. 16709	2. 677908
山　东	4824	182672. 6358	7495. 0039	180029. 8728	26. 052138	1. 38982	204058. 4411	38. 64788	5. 340658
河　南	3210	140324. 5647	0	134849. 5983	29. 765691	2. 571599	193655. 5934	32. 856097	4. 95246
湖　北	2068	91324. 0928	0	87594. 1381	14. 374622	1. 507828	174432. 65505	43. 195276	4. 95282
湖　南	3195	96395. 6901	0	88059. 3973	21. 558276	2. 396441	163882. 7152	63. 277769	6. 001994
广　东	9826	188843. 8868	7549. 0387	174376. 9592	21. 680817	1. 002941	498585. 2991	69. 442413	10. 511466
广　西	2539	161596. 4313	1597. 0146	153458. 3477	51. 881736	1. 397711	143910. 661	45. 748001	3. 406562
海　南	267	7031. 2967	2263. 388	6789. 4926	1. 155277	0. 055692	30486. 4425	8. 872428	0. 757429
重　庆	1638	65683. 845	0	61925. 1745	10. 029027	0. 725145	81384. 9087	13. 950681	1. 952718
四　川	4377	105909. 5519	0	101028. 7924	24. 514576	1. 348246	156799. 0988	50. 254051	4. 602404
贵　州	2050	13477. 6223	0	9569. 6254	1. 29882	0. 09327	45681. 5508	20. 297522	1. 616287
云　南	2088	32375. 2054	0	29990. 844	8. 531217	0. 321247	55215. 43015	18. 777053	1. 581091
西　藏	16	941. 5523		210. 3624	0. 073685	0. 000536	2513. 68	1. 466315	0. 146631
陕　西	1728	49136. 7627	0	47522. 8991	12. 63725	0. 720232	62081. 98	19. 169819	2. 469962
甘　肃	657	16363. 6072	0	13265. 6582	4. 867067	1. 186008	32906. 5562	11. 942651	1. 482184
青　海	155	8403. 9096		4691. 794	3. 925268	0. 170566	13767. 0241	3. 688018	0. 540177
宁　夏	353	21542. 3672	0	18834. 8363	9. 733382	0. 422782	19794. 0962	2. 784406	0. 388344
新　疆	877	24200. 7094	0	16152. 3074	15. 144663	0. 583442	52982. 9228	13. 525899	1. 970055

表 32－4　2009 年主要城市工业废水排放及处理情况

Table 32－4　Industrial Waste Water Discharge and Treatment Status of Major Cities in 2009

指标	工业废水排放总量（万吨）	工业废水排放达标量（万吨）	工业废水中化学需氧量排放量（吨）	工业废水中氨氮排放量（吨）	废水治理设施数（套）	废水治理设施本年运行费用（万元）
北　京	8712. 5326	8574. 1519	4898. 1073	452. 7949	524	65702. 5
天　津	19441. 0525	19440. 139	23469. 359	2915. 4927	848	73073. 1
石家庄	19045. 139	18991. 619	49738. 4056	4068. 5058	589	46852. 5
太　原	2483. 3878	2416. 85	4920. 0024	353. 7945	238	41219. 8
呼和浩特	2373. 9156	2372. 7403	2474. 0545	236. 3187	54	9061. 4
沈　阳	6259. 2573	5722. 3785	7947. 0281	1086. 4182	356	24178. 2
长　春	5489. 3273	5133. 0766	25946. 9786	468. 4983	104	5708
哈尔滨	3538. 5413	3445. 7679	13865. 0842	510. 8835	153	20847. 1
上　海	41192. 0286	40686. 6005	29030. 6183	1982. 8384	1730	264295. 6
南　京	36339. 4251	33698. 1146	26702. 4883	1098. 9958	677	85539. 5
杭　州	79958. 8214	76924. 7295	86449. 3141	2364. 4057	1221	90202. 8
合　肥	2035. 5977	1961. 4124	1380. 7407	80. 0687	166	9570. 4
福　州	4287. 8139	3859. 7034	3980. 0855	550. 2543	382	12994. 5
南　昌	10237. 7173	9554. 4604	20999. 9956	1085. 4986	207	12303. 4
济　南	5013. 7909	4954. 1251	7004. 5971	319. 8824	264	25822. 3
郑　州	11240. 4304	11155. 4304	9353. 7247	233. 1389	377	19533
武　汉	22531. 8013	22334. 3202	21808. 0778	1225. 8609	248	27054. 3
长　沙	3725. 7706	3354. 0923	4920. 8685	228. 6527	260	5283
广　州	26022. 7222	25115. 7649	23425. 2672	935. 7815	996	61140
南　宁	12346. 5861	11290. 7203	71981. 7463	1341. 5011	384	17523. 6
海　口	475. 1816	475. 1816	328. 5589	8. 7445	33	1439. 9
重　庆	65683. 845	61925. 1745	100290. 2713	7251. 448	1638	54792. 7
成　都	24553. 5832	24486. 977	41503. 9981	5755. 4998	1512	32644. 1
贵　阳	2356. 4051	2291. 5694	1790. 9573	96. 9757	298	8335. 3
昆　明	4256. 4331	4248. 9431	3276. 2642	209. 8326	538	20781. 8
拉　萨	986. 1524	210. 3624	489. 9445	2. 659	13	158. 2
西　安	14202. 9901	13280. 8633	40823. 4904	2230. 1972	383	13360. 2
兰　州	2945. 1777	2905. 4693	1834. 6179	134. 8943	78	15379. 2
西　宁	4387. 4786	3827. 9432	15990. 1134	970. 8375	127	2771. 7
银　川	4987. 3901	4945. 9701	12552. 645	1227. 7374	128	6317. 6
乌鲁木齐	5968. 3204	5250. 9893	8667. 4785	1668. 4997	107	7879. 8

表 32－5　2009 年工业按行业分废水排放及处理情况

Table 32－5　Trade－classified Industrial Waste Water Discharge and Treatment Status in 2009

指标	参与调查工业企业单位数（个）	工业废水排放总量（万吨）	工业废水排放达标量（万吨）	废水治理设施数（套）
行业合计	110903	2090299. 7855	1980074. 6375	77017

续表

指标	参与调查工业企业单位数（个）	工业废水排放总量（万吨）	工业废水排放达标量（万吨）	废水治理设施数（套）
煤炭开采和洗选业	4261	80235.5008	73665.1135	3435
石油和天然气开采业	233	10197.4901	10005.013	521
黑色金属矿采选业	1146	15546.2476	14649.9689	1024
有色金属矿采选业	1628	37307.3058	32729.9969	1739
非金属矿采选业	958	7718.5173	7291.5488	954
其他采矿业	74	573.6499	460.3877	37
农副食品加工业	7481	143837.6597	132050.2494	4827
食品制造业	3707	52699.0946	48365.3489	2517
饮料制造业	2748	69673.5881	65458.5013	1881
烟草制品业	158	3252.7877	3175.7161	133
纺织业	8070	239115.5989	230805.8211	6564
纺织服装、鞋、帽制造业	1451	14727.6014	14323.5814	686
皮革毛皮羽毛（绒）及其制品业	1695	24964.3198	23086.9055	1238
木材加工及木竹藤棕草制品业	1526	6137.2545	5702.9617	618
家具制造业	340	1855.8908	1835.7843	162
造纸及纸制品业	5771	392604.1394	367176.3453	5477
印刷业和记录媒介的复制	558	1783.4732	1742.9614	196
文教体育用品制造业	270	1238.5236	1108.6124	142
石油加工、炼焦及核燃料加工业	1139	66406.2781	63474.3194	1667
化学原料及化学制品制造业	9985	297061.6591	282137.906	9937
医药制造业	2939	52718.3901	51003.4747	2553
化学纤维制造业	350	43854.8108	41857.5245	355
橡胶制品业	918	6783.469	6704.4234	419
塑料制品业	1409	4387.4815	4005.375	597
非金属矿物制品业	21435	32777.0656	30433.9746	4987
黑色金属冶炼及压延加工业	3562	125978.3087	122036.1624	4388
有色金属冶炼及压延加工业	2643	28975.96	27642.7684	2172
金属制品业	6118	31345.7921	30165.6675	5088
通用设备制造业	3764	13452.2871	12953.3963	1365
专用设备制造业	1294	11006.4492	10746.9781	794
交通运输设备制造业	2539	27421.8168	26472.5398	2113
电气机械及器材制造业	1654	9324.4731	9001.7476	1149
通信计算机及其他电子设备制造业	1872	33513.2748	32776.4392	2047
仪器仪表及文化办公用机械制造业	511	5798.1608	5675.1402	470
工艺品及其他制造业	948	3587.4709	3362.3413	597
废弃资源和废旧材料回收加工业	189	959.4934	917.2488	103
电力、热力的生产和供应业	4132	149009.5609	144572.5837	3379

续表

指标	参与调查工业企业单位数（个）	工业废水排放总量（万吨）	工业废水排放达标量（万吨）	废水治理设施数（套）
燃气生产和供应业	88	2013.1319	1926.8711	56
水的生产和供应业	262	22918.9634	22420.9482	135
其他行业	1077	17536.845	16151.9897	495

表 32－6　工业废气排放量年度统计

Table 32－6　Yearly Statistics of Industrial Waste Gas Emission

单位：亿标立方米

	2005 年	2006 年	2007 年	2008 年	2009 年
全　国	268988	330992	388169.1921	403866	436063.8475
北京市	3532	4641	5146.0048	4316	4408.253
天津市	4602	6512	5505.7872	6005	5982.7565
河北省	26518	39254	48036.3617	37558	50779.4401
山西省	15142	18128	21429.2311	23180	23692.9067
内蒙古自治区	12071	18415	18199.6463	20190	24844.3649
辽宁省	20903	27195	23945.9177	40219	25211.1855
吉林省	4939	5352	5730.3715	6155	7123.8048
黑龙江省	5261	5991	7282.6717	7796	9977.067
上海市	8482	9428	9590.5041	10436	10058.5972
江苏省	20197	24881	23585.3703	25245	27431.7456
浙江省	13025	14702	17466.8643	17633	18860.3302
安徽省	6960	8677	13253.9733	15749	15272.5709
福建省	6265	6884	9153.4511	9150	10497.1038
江西省	4379	5096	6103.4669	7456	8286.0511
山东省	24129	25751	31341.3607	33505	35126.7039
河南省	15498	16770	18890.3878	20264	22185.5727
湖北省	9404	11015	10373.2651	11558	12522.6292
湖南省	6014	5986	8762.2001	9249	10972.5646
广东省	13447	13584	16938.5413	20510	22681.9537
广西壮族自治区	8339	8969	12723.5468	11643	13184.1989
海南省	910	860	1114.5677	1345	1353.2064
重庆市	3655	6757	7616.6177	7351	12586.5208
四川省	8140	10553	22970.0543	12997	13410.0185
贵州省	3852	8344	10356.4916	6842	7785.763
云南省	5444	6646	8081.7129	8316	9483.8008
西藏自治区	13	13	12.8151	13	15.4092
陕西省	4916	5535	6469.1844	9706	11031.9024
甘肃省	4250	4761	5817.9231	5685	6313.9457
青海省	1370	2099	2492.4629	3237	3307.9922
宁夏回族自治区	2844	3140	3981.225	4403	4700.6096
新疆维吾尔自治区	4485	5053	5797.2136	6154	6974.8786

表 32－7 2009 年各地区废气排放及处理情况

Table 32－7 Waste Gas Emission and Treatment Status of Different Regions in 2009

指标	工业废气排放量（亿标立方米）	废气治理设施数(套)	工业二氧化硫排放量（万吨）	生活二氧化硫排放量（万吨）	工业二氧化硫去除量（万吨）	工业烟尘排放量(万吨)	生活烟尘排放量(万吨)	工业烟尘去除量(万吨)	工业粉尘排放量(万吨)	工业粉尘去除量(万吨)	建成烟尘控制区数(个)	建成烟尘控制区面积(平方公里)
全　国	436063. 8475	241201. 4861	194862. 4312	176489	1865. 942396	348. 459356	2889. 860702	604. 363077	243. 28986	32848. 145775	523. 647237	8722. 647963
北京市	4408. 253	1777. 3707	2630. 8823	2603	5. 992186	5. 8872	11. 282947	1. 907694	2. 5295	181. 676874	1. 732078	80. 292568
天津市	5982. 7565	4039. 9449	1942. 8116	3157	17. 297981	6. 372	25. 651813	5. 868706	1. 2636	420. 324749	0. 794584	65. 146581
河北省	50779. 4401	24836. 4754	25942. 9647	11845	104. 267929	21. 0784	123. 773847	32. 981929	18. 8817	2355. 60249	42. 698778	561. 015861
山西省	23692. 9067	13065. 708	10627. 1987	9200	100. 958569	25. 8842	143. 421959	43. 823058	20. 8596	2137. 176643	42. 778673	319. 167115
内蒙古自治区	24844. 3649	14382. 4108	10461. 9541	4956	120. 401838	19. 4785	164. 64875	32. 052914	17. 3219	2458. 21147	16. 412704	234. 402233
辽宁省	25211. 1855	11283. 5173	13927. 6682	10067	91. 883747	13. 2582	102. 24648	40. 146672	21. 1218	1865. 502044	22. 690429	636. 484887
吉林省	7123. 8048	5094. 6273	2029. 1775	3037	30. 020024	6. 2805	12. 777176	27. 735378	10. 6936	868. 876742	6. 568283	374. 824429
黑龙江省	9977. 067	8260. 336	1716. 731	4313	41. 890548	7. 1478	7. 385126	31. 88936	11. 3648	1058. 733782	10. 145693	79. 842548
上海市	10058. 5972	3381. 7136	6676. 8836	3986	23. 934751	13. 956	38. 462354	3. 644988	6. 5358	513. 513232	0. 832044	95. 749159
江苏省	27431. 7456	17752. 2994	9679. 5162	11508	101. 182354	6. 2331	185. 55153	30. 14691	2. 8644	2279. 389595	16. 3712	323. 07572
浙江省	18860. 3302	11912. 9063	6947. 4239	14526	67. 700251	2. 4327	130. 563651	17. 97135	0. 9849	1234. 328973	16. 833978	869. 743779
安徽省	15272. 5709	7038. 1629	8234. 408	4733	48. 686397	5. 156	161. 852427	22. 962566	5. 0396	1363. 630334	28. 474493	344. 427508
福建省	10497. 1038	6580. 0664	3917. 0374	6931	39. 919235	2. 0464	35. 454558	7. 072899	4. 2604	606. 427841	15. 66055	249. 188055
江西省	8286. 0511	4710. 7339	3575. 3172	3953	49. 019998	7. 4022	142. 950383	13. 938746	2. 4988	741. 831633	26. 321392	567. 361305
山东省	35126. 7039	19539. 4024	15587. 3015	11727	136. 615046	22. 4151	273. 32868	30. 196797	11. 5221	3339. 775839	22. 060778	554. 550486
河南省	22185. 5727	12773. 3945	9412. 1782	8740	117. 596851	17. 9032	141. 417979	52. 05163	7. 6493	2685. 920523	24. 914459	572. 69309
湖北省	12522. 6292	5957. 7107	6564. 9185	5211	52. 743077	11. 633056	88. 46466	17. 935771	3. 73016	788. 967764	18. 308504	408. 255707
湖南省	10972. 5646	4373. 163	6599. 4016	5117	64. 942555	16. 2076	84. 762812	27. 554093	6. 5518	660. 215597	57. 516508	323. 486943

续表

指标	工业废气排放量（亿标立方米）	废气治理设施数（套）	工业二氧化硫排放量（万吨）	生活二氧化硫排放量（万吨）	工业二氧化硫去除量（万吨）	工业烟尘排放量（万吨）	生活烟尘排放量（万吨）	工业烟尘去除量（万吨）	工业粉尘排放量（万吨）	工业粉尘去除量（万吨）	建成烟尘控制区数（个）	建成烟尘控制区面积（平方公里）
广东省	22681. 9537	14033. 0002	8648. 9535	13011	101. 257915	5. 7909	187. 361308	24. 815895	5. 3206	988. 332189	10. 518952	464. 081446
广西壮族自治区	13184. 1989	7588. 292	5595. 9069	6006	83. 529578	5. 5199	87. 761947	24. 650134	1. 2974	372. 594778	46. 679755	323. 607228
海南省	1353. 2064	1023. 0279	330. 1785	397	2. 14332	0. 0598	6. 016681	0. 758894	0. 1521	55. 860586	0. 900491	12. 622962
重庆市	12586. 5208	6658. 655	5927. 8658	3309	58. 611662	15. 9975	78. 018101	10. 867377	8. 19	439. 048283	10. 774933	59. 000929
四川省	13410. 0185	7435. 7545	5974. 2638	6927	94. 642413	18. 8875	83. 800762	19. 569956	8. 9272	946. 257946	11. 358312	179. 379375
贵州省	7785. 763	4762. 819	3022. 944	2441	62. 374319	55. 1751	131. 784449	11. 780574	32. 2018	1396. 856465	9. 872041	177. 428105
云南省	9483. 8008	4758. 1216	4725. 6792	5432	41. 782078	8. 1486	141. 725599	12. 353524	5. 472	1105. 254763	10. 184334	370. 493312
西藏自治区	15. 4092	13. 8971	1. 5121	51	0. 166	0. 034	0	0. 071978	0. 1674	0. 083167	0. 117911	0. 013002
陕西省	11031. 9024	6822. 2205	4209. 6819	4035	74. 187263	6. 2535	74. 466299	15. 086473	5. 1528	779. 873666	14. 790071	221. 365298
甘肃省	6313. 9457	3374. 508	2939. 4377	2716	40. 092571	9. 938	191. 338556	9. 161825	7. 0188	304. 710478	8. 366753	89. 350944
青海省	3307. 9922	757. 0174	2550. 9748	752	12. 727526	0. 8423	1. 348499	5. 440956	2. 1826	107. 791535	6. 875656	63. 695984
宁夏回族自治区	4700. 6096	2556. 4583	2144. 1513	1503	27. 834607	3. 5899	29. 284619	7. 947248	1. 846	466. 039061	3. 625599	45. 581568
新疆维吾尔自治区	6974. 8786	4657. 7711	2317. 1075	4299	51. 539805	7. 4502	2. 95675	21. 976783	9. 6874	325. 336734	18. 467301	56. 319835

表 32－8　2009 年主要城市工业废气排放及处理情况

Table 32－8　Industrial Waste Gas Emission and Treatment Status of Major Cities in 2009

指标	工业二氧化硫排放量(吨)	工业烟尘排放量(吨)	工业粉尘排放量(吨)	工业二氧化硫去除量(吨)	工业烟尘去除量(吨)	工业粉尘去除量(吨)	废气治理设施数(套)	废气治理设施本年运行费用(万元)	脱硫设施(套)
北　京	59921. 864	19076. 935	17320. 783	112829. 469	1816768. 737	802925. 678	2603	1095	90121. 7
天　津	172979. 807	58687. 056	7945. 843	256518. 126	4203247. 493	651465. 811	3157	1631	168240
石家庄	143674. 931	31343. 34	15446. 647	277484. 028	3982449. 997	151841. 229	1810	365	72281. 2
太　原	90486. 541	38607. 282	30460. 557	173708. 219	3092713. 155	873954. 449	1040	407	89745. 7
呼和浩特	74040. 998	12731. 331	1309. 325	179785. 061	4193629. 794	71687. 169	726	16	40431. 6
沈　阳	82361. 647	52941. 364	4846. 717	44132. 333	1644313. 66	5468. 501	2140	513	46795. 7
长　春	54693. 004	68922. 367	25355. 108	14817. 01	1591061. 547	849075. 155	673	52	11922. 6
哈尔滨	51878. 795	51205. 496	17388. 414	27485. 945	1717866. 729	62672. 974	869	56	16620. 6
上　海	239347. 514	36449. 876	8320. 436	384623. 535	5135132. 324	957491. 589	3986	709	220887
南　京	134276. 574	31167. 567	41674. 918	494692. 119	2435625. 759	779644. 378	1144	130	88241. 9
杭　州	92925. 608	35417. 733	18349. 419	80713. 36	1210748. 483	935500. 833	1776	437	78786. 3
合　肥	30453. 002	10602. 043	4591. 128	17619. 882	1054716. 896	126714. 955	381	30	10252. 5
福　州	90277. 488	7026. 68	2126. 588	91876. 76	1462427. 938	15043. 372	558	62	67911. 6
南　昌	22514. 123	8103. 749	6115. 364	11547. 748	155627. 276	4172. 251	536	46	12350. 2
济　南	65944. 004	20145. 833	30021. 837	141691. 331	1633220. 028	697153. 254	940	158	113506. 9
郑　州	112167. 043	51410. 473	38210. 978	59190. 963	2785352. 221	400954. 352	1404	119	55391
武　汉	114579. 351	30890. 215	8162. 814	107380. 261	2160838. 47	409924. 925	628	42	45883. 6
长　沙	52052. 162	34703. 548	133529. 638	38484. 215	100414. 696	191883. 246	347	58	3353. 8
广　州	91722. 03	11943. 057	2237. 36	896167. 217	1806518. 381	95690. 231	1676	350	93455. 9
南　宁	58486. 786	26759. 677	8407. 25	20127. 021	513520. 563	400234. 586	916	171	13636. 9
海　口	103. 334	93. 619	1. 43	0	578. 657	16. 5	21		1731. 4
重　庆	586116. 624	108673. 773	107749. 334	780181. 013	4390482. 83	590009. 288	3309	328	149234. 1

续表

指标	工业二氧化硫排放量(吨)	工业烟尘排放量(吨)	工业粉尘排放量(吨)	工业二氧化硫去除量(吨)	工业烟尘去除量(吨)	工业粉尘去除量(吨)	废气治理设施数(套)	废气治理设施本年运行费用(万元)	脱硫设施(套)
成　都	93045. 412	36670. 267	5044. 781	153731. 468	1357526. 661	132141. 949	1607	64	59814. 1
贵　阳	84487. 682	13831. 369	14687. 907	187718. 436	1109016. 066	413900. 807	497	164	35510. 2
昆　明	88337. 235	8022. 698	5563. 153	593326. 175	1307392. 595	428396. 158	1407	124	86628. 6
拉　萨	1201. 116	570. 608	680. 445		505. 267	8. 192	17		1345
西　安	82864. 221	20409. 68	4235. 231	48743. 105	780445. 921	133130. 143	852	106	19073. 9
兰　州	70687. 002	11046. 488	8377. 917	207025. 377	1217512. 963	113681. 713	529	18	35426. 9
西　宁	70403. 188	24686. 936	32811. 319	13402. 4	798551. 708	383890. 905	421	4	33321. 6
银　川	20403. 259	6935. 655	2126. 295	40133. 956	143054. 89	94072. 704	281	59	10384. 2
乌鲁木齐	107971. 081	35058. 206	8524. 597	20205. 655	764050. 601	117003. 287	519	7	24411. 1

表 32－9 2009 年工业按行业分废气排放及处理情况

Table 32－9 Trade－classified Industrial Waste Gas Emission and Treatment Status in 2009

指标	废气治理设施数(套)	工业废气排放量(亿标立方米)	工业二氧化硫排放量(万吨)	工业二氧化硫去除量(万吨)	工业烟尘排放量(万吨)	工业烟尘去除量(万吨)	工业粉尘排放量(万吨)	工业粉尘去除量(万吨)	燃料燃烧过程中废气排放量(亿标立方米)	生产工艺过程中废气排放量(亿标立方米)
行业总计	176488	436062.727	241200.5456	194862.2512	1694.064465	2889.860639	544.624196	32848.145707	476.199308	8722.647946
煤炭开采和洗选业	4371	2333.9849	2005.4228	328.5621	14.986125	9.728403	9.82824	171.179148	18.783869	16.224063
石油和天然气开采业	328	1092.2138	1045.6438	46.57	3.530192	6.261235	1.105167	4.668247	0.000476	0
黑色金属矿采选业	700	1489.0855	914.0506	575.0349	5.453809	1.51285	1.840143	10.281818	3.781525	14.632132
有色金属矿采选业	639	359.0267	147.8008	211.2259	12.30279	58.333271	1.207192	12.626212	1.161355	4.95442
非金属矿采选业	548	834.4517	396.9282	437.5235	4.523044	2.623708	2.232532	25.813343	2.59052	5.661424
其他采矿业	33	18.8216	14.2179	4.6037	0.107858	0.007038	0.194581	0.124009	0.061658	0.02462
农副食品加工业	5018	3682.2986	3291.2402	391.0582	16.094383	6.652617	11.226023	123.426821	0.383527	6.177471
食品制造业	2958	3198.1059	3089.535	108.5709	10.758108	4.881828	5.718913	85.317663	0.131278	4.242188
饮料制造业	2373	2109.894	2082.3793	27.5147	10.57802	5.113284	6.747039	103.869536	0.110029	0.669354
烟草制品业	821	516.5106	166.1509	350.3597	1.161982	0.597041	0.505197	3.1097	0.100686	1.719115
纺织业	8366	3447.781	3203.4733	244.3077	25.611288	10.450743	12.663424	141.89742	0.170057	0.902526
纺织服装、鞋、帽制造业	851	226.6175	169.9341	56.6834	1.241669	0.69313	0.663383	7.275846	0.026481	0.052033
皮革毛皮羽毛(绒)及其制品业	1334	333.2868	256.8921	76.3947	1.783544	0.310889	1.027978	5.623524	0.034455	0.008631
木材加工及木竹藤棕草制品业	1905	1488.9297	884.0255	604.9042	3.213957	0.533811	3.226727	10.231266	2.482061	29.285957
家具制造业	870	120.0981	37.6468	82.4513	0.263391	0.08711	0.374268	1.254473	0.098391	0.223252
造纸及纸制品业	5287	6106.1342	5837.5419	268.5923	45.736559	23.422303	19.181849	344.336093	0.75012	1.438014
印刷业和记录媒介的复制	220	134.6669	31.3271	103.3398	0.267229	0.062013	0.153606	0.515432	0.002189	0.001333
文教体育用品制造业	322	82.681	14.8011	67.8799	0.114646	0.17451	0.06223	0.185997	0.028828	0.215471
石油加工、炼焦及核燃料加工业	2614	15803.7397	5834.4403	9969.2994	61.423491	230.753614	22.455524	279.734928	15.54576	117.596582
化学原料及化学制品制造业	16770	23174.4613	12133.8059	11040.6554	97.516036	112.358628	41.717687	764.710848	11.156988	104.903718

续表

指标	废气治理设施数(套)	工业废气排放量(亿标立方米)	工业二氧化硫排放量(万吨)	工业二氧化硫去除量(万吨)	工业烟尘排放量(万吨)	工业烟尘去除量(万吨)	工业粉尘排放量(万吨)	工业粉尘去除量(万吨)	燃料燃烧过程中废气排放量(亿标立方米)	生产工艺过程中废气排放量(亿标立方米)
医药制造业	3202	1286. 6704	1089. 6229	197. 0475	7. 761314	4. 956899	4. 507844	41. 857576	0. 080236	0. 146687
化学纤维制造业	781	3232. 9525	1896. 989	1335. 9635	11. 455321	8. 584983	2. 777292	113. 480755	0. 057126	1. 176983
橡胶制品业	1429	904. 6454	511. 4452	393. 2002	3. 779628	3. 566248	1. 840105	29. 073395	0. 062709	0. 916958
塑料制品业	1354	657. 4168	310. 0884	347. 3284	2. 339588	0. 623216	1. 014958	7. 486738	0. 029084	0. 291247
非金属矿物制品业	52848	78872. 7672	19233. 4385	59639. 3987	160. 52366	41. 792894	92. 50996	1114. 146139	309. 042285	5259. 284413
黑色金属冶炼及压延加工业	15019	103582. 5615	24098. 5514	79484. 0101	170. 183929	126. 29808	51. 841405	1244. 77565	84. 148857	2649. 427726
有色金属冶炼及压延加工业	7015	19456. 4685	4346. 0494	15110. 4191	66. 089028	783. 146032	12. 282095	420. 955293	8. 835261	370. 752649
金属制品业	4168	1934. 8018	842. 2304	1092. 5714	3. 837472	0. 801661	2. 179412	9. 49892	0. 762935	8. 995402
通用设备制造业	3952	1930. 8436	731. 1038	1199. 7398	4. 617954	1. 228607	3. 008324	10. 719914	2. 964861	43. 549174
专用设备制造业	1920	4177. 979	1181. 7039	2996. 2751	3. 831172	1. 541433	1. 302635	19. 475311	1. 665338	3. 394233
交通运输设备制造业	4755	3667. 889	773. 2303	2894. 6587	3. 684444	1. 655906	2. 931621	37. 421827	2. 692682	14. 05081
电气机械及器材制造业	2718	1101. 6614	339. 131	762. 5304	1. 133007	0. 470686	0. 643793	4. 883296	0. 064109	0. 184116
通信计算机及其他电子设备制造业	3602	3387. 3969	200. 4154	3186. 9815	1. 004072	0. 605884	0. 502153	5. 326572	0. 272705	5. 519801
仪器仪表及文化办公用机械制造业	646	541. 1506	28. 7127	512. 4379	0. 143455	0. 058603	0. 066907	0. 405306	0. 034009	0. 227524
工艺品及其他制造业	500	111. 5145	32. 7792	78. 7353	0. 365014	0. 057923	0. 162236	0. 779255	0. 906147	0. 636692
废弃资源和废旧材料回收加工业	104	64. 9852	47. 6503	17. 3349	0. 173932	0. 100106	0. 051805	0. 371466	0. 199382	0. 252445
电力、热力的生产和供应业	15135	143623. 7351	143478. 5034	145. 2317	932. 990443	1437. 89583	222. 148418	27665. 559312	0. 67337	1. 71317
燃气生产和供应业	406	764. 3009	309. 9415	454. 3594	2. 332863	0. 907654	1. 380558	19. 309353	0. 410091	0. 047715
水的生产和供应业	83	7. 3942	5. 5234	1. 8708	0. 045778	0. 002037	0. 049883	0. 165344	0. 000067	0. 000031
其他行业	523	202. 803	186. 1779	16. 6251	1. 104272	1. 007929	1. 291091	6. 271962	5. 897803	53. 147867

表 32－10　工业固体废物产生量年度统计

Table 32－10　Yearly Statistics of Industrial Solid Waste Generation

单位：万吨

	2005 年	2006 年	2007 年	2008 年	2009 年
全　国	134449	151541	175631. 603	190127	203943. 4136
北京市	1238	1356	1274. 8445	1157	1242. 4435
天津市	1123	1292	1399. 404	1479	1515. 6552
河北省	16279	14229	18688. 2903	19769	21975. 8085
山西省	11183	11817	13819. 136	16213	14742. 8515
内蒙古自治区	7363	8710	10972. 7788	10622	12108. 3241
辽宁省	10242	13013	14341. 8074	15841	17221. 4135
吉林省	2457	2802	3112. 5556	3415	3940. 5179
黑龙江省	3210	3914	4130. 1331	4472	5274. 7245
上海市	1964	2063	2165. 4038	2347	2254. 5883
江苏省	5757	7195	7354. 2185	7724	8027. 8134
浙江省	2514	3096	3613. 4526	3785	3909. 6637
安徽省	4196	5028	5960. 3475	7569	8470. 8317
福建省	3773	4238	4814. 8577	5371	6348. 9079
江西省	7007	7393	7777. 2787	8190	8898. 1797
山东省	9175	11011	11934. 7274	12988	14137. 9472
河南省	6178	7464	8850. 5849	9557	10785. 8162
湖北省	3692	4315	4682. 659	5014	5561. 4544
湖南省	3366	3688	4559. 7349	4520	5092. 7863
广东省	2896	3057	3852. 4325	4833	4740. 8511
广西壮族自治区	3489	3894	4543. 6418	5417	5693. 1042
海南省	127	147	157. 9343	220	200. 8575
重庆市	1777	1764	2086. 8441	2311	2551. 8474
四川省	6421	7600	9653. 8244	9237	8596. 8623
贵州省	4854	5827	5988. 5757	5844	7317. 366
云南省	4661	5972	7097. 5247	7986	8672. 8314
西藏自治区	8	9	5. 4869	6	11. 116
陕西省	4588	4794	5480. 018	6121	5546. 6679
甘肃省	2249	2591	3001. 4609	3199	3150. 2109
青海省	649	882	1129. 2675	1337	1347. 6239
宁夏回族自治区	719	799	1045. 7406	1143	1398. 2546
新疆维吾尔自治区	1295	1581	2136. 6368	2438	3206. 0831

表 32－11 工业固体废物排放量年度统计

Table 32－11 Yearly Statistics of Industrial Solid Waste Discharge

单位：吨

	2005 年	2006 年	2007 年	2008 年	2009 年
全 国	16546848	13020919	11967190. 58	7817522	7104521. 074
北京市	91403	1011	887	876	881. 4
天津市	44			0	0
河北省	424149	419579	389464. 12	608352	304551. 12
山西省	6046901	4357022	4142510. 606	2323858	1415687. 881
内蒙古自治区	624958	282161	89404. 525	395544	92170. 9164
辽宁省	93938	246918	44817. 28	11630	27510
吉林省	18642	20509	13070. 28	0	0
黑龙江省	300	10269	310	7098	9763
上海市	1233	2414	1514	0	42
江苏省	53	270	2585. 8	5	0
浙江省	56380	51800	14385. 41	16657	7824. 05
安徽省	450	468	68	15	15
福建省	57682	33707	27549. 573	26495	24334. 393
江西省	102819	83022	82367. 2	123088	139937. 88
山东省	1376	4139	659	803	144
河南省	36354	32217	22192. 8	26397	13104. 35
湖北省	167034	115310	79968. 01	58239	51169. 75
湖南省	567050	387231	318141. 51	290168	185972. 3499
广东省	138538	135771	115031. 81	119897	159839. 0453
广西壮族自治区	1104791	227434	103403. 679	96154	120978. 9768
海南省	368	200	1114	60	0
重庆市	1844955	1221841	1381292. 31	1488357	1498598. 32
四川省	1157015	848568	2045049. 5	134045	61158. 5
贵州省	1312832	1376600	818754. 75	555853	944883. 734
云南省	706602	995583	826555. 5	394239	606462. 018
西藏自治区	73000	73860	39299	42789	41243. 16
陕西省	348679	429068	422577. 28	263058	172657. 03
甘肃省	406347	337345	248185. 84	116261	123101. 84
青海省	27625	3154	8173. 5	17504	13904
宁夏回族自治区	41047	60178	47670	49908	36588
新疆维吾尔自治区	1094286	1263270	680188. 297	650174	1051998. 36

表 32－12　“三废”综合利用产品产值年度统计

Table 32－12　Early Statistics of Comprehensive Utilization of “Three Wastes”

单位：万元

	2005 年	2006 年	2007 年	2008 年	2009 年
全　国	7555064. 3	10267926	13512691. 9	16213662	16082440. 4
北京市	94715	110528. 6	91801. 6	117524. 5	71679. 6
天津市	77228. 9	171747. 8	110850. 9	113838. 2	187882. 1
河北省	428703	561234. 2	796650. 8	1339373. 3	936389. 8
山西省	249696. 5	225883. 8	403412	411687	342720. 7
内蒙古自治区	86505	83917. 1	198533. 2	218529	217935. 7
辽宁省	256174. 2	305735. 7	517482. 6	423803	443699. 2
吉林省	128649. 3	207413. 7	220825. 5	298590. 9	308741. 1
黑龙江省	138067	175473. 4	175738. 6	474109. 2	247210
上海市	91108. 9	97197. 9	152091. 3	169560	161409. 4
江苏省	931999. 9	1440509. 3	1834836. 2	2139397. 7	2014355. 9
浙江省	575110. 8	1497950. 1	2240597. 9	2409843. 1	2513210. 4
安徽省	187615. 8	232747. 6	362346. 3	576974	509654. 3
福建省	148978. 8	182826. 5	205186. 6	253523. 2	492685. 9
江西省	178811. 2	252891. 5	324863. 4	390934. 7	470276. 6
山东省	935661. 8	1074662. 4	1479858. 1	1650123. 5	1725360. 8
河南省	339139. 5	444607. 7	517888. 1	716717. 1	693260. 7
湖北省	462127. 9	559119. 3	613941. 9	648621. 7	699427. 9
湖南省	263513. 6	373538. 1	491844. 9	596282. 3	695003. 8
广东省	365616. 3	433236. 6	497618. 8	601479	509826. 8
广西壮族自治区	238923. 4	312363. 4	418233. 7	457778. 6	432197
海南省	14738. 3	12714. 2	24550. 5	41309. 1	24439. 8
重庆市	81493. 7	133994. 4	165138. 6	238383. 4	274132. 6
四川省	505656. 7	529852. 3	493074. 3	515723. 1	612685. 5
贵州省	121322. 2	152040. 3	157127. 6	197467. 4	165524
云南省	392999. 2	377929. 2	537403. 3	652348. 3	604415
西藏自治区		169. 3	558. 3	425. 7	238. 7
陕西省	49489. 4	98525. 2	160527	125382	222059
甘肃省	91664. 6	104990. 9	167268. 1	211219. 4	260170
青海省	5289. 7	10369. 2	20431. 2	46889. 7	26311. 1
宁夏回族自治区	23498. 2	36529. 2	40856. 9	58510	66672
新疆维吾尔自治区	90565. 5	67227. 1	91153. 7	117313. 9	152865

表 32－13　2009 年各地区工业固体废物产生及处理利用情况

Table 32－13　Industrial Solid Waste Generation and Treatment Status of Different Regions in 2009

指标	工业固体废物产生量（万吨）	工业固体废物综合利用量（万吨）	工业固体废物贮存量（万吨）	工业固体废物处置量（万吨）	工业固体废物排放量（吨）	“三废”综合利用产品产值（万元）	危险废物（万吨）	危险废物（吨）
全　国	203943.413647	1429.851696	138185.824385	20929.277443	47487.748987	7104521.0744	1.5	16082440.4
北京市	1242.443507	11.18807	910.448572	43.668292	754.635364	881.4	0	71679.6
天津市	1515.66516	7.837857	1498.304359	0.070729	25.672342	0	0	187882.1
河北省	21975.808457	27.703812	15693.324277	1168.439988	5259.878295	304551.12	0	936389.8
山西省	14742.851546	6.780205	8955.803462	718.673281	5099.440165	1415687.881	0	342720.7
内蒙古自治区	12108.324092	50.189538	6367.910333	1235.623108	4513.038603	92170.9164	0	217935.7
辽宁省	17221.413526	90.980815	8240.681619	1981.221514	7261.097777	27510	1.5	443699.2
吉林省	3940.517935	57.532248	2538.824836	1324.81611	88.522189	0	0	308741.1
黑龙江省	5274.724492	16.973913	3809.690848	993.710658	513.044403	9763	0	247210
上海市	2254.588323	47.623233	2171.595012	12.673351	85.657886	42	0	161409.4
江苏省	8027.813389	126.056609	7862.247091	168.300713	102.008586	0	0	2014355.9
浙江省	3909.663688	60.105256	3585.854268	74.698299	255.95925	7824.05	0	2513210.4
安徽省	8470.831669	11.583348	7227.037019	543.981585	922.701395	15	0	509654.3
福建省	6348.907864	8.038504	5425.812088	55.870204	874.470938	24334.393	0	492685.9
江西省	8898.179655	6.451636	3702.653579	786.039476	4416.128847	139937.88	0	470276.6
山东省	14137.9472	221.441892	13826.374037	237.94183	523.935459	144	0	1725360.8
河南省	10785.816245	17.895089	8064.285724	282.96391	2691.376069	13104.35	0	693260.7
湖北省	5561.454394	91.382946	4210.153299	251.428355	1164.442994	51169.75	0	699427.9
湖南省	5092.786298	53.946491	4010.409201	822.208165	379.75763	185972.3499	0	695003.8
广东省	4740.851108	99.214703	4321.562586	137.830052	313.48505	159839.0453	0	509826.8
广西壮族自治区	5693.10416	12.489471	3856.683338	461.393133	1434.4218	120978.9768	0	432197
海南省	200.857465	0.233027	167.946003	11.763955	21.207507	0	0	24439.8
重庆市	2551.847352	14.684162	2076.735085	262.30147	126.683114	1498598.32	0	274132.6
四川省	8596.862297	83.431625	4952.298072	1076.099881	2845.249684	61158.5	0	612685.5
贵州省	7317.36598	52.635844	3350.674137	1792.372141	2109.385918	944883.734	0	165524
云南省	8672.831413	50.439187	4264.802277	2001.567227	2615.389711	606462.018	0	604415
西藏自治区	11.116016		0.1984	6.7933	0	41243.16		238.7
陕西省	5546.667882	12.984137	2997.552181	1099.464095	1445.14365	172657.03	0	222059
甘肃省	3150.210871	20.937508	1072.650939	936.964779	1216.754299	123101.84	0	260170
青海省	1347.623903	83.305336	508.210585	850.176648	1.47117	13904	0	26311.1
宁夏回族自治区	1398.254636	0.376363	987.913193	169.67208	251.023283	36588	0	66672
新疆维吾尔自治区	3206.083122	85.40887	1527.187961	1420.549114	175.765611	1051998.36	0	152865

表 32－14　2009 年主要城市工业固体废物产生及处理利用情况

Table 32－14　Industrial Solid Waste Generation and Treatment Status of Major Cities in 2009

指标	工业固体废物产生量（万吨）	工业固体废物综合利用量（万吨）	工业固体废物排放量（吨）	工业固体废物综合利用率（%）	危险废物（吨）
北　京	1242. 443507	111880. 705	910. 448572	881. 4	68. 87
天　津	1515. 66516	78378. 5654	1498. 304359	0	98. 31
石家庄	1272. 744364	194148. 974	1208. 801685	0	92. 5
太　原	2410. 266612	19085. 94	1171. 635999	92911	48. 57
呼和浩特	591. 686126	2958. 09	158. 131857	20	64. 55
沈　阳	644. 788441	97157. 645	605. 631915	173. 5	17. 94
长　春	406. 19277	12653. 4	404. 17288	0	76. 31
哈尔滨	1329. 768028	14737. 89	1014. 79195	2793	70. 99
上　海	2254. 588323	476232. 3325	2171. 595012	42	98. 63
南　京	1442. 285545	182495. 702	1318. 767552	0	99. 26
杭　州	635. 189205	71953. 026	606. 194188	2651	93. 93
合　肥	276. 719121	11177. 132	273. 138439	0	60. 4
福　州	425. 131151	8283. 014	401. 162156	6100	94. 37
南　昌	117. 84228	1727. 877	114. 161048	3493. 05	93. 94
济　南	967. 880363	67138. 993	909. 264369	0	89. 2
郑　州	914. 877959	2840. 3	755. 750134	0	35. 19
武　汉	1215. 351422	27182. 307	1087. 071945	0	99. 85
长　沙	154. 623119	1058. 2	140. 106989	287	95. 61
广　州	641. 837657	225869. 8431	597. 949899	260	90. 06
南　宁	379. 724656	515. 288	344. 482861	1553	97. 05
海　口	4. 278592	561. 5	4. 222442	0	79. 8
重　庆	2551. 847352	146841. 62	2076. 735085	1498598. 32	19. 98
成　都	576. 427775	7822. 2353	569. 867174	0	74. 82
贵　阳	872. 642764	10354. 345	411. 32721	99. 5	39. 9
昆　明	2161. 425723	12358. 326	862. 352946	97022	18. 74
拉　萨	26. 46588	0	4. 9586	12060. 8	81. 12
西　安	246. 658117	8252. 443	241. 301483	4004. 66	29. 49
兰　州	485. 822324	157902. 66	364. 876506	0	22. 79
西　宁	250. 315002	106168. 36	242. 641016	5101. 5	91. 93
银　川	164. 08513	3053. 628	149. 894257	7200	90. 8
乌鲁木齐	550. 626137	16856	359. 557698	3127	65. 3

表 32 - 15 2009 年按行业分工业固体废物产生及处理利用情况

Table 32 - 15 Trade - classified Industrial Solid Waste Generation and Treatment Status in 2009

指标	工业固体废物产生量（万吨）	工业固体废物综合利用量（万吨）	工业固体废物贮存量（万吨）	工业固体废物置量（万吨）	工业固体废物排放量（吨）	“三废”综合利用产品产值（万元）	危险废物（万吨）	危险废物（吨）
行业总计	190673.512347	1429.839696	128607.633185	19391.017393	44940.198987	6315720.5744	1.5	16082440.4
煤炭开采和洗选业	23868.562784	0.063104	18409.943807	1490.345877	5069.665493	2613489.37	0	215935.1
石油和天然气开采业	175.482478	12.314435	57.582201	11.232548	106.650309	294.21	0	124425.3
黑色金属矿采选业	23441.799901	0.011135	5823.345816	3905.137726	13700.183405	255339.54	0	40443.9
有色金属矿采选业	25847.909888	158.079669	8953.515568	4713.700028	12164.683288	1215665.848	0	296452
非金属矿采选业	1454.586636	72.75251	764.987291	329.967864	353.269804	69916.77	0	33333.5
其他采矿业	46.012173	0	30.173623	10.07655	5.685	770	0	1679.1
农副食品加工业	2090.558793	0.141482	2055.715463	7.822383	27.532396	29231.6614	0	441430.5
食品制造业	531.940827	0.436544	522.366784	0.208578	8.263463	11586.78	0	175439.8
饮料制造业	930.033696	0.269623	914.26815	1.233503	7.172308	74779.44	0	322743.9
烟草制品业	40.68691	0.014637	33.896125	0.184974	6.348011	2751	0	4491.7
纺织业	732.451685	18.481517	685.428578	0.298002	45.823306	9571.324	0	258372.5
纺织服装、鞋、帽制造业	46.926873	0.1315	44.245119	0.12392	2.49611	616.94	0	6317.7
皮革毛皮羽毛（绒）及其制品业	80.444646	1.919619	65.929578	0.621964	13.743074	1500.331	0	15079.3
木材加工及木竹藤棕草制品业	170.025305	0.028416	167.600786	0.10388	2.071739	2639	0	134382.5
家具制造业	15.770158	0.180493	14.817731	0.06098	0.852617	399.3	0	4824.1
造纸及纸制品业	1938.667773	8.665057	1734.870994	28.68272	177.224476	47457.345	0	2045188.2
印刷业和记录媒介的复制	15.821066	1.693022	14.157351	0.042442	1.617872	36.6	0	17625.8
文教体育用品制造业	2.996852	0.193741	2.345386	0.00101	0.619406	310.5	0	1216.1
石油加工、炼焦及核燃料加工业	2994.396915	141.267025	2703.970221	110.233508	199.590901	28898.28	0	877195
化学原料及化学制品制造业	12595.463612	521.880096	8660.346744	1545.986972	2528.760214	151328.52	0	1352447.8
医药制造业	345.919562	31.93435	315.993701	1.473992	27.258555	16845.03	0	126705.5

续表

指标	工业固体废物产生量（万吨）	工业固体废物综合利用量（万吨）	工业固体废物贮存量（万吨）	工业固体废物置量（万吨）	工业固体废物排放量（吨）	“三废”综合利用产品产值（万元）	危险废物（万吨）	危险废物（吨）
化学纤维制造业	373.01709	41.488963	347.400652	9.372291	23.734526	914	0	87763.1
橡胶制品业	138.715775	0.543576	135.05703	0.14543	3.383792	2795.96	0	56085.5
塑料制品业	65.668914	2.32507	61.271485	0.004308	4.311652	818.7123	0	70474.1
非金属矿物制品业	4358.574589	2.718624	4514.019892	49.822478	115.364303	350510.9788	0	4067650.2
黑色金属冶炼及压延加工业	33893.703219	62.12623	29382.392937	1296.608657	3353.191206	499082.07	0	2447857.2
有色金属冶炼及压延加工业	7087.266833	169.867777	3209.301281	898.559266	2963.245895	400896.4799	0	1089261.1
金属制品业	505.837056	36.194056	471.61242	0.671732	33.455025	2237.8	0	132106.9
通用设备制造业	488.590654	6.492449	453.503774	0.658452	32.103487	23538.493	0	169775.3
专用设备制造业	186.809926	3.614702	160.607924	1.03782	23.234064	19315.331	0	98384.7
交通运输设备制造业	505.786747	15.009148	455.912133	0.163017	47.744096	20079.89	0	240684
电气机械及器材制造业	71.295098	10.835151	62.730355	0.182677	8.454357	834.62	0	85284.3
通信计算机及其他电子设备制造业	172.548203	79.563015	132.512043	0.519185	39.321071	2553.35	0	170357.9
仪器仪表及文化办公用机械制造业	26.104556	11.398102	18.27686	0.021892	7.781643	249.15	0	31346.1
工艺品及其他制造业	18.732723	0.332382	17.734539	0.011634	0.617078	3697.186	0	6044.8
废弃资源和废旧材料回收加工业	64.064438	0.088917	59.255252	0.045182	4.498478	2655.254	0	91021.6
电力、热力的生产和供应业	45131.205566	15.666149	36963.757635	4975.57887	3798.017519	451690.11	1.5	718419
燃气生产和供应业	55.60122	0.31065	35.36468	0.04	20.18254	140	0	1886.6
水的生产和供应业	19.423995	0.768951	8.928983	0.03093	10.437742	263.4	0	15869.8
其他行业	144.10721	0.03781	142.492293	0.00415	1.608767	20	0	6438.7

表 32－16 环境污染治理投资总额年度统计

Table 32－16 Yearly Statistics of Total Investment in Pollution Treatment

单位：亿元

	2005 年	2006 年	2007 年	2008 年	2009 年
全　国	2388	2566	3387.2801	4490.3	4525.2989
北京市	84.9	165.5	185.3066	152.9	208.68
天津市	71.4	40.8	59.7938	68.1	103.6885

续表

	2005年	2006年	2007年	2008年	2009年
河北省	121.4	132.2	170.2463	208.3	248.5772
山西省	48.5	63.2	97.0394	140.9	157.8136
内蒙古自治区	68	104.8	90.5024	135	155.2164
辽宁省	129	145.8	125.1348	163.7	204.9062
吉林省	34	42.3	50.9416	59.6	66.0618
黑龙江省	46.7	54.2	58.7155	98.8	107.786
上海市	88.1	94.3	123.0299	153.5	160.0777
江苏省	294.3	282.7	318.2073	395.9	369.8838
浙江省	160.3	140.3	177.4312	519.7	198.0342
安徽省	49.3	52	82.3536	139	139.2421
福建省	80.9	59.8	77.9631	83.1	87.2094
江西省	37.1	37.5	45.5107	39.2	70.4264
山东省	238.8	258.1	320.7981	432.2	459.4921
河南省	82.4	95.1	114.4035	109.9	121.3169
湖北省	62	67.7	64.3201	90.1	150.6052
湖南省	37.7	54	64.5603	91.4	146.3824
广东省	171.5	160.4	153.5833	164.6	240.1453
广西壮族自治区	41.4	41.2	65.4506	93	132.2614
海南省	6.3	8.3	14.8809	12.7	19.7265
重庆市	50.2	60.1	63.7121	67.3	109.7174
四川省	78.3	71.1	102.1586	100.7	103.5253
贵州省	14.1	19.8	22.4421	23.2	21.2072
云南省	28.4	29	29.9096	44.1	79.6256
西藏自治区	0.5	1.7	0.5722	0.2	2.7138
陕西省	36.5	41	63.7555	75.5	119.0551
甘肃省	20.4	27.8	38.0997	31.2	44.4275
青海省	5.3	6.1	10.645	18.1	12.257
宁夏回族自治区	12.1	21.3	33.445	30.9	28.742
新疆维吾尔自治区	33.5	23.3	35.1552	47.7	78.2147

表32-17 环境污染治理投资总额占国内生产总值年度统计

Table 32-17 Proportion of Investment in Pollution Treatment in Total GDP

单位:%

	2005年	2006年	2007年	2008年	2009年
全　国	1.3	1.22	1.3574	1.4934	1.329
北京市	1.23	2.1	1.9811	1.4579	1.7171
天津市	1.93	0.93	1.1846	1.0717	1.3785
河北省	1.2	1.13	1.2421	1.2867	1.4422
山西省	1.16	1.33	1.692	2.0306	2.1447

续表

	2005 年	2006 年	2007 年	2008 年	2009 年
内蒙古自治区	1.75	2.19	1.4863	1.7393	1.5936
辽宁省	1.61	1.58	1.1352	1.2161	1.347
吉林省	0.94	0.99	0.9635	0.9278	0.9076
黑龙江省	0.85	0.88	0.8304	1.1889	1.2552
上海市	0.96	0.91	1.0095	1.1206	1.0639
江苏省	1.61	1.31	1.2361	1.3061	1.0735
浙江省	1.19	0.89	0.9446	2.4187	0.8614
安徽省	0.92	0.84	1.118	1.5663	1.3837
福建省	1.23	0.79	0.8434	0.7678	0.7127
江西省	0.91	0.8	0.8269	0.6049	0.92
山东省	1.29	1.17	1.2355	1.391	1.3556
河南省	0.78	0.76	0.7621	0.597	0.6228
湖北省	0.95	0.89	0.6972	0.7952	1.162
湖南省	0.58	0.71	0.7012	0.8192	1.1209
广东省	0.77	0.61	0.494	0.4611	0.6082
广西壮族自治区	1.01	0.85	1.0998	1.2968	1.7046
海南省	0.7	0.79	1.2199	0.8703	1.1925
重庆市	1.64	1.72	1.5461	1.3205	1.6802
四川省	1.06	0.82	0.9729	0.8052	0.7316
贵州省	0.71	0.86	0.8174	0.696	0.542
云南省	0.82	0.72	0.6298	0.7737	1.2906
西藏自治区	0.19	0.59	0.1526	0.0505	0.6149
陕西省	0.99	0.91	1.1657	1.102	1.4573
甘肃省	1.05	1.22	1.4095	0.9823	1.3115
青海省	0.97	0.94	1.3526	1.8824	1.1336
宁夏回族自治区	2	3	3.7616	2.8129	2.1238
新疆维吾尔自治区	1.28	0.77	0.9992	1.1348	1.8287

表 32-18 环境污染治理城市环境基础设施建设投资完成投资年度统计

Table 32-18 Yearly Statistics of Implemented Investment in Urban Pollution Treatment Infrastructure

单位：亿元

	2005 年	2006 年	2007 年	2008 年	2009 年
全　国	1289.7	1314.9	1467.5002	1801.0293	2511.9744
北京市	66	126.7	135.9766	119.0258	180.1964
天津市	28.5	17.9	20.0768	35.3015	56.0518
河北省	75.2	84.6	87.2318	115.7037	158.9841
山西省	17.5	13.7	32.5823	42.6334	63.7727
内蒙古自治区	47.8	57.5	51.2855	81.8078	112.9743
辽宁省	75	77.3	79.1382	115.5044	152.9132

续表

	2005 年	2006 年	2007 年	2008 年	2009 年
吉林省	25.4	31.2	31.8853	26.1539	42.4319
黑龙江省	37.9	40.2	39.2564	63.7857	84.0851
上海市	50.4	50.6	68.3187	71.1973	63.9136
江苏省	182	161.4	160.9809	185.6277	232.7926
浙江省	84.3	57.9	65.6171	105.368	99.0716
安徽省	38.3	36.6	45.3478	70.7825	93.3185
福建省	26.1	24.9	40.206	33.8293	31.4899
江西省	20.8	24	18.213	23.9367	48.4864
山东省	156.6	160.5	173.9716	223.8444	296.0015
河南省	44.3	47.4	41.2962	49.8788	57.911
湖北省	36	38.5	38.1928	48.4002	70.5803
湖南省	15.6	28.7	34.4471	52.4664	90.3355
广东省	81.7	62.6	66.9813	86.9519	183.3176
广西壮族自治区	24.1	25.6	41.7055	46.3153	84.9997
海南省	3.8	3.5	9.0338	7.5523	15.2358
重庆市	34.7	32.7	28.333	33.5449	62.0851
四川省	43.9	31.5	53.2023	40.9549	51.5472
贵州省	4.1	5.4	8.1479	7.8978	5.1481
云南省	10.7	10.1	5.3194	8.6983	33.4954
西藏自治区	0.5	1.7		0.202	2.7138
陕西省	18.6	21.2	29.4115	43.1517	61.0463
甘肃省	9.8	11.2	14.4819	14.4569	18.5551
青海省	1.5	3.9	4.708	6.0645	3.6976
宁夏回族自治区	7.1	13.5	22.4213	12.9941	10.0273
新疆维吾尔自治区	21.7	12.5	19.7302	26.9972	44.795

表 32-19　环境污染治理工业污染源治理投资完成投资年度统计

Table 32-19　Yearly Statistics of Implemented Investment in Industrial Pollution Source Treatment

单位：亿元

	2005 年	2006 年	2007 年	2008 年	2009 年
全　国	458.2	483.9	552.3909	542.6404	442.6207
北京市	10.9	10.1	8.1207	7.8475	3.4421
天津市	18.6	15	15.0527	16.827	18.0054
河北省	25.2	19.1	21.5485	20.5675	13.2272
山西省	19.8	36.8	45.724	52.937	38.6711
内蒙古自治区	2.6	17.7	16.7487	21.9189	17.8258
辽宁省	36.9	52	23.7002	20.1645	19.6562
吉林省	5.2	4	8.0308	9.3865	7.9254

续表

	2005 年	2006 年	2007 年	2008 年	2009 年
黑龙江省	4. 6	5. 8	10. 211	9. 5079	9. 9318
上海市	8. 8	5. 9	16. 4318	10. 3901	6. 8357
江苏省	38. 9	28	53. 7032	39. 7126	27. 0554
浙江省	19. 9	25	21. 3773	14. 8006	19. 3574
安徽省	4. 5	5. 5	11. 3852	11. 5341	10. 8282
福建省	34. 5	19. 6	13. 8007	15. 5763	12. 8692
江西省	7. 2	6. 9	8. 2688	5. 0665	3. 954
山东省	60. 5	59. 7	67. 342	84. 4159	51. 5832
河南省	20. 7	24. 7	33. 8132	24. 611	15. 4242
湖北省	14. 8	14. 9	18. 8634	16. 1453	28. 1332
湖南省	14. 1	17. 3	13. 3641	14. 3905	13. 3806
广东省	37	31. 4	46. 2758	40. 3276	22. 7464
广西壮族自治区	10. 4	8. 7	18. 194	14. 9751	11. 7118
海南省	0. 4	2. 1	0. 3889	0. 3774	0. 3562
重庆市	3. 9	3. 7	10. 007	9. 7396	7. 0747
四川省	20	20. 3	20. 1033	19. 3808	9. 6191
贵州省	5. 9	10. 1	4. 5646	10. 2029	8. 9475
云南省	6. 8	9. 4	8. 6423	10. 2677	9. 488
西藏自治区				0	0
陕西省	12. 7	7. 4	9. 7045	10. 6582	20. 5999
甘肃省	6. 7	13. 6	14. 9087	11. 8436	12. 3302
青海省	0. 5	0. 8	0. 7913	1. 1165	2. 9439
宁夏回族自治区	1. 8	4	4. 6272	9. 0631	4. 3472
新疆维吾尔自治区	4. 4	4. 5	6. 6748	8. 8878	14. 3497

表 32－20　环境污染治理建设项目“三同时”环保投资完成投资年度统计

Table 32－20　Yearly Statistics of Implemented Investment in the “Three Simultaneity”

单位：亿元

	2005 年	2006 年	2007 年	2008 年	2009 年
全国	640. 1	767. 2	1367. 389	2146. 7	1570. 7038
北京市	8	28. 7	41. 2094	26. 1	25. 0414
天津市	24. 3	7. 9	24. 6643	16	29. 6313
河北省	21	28. 5	61. 466	72	76. 3659
山西省	11. 2	12. 7	18. 733	45. 4	55. 3698
内蒙古自治区	17. 6	29. 6	22. 4681	31. 3	24. 4163
辽宁省	17. 1	16. 5	22. 2964	28	32. 3368
吉林省	3. 4	7. 1	11. 0255	24	15. 7044
黑龙江省	4. 2	8. 2	9. 2481	25. 5	13. 7591
上海市	28. 9	37. 8	38. 2794	71. 9	89. 3284

续表

	2005 年	2006 年	2007 年	2008 年	2009 年
江苏省	73.4	93.3	103.5232	170.6	110.0358
浙江省	56.1	57.4	90.4368	399.5	79.6052
安徽省	6.5	9.9	25.6206	56.7	35.0954
福建省	20.3	15.3	23.9564	33.7	42.8503
江西省	9.1	6.6	19.0289	10.2	17.9861
山东省	21.7	37.9	79.4844	124	111.9075
河南省	17.4	23	39.2941	35.4	47.9817
湖北省	11.2	14.3	7.2639	25.6	51.8916
湖南省	8	8	16.7491	24.5	42.6662
广东省	52.8	66.4	40.3262	37.3	34.0813
广西壮族自治区	6.9	6.9	5.5511	31.7	35.5499
海南省	2.1	2.7	5.4582	4.7	4.1344
重庆市	11.6	23.7	25.3722	24.1	40.5577
四川省	14.4	19.3	28.853	40.3	42.359
贵州省	4.1	4.3	9.7296	5.1	7.1116
云南省	10.9	9.5	15.9478	25.1	36.6422
西藏自治区			0.55	0	0
陕西省	5.2	12.4	24.6395	21.6	37.4089
甘肃省	3.9	3	8.7091	4.9	13.5422
青海省	3.3	1.4	5.1458	10.9	5.6156
宁夏回族自治区	3.2	3.8	6.3965	8.8	14.3675
新疆维吾尔自治区	7.4	6.3	8.7501	11.8	19.07

表 32－21　2009 年各地区环境污染治理投资

Table 32－21　Investment in Pollution Treatment of Different Regions in 2009

指标	环境污染治理投资总额（亿元）	环境污染治理投资总额占国内生产总值比重（%）	城市环境基础设施建设投资（亿元）	工业污染源治理投资（亿元）	建设项目“三同时”环保投资（亿元）	燃气（亿元）	集中供热（亿元）	排水（亿元）	园林绿化（亿元）	市容环境卫生（亿元）
全　国	4525.2989	2511.9744	182.1692	368.6727	729.8043	914.8554	316.4728	442.62069	1570.70381	1.328989
北　京	208.67995	180.1964	10.5461	40.3911	24.8839	44.4321	59.9432	3.44213	25.04142	1.717102
天　津	103.68848	56.0518	10.7878	5.1851	19.5664	18.0727	2.4398	18.0054	29.63128	1.378497
河　北	248.57723	158.9841	12.2795	25.2891	34.2631	72.6126	14.5398	13.2272	76.36593	1.442241
山　西	157.81359	63.7727	1.6759	24.2858	18.9755	14.6352	4.2003	38.67109	55.3698	2.144699
内蒙古	155.21642	112.9743	4.4628	37.0495	13.4819	49.2523	8.7278	17.82581	24.41631	1.593557
辽　宁	204.90621	152.9132	3.7223	87.6139	26.6128	29.9006	5.0636	19.65622	32.33679	1.34696
吉　林	66.06179	42.4319	2.0264	14.632	12.1002	7.7625	5.9108	7.92545	15.70444	0.907598

续表

指标	环境污染治理投资总额（亿元）	环境污染治理投资总额占国内生产总值比重（%）	城市环境基础设施建设投资（亿元）	工业污染源治理投资（亿元）	建设项目“三同时”环保投资（亿元）	燃气（亿元）	集中供热（亿元）	排水（亿元）	园林绿化（亿元）	市容环境卫生（亿元）
黑龙江	107.78596	84.0851	3.0664	34.0865	19.8361	19.1764	7.9197	9.93178	13.76908	1.255223
上　海	160.07771	63.9136	19.0828	0	28.7717	7.7791	8.28	6.83569	89.32842	1.06389
江　苏	369.88381	232.7926	16.7828	2.728	57.6921	139.402	16.1877	27.0554	110.03581	1.073456
浙　江	198.034163	99.0716	9.8623	0.7523	43.9405	34.2968	10.2197	19.35738	79.605183	0.86138
安　徽	139.24213	93.3185	7.2243	2.7341	25.0066	52.834	5.5195	10.82821	35.09542	1.383729
福　建	87.20937	31.4899	3.4455	0	12.3486	6.9438	8.752	12.86918	42.85029	0.712697
江　西	70.42643	48.4864	3.5931	0	13.2432	29.6733	1.9768	3.95395	17.98608	0.919984
山　东	459.49214	296.0015	21.442	60.3513	76.054	125.1983	12.9559	51.58316	111.90748	1.355568
河　南	121.31687	57.911	7.3551	8.0391	21.0839	17.4264	4.0065	15.4242	47.98167	0.622762
湖　北	150.60518	70.5803	6.4964	0	29.1569	21.2316	13.6954	28.13323	51.89165	1.161978
湖　南	146.38236	90.3355	5.2116	0	42.7244	32.924	9.4755	13.38063	42.66623	1.120872
广　东	240.14532	183.3176	5.7686	0	76.3439	9.9019	91.3032	22.74643	34.08129	0.608231
广　西	132.26141	84.9997	3.0838	0	36.1738	41.0363	4.7058	11.71184	35.54987	1.704584
海　南	19.72647	15.2358	0.4803	0	11.5233	2.3837	0.8485	0.35625	4.13442	1.192501
重　庆	109.71745	62.0851	7.3627	0	11.4528	42.078	1.1916	7.07466	40.55769	1.680203
四　川	103.52528	51.5472	2.662	0	18.3607	27.5858	2.9387	9.61906	42.35902	0.731561
贵　州	21.2072	5.1481	0.5118	0	3.3648	0.6035	0.668	8.94746	7.11164	0.542012
云　南	79.62562	33.4954	1.3959	0	16.3566	7.0387	8.7042	9.48801	36.64221	1.290581
西　藏	2.7138	2.7138	0	0	0	2.6197	0.0941	0	0	0.614872
陕　西	119.05507	61.0463	3.0822	5.4332	15.9818	33.7178	2.8313	20.5999	37.40887	1.457258
甘　肃	44.42746	18.5551	2.3411	8.8776	4.191	2.7553	0.3901	12.33019	13.54217	1.311489
青　海	12.25705	3.6976	0.0168	0.68	1.5954	1.2349	0.1705	2.94388	5.61557	1.133579
宁　夏	28.742	10.0273	0.558	1.4244	3.1441	4.2568	0.644	4.34718	14.36752	2.12383
新　疆	78.21472	44.795	5.8429	9.1197	11.5743	16.0893	2.1688	14.34972	19.07	1.828707

表 32－22　2009 年工业污染治理投资完成情况

Table 32－22　Implemented Investment in Industrial Pollution Treatment in 2009

指标	工业污染治理项目本年完成投资（万元）	本年竣工项目数（个）	治理废水（万元）	治理废气（万元）	治理固体废物（万元）	治理噪声（万元）	治理其他（万元）
全　国	4426206.9	1494606	2324616	218535.7	14100	374349.2	8236
北　京	34421.3	1205.4	25718.2	0	12.4	7485.3	56
天　津	180054	40866.6	75920.9	545.8	398.8	62321.9	174
河　北	132272	35816.8	91106.1	0	1398.7	3950.4	226

续表

指标	工业污染治理项目本年完成投资（万元）	本年竣工项目数（个）	治理废水（万元）	治理废气（万元）	治理固体废物（万元）	治理噪声（万元）	治理其他（万元）
山　西	386710. 9	103324. 3	235573. 5	19686. 9	1742	26384. 2	707
内蒙古	178258. 1	33154. 6	123758. 5	10193. 6	120	11031. 4	247
辽　宁	196562. 2	36811	157287. 4	1519. 3	99. 2	845. 3	132
吉　林	79254. 5	26959. 2	48408. 9	435. 2	18. 3	3432. 9	95
黑龙江	99317. 8	45332. 7	50109. 3	755	0	3120. 8	117
上　海	68356. 9	9702. 9	40745. 7	11178	1598. 3	5132	237
江　苏	270554	145428	104852. 9	3363. 1	379. 1	16530. 9	600
浙　江	193573. 8	57671. 6	108609. 3	16756. 2	58	10478. 7	654
安　徽	108282. 1	19569. 5	65879. 8	80	471	22281. 8	212
福　建	128691. 8	51297. 5	61467. 8	9433	1059. 8	5433. 7	435
江　西	39539. 5	17427. 4	15339. 1	3780. 6	3	2989. 4	110
山　东	515831. 6	237316. 2	197083. 6	28790. 7	1045. 3	51595. 8	644
河　南	154242	66224. 4	61146. 8	5686. 9	1105. 2	20078. 7	329
湖　北	281332. 3	54001. 6	213918. 3	5969. 7	674. 8	6767. 9	273
湖　南	133806. 3	66314. 7	58965. 1	5375. 1	404. 7	2746. 7	286
广　东	227464. 3	84464. 6	65649. 4	11197. 3	302. 5	65850. 5	883
广　西	117118. 4	75922. 1	28285. 7	7089. 6	0	5821	223
海　南	3562. 5	3215. 6	142. 3	0	0	204. 6	14
重　庆	70746. 6	28813. 2	37521. 7	590. 4	675. 6	3145. 7	118
四　川	96190. 6	52685. 7	31900	6839. 7	201. 5	4563. 7	307
贵　州	89474. 6	6925. 8	38746. 9	35266. 7	212. 7	8322. 5	197
云　南	94880. 1	14807. 5	63768. 6	13398. 9	424. 3	2480. 8	381
西　藏							
陕　西	205999	75180. 8	106440	6675. 5	1003. 8	16698. 9	186
甘　肃	123301. 9	56648. 5	51172. 8	13313. 5	511	1656. 1	154
青　海	29438. 8	3885. 2	25543. 6	0	10	0	28
宁　夏	43471. 8	12890. 5	30197. 8	15	0	368. 5	84
新　疆	143497. 2	30742. 1	109356	600	170	2629. 1	127

大事记

（2010 年 10 月—2011 年 9 月）

浙江绍兴全国首创用电“诊断书”

2010 - 10 - 13

本月底，浙江省绍兴地区 32.66 万户居民都会收到一份由电力部门为他们量身定做的《居民家庭科学用电建议书》，此举乃全国首创。《建议书》对每户居民一年来每月高峰、低谷用电情况，以及每月电费、平均电价等作了详细介绍，还用曲线图形象地展示了全年每月总电量、峰电量、谷电量的使用情况，还给用户提出了“合理减少电费支出”的建议。

商务部发布流通业节能减排的指导意见

2010 - 10 - 13

为贯彻落实国务院节能减排工作电视电话会议上的讲话精神，商务部发布流通业节能减排的指导意见，从推动商业企业（市场）开展节能技术改造，加大汽车、家电以旧换新工作力度，抑制商品过度包装和限制塑料袋使用工作，餐饮服务业节能减排，推广使用散装水泥，发展低碳物流等七个方面进一步加强流通服务业节能减排工作。

中共中央十七届五中全会带来节能减排发展新契机

2010 - 10 - 18

中共中央十七届五中全会于 15 日—18 日在北京举行，会议公报和随后发布的“十二五”规划建议中均强调加快建设资源节约型、环境友好型社会，提高生态文明水平，积极应对全球气候变化，大力发展循环经济，加强资源节约和管理等内容。中央落实节能减排的决心和目标超出预期，环保、节能、新能源行业在政策和资金上将得到很大的支持，“十二五”将是相关行业的黄金发展时期。

“中国西部工业绿色发展高峰论坛”举行

2010 - 10 - 22

10 月 22 日，由四川省政府主办，国家工信部支持的“2010 中国西部工业绿色发展高峰论坛”在成都举办。该论坛以“节能减排，绿色发展”为主题，分析当前工业节能减排、清洁生产、资源综合利用和循环经济等面临的形势；介绍国外工业绿色发展、低碳经济建设的先进经验和做法，推介先进适用的技术和产品；探讨“十二五”工业节能减排、清洁生产、资源综合利用和循环经济的对策及措施，为西部发展绿色工业提出建议。

工信部发布首批工业节能服务公司名单

2010－10－22

10月22日，工信部发布了首批面向工业和通信业领域开展业务的节能服务公司推荐名单，上市公司中动力源、九洲电气、达实智能、智光电气等榜上有名。

工信部表示，这批名单是为贯彻落实《国务院办公厅转发发展改革委等部门关于加快推行合同能源管理促进节能服务产业发展意见的通知》精神，推进工业和通信业领域开展合同能源管理而制定的。通知为相关上市公司在获取节能补贴、绿色低息信贷、税收减免方面提供了制度保障，在主管部门的扶持下，入选工信部首批名单的上市公司有望作为专业化节能服务公司快速做强。

2010年全国关停小火电名单公布

2010－11－01

国家发改委公布我国2010年关停小火电机组的执行情况，并发布已关闭企业名单，接受社会各界监督。在首批公布的1—7月份关停名单中涉及山西企业42家，占到全国总数的28%，列全国第一；涉及发电量110.01万千瓦，占全国削减发电量的10%。

我国设备制造业首个节能自愿协议签署

2010－11－13

工业和信息化部与华为在北京签署了节能自愿协议，根据协议内容规定，双方未来将在通信业绿色环保节能减排领域展开深入沟通与合作，共同推进中国通信业的节能减排战略目标达成。此次双方节能自愿协议的签署，不仅将直接推进整个行业在节能减排上的行动自觉，促成战略目标的达成，更将纵深影响全球的绿色革命。

发改委发布《电力需求侧管理办法》

2010－11－18

国家发改委等部委日前联合发布《电力需求侧管理办法》。《办法》指出，为了节约用电和科学用电，明年起各地实施“电力需求侧管理”，规定各省制定年度电力节约指标，并将此纳入地区能源和经济发展规划。

我国首次承担节能减排领域国际标准化技术委员会的组织工作

2010－11－25

ISO技术管理局通过了由我国提出的“节能技术改造项目、工业企业和区域节能量确定方法新技术委员会”提案，并由我国承担该技术委员会秘书处。这是我国承担的第一个节能减排领域的国际标准化技术委员会，是我国节能减排标准化工作取得的一项重大突破。同时获得通过的还有我国提出的国际标准提案，这是一项包括改造项目、工业企业和区域三个层面节能量计算技术通则的国际标准。

2010 中国绿色产业和绿色经济高科技国际博览会在京举办

2010－11－27

11 月 24 日，经国务院批准，由商务部、中宣部等 12 个部委联合主办的 2010 中国绿色产业和绿色经济高科技国际博览会（简称“绿博会”）在北京展览馆开幕，历时 4 天，于 11 月 27 日圆满落下帷幕。本届绿博会集中展示了 25 个国家和地区的 212 家国际知名企业在清洁能源、节能减排、环境保护、低碳技术、循环经济等领域的一流企业、一流技术、一流产品、一流理念以及前沿科技研究成果。展会同期，还举办了 2010 中国绿色产业和绿色经济高科技高峰论坛，围绕“绿色经济引领世界未来发展”这一主题，邀请中外政要、专家学者、企业高管等进行讲演。

中国城市节能减排高峰论坛开幕

2010－11－29

由发改委、工信部、环保部、国资委等部门主办的第二届中国城市节能减排高峰论坛，于 11 月 29 日—30 日在北京举办。论坛以“加快转变经济发展方式”为主题，围绕低碳经济、低碳生活，垃圾分类、污水处理与资源化利用，节能减排与高新技术，节能减排体制机制，节能减排与公众参与，工业、电力、汽车、建筑等重点行业节能减排展开。

第四批节能汽车目录公布

2010－12－3

国家发改委、工信部、财政部三部委联合公布了第四批节能汽车推广目录，此批包括 17 家汽车厂商的 66 款车型。截至最新公布的第四批节能汽车推广目录，至此纳入四批节能推广汽车目录的汽车上升至 272 款。同时，三部委还联合发布了《“节能产品惠民工程”节能汽车推广专项核查办法》。

2010 年坎昆会议闭幕

2010－12－11

《联合国气候变化框架公约》第 16 次缔约方大会（COP16）和《京都议定书》第 6 次缔约方大会（CMP6）于 11 日在墨西哥坎昆闭幕。会议通过了两项应对气候变化的决议，推动气候谈判进程继续向前。会议坚持了《联合国气候变化框架公约》、《京都议定书》和“巴厘路线图”，坚持了“共同但有区别的责任”原则并就适应、技术转让、资金和能力建设等发展中国家关心问题的谈判取得了不同程度的进展。

2010 年中央经济工作会议明确节能减排产业发展方向

2010－12－12

12 日，2010 年中央经济工作会议闭幕，会议提出要强化节能减排和应对气候变化，大力发展绿色经济，完善政府节能减排目标责任考核评价体系，进一步发挥市场作用，健全激励和约束机制，增强企业和全社会节能减排内生动力。同时提出加强重点节能工程建设，大力发展循环经济和环保产业，加快低碳技术研发应用；加强重点流域、区域、行业污染治理，加快建立生态补偿机制；落实控制温室气体排放行动目标，积极开展应对全球气候变化国际合作。预计“十二五”期间，这一投资规模将达到3.1万亿元人民币，年均投资6200亿元，增幅超过100%，占全国GDP的比重约为1.35%。

《节能产品认证管理办法》发布

2010-12-15

15日，国务院法制办和国家质检总局联合发布了《节能产品认证管理办法（征求意见稿）》。办法中明确规定，取得节能产品认证的产品、设备，将优先列入政府采购名录，并对认证产品进行跟踪检查，不合格的将撤销证书并对外公布。同时，办法对认证申请与受理、产品检测和检验机构都进行了更为详细的规定。目前，节能产品认证在我国已涉及家用电器、照明产品、办公设备、机电产品、电力设备等50多个品种。

国内首个生物质炉VER自愿减排项目启动

2010-12-20

河北光磊炉业有限公司实施的“30万台生物质炉具VER自愿减排项目”在河北省故城县启动。这是国内首个在生物质炉具行业实施的VER自愿减排项目。VER自愿减排，是一种自发的、公益的、可认证的减排信用额度。自愿减排市场指的是一种碳排放交易市场。生物质炉具VER自愿减排项目的实施，实现了农村秸秆资源、能源化利用，杜绝秸秆露天焚烧，替代燃煤，节约能源，减少大气污染和二氧化碳等温室气体排放，对提高农民生活质量，促进农民增收节支具有积极的推动作用。

南方电网节能发电调度运行启动

2010-12-27

国家能源局今天在广州召开南方电网节能发电调度运行启动会议，南方电网广东、广西、云南、贵州和海南等五省区全网实施节能发电调度，这是我国电力行业加大节能减排力度的一项具体措施。

我国正式实施合同能源管理税收优惠政策

2011-01-01

财政部、国家税务总局2010年11月联合发布《关于促进节能服务产业发展增值税、营业税和企业所得税政策问题的通知》，明确了我国合同能源管理项目的具体税收优惠政策。从2011年起，我国节能服务公司实施合同能源管理项目将享有增值税、营业税和企业所得税等多项税收优惠政策。

国内首个工业建筑节能示范园开建

2011-01-07

1月7日，位于沈阳铁西现代建筑产业园的日本工业园二期工程通过专家评审，建设工作正式启动。该项目建成后，将成为全国第一个工业建筑节能示范园区。该项目总建筑面积30余万平方米，在设计上严格执行沈阳市《工业建筑节能设计导则》，在节能降耗方面采用了许多新技术、新材料，主体采用轻钢结构，外墙、屋面等围护结构采用聚氨酯夹芯板进行保温，供暖采用地源热泵系统，照明采用自然光和节能灯具，节能措施完备，实现资源的循环利用。据测算，二期项目 8 栋厂房 50 年可以节省标准煤 4.3 万吨，节电 20 万千瓦时。

中国节能减排标准化技术联盟启动

2011 - 01 - 18

1 月 18 日，由中国标准化研究院、国家节能中心、国家发改委培训中心、中国节能协会等 11 家机构联合发起的中国节能减排标准化技术联盟正式启动。联盟定位为“致力于推进中国节能减排标准化工作，集公益性、服务性和学术性为一体的非盈利性技术组织”，主要在节能减排标准化研究、技术服务、打造交流平台、组织公益活动、建设宣传渠道等方面开展工作，国内有多家高校、企业申请加入。

国管局明确中央国家机关“十二五”节能减排目标任务

2011 - 01 - 19

国务院机关事务管理局 19 日组织召开中央国家机关 2011 年节能减排工作会议，明确中央国家机关“十二五”节能减排目标任务。会议要求，“十二五”期间中央国家机关要加快推进重点领域节能减排，突出抓好节电，节水，节油工作。力争到“十二五”末中央国家机关新能源和可再生能源利用比例提高到 5% 以上；三年内实现中央国家机关全部达到北京市节水型单位标准；实施公务用车油耗定额管理等制度。

中国节能蓄能技术中心成立

2011 - 01 - 27

由国家发展和改革委员会下辖中国设备管理协会领导的中国节能蓄能技术中心（CEST）正式成立。中心的核心任务是发展、推广节能蓄能技术。主要包括组织宣传国家有关方针政策，组织制定节能蓄能行业标准，推广和发布节能蓄能新产品、新设备、新技术等。

18 部委联合发布《淘汰落后产能工作考核实施方案》

2011 - 02 - 01

工信部、国家发改委、能源局、环保部等 18 个部委联合发布的文件为《淘汰落后产能工作考核实施方案》，对于今后地方淘汰落后产能实施进度制订了完整的时间表。包括每年 2 月底前提出本地区重点行业淘汰落后产能年度目标任务和计划淘汰落后产能的企业名单；12 月底前，列入各地区本年度淘汰计划的落后产能全部拆除主体设备、生产线，使其不能恢复生产；次年 3 月底前 18 部委进行检查考核。

IFC 向杭州提供 50 亿元节能减排贷款

2011 - 02 - 09

2 月 9 日，国际金融公司（IFC）与杭州市政府签署谅解备忘录，建立一个损失分担机制，支持杭州地区最多7.7亿美元（约50亿人民币）的能源效率、可再生能源和水资源利用效率贷款，共同应对气候变化。这一损失分担机制是国际金融公司“中国节能减排融资项目”的一部分，根据这一项目，在银行发放贷款支持能效和可再生能源项目时，国际金融公司将承担银行的部分信贷损失。

我国主导的IEEE 1888绿色节能国际标准发布

2011－02－28

经美国电气和电子工程师协会标准协会（IEEE－SA）批准，由我国企业主导制定的IEEE 1888标准（Ubiquitous Green Community Control Network，泛在绿色控制网络协议）正式发布，此举标志着我国在绿色节能国际标准制定方面取得重大突破。IEEE 1888是首个以绿色节能为宗旨，信息通信技术与节能减排融合的创新型技术标准，该标准支持广域IPv4和IPv6网络。

“十一五”节能减排任务基本或超额完成

2011－03－01

3月1日，国家统计局发布“十一五”时期经济社会发展主要指标报告显示，2010年全国单位国内生产总值能耗比2005年累计下降19.1%，基本完成“十一五”规划中20%的任务；全国化学需氧量排放量比2005年累计下降12.45%，二氧化硫累计下降14.29%，双双超额完成“十一五”规划确定的减排任务。

《中国低碳城市发展绿皮书2011》在京发布

2011－03－02

中国社会科学院城市发展与环境研究所、湖南工业大学、经济杂志社等于3月2日在北京正式对外联合发布了《中国低碳城市发展绿皮书2011》，据绿皮书数据显示：西宁能源碳排放强度最低，深圳单位GDP能耗最低，上海能源消费总量最多，湛江人均CO_2排放量最少，北京居民碳消费结构最优，张家界市居住建筑单位面积能耗最低。

“十二五”规划纲要发布

2011－03－16

《中华人民共和国国民经济和社会发展第十二个五年（2011—2015年）规划纲要》经全国人大审议通过并发布。《纲要》明确提出健全节能减排激励约束机制的政策导向。在《纲要》中，节能环保、新能源、新能源汽车等产业被列为战略性新兴产业，强调推动能源生产和利用方式变革。同时，《纲要》明确了积极应对气候变化、加强资源节约和管理、大力发展循环经济的方向和重点工程。

“十二五”期间节能减排主要指标公布

2011－03－16

在《国民经济和社会发展第十二个五年规划纲要》中，我国节能减排相关的主要指标公布。六项指标均为约束性指标，分别是单位工业增加值用水量降低30%；非化石能源占一次能源消费比重达到 11.4%；单位国内生产总值能源消耗降低 16%；单位国内生产总值二氧化碳排放降低 17%；主要污染物排放总量显著减少，化学需氧量、二氧化硫排放分别减少 8%；氨氮、氮氧化物排放分别减少 10%。其中，非化石能源比重和氨氮、氮氧化物减排均为新增指标。

交通运输业“十二五”降耗目标公布

2011－03－23

23 日的交通运输部例行记者会上，交通运输部公布了我国交通运输行业“十二五”节能减排的相关指标：到 2015 年，在能源强度指标方面与 2005 年相比，营运车辆单位运输周转量能耗下降 10% 左右，其中营运客车、营运货车分别下降 6% 和 12% 左右。这为今后一个阶段，国内交通运输行业的节能减排工作提出了新标准和新要求。

中国首份“无悔减排”报告公布

2011－03－24

《中国的无悔减排政策：给力低碳城市发展》报告在深圳发布，报告提出：中国在节能减排和促进低碳城市发展的政策设计中，要建立地方政府实施无悔减排激励相容的政策；报告还提出从“汽车排放税”入手加快试行“碳税”、中国加快建立区域性的碳交易市场等建议。

“无悔减排”，就是不影响经济发展的减排，或者说是“无成本”的减排，这是发展中国家应对气候变化的重要理念。

我国“十二五”和 2011 年工业节能减排四大指标公布

2011－03－28

工业和信息化部 28 日公布“十二五”期间和今年我国工业节能减排四大约束性指标，明确 2015 年我国单位工业增加值能耗、二氧化碳排放量和用水量分别要比“十一五”末降低 18%、18% 以上和 30%，工业固体废物综合利用率要提高到 72% 左右；明确今年这四项指标同比要分别降低 4%、4% 以上和 7% 左右以及提高 2.2 个百分点。

我国首个自愿减排标准实现首笔交易

2011－03－29

2011 年 3 月 29 日，国内商业地产企业方兴地产（中国）有限公司通过北京环境交易所成功购买 16800 吨“熊猫标准”的自愿碳减排量。此次交易双方均为中国企业，减排量的核算采用本土标准，这在发展中国家尚属首次，对于中国碳交易领域具有里程碑意义的事件。

武汉成国内首个利用外资对公共建筑节能改造城市

2011－03－31

武汉市与法国合作的既有公共建筑节能改造示范工程贷款项目获得法国开发署批准，武汉市图书馆、市委大院等 6 处建筑将于今年上半年率先利用外资进行节能改造示范。这是全国第一个利用外国优惠贷款和赠款开展既有公共建筑节能改造的项目，在全国范围内具有一定的示范性。该项目估算总投资 1.99 亿元。

"十一五"期间北京万元GDP能耗绝对值全国最低

2011－04－01

统计数据显示，"十一五"期间，北京市万元GDP能耗由2005年的0.792吨标准煤下降到2010年的0.58吨标准煤左右，累计下降26.5%左右，下降幅度位于全国第一，万元GDP能耗绝对值全国最低，也是全国唯一连续五年完成年度目标的省级地区。

北京银行百亿元授信助力节能减排

2011－04－08

4月8日，北京银行与中国节能协会节能服务产业委员会（以下简称EMCA）签署战略合作协议。双方将充分发挥各自优势，共同搭建节能融资服务平台，以积极响应政策号召、加大对节能减排领域的融资支持力度。根据协议，未来五年内北京银行将为EMCA会员企业提供意向性授信100亿元人民币，约形成年节能能力560万吨标准煤，实现年减排二氧化碳1400万吨。

民航局发文加速行业节能减排

2011－04－12

民航总局出台了《关于加快推进行节能减排工作的指导意见》。《意见》明确，我国民航业要全面建立适应国际节能减排发展趋势的技术和管理体系，航空公司要通过技术节能以提高应对国际节能减排新形势的能力。此外，《意见》还提出积极推进机场建设和地面服务中的节能减排。《意见》提出，到2020年我国民航单位产出能耗和排放比2005年下降22%，达到航空发达国家水平。

春兰高能电池通过国家"863"节能重大专项验收

2011－04－12

12日，国家"863"计划节能与新能源汽车重大项目专家组对春兰清洁能源研究院有限公司研发的系列车用镍氢动力电源系统课题进行了验收。专家组认为春兰研发的系列车用镍氢动力电源系统，完全满足混合电动车规模化应用要求，相关课题组掌握了车用动力电源系统及其主要组成部件的关键核心技术、应用技术、集成技术和产业化制造与服务技术，而且为多个动力系统技术平台和整车项目提供了车用动力电源标准化、通用化的整体解决方案，并形成了规模化产品供应能力和服务能力，其主要性能指标处于国际先进水平。

国家电网首个能效服务小组在常州成立

2011－04－13

13日，国家电网公司在江苏常州启动了系统内的第一个能效服务活动小组。这意味着，中国的国家电网公司将利用自己庞大的用户网络资源，介入到力推节能减排的行动中来。此次成立的能效服务活动小组由供电公司牵头，12家企业组成，其中6家为中资企业，6家为外资企业。行业主要覆盖一些高耗能企业，用电量在200万千瓦时—6亿千瓦时。

中国碳减排第一案开庭

2011－04－14

中国碳减排行业的第一案在北京朝阳法院开庭。上海太比雅环保公司起诉国际权威认证机构挪威船级社北京分公司“涉嫌合同审定日期倒签”，导致自己通过出售碳减排额度获利目的落空，这是我国碳减排交易领域的第一起诉诸法律的案件。

按《京都议定书》要求之下的全球性减排协议，企业可以通过自愿减排项目，获得由相关认证机构签发的碳减排额度，然后通过出售此额度而获得收益。全世界有不到30家这样的机构，挪威船级社就是其中之一。

湖南发布国内首份电力减排通报

2011－04－28

国家电力监管委员会湖南电监办28日对外发布《湖南省电力工业“十一五”期间节能减排工作情况通报》。该《通报》显示，湖南电网经营企业在“十一五”期间大力加强输配电线路损耗管理，电网综合线损率由“十五”末的8.69%下降到“十一五”末的7.07%，累计节约电量38.49亿千瓦时，节约标准煤47万吨，减少二氧化硫排放1.4万吨。该报告是国内电力监管机构首份向社会发布的电力工业“十一五”期间节能减排工作情况通报。

两部委明确“十二五”公共建筑能耗标准

2011－05－04

财政部和住建部下发《关于进一步推进公共建筑节能工作的通知》，明确要争取在“十二五”期间实现公共建筑单位面积能耗下降10%，并将通过启动公共建筑节能改造重点城市等方式全面推进。《通知》提出，争取在“十二五”期间，实现公共建筑单位面积能耗下降10%，其中大型公共建筑能耗降低15%。

工信部公布2011年落后产能淘汰任务

2011－05－10

工业和信息化部10日发布公告，已于近期将工业行业2011年淘汰落后产能目标任务分解下达到各地。18个工业行业淘汰落后产能目标任务为：炼铁2653万吨，炼钢2627万吨，焦炭1870万吨，铁合金185.7万吨，电石137.5万吨，电解铝60万吨，铜冶炼29.1万吨，铅冶炼58.5万吨，锌冶炼33.7万吨，水泥13355万吨，平板玻璃2600万重量箱，造纸744.5万吨，酒精42.7万吨，味精8.3万吨，柠檬酸1.45万吨，制革397万标张，印染17.3亿米，化纤34.97万吨。

国内首例碳资产抵押获得授信

2011－05－10

福州市闽侯县兴源水力发电有限公司的水电站CDM项目获得了兴业银行提供的碳资产质押授信，即以水电站未来的应收账款（售碳收入），作为质押担保提供授信。这是中国首笔单纯以未来碳资产作为质押担保（无其他抵押担保条件）的授信，兴业银行成为我国CDM配套金融产品市场的先行者。

审计署发布节能减排资金审计报告

2011－05－13

中国审计署13日发布审计公告，披露在节能减排审计调查中发现的若干问题，其中重点指出，国家节能减排专项资金管理使用不够规范，被一些企业违规套取、挪用。审计公告同时称，违规建设高耗能、高污染项目的问题没有完全杜绝，淘汰落后产能工作还不够彻底。此次审计调查主要涉及20个省政府、相关主管部门和企业2007—2009年贯彻执行节能减排政策法规和管理使用节能减排资金等情况。

亚行和浦发银行共推建筑节能

2011－05－16

亚洲开发银行与上海浦东发展银行举行绿色建筑投融资合作签字仪式，亚行将向浦发银行提供3亿元人民币的部分损失分担，支持私营部门融资，在中国打造节能型建筑。亚行正在实施一项节能计划，旨在鼓励金融机构为致力于节能型建筑的企业提供融资支持，帮助企业改造既有建筑降低能耗；或建造新的绿色建筑，即对建筑进行设计、建造和维护，使其在使用期限内最大程度地达到节能和节水的目标。浦发银行则是亚行该项计划的首家中国合作伙伴。

空调节能补贴政策终结

2011－05－31

持续两年的节能空调惠民补贴政策于5月31日正式结束。2009年6月国家开始实施节能空调惠民补贴，对生产企业补贴标准为每台300元—850元，2010年6月对补贴额度做了调整，降为每台150元—250元。失去了节能补贴的空调企业成本压力将随之增大。目前，国内市场上的高能效定频空调已经占到定频总销量的80%。

央企“十二五”节能减排目标确定

2011－05－31

国资委在5月31日召开中央企业节能减排工作会议，明确央企“十二五”的节能减排目标：到“十二五”末，万元产值综合能耗（可比价）下降16%左右。国资委要求，到“十二五”末，中央企业二氧化碳、二氧化硫、化学需氧量、氨氮、氮氧化物等主要污染物排放总量降幅要高于全国平均水平。石油石化、电力、钢铁、有色、煤炭、建材、化工、交通运输等行业主要产品能耗指标达到国内领先水平。

第二届全球绿色经济峰会在沪召开

2011－06－08

第二届全球绿色经济峰会在上海举行。本次峰会由上海市政府、中国企业联合会、彭博商业周刊共同主办。会议围绕“绿色创新——为可持续发展的未来”这个议题，分别从城市、能源、水资源等领域探讨全球在可持续发展商所面临的主要机遇和挑战。逾四百位来自海内外的政府官员、专家学者及商业领袖，就未来5年—10年城市和企业在可持续发展商或将面临的挑战进行讨论，并分享行之有效的创新策略。

北京市公共机构节能目标公布

2011－06－14

北京市召开公共机构节能工作会议，“十二五”期间北京市公共机构节能目标发布，到2015年，全市公共机构单位建筑面积能耗计划比2010年降低12%，人均能耗计划比2010年降低16%以上，能源消费总量计划控制在200万吨标准煤左右。

首个工业建筑节能示范工程投入使用

2011－6－20

全国第一个工业建筑节能示范工程，沈阳贝卡尔特钢帘线有限公司厂房揭牌并投入使用。这座在节能减排方面投入2800万元的“绿色工厂”，该项目一年可节电4722兆瓦，节水1300吨，节约标准煤5462吨，废弃物排放量减至原来的1/8，使二氧化碳排放量减少18%，一年减少的二氧化碳排放量，相当于停用2000辆轿车。

APEC低碳城镇示范项目启动

2011－06－21

21日，天津于家堡金融区“APEC首例低碳城镇示范项目”正式启动。于家堡金融区作为天津金融改革重点项目之一，在2010年6月9日被确定为“APEC首例低碳城镇示范项目”。按我国有关规划，到2015年我国建成100座新能源城市，200个绿色能源示范县，1000座新能源示范区和10000个新能源示范村，形成新能源运用的局部优势示范区域。

环保部与新疆签署环保合作协议

2011－06－28

环境保护部与新疆维吾尔自治区人民政府签署《进一步促进新疆环境保护工作的协议》。根据协议，环境保护部将从治污减排、环境影响评价、污染防治和农村环境综合整治、环保基础能力、环境政策创新和试点、人才队伍建设等六个方面，对新疆环境保护工作进一步加大支持力度。“十二五”期间对新疆环保投资比“十一五”翻番，并于近期安排13亿元资金支持新疆环境保护的10项工程。

国际气候组织报告发布中国碳捕捉封存技术报告

2011－6－10

国际非营利机构气候组织发布了《CCUS在中国：18个热点问题》报告。报告介绍了CCUS（碳捕集、利用及封存）技术在中国进展的情况，认为应用CCUS技术能帮助中国有效的减少二氧化碳排放，从而达成国家减排目标、履行国际责任。并由于其对煤基工业（燃煤发电和煤化工）清洁化、低碳化发展的积极影响，对富煤贫油的中国具有特殊意义，有利于在保证能源安全的同时积极应对气候变化。

国家环境保护“十二五”科技发展规划公布

2011－06－28

环保部28日公布《国家环境保护“十二五”科技发展规划》。该《规划》称，为实施本规划12个领域的重点科技任务以及建设一批国家环境保护重点实验室、环境保护工程技术中心和野外观测研究站，预计需要国家在环境保护科技领域投入经费约220亿元。该《规划》还提出，到“十二五”末，要研发出一批具有核心竞争力的环境污染物控制与生态保护关键技术，提出重点领域和重点行业的绿色经济、循环经济发展模式与关键技术，同时引导和培育5家以上环保产学研联盟，创新流域、区域污染治理机制，促进产学研用有机结合。

《企业环境报告书编制导则》发布

2011－06－14

环保部发布《企业环境报告书编制导则》用以规范企业的环境信息公开行为。《导则》中规定，企业环境报告书中应包含高层致辞、企业概况及编制说明、环境管理状况、环保目标、降低环境负荷的措施及绩效、与社会及利益相关者的关系等内容，还应结合企业的行业特点和利益相关者关注的焦点，适时增加企业环境报告书的内容。《导则》还对企业环境报告书中所涉及的资源与能源消耗、污染排放和资源循环利用率的计算方法做了规定。对企业环境报告书的界限和时限、指标确定等也进行了相关规定。

我国首个自愿减排排行榜发布

2011－07－01

国内首个以温室气体自愿减排量为主要衡量指标的企业排行榜“中国企业自愿减排2010年度排行榜”发布。该项由北京环境交易所联合众多专业机构共同发布的榜单，汇集了过去一年中通过购买自愿碳减排指标（VER）实现碳抵消的企业与机构。这些企业和机构通过购买自愿碳减排指标的方式来抵消自身在运营企业或者组织活动过程中所产生的温室气体排放，共计减少温室气体排放量约21万吨。

机动车国Ⅳ标准实施

2011－07－01

根据国家环保部2011年第49号公告——《关于实施国家第四阶段轻型汽油车、两用燃料车和单一气体燃料车污染物排放标准的公告》的要求，2011年7月1日起，所有生产、进口、销售的轻型汽油车、两用燃料车、单一气体燃料车必须符合国Ⅳ标准的要求。与国Ⅲ标准相比，国Ⅳ标准对机动车排放的控制更严格，要在满足国Ⅲ标准上，再进一步降低30%—50%的排放才能达标。

中央财政支持交通运输节能减排

2011－07－05

财政部、交通部联合公布《交通运输节能减排专项资金管理暂行办法》及《交通运输节能减排专项资金申请指南（2011年度）》，明确交通运输企事业单位可依据相关政策申请节能减排专项资金。专项资金的使用原则上采取以奖代补方式，由财政部、交通运输部根据项目性质、投资总额、实际节能减排量以及产生的社会效益等综合测算确定补助额度。对节能减排量可以量化的项目，奖励资金原则上与节能减排量挂钩，对完成节能减排量目标的项目承担单位给予一次性奖励，对单个项目的补助额度原则上不超过1000万元。

工信部公告2011年淘汰落后产能企业名单

2011－07－11

工信部公告了2011年工业行业淘汰落后产能企业名单，18个工业行业淘汰落后产能共涉及2255家企业，其中河北291家、湖南226家、山西173家，成为淘汰落后产能涉及企业最多的三省份。其中，淘汰落后产能炼铁3122万吨、涉及96家企业，炼钢2794万吨、涉及58家企业，焦炭1975万吨、涉及87家企业，水泥15327万吨、涉及782家企业，平板玻璃2940.7万重量箱、涉及45家企业，造纸819.6万吨、涉及599家企业。此份名单是对工信部2011年淘汰落后产能任务的具体落实结果。

2011深圳节博会开幕

2011－07－14

2011年中国（深圳）国际节能减排和新能源科技成果产业化及投融资博览会在深圳开幕。本届博览会有300多家海内外企业，近百个国内外政府、企业和社团确认携采购项目参加，采购订单总额有望达百亿元。节博会上将展示国内外最先进的节能减排和新能源领域的相关产品和技术，同期还将举办多场规格高、专业性强的论坛活动。

“十二五”节能减排综合性工作方案获通过

2011－07－19

国务院召开国家应对气候变化及节能减排工作领导小组会议，审议并原则同意“十二五”节能减排综合性工作方案，以及节能目标分解方案、主要污染物排放总量控制计划。方案主要涉及重点领域节能减排、调整优化产业结构、节能减排重点工程、推广使用先进技术、加强节能减排管理、完善节能减排长效机制等方面。由此，中国“十二五”节能减排的大政方针确定，下一步能源结构调整的方向和步骤也就此定下基调。

国家发改委等支持餐厨废弃物炼制生物柴油

2011－08－02

国家发展和改革委员会、财政部、住房城乡建设部8月2日宣布，将安排循环经济发展专项资金6.3亿元，支持北京市朝阳区等33个试点市（区）开展餐厨废弃物资源化利用和无害化处理。试点工作的顺利开展将为全国其他城市的餐厨废弃物资源化利用和无害化处理提供经验和借鉴。目前，各试点城市（区）的实施方案规划了完善的餐厨废弃物收集、运输、处理和利用体系，提出了废弃油脂、固形物和液体的一体化统筹解决方案，探索餐厨废弃物资源化利用和无害化处理的治本之策。

大学生节能减排科技竞赛开赛

2011－08－08

第四届全国大学生节能减排社会实践与科技竞赛8日在哈尔滨工业大学拉开帷幕，共有73所高校的135件作品进入决赛环节。决赛期间，800余名师生将现场展示参赛作品。全国大学生节能减排社会实践与科技竞赛是教育部高等教育司主办的全国性大学生课外科技作品竞赛，竞赛以“节能减排、绿色能源”为主题，以“培养普及节能减排意识，提高科技创新能力”为宗旨，每年举办一届。

全国首个节能减排在线监测平台通过评审

2011－08－09

8月9日，湖南省两型办组织专家对湖南省节能减排在线监测运营平台进行评审。专家一致认为，该平台技术先进，设计规范，为全面推进节能减排全覆盖工程提供了重要技术支撑，一致同意通过评审。这也是全国首例全省统一兼容、共建共享的综合性监管平台。

该平台共包括节能减排监管信息、实时监测、统计申报、节能量碳减排量运营和考核评估等五大系统。采用GPRS无线传输技术实时采集能耗数据和排污数据，建立数据库，进行综合分析，为节能减排管理提供依据。从今年4月起，该平台也已在19个试点单位试运行，效果明显。

“十二五”节能减排将实施问责制 建评价考核体系

2011－08－10

国家发改委表示，“十二五”节能减排指标已下发到各省市，节能减排指标评价考核体系已建立，考核结果将向社会公告，并纳入政府绩效管理，对于未完成任务的地区将实施问责制。为完成“十一五”减排指标，国务院曾在2010年要求加强问责力度，对未完成任务的要追究主要领导责任，根据情节给予相应处分。一些省市也在2010年末突击公布问责办法，力促目标完成。

交通运输节能减排专项资金项目专家评审启动

2011－08－10

8月10日，交通运输节能减排专项资金项目专家评审工作在京启动。本次专家评审主要对申请项目逐一进行审核，淘汰不符合《交通运输节能减排专项资金管理暂行办法》和《交通运输节能减排专项资金申请指南（2011年度）》基本要求的项目；对符合条件项目从符合度、申请资料的完整性、基础数据的可信度、节能减排或投资额核算的科学性、技术方案的合理性、节能减排的促进和导向作用、在全行业推广应用的价值等八方面进行评价和打分，核实其节能减排量（替代燃料量）或投资额，并提出建议补助标准，为确定专项资金补助方案提供依据。

政策推动水泥行业整合

2011－08－25

8月25日，工信部印发了《水泥行业准入公告管理暂行办法》，对申请准入公告的企业作了一系列规定。《办法》延续了政策层面对于水泥行业的产能限制宗旨，旨在加强水泥行业准入管理，发挥先进企业的引导和示范作用，促进水泥行业结构调整和转型升级，进一步抑制新建产能的投放。《办法》对明确表示对准入公告企业在投融资、技术改造、兼并重组、节能减排、循环经济等方面优先予以支持。

2012年北京率先实施建筑节能75%标准

2011-09-01

北京市住建委和市发改委联合发布《北京市“十二五”时期民用建筑节能规划》。根据规划，北京市将在全国率先执行新建居住建筑节能75%的设计标准。规划中，北京市将通过8种节能方式节省620万吨标准煤，包括新建民用建筑2亿平方米；计划拆除原有非节能建筑1000万平方米，拆除农村非节能建筑4000万平方米并重建为节能建筑；总计6万平方米的居住建筑和公共建筑节能改造；可再生能源建筑应用以及太阳能光热系统应用等。

节能汽车补贴新政今起实施 三线城市影响最大

2011-09-07

财政部、国家发展改革委、工信部联合印发《关于调整节能汽车推广补贴政策的通知》。从10月1日起，中国将提高节能汽车补贴门槛，百公里平均油耗从6.9升降低到6.3升，补贴标准仍维持每辆3000元不变。对消费者购买节能汽车继续给予一次性3000元定额补助，由生产企业在销售时兑付给购买者。与之前只要排量在1.6升以下、百公里油耗低于6.9升不同，此次新出台的标准更详细、更严格。

国务院印发“十二五”节能减排方案

2011-09-07

国务院将《“十二五”节能减排综合性工作方案》印发给各省、自治区、直辖市人民政府，国务院各部委和直属机构。《工作方案》包括50条具体要求，涉及12个方面，包括：节能减排总体要求和主要目标、强化节能减排目标责任、调整优化产业结构、实施节能减排重点工程、加强节能减排管理、大力发展循环经济、加快节能减排技术开发和推广应用、完善节能减排经济政策、强化节能减排监督检查、推广节能减排市场化机制、加强节能减排基础工作和能力建设、动员全社会参与节能减排。

《方案》确定的总体目标包括：到2015年，我国万元国内生产总值能耗下降到0.869吨标准煤（按2005年价格计算），比2010年的1.034吨标准煤下降16%，比2005年的1.276吨标准煤下降32%。“十二五”期间，我国将实现节约能源6.7亿吨标准煤。

《火电厂大气污染物排放标准》出台

2011-09-21

环境保护部与质检总局联合发布了新的国家排放标准《火电厂大气污染物排放标准》，其相关标准被称为“世界最严”。《标准》规定：从2012年1月1日开始，所有新建火电机组氮氧化物污染物排放标准为100毫克/立方米；从2014年7月1日开始，现有火电机组氮氧化物污染物排放标准为100毫克/立方米（特殊规定的执行200毫克/立方米的标准）。而重点区域火电机组的氮氧化物污染物排放标准则统一为100毫克/立方米。

实施新标准后，到2015年，电力行业氮氧化物排放可减少580万吨、二氧化硫排放可减少618万吨。在大幅削减污染物排放的同时，标准将带动相关的环保技术和产业市场的发展，形成脱硝、脱硫和除尘等环保治理和设备制造行业约2600亿元的市场规模。

全国节能减排工作电视电话会议召开

2011－09－27

国务院9月27日召开全国节能减排工作电视电话会议，全面动员和部署“十二五”节能减排工作。会上，温家宝总理就全面落实“十二五”节能减排综合性工作方案提出五点要求：一是着力调整优化产业结构，促进节能减排；二是坚持以科技创新和技术进步推动节能减排；三是完善节能减排长效机制。深化资源性产品价格改革，完善价格形成机制；四是加强节能减排能力建设；五是推进重点领域节能减排。会上还宣布，国务院决定对“十一五”期间在节能工作和减排工作中表现突出的北京、山东等省份予以通报表扬。

附录1：节能减排政策清单（1980—2010）

年份	政策名称	发文单位
	《关于加快推行合同能源管理促进节能服务产业发展的意见》	发改委、财政部、央行、国税总局
	《私人购买新能源汽车试点财政补助资金管理暂行办法》	发改委、财政部、科技部、工信部
	《关于进一步加强中小企业节能减排工作的指导意见》	工信部
	《轻型汽车燃料消耗量标识管理规定》	工信部
	《家电以旧换新实施办法（修订稿）》	商务部、财政部
2010	《关于印发〈2010年工业节能与综合利用工作要点〉的通知》	工信部
	《中央企业节能减排监督管理暂行办法》	国务院国有资产监督管理委员会
	《关于组织申请国家机关办公建筑和大型公共建筑节能监管体系建设补助资金的通知》	财政部 、国家住房与城乡建设部
	《关于加强工业固定资产投资项目节能评估和审查工作的通知》	工信部
	《国务院关于进一步加大工作力度确保实现“十一五”节能减排目标的通知》	国务院
	《中华人民共和国循环经济促进法》	全国人大常委会
	《关于开展“节能产品惠民工程 ”的通知》	财政部、国家发改委
	《关于中国清洁发展机制基金及清洁发展机制项目实施企业有关企业所得税政策问题的通知》	财政部、国家税务总局
	《关于 <太阳能光电建筑应用财政补助资金管理暂行办法 >的通知》	财政部
2009	《废弃电器电子产品回收处理管理条例》	国务院
	《关于清理优惠电价有关问题的通知》	国家发展改革委、国家电监会、国家能源局
	《关于加强外商投资节能环保统计工作的通知》	商务部、环保部
	《国务院办公厅关于治理商品过度包装工作的通知》	国务院
	《关于开展节能与新能源汽车示范推广试点工作的通知》	财政部、科技部
	《关于实施成品油价格和税费改革的通知》	国务院
	《北方采暖地区既有居住建筑供热计量改造工程验收办法》	国家住房和城乡建设部
	《关于印发公路水路交通节能中长期规划纲要的通知》	交通部
	《关于进一步加强生物质发电项目环境影响评价管理工作的通知》	环保部、国家发改委
2008	《关于贯彻实施<中华人民共和国节约能源法>的通知》	国家发改委
	《公共机构节能条例》	国家发改委
	《关于进一步加强节油节电工作的通知》	国务院
	《民用建筑节能条例》	国家发改委
	《可再生能源发展“十一五”规划》	国家发改委

续表

年份	政策名称	发文单位
2007	《中华人民共和国节约能源法》	全国人大常委会
	《建设部关于落实〈国务院关于印发节能减排综合性工作方案的通知〉的实施方案》	国家建设部
	《国务院关于印发节能减排综合性工作方案的通知》	国务院
	《能源发展“十一五”规划》	国家发改委
2006	《“十一五”资源综合利用指导意见》	国家发改委
	《建筑门窗节能性能标识试点工作管理办法》	国家住房和城乡建设部
	《国务院关于加强节能工作的决定》	国务院
	《民用建筑节能管理规定》（修订）	国家建设部
	《中华人民共和国国民经济和社会发展第十一个五年规划纲要》	国家发改委
2005	《关于印发千家企业节能行动实施方案的通知》	国家发改委
	《关于做好建设节约型社会近期重点工作的通知》	国务院
	《关于发展节能省地型住宅和公共建筑的指导意见》	国家建设部
	《关于进一步推进墙体材料革新和推广节能建筑的通知》	国务院
	《关于加快发展循环经济的若干意见》	国务院
	《关于新建居住建筑严格执行节能设计标准的通知》	国家建设部
	《中国节水技术政策大纲》	国家发展和改革委员会、科技部、水利部、建设部、农业部
	《关于印发加快火电厂烟气脱硫产业化发展的若干意见的通知》	国家发展改革委
	《关于鼓励发展节能环保型小排量汽车意见的通知》	国务院
2004	《关于加强民用建筑工程项目建筑节能审查工作的通知》	国家建设部
	《关于加强城市照明管理促进节约用电工作的意见》	建设部、国家发改委
	《节能产品政府采购实施意见》	财政部、国家发改委
2002	《关于印发 < 建设部建筑节能“十五”计划纲要 > 的通知》	建设部、国家发改委
	《节约用电管理办法》	国家经济委员会
2001	《关于加强利用废塑料生产汽油、柴油管理有关规定的通知》	国家经济委员会
	《夏热冬冷地区居住建筑节能设计标准》	国家住房和城乡建设部
2000	《民用建筑节能管理规定》	国家建设部
	《交通行业实施节能法细则》	国家交通部
	《关于加强工业节水工作的意见》	国家经济委员会
1999	《中国节能产品认证管理办法》	中国节能产品认证管理委员会
	《重点用能单位节能管理办法》	国家经济贸易委员会
1995	《1996—2010 年新能源和可再生能源发展纲要》	国家发改委
1994	《中国电力法》	全国人大常委会
1986	《节约能源管理暂行条例》	国务院
1984	《国家经委关于开展资源综合利用若干问题的暂行规定》	国家经济委员会
1982	《节能技术政策大纲》	国家发改委
	《关于按省、市、自治区实行计划用电包干的暂行管理办法》	国务院
1980	《关于加强节约能源工作的报告》	国家经济委员会

附录2:“十二五”节能减排综合性工作方案

一、节能减排总体要求和主要目标

（一）总体要求。以邓小平理论和“三个代表”重要思想为指导，深入贯彻落实科学发展观，坚持降低能源消耗强度、减少主要污染物排放总量、合理控制能源消费总量相结合，形成加快转变经济发展方式的倒逼机制；坚持强化责任、健全法制、完善政策、加强监管相结合，建立健全激励和约束机制；坚持优化产业结构、推动技术进步、强化工程措施、加强管理引导相结合，大幅度提高能源利用效率，显著减少污染物排放；进一步形成政府为主导、企业为主体、市场有效驱动、全社会共同参与的推进节能减排工作格局，确保实现“十二五”节能减排约束性目标，加快建设资源节约型、环境友好型社会。

（二）主要目标。到2015年，全国万元国内生产总值能耗下降到0.869吨标准煤（按2005年价格计算），比2010年的1.034吨标准煤下降16%，比2005年的1.276吨标准煤下降32%；“十二五”期间，实现节约能源6.7亿吨标准煤。2015年，全国化学需氧量和二氧化硫排放总量分别控制在2347.6万吨、2086.4万吨，比2010年的2551.7万吨、2267.8万吨分别下降8%；全国氨氮和氮氧化物排放总量分别控制在238.0万吨、2046.2万吨，比2010年的264.4万吨、2273.6万吨分别下降10%。

二、强化节能减排目标责任

（三）合理分解节能减排指标。综合考虑经济发展水平、产业结构、节能潜力、环境容量及国家产业布局等因素，将全国节能减排目标合理分解到各地区、各行业。各地区要将国家下达的节能减排指标层层分解落实，明确下一级政府、有关部门、重点用能单位和重点排污单位的责任。

（四）健全节能减排统计、监测和考核体系。加强能源生产、流通、消费统计，建立和完善建筑、交通运输、公共机构能耗统计制度以及分地区单位国内生产总值能耗指标季度统计制度，完善统计核算与监测方法，提高能源统计的准确性和及时性。修订完善减排统计监测和核查核算办法，统一标准和分析方法，实现监测数据共享。加强氨氮、氮氧化物排放统计监测，建立农业源和机动车排放统计监测指标体系。完善节能减排考核办法，继续做好全国和各地区单位国内生产总值能耗、主要污染物排放指标公报工作。

（五）加强目标责任评价考核。把地区目标考核与行业目标评价相结合，把落实五年目标与完成年度目标相结合，把年度目标考核与进度跟踪相结合。省级人民政府每年要向国务院报告节能减排目标完成情况。有关部门每年要向国务院报告节能减排措施落实情况。国务院每年组织开展省级人民政府节能减排目标责任评价考核，考核结果向社会公告。强化考核结果运用，将节能减排目标完成情况和政策措施落实情况作为领导班子和领导干部综合考核评价的重要内容，纳入政府绩效和国有企业业绩管理，实行问责制和“一票否决”制，并对成绩突出的地区、单位和个人给予表彰奖励。

三、调整优化产业结构

（六）抑制高耗能、高排放行业过快增长。严格控制高耗能、高排放和产能过剩行业新上项目，进一步提高行业准入门槛，强化节能、环保、土地、安全等指标约束，依法严格节能评估审查、环境影响评价、建设用地审查，严格贷款审批。建立健全项目审批、核准、备案责任制，严肃查处越权审批、分拆审批、未批先建、边批边建等行为，依法追究有关人员责任。严格控制高耗能、高排放产品出口。中西部地区承接产业转移必须坚持高标准，严禁污染产业和落后生产能力转入。

（七）加快淘汰落后产能。抓紧制定重点行业“十二五”淘汰落后产能实施方案，将任务按年度分解落实到各地区。完善落后产能退出机制，指导、督促淘汰落后产能企业做好职工安置工作。地方各级人民政府要积极安排资金，支持淘汰落后产能工作。中央财政统筹支持各地区淘汰落后产能工作，对经济欠发达地区通过增加转移支付加大支持和奖励力度。完善淘汰落后产能公告制度，对未按期完成淘汰任务的地区，严格控制国家安排的投资项目，暂停对该地区重点行业建设项目办理核准、审批和备案手续；对未按期淘汰的企业，依法吊销排污许可证、生产许可证和安全生产许可证；对虚假淘汰行为，依法追究企业负责人和地方政府有关人员的责任。

（八）推动传统产业改造升级。严格落实《产业结构调整指导目录》。加快运用高新技术和先进适用技术改造提升传统产业，促进信息化和工业化深度融合，重点支持对产业升级带动作用大的重点项目和重污染企业搬迁改造。调整《加工贸易禁止类商品目录》，提高加工贸易准入门槛，促进加工贸易转型升级。合理引导企业兼并重组，提高产业集中度。

（九）调整能源结构。在做好生态保护和移民安置的基础上发展水电，在确保安全的基础上发展核电，加快发展天然气，因地制宜大力发展风能、太阳能、生物质能、地热能等可再生能源。到2015年，非化石能源占一次能源消费总量比重达到11.4%。

（十）提高服务业和战略性新兴产业在国民经济中的比重。到2015年，服务业增加值和战略性新兴产业增加值占国内生产总值比重分别达到47%和8%左右。

四、实施节能减排重点工程

（十一）实施节能重点工程。实施锅炉窑炉改造、电机系统节能、能量系统优化、余热余压利用、节约替代石油、建筑节能、绿色照明等节能改造工程，以及节能技术产业化示范工程、节能产品惠民工程、合同能源管理推广工程和节能能力建设工程。到2015年，工业锅炉、窑炉平均运行效率比2010年分别提高5个百分点和2个百分点，电机系统运行效率提高2—3个百分点，新增余热余压发电能力2000万千瓦，北方采暖地区既有居住建筑供热计量和节能改造4亿平方米以上，夏热冬冷地区既有居住建筑节能改造5000万平方米，公共建筑节能改造6000万平方米，高效节能产品市场份额大幅度提高。"十二五"时期，形成3亿吨标准煤的节能能力。

（十二）实施污染物减排重点工程。推进城镇污水处理设施及配套管网建设，改造提升现有设施，强化脱氮除磷，大力推进污泥处理处置，加强重点流域区域污染综合治理。到2015年，基本实现所有县和重点建制镇具备污水处理能力，全国新增污水日处理能力4200万吨，新建配套管网约16万公里，城市污水处理率达到95%，形成化学需氧量和氨氮削减能力280万吨、30万吨。实施规模化畜禽养殖场污染治理工程，形成化学需氧量和氨氮削减能力140万吨、10万吨。实施脱硫脱硝工程，推动燃煤电厂、钢铁行业烧结机脱硫，形成二氧化硫削减能力277万吨；推动燃煤电厂、水泥等行业脱硝，形成氮氧化物削减能力358万吨。

（十三）实施循环经济重点工程。实施资源综合利用、废旧商品回收体系、"城市矿产"示范基地、再制造产业化、餐厨废弃物资源化、产业园区循环化改造、资源循环利用技术示范推广等循环经济重点工程，建设100个资源综合利用示范基地、80个废旧商品回收体系示范城市、50个"城市矿产"示范基地、5个再制造产业集聚区、100个城市餐厨废弃物资源化利用和无害化处理示范工程。

（十四）多渠道筹措节能减排资金。节能减排重点工程所需资金主要由项目实施主体通过自有资金、金融机构贷款、社会资金解决，各级人民政府应安排一定的资金予以支持和引导。地方各级人民政府要切实承担城镇污水处理设施和配套管网建设的主体责任，严格城镇污水处理费征收和管理，国家对重点建设项目给予适当支持。

五、加强节能减排管理

（十五）合理控制能源消费总量。建立能源消费总量控制目标分解落实机制，制定实施方案，把总量控制目标分解落实到地方政府，实行目标责任管理，加大考核和监督力度。将固定资产投资项目节能评估审查作为控制地区能源消费增量和总量的重要措施。建立能源消费总量预测预警机制，跟踪监测各地区能源消费总量和高耗能行业用电量等指标，对能源消费总量增长过快的地区及时预警调控。在工业、建筑、交通运输、公共机构以及城乡建设和消费领域全面加强用能管理，切实改变敞开口子供应能源、无节制使用能源的现象。在大气联防联控重点区域开展煤炭消费总量控制试点。

（十六）强化重点用能单位节能管理。依法加强年耗能万吨标准煤以上用能单位节能管理，开展万家企业节能低碳行动，实现节能2.5亿吨标准煤。落实目标责任，实行能源审计制度，开展能效水平对标活动，建立健全企业能源管理体系，扩大能源管理师试点；实行能源利用状况报告制度，加快实施节能改造，提高能源管理水平。地方节能主管部门每年组织对进入万家企业节能低碳行动的企业节能目标完成情况进行考核，公告考核结果。对未完成年度节能任务的企业，强制进行能源审计，限期整改。中央企业要接受所在地区节能主管部门的监管，争当行业节能减排的排头兵。

（十七）加强工业节能减排。重点推进电力、煤炭、钢铁、有色金属、石油石化、化工、建材、造纸、纺织、印染、食品加工等行业节能减排，明确目标任务，加强行业指导，推动技术进步，强化监督管理。发展热电联产，推广分布式能源。开展智能电网试点。推广煤炭清洁利用，提高原煤入洗比例，加快煤层气开发利用。实施工业和信息产业能效提升计划。推动信息数据中心、通信机房和基站节能改造。实行电力、钢铁、造纸、印染等行业主要污染物排放总量控制。新建燃煤机组全部安装脱硫脱硝设施，现役燃煤机组必须安装脱硫设施，不能稳定达标排放的要进行更新改造，烟气脱硫设施要按照规定取消烟气旁路。单机容量30万千瓦及以上燃煤机组全部加装脱硝设施。钢铁行业全面实施烧结机烟气脱硫，新建烧结机配套安装脱硫脱硝设施。石油石化、有色金属、建材等重点行业实施脱硫改造。新型干法水泥窑实施低氮燃烧技术改造，配套建设脱硝设施。加强重点区域、重点行业和重点企业重金属污染防治，以湘江流域为重点开展重金属污染治理与修复试点示范。

（十八）推动建筑节能。制定并实施绿色建筑行动方案，从规划、法规、技术、标准、设计等方面全面推进建筑节能。新建建筑严格执行建筑节能标准，提高标准执行率。推进北方采暖地区既有建筑供热计量和节能改造，实施“节能暖房”工程，改造供热老旧管网，实行供热计量收费和能耗定额管理。做好夏热冬冷地区建筑节能改造。推动可再生能源与建筑一体化应用，推广使用新型节能建材和再生建材，继续推广散装水泥。加强公共建筑节能监管体系建设，完善能源审计、能效公示，推动节能改造与运行管理。研究建立建筑使用全寿命周期管理制度，严格建筑拆除管理。加强城市照明管理，严格防止和纠正过度装饰和亮化。

（十九）推进交通运输节能减排。加快构建综合交通运输体系，优化交通运输结构。积极发展城市公共交通，科学合理配置城市各种交通资源，有序推进城市轨道交通建设。提高铁路电气化比重。实施低碳交通运输体系建设城市试点，深入开展“车、船、路、港”千家企业低碳交通运输专项行动，推广公路甩挂运输，全面推行不停车收费系统，实施内河船型标准化，优化航路航线，推进航空、远洋运输业节能减排。开展机场、码头、车站节能改造。加速淘汰老旧汽车、机车、船舶，基本淘汰2005年以前注册运营的“黄标车”，加快提升车用燃油品质。实施第四阶段机动车排放标准，在有条件的重点城市和地区逐步实施第五阶段排放标准。全面推行机动车环保标志管理，探索城市调控机动车保有总量，积极推广节能与新能源汽车。

（二十）促进农业和农村节能减排。加快淘汰老旧农用机具，推广农用节能机械、设备和渔船。推进节能型住宅建设，推动省柴节煤灶更新换代，开展农村水电增效扩容改造。发展户用沼气和大中型沼气，加强运行管理和维护服务。治理农业面源污染，加强农村环境综合整治，实施农村清洁工程，规模化养殖场和养殖小区配套建设废弃物处理设施的比例达到50%以上，鼓励污染物统一收集、集中处理。因地制宜推进农村分布式、低成本、易维护的污水处理设施建设。推广测土配方施肥，鼓励使用高效、安全、低毒农药，推动有机农业发展。

（二十一）推动商业和民用节能。在零售业等商贸服务和旅游业开展节能减排行动，加快设施节能改造，严格用能管理，引导消费行为。宾馆、商厦、写字楼、机场、车站等要严格执行夏季、冬季空调温度设置标准。在居民中推广使用高效节能家电、照明产品，鼓励购买节能环保型汽车，支持乘用公共交通，提倡绿色出行。减少一次性用品使用，限制过度包装，抑制不合理消费。

（二十二）加强公共机构节能减排。公共机构新建建筑实行更加严格的建筑节能标准。加快公共机构办公区节能改造，完成办公建筑节能改造6000万平方米。国家机关供热实行按热量收费。开展节约型公共机构示范单位创建活动，创建2000家示范单位。推进公务用车制度改革，严格用车油耗定额管理，提高节能与新能源汽车比例。建立完善公共机构能源审计、能效公示和能耗定额管理制度，加强能耗监测平台和节能监管体系建设。支持军队重点用能设施设备节能改造。

六、大力发展循环经济

（二十三）加强对发展循环经济的宏观指导。研究提出进一步加快发展循环经济的意见。编制全国循环经济发展规划和重点领域专项规划，指导各地做好规划编制和实施工作。研究制定循环经济发展的指导目录。制定循环经济专项资金使用管理办法及实施方案。深化循环经济示范试点，推广循环经济典型模式。建立完善循环经济统计评价制度。

（二十四）全面推行清洁生产。编制清洁生产推行规划，制（修）订清洁生产评价指标体系，发布重点行业清洁生产推行方案。重点围绕主要污染物减排和重金属污染治理，全面推进农业、工业、建筑、商贸服务等领域清洁生产示范，从源头和全过程控制污染物产生和排放，降低资源消耗。发布清洁生产审核方案，公布清洁生产强制审核企业名单。实施清洁生产示范工程，推广应用清洁生产技术。

（二十五）推进资源综合利用。加强共伴生矿产资源及尾矿综合利用，建设绿色矿山。推动煤矸石、粉煤灰、工业副产石膏、冶炼和化工废渣、建筑和道路废弃物以及农作物秸秆综合利用、农林废物资源化利用，大力发展利废新型建筑材料。废弃物实现就地消化，减少转移。到2015年，工业固体废物综合利用率达到72%以上。

（二十六）加快资源再生利用产业化。加快“城市矿产”示范基地建设，推进再生资源规模化利用。培育一批汽车零部件、工程机械、矿山机械、办公用品等再制造示范企业，发布再制造产品目录，完善再制造旧件回收体系和再制造产品标准体系，推动再制造的规模化、产业化发展。加快建设城市社区和乡村回收站点、分拣中心、集散市场“三位一体”的再生资源回收体系。

（二十七）促进垃圾资源化利用。健全城市生活垃圾分类回收制度，完善分类回收、密闭运输、集中处理体系。鼓励开展垃圾焚烧发电和供热、填埋气体发电、餐厨废弃物资源化利用。鼓励在工业生产过程中协同处理城市生活垃圾和污泥。

（二十八）推进节水型社会建设。确立用水效率控制红线，实施用水总量控制和定额管理，制定区域、行业和产品用水效率指标体系。推广普及高效节水灌溉技术。加快重点用水行业节水技术改造，提高工业用水循环利用率。加强城乡生活节水，推广应用节水器具。推进再生水、矿井水、海水等非传统水资源利用。建设海水淡化及综合利用示范工程，创建示范城市。到2015年，实现单位工业增加值用水量下降30%。

七、加快节能减排技术开发和推广应用

（二十九）加快节能减排共性和关键技术研发。在国家、部门和地方相关科技计划和专项中，加大对节能减排科技研发的支持力度，完善技术创新体系。继续推进节能减排科技专项行动，组织高效节能、废物资源化以及小型分散污水处理、农业面源污染治理等共性、关键和前沿技术攻关。组建一批国家级节能减排工程实验室及专家队伍。推动组建节能减排技术与装备产业联盟，继续通过国家工程（技术）研究中心加大节能减排科技研发力度。加强资源环境高技术领域创新团队和研发基地建设。

（三十）加大节能减排技术产业化示范。实施节能减排重大技术与装备产业化工程，重点支持稀土永磁无铁芯电机、半导体照明、低品位余热利用、地热和浅层地温能应用、生物脱氮除

磷、烧结机烟气脱硫脱硝一体化、高浓度有机废水处理、污泥和垃圾渗滤液处理处置、废弃电器电子产品资源化、金属无害化处理等关键技术与设备产业化，加快产业化基地建设。

（三十一）加快节能减排技术推广应用。编制节能减排技术政策大纲。继续发布国家重点节能技术推广目录、国家鼓励发展的重大环保技术装备目录，建立节能减排技术遴选、评定及推广机制。重点推广能量梯级利用、低温余热发电、先进煤气化、高压变频调速、干熄焦、蓄热式加热炉、吸收式热泵供暖、冰蓄冷、高效换热器，以及干法和半干法烟气脱硫、膜生物反应器、选择性催化还原氮氧化物控制等节能减排技术。加强与有关国际组织、政府在节能环保领域的交流与合作，积极引进、消化、吸收国外先进节能环保技术，加大推广力度。

八、完善节能减排经济政策

（三十二）推进价格和环保收费改革。深化资源性产品价格改革，理顺煤、电、油、气、水、矿产等资源性产品价格关系。推行居民用电、用水阶梯价格。完善电力峰谷分时电价政策。深化供热体制改革，全面推行供热计量收费。对能源消耗超过国家和地区规定的单位产品能耗（电耗）限额标准的企业和产品，实行惩罚性电价。各地可在国家规定基础上，按程序加大差别电价、惩罚性电价实施力度。严格落实脱硫电价，研究制定燃煤电厂烟气脱硝电价政策。进一步完善污水处理费政策，研究将污泥处理费用逐步纳入污水处理成本问题。改革垃圾处理收费方式，加大征收力度，降低征收成本。

（三十三）完善财政激励政策。加大中央预算内投资和中央财政节能减排专项资金的投入力度，加快节能减排重点工程实施和能力建设。深化“以奖代补”、“以奖促治”以及采用财政补贴方式推广高效节能家用电器、照明产品、节能汽车、高效电机产品等支持机制，强化财政资金的引导作用。国有资本经营预算要继续支持企业实施节能减排项目。地方各级人民政府要加大对节能减排的投入。推行政府绿色采购，完善强制采购和优先采购制度，逐步提高节能环保产品比重，研究实行节能环保服务政府采购。

（三十四）健全税收支持政策。落实国家支持节能减排所得税、增值税等优惠政策。积极推进资源税费改革，将原油、天然气和煤炭资源税计征办法由从量征收改为从价征收并适当提高税负水平，依法清理取消涉及矿产资源的不合理收费基金项目。积极推进环境税费改革，选择防治任务重、技术标准成熟的税目开征环境保护税，逐步扩大征收范围。完善和落实资源综合利用和可再生能源发展的税收优惠政策。调整进出口税收政策，遏制高耗能、高排放产品出口。对用于制造大型环保及资源综合利用设备确有必要进口的关键零部件及原材料，抓紧研究制定税收优惠政策。

（三十五）强化金融支持力度。加大各类金融机构对节能减排项目的信贷支持力度，鼓励金融机构创新适合节能减排项目特点的信贷管理模式。引导各类创业投资企业、股权投资企业、社会捐赠资金和国际援助资金增加对节能减排领域的投入。提高高耗能、高排放行业贷款门槛，将企业环境违法信息纳入人民银行企业征信系统和银监会信息披露系统，与企业信用等级评定、贷款及证券融资联动。推行环境污染责任保险，重点区域涉重金属企业应当购买环境污染责任保险。建立银行绿色评级制度，将绿色信贷成效与银行机构高管人员履职评价、机构准入、业务发展相挂钩。

九、强化节能减排监督检查

（三十六）健全节能环保法律法规。推进环境保护法、大气污染防治法、清洁生产促进法、建设项目环境保护管理条例的修订工作，加快制定城镇排水与污水处理条例、排污许可证管理条例、畜禽养殖污染防治条例、机动车污染防治条例等行政法规。修订重点用能单位节能管理办法、能效标识管理办法、节能产品认证管理办法等部门规章。

（三十七）严格节能评估审查和环境影响评价制度。把污染物排放总量指标作为环评审批的前置条件，对年度减排目标未完成、重点减排项目未按目标责任书落实的地区和企业，实行阶段性环评限批。对未通过能评、环评审查的投资项目，有关部门不得审批、核准、批准开工建设，不得发放生产许可证、安全生产许可证、排污许可证，金融机构不得发放贷款，有关单位不得供水、供电。加强能评和环评审查的监督管理，严肃查处各种违规审批行为。能评费用由节能审查机关同级财政部门安排。

（三十八）加强重点污染源和治理设施运行监管。严格排污许可证管理。强化重点流域、重点地区、重点行业污染源监管，适时发布主要污染物超标严重的国家重点环境监控企业名单。列入国家重点环境监控范围的电力、钢铁、造纸、印染等重点行业的企业，要安装运行管理监控平台和污染物排放自动监控系统，定期报告运行情况及污染物排放信息，推动污染源自动监控数据联网共享。加强城市污水处理厂监控平台建设，提高污水收集率，做好运行和污染物削减评估考核，考核结果作为核拨污水处理费的重要依据。对城市污水处理设施建设严重滞后、收费政策不落实、污水处理厂建成后一年内实际处理水量达不到设计能力60%，以及已建成污水处理设施但无故不运行的地区，暂缓审批该城市项目环评，暂缓下达有关项目的国家建设资金。

（三十九）加强节能减排执法监督。各级人民政府要组织开展节能减排专项检查，督促各项措施落实，严肃查处违法违规行为。加大对重点用能单位和重点污染源的执法检查力度，加大对高耗能特种设备节能标准和建筑施工阶段标准执行情况、国家机关办公建筑和大型公共建筑节能监管体系建设情况，以及节能环保产品质量和能效标识的监督检查力度。对严重违反节能环保法律法规，未按要求淘汰落后产能、违规使用明令淘汰用能设备、虚标产品能效标识、减排设施未按要求运行等行为，公开通报或挂牌督办，限期整改，对有关责任人进行严肃处理。实行节能减排执法责任制，对行政不作为、执法不严等行为，严肃追究有关主管部门和执法机构负责人的责任。

十、推广节能减排市场化机制

（四十）加大能效标识和节能环保产品认证实施力度。扩大终端用能产品能效标识实施范围，加强宣传和政策激励，引导消费者购买高效节能产品。继续推进节能产品、环境标志产品、环保装备认证，规范认证行为，扩展认证范围，建立有效的国际协调互认机制。加强标识、认证质量的监管。

（四十一）建立"领跑者"标准制度。研究确定高耗能产品和终端用能产品的能效先进水平，制定"领跑者"能效标准，明确实施时限。将"领跑者"能效标准与新上项目能评审查、节能产品推广应用相结合，推动企业技术进步，加快标准的更新换代，促进能效水平快速提升。

（四十二）加强节能发电调度和电力需求侧管理。改革发电调度方式，电网企业要按照节能、经济的原则，优先调度水电、风电、太阳能发电、核电以及余热余压、煤层气、填埋气、煤矸石和垃圾等发电上网，优先安排节能、环保、高效火电机组发电上网。研究推行发电权交易。电网企业要及时、真实、准确、完整地公布节能发电调度信息，电力监管部门要加强对节能发电调度工作的监督。落实电力需求侧管理办法，制定配套政策，规范有序用电。以建设技术支撑平台为基础，开展城市综合试点，推广能效电厂。

（四十三）加快推行合同能源管理。落实财政、税收和金融等扶持政策，引导专业化节能服务公司采用合同能源管理方式为用能单位实施节能改造，扶持壮大节能服务产业。研究建立合同能源管理项目节能量审核和交易制度，培育第三方审核评估机构。鼓励大型重点用能单位利用自身技术优势和管理经验，组建专业化节能服务公司。引导和支持各类融资担保机构提供风险分担服务。

（四十四）推进排污权和碳排放权交易试点。完善主要污染物排污权有偿使用和交易试点，建立健全排污权交易市场，研究制定排污权有偿使用和交易试点的指导意见。开展碳排放交易试点，建立自愿减排机制，推进碳排放权交易市场建设。

（四十五）推行污染治理设施建设运行特许经营。总结燃煤电厂烟气脱硫特许经营试点经验，完善相关政策措施。鼓励采用多种建设运营模式开展城镇污水垃圾处理、工业园区污染物集中治理，确保处理设施稳定高效运行。实行环保设施运营资质许可制度，推进环保设施的专业化、社会化运营服务。完善市场准入机制，规范市场行为，打破地方保护，为企业创造公平竞争的市场环境。

十一、加强节能减排基础工作和能力建设

（四十六）加快节能环保标准体系建设。加快制（修）订重点行业单位产品能耗限额、产品能效和污染物排放等强制性国家标准，以及建筑节能标准和设计规范，提高准入门槛。制定和完善环保产品及装备标准。完善机动车燃油消耗量限值标准、低速汽车排放标准。制（修）订轻型汽车第五阶段排放标准，颁布实施第四、第五阶段车用燃油国家标准。建立满足氨氮、氮氧化物控制目标要求的排放标准。鼓励地方依法制定更加严格的节能环保地方标准。

（四十七）强化节能减排管理能力建设。建立健全节能管理、监察、服务“三位一体”的节能管理体系，加强政府节能管理能力建设，完善机构，充实人员。加强节能监察机构能力建设，配备监测和检测设备，加强人员培训，提高执法能力，完善覆盖全国的省、市、县三级节能监察体系。继续推进能源统计能力建设。推动重点用能单位按要求配备计量器具，推行能源计量数据在线采集、实时监测。开展城市能源计量建设示范。加强减排监管能力建设，推进环境监管机构标准化，提高污染源监测、机动车污染监控、农业源污染检测和减排管理能力，建立健全国家、省、市三级减排监控体系，加强人员培训和队伍建设。

十二、动员全社会参与节能减排

（四十八）加强节能减排宣传教育。把节能减排纳入社会主义核心价值观宣传教育体系以及基础教育、高等教育、职业教育体系。组织好全国节能宣传周、世界环境日等主题宣传活动，加强日常性节能减排宣传教育。新闻媒体要积极宣传节能减排的重要性、紧迫性以及国家采取

的政策措施和取得的成效，宣传先进典型，普及节能减排知识和方法，加强舆论监督和对外宣传，积极为节能减排营造良好的国内和国际环境。

（四十九）深入开展节能减排全民行动。抓好家庭社区、青少年、企业、学校、军营、农村、政府机构、科技、科普和媒体等十个节能减排专项行动，通过典型示范、专题活动、展览展示、岗位创建、合理化建议等多种形式，广泛动员全社会参与节能减排，发挥职工节能减排义务监督员队伍作用，倡导文明、节约、绿色、低碳的生产方式、消费模式和生活习惯。

（五十）政府机关带头节能减排。各级人民政府机关要将节能减排作为机关工作的一项重要任务来抓，健全规章制度，落实岗位责任，细化管理措施，树立节约意识，践行节约行动，作节能减排的表率。

附件1：

“十二五”各地区节能目标

地区	单位国内生产总值能耗降低率（%）		
	“十一五”时期	“十二五”时期	2006—2015年累计
全国	19.06	16	32.01
北京	26.59	17	39.07
天津	21.00	18	35.22
河北	20.11	17	33.69
山西	22.66	16	35.03
内蒙古	22.62	15	34.23
辽宁	20.01	17	33.61
吉林	22.04	16	34.51
黑龙江	20.79	16	33.46
上海	20.00	18	34.40
江苏	20.45	18	34.77
浙江	20.01	18	34.41
安徽	20.36	16	33.10
福建	16.45	16	29.82
江西	20.04	16	32.83
山东	22.09	17	35.33
河南	20.12	16	32.90
湖北	21.67	16	34.20
湖南	20.43	16	33.16
广东	16.42	18	31.46
广西	15.22	15	27.94
海南	12.14	10	20.93
重庆	20.95	16	33.60
四川	20.31	16	33.06

续表

地区	单位国内生产总值能耗降低率（%）		
	“十一五”时期	“十二五”时期	2006—2015 年累计
贵州	20.06	15	32.05
云南	17.41	15	29.80
西藏	12.00	10	20.80
陕西	20.25	16	33.01
甘肃	20.26	15	32.22
青海	17.04	10	25.34
宁夏	20.09	15	32.08
新疆	8.91	10	18.02

注：“十一五”各地区单位国内生产总值能耗降低率除新疆外均为国家统计局最终公布数据，新疆为初步核实数据。

附件 2：

“十二五”各地区化学需氧量排放总量控制计划

单位：万吨

地区	2010 年		2015 年		2015 年比 2010 年（%）	
	排放量	其中：工业和生活	控制量	其中：工业和生活	增加或减少	其中：工业和生活
北京	20.0	10.9	18.3	9.8	-8.7	-9.8
天津	23.8	12.3	21.8	11.2	-8.6	-9.2
河北	142.2	45.6	128.3	40.7	-9.8	-10.8
山西	50.7	31.2	45.82	7.9	-9.6	-10.6
内蒙古	92.1	27.5	95.9	25.4	-6.7	-7.5
辽宁	137.3	47.0	124.7	42.1	-9.2	-10.4
吉林	83.4	28.8	76.1	26.1	-8.8	-9.4
黑龙江	161.2	47.8	147.3	43.4	-8.6	-9.3
上海	26.6	22.5	23.9	20.1	-10.0	-10.5
江苏	128.0	86.3	112.8	75.3	-11.9	-12.8
浙江	84.2	61.4	74.6	53.7	-11.4	-12.5
安徽	97.3	55.6	90.3	52.0	-7.2	-6.5
福建	69.6	45.8	65.2	43.1	-6.3	-6.0
江西	77.7	51.9	73.2	48.3	-5.8	-7.0
山东	201.6	62.7	177.4	54.6	-12.0	-12.9
河南	148.2	62.01	33.555.8	-9.9	-10.0	
湖北	112.4	62.1	104.1	59.0	-7.4	-5.0
湖南	134.1	71.8	124.4	66.8	-7.2	-7.0
广东	193.3	130.6	170.1	113.8	-12.0	-12.9
广西	80.7	58.1	74.6	53.6	-7.6	-7.8
海南	20.4	9.2	20.4	9.2	0	0

续表

地区	2010年		2015年		2015年比2010年（%）	
	排放量	其中：工业和生活	控制量	其中：工业和生活	增加或减少	其中：工业和生活
重庆	42.6	29.4	39.5	27.5	-7.2	-6.5
四川	132.4	75.0	123.1	71.3	-7.0	-5.0
贵州	34.8	28.1	32.7	26.4	-6.0	-6.1
云南	56.4	48.0	52.9	45.0	-6.2	-6.2
西藏	2.72.3	2.7	2.3	0	0	
陕西	57.0	36.4	52.7	33.5	-7.6	-7.9
甘肃	40.2	25.5	37.6	23.7	-6.4	-6.9
青海	10.4	8.1	12.3	9.6	18.0	18.0
宁夏	24.0	13.3	22.6	12.5	-6.0	-6.3
新疆	56.9	26.2	56.9	26.2	0	0
新疆生产建设兵团	9.5	4.7	9.5	4.7	0	0
合计	2551.7	1328.1	2335.2	1214.6	-8.5	-8.5

注：全国化学需氧量排放量削减8%的总量控制目标为2347.6万吨（其中工业和生活1221.9万吨），实际分配给各地区2335.2万吨（其中工业和生活1214.6万吨），国家预留12.4万吨，用于化学需氧量排污权有偿分配和交易试点工作。

附件3：

“十二五”各地区氨氮排放总量控制计划

单位：万吨

地区	2010年		2015年		2015年比2010年（%）	
	排放量	其中：工业和生活	控制量	其中：工业和生活	增加或减少	其中：工业和生活
北京	2.20	1.64	1.98	1.47	-10.1	-10.2
天津	2.79	2.18	2.50	1.95	-10.5	-10.4
河北	11.61	6.98	10.14	6.10	-12.7	-12.6
山西	5.93	4.66	5.21	4.08	-12.2	-12.4
内蒙古	5.45	4.19	4.92	3.79	-9.7	-9.5
辽宁	11.25	7.56	10.01	6.69	-11.0	-11.5
吉林	5.87	3.92	5.25	3.49	-10.5	-10.9
黑龙江	9.45	6.14	8.47	5.49	-10.4	-10.6
上海	5.21	4.83	4.54	4.21	-12.9	-12.9
江苏	16.12	11.98	14.04	10.40	-12.9	-13.2
浙江	11.84	8.96	10.36	7.84	-12.5	-12.5
安徽	11.20	7.07	10.09	6.38	-9.9	-9.8
福建	9.72	6.16	8.90	5.67	-8.4	-8.0
江西	9.45	6.18	8.52	5.57	-9.8	-9.8
山东	17.64	10.06	15.29	8.70	-13.3	-13.5
河南	15.57	8.80	13.61	7.66	-12.6	-12.9

续表

地区	2010 年		2015 年		2015 年比 2010 年（%）	
	排放量	其中：工业和生活	控制量	其中：工业和生活	增加或减少	其中：工业和生活
湖北	13.29	8.25	12.00	7.43	-9.7	-9.9
湖南	16.95	10.15	15.29	9.16	-9.8	-9.8
广东	23.52	17.53	20.39	15.16	-13.3	-13.5
广西	8.45	5.63	7.71	5.13	-8.7	-8.9
海南	2.29	1.36	2.29	1.37	0	1.0
重庆	5.59	4.19	5.10	3.81	-8.8	-9.0
四川	14.56	8.50	13.31	7.78	-8.6	-8.5
贵州	4.03	3.19	3.72	2.94	-7.7	-7.8
云南	6.00	4.66	5.51	4.29	-8.1	-8.0
西藏	0.33	0.28	0.33	0.28	0	0
陕西	6.44	4.80	5.81	4.34	-9.8	-9.6
甘肃	4.33	3.70	3.94	3.38	-8.9	-8.7
青海	0.96	0.87	1.10	1.00	15.0	15.0
宁夏	1.82	1.60	1.67	1.47	-8.0	-8.0
新疆	4.06	3.08	4.06	3.08	0	0
新疆生产建设兵团	0.51	0.25	0.51	0.25	0	0
合计	264.4	179.4	236.6	160.4	-10.5	-10.6

注：全国氨氮排放量削减10%的总量控制目标为238.0万吨（其中工业和生活161.5万吨），实际分配给各地区236.6万吨（其中工业和生活160.4万吨），国家预留1.4万吨，用于氨氮排污权有偿分配和交易试点工作。

附件4：

“十二五”各地区二氧化硫排放总量控制计划

单位：万吨

地区	2010 年排放量	2015 年控制量	2015 年比 2010 年（%）
北京	10.4	9.0	-13.4
天津	23.8	21.6	-9.4
河北	143.8	125.5	-12.7
山西	143.8	127.6	-11.3
内蒙古	139.7	134.4	-3.8
辽宁	117.2	104.7	-10.7
吉林	41.7	40.6	-2.7
黑龙江	51.3	50.3	-2.0
上海	25.5	22.0	-13.7
江苏	108.6	92.5	-14.8
浙江	68.4	59.3	-13.3
安徽	53.8	50.5	-6.1

续表

地区	2010 年排放量	2015 年控制量	2015 年比 2010 年（%）
福建	39.3	36.5	-7.0
江西	59.4	54.9	-7.5
山东	188.1	160.1	-14.9
河南	144.0	126.9	-11.9
湖北	69.5	63.7	-8.3
湖南	71.0	65.1	-8.3
广东	83.9	71.5	-14.8
广西	57.2	52.7	-7.9
海南	3.1	4.2	34.9
重庆	60.9	56.6	-7.1
四川	92.7	84.4	-9.0
贵州	116.2	106.2	-8.6
云南	70.4	67.6	-4.0
西藏	0.4	0.4	0
陕西	94.8	87.3	-7.9
甘肃	62.2	63.4	2.0
青海	15.7	18.3	16.7
宁夏	38.3	36.9	-3.6
新疆	63.1	63.1	0
新疆生产建设兵团	9.6	9.6	0
合计	2267.8	2067.4	-8.8

注：全国二氧化硫排放量削减8%的总量控制目标为2086.4万吨，实际分配给各地区2067.4万吨，国家预留19.0万吨，用于二氧化硫排污权有偿分配和交易试点工作。

附件5：

"十二五"各地区氮氧化物排放总量控制计划

单位：万吨

地区	2010 年排放量	2015 年控制量	2015 年比 2010 年（%）
北京	19.8	17.4	-12.3
天津	34.0	28.8	-15.2
河北	171.3	147.5	-13.9
山西	124.1	106.9	-13.9
内蒙古	131.4	123.8	-5.8
辽宁	102.0	88.0	-13.7
吉林	58.2	54.2	-6.9
黑龙江	75.3	73.0	-3.1
上海	44.3	36.5	-17.5
江苏	147.2	121.4	-17.5

续表

地区	2010年排放量	2015年控制量	2015年比2010年（%）
浙江	95.3	69.9	-18.0
安徽	95.9	82.0	-9.8
福建	44.8	40.9	-8.6
江西	58.2	54.2	-6.9
山东	174.0	146.0	-16.1
河南	159.0	135.6	-14.7
湖北	63.1	58.6	-7.2
湖南	60.4	55.0	-9.0
广东	132.3	109.9	-16.9
广西	45.1	41.1	-8.8
海南	8.0	9.8	22.3
重庆	38.2	35.6	-6.9
四川	62.0	57.7	-6.9
贵州	49.3	44.5	-9.8
云南	52.0	49.0	-5.8
西藏	3.8	3.8	0
陕西	76.6	69.0	-9.9
甘肃	42.0	40.7	-3.1
青海	11.6	13.4	15.3
宁夏	41.8	39.8	-4.9
新疆	58.8	58.8	0
新疆生产建设兵团	8.8	8.8	0
合计	2273.6	2021.6	-11.1

注：全国氮氧化物排放量削减10%的总量控制目标为2046.2万吨，实际分配给各地区2021.6万吨，国家预留24.6万吨，用于氮氧化物排污权有偿分配和交易试点工作。

附录3：工业企业能源审计和评估：对全球相关项目的调研①

引言

从全球范围内来看，在不同国家对不同行业的分析都表明工业行业领域存有大量的能效提升机会，其中很多机会从成本的角度来说也都是划算的，要挖掘出这些能效机会不仅需要采取跨行业的措施，有时候还需要采取一些与行业工艺相关的节能措施。很多时候工业设施却并未意识到自身能效提升的潜力，或不清楚达到这一目的需要采用哪些具体的技术和措施。因此进行工业能源审计是全球范围内应对上述问题最为广泛采用的方法之一。

能源审计旨在审查相关工业设施的能源消耗状况，并“发现在设备正常运行的条件下，设备所消耗的额外能耗”。通常情况下，能源审计的相关结果都以一份正式报告的形式呈现出来，内容包括对能效提升潜力的相关发现、对节约能源和成本的预估以及对实施成本的预估等（Russell，2010）。能源审计模式各不相同，一般来说大致上可以分为两类，即简单能源审计和复杂的能源审计。其中，简单能源审计也叫做筛选/初级或是“走过式”能源审计，即通过分析工业设施相关的能耗账单及其他运行数据，从而获得对审计工业设施能源消耗状况的基本了解，从而找出设施日常运营和维护方面能够进行简单改善的地方，并帮助确定是否有必要进行一次更加全面的能源审计。全面的或详细的能源审计可以以具体能耗系统为目标，涵盖更多的程序、设备或设施，从而找到更全面的能效提升措施。

开展工业企业能源审计是确定终端能源消耗情况并找出工业运行设施中能源节约关键领域的必要第一步。没有能源消耗的相关数据作为基础，我们就不能对当前的状况有一个清楚的认识，也难以在能效提升战略方面做出明智的决定。此外，对于工业企业来说，进行能效审计也是一个重要的推动力，可以鼓励他们采用那些最具有节能潜力的能效措施和技术。通常来说，简单能源审计能够就能耗情况、能源节约的潜在成本和节约潜能，以及简单的投资回报周期做出快速预估。另一方面，全面的能源审计就能够根据具体工厂的运行环境，对提出的所有节能措施和技术进行详细的性价比分析。高质量的能源审计还能够为工厂提供一揽子个性化的节能建议。进行能源审计通常是工业企业能效提升项目的一个关键组成部分，而且能源审计通常也作为自愿协议或总量管制与排放交易等政策的支撑政策工具。

为了更好地挖掘工业领域能源节约的潜力，并克服诸如对进行能源审计缺乏兴趣之类的障

---

① 本文由劳伦斯·伯克利国家实验室环境能源科技研究室能源分析项目组中国能源小组提供，作者为Lynn Price，鲁虹佑。

碍问题，世界范围内各个国家经常将工业的能源审计纳入到具体的政策或项目中，以便做到机制化。这类项目，在本报告中我们称之为独立型能源审计项目，主要关注能源审计本身，要求参与企业进行能源审计（这类项目通常向参与企业提供相关支持或要求企业必须参加）。另一种项目，在本报告中我们称之为一体化的能源审计项目，将能源审计与其他政策措施相结合，旨在更好地激励参与企业，帮助相关决策者设定一个合理但又宏大的节能目标，从而实现更广泛的目标要求。由于能效技术和措施都是随着时间的推移而不断进步的，所以能源审计也不应该是一次性的事情。相反，能源审计应该经常地开展，或其他政策机制相结合，以便不断提高工业能效。

完善的工业能源审计项目可以是独立型的能源审计项目，也可以是一体化的能源审计项目，这些能源审计项目都应该为开展标准化的能源审计提供规范、标准和指引，对能源审计的结果进行收集、分析和评估，同时也对项目的激励和支持措施进行评价。工业能源审计项目还应该对能源审计人员进行培训和认证，因为能源审计人员的素质对于能源审计项目的质量和最终结果具有重要影响。能源审计项目的质量控制和监控以及信息发布对于能源审计项目最终能否取得成功也十分重要。

成熟的工业能源审计项目不仅应该寻求将能源审计长期化、机制化，对能源审计人员的绩效进行跟踪，对相关建议的实施情况进行监测，为工业领域的参与方提供反馈，并根据对工业领域和能源使用系统一体化能源审计报告的分析提出相关的政策建议。

本文通过对15个国家和1个地区的22项工业能源审计项目进行调研，旨在提供这项能源审计项目设立的相关信息，其中包括不同的项目类型、激励政策、能源审计手册和工具、能源审计人员的培训和认证、能源审计结果的数据库建设、能源审计后续跟踪以及其他内在政策措施。随后，本文对独立的能源审计项目和一体化的政策项目进行了对比，一体化的政策项目中包括了根据前述关键元素设计的能源审计分项目。最后，本文就如何建立起强有力的能源审计或能源评估项目提出了建议，其目的就在于为决策者和项目设计人员，特别是目前还没有实施过能源审计项目、或是正在实施并进一步改进现有能源审计项目的国家和相关人员提供信息和方向。

方法论

我们对22项工业能源审计项目进行了调研，其中一个项目位于欧盟内，其他21个项目位于特定的15个国家，其中包括加拿大、丹麦、芬兰、法国、印度、爱尔兰、日本、荷兰、挪威、葡萄牙、瑞典、瑞士、英国和美国。表1就按照国别对上述22个项目进行了一一列举。

表1　工业能源审计项目

国别	能源审计项目名称	审计要求和项目类型（独立型或一体化）
澳大利亚	能效机会项目	法律规定的强制能源审计（I）
加拿大	工业生态能源项目	自愿协议中的自愿能源审计（I）
丹麦	附带温室气体排放税的自愿协议项目	附带税务减免自愿协议中的自愿能源审计（I）
芬兰	芬兰工业能源审计项目	自愿协议中的自愿能源审计（I）
法国	中小企业能源审计项目	免费能源审计（S）
	附带有税务威慑的自愿协议	附带税务威慑自愿协议中的自愿能源审计（I）

续表

国别	能源审计项目名称	审计要求和项目类型（独立型或一体化）
印度	能源经理培训	能源审计人员培训和认证项目（I）
爱尔兰	中小企业能源咨询、辅导和评估	免费能源审计（S）
	大型工业企业能源网络能源协议项目	附带能源管理体系要求自愿协议中的强制能源审计（I）
日本	能源节约审计	免费能源审计（S）
	能源经理认证项目（按照2005年能源节约法案修正案的相关要求）	附带能源管理要求法律规定的强制能源审计（I）
荷兰	长期自愿协议项目	附带能效目标要求自愿协议的强制能源审计（I）
挪威	工业企业能效网络	附带二氧化碳排放税自愿协议的自愿能源审计（I）
	能源管理网络	附带能效目标要求自愿协议的自愿能源审计（I）
葡萄牙	能源密集型企业的管理体系	附带能源管理体系和能效目标要求的强制能源审计（I）
瑞典	重点能耗行业能效提升项目	附带税务减免自愿协议中的强制审计（I）
	公司能源审计	免费能源审计（S）
瑞士	二氧化碳排放目标协议	附带税务减免自愿协议中的强制审计（I）
英国	英国碳信用基金的碳调查	免费能源/碳审计
美国	“现在开始节能”领导者计划	附带能效目标要求自愿协议的强制审计（I）
	工业评估中心	免费能源评估（S）
欧盟	欧盟生态管理和审计项目	附带能效目标和环境管理体系要求自愿协议的强制审计（I）

注：（S）独立型的能源审计项目；（I）一体化的能源审计项目。
资料来源：国际能源署政策和措施数据库，MURE II 数据库，世界能源理事会能源效力政策和措施。

如前所述，工业能源审计项目可以分类为：（1）独立型能源审计项目；（2）将能源审计作为关键组成部分的一体化政策项目。在本文中，我们对6个国家的6个独立型能源审计项目以及9个国家和1个地区的16个一体化政策项目进行了调研，相关调研信息主要来自于有关国际能效政策的数据库、相关项目的网页、政府发布的声明和法律法规文件、项目评估报告、相关研究人员发布的研究文献以及网络调查的结果。

在本文下面部分我们就来介绍独立型能源审计项目的共同特征，并选取特定国家的某些项目作为研究案例。随后我们对一体化工业能源审计的项目进行介绍，挖掘出这类能源审计项目的关键特征，并选取包括丹麦、爱尔兰、日本、瑞典和美国等发达工业国家在内的相关能源审计项目作为案例。根据这类国际经验和实践，我们揭示了实施能源审计项目的关键领域，以供能源审计项目开发人员参考。文章结尾部分，我们对还没有在全国范围内实施工业能源审计项目的国家提出了一些意见和建议。

独立型的能源审计项目

在本文中，我们对6个国家的独立型能源审计项目进行了调研，所有这些能源审计项目都由所在国政府发起，旨在刺激工业企业，特别是中小企业对能源审计的需求。进一步细分，独立型能源审计项目又可分为两类（见表2）：（1）为企业免费提供能源审计的项目；（2）为公司提供补贴以部分抵消其能源审计开支的项目。

表2　独立型能源审计项目分类

类型	项目示例
免费能源审计	美国工业评估中心（美国能源部）
	英国碳调查（英国碳信用基金）
	日本工业能源审计（日本节能中心）
	爱尔兰中小企业能源建议（爱尔兰可持续能源局）
补贴型能源审计	瑞典公司能源审计项目（瑞典国家能源局）
	法国中小企业能源审计项目（法国经济、财政和工业部）

独立型能源审计项目有一些共同的特征：第一，所有这些国家的项目都是自愿性质的，并对所有有兴趣的参与企业开放；第二，除了瑞典实施的公司能源审计项目外，所有其他国家的能源审计项目都集中于中小企业。同大型企业相比，中小企业缺少能源效率方面的专长和信息等资源（Nagesha and Balachandra，2006；Worrell and Price，2001；Reddy，1991）。通常来说，由于经济上面临紧迫的生存压力以及管理能力有限，中小企业要么就是对能源审计不感兴趣，要么就是缺少开展专业能源审计的财力资源，因此对于中小企业来说，政府发起的能源审计项目就是一个便利的载体，以便他们寻求能源审计方面的专长或财务支持。

美国通过能源部下属的工业评估中心进行的能源审计项目重点关注以下类型的工业设施：（1）年销售总额不超过1亿美元；（2）雇员总数不超过500人；（3）年度能源账单支出额超过10万美元但不足200万美元（US DOE ITP，2010a）。在英国开展的碳调查项目中，有资格获得免费能源审计的公司，其位于英国的工厂年度能源支出需超过5万英镑（7.73万美元）但不足50万英镑（77.3万美元）①（英国碳信用基金，2010）②。除了在碳调查项目项下提供的能源审计外，其母项目——英国碳信用基金本身还提供一系列的碳审计，通过免费提供碳足迹指标和碳足迹计算器，碳信用基金还能帮助工业企业设施度量和测算出自己的碳足迹。

对于独立型的能源审计项目来说，要想了解其是否实施有效关键要看八项指标，下面我们就将对这八项指标进行深入评价③，这些指标特征对于后面我们将要讨论的一体化能源审计项目来说也是重要的，这八项指标特征分别是：

- 能源审计的成本；
- 标准化的能源审计手册；
- 能源审计人员的培训；
- 能源审计人员的认证；
- 能源审计结果数据库；
- 能源审计之后的跟踪回访；
- 标准化的工具；

① 本文中采用的英镑对美元汇率为1英镑=1.546美元（2010年年均汇率），http：//www.oanda.com/currency/historical-rates。

② 自2011年起，该项能源审计项目仅对苏格兰、爱尔兰和威尔士地区的公司提供能源审计。

③ 资金困难和投资回报周期过长通常是工业企业设施未实施相关能效措施最常见的两个原因，但由于各个能源审计项目本身的资金和资源也是有限的，在我们所调研的能源审计项目中直接提供投资支持的项目也不是很常见。另一方面，能效融资先也变得越来越流行，需求也越来越大，各种类型的融资模式和项目也被开发出来，以促进企业工厂充分挖掘能效方面的潜能。因此，在很多国家，在工业设施进行能源审计之后，他们可以通过其他公共或私营融资渠道申请融资支持。

- 案例研究。

表3对上述指标特征在6个独立型能源审计项目中的实施应用进行了总结，在本文下面部分我们还会这些指标特征进行深入讨论。

能源审计成本

通过补贴措施可以减少能源审计的附带成本，或者也可以向企业提供免费的能源审计。政府可以为补贴设定上限，具体形式可以是能源审计成本的一定比例，或是一个绝对的数额，或是两者的结合。

例如，日本节能中心在日本政府（主要来自日本经济贸易产业省）和日本私营行业提供的资金支持下，自1978年起就一直为日本的工厂进行工业能源审计工作（ECCJ, 2009a），对于注册资本不足1亿日元或雇员人数不足300人的公司来说，这些能源审计服务都是免费提供的。①。

表3　选定独立型能源审计项目的关键指标特征状况

<table>
<tr><th>项目名称</th><th>审计成本</th><th>标准化手册</th><th>审计人员培训</th><th>审计人员认证</th><th>能源审计结果数据库</th><th>能源审计后续跟踪</th><th>标准化工具</th><th>典型案例</th></tr>
<tr><td>美国工业评估中心（IACs）</td><td rowspan="4">免费</td><td>有</td><td>有</td><td rowspan="5">无</td><td>有</td><td rowspan="5">有</td><td>有</td><td rowspan="6">有</td></tr>
<tr><td>英国碳调查（英国碳信用基金）</td><td rowspan="2">不适用</td><td>无[1]</td><td rowspan="5">无</td><td rowspan="2">无</td></tr>
<tr><td>日本（日本节能中心）</td><td>有</td></tr>
<tr><td>爱尔兰（对中小企业的能源建议）</td><td rowspan="3">无</td><td>无</td><td>有</td></tr>
<tr><td>瑞典（公司能源审计）</td><td rowspan="2">成本共担</td><td>无[2]</td><td>无</td></tr>
<tr><td>法国（中小企业能源审计）</td><td>有</td><td>有</td><td>无</td><td>有</td></tr>
</table>

注：1. 尽管没有专门提供有关能源审计的特别培训，但每一次碳调查都由一位独立且经过认证的咨询人士执行，在英国碳信用基金的网站上，对于如何完成一个个性化的“能源节约行动计划”有在线培训。
2. 瑞典能源局提供了能够为公司提供咨询和建议的能源专家和专业人士列表。

自2000年以来，在法国开展的中小企业能源审计项目项下，法国环境和能源管理署（ADEME）对于工业企业和建筑行业开展的能源审计成本进行了部分补贴，但不包括独立的个人住房②。补贴主要针对调研和预可行性研究，具体情况如下（IEA, 2010）：

- 调研（预诊断）主要用来识别资源不当管理的可能解决方案并对其按照缓急轻重进行排序。在为期两天的调研中，经过认证的专家将对企业当前情形进行审计和诊断，总费用为2300欧元；。
- “指导建议”主要用于一些设施比较复杂的场所，相关费用可高达75000欧元（住房和交通项目的封顶费用由各地自行制定），法国环境和能源管理署最高可以提供技术研究成本70%的补贴。
- 诊断和可行性研究就在具体细节上更近一步，工业和农业项目诊断方面的费用可以高达

---

① 超过上述门槛的公司被纳入到一体化的能源审计项目中，即日本政府实行的能源经理认证项目。
② 对独立个人住房，法国环境和能源管理署的网站上有一系列的工具可供其开展能源审计使用。

30000欧元，住房和交通项目的封顶费用由各地自行制定。工业领域可行性研究方面的费用最高可达75000欧元，政府提供补贴的最高比率为成本支出的50%。

能源审计项目的设计者在设计相关激励政策，特别是补贴政策时必须要将搭便车的因素考虑在内，所谓搭便车者指的是“能源节约项目中的那些哪怕没有相关项目规划也得采取相关能越节约措施的参与方”（Vine，1992）。搭便车的公司可能就不会将能源审计看成是最紧迫的事情，或是仅仅利用补贴资金进行一下初步的能源审计，有些公司还可能将原本应该用于能源审计或其他能效措施的补贴用于进行可行性研究（Vaisanen，et al.，2002）。尽管这类研究项目可能会对能源审计产生一定的积极贡献，但由于这类研究项目程序覆盖的范围较小，其对能效提升的影响有限，这类研究项目不能取代也不会获得能源审计项目所能取得的成果。一份芬兰的项目评估报告指出，能源审计项目的搭便车因素在全部能源审计行为的占比大约为10%—15%（Thomas，2008；Khan，2006）。在丹麦，人们估计大约有50%实施的能效项目在没有能源审计补贴的情况下也会实施（Gynther and Suomi，2009）。

标准化的手册和标准化的工具

标准化的手册指的是提供标准化程度供能源审计人员开展能源审计工作使用的指南和模板。标准化手册能够极大提高能源审计的项目质量，确保能源审计项目以一种连贯的方式实施，确保整体数据能够对接，从而产生富有见地的能源审计结果。标准化的指南和模板还可以被用来作为对能源审计人员就行培训的基础。

标准化的手册应该包括：（1）能源审计项目的一般原则和要求；（2）怎样开展工业企业能源所审计项目，特别是在机械和电力系统内；（3）能源审计领域的最佳实践；（4）最经常提出的能源节约措施（Vaisanen and Reinikainen，2002）。考虑到不同的程序或不同的工业领域内具体的能源审计项目也各不相同，因此我们可以编制出一些专门行业、程序或设备的指南或模本。在我们调研的独立型能源审计项目中，6个项目中有4个项目提供了标准化的能源审计手册。

标准化的工具可以是清单列表、基于Excel表格或以项目为基础的计算工具（通常和数据收集表格一起使用）、各种在线工具以及可下载的或CD储存的软件等。清单列表可以在能源经理对自己工厂设施以“走过式”形式开展的能源审计中加以应用，一些简单的计算工具可以用于进行自评，或是在更广范围内作为一种提升能效意识的工具。软件工具应便于携带，并具备多种功能，但要想对相关结果进行很好的分析，高质量的数据也是必需的。

标准化的工具不仅会极大地减少能源审计的成本，而且还会最大程度地提升能源审计的质量。通过使用能源审计评估工具，能源审计人员可以找出潜在的问题，迅速就能源节约的潜力进行评估并提出意见建议，并对各种意见建议的组合进行分析，通过对潜在影响进行分析拿出最优化的能源节约方案，然后对相关成果进行确认（Muller，2010a）。

人们还为能源审计开发出了各种各样的工具，其中有些工具需要购买才能得到，而有些工具是免费对公众开放的，需要花钱购买的能源审计工具在功能上都很强大，审计准确度上也有保证，当然价格方面也就不菲了，而且还有许可证方面的限制。免费对公众开放的能源审计工具在质量和功能方面不一定都能保证，这主要取决于工具到底来源于何方（Muller，2010a）。

美国能源部开发出了一系列用于不同行业对工业企业进行能源评估的软件，这些软件都可以从美国能源部的网站上免费下载，这些用于不同行业的工具在美国工业评估中心（IAC）项目中就被当成是标准化的能源审计工具来使用（如表4）。这些工具由能源专家开发，其设计理念主要是为了识别不同行业领域的节能机会，其中每一项工具都附带有用户手册和案例研究，而且美国能源部还提供使用这些工具的专门培训。

表4　美国工业评估中心进行能源审计时采用的标准化工具

使用范围	能源审计工具
全二厂范围内	工业企业设施工具包 工厂能源状况快速诊断工具/一体化的工具包（Quick PEP）
蒸汽	蒸汽系统范围界定工具 蒸汽系统评估工具 3E +
过程加热	热电联营应用工具 氮氧化物和能源评估工具 过程加热和调研评估工具（PHAST）
电机驱动	AirMaster + 风扇系统评估工具 MotorMaster + 水冷却系统分析工具 泵系统评估工具
数据中心	数据中心专业软件工具包

资料来源：Muller，2010a，美国能源部，2010b。

能源审计人员的培训和认证

对于能源审计项目来说，聘用高质量、技术娴熟的能源审计人员是至关重要的，能源审计人员提供的评估和相关建议和能源审计的最终结果直接相关。能源审计方面的培训可以以多种形式展开，如关于能源审计程序的培训、关于技术方法论和实践方法的培训等等。根据能源审计项目的不同要求，培训项目可以是自愿的，也可以是建议或是强制的。

在能源审计培训完成之后，培训组织者还要安排一次考试或测验，以便确定受训人员是否可以通过认证，有些情况下，如果没有安排相关培训项目，可以通过一些能源审计的要求或资格来进行审计人员的认证。通过考试的受训人员将会被授予资质或“证书”，从而可以从事能源审计工作。对能源审计人员进行资格认证的目的就是要培养一批熟练的能源审计人员，或是对能源审计人员的最低审计能力进行控制，被认证的对象可以是个人、公司或其他组织，如能源服务公司（ESCO）等。根据不同项目的不同需求，能源审计项目的设计者可以自行决定是否对能源审计人员的认证是强制的、建议的还是自愿的。

通过美国能源部下属的工业评估中心以及“现在开始节能”项目进行的能源评估，都是由能源专家（Energy Experts）或最佳实践资深专业人士（Best Practice Qualified Specialists）来完成的。对这些专业人士的培训涵盖能源审计工具和能源审计实践，具体包括蒸汽系统工具包，过程加热和调研评估工具、泵系统评估工具、风扇系统评估工具以及 Airmaster + 等。通常情况下，培训周期大约为3—5天，想成为合格专业人士的受训者不仅应该满足一些基本条件并参加相关培训项目，而且他们还需要通过实践和笔头两项考试。若他们能够成功通过这些考试，美国能源部就会对他们进行认证，并在能源部的网站上将他们的名字作为合格专业人士公布出来，他们也就可以从事一些跨能源消耗系统的能源审计工作，例如，空气压缩、风扇、过程加热、泵和蒸汽系统。

日本节能中心（ECCJ）为负责工业设施能源节约的工程师举办各种以教育为目的的研讨会，这种以教育为目的研讨会能让参与者了解与日本能源节约法案的相关知识并了解最新的节能科技。此外日本节能中心还通过设施与设备的示范项目，举办教授具体节能方法的研讨会（ECCJ，2009c）。如果某个工厂或组织想要个性化的培训，他们也可以向日本节能中心提出这种培训研讨会的要求。

在法国的能源审计项目中，为提高能源审计人员的素质和能源审计的质量，法国环境和能源管理署在法国境内开展了一系列名为“培训巡演”的培训活动。根据能源审计人员所关注的不同行业（如建筑行业或工业行业），培训的内容也不尽相同，但一般来说一个2天半—3天的培训，其基本内容包括（Despretz，2002）：

- 法国环境和能源管理署规划和审计项目简介；
- 能源审计模式和要求简介；
- 介绍日常审计工作中最为常见的错误和容易出现的遗漏；
- 能源审计软件和工具简介和培训；
- 根据两个真实案例进行演示。

法国环境和能源管理署任命能源领域合格的独立专业人士为能源审计人员，并通过与能源咨询人员签订技术性和非技术性协议的形式公布得到认证的审计人员名单。在法国环境和能源署进行的能源审计项目下，能源审计必须由认证的能源审计人员进行。在英国，碳信用基金通过指南、电话热线、具体的行业建议、针对中小企业的在线培训等为工业设施提供各种建议和技术支持，帮助这些工业设施去测量和检测自己的能源使用情况（如能量计量技术的入门指南和高级指南）。此外，碳信用基金还开发出各类能源审计工具，比如能源顾问工具、碳足迹指标和碳足迹计算器等，这些工具都可以为能源审计人员和能源经理提供帮助。

能源审计结果数据库

在我们调研的独立性能源审计项目中，只有美国的工业评估中心建立了能源审计结果的数据库。美国工业评估中心的数据库开始于1981年，到现在为止，这个数据库包括了大约15000个评估项目和110000多条建议，这个可供下载的数据库可像地图一样显示美国各地实施的能源审计项目，而且还能对项目评估项目进行筛选过滤，这个数据库不仅记录项目评估和建议，而且还可以分析有关所有项目数据并生成相关报告。例如，根据该数据库的相关数据，每一次能源审计找出的节能价值大约为7万美元，每一次评估中实现的能源节约量价值大约为2.3万美元。通过美国工业评估中心进行的能源评估，所有能源评估项目平均每年建议的能源节约量总价值为8400万美元，或125000百万英热单位（6462 TJ）。自美国工业中心开始进行第一个能源评估项目开始，在相关能源节约措施实施后，到目前为止实现的实际节能量价值达到52亿美元，或4.04百万英热单位（426242 TJ）（Muller，2010b）。这一数据库还根据行业、企业规模或产品对能源审计的结果进行检索，或是对有关建议按照节能量和成本来进行检索。此外，该数据库还可以作为对审计人员进行培训训”的好帮手。另外，其他企业也可通过该数据库获取相关评估结果，进行能效对标，从而对自己企业内部的节能潜力和机会进行评估。

在为能源审计项目设计数据库的时候，项目开发人员和工程师需要考虑以下因素，如数据库的使用者、数据库的功能和特征、数据库容易接入等等。通常情况下，一些必须的数据输入包括（Muller, 2010b）：

（1）能源审计活动。其类型、日期、参与方等。

（2）工厂汇总数据。联系信息、能源消耗情况、生产企业类型、特定产品信息、生产时间和生产率水平。

（3）建议数据。每一单位能源节约情况、成本节约情况、实施成本和资源类型。

（4）参考数据。工业类型、能源审计人员细节信息、标准行业代码、未实施某项建议的原因。

相关数据库可以对能源审计人员、工业工厂和其他相关方开放，让更多的人进入数据库可以让这些数据在其他的节能项目中得到更多的应用，从而实现更多的节能目标。但同时对一些保密或敏感的数据也应该有所限制，一般人员不应直接从数据库上获得上述信息（Muller, 2010b）。在荷兰实行的能源效率长期协议项目中，为了认真应对敏感数据问题，政府要求必须在不损害公司敏感数据保密性的前提下，有关项目实际进展的相关信息方才可以公布（Nuijen and Booij, 2002）。

数据库还可以被用来对能源审计项目进行监测和检查。在欧盟开展的第二期能源审计项目中，项目设计人员就为整个能源审计项目设计了六个不同层次的监测体系：对补贴开支的监测、能源审计容量、能源节约潜力、建议节能措施的理论节能量、实施的相关措施在现场测量到的节能量以及确认的节能结果（Vaisanen, et al. , 2002）。在美国工业评估中心实行的项目中，在相关评估数据和能源审计报告提交给数据库之后，有关人员会对报告进行审议并对相关数据进行误差核查，通过使用该中心的数据库，可以识别并解决存在的问题，而且这一数据库还为美国工业评估中心进行跟踪调查和收集相关研究案例提供了一个平台。

能源审计后的跟踪回访和案例研究

对于了解能源审计后建议的相关节能措施是如何得到实施的，能源审计后的跟踪回访就很重要，只有在对为什么工厂在采取某些节能措施方面存有困难进行分析评估后，我们才能为工业设施提出更加具有针对性的服务。在我们调研的独立型能源审计项目中，有好几个项目都在进行能源审计后对工厂进行了跟踪回访。

在爱尔兰，中小企业能源建议项目通常会通过电话和邮件的形式对工厂进行为期95天的跟踪回访，以便为他们提供一些额外的支持和咨询（爱尔兰可持续能源局，2010）。在英国，碳调查项目通过电话或邮件的形式为客户提供反馈和跟踪回访。在美国工业企业评估中心实行的能源审计项目中，在对工厂进行能源评估后6个月内，他们还通过电话的形式与被审计的工厂保持联系，并准备了一些调查表格，上面列出了具体的建议、建议实施日期以及未实施某些建议的原因。此外，他们还会对被审计公司进行一次跟踪回访，以便确认被审计公司是否在公司其他设施上通过复制类似节能措施实现了另外的能源节约，或是通过实施其他建议措施实现了额外的能源节约（Gopalakrishnan, 2010）。

所有独立型的能源审计项目都为公众提供案例研究，案例研究和其他类似的手册都是非常有用的信息工具，他们可以吸引到潜在的参与方，并为相关培训项目提供支持。同其他支持措施一起，案例研究可以被用来提升公众意识，让公众对能源审计树立起积极的认知，他们还可以被用来说服潜在的参与方（如企业和工厂）参加到能源审计项目中来，或是对相关问题进行自我诊断或是实行相关能效提升措施。

日本节能中心就通过其网站提供研究案例和其他信息资源，日本节能中心还通过电子邮件提供有关节能设备、日本能源节约法案、能源税务改革等方面的咨询服务，并开展相关能源节约的宣传工作（Galitsky, et al. , 2004）。此外，日本节能中心还携手相关公司和组织通过以教育为目的的研讨会、技术课程、月刊杂志和手册等形式公布一些好的能源节约做法（日本节能中心，2009c）。

附带能源审计的一体化工业能效政策项目

很多的工业能效政策项目都将工业能源审计当成是关键组成部分，并将其与其他政策措施一道用来更好地激励相关参与方，以便在更广范围内实现能源节约的目标。在本文中，我们对14个国家以及欧盟地区的16个项目进行了调研，所有这些项目都是附带能源审计的一体化工业能效政策项目。

一体化的政策项目又包括自愿协议和强制要求两种，自愿协议（即工业企业和政府之间签订的协议）的使用范围很广（Price, 2005），在很多情况下自愿协议都要求参与方进行能源审计。强制要求指的是政府制定的相关法律或法律规定，其要求工业设施进行能源审计，或是达到能效提升的特定目标，或是建立起经认证的能源/环境管理体系。通常情况下，能源审计都被当成是一个非常有用的工具，以便在更广范围内实现政府法规的节能目标。

附带能源审计的一体化工业能效政策项目在能源审计之外，还使用一系列的其他政策和项目，其中包括使用经认证的能源或环境管理系统、要求建立起具体的能效提升目标或指标、要求建立起能源行动计划、免于征收能源和/或二氧化碳税、若未能实现预先设定的目标或指标就威胁征收能源或二氧化碳税、对相关投资的财务支持、对能源审计的补贴、认可标识和奖励。根据各国不同的国情，能源审计项目开发人员（如相关政策决策者）可以自行决定采用什么样的项目类型（如自愿协议还是强制要求）以及具体包含哪些措施。

在我们所调研的16个能源审计项目中，其中4个项目强制要求工业设施进行能源审计并要求其符合其他的特定要求，如建立起经认证的能源管理体系，另外12个项目都采取了自愿协议的形式，但其中7个自愿协议项目都要求所有参与方必须进行能源审计。表5列出了附带能源审计的一体化工业能效政策项目中使用到的其他政策措施。

能源或环境管理体系

能源或环境管理体系的推广和采用对于工业企业提高能效来说可谓是关键的基础性工作，可以帮助他们不断采用严格的程序来对自己的能源使用形态进行评估，制定出能源节约的目标，发现并实施相关节能措施（Taylor, 2010）。好几个国家都在能源管理方面建立起了地区性的和国家性的标准，例如欧盟的能源管理标准（EN 1601）以及各国自己开发的标准，如丹麦的（DS 2403）、瑞典的（SS 627750）、爱尔兰的（IS 39）和美国的（MSE 2000）。目前，国际标准组织（ISO）正在开发能源管理体系的国际标准，也就是ISO50001体系，该标准将于2011年公布，ISO50001体系将帮助工业企业公司建议其检测、控制和节约能源的框架。

爱尔兰、日本、葡萄牙和欧盟都将能源审计与经过认证的能源管理体系相结合，爱尔兰实施了两类不同的自愿协议项目，即大型工业企业能源网络项目（LIEN）和能源协议项目（EAP）。能源协议项目可以看成是大型工业企业能源网络项目的子项目，针对的也是大型工业企业公司，其目的旨在“通过预订的程序为大型工业企业提供能源管理最佳实践的相关机会和方法，避免大型工业企业一而再再而三地实施一些低质量的能效项目”（Price，et al.，2010）。加入爱尔兰能源协议项目的公司都被强制要求进行能源审计，以便获得爱尔兰能源管理体系IS393的认证，并通过与爱尔兰可持续能源局（SEAI）签订为期三年的合约来实施这一标准体系，以便做到节能潜力的最大化。此外，加入能源协议项目的公司还应致力于开展并完成三项特定调研任务，其重点就是在重点能耗领域实施能效提升技术和/或关键程序（Price，et al.，2010）。上面提到的三项特定调研任务必须在合约开始的头三年里完成，每年能效方面的数据都必要向爱尔兰可持续能源局报告（Gudbjerg，2009）。作为交换，参加能源协议项目的参与方可从爱尔兰可持续能源局获得相关技术支持、量身定制的培训以及识别能源节约机会的辅导和顾问服务（Price et al.，2010）。

在日本，根据修订的2005年《能源节约法案》，1类①和2类②指定能源管理工厂都被要求雇用或指定专人担任能源经理或能源管理人员，并通过日本节能中心每年向日本经济贸易产业省提交有关其工业设施能源消耗和使用状况的报告（Kawano，2007）。1类和2类指定能源管理工厂都被必须聘用能源经理或能源管理人员开展能源审计工作，这类能源审计项目是能源经理或能源管理人员履行其职责的一个重要途径，其主要职责就包括起草能源消耗状况包括、对能源使用情况进行现场检查和监测并提升设施能效。日本还要求所有能源经理或能源管理人员都是经日本政府批准和认证过的。要获得相关许可成为合格的能源经理，申请人士必须要通过全国性的资格考试或是完成一个培训研讨会。申请参加每年全国资格考试的申请人必须至少具备一年相关工作经验。2005年，日本全国共有8950人申请参加了这一资格考试，通过率仅为22.5%。每年为能源经理举行的培训研讨会为期7天，其中包括6天的培训课程和1天的考试，受训人员应该至少具有三年的从业经验。2005年，全国共有2700人参加了这一培训，通过率大约为65%。日本对能源管理人员也是通过培训的形式进行认证，日本要求这些能源管理人员每隔三年参加一次培训，以确保他们能够不断更新其专业特长和技术知识，这项为期1天的培训项目主题包括能源管理、能源效率政策和规定以及能源管理实践。

---

① 每年燃料消耗量少于1000立升原油当量但不超过3000立升原油当量的工厂就属于“2类指定能源管理工厂”。

② 截止2006年8月，日本共有1类指定能源管理工厂7433家，2类指定能源管理工厂5860家（Kawano，2007）。

表 5　能源审计作为关键组成部分的国家一体化能效政策项目中运用的政策措施

<table>
<tr><th>政策类型</th><th>项目名称</th><th>自愿或强制要求的能源审计</th><th>能源或环境管理体系</th><th>能源审计补贴</th><th>能源或二氧化碳税</th><th>能效提升</th></tr>
<tr><td rowspan="12">自愿协议</td><td>加拿大（工业企业生态能源）</td><td rowspan="5">自愿审计</td><td rowspan="6">无强制要求</td><td rowspan="5">有</td><td rowspan="2">不适用</td><td rowspan="3">无强制要求</td></tr>
<tr><td>芬兰（工业企业能源管理项目）</td></tr>
<tr><td>法国（AERES）</td><td>威胁使用</td></tr>
<tr><td>挪威（能源管理网络）</td><td>不适用</td><td>强制要求达标</td></tr>
<tr><td>挪威（工业企业能效网络）</td><td>适用税务</td><td>无强制要求</td></tr>
<tr><td>“现在开始节能”领导者计划</td><td rowspan="11">强制性审计</td><td rowspan="10">无</td><td rowspan="4">不适用</td><td>强制要求达标</td></tr>
<tr><td>欧盟生态管理和审计项目</td><td rowspan="2">强制要求</td><td rowspan="2">无强制要求</td></tr>
<tr><td>爱尔兰（大型工业企业能源网络和能源协议项目）</td></tr>
<tr><td>荷兰长期自愿协议项目</td><td rowspan="6">无强制要求</td><td>强制要求达标</td></tr>
<tr><td>瑞典（重点能耗行业能效提升项目）</td><td rowspan="3">税务减免</td><td rowspan="6">无强制要求</td></tr>
<tr><td>丹麦（附带温室气体排放税的自愿协议）</td></tr>
<tr><td>瑞士（二氧化碳排放目标协议）</td></tr>
<tr><td rowspan="4">法规强制要求</td><td>澳大利亚（能效机会项目）</td><td rowspan="3">不适用</td></tr>
<tr><td>印度（能源经理培训项目）</td></tr>
<tr><td>日本（能源经理认证项目）</td><td rowspan="2">强制要求</td></tr>
<tr><td>葡萄牙（能源密集型企业的管理体系）</td><td>有</td><td>税务减免</td><td>强制要求</td></tr>
</table>

能源审计激励机制

为吸引并鼓励工业企业积极参与能源审计项目，很多政府发起的能源审计项目都采用补贴或其他形式的激励来对能源审计提供支持，这类激励措施不仅仅局限于一体化的能源审计项目中，他们在独立型的能源审计中也被广泛采用。

在设计相关政策或机制时，具体补贴的水平以及补贴的发放方式应该与能源审计项目的项目目标、范围、经济发展状况以及能源审计项目希望取得的经济效益等紧密相连。政府可以为补贴设置一定的上限，具体可以是能源审计成本的一定比率或是一绝对数额，或是两者的结合。正如我们在独立型能源审计项目中看到的那样，能源审计通常都是由政府免费提供的，或是政府补贴了能源审计项目的部分支出。此外，补贴即可以发放给参与能源审计项目的所有公司，或是在发放具体补贴时设置一定的前提条件，或是要求参与能源审计项目的公司在能源节约方面取得一定的成绩。

除了补贴之外，全球各个国家还采用其他各种不同的措施来鼓励工业企业参与并实施能源审计或能源评估，例如提供贷款、税务减免和投资资助等。例如，在欧盟实行的欧盟生态管理和审计项目中，很多欧盟成员国家就为组织或公司参与到该项目中来而制定了各种激励政策，参与到欧盟生态管理和审计项目中的组织可以“从减少检查频率、减少许可费用、快速获得许可和更容易获得资金等方面获益”（European Commission，2010）。在葡萄牙政府实行的能源密集型企业的管理体系项目中，参与公司可以获得在能效设备和能效管理体系方面投入25%的补贴，补贴额最高不超过1万欧元（或1.55万美元）（SGCIE，2010）。

能源或二氧化碳税减免

某些国家实施的一体化工业能效政策项目中还允许对开展能源审计的参与企业免征能源税或二氧化碳税，这方面的例子包括丹麦实施的附带自愿协议的绿色税务一揽子计划、瑞典实施的二氧化碳目标协议项目以及葡萄牙实施的能源密集型企业的管理体系，参与企业在达到预先同意设定的目标后将获得回报，具体形式就是可以少缴强制性的能源税（DEFRA，2004；Togeby et al.，1999）。对于不遵守约定的公司，还可以运用能源税或二氧化碳税作为一项惩罚或威胁措施。在法国开展的自愿协议项目中（法国减少温室气体企业协会，AERES），如果一家公司未能遵守其承诺，其中包括开展能源审计的承诺，该公司将会受到二氧化碳税的惩罚（AERES，2009）。1995年，丹麦政府提出了到2005年将丹麦二氧化碳排放量在1988年的基础上降低20%的目标，此外按照《京都议定书》以及欧盟内部制定的相关减排分担协议规定，丹麦需要在2008—2012年间将其温室气体排放量在1995年排放水平上降低21%。为了在全国范围内实现上述二氧化碳减排目标，丹麦政府分别于1992年5月15日和1993年1月1日为家庭和工业企业引入了二氧化碳税，1996年丹麦政府更是制定了绿色税务的一揽子计划，其中就包括加码的二氧化碳税、新引入的二氧化硫税和取暖的新能源税（Price，et al.，2010）。同丹麦政府签订相关协议的公司企业在实现特定的节能目标后可以获得二氧化碳税的退税，要求之一就包括强制进行能源审计的要求。2000年后，丹麦政府不再强制要求公司企业开展能源审计活动，但仍要求公司企业根据其能源管理体系建立起相应的能源行动计划（Price，et al.，2010）。

瑞典于1991年引入碳税，2004年瑞典工业企业生产用电的税率提高到0.5欧元/兆瓦时，这项新实施的税务规定影响到了大部分的瑞典工业企业（Zhou，et al.，将要发表），作为对这项税令的补偿，瑞典政府于2005年实施了一项名为提升重点能耗行业能效的项目，参与该项目的公司企业可以获得一定的税务减免，但要求这些公司企业采取实质性措施提升其能效，并采用相关节能措施，如强制性的能源审计要求和建立起标准化的能源管理体系（瑞典能源局，2007）。

强制要求的能效提升

在我们调研的一体化工业能效政策项目中有4个项目强制要求参与企业需达到一定的能效提升目标，还有3个项目强制要求参与企业开展能源审计，并将其作为实现能效提升目标的机制之一。附带强制能效提升目标要求的能源审计项目不仅会极大地提高相关能效提升建议措施的实施效率，它还会帮助解决其他一些重要的问题，如公司企业高层对能源节约缺乏承诺、在能效提升方面缺乏持久的动力或是对能效提升未予以足够重视等。

美国实行的“现在开始节能”领导者计划针对的主要是这些主动承诺在未来10年要将自己企业能源强度降低25%的公司企业。作为回报，这些公司企业在进行能源评估时，政府将分担其一部分成本，而且这些企业公司也将优先获得个性化的支持、个性化的资源和融资支持（美国能源部ITP，2009a）。

荷兰政府实行的长期协议项目（LTA）就为参与项目的企业设定了明确的目标和承诺，如在第一期的长期协议项目中（1989—2000），全行业的整体承诺目标是到2000年，全行业的能效水平要比1989年的能效水平提高20%。参与公司企业通过开展能源评估来建立起所要求的长期计划，并用来设定能效提升方面的具体目标（Kerssemeeckers，2002；Rietenbergen et al.，1998）。在开展完能源评估并找到潜在的节能机会后，荷兰的长期协议项目还要求参与工业企业对投资回报期低于五年的"所有合适的能效提升措施"进行投资（MURE II Database，2010）。

结论部分

作为帮助工业企业识别能源提升机会的手段之一，工业能源审计在很多国家都得到了广泛采用。希望开展能源效率审计项目的国家，或是希望对现有能源审计项目进行改进的国家都可以从本文调研到的22个能源审计项目中获益。

通常来说，独立型的能源审计项目更加关注中小企业，这类能源审计项目有时候是免费提供的，或是由工业企业和政府共同分担审计成本。独立型的能源审计项目经常强调怎样建立起一套有效的、标准化的实用体系，在项目设计过程中主要应确保参与的工业企业能够比较经济地实施所建议的节能和能效措施，应确保能源审计以一种可比、连贯的方式进行，还应确保审计结果是可度量的、可核实的，并能对其他企业有所帮助。要想开展一项强有力的独立型能源审计项目，对能源审计的补贴、对能源审计人员的培训和认证、标准化的工具和指南、能源审计数据库建设、审计后的跟踪回访以及对研究案例的宣传等等都是非常重要的因素。

在附带能源审计的一体化政策项目中，能源审计或是包含到自愿协议的框架中或是由政府相关法律法规强制规定。如果通过法律形式强制要求进行能源审计是难以实现的的，那么能源审计项目的开发人员就可以考虑通过自愿协议的形式将能源审计与其他政策措施相结合，通常来说其他互补的政策措施主要包括对能源审计提供补贴、经过认证的能源管理体体系、使用能源税或碳税（或税务减免）、对能效提升投资提供资金支持、设定能效提升目标和强制要求提升能效等。

能源审计项目开发人员在设计自己国家的能源审计项目时应将本国特定国情考虑在内，但是正如相关国际经验所展示的那样，很多不同国家的独立型能源审计项目都有一些共同的关键指标特征。为了更好地激励工业企业参与能源审计，并在更广范围内实现项目目标，如能源管理体系的推广或是更多地采用能效提升方面的措施，附带能源审计的一体化政策项目将是一个不错的选择。

参考文献

1. 王绍武，罗勇，赵宗慈，闻新宇，黄建斌．全球气候变暖原因的争议［J］．气候变化研究进展，2011，7（2）：79－84.

2. 何英．希望能在最后时刻达成共识——专访 IPCC 主席拉金德拉·帕乔里［N］．中国能源报，2011－08－08.

3. 冯迪凡．后《京都议定书》之问：难道要等到 2015？［N］．第一财经日报，2011－07－06.

4. 文雯．《京都议定书》为何陷入停摆？［N］．中国环境报，2011－07－08.

5. 陈晓晨．印度：《京都议定书》不可废［N］．第一财经日报，2010－10－12.

6. 解振华．强调《京都议定书》必须坚持［N］．中国气象报，2010－12－08.

7. 李学华．延长《京都议定书》至关重要［N］．科技日报，2011－04－23.

8. 冯迪凡．抛开《京都议定书》并不明智［N］．第一财经日报，2009－12－10.

9. 辛本健．坎昆气候大会的危机与希望［N］．人民日报海外版，2010－12－07.

10. 王东．德班会议难达最终目标［N］．人民日报，2011－01－26.

11. 邹志鹏．量化减排成坎昆难解之题［N］．人民日报，2010－12－02.

12. 李学华．需要的是立即行动［N］．科技日报，2011－05－07.

13. 张守营．中国在碳排放问题上将变被动为主动［N］．中国经济导报，2011－03－26.

14. 迈向绿色经济．联合国环境署．

15. 张传国，陈晓庆．国外碳金融研究的新进展［J］．审计与经济研究，Journal of Audit & Economics，2011，（05）.

16. 大卫·霍恩，编译 王林．澳大利亚碳税之路在何方［N］．中国能源报，2011－03－07.

17. 朱其忠．澳碳税方案因何“凤凰变鸡”［N］．中国石化报，2011－05－13.

18. 傅云威．“碳税”方案因何凤凰变鸡［N］．经济参考报，2011－04－28.

19. 李慧．2010 年全球能源消费量再创新高［N］．中国能源报，2011－07－04.

20. 欧盟发布 2050 年减排路线图：最高减排 90% 时间［N］．21 世纪经济报道，2011－03－02.

21. 于晓玲．国际贸易中的“碳壁垒”［J］．合作经济与科技，2010，（23）.

22. 黄文旭．碳关税相关概念辨析［J］．岭南学刊，2011，（01）.

23. 王志华．碳壁垒合法性的国际法考量［J］．山东师范大学学报（人文社会科学版），2010，（06）.

24. 周生贤部长在主要污染物总量减排核查核算视频会议上的讲话．

25. 李毅 水电行业加速跑将遇资源瓶颈［N］. 第一财经日报，2011－06 －21.

26. 北京清洁能源推动绿色电力业务发展［N］. 明报，2011－06－28.

27. “十二五”将推动工业节能降耗减排迈上新台阶.

28. 韩晓平. 能源改革的“决心和勇气”［J］. 商务周刊，2011，（1）.

29. 盘点2010年二十个最“给力”能源事件［J］. 农业工程技术·新能源产业，2011，（2）.

30. 黄淑和在中央企业节能减排工作会议上的讲话.

31. 姜克隽. 中国节能应转向总量控制［J］. 新世纪周刊，2010，（35）.

32. 侯艳丽，李晶晶，杨富强. 中国“十二五”节能减碳应坚持高目标［J］. 中国能源，2011，（2）.

33. 廖华，魏一鸣. “十二五”中国能源和碳排放预测与展望［J］. 中国科学院院刊，2011，（2）.

34. 石油和化学工业“十二五”发展指南.

35. “十一五”经济社会发展成就系列报告. http：//www. stats. gov. cn/tjfx/ztfx/sywcj/.

36. 中国统计摘要2011. 中国统计出版社.

37. 吴力波. “十二五”期间中国能源供应结构的高效化［J］. 电力与能源，2011，（2）：90－95.

38. 宏观经济研究院国地所课题组. “十一五”我国节能减排工作回顾及“十二五”政策建议［J］. 宏观经济管理，2011，（1）：14－16.

39. 李迪. 黑龙江省节能减排实施现状的分析及战略选择［J］. 中小企业管理与科技，2010，（19）.

40. 唐方方. 节能减排的市场化手段［J］. Energy，2011（1）.

41. 李琪，赵春雨，安树伟. “十一五”以来我国四大区域经济发展回顾及“十二五”展望［J］. 发展研究，2011，（5）：13－17.

42. 刘晓蓉，安树伟. “十二五”时期我国区域经济发展环境分析［J］. 西安财经学院学报，2010，（3）：11－15.

43. 张有生. 能源规划：“十一五”回顾与“十二五”展望［J］. 国际石油经济，2010，（12）：9－19.

44. 各省市“十二五”规划纲要. 各省市2011年政府工作报告. 各省政府、发改委网站.

45. 中国节能环保集团公司，中国工业节能与清洁生产协会. 2010中国节能减排产业发展报告［M］. 北京：水利水电出版社，2010.

46. 王伟光. 应对气候变化报告（2010版坎昆的挑战与中国的行动）/气候变化绿皮书［M］. 北京：社会科学文献出版社，2010.

47. 国家环境保护总局. 中国环境统计年报2005—2010，北京：中国环境科学出版社，历年.

48. 蔡守秋，莫神星. 节能减排，机制先行——写在《节能减排机制法律政策研究》出版之际. 前沿，2009，（2）.

49. 李广杰. 转变经济发展方式促进节能减排. http：//www. sdetn. gov. cn/pxc/gzdt/webinfo/

2008/06/1214299921949390. htm.

50. 国家统计局．中国统计摘要 2011 [M]. 北京：中国统计出版社，2011.

51. 叶翠娟，张冰莹，刘俊勋．浙江省节能减排的成效、问题与对策 [J]. 北方经济，2010，(1)：76 -78.

52. 曾凡银．中国节能减排政策：理论框架与实践分析 [J]. 财贸经济，2010，(7)：111 -115.

53. 孙海彬，苏迪．节能减排现状分析及对策 [J]. 上海电力学院学报，2010，(3)：233 -236.

54. 米强，米娟．中国节能降耗指标体系初探 [J]. 生态经济，2010，(2)：58 -61.

55. 李琪，赵春雨，安树伟．"十一五" 以来我国四大区域经济发展回顾及 "十二五" 展望 [J]. 发展研究，2011，(5)：13 -17.

56. 唐志军，谌莹．为什么中国的城市化进程滞后？———基于比较和历史的视角 [J]. 云南财经大学学报，2011，(1)：12 -19.

57. 童中贤，肖琳子，熊柏隆，佘纪国．中部地区城市群整合及发展战略研究．中国网，http：//www. china. com. cn/economic/txt/2009 -09/16/content_ 18538416. htm.

58. 任超然．节能减排亟待突破产业结构瓶颈 [EB/OL]. 证券之星网，2010. http：//news. stockstar. com/info/darticle. aspx？ id = JL，20100702，00000995.

59. 洪流．洪流同志在全省财政经济建设暨节能减排工作会议上的讲话 [R/OL]. 湖北省财政厅公众网，2009. http：//www. ecz. gov. cn/structure/zwdt/ldhdzw_ 1261_ 1. htm.

60. 赵秉栋，焦士兴，赵军凯．我国西部地区可持续发展战略思考 [J]. 河南大学学报（自然科学版），2003，(2)：50 -53.

61. 黄松龄，昌苗苗，韦彦．广西安排 2. 1 亿元资金以奖代补支持节能减排 财政奖励与节能减排量挂钩 [R/OL]. 中国环境与发展合作委员会，2008. http：//www. china. com. cn/tech/zhuanti/wyh/2008 -09/08/content_ 16409729. htm.

62. 阿尔泰地区经贸委．新疆维吾尔自治区节能减排专项资金项目申报指南 [R/OL]. 阿尔泰地区经济贸易委员会，2009. http：//www. xjalt. gov. cn/10220/10958/2009/20296. htm.

63. 环境保护部自然生态司．生态省建设理论与实践 [M]. 北京：中国环境科学出版社，2008.

64. 中国的能源状况与政策 http：//www. gov. cn/zwgk/2007 -12/26/content_ 844159. htm.

65. 李广杰．转变经济发展方式促进节能减排．http：//www. sdetn. gov. cn/pxc/gzdt/webinfo/2008/06/1214299921949390. htm.

66. 国家统计局．中国统计摘要 2011 [M]. 北京：中国统计出版社，2011. 5.

67. 包婧文，丽珍．可再生能源：回顾 "十一五" 展望 "十二五" [J]. 太阳能，2011，(04).

68. 王世江，高宏玲．2010 年全球太阳能电池产业发展总结 [J]. 太阳能，2011，(13).

69. 何晓亮．传统能源继续发力 新兴能源起步迅猛 [N]. 科技日报，2011 -01 -25.

70. 罗清启．太阳能热水器产业需要明确战略模式 [N]. 经济参考报，2011 -05 -10.

71. 邱江勇．太阳能热水器：门槛将抬高 预备 "进城" [N]. 中国电子报，2011 -04 -01.

72. 周涛. 光热版块崛起指日可待［N］. 长城证券，2010－06－24.

73. 李纬娜，王奇华. 光热发电蹒跚起步［J］. 财经，2010，(11).

74. 易蓉蓉. 聚焦光热发电商业化［N］. 中国日报，2011－08－29.

75. BP Stats Review 2009、2010.

76. 谭倩，张晓霞. 光伏行业：政策趋向明朗，长期发展有望提速［J］. 太阳能，2011，(10).

77. SEMI PV Group. 2011 中国光伏产业发展报告.

78. 俞文杰. 2010 年太阳能光伏行业分析与 2011 年展望，国投资咨询网.

79. 东阳. 太阳能热利用产业路在何方［N］. 太阳能信息报，2011－04－01.

80. 谢丹. 太阳能闪耀中国［J］. 环球企业家，2011，(06).

81. 郭力方. "十二五"光热发电规模有望超千亿 8 股蓄势待发［N］. 中国证券报，2011－08－24.

82. 中投信德产业研究中心. 2010—2013 年生物柴油行业竞争格局与投资战略研究咨询报告，2010.

83. 中投信德产业研究中心. 2010—2013 年中国沼气行业市场调研投资分析预测报告，2010.

84. 张庆丰，渡边真树子等. 中国农村生物质能 2020［D］. 亚洲开发银行，2010.

85. 颜新华. 2010 年度新兴能源行业十大新闻［N］. 中国电力报，2011－01－11.

86. 郭周芳. 城市污水生物脱氮除磷新技术［J］. 水科学与工程技术，2011，(03).

87. 韩晶晶. 新型生物脱氮除磷技术的研究进展［J］. 广州化工，2011，(10).

88. 王永林. 干熄焦技术的节能减排及环保分析［J］. 煤化工，2011，(06).

89. 廖继勇. 烧结烟气脱硫脱硝技术的发展与应用前景［J］. 烧结球团，2008，(08).

90. 戈志华. 能量梯级利用在热电联产中的应用［J］. 华北电力大学学报，2010，(01).

91. 沈纲轮. 煤气化技术现状和展望［J］. 上海煤气，2009，(01).

92. 黄庆军. 中国换热器产业现状及发展趋势［J］. 石油与化工设备，2010，(13).